Gekeler
Mathematical Methods for Mechanics

Eckart W. Gekeler

Mathematical Methods for Mechanics

A Handbook with MATLAB Experiments

Springer

Prof. Dr. Eckart W. Gekeler
Former affiliation:
Institut für Angewandte Analysis
und Numerische Simulation (IANS)
Universität Stuttgart
Pfaffenwaldring 57
70550 Stuttgart
Germany
Eckart.Gekeler@mathematik.uni-stuttgart.de
eckart.gekeler@t-online.de
http://www.ians.uni-stuttgart.de/AbNumMath/Gekeler

Originally published in German as *Mathematische Methoden zur Mechanik*,
ISBN 978-3-540-30267-4, © 2006
ISBN: 978-3-540-69278-2 e-ISBN: 978-3-540-69279-9

Library of Congress Control Number: 2008929625

Cover design: WMXDesign, Heidelberg, Germany

Printed on acid-free paper

9 8 7 6 5 4 3 2 1

springer.com

Preface

Mathematics is undoubtedly the key to state-of-the-art high technology. It is an international technical language and proves to be an eternally young science to those who have learned its ways. Long an indispensable part of research thanks to modeling and simulation, mathematics is enjoying particular vitality now more than ever. Nevertheless, this stormy development is resulting in increasingly high requirements for students in technical disciplines, while general interest in mathematics continues to wane at the same time. This book and its appendices on the *Internet* seek to deal with this issue, helping students master the difficult transition from the receptive to the productive phase of their education.

The author has repeatedly held a three-semester introductory course entitled *Higher Mathematics* at the University of Stuttgart and used a series of *"handouts"* to show further aspects, make the course contents more motivating, and connect with the mechanics lectures taking place at the same time. *One* part of the book has more or less evolved from this on its own. True to the original objective, this part treats a variety of separate topics of varying degrees of difficulty; nevertheless, all these topics are oriented to mechanics.

Another part of this book seeks to offer a selection of understandable realistic models that can be implemented directly from the multitude of mathematical resources. The author does not attempt to hide his preference of *Numerical Mathematics* and thus places importance on careful theoretical preparation. Proofs are only shown when they are necessary for comprehension; additional proofs are available on the *Web pages* that are part of this book. Overall, the book is divided into four parts of varying lengths:

- a summary of the aids used in the book,
- general methods and mathematical methods in particular,
- the fascinating topic of mechanics from a mathematician's point of view,
- a survey of tensor calculus.

Physics programs at the university levels have always focused on the real experiment and continue to do so in the age of virtual worlds while experiments have been a rare exception in lectures on numerical and applied mathematics. This was at the great expense of comprehension, much like the way that reading music is a sorry substitute for the sound produced or even for one's own intonation. However, the situation has changed fundamentally in the last two decades. Problems that no one would have dared to imagine in the past can now be solved on a *laptop*, and these days anyone can test numerical methods themselves. Taking into account this development, the book provides experiments for *all* of the numerical methods that it treats. If more complex case studies are involved, bear in mind that it is a difficult road from the formula to the illustration and that detailed instructions are required here. Many well-established algorithms are known not to develop the performance required of them until after they have been subjected to careful *parameter tuning*, which is something that the computer cannot do. Sheer lack of space does not permit simultaneous treatment of theory and numerics in their entirety, and still maintain clarity. In order to deal with this dilemma, the author has relegated many proofs and all the algorithms to the internet. An extensive library of MATLAB programs are available there that is continually supplemented and expanded. The development environment of MATLAB allows algorithms to be displayed without all the unnecessary ballast in a clearly structured form that is often more impressive than a longwinded description and that — by the way – unconvers all errors without shred of mercy. These programs and the way the materials are presented are not only meant to unite reader's desire to play with their inherent curiosity but also to break a lance for the experiment in mathematics as the way to indings. In extreme cases, the user can even experience a certain esthetic charm from *the algorithm itself* and start viewing the problem from an entirely different angle.

All the illustrations and charts can be reproduced by readers themselves if they download the appropriate programs.

The *first* chapter introduces the mathematical toolbox. Terms are defined for later use, and some calculus topics are reviewed and worked through.

The *second* chapter provides concise treatment of the classic problems of numerical mathematics, in particular inital and boundary value problems for ordinary differential systems .

The *third and fourth* chapters focus on optimization and its sister in the continuum, control theory. Many qucstions in technical disciplines and business alike lead quite naturally to an extremal problem with constraints. Formulation as an optimization problem is often the last resort when a numerical approximation problem refuse to be solved in the usual way. Incidentally, optimization under constraints follows its own laws, which can be tracked throughout the whole of theoretical mechanics. In control theory, the calculus of variations and the LAGRANGE theory enter into a fruitful symbiosis. Concrete problems are solved numerically after discretization using the discrete optimization methods treated.

The *fifth* chapter attempts to make theory of branching and path continuation understandable for practically oriented users. This basically involves the task of calculating non-trivial solutions of a system when they exist simultaneously alongside the trivial solution, as in the case of the classical eigenvalue problem. The methods discussed provide accesss to new, somewhat exotic types of solutions that remain unattainable under the usual assumptions on existence and uniqueness. For numerical approximation, the trivial solution has an almost irresistible power of attraction that must be overcome. The usefulness of the numerical methods presented here is tested by means of some benchmark problems from the literature.

Chapters *six* and *seven* treat problems known from introductory mechanics such as planetary orbits, n-body problem, gyroscope theory, and framing. These chapters also place particular value on visualization that readers can comprehend.

Chapter *eight* presents the fundamentals of continuum theory for later use, allowing readers to understand how differential systems such as the POISSON equation and the NAVIER-STOKES equations evolved from conservation laws.

Chapter *nine* uses numerical methods to solve problems in continuum mechanics for solids and *incompressible* flow. At first, some general aspects of the finite element method are considered. Then its implementation by fundamental matrices and by shape functions is elaborated more thoroughly, and some special examples from the family of finite elements are presented in detail. For the numerical treatment of the NAVIER-STOKES equations, stream-function vorticity form with the most simple triangular elements is chosen for the most part; although it places high demands on the smoothness of the solution, it allows easier access to related problems such as convection and mass transport (and is also not as bad as its reputation would suggest, by the way). A selection of algorithm examples available on the Internet provides further explanations.

Everone lives with at least two sets of coordinate systems, the absolute coordinate system and one's own relative system; other systems are added constantly. Switching back and forth between coordinate systems has to be described mathematically, leading to tensor calculus and differential geometry. The *tenth* chapter introduces the disciplines, applying dual pairing of vector spaces.

The *eleventh* chapter includes, in addition to a model-like example from gas dynamics and three examples of multibody problems, a section on rolling discs and cogwheels with its numeric implementation on the corresponding Web page.

Many innovative impulses for Numerical Mathematics originate from the engineering fields. The author hopes that this book will raise interest in the numerical components of technical and physical problems and encourage readers to do some experimenting of their own.

The book is strictly divided into topics, and the individual parts are presented as separately from each other as possible to allow the book to serve as a reference volume and maintain clarity despite the multitude of material. It should more or less build a bridge from introductory studies to the requirements of advanced studies that paves the way to the "upper class" of the community.

The finite elements presented for the static problems are for the major part MATLAB versions of FORTRAN programs from the books by H.R. SCHWARZ. Their reliability has made them a great help over the course of developing and testing further elements, and they should also be made accessible to the MATLAB community in this way.

Warning: As the interest in experimenting grows, readers will observe that every numerical method can be undermined by a suitable example. For many problems such as non-convex optimization problems, convergence is not a matter of course and extreme caution is advised; even for bifurcation problems, iteration can quickly lead to non-realistic regions or go back to zero; spectacular accidents have been known to happen when finite elements are used in a wrong way. This is why results always need to be checked carefully.

Hint for using the MATLAB programs: make first the directory `AAMESH` permanent.

Reutlingen, July 2008 *Eckart W. Gekeler*

Contents

1

Mathematical Auxiliaries

This chapter is meant as a brief, independent compendium of the mathematical tools used in the following chapters, so that the reader may easily refer to them. It contains many well-known results from calculus and other results which are less familiar or lead a somewhat hidden life in literature. According to the intentions of this book, not only a rather comprehensive collection of formulas of vector and tensor analysis in $\mathbb{R}^n$ is contained but also linear differential equations of any "couleur" since they occupy a large area of the engineering sciences. Some general properties of vector fields are briefly studied and modern notations and concepts of functional analysis are introduced in compact form. Weak and strong *derivatives* play a fundamental role in technical applications from variational calculus to numerical approaches in continuum theory. *Convex* sets and functions are an equally important tool in any form of optimization. On the other side, numerical devices for solving linear systems of equations are almost entirely omitted here and in the next chapter because the user may consult the MATLAB suite here directly. A final section is dedicated to *quadratic forms* in HILBERT space deserving particular interest as indispensable preparation for the finite element method, which is widely used in today's technical applications.

1.1 Matrix Computations

Notations

Greek letters:	scalars or scalar fields
lowercase letters:	coefficients, points (position vectors)
underlined lowercase:	vectors and vector fields
capitals or bold-faced letters:	matrices or tensor fields

Points resp. their associated position vectors are *not* underlined when a distinction has to be made between points and vectors; cf. Chap. 10.

(a) Vector and Matrix Products

(a1) To conform with the notations in general tensor calculus — and with the view to MATLAB implementations — we distinguish clearly between **row vectors** and **column vectors**: For a (m,n)-matrix $A \in \mathbb{R}^m{}_n$ we write

$$A = \begin{bmatrix} \underline{a}^1 \\ \vdots \\ \underline{a}^m \end{bmatrix} = [\underline{a}_1, \ldots, \underline{a}_n] = [a^i{}_j]\,, \quad \text{i row index}\,,$$

with m rows $\underline{a}^i$ (index above) and n columns $\underline{a}_j$ (index below). By this way,

a *column* vector $\underline{a} \in \mathbb{R}^m{}_1 =: \mathbb{R}^m$ has the *"rows"* $a^i \in \mathbb{R}$ for elements
a *row* vector $\underline{a} \in \mathbb{R}^1{}_n =: \mathbb{R}_n$ has the *"columns"* $a_i \in \mathbb{R}$ for elements

.

(a2) The **scalar product** of two vectors $\underline{a}$, $\underline{b}$ is geometrically, i.e., without using a coordinate system, defined by

$$\underline{a} \cdot \underline{b} := |\underline{a}|\,|\underline{b}| \cos\varphi(\underline{a}, \underline{b}) \in \mathbb{R}\,, \tag{1.1}$$

where φ is the smaller of the two positive taken angles between $\underline{a}$ and $\underline{b}$; then $-|\underline{a}|\,|\underline{b}| \le \underline{a} \cdot \underline{b} \le |\underline{a}|\,|\underline{b}|$. For two vectors in the space of coordinates, $\underline{a}$, $\underline{b} \in \mathbb{R}^n$, we have $\underline{a} \cdot \underline{b} = \underline{a}^T \underline{b} = \sum_{i=1:n} a^i\, b^i \in \mathbb{R}$.

(a3) The **matrix product** of two matrices

$$A = [a^i{}_j] = \begin{bmatrix} \underline{a}^1 \\ \vdots \\ \underline{a}^m \end{bmatrix} \in \mathbb{R}^m{}_p \quad \text{and} \quad B = [b^j{}_k] = [\underline{b}_1, \ldots, \underline{b}_n] \in \mathbb{R}^p{}_n$$

is defined by

$$\boxed{C = [c^i{}_k] = A\,B \in \mathbb{R}^m{}_n\,, \quad c^i{}_k = \underline{a}^i\,\underline{b}_k := \sum_{j=1}^{p} a^i{}_j b^j{}_k}\,.$$

Note that the multiplication point is reserved for the *scalar* product and not used in any matrix product. The requirement of *compatibility*, namely the equality of the column number of the left factor with the row number of the right factor, must always be observed. The matrix product is *not commutative*; note in particular, for $\underline{a}$, $\underline{b} \in \mathbb{R}^3$, the difference between

$$\underline{a} \cdot \underline{b} = \underline{a}^T \underline{b} = \underline{b}^T \underline{a} \in \mathbb{R} \quad \text{and} \quad \underline{a}\,\underline{b}^T = \begin{bmatrix} a^1b^1 & a^1b^2 & a^1b^3 \\ a^2b^1 & a^2b^2 & a^2b^3 \\ a^3b^1 & a^3b^2 & a^3b^3 \end{bmatrix} \in \mathbb{R}^3{}_3$$

where the second special matrix product carries the name *dyadic product* or simply *dyade* because of its importance. A *quadratic* matrix A is *regular* if it has an inverse A^{-1}, $A\,A^{-1} = I$ (I unit matrix), otherwise it is *singular*. The inverse is uniquely determined in case of existence, in particular, *left inverse equals right inverse.*

(a4) The **trace** of a (quadratic) matrix is the sum of its diagonal elements (trace(AB) = trace(BA)). The **scalar product** or **tensor product** of two matrices of the same dimension is defined by

$$A : B = B : A := \text{trace}(A^T B) = \text{trace}(B^T A) = \sum_{i,k} a^i{}_k b^i{}_k \in \mathbb{R}\,, \tag{1.2}$$

cf. e.g., (1.21) for application; note that $A : B = A^T : B$ if B is symmetric.

(a5) The **vector space** $\mathbb{R}^n$ of n-tuples is called the *space of coordinates*. Its *canonical basis* is formed by the unit vectors $\underline{e}_i$, $i = 1 : n$, i.e., the columns of the unit matrix I. In general, a coordinate system in $\mathbb{R}^n$ is called *cartesian* (COS) if the vectors of its basis are normed to have length unity, are perpendicular to each other (both in the sense of the canonical scalar product), and form a *right oriented* system with a positive determinant.

Let $F = [\underline{f}_1, \ldots, \underline{f}_n]$ and $G = [\underline{g}_1, \ldots, \underline{g}_n]$ now be *arbitrary* regular matrices; then their respective columns also form bases of $\mathbb{R}^n$, and we have $\underline{g}_k = \sum_{i=1:n} \underline{f}_i d^i{}_k$, $k = 1 : n$, or $G = F\,D$ with a regular transformation matrix $D \in \mathbb{R}^n{}_n$. If both bases are cartesian, D is orthogonal with positive determinant. Of course every vector $\underline{v} \in \mathbb{R}^n$ can be written in both bases with different components:

$$\underline{v} = \sum_{k=1}^{n} \underline{g}_k y^k \equiv G\underline{y} = F\underline{x} = F\,D\,D^{-1}\underline{x} = G\,D^{-1}\underline{x} \implies \underline{y} = D^{-1}\underline{x}\,.$$

The contrary behavior of basis and components under transformation, namely $G = F\,D$ and $\underline{y} = D^{-1}\underline{x}$, is a fundamental aspect of general tensor calculus and justifies, in addition to other aspects, the different position of the indices. For a more general treatment of scalar product spaces we refer to Chap. 10.

(a6) The **vector product** or **cross product** $\underline{a} \times \underline{b}$ of two (column) vectors $\underline{a}\,,\,\underline{b} \in \mathbb{R}^3$ is a *vector* perpendicular to $\underline{a}$ and $\underline{b}$ with the geometric properties

$$\{\underline{a}\,,\,\underline{b}\,,\,\underline{a} \times \underline{b}\} \text{ form a right oriented system and } |\underline{a} \times \underline{b}| = |\underline{a}|\,|\underline{b}| \sin(\varphi)$$

where the angle φ has the same meaning as in (1.1). For a representation in coordinate space (with canonical basis $\{\underline{e}_i\}$), it follows that

$$\underline{a} \times \underline{b} = \begin{bmatrix} a^2b^3 - a^3b^2 \\ a^3b^1 - a^1b^3 \\ a^1b^2 - a^2b^1 \end{bmatrix} = \begin{vmatrix} \underline{e}_1 & a^1 & b^1 \\ \underline{e}_2 & a^2 & b^2 \\ \underline{e}_3 & a^3 & b^3 \end{vmatrix}\,.$$

The second rule is to be understood as memo rule with the formal determinant $|\ldots|$ and the canonical basis $\{\underline{e}_1, \underline{e}_2, \underline{e}_3\}$ in $\mathbb{R}^3$ (more suggestive than SARRUS' rule). The vector product is linear in each argument but skew-symmetric, i.e., $\underline{b} \times \underline{a} = -\underline{a} \times \underline{b}$; in particular, we have $\underline{a} \times \underline{a} = \underline{0}$ hence $(\underline{a} + \underline{b}) \times \underline{b} = \underline{a} \times \underline{b}$.

Furthermore, the *expansion theorem* (GRASSMANN *identity*)

$$\begin{aligned} \underline{a} \times (\underline{b} \times \underline{c}) &= (\underline{a} \cdot \underline{c})\, \underline{b} - (\underline{a} \cdot \underline{b})\, \underline{c} \\ (\underline{a} \times \underline{b}) \times \underline{c} &= (\underline{a} \cdot \underline{c})\, \underline{b} - (\underline{b} \cdot \underline{c})\, \underline{a} \end{aligned} \tag{1.3}$$

has to be mentioned and the sometimes useful formula

$$|\underline{a} \times \underline{b}|^2 = |\underline{a}|^2 |\underline{b}|^2 - (\underline{a} \cdot \underline{b})^2 .$$

Example 1.1. A force $\underline{k}$, attacking at a point P of a rigid body, yields the *moment (of force)* $\underline{m}_Q = \overrightarrow{QP} \times \underline{k}$ at each point Q of the body, and

$$\underline{m}_Q = \left(\overrightarrow{QP} + t\,\underline{k}\right) \times \underline{k} = \overrightarrow{QP} \times \underline{k} + t\,\underline{k} \times \underline{k} = \overrightarrow{QP} \times \underline{k}$$

for all $t \in \mathbb{R}$. The vector (of force) $\underline{k}$ is sometimes called *aligned* in this context because it may be shifted arbitrarily along its "action line".

If we operate in the plane, the third components of $\underline{k}$ and of $\overrightarrow{QP}$ are zero; hence the first both components of $\underline{m}_Q$ are likewise zero. In this case the remaining nontrivial third component of $\underline{m}_Q$,

$$\underline{e}_3 \cdot \underline{m}_Q = [\underline{m}_Q]_3 = [\overrightarrow{QP}]^1\, k^2 - [\overrightarrow{QP}]^2\, k^1 \in \mathbb{R} ,$$

is frequently called moment of the force $\underline{k}$ with a slight abuse of notation.

Example 1.2. For two vectors $\underline{a}$, $\underline{b} \in \mathbb{R}^n$, the vector

$$\underline{a}_{\underline{b}} = \frac{\underline{a} \cdot \underline{b}}{\underline{b} \cdot \underline{b}}\, \underline{b}$$

is the *projection* of $\underline{a}$ onto $\underline{b}$ (without using the angle φ explicitly), and $\underline{a} = \underline{a}_{\underline{b}} + (\underline{a} - \underline{a}_{\underline{b}}) =: \underline{a}_{\underline{b}} + \underline{a}_{\underline{n}}$ is the *orthogonal decomposition* of $\underline{a}$ into direction of $\underline{b}$ (called HUYGENS' decomposition in the case of the acceleration vector). By (1.3) we have $\underline{b} \times (\underline{a} \times \underline{b}) = (\underline{b} \cdot \underline{b})\underline{a} - (\underline{a} \cdot \underline{b})\underline{b}$; hence

$$\underline{a}_n = \underline{a} - \underline{a}_{\underline{b}} = \frac{1}{\underline{b} \cdot \underline{b}}\, \underline{b} \times (\underline{a} \times \underline{b}) .$$

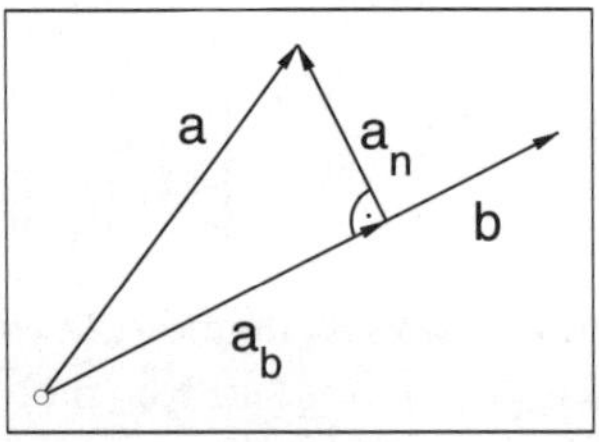

Figure 1.1. Projection of $\underline{a}$ onto $\underline{b}$

(b) Determinants and Cofactors Let $A = [a^i{}_k] = [\underline{a}_1, \underline{a}_2, \underline{a}_3] \in \mathbb{R}^3{}_3$ be a matrix with columns $\underline{a}_i$ then

$$\det(A) = <\underline{a}_1, \underline{a}_2, \underline{a}_3> = \underline{a}_1 \cdot (\underline{a}_2 \times \underline{a}_3) = (\underline{a}_1 \times \underline{a}_2) \cdot \underline{a}_3$$

is the *determinant* of A and $|\det(A)|$ is the volume of the *parallelepiped* (*spat*) spanned by the vectors $\underline{a}_i$, $i = 1:3$. More generally, let $A = [a^i{}_k] \in \mathbb{R}^n{}_n$ be a quadratic matrix of order n and let A'_{ik} be the matrix of dimension $n-1$ obtained by deleting the i-th row and k-th column of A. Then cof $A = [d^i{}_k] \in \mathbb{R}^n{}_n$ with the components

$$d^i{}_k = (-1)^{i+k} \det(A'_{ik})$$

is called the *cofactor matrix* of A. By CRAMER's rule we obtain the *explicit* representation of the inverse of a regular matrix A:

$$A^{-1} = \frac{1}{\det(A)}\,[\text{cof } A]^T\,;$$

hence cof $A = \det(A)A^{-T}$ with the notation $A^{-T} = [A^{-1}]^T = [A^T]^{-1}$. In particular,

$$\boxed{\begin{bmatrix} a & b \\ c & d \end{bmatrix}^{-1} = \frac{1}{ad-bc}\begin{bmatrix} d & -b \\ -c & a \end{bmatrix}}\,,$$

and in the likewise important case $n = 3$,

$$d^i{}_k = a^{i+1}{}_{k+1}a^{i+2}{}_{k+2} - a^{i+1}{}_{k+2}a^{i+2}{}_{k+1}\,, \quad i,k = 1:3$$

(counting indices modulo 3, no summation).

Rules for computation (cf. (1.2)):

$$\begin{aligned} \det(A) &= \det(A^T)\,, \ \det(A\,B) = \det(A)\det(B)\,, \\ \text{cof}(A^T) &= \text{cof}(A)^T, \ \text{cof}(A\,B) = \text{cof}(A)\,\text{cof}(B)\,, \end{aligned}$$

$$\boxed{\frac{d}{dt}\det\, A(t) = \left[\frac{d}{dt}A(t)\right] : \text{cof } A(t)}\,. \tag{1.4}$$

To prove the last rule for $A(t) \in \mathbb{R}^3{}_3$, let $B = [\underline{b}_1, \underline{b}_2, \underline{b}_3] = \text{cof } A(t)$ then

$$\begin{aligned} \frac{d}{dt}\det(A(t)) &= \det[\dot{\underline{a}}_1, \underline{a}_2, \underline{a}_3] + \det[\underline{a}_1, \dot{\underline{a}}_2, \underline{a}_3] + \det[\underline{a}_1, \underline{a}_2, \dot{\underline{a}}_3] \\ &= \dot{\underline{a}}_1^T\underline{b}_1 = \dot{\underline{a}}_2^T\underline{b}_2 + \dot{\underline{a}}_3^T\underline{b}_3 = \text{trace}(\dot{A}^T B)\,. \end{aligned}$$

(c) Eigenvalues and Eigenvectors

(c1) A pair $(\lambda, \underline{a})$ with $\underline{a} \neq \underline{0}$ and $A\underline{a} = \lambda\underline{a}$ is called a *characteristic pair* of the (quadratic) matrix A with *eigenvalue* λ and *eigenvector* $\underline{a}$, the null vector *never* being an eigenvector. The eigenvalues of $A \in \mathbb{R}^n{}_n$ are exactly the n roots of the *characteristic polynomial* $p(\lambda) := \det(A - \lambda I)$ (counting multiplicities). Because of $\det(Q\,A\,Q^{-1}) = \det(A)$ for every regular matrix Q, the coefficients of $p(\lambda)$ and thus also the eigenvalues of a matrix are invariant under transformations of the coordinate system; therefore the components are sometimes called *principal invariants* of A. Using cofactor and determinant we can write them explicitly for $A \in \mathbb{R}^3{}_3$:

$$\det(A - \lambda I) = -\lambda^3 + p_1\lambda^2 - p_2\lambda + p_3\,,$$

$$\begin{aligned} p_1 &= \operatorname{trace}(A) &&= \lambda_1 + \lambda_2 + \lambda_3\,,\\ p_2 &= \operatorname{trace}\operatorname{cof}(A) &&= [(\operatorname{trace} A)^2 - \operatorname{trace}(A^2)]/2 = \lambda_1\lambda_2 + \lambda_1\lambda_3 + \lambda_2\lambda_3\,,\\ p_3 &= \det(A) &&= [(\operatorname{trace} A)^3 - 3\operatorname{trace}(A)\operatorname{trace}(A^2) + 2\operatorname{trace}(A^3)]/6\\ &= \lambda_1\lambda_2\lambda_3\,. \end{aligned}$$

(c2) A matrix $A \in \mathbb{R}^n{}_n$ is *diagonalizable* if it can be changed into diagonal form by a similarity transformation, i.e., if there exists a regular matrix U such that $U^{-1}AU = \Lambda$ is a diagonal matrix. This implies, on the one hand, that there exists a (not necessarily cartesian) system of coordinates in which all off-diagonal elements of A disappear, and on the other side that the matrix $A \in \mathbb{R}^n{}_n$ has n linear independent *column* eigenvectors, namely, the columns of U. The rows of U^{-1} are *row* or left-eigenvectors, and the respective eigenvalues are the diagonal elements of Λ. The (canonical) decomposition

$$\boxed{A = U\Lambda V\,,\quad V = U^{-1}\,,\quad V\,A\,U = \Lambda}$$

is called *Jordan decomposition* of the matrix A. In other words, a diagonalizable matrix can be written as a sum of dyades, $A = \sum_{i=1:n}\lambda_i\,\underline{u}_i\,\underline{v}^i$, with eigenvalues λ_i, column (or right-)eigenvectors $\underline{u}_i$ and row (or left-)eigenvectors $\underline{v}^i$ of A, yet both have to satisfy $\underline{v}^i\,\underline{u}_i = 1\,,\ i = 1:n$. Exactly *normal* matrices with the property $A^TA = AA^T$ are diagonalizable with *orthogonal* eigenvectors, in particular symmetric matrices (Stoer).

(c3) If a matrix $A \in \mathbb{R}^n{}_n$ is *not* diagonalizable, there exist less than n linearly independent eigenvectors, and they have to be completed to a full basis of $\mathbb{R}^n$ with *generalized eigenvectors* or *principal vectors*. These principal vectors $\underline{u}_k$ with the property

$$(A - \lambda I)^k\underline{u}_k = \underline{0}\,,\quad (A - \lambda I)^{k-1}\underline{u}_k \neq \underline{0}\,,\ k = 1,\ldots,\quad \underline{u}_1 = \underline{u}\,,$$

have to be determined for every characteristic pair $(\lambda, \underline{u})$. They are obtained by solving successively the linear systems of equations

$$(A - \lambda I)\underline{u}_{k+1} = \underline{u}_k,\ k = 1,\ldots,\quad \underline{u}_1 = \underline{u}\,,$$

(I unit matrix), the leading matrix $A - \lambda I$ being singular. Therefore the elements of such a *chain of principal vectors* are never determined uniquely and the chain breaks down if the system becomes unsolvable. Now the JORDAN decomposition may be written in the form

$$\boxed{A = U(\Lambda + T)V\,, \quad V = U^{-1}} \tag{1.5}$$

with a matrix T having zeros everywhere apart from the first upper diagonal where the elements are zero or one. Consequently the matrix $\Lambda + T$ is now a block-diagonal matrix with so-called JORDAN *blocks* in the main diagonal, every single block corresponding to a chain of principal vectors. If no chain of principal vectors exists for a characteristic pair, the associated JORDAN block consists of the scalar eigenvalue λ only. Because every eigenvector generates a JORDAN block, possibly of dimension one only, the number of JORDAN blocks equals the number of linearly independent eigenvectors appertaining to the eigenvalue λ. This number is called the *geometric multiplicity* of λ in contrast to the algebraic multiplicity, namely the multiplicity of λ as a root of the characteristic polynomial. Besides, the JORDAN decomposition is uniquely determined up to *permutations* of columns of U and corresponding rows of V so that *different* eigenvalues in Λ can be ordered in an arbitrary way.

Example 1.3. We suppose that the matrix $A \in \mathbb{R}^5{}_5$ has the double eigenvalue λ and the triple eigenvalue μ and that both eigenvalues possess only one eigenvector. Then there exists one chain of principal vectors of length two for λ and one chain of length three for μ. Accordingly, the JORDAN decomposition of A has up to permutations the form

$$A = [\underline{u}_1, \underline{u}_2, \underline{v}_1, \underline{v}_2, \underline{v}_3] \begin{bmatrix} \lambda & 1 & 0 & 0 & 0 \\ 0 & \lambda & 0 & 0 & 0 \\ 0 & 0 & \mu & 1 & 0 \\ 0 & 0 & 0 & \mu & 1 \\ 0 & 0 & 0 & 0 & \mu \end{bmatrix} \begin{bmatrix} \underline{u}^2 \\ \underline{u}^1 \\ \underline{v}^3 \\ \underline{v}^2 \\ \underline{v}^1 \end{bmatrix}, \quad \underline{u}_i\,,\ \underline{v}_k \text{ columns}\,, \quad \underline{u}^i\,,\ \underline{v}^k \text{ rows}$$

where

$$\begin{array}{lll} \{\underline{u}_1, \underline{u}_2\} & \text{column chain for} & (\lambda, \underline{u}_1) \\ \{\underline{v}_1, \underline{v}_2, \underline{v}_3\} & \text{column chain for} & (\mu, \underline{v}_1) \\ \{\underline{u}^1, \underline{u}^2\} & \text{row chain for} & (\lambda, \underline{u}^1) \\ \{\underline{v}^1, \underline{v}^2, \underline{v}^3\} & \text{row chain for} & (\mu, \underline{v}^1)\,. \end{array}$$

The *nilpotent* matrix T satisfies $T^n = 0 \in \mathbb{R}^n{}_n$ and has in this example the form

$$T = \begin{bmatrix} T_1 & O \\ O & T_2 \end{bmatrix}, \ T_1 = \begin{bmatrix} 0 & 1 \\ 0 & 0 \end{bmatrix}, \ T_1^2 = 0 \in \mathbb{R}^2{}_2\,, \ T_2 = \begin{bmatrix} 0 & 1 & 0 \\ 0 & 0 & 1 \\ 0 & 0 & 0 \end{bmatrix}, \ T_2^3 = 0 \in \mathbb{R}^3{}_3\,.$$

If $\dot{\underline{x}} = A\underline{x}$ is now a differential system with the above matrix A, then a *fundamental system* of solutions is found immediately:

$$\underline{x}_1(t) = \underline{u}_1 e^{\lambda t}\,,\ \underline{x}_2(t) = e^{\lambda t}(\underline{u}_2 + t\,\underline{u}_1)$$

$$\underline{x}_3(t) = \underline{v}_1 e^{\mu t}\,,\ \underline{x}_4(t) = e^{\mu t}(\underline{v}_2 + t\,\underline{v}_1)\,,\ \ \underline{x}_5(t) = e^{\mu t}\left(\underline{v}_3 + t\,\underline{v}_2 + \frac{t^2}{2}\,\underline{v}_1\right)\,;$$

cf. Sect. 1.5**(b)**.

(c4) By means of the JORDAN decomposition it can be shown that the eigenvalues of a matrix A essentially govern the convergence $\lim_{n\to\infty} A^n$ of the important sequence $\{A^n\}_{n=1}^\infty$. To this end we introduce some

Further Notations:

(1°) An arbitrary vector norm $\|\circ\|$ generates the *operator norm* $\|A\| = \max\{\|Ax\|\,;\ \|x\|=1\}$; other matrix norms are not necessarily submultiplicative, $\|A\,B\| \le \|A\|\|B\|$.

(2°) The set $\sigma(A)$ of eigenvalues of a quadratic matrix A is called the *spectrum* of (A) .

(3°) The radius $\varrho(A)$ of the smallest circle with the center at the origin containing all eigenvalues of A is called the *spectral radius* of A .

(4°) An eigenvalue is called *semi-simple* if it has only eigenvectors and no principal vectors. Then all *associated* JORDAN blocks consist of only one element.

(5°) A quadratic matrix A is called an *M-matrix* if all eigenvalues λ with a maximum absolute value $|\lambda| = \varrho(A)$ are semi-simple. (This notion is, however, also used in a different context; cf. (Ortega), Def. 2.4.7.)

Theorem 1.1. *Let A be a quadratic matrix. Then*

(1°) $\varrho(A) \le \|A\|$.

(2°) *For every $\varepsilon > 0$ there exists an operator norm $\|\circ\|_{A,\varepsilon}$ depending on A and ε such that*

$$\varrho(A) \le \|A\|_{A,\varepsilon} \le \varrho(A) + \varepsilon\,.$$

(3°) $\lim_{n\to\infty} \|A\|^n = 0 \iff \varrho(A) < 1$.

(4°) $\exists\, K > 0\ \ \forall\, n \in \mathbb{N} : \|A\|^n \le K \iff \varrho(A) < 1$ *or* ($\varrho(A) = 1$ *and A is a M-matrix).*

For the proof cf., e.g., (Stoer), Sect. 6.9.

Because of $\|A^n\| \le \|A\|^n$, the sequence $\underline{x}^{n+1} = A\underline{x}^n = A^{n+1}\underline{x}^0$, $n = 0, 1, \ldots$, converges to the null vector for *arbitrary* $\underline{x}^0$ if the first condition in (4°) is fulfilled, and it always remains bounded if the second condition in (4°) is fulfilled.

Solving linear systems of equations $A\underline{x} = \underline{b}$ with a regular matrix A has always been and is up to this day one of the main subjects dealt with in numerical mathematics and there is a grat number of communications and monographs on this field. The reason for these permanent activities lies in the fact that "regularity" is an algebraic property of a matrix which cannot be transposed simply to numerical computations. Let, e.g., **eps** be the smallest number that $1 + \textsf{eps} > 1$ on the computer; then the MATLAB matrix [1, 0; 0, eps] is regular in the algebraic sense but the use of its inverse is rather daring. Instead, the *condition* of a matrix A is revealed to be the crucial criterium for an acceptable result in numerical computations — and it can be arbitrarily bad in dependence of the desired accuracy of the underlying analytical problem. Usually it is defined by the ratio of the largest absolute value and the smallest absolute value of all eigenvalues of that matrix A. Every algorithm for solving linear systems breaks down for a properly chosen counterexample. In ill-conditioned systems, direct methods such as the GAUSS algorithm and its related, on the on hand, and iterative methods, on the other hand, have been standing forever in noble contest with each other. Fortunately, the direct methods presented by MATLAB have a high degree of perfection and yield respectable results over a large range. If they break down one has to switch to iterative methods, some of which are offered by MATLAB, too. These latter methods are frequently of the form

$$\underline{x}^{n+1} = \underline{x}^n - C(A\underline{x}^n - \underline{b}), \quad n = 0, 1, \ldots,$$

and have the advantage that, in case of convergence, arising rounding errors etc. are ruled out again during computation. By Theorem 1.1, the method is convergent to the solution $A^{-1}\underline{b}$ (for arbitrary initial vector) if and only if the spectral radius of the matrix $I - C\,A$ is less than one. Essentially, the individual iterative methods differ from each other by the choice of the *preconditioner* C or a sequence of matrices C in order to satisfy this convergence criterium under more or less vague knowledge of the leading matrix A.

(d) Decompositions of a Matrix For an arbitrary, sufficiently smooth vector field we have by TAYLOR

$$\underline{w}(x + \underline{h}) = \underline{w}(x) + \operatorname{grad} \underline{w}(x)\underline{h} + o(|\underline{h}|);$$

hence the gradient of $\underline{w}$ describes the local or *infinitisimal variation* of $\underline{w}$. Three decompositions of $A := \operatorname{grad} \underline{w}(x) =: \nabla \underline{w}(x) \in \mathbb{R}^3{}_3$ play a particular role in the mechanics of continua.

(d1) In the decomposition into a symmetric and a skew-symmetric component,

$$A = D + S, \quad D = \frac{1}{2}[\nabla \underline{w}(x) + \nabla \underline{w}(x)^T], \quad S = \frac{1}{2}[\nabla \underline{w}(x) - \nabla \underline{w}(x)^T],$$

the symmetric matrix D describes the local dilatation, or rather, compression of the vector field $\underline{w}$ in the direction of the principal axes of D. On the other hand, the skew-symmetric matrix S describes the local rotation of $\underline{w}$ because

$$\operatorname{rot} \underline{w}(x) \times (y - x) = [\nabla \underline{w}(x) - \nabla \underline{w}(x)^T](y - x).$$

(d2) For $A = I + \nabla \underline{v}(x) \in \mathbb{R}^3{}_3$ an expansion of $\det(A)$ yields

$$\det(A) = 1 + \operatorname{trace}(A) + \text{ h.o.t.}$$

(h.o.t. = higher order terms = terms of higher order in the components of A).

Therefore, in the decomposition

$$A = A' + \frac{1}{3}\operatorname{trace}(A)I\,,\quad A' = A - \frac{1}{3}\operatorname{trace}(A)I\,,\quad I \in \mathbb{R}^3{}_3 \text{ unit matrix,}$$

($\operatorname{trace}(A') = 0$) the matrix A' describes the local variation of the *shape* of A and $\operatorname{trace}(A)I/3$ the local variation of the *volume* of A.

(d3) Besides the above additive decompositions of a quadratic matrix — and triangular factorizations not mentioned here — there is a further factorization originally due to FINGER 1892, see also (Ciarlet93), that plays a crucial role in the characterization of material frame-indifference in stress tensors.

Lemma 1.1. *(Polar Factorization) Let A be a real and regular matrix then there exist uniquely orthogonal matrices P, Q and unique symmetric positive definite matrices U, V so that*

$$A = PU = VQ\,.$$

It can be shown that $\det(P) = 1$ for $\det(A) > 0$. Then the result says that a regular matrix A can always be considered as a dilatation followed by a rotation or vice versa.

(e) Linear Systems of Equations

(e1) Let $A \in \mathbb{R}^m{}_n$ be a matrix of *arbitrary* dimension, let

Range $A = \{\underline{y} \in \mathbb{R}^m,\ \exists\, \underline{x} \in \mathbb{R}^n : \underline{y} = A\underline{x}\} \in \mathbb{R}^m$ the *range or image* of A,
Ker $A = \{\underline{x} \in \mathbb{R}^n,\ A\underline{x} = 0\} \in \mathbb{R}^n$ the *null space* or *kernel* of A,

(both are vector spaces), and let

$$\mathcal{U}^\perp := \{\underline{v} \in \mathbb{R}^n,\ \forall\, \underline{u} \in \mathcal{U} : \underline{v}^T\underline{u} = 0\}$$

be the *orthogonal complement* of a set $\mathcal{U} \subset \mathbb{R}^n$; $\mathcal{U}^\perp$ is a vector space if $\mathcal{U}$ is a vector space. Obviously, a linear system $A\underline{x} = \underline{b}$ is solvable if and only if the right side $\underline{b}$ is contained in the range of A. The set of all these right sides is characterized by the following important *Range Theorem.*

Theorem 1.2. *Let $A \in \mathbb{R}^m{}_n$ be an arbitrary matrix, then*

$$\boxed{[\operatorname{Range} A]^\perp = \operatorname{Ker} A^T}\,.$$

Proof.

$$\underline{0} \neq \underline{y} \in [\operatorname{Range} A]^\perp \iff \underline{y}^T A\underline{x} = 0\ \forall\, \underline{x} \in \mathbb{R}^n$$

$$\iff \underline{y}^T A = \underline{0} \iff A^T\underline{y} = \underline{0} \iff \underline{y} \in \operatorname{Ker} A^T.$$

□

Formulating the result for the orthogonal complement immediately yields

$$\operatorname{Range} A = \left[\operatorname{Ker} A^T\right]^\perp.$$

In other words, the range of a matrix is the orthogonal complement of the null space of the transposed matrix; see also generalizations in Sect. 1.9**(a)**. Some fundamental inferences follow at once:

Corollary 1.1. (1°)

$$\operatorname{Range} A = \mathbb{R}^m \Longleftrightarrow \operatorname{Ker} A^T = \{\underline{0}\} \in \mathbb{R}^m \,. \tag{1.6}$$

(2°) *For quadratic* A, *the system* $A\underline{x} = \underline{b}$ *has by (1.6) a unique solution if and only if the system* $A\underline{x} = \underline{0}$ *has only the trivial solution (*FREDHOLM's alternative*).*

(3°)

$$\dim \operatorname{Ker} A + \dim \operatorname{Range} A = n \,. \tag{1.7}$$

(4°) *For quadratic* A,

$$\operatorname{Ker} A \cap \operatorname{Range} A = \{\underline{0}\} \Longleftrightarrow A \text{ diagonalizable}\,. \tag{1.8}$$

Proof of (3°). It follows from Theorem 1.2 that $[\operatorname{Range} A^T]^\perp = \operatorname{Ker} A$ hence

$$\dim \operatorname{Ker} A = \dim[\operatorname{Range} A^T]^\perp = \operatorname{codim} \operatorname{Range} A^T = n - \dim \operatorname{Range} A^T \,.$$

But as row rank equals column rank in a matrix, we have $\dim \operatorname{Range} A^T = \dim \operatorname{Range} A$. □

(e2) In optimization and mechanics of continua one frequently encounters linear systems of equations with a LAGRANGE matrix

$$L = \begin{bmatrix} A & B^T \\ B & 0 \end{bmatrix}, \tag{1.9}$$

which is often indefinite — also if A is symmetric positive definite.

Lemma 1.2. *Let* $A \in \mathbb{R}^n{}_n$ *be symmetric positive semi-definite. The* LAGRANGE *matrix (1.9) is regular if and only if* $B \in \mathbb{R}^m{}_n$ *satisfies the following three conditions:*

(1°) $m \leq n$;

(2°) $\operatorname{Rank}(B) = m$, *i.e.* B *has maximum rank;*

(3°) A *is positive definite on the kernel of* B, *i.e.*

$$B\underline{x} = \underline{0}\,,\ \underline{x} \neq \underline{0} \Longrightarrow \underline{x}^T A \underline{x} > 0\,.$$

Proof. We show that $L\underline{z} \neq \underline{0}$ for $\underline{z} \neq \underline{0}$. Necessity of (1°) and (2°) follows in a simple way. Let now $\underline{z} = [\underline{x}, \underline{y}]^T$ then $L\underline{z} \neq \underline{0}$ if $B\underline{x} \neq \underline{0}$. By $B\underline{x} = \underline{0}$, $\underline{x} \neq \underline{0}$ and $L\underline{z} = \underline{0}$, it follows that

$$A\underline{x} + B^T\underline{y} = \underline{0} \Longrightarrow \underline{x}^T A \underline{x} + \underline{x}^T B^T \underline{y} = 0\,,$$

yielding $\underline{x}^T B \underline{y} < 0$ because $\underline{x}^T A \underline{x} > 0$. This is a contradiction of the assumption that $B\underline{x} = \underline{0}$ and $\underline{x} \neq \underline{0}$. The same argumentation shows that (3°) is necessary. □

(e3) Sometimes, as for instance in the complete cubic element of Sect. 9.4**(a)**, a linear system is to be reduced by eliminating some components of the solution. This procedure is called *condensation*. Consider the system

$$\begin{bmatrix} A & B \\ C & D \end{bmatrix} \begin{bmatrix} \underline{x} \\ \underline{y} \end{bmatrix} = \begin{bmatrix} \underline{f} \\ \underline{g} \end{bmatrix}$$

and suppose that D is regular. Then we may insert the solution $\underline{y} = D^{-1}[\underline{g} - C\underline{x}$ of the second row into the first row and obtain a smaller system where, however, $\underline{y}$ has been cancelled out completely:

$$[A - BD^{-1}C]\underline{x} = \underline{f} - BD^{-1}\underline{g}\,.$$

(f) Projectors and Reflectors

(f1) Projectors For $\underline{a}\,,\ \underline{b} \in \mathbb{R}^n$, the projection of $\underline{a}$ onto $\underline{b}$ can be written as a matrix-vector product,

$$\underline{a}_b = \frac{\underline{a}\cdot\underline{b}}{\underline{b}\cdot\underline{b}}\,\underline{b} = \frac{\underline{a}^T\underline{b}}{\underline{b}^T\underline{b}}\,\underline{b} = \frac{\underline{b}\,\underline{b}^T}{\underline{b}^T\underline{b}}\,\underline{a} =: P\underline{a}\,;$$

cf. **(a)**, Example 1.2. Let more generally $\mathcal{U} \subset \mathbb{R}^n$ be a subspace with basis $\{\underline{b}_1, \ldots, \underline{b}_m\}$ and let $B = [\underline{b}_1, \ldots, \underline{b}_m] \in \mathbb{R}^n{}_m$ be the matrix with columns $\underline{b}_i$ then B has maximum rank. The matrix

$$P = B(B^TB)^{-1}B^T$$

projects every vector $\underline{x} \in \mathbb{R}^n$ onto $\mathcal{U}$ and is, moreover, symmetric and idempotent, $P^2 = P$. Every matrix with these two properties is called a *projector.*

(f2) Reflectors Let $\underline{0} \neq \underline{u} \in \mathbb{R}^n$, let $\mathcal{H} := \{\underline{x} \in \mathbb{R}^n,\ \underline{u}^T\underline{x} = 0\}$ be a hyperplane through the origin and P a point with position vector $\underline{p}$. By subtracting the projection $\underline{p}_u$ of $\underline{p}$ onto $\underline{u}$ from $\underline{p}$ twice , the point P is reflected at the plane $\mathcal{H}$ into the point Q with position vector $\underline{q}$:

$$\underline{q} = \underline{p} - 2\underline{p}_u = \left(I - 2\frac{\underline{u}\,\underline{u}^T}{\underline{u}^T\underline{u}}\right)\underline{p} =: S\underline{p}\,;$$

cf. Figure 1.2. The reflection matrix S is also called a *reflector.* On the other hand, if $\underline{p}$ and $\underline{q}$ are given and the reflection plane H is sought then we have to set $\underline{u} = \pm(\underline{p} - \underline{q})$; the sign is cancelled in the reflection matrix.

(f3) The QR algorithm Let $A = [\underline{a}_1, \ldots, \underline{a}_m] \in \mathbb{R}^n{}_m$ be a matrix which is to be factorized into $A = Q\,R$ with orthogonal matrix Q and upper triangular R. In the first step of the QR algorithm we choose $\underline{p} = \underline{a}_1$ and $\underline{q} = \pm|\underline{a}_1|\underline{e}_1$, with $\underline{e}_1$ being the first unit vector, and form again $A^{(1)} = S\,A$. For reasons of numerical stability the sign of $\underline{q}$ is chosen such that no subtraction appears in $\underline{u} = \pm(\underline{p} - \underline{q})$. The same operation is then repeated on the remaining matrices (of lower dimension) until complete factorization is attained.

A QR decomposition can be obtained also by means of simple rotations preserving the orientation of the columns in A. The MATLAB algorithm `[Q,R] = qr(A)`, however, uses the above reflection matrices S with $\det(S) = -1$

$$\det(A) = \det(Q) \cdot \det(R) = \pm \det(R)$$

which has to be observed, e.g., in *continuation methods*; cf. Sect. 5.8.

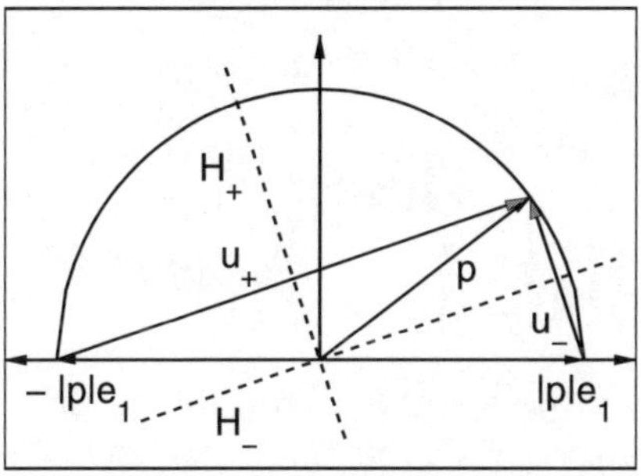

Figure 1.2. Reflection

(g) The Moore-Penrose Inverse (Pseudo-inverse) For an arbitrary matrix $A \in \mathbb{R}^m{}_n$, the (non-negative) *roots* of the eigenvalues of the semi-definite matrix $A^TA \in \mathbb{R}^n{}_n$ are called *singular values* of A .

Theorem 1.3. *For every matrix $A \in \mathbb{R}^m{}_n$ there exist orthogonal matrices $U \in \mathbb{R}^m{}_m$ and $V \in \mathbb{R}^n{}_n$ such that*

$$U^TAV = S \in \mathbb{R}^m{}_n \tag{1.10}$$

is a diagonal matrix. The diagonal of S contains the singular values $\sigma_1, \ldots, \sigma_n$ of A .

Proof. Without loss of generality let $m \geq n$; otherwise consider A^T instead of A . Let $V^* \in \mathbb{R}^n{}_n$ be an orthogonal matrix transforming A^TA into diagonal form and let $AV^* = QR$ be a QR decomposition of AV^* . Then one can choose $V = V^*$ and $U = Q$ in (1.10). For the proof consider the identity $U^TAV = Q^TAV^* = R$. By definition of V we have

$$V^TA^TAV = V^TA^TUU^TAV = R^TR = \mathrm{diag}(\sigma_1^2, \ldots, \sigma_n^2)\,,$$

hence R must be a diagonal matrix and thus $R = S$. □

If A has rank $r \leq \min\{m, n\}$ then, without loss of generality, S is a diagonal matrix of the form

$$S = \begin{bmatrix} D & 0 \\ 0 & 0 \end{bmatrix} \in \mathbb{R}^m{}_n$$

where either D empty or $D = [\sigma_1, \ldots, \sigma_p]$ is a *regular* diagonal matrix with $p \leq n$. Define

$$S^+ := \begin{bmatrix} D^{-1} & 0 \\ 0 & 0 \end{bmatrix} \in \mathbb{R}^n{}_m$$

then the matrix $A^+ = VS^+U^H \in \mathbb{R}^n{}_m$ is called the MOORE-PENROSE *inverse* of A, being uniquely determined up to representation.

Lemma 1.3. *(Properties) For* $A \in \mathbb{R}^m{}_n$

$$\begin{aligned} A^{++} &= A, & (A^+)^T &= (A^T)^+, \\ A^+A &= (A^+A)^T, & AA^+ &= (AA^+)^T, \\ AA^+A &= A, & A^+AA^+ &= A^+, \\ (A^TA)^+A^T &= A^+. \end{aligned}$$

$P := AA^+$ *is projector of* $\mathbb{R}^m$ *onto* $\mathrm{Range}(A)$,
$Q := A^+A$ *is projector of* $\mathbb{R}^n$ *onto* $\mathrm{Ker}(A)^\perp = \mathrm{Range}(A^T)$.

Lemma 1.4. *Letting* $\|\underline{x}\|^2 = \underline{x}^T\underline{x}$, *the vector* $\underline{x}^* = A^+\underline{b}$ *satisfies*
(1°) $\forall\, \underline{x} \in \mathbb{R}^n : \|A\underline{x}^* - \underline{b}\| \le \|A\underline{x} - \underline{b}\|$,
(2°) $\|A\underline{x}^* - \underline{b}\| = \|A\underline{x} - \underline{b}\| \implies \|\underline{x}^*\| \le \|\underline{x}\|$.

Proof (Stoer), Sect. 4.8.5 and `SUPPLEMENT\chap01a`.

In other words, $\underline{x}^* = A^+\underline{b}$ is the solution to the problem $\|A\underline{x} - \underline{b}\| =$ min! with *minimum norm property* (2°); MATLAB computes this solution by `x = pinv(A)*b`; the solution is unique for $m \ge n$, A having maximum rank.

(h) Over- and Underdetermined Systems Look for a "solution" of $A\underline{x} = \underline{b}$ where $A \in \mathbb{R}^m{}_n$ is a matrix of arbitrary dimension.

(h1) If A is quadratic and singular, the MATLAB algorithms `x = A\b` and

```
[Q,R] = qr(A); y = Q'*b; x = R\y
```

do *not* provide a solution and an estimation of the condition of A is put out.

(h2) If A is not quadratic, both algorithms supply the same solution of $\|A\underline{x} - \underline{b}\| =$ min!, also in the case that the rank of A is not maximum. But, in the normal case, the solution does not have the minimum norm property of Lemma 1.4.

(h3) In continuation methods and in the solution of periodic boundary value problems, the QR decomposition of a maximum rank matrix $A \in \mathbb{R}^n{}_{n+1}$ is of particular interest, but the application of `x = pinv(A)*b` is mostly too laborious. The following MATLAB algorithm uses the QR decomposition of A without storage of the matrix Q being fully occupied in general; cf. (Hoellig), Sect. 3.2.

```
function x = mpsolv(A,b);
% computes solution x = mpinv(A)*b
% for (n,n+1)-matrix A of maximum rank
[m,n] = size(A);
if n ~= m+1 disp('A no (n,n+1)-matrix'), return
end
R = triu(qr([A,b]));
C = R(1:m,m+2); T = R(1:m,m+1); S = R(1:m,1:m);
G = [-T,C]; U = S\G;
V = -[U(:,1);1]\[U(:,2);0];
x = [U(:,1);1]*V + [U(:,2);0];
```

(i) Rotations in $\mathbb{R}^3$. We consider the rotation of a point $\underline{x}$ about a normed rotation axis $\underline{a}$, $|\underline{a}| = 1$, (both are position vectors with common tail in origin). Recall that $\underline{x}_a = (\underline{x} \cdot \underline{a})\underline{a} = [\underline{a}\,\underline{a}^T]\underline{x}$ is the projection of $\underline{x}$ onto $\underline{a}$ and that the axis $\underline{a}$ is the only vector unchanged under rotation. Then $\underline{x}_a$, $\underline{x} - \underline{x}_a$ and $\underline{a} \times \underline{x}$ form a local cartesian coordinate system and rotation about the angle φ has the form

$$\begin{aligned} \underline{y} &= \underline{x}_a + \cos(\varphi)(\underline{x} - \underline{x}_a) + \sin(\varphi)(\underline{a} \times \underline{x}) \\ &= \cos(\varphi)\underline{x} + (1 - \cos(\varphi))\underline{a}\,\underline{a}^T\underline{x} + \sin(\varphi)(\underline{a} \times \underline{x}) . \end{aligned} \tag{1.11}$$

Now, introduce the skew-symmetric matrix C,

$$C_a\underline{x} = \underline{a} \times \underline{x} = \begin{bmatrix} 0 & -a^3 & a^2 \\ a^3 & 0 & -a^1 \\ -a^2 & a^1 & 0 \end{bmatrix} \underline{x} .$$

If, conversely, $C = [c^i{}_k] \in \mathbb{R}^3{}_3$ is a skew-symmetric matrix then

$$C\,\underline{x} = \underline{a} \times \underline{x}, \quad \underline{a} = \begin{bmatrix} -c^2{}_3 \\ c^1{}_3 \\ -c^1{}_2 \end{bmatrix} .$$

Because of $\underline{a} \times \underline{a} = 0$ and the representation formula,

$$\underline{a} \times (\underline{b} \times \underline{c}) = (\underline{a} \cdot \underline{c})\underline{b} - (\underline{a} \cdot \underline{b})\underline{c}, \quad \underline{a},\, \underline{b},\, \underline{c} \in \mathbb{R}^3 ,$$

applied to $C_a^2\,\underline{x} = C_a C_a\,\underline{x} = \underline{a} \times (\underline{a} \times \underline{x})$ we have

$$\underline{a}^T C_a = 0\,, \quad C_a\,\underline{a} = 0, \quad C_a^2 = \underline{a}\,\underline{a}^T - |\underline{a}|^2 I \tag{1.12}$$

(I unit matrix).

Suppose that $\varphi(t) = \omega t$ is the rotation angle with angular velocity ω then the rotated vector $\underline{x}_0 = \underline{x}(0)$ has, by (1.11), the form $\underline{x}(t) = D(\underline{a}, \omega t)\underline{x}_0$ with the *rotation matrix* $D(\underline{a}, \omega t)$:

$$\begin{aligned} D(\underline{a},\omega t) &= \cos(\omega t)I + (1-\cos(\omega t))\underline{a}\,\underline{a}^T + \sin(\omega t)C_a\,,\\ \dot{D}(\underline{a},\omega t) &= \omega\big[-\sin(\omega t)I + \sin(\omega t)\underline{a}\,\underline{a}^T + \cos(\omega t)C_a\big]\,. \end{aligned} \tag{1.13}$$

Using (1.12) one computes $\dot{D} = \omega CD = \omega DC$; therefore, it follows that

$$\boxed{\dot{\underline{x}}(t) = \omega\, C_a D(\underline{a},\omega t)\underline{x}_0 = \omega\,\underline{a}\times\underline{x}(t)}\,. \tag{1.14}$$

In other words, the vector field $\omega\,\underline{a}\times\underline{x}(t)$ is the velocity field of rotation about $\underline{a}$, $|\underline{a}| = 1$, with constant angular velocity ω.

Conversely, if $D = [d^i{}_k] \neq I \in \mathbb{R}^3{}_3$ is an orthogonal matrix with positive determinant, it is a rotation matrix. From the representation (1.13) we obtain the following formulas for the rotation angle $0 \leq \varphi$, $\varphi \neq \pi$, and the rotation axis $\underline{a}$

$$\boxed{\cos\varphi = \frac{1}{2}(\operatorname{trace} D - 1)\,,\quad \underline{a} = \frac{1}{2\sin\varphi}\left[d^3{}_2 - d^2{}_3,\, d^1{}_3 - d^3{}_1,\, d^2{}_1 - d^1{}_2\right]^T}\,;$$

cf., e.g., (Meyberg), Sect. 6.6.4. Furthermore, from $\dot{D} = \omega CD = \omega DC$, the relations

$$\omega C = \dot{D}\,D^T = -D\,\dot{D}^T = D^T\dot{D} = -\dot{D}^T D \tag{1.15}$$

are derived, being used in the theory of top as well as the following result.

Lemma 1.5. *Let D be a rotation matrix; then $D\underline{b}\times D\underline{c} = D(\underline{b}\times\underline{c})$ for $\underline{b}, \underline{c} \in \mathbb{R}^3$.*

Proof. For $\underline{a}$, $\underline{b}$, $\underline{c} \in \mathbb{R}^3$ the determinant is the parallepiped product, $\det(\underline{a},\underline{b},\underline{c}) = \underline{a}^T(\underline{b}\times\underline{c})$. For a matrix $C \in \mathbb{R}^3{}_3$ it follows that

$$\begin{aligned} \det(\underline{a}, C\underline{b}, C\underline{c}) &= \det(CC^{-1}\underline{a}, C\underline{b}, C\underline{c}) = \det(C)\det(C^{-1}\underline{a},\underline{b},\underline{c})\\ &= <\det(C)C^{-1}\underline{a},\underline{b},\underline{c}> = <(\operatorname{cof} C^T)\underline{a},\underline{b},\underline{c}>\\ &= \underline{a}^T(\operatorname{cof} C)(\underline{b}\times\underline{c})\,. \end{aligned}$$

cf. **(b)**. Substituting the rotation matrix D with $\operatorname{cof} D = (\det D)D^{-T} = D$ for C, we obtain

$$\underline{a}\cdot(D\underline{b}\times D\underline{c}) = \underline{a}\cdot D(\underline{b}\times\underline{c})\,.$$

This proves the assertion because $\underline{a} \in \mathbb{R}^3$ has been chosen arbitrarily. □

Lemma 1.5 describes the geometrically obvious fact that a rotated vector product is equal to the product of the rotated vectors.

Remember once again that exactly all orthogonal matrices D with positive determinant are rotation matrices and that their product yields again a rotation. But the angle and axis of the composed rotation are difficult to calculate by using matrices of the above form. To this end we refer the reader to Sect. 12.7 on *quaternions*.

(j) Matrices with Definite Real Part For *symmetric* matrices B, $C \in \mathbb{R}^n{}_n$ one usually writes $B \geq C$ in the case where $\underline{x}^T(B-C)\underline{x} \geq 0$ holds for all $\underline{x} \in \mathbb{R}^n$.

Lemma 1.6. *Let* $A \in \mathbb{R}^n{}_n$, $\alpha > 0$, *and let* $\mathrm{Re}(A) := \dfrac{1}{2}(A + A^T) \geq \alpha I$, *then* A *is regular and* $\|A^{-1}\|_2 \leq \alpha^{-1}$. *Moreover, the sequence*

$$\underline{y}_{n+1} = \underline{y}_n - \frac{h}{2}(A\underline{y}_{n+1} + A\underline{y}_n),\ n = 0, 1, \ldots,\ \ \underline{y}_0 \in \mathbb{R}^n,\ h > 0,$$

is bounded.

Cf. the trapezoidal rule for the differential system $\dot{\underline{x}} + A\underline{x} = \underline{0}$ in Sect. 2.4, *Example* 2.12. Note that A is not supposed to be diagonalizable; convergence instead of boundedness is, however, difficult to verify under the above weak assumptions.

Proof. (1°) Let $A \in \mathbb{R}^n{}_n$ and $\underline{x} \in \mathbb{R}^n$; then $\underline{x}^T \mathrm{Re}(A)\underline{x} = \dfrac{1}{2}\underline{x}^T(A + A^T)\underline{x} = \underline{x}^T A\underline{x}$. The assumption $0 < \alpha\,\underline{x}^T\underline{x} \leq \underline{x}^T A\underline{x}$, $\underline{x} \neq 0$, yields a contradiction if A is singular and an eigenvector belonging to the eigenvalue null is substituted for $\underline{x}$.
(2°) Using $\underline{x} = A^{-1}\underline{y}$ and the Cauchy-Schwarz inequality we obtain

$$\begin{aligned}\alpha\,\underline{x}^T\underline{x} &= \alpha\|A^{-1}\underline{y}\|_2^2 \leq (A^{-1}\underline{y})^T \mathrm{Re}(A)(A^{-1}\underline{y}) = (A^{-1}\underline{y})^T A(A^{-1}\underline{y}) \\ &= (A^{-1}\underline{y})^T\underline{y} \leq \|A^{-1}\underline{y}\|_2\|\underline{y}\|_2\,,\end{aligned}$$

hence $\forall\ \underline{y} : \|A^{-1}\underline{y}\|_2 \leq \alpha^{-1}\|\underline{y}\|_2$.
(3°) Let $h/2 = 1$ without loss of generality, then $A(\underline{y}_{n+1} + \underline{y}_n) = \underline{y}_n - \underline{y}_{n+1}$ and, by assumption,

$$0 \leq (\underline{y}_n + \underline{y}_{n+1})^T A(\underline{y}_n + \underline{y}_{n+1}) = (\underline{y}_n + \underline{y}_{n+1})^T(\underline{y}_n - y_{n+1}) = \|\underline{y}_n\|_2^2 - \|\underline{y}_{n+1}\|_2^2\,.$$

1.2 Brief on Vector Analysis

In this section we compile a collection of formulas of vector analysis in coordinate space $\mathbb{R}^3$. All these formulas hold also in n dimensions as long as the rotation operator does not occur. The computational rules are preferably written here with matrix multiplication and not with scalar product because the latter does not apply in Matlab. For a more general treatment of the matter discussed here, we refer to Chap. 10.

(a) Notations and Definitions Let $\underline{x} \in \mathbb{R}^3$ be a point or position vector and let $\partial_i = \partial/\partial x^i$ be the operator of partial derivation w.r.t. the i-th component. Let φ, χ, ψ be scalar fields, e.g., $\varphi : \mathbb{R}^3 \ni \underline{x} \mapsto \varphi(\underline{x}) \in \mathbb{R}$, let $\underline{u}$, $\underline{v}$, $\underline{w}$ column vector fields, e.g.

$$\underline{v} : \mathbb{R}^3 \ni \underline{x} \mapsto \underline{v}(\underline{x}) = \begin{bmatrix} v^1(\underline{x}) \\ v^2(\underline{x}) \\ v^3(\underline{x}) \end{bmatrix} = [v^1(\underline{x}),\, v^2(\underline{x}),\, v^3(\underline{x})]^T \in \mathbb{R}^3\,,$$

and let R, S, T tensor fields, e.g. $R : \mathbb{R}^3 \ni \underline{x} \mapsto R(\underline{x}) = [r^i{}_j(\underline{x})] \in \mathbb{R}^3{}_3$; frequently r_{ij} is written instead of $r^i{}_j$.

(a1) *Gradient, Gradient Field:*

$$\begin{aligned}
\operatorname{grad}\varphi \quad &: \quad \underline{x} \mapsto \operatorname{grad}\varphi(\underline{x}) := [\partial_1\varphi(\underline{x}), \partial_2\varphi(\underline{x}), \partial_3\varphi(\underline{x})] \in \mathbb{R}^3 \\
&\quad\ \text{row vector field,} \\
\operatorname{grad}\underline{v} \quad &: \quad \underline{x} \mapsto \operatorname{grad}\underline{v}(\underline{x}) \in \mathbb{R}^3{}_3 \ \text{ tensor field,}
\end{aligned}$$

$$\operatorname{grad}\underline{v}(\underline{x}) := \begin{bmatrix} \partial_1 v^1 & \partial_2 v^1 & \partial_3 v^1 \\ \partial_1 v^2 & \partial_2 v^2 & \partial_3 v^2 \\ \partial_1 v^3 & \partial_2 v^3 & \partial_3 v^3 \end{bmatrix}(\underline{x}) = \begin{bmatrix} \operatorname{grad} v^1(\underline{x}) \\ \operatorname{grad} v^2(\underline{x}) \\ \operatorname{grad} v^3(\underline{x}) \end{bmatrix} = [\partial_1\underline{v}, \partial_2\underline{v}, \partial_3\underline{v}]\,(\underline{x}),$$

$$\operatorname{grad} T(\underline{x}) = [\partial_1 T(\underline{x}), \partial_2 T(\underline{x}), \partial_3 T(\underline{x})] \in \mathbb{R}^3{}_9\,.$$

> The gradient of a mapping $\underline{v} : \mathbb{R}^n \to \mathbb{R}^m$ at point $\underline{x}$ is a (m,n)-matrix, in particular a *row vector* for $m = 1$, but note that sometimes (not in this volume) the transposed matrix $[\operatorname{grad}\underline{v}(\underline{x})]^T$ is defined for "gradient".

If $f : \mathbb{R}^3{}_3 \ni S \mapsto f(S) \in \mathbb{R}$ is a scalar function with tensor argument $S = [s^i{}_j]$,

$$\operatorname{grad} f(S) = \frac{\partial f}{\partial S}(S) = \left[\frac{\partial f}{\partial s^i{}_j}(S)\right] \in \mathbb{R}^3{}_3\,,$$

and the directional derivative is a tensor product,

$$\left.\frac{d}{dt} f(S + tT)\right|_{t=0} = [\operatorname{grad} f(S)] : T\,.$$

(a2) *Divergence:*

$$\begin{aligned}
\operatorname{div}\underline{v} \quad &: \quad \underline{x} \mapsto \operatorname{div}\underline{v}(\underline{x}) := \partial_1 v^1(\underline{x}) + \partial_2 v^2(\underline{x}) + \partial_3 v^3(\underline{x}) \in \mathbb{R} \ \text{ scalar field,} \\
\operatorname{div} R \quad &: \quad \underline{x} \mapsto \operatorname{div} R(\underline{x}) \in \mathbb{R}^3 \ \text{ column vector field,}
\end{aligned}$$

$$\begin{aligned}
\operatorname{div} R(\underline{x}) &:= \begin{bmatrix} \partial_1 r_{11} + \partial_2 r_{12} + \partial_3 r_{13} \\ \partial_1 r_{21} + \partial_2 r_{22} + \partial_3 r_{23} \\ \partial_1 r_{31} + \partial_2 r_{32} + \partial_3 r_{33} \end{bmatrix}(\underline{x}) = \begin{bmatrix} \operatorname{div}\underline{r}^1(\underline{x}) \\ \operatorname{div}\underline{r}^2(\underline{x}) \\ \operatorname{div}\underline{r}^3(\underline{x}) \end{bmatrix} \\
&= [\partial_1\underline{r}_1 + \partial_2\underline{r}_2 + \partial_3\underline{r}_3]\,(\underline{x})\,.
\end{aligned}$$

Accordingly, the divergence is applied *row by row* in tensor divergence.

(a3) *Vorticity, Curl, Rotation of a Vector Field,* cf. also Sect. 1.1**(a6)**:

$$\operatorname{rot}\underline{v} \quad : \quad \underline{x} \mapsto \operatorname{rot}\underline{v}(\underline{x}) \in \mathbb{R}^3 \ \text{ column vector field,}$$

$$\operatorname{rot}\underline{v}(\underline{x}) := \begin{bmatrix} \partial_2 v^3(\underline{x}) - \partial_3 v^2(\underline{x}) \\ \partial_3 v^1(\underline{x}) - \partial_1 v^3(\underline{x}) \\ \partial_1 v^2(\underline{x}) - \partial_2 v^1(\underline{x}) \end{bmatrix} = \begin{vmatrix} \underline{e}_1 & \partial_1 & v^1 \\ \underline{e}_2 & \partial_2 & v^2 \\ \underline{e}_3 & \partial_3 & v^3 \end{vmatrix}(\underline{x})\,.$$

As already mentioned, the second formula is to be understood as a memo rule. If $\underline{v}$ is a vector field in the plane, the first two components of its vorticity field disappear and then one often writes often with a slight abuse of notation

$$\operatorname{rot} \underline{v} := \underline{e}_3^T \operatorname{rot} \underline{v} = v_x^2 - v_y^1 \in \mathbb{R}. \tag{1.16}$$

The notion "rotation of a vector field" stems from the important example

$$\boxed{\underline{w}(x) = \underline{a} \times \underline{x} \Longrightarrow \operatorname{rot} \underline{w} = 2\underline{a}}\ ;$$

cf. (1.14). Sometimes also the notation

$$\operatorname{curl} \varphi(x, y) = [\varphi_y, -\varphi_x]^T \in \mathbb{R}^2$$

is used for a *tangent vector* of an implicitely defined curve $\varphi(x, y) = c$ in the plane where the sign is not fixed.

Formal notations with the nabla operator ∇:

$$\begin{array}{lll}
\nabla & := (\partial_1, \partial_2, \partial_3) & \text{formal row vector (!)},\\
\nabla\varphi & := \operatorname{grad}\varphi\,, & \nabla\underline{v} := \operatorname{grad}\underline{v}\,,\\
\nabla \cdot \underline{v} & := \operatorname{div}\underline{v}\,, & \\
\nabla \times \underline{v} & := \operatorname{rot}\underline{v}\,, & (\text{also } \nabla \wedge \underline{v}),\\
\underline{v} \cdot \nabla & : \quad \varphi \mapsto (\underline{v} \cdot \nabla)\varphi := (\nabla\varphi)\underline{v}\,, & \\
\underline{v} \cdot \nabla & : \quad \underline{w} \mapsto (\underline{v} \cdot \nabla)\underline{w} := (\nabla\underline{w})\underline{v}\,. &
\end{array}$$

The notation for the operator $\underline{v} \cdot \nabla$ is somewhat confusing, one should prefer instead $\partial_v \varphi = (\operatorname{grad} \varphi)\underline{v}$:

$$\begin{aligned}
\underline{v} \cdot \nabla &:= \partial_v := v^1\partial_1 + v^2\partial_2 + v^3\partial_3\\
\partial_v \underline{w} &= v^1\partial_1\underline{w} + v^2\partial_2\underline{w} + v^3\partial_3\underline{w} = (\nabla\underline{w})\underline{v}\,.
\end{aligned}$$

Representations with the JACOBI *matrix* $\nabla\underline{v}$:

$$\begin{aligned}
\operatorname{div} \underline{v}(\underline{x}) &= \operatorname{trace}(\nabla\underline{v})(\underline{x})\,,\\
(\operatorname{rot} \underline{v}(\underline{x})) \times \underline{w} &= [\nabla\underline{v}(\underline{x}) - (\nabla\underline{v}(\underline{x}))^T]w\,.
\end{aligned}$$

(a4) LAPLACE *Operator*:

$$\begin{aligned}
\Delta\varphi : \ \underline{x} \mapsto \Delta\varphi(\underline{x}) &:= \partial_1^2\varphi(\underline{x}) + \partial_2^2\varphi(\underline{x}) + \partial_3^2\varphi(\underline{x})\\
&= \operatorname{div}\operatorname{grad}\varphi(\underline{x}) \in \mathbb{R} \ \text{ scalar field,}
\end{aligned}$$

$$\begin{aligned}
\Delta\underline{v} : \ \underline{x} \mapsto \Delta\underline{v}(\underline{x}) &:= \begin{bmatrix} \Delta v^1(x)\\ \Delta v^2(\underline{x})\\ \Delta v^3(\underline{x}) \end{bmatrix}\\
&= \operatorname{div}\operatorname{grad}\underline{v}(\underline{x}) \in \mathbb{R}^3 \ \text{ vector field.}
\end{aligned}$$

(b) Differential Rules Note that all arguments have to be sufficiently smooth in the following formulas!

(b1) The operators grad, div and rot are *linear*, e.g.

$$\mathrm{rot}(\alpha\,\underline{v}+\beta\,\underline{w}) = \alpha\ \mathrm{rot}\,\underline{v}+\beta\ \mathrm{rot}\,\underline{w}\,,\quad \alpha,\ \beta\in\mathbb{R}\,.$$

(b2) *Chain rule:*

$$\mathrm{grad}(\underline{w}\circ\underline{v})(\underline{x}) = \mathrm{grad}\,\underline{w}(\underline{v}(\underline{x}))\ \mathrm{grad}\,\underline{v}(\underline{x})\,.$$

(b3) *Further Computational Rules:* As already announced, matrix multiplication is used in the subsequent rules instead of the scalar product because the latter does not apply in MATLAB.

Product Rules:

$\mathrm{grad}(\varphi\,\psi)$	$= \varphi\ \mathrm{grad}\,\psi + \psi\ \mathrm{grad}\,\varphi$	row vector field
$\mathrm{grad}(\underline{v}^T\underline{w})$	$= \underline{v}^T\ \mathrm{grad}\,\underline{w} + \underline{w}^T\ \mathrm{grad}\,\underline{v}$	row vector field
$\mathrm{grad}(\varphi\,\underline{v})$	$= \underline{v}\ \mathrm{grad}\,\varphi + \varphi\ \mathrm{grad}\,\underline{v}$	tensor field
$\mathrm{div}(\varphi\,\underline{v})$	$= \varphi\ \mathrm{div}\,\underline{v} + (\mathrm{grad}\,\varphi)\underline{v}$	scalar field
$\mathrm{rot}(\varphi\,\underline{v})$	$= \varphi\ \mathrm{rot}\,\underline{v} + (\mathrm{grad}\,\varphi)^T\times\underline{v}$	column vector field
$\mathrm{div}(\varphi\,T)$	$= \varphi\ \mathrm{div}\,T + T(\mathrm{grad}\,\varphi)^T$	column vector field
$\mathrm{div}(S\underline{v})$	$= [\mathrm{div}(S^T)]^T\underline{v} + S^T : \mathrm{grad}\,\underline{v}$	scalar field
$\mathrm{div}(\underline{v}^T S)$	$= \mathrm{div}(S^T\underline{v}) = \underline{v}^T\,\mathrm{div}\,S + \mathrm{grad}\,\underline{v} : S$	scalar field
$\mathrm{div}(S\,T)$	$= [\mathrm{grad}\,([s^i]^T) : T]_{i=1}^3 + S\ \mathrm{div}\ T$	column vector field
$\mathrm{div}(\underline{v}\,\underline{w}^T)$	$= (\mathrm{grad}\,\underline{v})\underline{w} + \underline{v}\,\mathrm{div}\,\underline{w}$	column vector field

$$\begin{aligned}
\mathrm{div}(\underline{v}\times\underline{w}) &= \underline{w}^T\,\mathrm{rot}\,\underline{v} - \underline{v}^T\,\mathrm{rot}\,\underline{w}\\
\mathrm{rot}(\underline{v}\times\underline{w}) &= (\mathrm{div}\,\underline{w})\underline{v} - (\mathrm{div}\,\underline{v})\underline{w} + (\mathrm{grad}\,\underline{v})\underline{w} - (\mathrm{grad}\,\underline{w})\underline{v}\,.
\end{aligned}$$

Twofold Applications:

$$(\mathrm{grad}\,\mathrm{div}\,\underline{v})^T = \begin{bmatrix} \partial_{11}^2 v^1 + \partial_{12}^2 v^2 + \partial_{13}^2 v^3 \\ \partial_{21}^2 v^1 + \partial_{22}^2 v^2 + \partial_{23}^2 v^3 \\ \partial_{31}^2 v^1 + \partial_{32}^2 v^2 + \partial_{33}^2 v^3 \end{bmatrix} = \begin{bmatrix} \partial_1\,\mathrm{div}\,\underline{v} \\ \partial_2\,\mathrm{div}\,\underline{v} \\ \partial_3\,\mathrm{div}\,\underline{v} \end{bmatrix},$$

$$\begin{aligned}
\mathrm{div}\,\mathrm{rot}\,\underline{v} &= 0\,, && \text{smooth rotational field has no sources}\\
\mathrm{rot}\,\mathrm{grad}\,\varphi &= \underline{0}\,, && \text{smooth gradient field has no rotations}\\
(\mathrm{grad}\,\mathrm{div}\,\underline{v})^T &= \Delta\underline{v} + \mathrm{rot}\,\mathrm{rot}\,\underline{v}\,,\\
\mathrm{div}(\varphi\ \mathrm{grad}\,\psi) &= \varphi\,\Delta\psi + (\mathrm{grad}\ \varphi)\ \mathrm{grad}\ \psi^T\\
\mathrm{div}([\mathrm{grad}\,\underline{v}]^T) &= [\mathrm{grad}\,\mathrm{div}\,\underline{v}]^T\\
\mathrm{rot}(\Delta\underline{v}) &= \Delta(\mathrm{rot}\,\underline{v})\,.
\end{aligned}$$

Observe that the nabla operator (gradient) " ∇ " is *not* a vector and is used only for abbreviation hence e.g.

$$[\nabla(\underline{v}^T\underline{w})]^T = (\nabla\underline{v})^T\underline{w} + (\nabla\underline{w})^T\underline{v}\,.$$

Furthermore, using $(\underline{w}\cdot\nabla)\underline{v} = (\nabla\underline{v})\underline{w} = [\operatorname{grad}\underline{v}]\underline{w}$, we have

$$\begin{aligned}(\nabla\underline{v})^T\underline{w} &= [\nabla\underline{v})^T - \nabla\underline{v}]\underline{w} + (\nabla\underline{v})\underline{w} = -[\nabla\underline{v} - (\nabla\underline{v})^T]\underline{w} + (\nabla\underline{v})\underline{w}\\ &= -(\operatorname{rot}\underline{v})\times\underline{w} + (\underline{w}\cdot\nabla)\underline{v} = \underline{w}\times\operatorname{rot}\underline{v} + (\underline{w}\cdot\nabla)\underline{v}\,;\end{aligned}$$

hence

$$(\nabla(\underline{v}^T\underline{w}))^T = (\underline{v}\cdot\nabla)\underline{w} + (\underline{w}\cdot\nabla)\underline{v} + \underline{v}\times\operatorname{rot}\underline{w} + \underline{w}\times\operatorname{rot}\underline{v}\,.$$

We especially obtain hereby the formula

$$\frac{1}{2}[\nabla(\underline{u}^T\underline{u})]^T = \underline{u}\times\operatorname{rot}\underline{u} + (\underline{u}\cdot\nabla)\underline{u}\,. \tag{1.17}$$

(c) Integral Rules Let $V\subset\mathbb{R}^3$ be an open, bounded set with sufficiently smooth boundary ∂V and let $F\in\mathbb{R}^3$ be a *regular* piece of area (surface) with piecewise smooth boundary ∂F.

(c1) *Elementary Differentials:* Cf. Chap. 10. For the substitution $(x^1,x^2) = (f(u,v),g(u,v))$, let

$$\frac{\partial(x^1,x^2)}{\partial(u,v)} = \begin{vmatrix} f_u & f_v\\ g_u & g_v\end{vmatrix}(u,v) := \det\begin{bmatrix} f_u & f_v\\ g_u & g_v\end{bmatrix}(u,v)\,.$$

The term $dx^1\wedge dx^2$ then stands for the computational device: Integrate over x^1 and x^2 if x^1 and x^2 are the independent variables, otherwise set

$$dx^1\wedge dx^2 = \frac{\partial(x^1,x^2)}{\partial(u,v)}(u,v)\,du\wedge dv\,;$$

the same device holds for

$$dx^1\wedge dx^2\wedge dx^3 = \frac{\partial(x^1,x^2,x^3)}{\partial(u,v,w)}(u,v,w)\,du\wedge dv\wedge dw\,;$$

in both cases the determinant must be *positive*. Then there is

$dV = dx^1\wedge dx^2\wedge dx^3$	the *volume element*
$d\underline{O} = (dx^2\wedge dx^3, dx^3\wedge dx^1, dx^1\wedge dx^2)^T$	the *vector area (surface) element*
$dO = \lvert d\underline{O}\rvert$	the *scalar area (surface) element*
$d\underline{x} = (dx^1, dx^2, dx^3)^T$	the *vector line element*
$\underline{n} = d\underline{O}/\lvert d\underline{O}\rvert$	the normalized *normal vector*

If $D\subset\mathbb{R}^2$ open and $F := \{\underline{x}(u,v),\ (u,v)\in D\}\subset\mathbb{R}^3$ is a piece of area, we have in explicit form with *unnormalized* normal vector of area (surface)

$$\underline{\widetilde{n}}(u,v) = \left(\frac{\partial\underline{x}}{\partial u}\times\frac{\partial\underline{x}}{\partial v}\right)(u,v) \equiv \left[\frac{\partial(x^2,x^3)}{\partial(u,v)},\frac{\partial(x^3,x^1)}{\partial(u,v)},\frac{\partial(x^1,x^2)}{\partial(u,v)}\right]^T,$$

$$dO = \lvert\underline{\widetilde{n}}(u,v)\rvert\,dudv\,,\quad d\underline{O} = \frac{\underline{\widetilde{n}}(u,v)}{\lvert\underline{\widetilde{n}}(u,v)\rvert}\lvert\underline{\widetilde{n}}(u,v)\rvert\,dudv = \underline{n}(u,v)dO\,.$$

(c2) The **Integral Theorems**,

Divergence Theorem of GAUSS in $\mathbb{R}^3$	$\int_V \operatorname{div} \underline{v}\, dV$	$=$	$\oint_{\partial V} \underline{v} \cdot d\underline{O}$
Rotation Theorem of STOKES in $\mathbb{R}^3$	$\int_F (\operatorname{rot} \underline{v}) \cdot d\underline{O}$	$=$	$\oint_{\partial F} \underline{v} \cdot d\underline{x}$

,

suppose always that $\underline{v}$ is a continuously differentiable vector field and V resp. F are "regular" volumes resp. surfaces. Of course, both theorems hold also for *planar* vector fields. However, one uses here the notation (1.16) for vorticity in plane, and

$$\underline{t} = \left(dx^1, dx^2\right), \quad \underline{n} = \left(-dx^2, dx^1\right)$$

for tangent and normal vector of the positive oriented planar *boundary curve* ∂F of $F \subset \mathbb{R}^2$. The *normal vector of area* is now $\underline{n} = [0, 0, 1]^T$, and we obtain *Divergence Theorem of* GAUSS in $\mathbb{R}^2$:

$$\boxed{\int_F \left(\partial_1 v^1 + \partial_2 v^2\right) dF = \oint_{\partial F} \left(-v^1 dx^2 + v^2 dx^1\right) ds},$$

Rotation Theorem of STOKES in $\mathbb{R}^2$:

$$\boxed{\int_F \left(\partial_1 v^2 - \partial_2 v^1\right) dF = \oint_{\partial F} \left(v^1 dx^1 + v^2 dx^2\right) ds}.$$

The subsequent two formulas follow easily from the Theorem of GAUSS:

$$\begin{aligned} \int_V (\operatorname{grad} \varphi)^T dV &= \oint_{\partial V} \varphi\, d\underline{O} = \oint_{\partial V} \varphi\, \underline{n} dO \in \mathbb{R}^3, \\ \int_V \operatorname{rot} \underline{v}\, dV &= \oint_{\partial V} d\underline{O} \times \underline{v} = -\oint_{\partial V} \underline{v} \times \underline{n}\, dO \in \mathbb{R}^3. \end{aligned} \tag{1.18}$$

The first one follows from the Divergence Theorem by substituting $\underline{v} = \varphi\ \underline{a}$ with arbitrary $\underline{a}$ and the second if $\underline{a} \times \underline{v}$ is substituted for $\underline{v}$.

The Theorem of STOKES yields

$$\int_F d\underline{O} \times (\operatorname{grad} \varphi)^T = \oint_{\partial F} \varphi\, d\underline{x} \in \mathbb{R}^3. \tag{1.19}$$

It follows by substituting $\underline{v} = \varphi\, \underline{a}$ with arbitrary $\underline{a}$ and observing the relation $(\underline{u} \times \underline{v}) \cdot \underline{w} = \underline{u} \cdot (\underline{v} \times \underline{w})$. See `SUPPLEMENT\chap01a`.

By using $\operatorname{div}(\varphi\, \underline{v}) = \varphi \operatorname{div} \underline{v} + \operatorname{grad} \varphi \cdot \underline{v}$, the Theorem of GAUSS leads to

$$\oint_{\partial V} \varphi\, \underline{v} \cdot d\underline{O} = \int_V [\varphi \operatorname{div} \underline{v} + \operatorname{grad} \varphi \cdot \underline{v}]\, dV. \tag{1.20}$$

If now T is a tensor field with rows $\underline{t}^i$, an elementwise application of (1.20) with $\varphi = v^i$ and $\underline{v} = \underline{t}^i$ yields the crucial relation

$$\boxed{\int_V \left[\underline{v} \cdot \operatorname{div} T + \operatorname{grad} \underline{v} : T\right] dV = \oint_{\partial V} \underline{v} \cdot T \, \underline{n} \, dO} \, . \tag{1.21}$$

Because of $\operatorname{div}(\varphi \operatorname{grad} \psi) = \varphi \Delta\psi + \operatorname{grad}\varphi \cdot \operatorname{grad}\psi$, an application of the Theorem of GAUSS yields the *First Theorem of* GREEN:

$$\oint_{\partial V} \varphi \frac{\partial \psi}{\partial n} dO = \int_V \left[\varphi \, \Delta\psi + \operatorname{grad}\varphi \cdot \operatorname{grad}\psi\right] dV \, , \quad \frac{\partial \psi}{\partial n} := \operatorname{grad}\psi \cdot \underline{n} \, . \tag{1.22}$$

Exchanging φ and ψ and subtracting the new result from the old yields the *Second Theorem of* GREEN:

$$\oint_{\partial V} \left[\varphi \frac{\partial \psi}{\partial n} + \psi \frac{\partial \varphi}{\partial n}\right] dO = \int_V [\varphi \Delta\psi - \psi \Delta\varphi] \, dV \tag{1.23}$$

from which a further interesting result is derived by inserting $\psi \equiv 1$.

(d) Coordinate-Free Definitions Let $K_r(\underline{x})$ be a ball with center $\underline{x}$, radius r and surface S_r , and let $\underline{n}(\underline{y})$, $|\underline{n}(\underline{y})| = 1$, be the normal vector of S_r in $\underline{y}$ pointing outward. Then

$$\operatorname{div} \underline{v}(\underline{x}) = \lim_{r \mapsto 0} \frac{1}{|K_r|} \oint_{S_r} (\underline{v} \cdot \underline{n}) \, dO \, ;$$

hence $\operatorname{div} \underline{v}$ is also called *source intensity* of $\underline{v}$.

On the other side, let $S \subset \mathbb{R}^3$ be a surface with $\underline{x} \in S$ and let S_r be cut out from S by $K_r(\underline{x})$ then, applying the Theorem of STOKES,

$$(\operatorname{rot} \underline{v} \cdot \underline{n})(\underline{x}) = \lim_{r \mapsto 0} \frac{1}{|S_r|} \oint_{\partial S_r} \underline{v} \cdot d\underline{x} = \lim_{r \mapsto 0} \frac{1}{|S_r|} \oint_{S_r} (\operatorname{rot} \underline{v}) \cdot \underline{n} \, dO \, .$$

The right side is called *vorticity* $\underline{v}$ about $\underline{n}$ and becomes maximum for $\operatorname{rot} \underline{v}$ parallel to $\underline{n}$, and $\underline{u} = \operatorname{rot} \underline{v}$ is again the *vorticity field* or *rotational field* of $\underline{v}$. Conversely, $\underline{v}$ with $\operatorname{rot} \underline{v} = \underline{u}$ is called a *vector potential* of $\underline{u}$ (not uniquely determined).

(e) Potentials and Vector Fields An open, connecting set $\Omega \subset \mathbb{R}^n$ is called *domain*, and a domain Ω is called *simply connected* if every closed curve $C \subset \Omega$ can be contracted to a point in Ω without leaving Ω. A domain Ω is *star-shaped* if

$$\exists \, \underline{z} \in \Omega \;\; \forall \, \underline{x} \in \Omega : [\underline{z}, \underline{x}] \subset \Omega$$

($[\underline{z}, \underline{x}]$ straight line segment between $\underline{z}$ and $\underline{x}$), the point $\underline{z}$ is then called a center of Ω. For instance, a disc in $\mathbb{R}^2$ *without* center is not simply connected whereas a ball in $\mathbb{R}^3$ *without* center is simply connected.

Theorem 1.4. *(Potential Criterium) Let $\Omega \subset \mathbb{R}^n$ be a simply connected domain and $\underline{v} : \Omega \to \mathbb{R}^n$ a continuously differentiable vector field. Then there exists a* potential *$\varphi : \Omega \to \mathbb{R}$ with $\underline{v} = \operatorname{grad} \varphi$ if and only if $\operatorname{grad} \underline{v}(\underline{x})$ is symmetric,*

$$\forall\, \underline{x} \in \Omega : \operatorname{grad} \underline{v}(\underline{x}) = [\operatorname{grad} \underline{v}(\underline{x})]^T ;$$

this condition is equivalent to $\operatorname{rot} \underline{v}(\underline{x}) = 0$ for $n = 2, 3$.

Proof (Meyberg) Vol. I, Sect. 8.2 and `SUPPLEMENT\chap01a`.

In the present case, $\underline{v}$ is called *potential field* or *conservative vector field* and exactly the rotation-free vector fields are potential fields in simply connected domains of $\mathbb{R}^3$.

Theorem 1.5. *Let $\Omega \subset \mathbb{R}^3$ be a star-shaped domain and $\underline{v} : \Omega \to \mathbb{R}^3$ continuously differentiable. Then the vector field $\underline{v}$ has a vector potential $\underline{w}$ with $\underline{v} = \operatorname{rot} \underline{w}$ if and only if $\operatorname{div} \underline{v} = 0$ in Ω.*

In other words, exactly the divergence-free vector fields have a vector potential in star-shaped domains of $\mathbb{R}^3$. (A divergence-free vector field is also called *solenoidal vector field.*)

If $\underline{z}$ is a center of Ω, a vector potential $\underline{w}$ of $\underline{v}$ can be given as

$$\underline{w}(\underline{x}) = \int_0^1 t\underline{v}(\underline{z} + t(\underline{x} - \underline{z})) \times (\underline{x} - \underline{z})\, dt .$$

Two vector potentials of a vector field differ only by an additive gradient field or, in other words, a vector potential of a divergence-free vector field is uniquely determined only up to an additive gradient field. Therefore we have in star-shaped domains

$$\underline{v} = \operatorname{rot}(\underline{w} + \operatorname{grad} \varphi)$$

with arbitrary φ for divergence-free vector fields $\underline{v}$. If the domain Ω is "regular" in some sense to be specified more exactly, then φ may be chosen so that $\operatorname{div}(\underline{w} + \operatorname{grad} \varphi) = 0$:

Theorem 1.6. *Let Ω be "regular" and let $\underline{v} : \Omega \to \mathbb{R}^3$ be a divergece-free vector field then $\underline{v}$ has a divergence-free vector potential.*

The next result says that, in normal case, a vector field can be decomposed into the sum of a divergence-free and a vorticity-free vector field.

Theorem 1.7. *(*HELMHOLTZ*' Decomposition Theorem) Let $\Omega \subset \mathbb{R}^3$ be "regular" and let $\underline{v} : \Omega \to \mathbb{R}^3$ be continuously differentiable then there exists a scalar field φ and a vector field $\underline{w}$ so that*

$$\boxed{\underline{v} = \operatorname{grad} \varphi + \operatorname{rot} \underline{w}} \, .$$

The vector field $\underline{w}$ can be chosen here that $\operatorname{div} \underline{w} = 0$ and $\operatorname{rot} \underline{w} \cdot \underline{n} = 0$ on the boundary $\partial\Omega$. The decomposition is then unique up to an additive constant in φ. As $\operatorname{div}\operatorname{rot} \underline{w} = 0$, the first Theorem of GREEN yields now, substituting $\underline{u} = \operatorname{rot} \underline{w}$,

$$\int_\Omega \operatorname{grad} \varphi \cdot \underline{u}\, dV = -\int_\Omega \varphi \operatorname{div} \underline{u}\, dV + \oint_{\partial\Omega} \varphi \underline{u} \cdot \underline{n}\, dO = 0\,.$$

By this way a vector field $\underline{v}$ is decomposed into two parts being "orthogonal" to each other w.r.t. the scalar product $(\underline{u}, \underline{v}) = \int_\Omega \underline{u}(\underline{x})^T \underline{v}(\underline{x})\, dV$. This decomposition is somctimes applied in numerical approximations of the NAVIER-STOKES equations.

Proof of the results in **(e)** see `SUPPLEMENT\chap01a`; cf. also (Burg), Bd. IV.

1.3 Curves in $\mathbb{R}^3$

Let $\underline{x} : \mathbb{R} \supset [a\,,\,b] \ni t \mapsto \underline{x}(t) \in \mathbb{R}^3$ be a *curve* with *graph* $\{(t\,,\,\underline{x}(t)),\ t \in [a\,,\,b]\}$ being likewise called *curve*. The curve $\underline{x}$ is *regular* if the function $\underline{x}$ is continuously differentiable and if *always* $\dot{\underline{x}}(t) \neq \underline{0}$. But note that there may exist regular and irregular parametrizations for the same graph.

Notations:

$s(t) = \displaystyle\int_a^t \lvert\dot{\underline{x}}(t)\rvert\, dt$	*arc length*	
$ds := \lvert\dot{\underline{x}}(t)\rvert\, dt$	*scalar differential (or element) of arc length*	
$d\underline{x} := \dot{\underline{x}}(t)\, dt$	*vectorial differential (or element) of arc length*	

$\underline{t}(t) := \dfrac{1}{\lvert\dot{\underline{x}}(t)\rvert}\, \dot{\underline{x}}(t)$	*tangent vector* (normed)
$\underline{n}(t) := \dfrac{1}{\lvert\dot{\underline{t}}(t)\rvert}\, \dot{\underline{t}}(t)$	*normal vector* (normed)
$\underline{b}(t) := \underline{t}(t) \times \underline{n}(t)$	*binormal vector* (normed)

.

The sign of $\underline{n}$ is handled differently in mechanics! Because

$$\underline{t}(t)^T \underline{t}(t) = 1 \Longrightarrow \dot{\underline{t}}(t)^T \underline{t}(t) + \underline{t}(t)^T \dot{\underline{t}}(t) = 2\,\dot{\underline{t}}(t)^T \underline{t}(t) = 0\,,$$

the normal vector $\underline{n}(t)$ is perpendicular to the tangent vector $\underline{t}(t)$. The right oriented system $\{\underline{t}\,,\,\underline{n}\,,\,\underline{b}\}(t)$ is called *accompanying trihedron* of the curve $\underline{x}$ in point $\underline{x}(t)$. Curvature vector and torsion vector are the variation of $\underline{t}$ and $\underline{b}$ relative to the arc length, respectively:

$$\boxed{\begin{array}{lll} \textit{vector of curvature}: & \displaystyle\lim_{h\to 0}\frac{1}{s(t+h)-s(t)}[\underline{t}(t+h)-\underline{t}(t)] = \frac{1}{\dot{s}(t)}\,\dot{\underline{t}}(t) \\ \textit{vector of torsion:} & \displaystyle\lim_{h\to 0}\frac{1}{s(t+h)-s(t)}[\underline{b}(t+h)-\underline{b}(t)] = \frac{1}{\dot{s}(t)}\,\dot{\underline{b}}(t) \end{array}}\,.$$

The *curvature* $\kappa(t)$ is the length of the curvature vector (here without sign) and the *absolute value* of the *torsion* $\tau(t)$ is the length of the torsion vector. The *radius of the circle of curvature* in point $\underline{x}(t)$ is the reciprocal of the curvature.

By the chain rule we obtain for $\underline{y}(s) := \underline{x}(t(s))$

$$\frac{d}{dt}\,\underline{y}(s) = \dot{\underline{x}}(t(s))\,\frac{dt}{ds} = \dot{\underline{x}}(t)/\frac{ds}{dt} = \dot{\underline{x}}(t)/|\dot{\underline{x}}(t)|\,,$$

therefore the tangent vector has always *unit length* if *arc length* is chosen for parameter (!) hence it offers itself for parametrization in canonical way.

By means of tangent and normal vector, the *osculating plane* S is described at the point $\underline{x}(t)$ of the curve:

$$S = \{\underline{y} \in \mathbb{R}^3,\ \underline{y} = \underline{x}(t) + \lambda\,\underline{t}(t) + \mu\,\underline{n}(t)\,,\ \lambda,\mu \in \mathbb{R}\}\,,$$

and its normal vector is the binormal vector $\underline{b}(t)$. Using $\dot{\underline{x}}(t) = |\dot{\underline{x}}(t)|\,\underline{t}(t) = \dot{s}(t)\,\underline{t}(t)$, one finds

$$\begin{aligned} \ddot{\underline{x}}(t) &= \ddot{s}(t)\,\underline{t}(t) + \dot{s}(t)\,\dot{\underline{t}}(t) = \ddot{s}(t)\,\underline{t}(t) + \dot{s}(t)^2\,\frac{|\dot{\underline{t}}(t)|}{\dot{s}(t)}\,\frac{\dot{\underline{t}}(t)}{|\dot{\underline{t}}(t)|} \\ &= \ddot{s}(t)\,\underline{t}(t) + \dot{s}(t)^2\kappa(t)\,\underline{n}(t)\,. \end{aligned} \tag{1.24}$$

As $\dot{\underline{x}}(t)$ is parallel to $\underline{t}$, there follows $\dot{\underline{x}}(t) \times \ddot{\underline{x}}(t) = 0 + \dot{s}(t)^3\kappa(t)\,(\underline{t} \times \underline{n})(t)$. Hence, because $\underline{b} = \underline{t} \times \underline{n}$ and $|\underline{b}| = 1$, we have

$$\kappa(t) = \frac{|\dot{\underline{x}}(t) \times \ddot{\underline{x}}(t)|}{|\dot{\underline{x}}(t)|^3} \geq 0\,.$$

For computation, the following formulas are chosen in case the denominator is not zero:

$$\boxed{\begin{array}{c} \underline{t}(t) = \dfrac{\dot{\underline{x}}(t)}{|\dot{\underline{x}}(t)|}\,,\quad \underline{b}(t) = \dfrac{\dot{\underline{x}}(t) \times \ddot{\underline{x}}(t)}{|\dot{\underline{x}}(t) \times \ddot{\underline{x}}(t)|}\,,\quad \underline{n}(t) = \underline{b}(t) \times \underline{t}(t)\,, \\ \kappa(t) = \dfrac{|\dot{\underline{x}}(t) \times \ddot{\underline{x}}(t)|}{|\dot{\underline{x}}(t)|^3}\,,\quad \tau(t) = \dfrac{\det(\dot{\underline{x}}, \ddot{\underline{x}}, \dddot{\underline{x}})(t)}{|\dot{\underline{x}}(t) \times \ddot{\underline{x}}(t)|^2} \end{array}}\,; \tag{1.25}$$

cf. `SUPPLEMENT\chap01a`. If the denominator disappears for some t then the limit values are to be considered at this point. The torsion disappears in the plane as well as the first and second component of $\dot{\underline{x}} \times \ddot{\underline{x}}$ whereas the third component is the curvature *with sign* in this case. Choosing the cartesian

x-coordinate for parameter of the curve in $\mathbb{R}^2$, we obtain the well-known relation

$$\kappa(x) = \frac{f''(x)}{(1+f'(x)^2)^{3/2}}\,.$$

If arc length s is chosen for parameter, of course $s' = 1$ and $s'' = 0$. Writing $\underline{y}(s) = \underline{x}(t(s))$ there follows then $\underline{t} = y'$, $\underline{t}' = \kappa\,\underline{n}$, $\underline{b}' = -\tau\,\underline{n}$, and thus, recalling $\underline{n} = \underline{b} \times \underline{t}$,

$$\underline{n}' = \underline{b}' \times \underline{t} + \underline{b} \times \underline{t}' = -\tau\,(\underline{n} \times \underline{t}) + \kappa\,(\underline{b} \times \underline{n}) = \tau\,\underline{b} - \kappa\,\underline{t}\,.$$

The *formulas of* FRENET

$$\boxed{\begin{array}{rlll} \underline{t}' = & & \kappa\,\underline{n} & \\ \underline{n}' = & -\kappa\,\underline{t} & & +\tau\,\underline{b} \\ \underline{b}' = & & -\tau\,\underline{n} & \end{array}} \tag{1.26}$$

constitute a differential system for the computation of the accompanying trihedron of a curve if curvature and torsion are prescribed in dependence of s (Fig. 1.3).

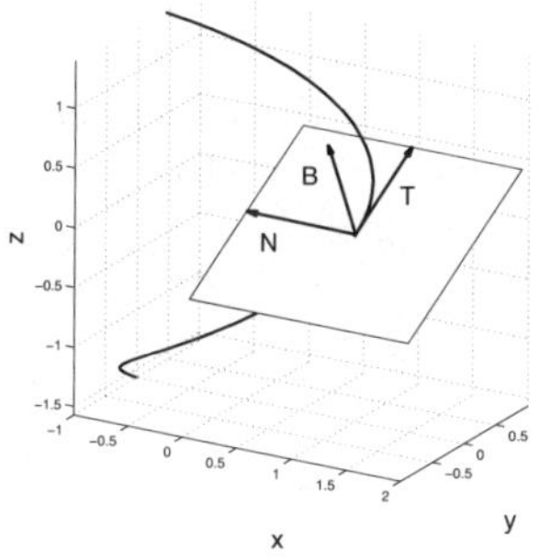

Figure 1.3. Osculating plane

1.4 Linear Differential Equations

Hint: $a, b, c, d, e, p, q, r, s$ are always real numbers in this section.

(a) Homogenous Linear Differential Equations with Constant Coefficients

(a1) Find the general real solution (GRSH) $y(x)$ of the homogenous differential equation

$$\boxed{ay'' + by' + cy = 0\,, \quad a \neq 0}$$

where "homogenous" says that the right side is zero. Compute the zeros α, β of the *characteristic polynomial* $p(\lambda) = a\lambda^2 + b\lambda + c$ then GRSH for

$$\begin{array}{ll} \alpha\,,\,\beta \in \mathbb{R}\,,\ \alpha \neq \beta : & y(x) = p\,e^{\alpha x} + q\,e^{\beta x} \\ \alpha\,,\,\beta \in \mathbb{R}\,,\ \alpha = \beta : & y(x) = p\,e^{\alpha x} + q\,x\,e^{\alpha x} \\ \alpha = \overline{\beta} = \mu + i\,\nu \in \mathbb{C} : & y(x) = e^{\mu x}\big[p\,\cos(\nu x) + q\,\sin(\nu x)\big]\,,\quad p\,,\,q \in \mathbb{R}\,. \end{array}$$

(a2) Find GRSH $y(x)$ of the homogenous equation

$$\boxed{ay''' + by'' + cy' + dy = 0\,,\quad a \neq 0}\,.$$

Compute the zeros α, β, γ of the characteristic polynomial $p(\lambda) = a\lambda^3 + b\lambda^2 + c\lambda + d$, then GRSH for $p, q, r \in \mathbb{R}$:

$$\begin{array}{ll} \alpha, \beta, \gamma \in \mathbb{R}\,,\ \text{all distinct} : & y(x) = p\,e^{\alpha x} + q\,e^{\beta x} + r\,e^{\gamma x} \\ \alpha, \beta, \gamma \in \mathbb{R}\,,\ \beta = \gamma : & y(x) = p\,e^{\alpha x} + q\,e^{\beta x} + r\,x\,e^{\beta x} \\ \beta = \overline{\gamma} = \mu + i\,\nu \in \mathbb{C} : & y(x) = p\,e^{\alpha x} + e^{\mu x}\big[q\,\cos(\nu x) + r\,\sin(\nu x)\big]\,. \end{array}$$

(a3) Find GRSH $y(x)$ of the homogenous equation

$$\boxed{ay^{(4)} + by''' + cy'' + dy' + ey = 0\,,\quad a \neq 0}\,.$$

Compute the zeros α_i, $i = 1:4$ of the characteristic polynomial $p(\lambda) = a\lambda^4 + b\lambda^3 + c\lambda^2 + d\lambda + e$, then GRSH for $\alpha_i \in \mathbb{R}$, all distinct:

$$y(x) = p_1\,e^{\alpha_1 x} + p_2\,e^{\alpha_2 x} + p_3\,e^{\alpha_3 x} + p_4\,e^{\alpha_4 x}$$

$\alpha_i \in \mathbb{R}$, $\alpha_3 = \alpha_4$:

$$y(x) = p_1\,e^{\alpha_1 x} + p_2\,e^{\alpha_2 x} + p_3\,e^{\alpha_3 x} + p_4\,x\,e^{\alpha_4 x}$$

$\alpha_i \in \mathbb{R}$, $\alpha_2 = \alpha_3 = \alpha_4$:

$$y(x) = p_1\,e^{\alpha_1 x} + p_2\,e^{\alpha_2 x} + p_3\,x\,e^{\alpha_2 x} + p_4\,x^2\,e^{\alpha_2 x}$$

$\alpha_i \in \mathbb{R}$, all equal α:

$$y(x) = p_1\,e^{\alpha x} + p_2\,x\,e^{\alpha x} + p_3\,x^2\,e^{\alpha x} + p_4\,x^3\,e^{\alpha x}$$

$\alpha_1 \neq \alpha_2 \in \mathbb{R}$, $\alpha_3 = \overline{\alpha}_4 = \mu + i\,\nu \in \mathbb{C}$:

$$y(x) = p_1\,e^{\alpha_1 x} + p_2\,e^{\alpha_2 x} + e^{\mu x}\,\big[p_3\,\cos(\nu x) + p_4\,\sin(\nu x)\big]$$

$\alpha_1 = \alpha_2 \in \mathbb{R}$, $\alpha_3 = \overline{\alpha}_4 = \mu + i\,\nu \in \mathbb{C}$:

$$y(x) = p_1\,e^{\alpha_1 x} + p_2\,x\,e^{\alpha_1 x} + e^{\mu x}\,\big[p_3\,\cos(\nu x) + p_4\,\sin(\nu x)\big]$$

$\alpha_1 = \alpha_2 = \mu + i\nu \in \mathbb{C}$:

$$y(x) = e^{\mu x}\big[(p_1 + p_2\,x)\cos(\nu x) + (p_3 + p_4\,x)\sin(\nu x)\big]\,;$$

and so on.

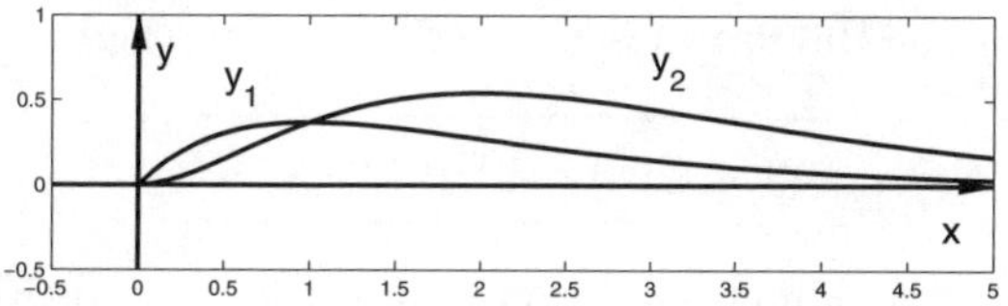

Figure 1.4. $y_1 = xe^{-x}$ and $y_2 = x^2e^{-x}$

(b) Inhomogenous Linear Differential Equations with Constant Coefficients and Special Right Sides Let the differential operator D be defined by $Dy(x) = y'(x)$, let $D^k y(x) = y^{(k)}(x)$, and let $p(\lambda) \in \Pi_m$ be a real polynomial of exact degree m. Then a general linear differential equation of order m with constant coefficients and right side $f(x)$ has the form

$$\boxed{p\,(D)y = f(x)}\,.$$

A **p**articular **s**olution (PSI) $z(x)$ of this **i**nhomogenous equation can be found in the following cases:

(b1) $\alpha = \mu + i\,\nu,\ \mu, \nu \in \mathbb{R}$ is a k-fold zero of $p(\lambda)$ and

$$\boxed{f(x) = a\,e^{\alpha x} = a\,e^{(\mu+i\nu)x} = a\,e^{\mu x}[\cos(\nu x) + i\sin(\nu x)]\,,\ a \in \mathbb{R}}\,.$$

Then PSI (possibly complex)

$$\boxed{z(x) = b\,x^k e^{\alpha x},\ \ b = \frac{a}{p^{(k)}(\alpha)}}\,.$$

Case 1: α real then $z(x)$ real.
Case 2: $g(x) = \text{Re}\ f(x) = a\,e^{\mu x}\cos(\nu x)$ then

Re $z(x)$ is a solution of $P(D)y = g(x) \equiv \text{Re}\ f(x) = a\,e^{\mu x}\cos(\nu x)$.

Case 3: $h(x) = \text{Im}\ f(x) = a\,e^{\mu x}\sin(\nu x)$ then

Im $z(x)$ is a solution of $P(D)y = h(x) = \text{Im}\ f(x) = a\,e^{\mu x}\sin(\nu x)$.

Of course the cases $\mu = 0$ and/or $\nu = 0$ are admitted here, too.

(b2) $\alpha = \mu + i\,\nu$, $\mu, \nu \in \mathbb{R}$ is a k-fold zero of $p(\lambda)$, $q(x) \in \Pi_n$ is a real polynomial of exact degree n, and

$$\boxed{f(x) = q(x)\, e^{\alpha x} = q(x)\, e^{(\mu + i\nu)x} = q(x)\, e^{\mu x}[\cos(\nu x) + i\sin(\nu x)]}\ .$$

Then a PSI $z(x)$ has the form ("ansatz of type of the right side"):

$$\boxed{z(x) = Q(x)\, x^k e^{\alpha x}, \quad Q(x) \in \Pi_n}\ .$$

Accordingly, the polynomial $Q(x)$ must have the same degree as $q(x)$. The polynomial $R(x) := Q(x)x^k$ (!) must be a solution of the *algebraic* equation

$$\boxed{\sum_{j=k}^{M} \frac{p^{(j)}(\alpha)}{j!} R^{(j)}(x) = q(x)\,, \quad M = \min\{m,\, n+k\}}\ .$$

The unknown coefficients of $Q(x)$ are computed from this equation by comparison of coefficients.

Hint: If α is complex then the polynomial $Q(x)$ has to be substituted with complex coefficients.

Case 1: α real, then $z(x)$ is real and solution of the inhomogenous equation.

Case 2: α complex, then

$$\text{Re } z(x) = \text{Re } R(x)e^{\mu x}\cos(\nu x) - \text{Im } R(x)e^{\mu x}\sin(\nu x)$$

is solution of $P(D)y = \text{Re } f(x) = q(x)\, e^{\mu x}\cos(\nu x)$ and

$$\text{Im } z(x) = \text{Re } R(x)e^{\mu x}\sin(\nu x) + \text{Im } R(x)e^{\mu x}\cos(\nu x)$$

is solution of $P(D)y = \text{Im } f(x) = q(x)\, e^{\mu x}\sin(\nu x)$.

By this way, solutions of the equation

$$P(D)y = g(x) \text{ with } g(x) = q(x)\, e^{\mu x}\cos(\nu x) \text{ resp. with } g(x) = q(x)\, e^{\mu x}\sin(\nu x)$$

can be computed as well as arbitrary linear combinations of these both types.

(c) The **general solution of** $p(D)y = f(x)$**.** Let $u(x)$ be the general solution of $p(D)y = 0$ and let $v(x)$ be a particular solution of $p(D)y = f(x)$; then $y(x) = u(x) + v(x)$ is the general solution of the affine linear equation $p(D)y = f(x)$.

(d) Example The *oscillator*,

$$\ddot{x} + a\,\dot{x} + b^2 x = 0\,, \quad \boxed{a \geq 0\,,\ 0 < b \ \text{ fixed}}\,,$$

has a quadratic characteristic polynomial with zeros

$$\lambda_{1,2}(a) = \frac{1}{2}\left(-a \pm (a^2 - 4b^2)^{1/2}\right).$$

This yields four possibilities:

$$\begin{array}{lll} a = 0: & \lambda_1 = \overline{\lambda}_2 \in i\,\mathbb{R} & \text{oscillation without damping} \\ 0 < a < 2b: & \lambda_1 = \overline{\lambda}_2 \in \mathbb{C} & \text{damped oscillation} \\ a > 2b: & \lambda_2 < \lambda_1 < 0 & \text{strong damping without oscillation.} \end{array}$$

In the critical *fourth case* $a = 2b$ we have $\lambda := \lambda_{1,2} = -a/2 < 0$ and the general solution has the form

$$x(t) = e^{\lambda t}(\alpha + \beta t)\,, \quad \alpha, \beta \in \mathbb{R}\,.$$

It is illustrated in Figure 1.5 that this limit case yields the most damped solution *without* oscillation ($b = 1/2$):

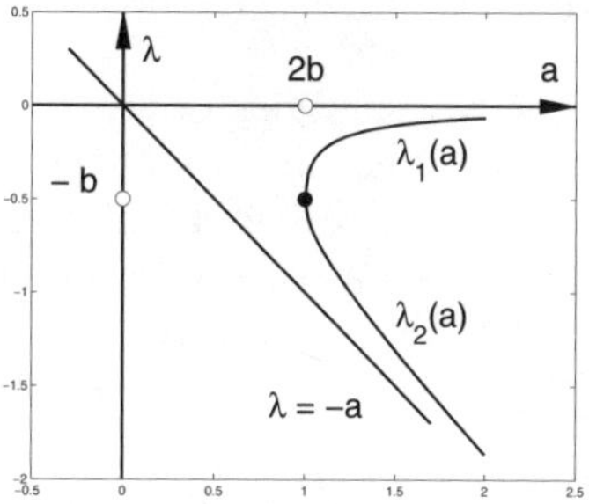

Figure 1.5. Damped oscillation, eigenvalues

1.5 Linear Differential Systems of First Order

Let $\mathcal{I}$ be an open interval, let $A : \mathcal{I} \ni t \mapsto A(t) \in \mathbb{R}^n{}_n$ be a continuous tensor field and $\underline{c} : \mathcal{I} \ni t \mapsto \underline{c}(t) \in \mathbb{R}^n$ a continuous vector field. We seek the general solution $\underline{x} : t \mapsto \underline{x}(t)$ of

$$\boxed{\dot{\underline{x}}(t) = A(t)\,\underline{x}(t) + \underline{c}(t)}\,, \tag{1.27}$$

resp. a solution with given inital or boundary conditions.

(a) Autonomous Homogenous Systems with Diagonalizable Matrix Here, the matrix A is constant and $\underline{c} = \underline{0}$; hence

$$\dot{\underline{x}}(t) = A\,\underline{x}(t)\,. \tag{1.28}$$

For every characteristic pair $(\lambda, \underline{u})$ of the matrix A, there exists a solution of (1.28) with four different possibilities:

Case 1: $\lambda \in \mathbb{R}$ is a simple eigenvalue, then $\underline{x}(t) = e^{\lambda t}\underline{u}$ is a solution of (1.28).

Case 2: $\lambda \in \mathbb{R}$ is a m-fold eigenvalue with linearly independent eigenvectors $\underline{u}_1, \ldots, \underline{u}_m$, then $\underline{x}_i(t) = e^{\lambda t}\underline{u}_i$, $i = 1 : m$, are linearly independent solutions of (1.28).

Case 3: $\lambda = \mu + i\,\nu \in \mathbb{C}$, $\nu > 0$, is a simple eigenvalue with eigenvector $\underline{u} = \underline{v} + i\,\underline{w} \in \mathbb{C}^n$, then $(\overline{\lambda}, \overline{\underline{u}})$ is also a characteristic pair and $\underline{x}(t) = e^{\lambda t}\underline{u}$, $\overline{\underline{x}}(t) = e^{\overline{\lambda} t}\overline{\underline{u}}$ are two *conjugate complex* solutions of (1.28). Their *real and imaginary part*,

$$\boxed{\text{Re}\ \underline{x}(t) = e^{\mu t}\big[\underline{v}\cos(\nu t) - \underline{w}\sin(\nu t)\big],\ \ \text{Im}\ \underline{x}(t) = e^{\mu t}\big[\underline{v}\sin(\nu t) + \underline{w}\cos(\nu t)\big]}$$

constitute two linearly independent *real* solutions of (1.28).

Case 4: As *Case 3* but λ is an m-fold eigenvalue with m (complex) linearly independent eigenvectors $\underline{u}_i$ then, accordingly,

$$\text{Re}\ \underline{x}_i(t) = \text{Re}\left(e^{\lambda t}\underline{u}_i\right),\ \ \text{Im}\ \underline{x}_i(t) = \text{Im}\left(e^{\lambda t}\underline{u}_i\right),\ \ i = 1 : m\,,$$

are $2m$ linearly independent *real* solutions of (1.28).

Repeating this process for all eigenvalues of the matrix A yields together a *fundamental system* $\underline{x}_1, \ldots, \underline{x}_n$ of n linearly independent real solutions, and the general solution of (1.28) is an arbitrary linear combination

$$\underline{x}(t) = a^1\,\underline{x}_1(t) + \ldots + a^n\,\underline{x}_n(t)\,,\ \ a^i \in \mathbb{R}\,.$$

(b) Autonomous Homogenous Systems with Undiagonalizable Matrix Here we need the JORDAN decomposition of the leading matrix A and, using the same notations as in Sect. 1.1**(c3)**, we write

$$\dot{\underline{x}}(t) = A\underline{x}(t) = U(\Lambda + T)V\underline{x}(t)\,,\ \ V = U^{-1}\,. \tag{1.29}$$

Let now $\exp(\Lambda t) = \text{diag}\left(\exp(\lambda_1 t), \ldots, \exp(\lambda_n t)\right)$ be a diagonal matrix and

$$e^A = \lim_{n\to\infty}\sum_{k=1}^{n}\frac{A^k}{k!}\,.$$

Then we obtain by the nilpotence of the matrix T, i.e., $T^n = 0$,

$$e^{(\Lambda+T)t} = e^{\Lambda t}\sum_{k=0}^{n}\frac{t^k}{k!}T^k\,,\ \ e^{At} = Ue^{(\Lambda+T)t}V\,,\ \ \frac{d}{dt}e^{At} = U(\Lambda+T)e^{(\Lambda+T)t}V\,.$$

The general solution of the homogenous system (1.28) thus has the form

$$\boxed{\underline{x}(t) = Ue^{\Lambda t}\sum_{k=0}^{n}\frac{t^k}{k!}T^k\underline{a}\,,\ \ \underline{a} \in \mathbb{R}^n}\ . \tag{1.30}$$

Example 1.4. We choose the matrix $A \in \mathbb{R}^5{}_5$ as in the example of Sect. 1.1**(c3)** but with one eigenvalue $\lambda = \mu$ and two eigenvectors $\underline{u}_1$ and $\underline{v}_1$. Then $T^3 = 0$,

$$T = \begin{bmatrix} 0\,1\,0\,0\,0 \\ 0\,0\,0\,0\,0 \\ 0\,0\,0\,1\,0 \\ 0\,0\,0\,0\,1 \\ 0\,0\,0\,0\,0 \end{bmatrix}, \quad T^2 = \begin{bmatrix} 0\,0\,0\,0\,0 \\ 0\,0\,0\,0\,0 \\ 0\,0\,0\,0\,1 \\ 0\,0\,0\,0\,0 \\ 0\,0\,0\,0\,0 \end{bmatrix},$$

and

$$S(t) := \sum_{k=0}^{2} \frac{t^k}{k!} T^k = \begin{bmatrix} 1\;t\;0\;0\;\;0 \\ 0\;1\;0\;0\;\;0 \\ 0\;0\;1\;t\;t^2/2 \\ 0\;0\;0\;1\;\;t \\ 0\;0\;0\;0\;\;1 \end{bmatrix}, \quad e^{\Lambda t} S(t)\underline{a} = e^{\lambda t} \begin{bmatrix} a^1 + a^2 t \\ a^2 \\ a^3 + a^4 t + a^5 t^2/2 \\ a^4 + a^5 t \\ a^5 \end{bmatrix}.$$

Using the same notations as in Sect. 1.1**(c3)** for the eigen- and principal vectors, the general solution of the homogenous system now reads as follows:

$$\begin{aligned} &\underline{x}(t) \\ &= e^{\lambda t}\left[\underline{u}_1\left(a^1 + a^2 t\right) + \underline{u}_2 a^2 + \underline{v}_1\left(a^3 + a^4 t + a^5 \frac{t^2}{2}\right) + \underline{v}_2\left(a_4 + a_5 t\right) + \underline{v}_3 a_5\right] \\ &= e^{\lambda t}\left[a^1 \underline{u}_1 + a^2\left(\underline{u}_2 + t\underline{u}_1\right) + a^3 \underline{v}_1 + a^4\left(\underline{v}_2 + t\underline{v}_1\right) + a^5\left(\underline{v}_3 + t\underline{v}_2 + \frac{t^2}{2}\underline{v}_1\right)\right]. \end{aligned}$$

(c) On Stability The system (1.28) is called *stable* if the absolute difference of every two solutions remains bounded for every t-interval, and it is called *asymptotically stable* if the difference of every two solutions tends to zero for $t \to \infty$. Then the following result is an immediate inference of (1.30) and Theorem 1.1.

Theorem 1.8. (1°) *The system (1.28) is stable if all eigenvalues of A have non-positive real part and all purely imaginary eigenvalues as well as the possible eigenvalue null are semi-simple, i.e., have no principal vectors.*
(2°) *The system (1.28) is asymptotically stable if all eigenvalues of A have negative real part.*

(d) General Linear Systems If the system $\dot{\underline{x}} = A(t)\underline{x}$ has dimension n then arbitrary n linearly independent solutions $\underline{x}_1, \ldots, \underline{x}_n$ are called *fundamental system* constituting a basis of the vector space of solutions. In order to show the independence, it is sufficient by a famous result of Liouville to verify the regularity of the *fundamental matrix* $X(t) = [\underline{x}_1(t), \ldots, \underline{x}_n(t)]$ in *only* one point t_0 of the interval $\mathcal{I}$. The general solution is then $\underline{x}(t) = X(t)\underline{a}$ with arbitrary vector $\underline{a}$, and a change of basis is obtained in the same way as in $\mathbb{R}^n$ by multiplication $X(t)$ from right with a constant regular matrix.

Theorem 1.9. *(Unique Existence) Let $B \in \mathbb{R}^n{}_n$ be an arbitrary regular matrix then, for every $t_0 \in \mathcal{I}$, there exists a unique continuously differentiable fundamental system $X(t)$ of (1.27) such that $X(t_0) = B$.*

The WRONSKI matrix

$$W : \mathcal{I} \times \mathcal{I} \ni (s,t) \mapsto W(s,t) = X(s)X(t)^{-1} \in \mathbb{R}^n{}_n ,$$

is continuously differentiable in both arguments and has the properties

$$\begin{aligned} W_t(t,t_0) &= A(t)W(t,t_0) \\ W(s,t)W(t,t_0) &= W(s,t_0) \\ W(t,t_0) &\quad \text{regular for all } t, t_0 \in \mathcal{I} , \end{aligned}$$

in particular, $W(s,t)W(t,s) = W(s,s) = I$ identity. By means of the WRONSKI matrix, the general solution (or also the *flux*) of (1.27) can be written explicitly depending on all variables:

$$\underline{x}(t;t_0,\underline{x}_0) = W(t,t_0)\underline{x}_0 + \int_{t_0}^{t} W(t,s)\underline{c}(s)\,ds . \tag{1.31}$$

Accordingly, the set of solutions of the initial value problem for (1.27) constitutes an affine vector space of dimension n, and the solution $\underline{x}$ depends continuously differentiable on the initial values $(t_0, \underline{x}_0)$. If a fundamental matrix X with $X(t_0) = I$ is known, the solution (1.31) has the somewhat simpler form

$$\boxed{\underline{x}(t;t_0,\underline{x}_0) = X(t)\underline{x}_0 + X(t)\int_{t_0}^{t} X(s)^{-1}\underline{c}(s)\,ds\,, \quad X(t_0) = I} . \tag{1.32}$$

(e) Special Right Sides Inhomogenous systems (1.27) with *constant* matrix A allow the computation of particular solutions $\underline{z}(t)$ for a large class of right sides as in the scalar case but here the JORDAN decomposition $A = U(\Lambda + T)V$ is a necessary tool. Using (1.30) we obtain by (1.32) the general representation

$$\underline{z}(t) = Ue^{(\Lambda+T)t}\int_{t_0}^{t} e^{-(\Lambda+T)s}V\underline{c}(s)\,ds = U\int_{t_0}^{t} e^{(\Lambda+T)(t-s)}V\underline{c}(s)\,ds .$$

(e1) Let the right side be $\underline{c}(t) = \underline{c}\,e^{\alpha t}$ then

$$\underline{z}(t) = Ue^{(\Lambda+T)t}\int_{t_0}^{t} e^{-(\Lambda+T)s}V\underline{c}(s)\,ds = Ue^{(\Lambda+T)t}\left(\int_{t_0}^{t} e^{(\alpha I-\Lambda-T)s}\,ds\right)V\underline{c}$$

where I is the unit matrix again. This result holds also in case of *resonance*, i.e. if α is single or multiple eigenvalue of the matrix A.

(e2) Let the right side be $\underline{c}(t) = \underline{c}\, e^{\mu t}\cos(\nu t)$ resp. $\underline{c}(t) = \underline{c}\, e^{\mu t}\sin(\nu t)$ with real vector $\underline{c}$ then we proceed as in **(e1)** but substituting a *complex* right side $\underline{c}\, e^{(\mu + i\,\nu)t}$. Ensuing Re $\underline{z}(t)$ resp. Im $\underline{z}(t)$ is a particular *real* solution of (1.27).

(e3) Let the right be $\underline{c}(t) = \underline{c}\, t^k\, e^{\alpha t}$ with real vector $\underline{c}$ and real α or — after complexification — complex α , then

$$\underline{z}(t) = Ue^{(\Lambda+T)t}\int_{t_0}^{t} e^{-(\Lambda+T)s}V\underline{c}(s)\, ds = Ue^{(\Lambda+T)t}\left(\int_{t_0}^{t} s^k\, e^{(\alpha I-\Lambda-T)s}\, ds\right)V\underline{c}\,.$$

The appearing integrals can be found explicitly in each case by multiple partial integration. Ensuing, Re $\underline{z}(t)$ resp. Im $\underline{z}(t)$ is a particular *real* solution of (1.27) again.

Example 1.5. Let the constant matrix A be diagonalizable and let $\underline{c}(t) = \underline{c}\, e^{\alpha t}$. If α is not an eigenvalue of A (no resonance), then

$$z(t) = U\, D\, V\underline{c}\,,\quad D = (\alpha I - \Lambda)^{-1}\left[e^{(\alpha I-\Lambda)t}\right] \text{ diagonal matrix}$$

is a particular solution. If α is single or multiple eigenvalue (resonance), all those diagonal elements of D must be replaced by $t\, e^{\alpha\, t}$ whose denominator is zero.

(f) Boundary Value Problems We consider the general linear boundary value problem

$$L\,\underline{x} := \dot{\underline{x}} - A(t)\underline{x} = \underline{c}(t),\quad R_0\underline{x}(0) + R_1\underline{x}(1) = \underline{d} \in \mathbb{R}^m\,,\quad m \leq n\,, \tag{1.33}$$

in interval $\mathcal{I} = [0,1]$ where R_0 , $R_1 \in \mathbb{R}^m{}_n$ are arbitrary matrices for the present. The general solution of $L\underline{x} = \underline{c}$ has the form

$$\underline{x}(t) = X(t)\underline{a} + \underline{z}(t)\,,\quad \underline{a} \in \mathbb{R}^n,$$

with an arbitrary fundamental matrix $X(t)$, a particular solution $\underline{z}(t)$ of the inhomogenous system, and an arbitrary vector $\underline{a}$. Substitution into the boundary conditions yields a linear system of equations for the constant vector $\underline{a}$,

$$R_0\underline{z}(0) + R_1\underline{z}(1) + [R_0X(0) + R_1X(1)]\underline{a} = \underline{d} \in \mathbb{R}^m\,. \tag{1.34}$$

The *rank* of the *characteristic matrix*

$$\boxed{C := R_0X(0) + R_1X(1) \in \mathbb{R}^m{}_n} \tag{1.35}$$

is determined uniquely and the system (1.34) has a unique solution $\underline{a}$ for regular C which implies that m must be equal to n .

Let now $\underline{w}$ be a continuous differentiable function satisfying the boundary conditions,

$$R_0\underline{w}(0) + R_1\underline{w}(1) = \underline{d}\,,$$

and let $\underline{v}$ be a solution of the *semi-homogenous* boundary value problem

$$L\,\underline{v} = \underline{c} - L\,\underline{w}\,,\quad R_0\underline{v}(0) + R_1\underline{v}(1) = \underline{0} \tag{1.36}$$

with homogeneous boundary conditions and inhomogenous right side, then $\underline{x} = \underline{v} + \underline{w}$ is a solution of (1.33). If $\underline{a} \in \mathrm{Ker}(C)$, then $\underline{x} : t \mapsto X(t)\underline{a}$ is a solution of the full homogenous boundary problem and, conversely, every solution $\underline{x}(t) = X(t)\underline{a}$ of the full homogenous problem must satisfy $C\underline{a} = 0$. Because $\dim \mathrm{Ker}(C) + \mathrm{rank}(C) = n$, we thus obtain the following result:

Theorem 1.10. (1°) *The boundary value problem (1.33) has a unique solution for all right sides $\underline{c}(t)$ if an only if the characteristic matrix C is regular.* (2°) *The full homogenous problem has $n - \mathrm{rank}(C)$ linearly independent solutions.*

Example 1.6. The scalar boundary problem $\ddot{x} + \lambda^2 x = 0\,,\ x(0) = 0\,,\ x(1) = d$ is equivalent to the system

$$\begin{array}{l}\dot{u} = v\\ \dot{v} = -\lambda^2 u\end{array}\,,\quad u(0) = 0\,,\ u(1) = d\,,$$

with solution $u(t) = \gamma\sin(\lambda t)\,,\ v = \dot{u}$ ($\mathbb{Z}$ entire numbers).

$$\begin{array}{lll}\lambda \notin \pi\mathbb{Z} \Longrightarrow \sin\lambda \neq 0\,, & \gamma = d/\sin\lambda & \text{unique solution}\\ \lambda \in \pi\mathbb{Z} \Longrightarrow \sin\lambda = 0\,, & d \neq 0 & \text{no solution}\\ & d = 0 & \text{infinitely number of solutions.}\end{array}$$

(g) Periodic Solutions A linear system

$$\dot{\underline{x}}(t) = A(t)\underline{x}(t) + \underline{c}(t) \tag{1.37}$$

is called T*-periodic*, if both A and $\underline{c}$ are T-periodic and everywhere continuous. By means of (1.31) we obtain immediately:

Lemma 1.7. *The system (1.37) has a T-periodic solution if and only if the mapping*

$$\underline{\xi} \mapsto W(T,0)\underline{\xi} + \int_0^T W(t,s)\underline{c}(s)\,ds$$

has a fixed point $\underline{\xi}^*$.

If $\underline{y} : t \mapsto \underline{y}(t)$ is a T-periodic solution, $\underline{y} : t \mapsto \underline{y}(t + \alpha)$ is also a T-periodic solution for every $\alpha \in \mathbb{R}$; hence T-periodic solutions are never determined uniquely which leads to apparent difficulties in their numerical computation.

(g1) The real system (1.37) with *constant* matrix A and $\underline{c}(t) \equiv \underline{0}$ has a T-periodic solution by **(a)** if and only if $\lambda = 2\pi\, i/T$ is eigenvalue of A. More general, let again $\sigma(A)$ be the spectrum of A in the complex plane and let the *neutral spectrum* be $\sigma_n(A) := \sigma(A) \cap i\,\mathbb{R}$, i.e., the intersection of the spectrum with the imaginary axis. If A is real then $\sigma_n(A)$ may contain the eigenvalue zero as well as conjugate complex, purely imaginary eigenvalues, or it can be empty. By **(a)** and Fredholm's alternative we obtain

Lemma 1.8. (1°) *If $\sigma_n(A) \neq \emptyset$, the homogenous T-periodic system (1.37) has periodic solutions.*
(2°) *If*

$$\sigma(A) \cap \frac{2\pi\, i}{T}\mathbb{Z} = \emptyset\,,$$

the inhomogenous T-periodic system (1.37) has at least one T-periodic solution.
(3°) *If $\sigma_n(A) = \emptyset$, the inhomogenous T-periodic system (1.37) has exactly one T-periodic solution.*

Proof see e.g. (Amann), Sect. 20, 22.

The solution $\underline{x}(t)$ in Lemma 1.8(3°) can be displayed explicitly if the matrix A is *diagonalizable.* Let

$$\begin{aligned}
\sigma_s(A) &= \{\lambda \in \sigma(A)\,,\ \mathrm{Re}\,\lambda < 0\} \quad \text{the stable spectrum}\\
\sigma_u(A) &= \{\lambda \in \sigma(A)\,,\ \mathrm{Re}\,\lambda > 0\} \quad \text{the unstable spectrum}\\
\{\underline{u}_1, \ldots, \underline{u}_p\} &= \text{a basis of the subspace } \mathcal{U} \text{ of all eigenvectors}\\
&\ \text{belonging to eigenvalues of } \sigma_s(A)\\
\{\underline{v}_1, \ldots, \underline{v}_q\} &= \text{a basis of the subspace } \mathcal{V} \text{ of all eigenvectors}\\
&\ \text{belonging to eigenvalues of } \sigma_u(A)
\end{aligned}$$

and let

$$\begin{aligned}
U &= [\underline{u}_1, \ldots \underline{u}_p]\,,\ P = U[U^TU]^{-1}U^T \ \text{projector onto } \mathcal{U}\\
V &= [\underline{v}_1, \ldots, \underline{v}_]\,,\ Q = V[V^TV]^{-1}V^T \ \text{projector onto } \mathcal{V}\,.
\end{aligned}$$

By assumption, $p + q$ must be equal to the dimension n of the system. With these notations it follows that

$$\underline{x}(t) = \int_{-\infty}^{t} e^{(t-s)A}P\,\underline{c}(s)\,ds - \int_{t}^{\infty} e^{(t-s)A}Q\underline{c}(s)\,ds\,.$$

(g2) Periodic solutions of the non-autonomic T-periodic system (1.37) can be obtained explicitly only if *all* solutions of the homogenous system $\dot{\underline{x}} = A(t)\underline{x}$ are known explicitly. Let $W(s,t)$ be the WRONSKI matrix of **(b)** then, by the property $W(r,s)W(s,t) = W(r,t)\,,\ W(r,r) = I$, there follows at first $W(t+T,T) = W(t+T,0)W(0,T)$ and then, by the periodicity of A, on the one side,

$$\begin{aligned}
\frac{d}{dt}W(t+T,T) &= \frac{d}{dt}W(t+T,0)W(0,T) = A(t+T)W(t+T,0)W(0,T)\\
&= A(t)W(t+T,0)W(0,T) = A(t)W(t+T,T)\,,
\end{aligned}$$

$W(0+T,T) = I$, and on the other side

$$\frac{d}{dt}W(t,0) = A(t)W(t,0)\,,\ \ W(0,0) = I\,.$$

Because of the unique existence of a solution of the initial value problem, we thus find that

$$W(t+T,0)W(0,T) = W(t+T,T) = W(t,0)\,;$$

hence with inversion

$$\boxed{W(t+T,0) = W(t,0)W(T,0)}\,.$$

The eigenvalues of the T-*translation* or *monodromy matrix* $W(T,0)$ are called FLOQUET *multipliers*; they are *independent* of the variable t! If $\lambda = 1$ is a FLOQUET multiplier, there exists a vector $\underline{a} \neq \underline{0}$ with $W(T,0)\underline{a} = W(0,0)\underline{a} = \underline{a}$. Accordingly,

$$\underline{y}(t+T) := W(t+T,0)\underline{a} = W(t,0)\underline{a} = \underline{y}(t)\,,$$

and thus $\underline{y}(t)$ is a T-periodic solution of the homogenous system. In inhomogenous systems FREDHOLM's alternative may be applied again and yields:

Lemma 1.9. (1°) *The homogenous T-periodic system $\dot{\underline{x}} = A(t)\underline{x}$ has a nontrivial T-periodic solution if and only if unity is a* FLOQUET *multiplier.*
(2°) *The inhomogenous T-periodic system $\dot{\underline{x}} = A(t)\underline{x} + \underline{c}(t)$ has a T-periodic solution if and only if unity is not a* FLOQUET *multiplier.*

References: (Amann).

1.6 The Flux Integral and its Vector Field

Hint: The notation "flux integral" is introduced here for abbreviation and does appear seldom in the literature with the same meaning.

(a) The Flux Integral Let $\mathcal{I} \subseteq \mathbb{R}$ be an open interval, let $\Omega \subseteq \mathbb{R}^n$ be an open, connected set (*domain*) and $\underline{v} : \mathcal{I} \times \Omega \to \mathbb{R}^n$ a C^r vector field, $\underline{v} \in C^r(\mathcal{I} \times \Omega; \mathbb{R}^n)$, with $r \geq 1$. (See Sect. 1.7 for the notation C^r).

Theorem 1.11. *(Unique Existence) For every pair $(t_0, \underline{x}_0) \in \mathcal{I} \times \Omega$ there exists a unique solution $\Phi(t; t_0, \underline{x}_0)$ of the initial value problem*

$$\dot{\underline{x}} = \underline{v}(t, \underline{x})\,, \quad \underline{x}(t_0) = \underline{x}_0\,, \tag{1.38}$$

with the following properties:

(1°) $\Phi(t_0; t_0, \underline{x}_0) = \underline{x}_0$;
(2°) Φ approaches the boundary of $\mathcal{I} \times \Omega$ arbitrary close — but not necessarily the ends of the interval $\mathcal{I}$ — ("cannot be continued further");
(3°) Φ is $(r+1)$-times continuously differentiable w.r.t. t;
(4°) Φ is r-times continuously differentiable w.r.t. t_0 and w.r.t. $\underline{x}_0$.

Proof see, e.g., (Arnold80).

The mapping

$$\Phi : \mathcal{I}_{t_0} \times \mathcal{I} \times \Omega \ni (t; t_0, \underline{x}_0) \mapsto \Phi(t; t_0, \underline{x}_0) \in \mathbb{R}^n\,, \quad \mathcal{I}_{t_0} \subset \mathcal{I} \text{ maximum}$$

is called *flux integral* of the vector field $\underline{v}$ in the sequel. It describes *all* solutions of the initial value problem (1.38) together and thus contains its complete information. One writes also briefly

$$\Phi(t; t_0, \underline{x}_0) = \underline{x}_0 + \int_{t_0}^{t} \underline{v}(\tau, \underline{x})\, d\tau \;\text{ for }\; \Phi(t; t_0, \underline{x}_0) = \underline{x}_0 + \int_{t_0}^{t} \underline{v}(\tau, \Phi(\tau; t_0, \underline{x}_0)\, d\tau\,.$$

(b) A **Stationary Vector Field** does not depend explicitly on the independent variable t. Then initial time is frequently set equal to zero, $t_0 = 0$, and this argument is entirely dropped in the integral Φ which is briefly called *flux* in this case. The mapping $\Phi(\circ, \underline{x}) : \mathcal{I} \ni t \mapsto \Phi(t, \underline{x})$ is called *phase curve* through $\underline{x}$ and $\Phi(\mathcal{I}, \underline{x})$ *orbit* of $\underline{x}$ (set of points of a phase curve) where $\underline{x}$ is supposed to be fixed both times.

Some properties:

(1°) Because

$$\underline{v}(\underline{x}) := \frac{\partial \Phi}{\partial t}(0, \underline{x}) \quad \Longrightarrow \quad \frac{\partial \Phi}{\partial t}(t, \underline{x}) = \underline{v}(\Phi(t, \underline{x}))\,,$$

$\underline{v}(\underline{x})$ is the phase velocity vector (tangent vector) of Φ.

(2°) For s, t with $s + t \in \mathcal{I}$, the following holds

$$\Phi(s + t, \;\cdot\;) = \Phi(t + s, \;\cdot\;) = \Phi(s, \;\cdot\;) \circ \Phi(t, \;\cdot\;) \equiv \Phi(s, \Phi(t, \;\cdot\;))\,,$$

$\Phi(0, \;\cdot\;) =$ identity, and Φ reveals to be a (local) *transformation group* in mathematical sense; see also Sect. 10.6.

(3°) Let $F : \Phi(\mathcal{I} \times \Omega) \cap \Omega \to \mathbb{R}^n$ be a reversible mapping being smooth in both directions (*diffeomorphism*) then $\Psi = F \circ \Phi$ is a flux with vector field $\underline{w}$,

$$\frac{\partial}{\partial t}\Psi(t, \underline{x})\Big|_{t=0} = \nabla F(\Phi(t, \underline{x}))\frac{\partial}{\partial t}\Phi(t, \underline{x})\Big|_{t=0} = \nabla F(\underline{x}) v(\underline{x})\,,$$

because $\Phi(0, \underline{x}) = \underline{x}$, and thus

$$\boxed{\forall\, \underline{x} \in \Omega : \underline{w}(\underline{x}) = \nabla F(\underline{x})\underline{v}(\underline{x})}\;.$$

(4°) A *fixed point* $\underline{x}$ of Φ with $\underline{x} = \Phi(t, \underline{x})\ \forall\, t \in \mathcal{I}$ yields $\underline{v}(\underline{x}) = \underline{0}$ by (1.38). All points with this property are *constant solutions* and are called *equilibrium points* or *singular points* of the vector field; points with $\underline{v}(\underline{x}) \neq \underline{0}$ are regular points of $\underline{v}$. Since exactly one phase curve passes through every point of the domain of $\underline{v} \in \mathcal{C}^1$ by Theorem 1.11 on unique existence, singular points are *isolated solutions* which can never be reached by other solutions in finite time; nevertheless they are sometimes called *bifurcation points*. A solution passing to a singular point in the limit $|t| \to \infty$ is called *separatrix* if that singular point is not a center. The next result shows that the global behavior of a vector field is determined *qualitatively* by its singular points and their topological character.

(c) Straightening of Vector Fields A nonlinear vector field $\underline{v}$ in a canonical coordinate system of coordinate space $\mathbb{R}^n$ has curvilinear phase curves for solutions. All these solutions together constitute the flux Φ of $\underline{v}$. Of course, a fixed vector $\underline{v}(x_0)$ can always be moved into, e.g., the first unit vector $\underline{e}_1$ by rotation or reflection. The question is now whether there exists *locally* a unique mapping F of which the gradient $\nabla F(\underline{x})$ has this property for all $\underline{x}$ in a *full* neighborhood of $\underline{x}_0$. The columns of the (regular) matrix $\nabla F(\underline{x})$ then constitute a curvilinear coordinate system (moving frame) along of which $\underline{v}(\underline{x}) = \underline{e}_1$ is *constant* in a suitable neighborhood $\mathcal{U}$ of $\underline{x}_0$. If such a mapping exists between two general vector fields, they are called *local similar.* The following result is sometimes called *Main Theorem of Ordinary Differential Systems* (Arnold80).

Theorem 1.12. *(Straightening Theorem) Let $\underline{v} : \Omega \to \mathbb{R}^n$ be a conservative vector field, let $\underline{x}_0 \in \Omega$, $\underline{v}(\underline{x}_0) \neq \underline{0}$, and let $\underline{e}_1 = [1, 0, \ldots, 0]^T \in \mathbb{R}^n$. Then there exists an open set $\mathcal{U}$ with $\underline{x}_0 \in \mathcal{U} \subset \mathbb{R}^n$ and a diffeomorphism $F : \mathcal{U} \to \mathcal{U}$ so that*

$$\forall\, \underline{x} \in \mathcal{U} : \underline{e}_1 = \nabla F(\underline{x})\underline{v}(\underline{x}) \,. \tag{1.39}$$

Proof see `SUPPLEMENT\chap01a` for $n = 3$; cf. also (Arnold80).

Corollary 1.2. *All smooth, conservative vector fields $\underline{v}$ are local similar in regular points $\underline{x}$ with $\underline{v}(\underline{x}) \neq \underline{0}$.*

In other words, if $\underline{v}$ and $\underline{w}$ are stationary, continuously differentiable vector fields and $\underline{v}(\underline{x}_0) \neq \underline{0}$, $\underline{w}(\underline{x}_0) \neq \underline{0}$, there exists a an open neighborhood $\mathcal{U}$ of $\underline{x}_0$ and a diffeomorphism H so that $\underline{w}(\underline{x}) = \nabla H(\underline{x})\underline{v}(\underline{x})$ for all $\underline{x} \in \mathcal{U}$.

If $\underline{v} : (t, \underline{x}) \mapsto \underline{v}(t, \underline{x})$ is a non-conservative vector field, we introduce the independent variable t for new *dependent* variable and consider the system

$$\dot{\underline{y}} = \begin{bmatrix} 1 \\ \underline{v}(\underline{y}) \end{bmatrix} , \quad \underline{y} = (t, \underline{x}) \,.$$

Corollary 1.3. *Let $\underline{v}(t_0, \underline{x}_0) \neq \underline{0} \in \mathbb{R}^n$ then there exists an open set $(t_0, \underline{x}_0) \in \mathcal{V} \subset \mathbb{R}^{n+1}$ and a diffeomorphism $G : \mathcal{V} \to \mathcal{V}$ so that*

$$\forall\, (t, \underline{x}) \in \mathcal{V} : \underline{e}_1 = \nabla G(t, \underline{x}) \begin{bmatrix} 1 \\ \underline{v}(t, \underline{x}) \end{bmatrix} \in \mathbb{R}^{n+1} ,$$

hence $\dot{t} = 1$, $\dot{\underline{x}} = \underline{v}(t, \underline{x})$ is local similar to the system $\dot{y}_1 = 1$, $\dot{y}_i = 0$, $i = 2 : n+1$ in every point $(t_0, \underline{x}_0)$ and, besides, t remains unchanged.

(d) Invariants A mapping $F : \Phi(\mathcal{I} \times \Omega) \to \mathbb{R}$ is called an *invariant* of Φ or a *first integral* if

$$\forall\, t \in \mathcal{I}\ \forall\, \underline{x} \in \Omega : F(\Phi(t, \underline{x})) = F(\underline{x}) \quad \Big(= F(\Phi(0, \underline{x}))\Big) .$$

In other words, F ist constant on every orbit of Φ. A mapping F is called *invariant* of a conservative vector field $\underline{v}$ if

$$\forall\, \underline{x} \in \Omega : \nabla F(\underline{x})\underline{v}(x) = 0 \in \mathbb{R} .$$

($\nabla F(\underline{x})$ is a row vector again.)

Theorem 1.13. *Let Φ be a flux with conservative vector field $\underline{v}$ then F is invariant of Φ if and only if F is invariant of $\underline{v}$.*

Proof. "$\Longrightarrow$" clear!. "$\Longleftarrow$" with ($g'(t) = 0 \Longrightarrow g(t) = c$). □

Let for instance

$$\dot{x} = H_y(x, y) \in \mathbb{R}\,,\ \dot{y} = -H_x(x, y) \in \mathbb{R}\,,$$

be a HAMILTON *system* then H is an invariant (implicit representation of the solution). Of course, *one* invariant suffices for implicit representation of the solution of a differential system with *two* equations and two dependent variables. More generally, $n - 1$ invariants are necessary for the implicit representation in a system $\dot{\underline{x}} = \underline{v}(\underline{x})$ with $\underline{x}(t) \in \mathbb{R}^n$.

Definition 1.1. *Let $\Omega \subset \mathbb{R}^n$ be open and let $\underline{f} \in \mathcal{C}^1(\Omega; \mathbb{R}^m)$. Then the components f^i of $\underline{f}$ are* functionally dependent *if*

$$\forall\, \underline{x} \in \Omega\ \exists\, \mathcal{U} \subset \mathbb{R}^n \text{ open } \exists\, 0 \neq G \in \mathcal{C}^1(\underline{f}(\Omega); \mathbb{R})\ \forall\, \underline{x} \in \mathcal{U} : G(\underline{f}(\underline{x})) = 0\,;$$

otherwise they are functionally independent.

Lemma 1.10. *Let $\underline{f} \in \mathcal{C}^1(\Omega; \mathbb{R}^m)$ then the components f^i of $\underline{f}$ are functionally independent if and only if $\nabla \underline{f}(\underline{x})$ has rank m for all $\underline{x} \in \Omega$.*

Let $\underline{v} : \Omega \to \mathbb{R}^n$ be a vector field and let $f^1, \ldots, f^k$ be the maximum number of functionally independent invariants then the most general invariant is given by $\Psi(f^1, \ldots, f^k)$ with arbitrary $\Psi \in \mathcal{C}^1$.

The following lemma supplies the desired information about the *local existence* of invariants.

Lemma 1.11. *Let $\underline{v} : \mathbb{R}^n \supset \Omega \to \mathbb{R}^n$ be a conservative $\mathcal{C}^1$ vector field and let $\underline{v}(\underline{x}_0) \neq \underline{0}$. Then there exist exactly $n - 1$ functionally independent invariants $F^i \in \mathcal{C}^1(\mathcal{U}; \mathbb{R})$ in a suitable open neighborhood $\underline{x}_0 \in \mathcal{U} \subset \Omega$.*

Proof. (1°) Let $\underline{v}(\underline{x}) = \underline{e}_1$. Then the simple mappings $F^i : \underline{x} \mapsto x_i$, $i = 2 : n$, are invariants and functionally independent.
(2°) Let $\underline{v}$ be general with $\underline{v}(x_0) \neq \underline{0}$, let F be the local straightening mapping so that (1.39) holds. Then the i-th component F^i of F is an invariant for $i = 2 : n$. Namely, if $v^1(x_0) \neq 0$ without loss of generality,

$$F^i(\Phi(t, \underline{X})) = \text{ const } \Longleftrightarrow (F^i \circ \Phi)_t(t, \underline{X}) = 0$$
$$\Longleftrightarrow \nabla F^i(\Phi(t, \underline{X})) \cdot \Phi_t(t, \underline{X}) = 0 \Longleftrightarrow [\nabla F(x)]^i \cdot \underline{v}(\underline{x}) = 0 , \; i = 2 : n ,$$

for $\underline{x} = \Phi(t, \underline{X}) \in \mathcal{U}(\underline{x}_0)$. But we have

$$\nabla F(\underline{x})\underline{v}(\underline{x}) = \underline{e}_1 , \quad \underline{x} \in \mathcal{U}(\underline{x}_0)$$

by assumption, hence the assertion is proved. □

(e) Transformation Let once more $F : \mathbb{R}^n(\underline{x}) \supset \Omega \to \mathbb{R}^n(\underline{y})$ be a diffeomorphism writing $\underline{y} = F(\underline{x})$, $\underline{x} = F^{-1}(\underline{y})$, and let $\Psi(t, \underline{x}) = F(\Phi(t, \underline{x}))$ be the flux Ψ with vector field $\underline{w}(\underline{x}) = \nabla F(\underline{x})\underline{v}(\underline{x})$ under the mapping F , then $\underline{y} = \Psi(0, \underline{x}) = F(\Phi(0, \underline{x})) = F(\underline{x})$. What does the flux Ψ look like in $\underline{y}$-coordinates, i.e., if the $\underline{x}$-coordinates are transformed by F, too? Of course it is the flux $\widetilde{\Psi}(t, \underline{y}) = \Psi(t, F^{-1}(\underline{y}))$ with vector field $\widetilde{\underline{w}}(\underline{y}) = \nabla F(\underline{x})\underline{v}(\underline{x})$, $\underline{x} = F^{-1}(\underline{y})$. One verifies easily the flux properties:

$$\widetilde{\Psi}(0, \underline{y}) = \Psi(0, F^{-1}(\underline{y})) = F(\Phi(0, F^{-1}(\underline{y}))) = F(F^{-1}(\underline{y})) = \underline{y}$$

and, using $\underline{x} = F^{-1}(\underline{y})$,

$$\begin{aligned}\widetilde{\Psi}(s + t, \underline{y}) &= \Psi(s + t, \underline{x}) = F(\Phi(s + t, \underline{x})) = F(\Phi(s, \Phi(t, \underline{x}))) \\ &= \Psi(s, F^{-1}(F(\Phi(t, \underline{x}))) = \widetilde{\Psi}(s, F(\Phi(t, \underline{x}))) = \widetilde{\Psi}(s, \Psi(t, \underline{x})) \\ &= \widetilde{\Psi}(s, \Psi(t, F^{-1}(\underline{y})) = \widetilde{\Psi}(s, \widetilde{\Psi}(t, \underline{y})) .\end{aligned}$$

(f) Examples

Example 1.7. The system $\dot{\underline{x}} = \lambda \underline{x}$ has the general solution $\underline{x}(t) = \underline{c} e^{\lambda t}$. A substitution of $\underline{x}(t_0) = \underline{x}_0$ yields the flux

$$\Phi(t; t_0, \underline{x}_0) = \underline{x}_0 e^{\lambda(t - t_0)} .$$

Example 1.8. The autonomeous system $\dot{x}_1 = x_2(1 - x_1)$, $\dot{x}_2 = -x_1(1 - x_2)$ has the critical points $\underline{x} = (0, 0)$ and $\underline{x} = (1, 1)$ and the first integral (implicit solution)

$$F(x, y) = \ln |y - 1| + \ln |1 - x| + y + x = c .$$

Example 1.9. The differential equation $\dot{x} = e^x \sin t$ has the general solution $x(t) = -\ln(\cos t + c)$. By substitution of (t_0, x_0) we obtain again the flux integral

$$\Phi(t; t_0, \underline{x}_0) = -\ln(\cos t + \exp(-x_0) - \cos t_0) .$$

However, the solutions exist for all $t \in \mathbb{R}$ only on and below the bold curve in Fig. 1.7.

Example 1.10. For displaying the flux integral of $\dot{x} = (t - x + 3)^2$, we have to decompose the (t, x)-plane into
$\Omega_1 = \{(t, x),\ |t - x + 3| < 1\}$, $\Omega_2 = \{(t, x),\ t - x + 3 < -1\}$,
$\Omega_3 = \{(t, x),\ t - x + 3 > 1\}$. One computes then

$$\Phi(t; t_0, x_0) = t + 3 - \frac{(t_0 - x_0 + 4)e^{2t} + (t_0 - x_0 + 2)e^{2t_0}}{(t_0 - x_0 + 4)e^{2t} - (t_0 - x_0 + 2)e^{2t_0}}$$

for $(t_0, x_0) \in \Omega_2 \cup \Omega_3$ and

$$\Phi(t; t_0, x_0) = t + 3 - \frac{(t_0 - x_0 + 4)e^{2t} - (t_0 - x_0 + 2)e^{2t_0}}{(t_0 - x_0 + 4)e^{2t} + (t_0 - x_0 + 2)e^{2t_0}}$$

for $(t_0, x_0) \in \Omega_1$; the solutions exist only in Ω_1 for all $t \in \mathbb{R}$.

Example 1.11. The HAMILTON system $\dot{x} = y(1 - x^2)$, $\dot{y} = x(y^2 - 1)$ has the critical points

$$\underline{x}_0 = (0, 0),\ \underline{x}_1 = (1,\ 1),\ \underline{x}_2 = (1,\ -1),\ \underline{x}_3 = (-1,\ 1),\ \underline{x}_4 = (-1,\ -1),$$

and for invariant the HAMILTON function

$$H(x, y) = \frac{1}{2}\left(x^2 + y^2 - x^2 y^2\right) = d.$$

All separatrices are straight line segments and can be found explicitly. For instance, inserting $x = 1$ yields

$$\dot{x} = 0, \qquad \dot{y} = y^2 - 1 \iff \frac{\dot{y}}{y^2 - 1} = 1.$$

Separation of variables yields with partial fraction expansion

$$\frac{\dot{y}}{y - 1} - \frac{\dot{y}}{y + 1} = 2.$$

Therefore

$$\ln|y - 1| - \ln|y + 1| = \ln\frac{|y - 1|}{|y + 1|} = 2t + c_1 \implies \frac{|y - 1|}{|y + 1|} = c\,e^{2t}$$

where $c = 1$ without loss of generality.

$$\begin{aligned}
y > 1 : y(t) &= \frac{1 + e^{2t}}{1 - e^{2t}}, \quad -\infty < t < 0, \lim_{t\to-\infty} y(t) = 1, \lim_{t\to 0-} y(t) = \infty \\
y < -1 : y(t) &= \frac{1 + e^{2t}}{1 - e^{2t}}, \quad 0 < t < \infty, \lim_{t\to\infty} y(t) = -1, \lim_{t\to 0+} y(t) = -\infty \\
|y| < 1 : y(t) &= \frac{1 - e^{2t}}{1 + e^{2t}}, \quad -\infty < t < \infty, \lim_{t\to\infty} y(t) = -1, \lim_{t\to-\infty} y(t) = 1.
\end{aligned}$$

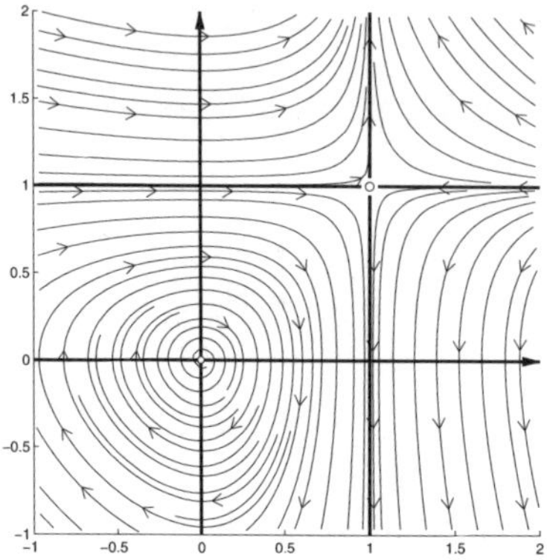

Figure 1.6. Example 1.8

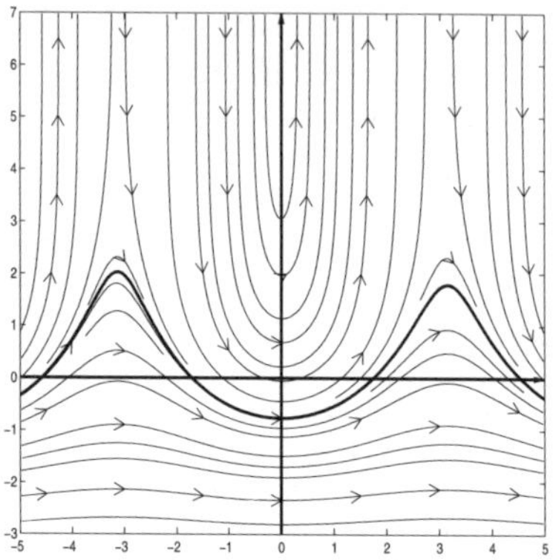

Figure 1.7. Example 1.9

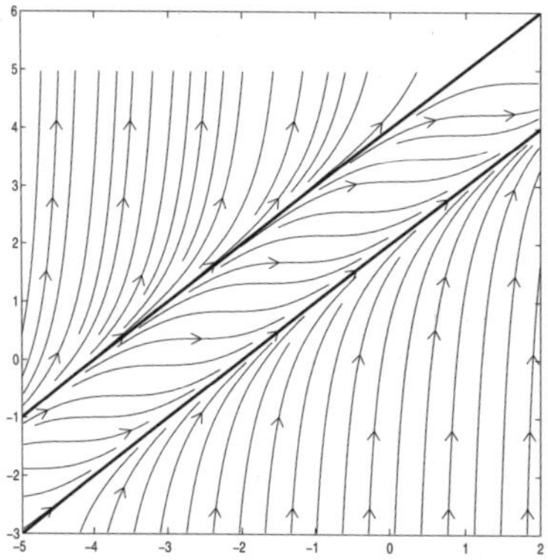

Figure 1.8. Example 1.10

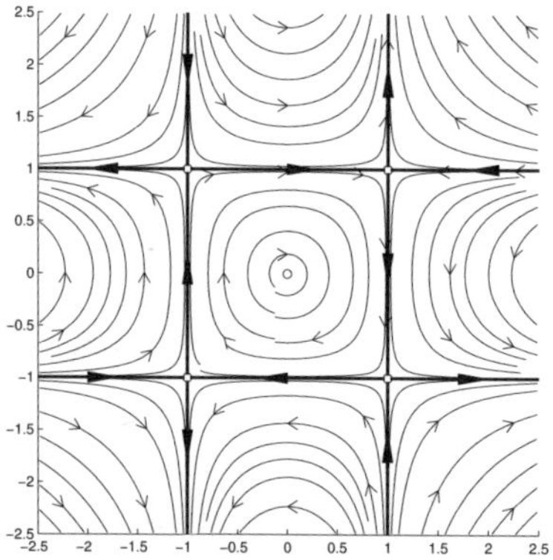

Figure 1.9. Example 1.11

Singular values for the restricted three-body problem are computed in Sect. 6.5.

References: (Amann), (Arnold80).

1.7 Vector Spaces

Let $\Omega \subset \mathbb{R}^n$ be a bounded, open and connected set (bounded *domain*) with continuous, piecewise smooth boundary $\Gamma = \partial\Omega$, and let $\overline{\Omega} = \Omega \cup \partial\Omega$ be the closure of Ω.

(a) Spaces of Continuous Functions

$$\begin{aligned}
\mathcal{C}(a,b) &:= \mathcal{C}((a,b),\mathbb{R}) := \text{vector space of continuous functions}\\
&\qquad f:(a,b)\to\mathbb{R}\\
\mathcal{C}[a,b] &:= \mathcal{C}([a,b],\mathbb{R}) := \text{vector space of continuous functions } f:[a,b]\to\mathbb{R}\\
\mathcal{C}^s(a,b) &:= \text{subspace of functions in } \mathcal{C}(a,b)\\
&\qquad \text{with } \textit{continuous} \text{ derivatives up to order } s\ge 1\\
\mathcal{C}^s[a,b] &:= \text{subspace of functions in } \mathcal{C}[a,b]\\
&\qquad \text{which together with their derivatives up to order } s\\
&\qquad \text{can be continued on } [a,b] \text{ to continuous functions}\\
\mathcal{C}^\infty(a,b) &:= \text{subspace of functions in } \mathcal{C}(a,b)\\
&\qquad \text{having derivatives of arbitrary order}\,.
\end{aligned} \tag{1.40}$$

In these examples, the null element is the constant function $u(x) = 0$ for $x \in (a,b)$ or $x \in [a,b]$, respectively. The spaces $\mathcal{C}(\Omega)$, $\mathcal{C}(\overline{\Omega})$, $\mathcal{C}^s(\Omega)$, $\mathcal{C}^s(\overline{\Omega})$ $\mathcal{C}^\infty(\Omega)$ are defined as in (1.40) with "derivative" replaced by "partial derivative" and "order" by "total order" in the sense of the subsequent Example 1.12.

(b) Banach Spaces As well known, a sequence $\{x_n\}$ in a (real) *normed vector space* $\mathcal{X}$ is called CAUCHY sequence if $\forall\ \varepsilon > 0\ \ \exists\ N_\varepsilon\ \ \forall\ m,n > N_\varepsilon$: $\|x_m - x_n\| < \varepsilon$. By the following result, every CAUCHY sequence is convergent to an element *in* $\mathcal{X}$ independent of the just applied norm in case $\mathcal{X}$ is *finite-dimensional.*

Theorem 1.14. *(Norm Equivalence Theorem) Let* $\dim(\mathcal{X}) < \infty$ *and let* $\|\circ\|_\alpha$, $\|\circ\|_\beta$ *be two norms in* $\mathcal{X}$ *then there exist uniform constants* $0 < m \le M$ *such that*

$$\forall\, x \in \mathcal{X} : m\,\|x\|_\alpha \le \|x\|_\beta \le M\,\|x\|_\alpha\,.$$

Proof see (Stoer), Sect. 4.4.

However, this result is no longer valid in spaces $\mathcal{X}$ of infinite dimension as, e.g., $\mathcal{C}(a,b)$ with the maximum norm. A normed vector space $\mathcal{X}$ is called BANACH *space* if every CAUCHY sequence with elements in $\mathcal{X}$ does converge to an element *belonging to* $\mathcal{X}$. In this case, a subset $\mathcal{M} \subset \mathcal{X}$ as well as $\mathcal{X}$ is called *closed* or *complete* w.r.t. the present norm. The two notions differ from each other only in more general topological vector spaces.

Example 1.12. The spaces of functions $\mathcal{C}(\overline{\Omega})$ and $\mathcal{C}^s(\overline{\Omega})$ are BANACH spaces with the supremum norm

$$\|f\|_{\mathcal{C}} := \sup_{x\in\Omega} |f(x)|\,,\ \text{and}\ \|f\|_{\mathcal{C}^s} := \sup\{|D^\sigma f(x)|,\ x\in\Omega,\ |\sigma|\le s\}\,,$$

respectively, where $x = (x^1, \ldots, x^n)$, $\sigma_i \in \mathbb{N}_0 := \mathbb{N} \cup \{0\}$, $\sigma = (\sigma_1, \ldots, \sigma_n)$, $|\sigma| = \sigma_1 + \cdots + \sigma_n$, and

$$D^\sigma f = \frac{\partial^{|\sigma|} f}{(\partial x^1)^{\sigma_1} \cdots (\partial x^n)^{\sigma_n}}.$$

(c) Linear Mappings Let $\mathcal{X}$ and $\mathcal{Y}$ be normed vector spaces, then a linear mapping $A : \mathcal{X} \to \mathcal{Y}$ is *continuous* if and only if it is *bounded*, i.e.,

$$\exists\, \kappa > 0 \;\; \forall\, x \in \mathcal{X} : \|Ax\| \leq \kappa \|x\| .$$

If the *image space* $\mathcal{Y}$ is a BANACH-Raum, the vector space $\mathcal{L}(\mathcal{X}, \mathcal{Y})$ of *continuous linear* mappings $\mathcal{X}$ in $\mathcal{Y}$ with the *operator norm*

$$\|A\| = \sup_{\|x\|=1} \|Ax\| \quad \left(= \sup_{\|x\|\leq 1} \|Ax\| = \sup_{x \neq 0} \frac{\|Ax\|}{\|x\|} \right)$$

is also a BANACH space.

(d) Linear Functionals and Hyperplanes Of course, $\mathbb{R}$ with the absolute value for norm is a BANACH space hence the vector space $\mathcal{L}(\mathcal{X}, \mathbb{R})$ of continuous linear *functionals* $f : \mathcal{X} \to \mathbb{R}$ is a BANACH space by **(c)**.

Example 1.13. Let $f(x) = \int_a^b x(t)\, dt$, x continuous and let $\|x\| = \sup_{a \leq t \leq b} |x(t)|$, then

$$\|f\| = \sup_{\|x\|=1} \left| \int_a^b x(t)\, dt \right| \leq \sup_{\|x\|=1} \|x\|(b-a) = b - a .$$

Let now $\mathcal{U} \subset \mathcal{X}$ be a linear subspace and let $0 \neq v \in \mathcal{X}$ be arbitrary then

$$\mathcal{V} = v + \mathcal{U} := \{x \in \mathcal{X},\; x - v \in \mathcal{U}\}$$

is called an *affine (linear) subspace.* Here, $\mathcal{U}$ is determined uniquely whereas v can be chosen arbitrarily in $\mathcal{V}$. The *maximum* affine subspaces $\mathcal{V}$ with the property

$$\exists\, w \in \mathcal{X} : (w \notin \mathcal{V}) \wedge (\mathcal{X} = \operatorname{span}\{w, \mathcal{V}\})$$

are called *hyperplanes.* In other words, there is no "place" for a further subspace $\mathcal{W}$ "between" a hyperplane $\mathcal{V} \subset \mathcal{X}$ and the vector space $\mathcal{X}$ itself:

$$\mathcal{V} \subset \mathcal{W} \subset \mathcal{X} \implies (\mathcal{V} = \mathcal{W}) \wedge (\mathcal{W} = \mathcal{X}) .$$

The following result displays the strong connection between linear functionals and hyperplanes.

Lemma 1.12. (1°) *$\mathcal{H} \subset \mathcal{X}$ is a hyperplane if and only if*

$$\exists\, 0 \neq f : \mathcal{X} \to \mathbb{R} \text{ linear } \exists\, c \in \mathbb{R} : \mathcal{H} = \{x \in \mathcal{X},\; f(x) = c\} .$$

(2°) *If $0 \notin \mathcal{H} \subset \mathcal{X}$ is a hyperplane, there exists a unique linear functional $f : \mathcal{X} \to \mathbb{R}$ with $\mathcal{H} = \{x \in \mathcal{X},\; f(x) = 1\}$.*
(3°) *If $0 \neq f : X \to \mathbb{R}$ is linear and $\mathcal{H} = \{x \in \mathcal{X},\; f(x) = c\}$ is a hyperplane, then $\mathcal{H} = \overline{\mathcal{H}}$ closed if and only if the mapping f is continuous.*

Proof (Luenberger) and `SUPPLEMENT\chap01a`.

Let $\mathcal{H} \subset \mathcal{X}$ be a hyperplane then either $\mathcal{H}$ closed, $\mathcal{H} = \overline{\mathcal{H}}$, or $\mathcal{H}$ *dense* in $\mathcal{X}$, $\overline{\mathcal{H}} = \mathcal{X}$; hence, e.g., the set of continuous functions is a hyperplane in the space of functions being quadratic integrable in LEBESGUE sense.

(e) Dual Spaces The vector space $\mathcal{L}(\mathcal{X}, \mathbb{R})$ of continuous linear functionals on $\mathcal{X}$ is called (topological) *dual space* $\mathcal{X}_d$ of $\mathcal{X}$. $\mathcal{X}_d$ is always a BANACH space also if $\mathcal{X}$ is only a normed vector space. Roughly spoken we have $\mathcal{X} \subset (\mathcal{X}_d)_d$ and, again roughly spoken, BANACH spaces with the property $\mathcal{X} = (\mathcal{X}_d)_d$ are called *reflexive*.

Example 1.14. Let Π_n be the space of real polynomials of degree $\leq n$ with dimension $(n+1)$ and let $x_i \in \mathbb{R}$, $i = 1 : n+1$, be $n+1$ arbitrary *different* numbers. Then the dual space $[\Pi_n]_d$ of Π_n is defined by an arbitrary basis $\mathcal{F}$ of functionals, e.g.,

$$\mathcal{F} := \{f_i(p) := p(x_i)\,,\ i = 1 : n+1\}\,.$$

Choose for Π_n the basis $\mathcal{E}$ of the LAGRANGE fundamental polynomials $p_1, \ldots, p_{n+1}$ with the property

$$p_i(x_k) = \delta^i{}_k \quad \text{(KRONECKER symbol)}$$

then $f_i(p_k) = \delta^i{}_k$ and $\mathcal{F}$ reveals to be a *dual basis* to $\mathcal{E}$; see also Chap. 10.

(f) Hilbert Spaces A *scalar product* in $\mathcal{X}$ is a positive definite symmetric bilinear form, i.e., a *bilinear mapping* $(\diamond, \circ) : \mathcal{X} \times \mathcal{X} \to \mathbb{R}$ with the properties

$$\begin{gathered}
\forall\, x \in \mathcal{X} : (x, \circ) : \mathcal{X} \to \mathbb{R} \text{ linear} \\
\forall\, y \in \mathcal{X} : (\diamond, y) : \mathcal{X} \to \mathbb{R} \text{ linear} \\
\forall\, x, y \in \mathcal{X} : \ (x, y) = (y, x) \\
(x, x) = 0 \Longleftrightarrow x = 0;\ \forall\, 0 \neq x \in \mathcal{X} : \ (x, x) > 0\,.
\end{gathered}$$

By means of the CAUCHY-SCHWARZ *inequality*,

$$\forall\, x, y \in \mathcal{X} : \ 0 \leq (x, y)^2 \leq (x, x)\,(y, y)\,,$$

it can be shown easily that a scalar product defines a norm on $\mathcal{X}$ by $\|x\|^2 = (x, x)$.

In general, there are arbitrary many scalar products in a vector space. If the norm is defined by special selected *canonical* scalar product, the space $\mathcal{X}$ is called *scalar product space* or pre-HILBERT space. If, additionally, the space is complete w.r.t. *this canonical* norm, it is called a HILBERT space.

Conversely, suppose that the vector space is normed and that the parallelogram identity $\|x+y\|^2 + \|x-y\|^2 = 2(\|x\|^2 + \|y\|^2)$ is valid for all elements x, y then a scalar product can be defined in that space by $(x, y) := \left(\|x+y\|^2 - \|x-y\|^2\right)/4$.

(g) Sobolev Spaces In this subsection we are forced to make use of the LEBESGUE-Integral (L-Integral). The main reason that this integral is preferred over the RIEMANN integral in theoretical investigations is a property of *completeness* namely that appropriate limits of integrable functions are integrable again. However, we do not pursue this concept further but refer, e.g., to (Halmos) for details. The L-integral may be conceived as generalization of the RIEMANN-Integral (R-Integral) to a larger class of functions (with less smoothness). A set $\Omega \subset \mathbb{R}^n$ with L-Integral $\int_\Omega dx = 0$ has per definitionem the LEBESGUE *measure* zero; see also Sect. 12.5. LEBESGUE integrals do not make a difference between functions which are identical *almost everwhere* (a.e.), i.e., which differ only on sets of L-measure zero. Every *proper* R-Integral is also a L-integral but the converse is not true. For instance, let f be the DIRICHLET function satisfying $f(x) = 1$ for rational $x \in (0,1)$ and $f(x) = 0$ for irrational $x \in (0,1)$ then

$$\text{R-integral} \int_0^1 f(x)\,dx \text{ not defined, L-Integral} \int_0^1 f(x)\,dx = 0\,.$$

However, sets of L-measure zero can be much larger than the countable set of rational numbers in $[0, 1]$.

In the below described properties of some commonly used HILBERT spaces, the continuous and piecewise smooth boundary of the bounded domain Ω must satisfy an additional regularity condition ("cone condition") cf., e.g. (Agmon), on which we refer simply by the adjective "regular". The *boundary* of the domain Ω is always denoted by $\partial\Omega$ or simply by Γ.

Further notations:

$$\begin{aligned}
\widetilde{\mathcal{L}}_2(\Omega) &= \{f : \Omega \to \mathbb{R},\ \int_\Omega |f(x)|^2\,dx \text{ exists finitely in L-sense}\}\\
&\quad\ \text{set of quadratically integrable functions on } \Omega \text{ in L-sense}\\
\mathcal{M}_2(\Omega) &= \{f \in \widetilde{\mathcal{L}}_2(\Omega),\ \int_\Omega |f(x)|^2\,dx = 0\} \text{ set of functions}\\
&\quad\ \text{being zero almost everywhere (a.e.) on } \Omega \text{ in L-sense}\\
\mathcal{L}_2(\Omega) &= \widetilde{\mathcal{L}}_2(\Omega)/\mathcal{M}_2(\Omega)\ \textit{classes} \text{ of functions being a.e. identical on } \Omega\\
\mathrm{Sup}(f) &= \overline{\{x \in \mathbb{R}^n,\ f(x) \neq 0\}} \text{ support of } f : \mathbb{R}^n \to \mathbb{R}\\
\mathcal{C}_0^s(\Omega) &= \{f \in \mathcal{C}^s(\Omega)\,, D^\sigma f = 0 \text{ on } \Gamma \text{ for all } |\sigma| \leq s\}\\
\mathcal{C}_0^\infty(\Omega) &= \{f \in \mathcal{C}^\infty(\Omega),\ \mathrm{Sup}(f) \subset \Omega\}
\end{aligned}$$

By $\mathrm{Sup}(f) \subset \Omega$, it follows that $f = 0$ on Γ. The bilinear form

$$(f,g)_0 := \int_\Omega f(x)g(x)\,dx \tag{1.41}$$

is a scalar product on $\mathcal{L}_2(\Omega)$, and the space is complete w.r.t. the norm $\|f\| = (f,f)_0^{1/2}$, hence a HILBERT space. However, the space $\mathcal{C}(\Omega)$ is *not* complete w.r.t. this norm, hence not a HILBERT space; equally,

$$(f,g)_s := (f,g)_0 + \sum_{1\le|\sigma|\le s} (D^\sigma f, D^\sigma g)_0$$

is a scalar product on $\mathcal{C}^s(\Omega)$ but, again, the space is not HILBERT space. We write

$$|f|_s := \Big[\sum_{|\sigma|=s} (D^\sigma f, D^\sigma f)_0\Big]^{1/2}, \quad \|f\|_s := (f,f)_s^{1/2} \quad (\text{"SOBOLEV norm"}). \tag{1.42}$$

Then, of course, the closures w.r.t. $\|\circ\|_s$, i.e.,

$$\mathcal{H}^s(\Omega) := \overline{\mathcal{C}^s(\Omega)}^{\|\cdot\|_s}, \quad \mathcal{H}_0^s(\Omega) := \overline{\mathcal{C}_0^s(\Omega)}^{\|\cdot\|_s}, \tag{1.43}$$

are HILBERT spaces with the scalar product $(\diamond,\circ)_s$ (SOBOLEV spaces, the letter H in honor of HILBERT). All these spaces need the LEBESGUE integral in the bilinear form (1.41) for completeness, therefore this concept is an indispensable tool in the theory of elliptic boundary value problems.

(h) On Boundary Values The elements of $\mathcal{H}^1(\Omega)$ are not necessarily continuous and, by the way, only defined a.e. in LEBESGUE sense. It is therefore to be explained which values they take on the boundary Γ of Ω.

Theorem 1.15. *(Trace Theorem) Let Ω be a bounded, regular domain. Then* (1°) $\exists\, 0 < M(\Omega)\ \forall\, f \in \mathcal{C}^s(\overline{\Omega})\ \forall\, |\sigma| \le s-1$*:*

$$\Big[\int_\Gamma |D^\sigma f|^2\, ds\Big]^{1/2} \le M(\Omega)\|D^\sigma f\|_{1,\Omega}\,.$$

(2°) *There exists a unique continuous linear mapping (*trace operator*)*

$$\mathrm{Tr}_\sigma : \mathcal{H}^s(\Omega) \to \mathcal{L}_2(\Gamma)\,, \quad |\sigma| \le s-1\,,$$

with the property

$$\forall\, f \in \mathcal{C}^s(\overline{\Omega}) : \mathrm{Tr}_\sigma f = D^\sigma f\Big|_\Gamma\,.$$

(3°) *If $M(\Omega)$ is the constant of* (1°),

$$\forall\, f \in \mathcal{C}^s(\overline{\Omega})\ \forall\, |\sigma| \le s-1 : \|\mathrm{Tr}_\sigma(f)\|_{0,\Gamma} \le M(\Omega)\|D^\sigma f\|_{1,\Omega}\,.$$

Accordingly, at first the trace operator Tr_σ is defined by means of functions $f \in \mathcal{C}^s(\overline{\Omega})$ and then the boundary values of functions $f \in H^s(\Omega)$ are defined by means of the trace operator, e.g., $f\Big|_\Gamma = \mathrm{Tr}_0(f)$.

Corollary 1.4.

$$f \in \mathcal{H}_0^s(\Omega) \iff f \in \mathcal{H}^s(\Omega) \text{ and } \forall\, |\sigma| \le s-1 : \mathrm{Tr}_\sigma(f) = 0\,.$$

(i) Properties of $\mathcal{H}_0^s(\Omega)$ and $\mathcal{H}^s(\Omega)$. Up to now, the spaces (1.43) are only known as closures w.r.t. some norm. For a better understanding we need the concept of *weak derivative*:

$$\boxed{\begin{array}{l} f^{(\sigma)} \text{ is the weak } \sigma\text{-derivative of } f \text{ if, in L-sense,} \\ \forall\, \varphi \in \mathcal{C}_0^{|\sigma|}(\Omega) : \displaystyle\int_\Omega f^{(\sigma)} \varphi \, dx = (-1)^{|\sigma|} \int_\Omega f D^\sigma \varphi \, dx \end{array}} .$$

By using the vector spaces

$$\begin{aligned} \mathcal{W}^s(\Omega) &:= \{f \in \mathcal{L}_2(\Omega),\ f^{(\sigma)} \in L_2(\Omega) \text{ for } |\sigma| \le s\} , \\ \mathcal{W}_0^s(\Omega) &:= \{f \in W^s(\Omega),\ f^{(\sigma)} = 0 \text{ on } \Gamma \text{ for } |\sigma| \le s\} , \end{aligned}$$

we can now present a result of (Meyers) which supplies the crucial information about the SOBOLEV spaces $\mathcal{H}^s$ and $\mathcal{H}_0^s$:

Theorem 1.16. *Let $\Omega \subset \mathbb{R}^n$ be an arbitrary open set and $s \in \mathbb{N}_0 = \mathbb{N} \cup \{0\}$, then*

$$\mathcal{W}^s(\Omega) = \mathcal{H}^s(\Omega), \quad \mathcal{W}_0^s(\Omega) = \mathcal{H}_0^s(\Omega) .$$

Further information on the smoothness of the elements of these spaces is supplied by the famous *imbedding theorem of* SOBOLEV, cf. (Agmon):

Theorem 1.17. (1°) *Let $\Omega \subset \mathbb{R}^n$ be a bounded set and $s > n/2$ then $\mathcal{H}_0^s(\Omega) \subset \mathcal{C}(\overline{\Omega})$ and*

$$\exists\, \kappa(\Omega) > 0 \ \forall\, f \in \mathcal{H}_0^s(\Omega) : \|f\|_C \le \kappa(\Omega)\, \|f\|_s .$$

(2°) *Let Ω be regular and $s > n/2$ then $\mathcal{H}^s(\Omega) \subset \mathcal{C}(\overline{\Omega})$ and*

$$\exists\, \kappa(\Omega) > 0 \ \forall\, f \in \mathcal{H}^s(\Omega) : \|f\|_C \le \kappa(\Omega)\, \|f\|_s .$$

(3°) *Let Ω be regular and $r - s > n/2$ then $\mathcal{H}^r(\Omega) \subset \mathcal{C}^s(\overline{\Omega})$ and*

$$\exists\, \kappa(\Omega) > 0 \ \forall\, f \in \mathcal{H}^s(\Omega) : \|f\|_{C^s} \le \kappa(\Omega)\, \|f\|_s .$$

More precisely, all these three inclusions $A \subset B$ are valid only w.r.t. to a suitable imbedding. For $n = 2$, $\mathcal{H}^1(\Omega) \subset \mathcal{C}(\overline{\Omega})$ is not correct but only $\mathcal{H}^1(\Omega) \subset \mathcal{L}_2(\Omega)$ (Theorem of RELLICH).

Also a further connection between the spaces $\mathcal{H}^s(\Omega)$ und $\mathcal{H}_0^s(\Omega)$ can be realized by means of weak derivatives. For this, let $\underline{n}$ be the normed normal vector a.e. on the boundary Γ showing in "outer" space, and let $(\partial_n)^r f$ denote the r-th directional derivative in weak sense, $(\partial_n)^0 f = f$. Then

$$\mathcal{H}_0^s(\Omega) = \{f \in \mathcal{H}^s(\Omega),\ (\partial_n)^r f = 0,\ r = 0 : s-1, \text{ on } \Gamma\} , \tag{1.44}$$

hence, in particular,

$$\mathcal{H}_0^2(\Omega) = \{f \in \mathcal{H}^2(\Omega),\ f = \partial_n f = 0 \text{ on } \Gamma\} .$$

(j) Equivalent Norms on $\mathcal{H}_0^s(\Omega)$ and $\mathcal{H}^s(\Omega)$. Note that $|p|_s = 0$ for all polynomials $p \in \Pi_{s-1}$ of degree less than or equal $s-1$ hence $|f|_s$ is only a *semi-norm* on $\mathcal{H}^s(\Omega)$. For the considerations in Sect. 9.1 we have to modify the functional $|\circ|_s$ such that it becomes a *norm* on $\mathcal{H}^s(\Omega)$ with the equivalence relation

$$\boxed{\exists\, 0 < m(\Omega) < M(\Omega)\ \forall\, f \in \mathcal{H}^s(\Omega):\ m(\Omega)\|f\|_s \le |f|_s \le M(\Omega)\|f\|_s}\,. \tag{1.45}$$

In the case of the space $\mathcal{H}_0^s(\Omega)$, these inequalities can be verified in a rather simple way by partial integration because all boundary terms disappear; cf., e.g., (Braess), Sect. 2.1:

Lemma 1.13. *(*POINCARÉ-FRIEDRICHS *Inequality) Let Ω be a bounded domain then (1.45) holds for $\mathcal{H}_0^s(\Omega)$ in place of $\mathcal{H}^s(\Omega)$.*

For $\mathcal{H}^s(\Omega)$, the right inequality follows at once by definition of the SOBOLEV norm but the left inequality reveals to be less accessible like all estimations from below. The following result can be proved only by applying some deeper auxiliaries of functional analysis; see e.g. (Velte), Sect. 2.2.

Theorem 1.18. *Let $\Omega \subset \mathbb{R}^n$ be a bounded regular domain and let $|\circ|_\Pi$ be an arbitrary norm on Π_{s-1} being at least a semi-norm on $\mathcal{H}^s(\Omega)$ then*

$$\exists\, 0 < m(\Omega)\ \forall\, f \in \mathcal{H}^s(\Omega):\ m(\Omega)\|f\|_s \le \|f\|$$

for the norm on $\mathcal{H}^s(\Omega)$ defined by $\|f\|^2 = |f|_\Pi^2 + |f|_s^2$.

Example 1.15. (1°) The choice $|f|_\Pi^2 = \|f\|_0^2 = \int_\Omega f^2\,dV$ yields a norm on $\mathcal{H}^s(\Omega)$ therefore

$$\exists\, 0 < m(\Omega)\ \forall\, f \in \mathcal{H}^s(\Omega): m(\Omega)\|f\|_s^2 \le \|f\|_0^2 + |f|_s^2\,. \tag{1.46}$$

(2°) Let $\Gamma_D \subset \Gamma$ be a part of the boundary satisfying $\int_{\Gamma_D} dO > 0$ then $|f|_\Pi^2 = \int_{\Gamma_D} f^2\,dO$ is a norm on Π_0 which is a semi-norm on $\mathcal{H}^1(\Omega)$, hence

$$\exists\, 0 < m(\Omega)\ \forall\, f \in \mathcal{H}^s(\Omega): m(\Omega)\|f\|_1^2 < \int_{\Gamma_D} f^2\,dO + |f|_1^2\,. \tag{1.47}$$

(3°) By

$$|f|_\Pi^2 = \sum_{|\sigma|\le s-1}\left(\int_\Omega D^\sigma f\,dx\right)^2,$$

a norm is defined on Π_{s-1} being a semi-norm on $\mathcal{H}^s(\Omega)$, hence

$$\exists\, 0 < m(\Omega)\ \forall\, f \in \mathcal{H}^s(\Omega): m(\Omega)\|f\|_s^2 \le \sum_{|\sigma|\le s-1}\left(\int_\Omega D^\sigma f\,dx\right)^2 + |f|_s^2\,. \tag{1.48}$$

Corollary 1.5. *Let $\Omega \subset \mathbb{R}^n$ be a bounded, regular domain with piecewise smooth boundary $\Gamma = \partial\Omega$.*
(1°) *The set*

$$\{f \in \mathcal{H}^s(\Omega)\,,\ \|f\|_0 = 0\} \subset \mathcal{H}^s(\Omega)$$

is a closed subspace, hence a HILBERT *space.*
(2°) *Let Γ_D be a subset of the boundary such that $\int_{\Gamma_D} dO > 0$ then the set*

$$\{f \in \mathcal{H}^1(\Omega)\,,\ f = 0 \text{ on } \Gamma_D \subset \Gamma\} \subset \mathcal{H}^s(\Omega)$$

is a closed subspace, hence a HILBERT *space.*
(3°) *The set*

$$\{f \in \mathcal{H}^s(\Omega),\ \int_\Omega D^\sigma f\,dx = 0,\ |\sigma| \leq s-1\} \subset \mathcal{H}^s(\Omega)$$

is a closed subspace, hence a HILBERT *space.*

The proof of closure is managed in a simple way by means of SCHWARZ' inequality, and in (2°) by an additional application of the trace theorem; cf. (Velte).

References: (Agmon), (Braess), (Brenner), (Evans), (Michlin), (Taylor), (Velte), (Wloka).

1.8 Derivatives

(a) Gâteaux and Fréchet Derivative Let $\mathcal{X}$, $\mathcal{Y}$ be real vector spaces and $\mathcal{D} \subset \mathcal{X}$ an arbitrary subset.

Notations:

(1°) A point $x \in \mathcal{D}$ is called *radial point* of $\mathcal{D}$ in direction $h \in \mathcal{X}$ if

$$\exists\, \varepsilon(h) > 0\ \ \forall\, 0 \leq \varepsilon < \varepsilon(h):\ x + \varepsilon h \in \mathcal{D};$$

x is called *interior point* of $\mathcal{D}$ in direction $h \in \mathcal{X}$ if

$$\exists\, \varepsilon(h) > 0\ \ \forall\, 0 \leq |\varepsilon| < \varepsilon(h):\ x + \varepsilon h \in \mathcal{D}.$$

(2°) Let $\mathcal{Y}$ be normed, let $f : \mathcal{D} \to \mathcal{Y}$ be a mapping, x *radial point* of $\mathcal{D}$ in direction h, and let the limit

$$\delta f(x;h)_+ := \frac{d}{d\varepsilon} f(x+\varepsilon h)|_{\varepsilon=0+} := \lim_{\varepsilon \to 0+} \varepsilon^{-1}[f(x+\varepsilon h) - f(x)] \qquad (1.49)$$

exist finitely, then $\delta f(x;h)_+$ is called *one-sided first variation* (one-sided GÂTEAUX variation) of f at the point x in direction h.

(3°) Let $\mathcal{Y}$ be normed, let $f : \mathcal{D} \to \mathcal{Y}$ be a mapping, x *interior point* of $\mathcal{D}$ in direction h, and let

$$\boxed{\delta f(x;h) := \lim_{\varepsilon\to 0}\varepsilon^{-1}[f(x+\varepsilon h) - f(x)]} \tag{1.50}$$

exist finitely, then $\delta f(x;h)$ is called *first variation* (GÂTEAUX variation) of f at the point x in direction h.

(4°) In the same way, variations of higher order are defined by

$$\delta^k f(x;h) := \frac{d^k}{d\varepsilon^k} f(x+\varepsilon h)|_{\varepsilon=0}\,, \quad k \in \mathbb{N}\,.$$

If $\mathcal{X}$ is normed and $\|h\| = 1$, then (1.50) is the *directional derivative*; besides, (1.50) is equivalent to

$$\lim_{\varepsilon\to 0}\varepsilon^{-1}\|f(x+\varepsilon h) - f(x) - \varepsilon\delta f(x;h)\| = 0\,.$$

The limit $\delta f(x;h)$ is not necessarily *additive* in the second argument but *homogeneous* because, for $\lambda \neq 0$,

$$\begin{aligned}\delta f(x;\lambda h) &= \lim_{\varepsilon\to 0}\varepsilon^{-1}[f(x+\varepsilon\lambda h) - f(x)]\\ &= \lambda\lim_{\varepsilon\to 0}(\lambda\varepsilon)^{-1}[f(x+\varepsilon\lambda h) - f(x)]\\ &= \lambda\lim_{\tau\to 0}\tau^{-1}[f(x+\tau h) - f(x)] = \lambda\delta f(x;h)\,.\end{aligned}$$

Therefore (1.50) is also equivalent to

$$\lim_{\varepsilon\to 0+}\varepsilon^{-1}\|f(x+\varepsilon h) - f(x) - \delta f(x;\varepsilon h)\| = 0\,.$$

The introduction of one-sided directional derivatives is necessary because $\mathcal{D}$ is not always open and extremals appear frequently on the boundary of the feasible domain.

The GÂTEAUX variation is the weakest form of a derivative, in particular, no norm is required on the space $\mathcal{X}$. The well-known *necessary* conditions for stationary points present themselves now in the following form:

Lemma 1.14. *Let $f : \mathcal{X} \supset \mathcal{D} \to \mathbb{R}$ be sufficiently smooth, $x^* \in \mathcal{D}$ and $\forall\, x \in \mathcal{D} : f(x^*) \leq f(x)$.*

(1°) *Let x^* be interior point of $\mathcal{D}$ in direction h then $\delta f(x^*;h) = 0$ and $\delta^2 f(x^*;h) \geq 0$.*
(2°) *Let x^* be radial point of $\mathcal{D}$ in direction h then $\delta f(x^*;h)_+ \geq 0$.*
(3°) *Let $\mathcal{U} \subset \mathcal{X}$ be a subspace, $\mathcal{V} = w + \mathcal{U}$ an affine subspace, $\mathcal{D} \subset \mathcal{V}$ open in $\mathcal{V}$, and let $\delta f(x^*;h)$ exists for all $h \in \mathcal{U}$ then $\forall\, h \in \mathcal{U} : \delta f(x^*;h) = 0$, i.e., $\delta f(x^*;\,\cdot\,) = 0$ in $\mathcal{U}$.*

Differentiating a mapping $f : \mathcal{X} \supset \mathcal{D} \to \mathcal{Y}$ at a point $x \in \mathcal{X}$ means that the linear part of f is filtered out at x. But this part may attain different values on different curves through x if we do not require some conditions on uniform approximation.

Let $\mathcal{X}$, $\mathcal{Y}$ be normed vector spaces.
(1°) Let the variation $\delta f(x;\circ) : \mathcal{X} \to \mathcal{Y}$ be linear and continuous in the second argument for all $h \in \mathcal{X}$ then f is called GÂTEAUX *differentiable* at the point x, $\delta f(x;\circ)$ is the GÂTEAUX *derivative* of f in x, and we write $\partial f(x;h) = \partial f(x)h$.
(2°) Let the limit $\delta f(x;h)$ exist *uniformly* for all $h \in \mathcal{X}$, i.e.

$$\lim_{\|h\|\to 0} \|h\|^{-1}\|f(x+h) - f(x) - \delta f(x;h)\| = 0$$

then $\delta f(x;\circ) : \mathcal{X} \to \mathcal{Y}$ is called FRÉCHET *variation* of f at the point x. If, in addition, $\delta f(x;\circ)$ is linear and continuous in the second argument, f is FRÉCHET *differentiable* or briefly differentiable in x, $\delta f(x;\circ)$ is called FRÉCHET *derivative* of f in x, and we write $\partial f(x;h) = \nabla f(x)h$ or $\partial f(x;h) = f'(x)h$.

Let f be continuous in x and let the FRÉCHET variation be linear, then it is also continuous in the second argument; cf. (Dieudonné), Sect. 8.1.1.

If G-derivatives or F-derivatives exist, they are determined uniquely and, of course, a F-derivative is also G-derivative; for $\mathcal{X} = \mathcal{Y} = \mathbb{R}$ both notations are equivalent. Sometimes the notion "G-differential" is used instead of "G-variation" but then we have to distinguish between G-differential and G-differentiability.

Example 1.16. Note that the derivatives are defined as linear mappings here, hence $f = \nabla f$ if $f : \mathcal{X} \to \mathcal{Y}$ is linear and continuous.

Example 1.17. Let $\underline{f} : \mathbb{R}^m \supset \mathcal{D} \to \mathbb{R}^n$ and let $\operatorname{grad} \underline{f}(x)$ be the gradient of $\underline{f}$ (matrix of partial derivatives, JACOBI matrix). Then $\underline{f}$ is G-differentiable in x if and only if $\forall\, \underline{h} \in \mathbb{R}^m \;\; \forall\, 0 < \varepsilon \ll 1$:

$$\underline{f}(x + \varepsilon\,\underline{h}) = \underline{f}(x) + \operatorname{grad}\underline{f}(x)\varepsilon\,\underline{h} + \underline{r}(\varepsilon,\underline{h})\,, \quad \lim_{\varepsilon\to 0+} \varepsilon^{-1}\underline{r}(\varepsilon,\underline{h}) = 0\,;$$

$\underline{f}$ is F-differentiable in x if and only if

$$\forall\, \underline{h} \in \mathbb{R}^m : \underline{f}(x+\underline{h}) = \underline{f}(x) + \operatorname{grad}\underline{f}(x)\underline{h} + \underline{r}(\underline{h})\,, \quad \lim_{\|\underline{h}\|\to 0} \underline{r}(\underline{h})/\|\underline{h}\| = 0\,.$$

(b) Properties and Examples $f : \mathbb{R}^m \to \mathbb{R}^n$, $(m,n) \neq (1,1)$.

$\underline{f}$ G-differentiable in x	$\Longrightarrow$	$\operatorname{grad}\underline{f}$ exists in x
$\operatorname{grad}\underline{f}$ exists in x	$\not\Longrightarrow$	$\underline{f}$ G-differentiable in x
$\operatorname{grad}\underline{f}$ exists continuously in x	$\Longrightarrow$	$\underline{f}$ F-differentiable in x
$\underline{f}$ F-differentiable in x	$\Longrightarrow$	$\underline{f}$ continuous in x.

The last property remains valid in BANACH spaces.

Lemma 1.15. *$f : \mathcal{X} \to \mathcal{Y}$. (Chain Rule, Mean Value Theorem)*
(1°) Let f be G-differentiable, g F-differentiable, and let $g \circ f$ exist then $g \circ f$ is G-differentiable and

$$\delta(g \circ f)(x) = \nabla g(f(x))\delta f(x) .$$

(2°) Let f be F-differentiable then, using the operator norm,

$$\|f(x+h) - f(x)\| \leq \|h\| \sup_{0<t<1} \|\nabla f(x+th)\| .$$

Lemma 1.16. *$f : \mathcal{X} \to \mathcal{Y}$.*
(1°) f has G-variation in $x \quad \not\Longrightarrow \quad f$ has G-derivative in x;
(2°) f has G-derivative in $x \not\Longrightarrow \quad f$ is continuous in x;
(3°) f has G-derivative in $x \quad \not\Longrightarrow \quad f$ has F-derivative in x;
(4°) f and g have both a G-derivative in $x \quad \not\Longrightarrow \quad g \circ f$ has a G-derivative in x.

In other words, the chain rule does not hold for G-derivatives.

Definition 1.2. *Let $\mathcal{X}$, $\mathcal{Y}$ be vector spaces and let $\mathcal{Y}$ be normed then $f : \mathcal{X} \supset \mathcal{D} \to \mathcal{Y}$ is h-continuous in x (hemi-continuous) if $\forall\, h \in \mathcal{X}\ \forall\, \varepsilon > 0\ \exists\, \delta(\varepsilon, \eta) > 0 : |t| \leq \delta$ and $x + th \in \mathcal{D} \Longrightarrow \|f(x+th) - f(x)\| < \varepsilon$, i.e., f is continuous on all straight lines through x.*

Lemma 1.17. *Using the notations of Definition 1.2 we have:*
(1°) Let f be G-differentiable in x. Then f is h-continuous in x.
(2°) Let $\mathcal{D}$ open and f F-differentiable in x. Then f is continuous in x (see above).

Example 1.18. (Ortega) Let $\underline{x} = (x, y)$ and

$$f(\underline{x}) = \frac{xy^2}{x^2 + y^4} \quad \text{for } \underline{x} \neq \underline{0}, \quad f(\underline{0}) = 0 ,$$

then f has a G-variation in $\underline{0}$ but no G-derivative, moreover f is not continuous in $\underline{0}$ (Fig. 1.10).

Example 1.19. (Ortega) Let

$$g(\underline{x}) = \frac{2y\, e^{-1/x^2}}{y^2 + e^{-2/x^2}} \quad \text{for } x \neq 0, \quad g(\underline{0}) = 0 ,$$

then g has a G-derivative in $\underline{0}$, but g is not continuous in $\underline{0}$.

Example 1.20. (Ortega) Consider the function

$$h(\underline{x}) = \frac{y(x^2+y^2)^{3/2}}{(x^2+y^2)^2+y^2} \quad \text{for } \underline{x} \neq \underline{0}\,, \quad h(\underline{0}) = 0\,,$$

then $h(x,0) = 0$, $h(0,y) = 0 \Longrightarrow h_x(0,0) = 0$, $h_y(0,0) = 0$, hence

$$h(\underline{x}) = h(\underline{0}) + \nabla h(\underline{0})(\underline{x} - \underline{0}) + r(|\underline{x}|) = 0 + 0 + 0 + r(|\underline{x}|)\,.$$

Applying polar coordinates $(x,y) = (r\cos\varphi, r\sin\varphi)$, we obtain

$$\frac{r(|\underline{x}|)}{|\underline{x}|} = \frac{r^4\sin\varphi}{r^5 + r^3\sin^2\varphi} \underset{r\longrightarrow 0}{\longrightarrow} 0$$

for any *fixed* angle φ hence h has the G-derivative $\delta h = [0,\,0]$ in $\underline{0}$. But, substituting $\varphi = \arcsin r$ for sufficiently small r, we obtain

$$\frac{r(|\underline{x}|)}{|\underline{x}|} = \frac{r^5}{r^5 + r^5} = \frac{1}{2}\,.$$

Accordingly, h does not have a F-derivative in $\underline{0}$. For $a \neq 0$ and the straight line $t \mapsto (t, at)$, we have

$$h(t,at) = \frac{at(t^2+a^2t^2)^{3/2}}{(t^2+a^2t^2)^2+a^2t^2} = \frac{t^2}{a}\,\frac{(1+a^2)^{3/2}}{1+\left(\dfrac{t+a^2t}{a^2}\right)^2} \underset{t\longrightarrow 0}{\longrightarrow} 0\,,$$

hence dh/dt exists in $t = 0$ on all straight lines through $h(0)$ but the limiting value zero does *not* exist *uniformly* for $a \neq 0$ (Fig. 1.11).

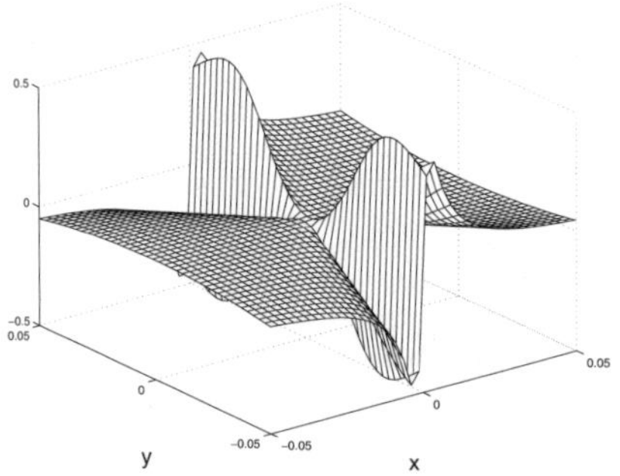

Figure 1.10. Example 1.18

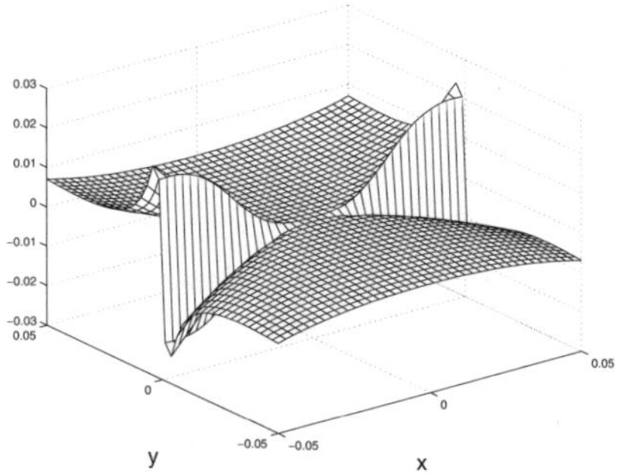

Figure 1.11. Example 1.20

Example 1.21. Let $\mathcal{X} = C[0,1]$, $f(x) = \int_0^1 g(t, x(t))\,dt$ with fixed g and let $\partial g/\partial x$ exist continuously in t and x. Then $\delta f(x)h = \int_0^1 g_x(x,t)h(t)\,dt$ and the G-derivative $\delta f(x)$ is F-derivative of f at the point x, too.

Example 1.22. (Craven95) Let $\mathcal{X} = \{x \in \mathcal{L}_2(0,1)\,,\ x \text{ bounded}\}$ and $f : \mathcal{X} \ni x \mapsto \int_0^1 x(t)^3\,dt$, then, for $h \in \mathcal{X}$,

$$f(x+h) - f(x) \ = \int_0^1 [3x(t)^2 h(t) + 3x(t)h(t)^2 + h(t)^3]\,dt\,,$$
$$f(x+\alpha\, h) - f(x) = \int_0^1 [3\alpha x(t)^2 h(t) + 3\alpha^2 x(t)h(t)^2 + \alpha^3 h(t)^3]\,dt$$

$\forall\ \alpha \in \mathbb{R}$. Accordingly, f has the G-derivative $\delta f(x)h = 3\displaystyle\int_0^1 x(t)^2 h(t)\,dt$ in $x \in \mathcal{X}$. But there does *not* hold in general

$$\int_0^1 h(t)^3\,dt = o(\|h\|) \ \text{ for } \ \|h\| \to 0\,, \tag{1.51}$$

hence f does not have a F-derivative in x. For proving the negation of (1.51) choose for instance $h(t) = n^{3/8}$ for $0 \le t \le n^{-1}$ and $h(t) = 0$ for $n^{-1} < t \le 1$, and consider the limit $n \to \infty$.

References: (Clarke), (Craven95), (Dieudonné), (Luenberger), (Ortega).

1.9 Mappings in Banach Spaces

Let $\mathcal{X}$ and $\mathcal{Y}$ be real BANACH spaces.

(a) Linear Operators. Notations:

(1°) Let $\mathcal{L}(\mathcal{X},\mathcal{Y})$ be the vector space of *linear continuous* mappings $L : \mathcal{X} \to \mathcal{Y}$. $\mathcal{L}(\mathcal{X},\mathcal{Y})$ is complete w.r.t. the operator norm $\|L\| = \sup_{\|x\|=1}\|Lx\|$, hence a BANACH space.

(2°) Let $\mathcal{U}$, $\mathcal{V}$, $\mathcal{W} \subset \mathcal{Y}$ be linear subspaces then $\mathcal{W} = \mathcal{U} \oplus \mathcal{V}$ is the *direct sum* or *direct decomposition* if

$$\forall\, w \in \mathcal{W}\ \ \exists!\, u \in \mathcal{U}\ \ \exists!\, v \in \mathcal{V} : w = u + v\,.$$

(3°) See Sect. 1.7**(e)**. Let $\mathcal{X}_d := \mathcal{L}(\mathcal{X},\mathbb{R})$ be the *dual space* of $\mathcal{X}$, let $\mathcal{Y}_d$ the dual space of $\mathcal{Y}$, and let $L \in \mathcal{L}(\mathcal{X},\mathcal{Y})$ then

$$x_d\,(x) := y_d\,(Lx)\,,\ \ x \in \mathcal{X}\,,\ y_d \in \mathcal{Y}_d\,,$$

is an element $x_d \in \mathcal{X}_d$. By this way the operator L_d *adjoint* to L is uniquely defined,

$$\boxed{L_d \circ y_d = y_d \circ L\,,\ \ L_d\, y_d\,(x) = y_d\,(L\,x)}\,,$$

$$\boxed{\begin{array}{ccccc} \mathcal{X}_d & \longleftarrow & L_d & \longleftarrow & \mathcal{Y}_d \\ \uparrow & & & & \uparrow \\ \mathcal{X} & \longrightarrow & L & \longrightarrow & \mathcal{Y} \end{array}}\ ;$$

in particular, we have $\|L\| = \|L_d\|$ and $0 = L_d y = yL \in \mathcal{X}_d$ for $y \in \operatorname{Ker}(L_d)$.

(4°) To emphasize the *dual pairing* but also to remember the notations in HILBERT spaces, we write in BANACH spaces also $\langle y, x\rangle := y(x)$ for $y \in \mathcal{X}_d$ and $x \in \mathcal{X}$ and, moreover,

$$\begin{aligned} \mathcal{S}^\perp &:= \{y \in \mathcal{X}_d,\ \forall\, x \in \mathcal{S} : \langle y, x\rangle = 0\} \\ \mathcal{S}_d^\perp &:= \{x \in \mathcal{X},\ \ \forall\, y \in \mathcal{S}_d : \langle y, x\rangle = 0\}\,. \end{aligned} \tag{1.52}$$

(5°) Let $\mathcal{U} \subset \mathcal{X}$ be a closed subspace then $\mathcal{X}/\mathcal{U}$ is the *quotient space* of $\mathcal{U}$ in $\mathcal{X}$ of equivalence classes with the equivalence relation $x \sim y \iff x - y \in \mathcal{U}$. If $\mathcal{X}/\mathcal{U}$ is the *quotient space*, $\dim(\mathcal{X}/\mathcal{U})$ is called *codimension* of $\mathcal{U}$.

(6°) In HILBERT spaces $\mathcal{X}$, $\mathcal{Y}$ the adjoint operator L_d is defined in a slightly different way by

$$(x,\, L_d y)_\mathcal{X} = (y,\, Lx)_\mathcal{Y}$$

using the both scalar products, see also Sect. 1.11**(b)**. For the sake of accuracy we mention that both definitions do no exactly coincide but only after some canonical identifications by means of RIESZ' representation theorem; cf. (Taylor), p. 249.

Example 1.23. Let $\mathcal{X} = \mathbb{R}^n$, $\mathcal{Y} = \mathbb{R}^m$ (columns).

(1°) Adopt that $\mathcal{X}_d = \mathbb{R}_n$, $\mathcal{Y}_d = \mathbb{R}_m$ (row vectors) with the usual matrix multiplication. Then $L = A \in \mathbb{R}^m{}_n$ and

$$L_d : \mathcal{Y}_d = \mathcal{R}_m \ni y \mapsto yA \in \mathbb{R}_n = \mathcal{X}_d\,,\quad (L_d y)(x) = yAx \in \mathbb{R}$$

thus L_d is simply the matrix A with left-multiplication.

(2°) Let $\mathcal{X}_d = \mathbb{R}^n$ after canonical identification (RIESZ mapping). Then $L_d = A^T$ and $(L_d y)(x) = (A^T y) \cdot x$ with the canonical *scalar product* of vectors $x, y \in \mathbb{R}^n$.

In the following theorem we collect general properties of linear continuous "operators" $L \in \mathcal{L}(\mathcal{X}, \mathcal{Y})$. Some of them have far-reaching consequences in many fields of applications. Remember that the *graph* of L is the subset

$$\mathcal{G}(L) = \{(x, Lx)\,,\ x \in \mathcal{X}\} \subset \mathcal{X} \times \mathcal{Y}\,.$$

Theorem 1.19. *(Properties of $L \in \mathcal{L}(\mathcal{X}, \mathcal{Y})$)*

(1°) *$L \in \mathcal{L}(\mathcal{X}, \mathcal{Y})$ (continuous) if and only if L (linear) and bounded,*

$$\exists\, \kappa > 0\ \ \forall\, x \in \mathcal{X} : \|Lx\| \le \kappa\|x\|\,.$$

(2°) *(Principle of Open Mapping) Let L be surjective. Then L is open, i.e., the image of every open set is open; in particular* $\text{Range}\, L$ *is open.*
(3°) *(Inverse Operator Theorem) Let L be bijective. Then L^{-1} (exists) and is continuous.*
(4°) *(Closed Graph Theorem) Let $L : \mathcal{X} \to \mathcal{Y}$ be linear. Then L is continuous if and only if the graph $\mathcal{G}(L)$ is closed, i.e., is a closed subset of $\mathcal{X} \times \mathcal{Y}$.*
(5°) *(Range Theorem)* $\overline{\text{Range}\, L} = [\text{Ker}\, L_d]^\perp$, $[\overline{\text{Range}\, L}]^\perp = \text{Ker}\, L_d$.
(6°) *(Range Theorem)* $\overline{\text{Range}\, L_d} = [\text{Ker}\, L]^\perp$, $\overline{\text{Range}\, L_d}^\perp = \text{Ker}\, L$.
(7°) *(Closed Range Theorem) The following statements are equivalent:*
(7.1°) $\text{Range}\, L$ *is closed;* (7.2°) $\text{Range}\, L_d$ *is closed;*
(7.3°) $\text{Range}\, L = [\text{Ker}\, L_d]^\perp$, (7.4°) $\text{Range}\, L_d = [\text{Ker}\, L]^\perp$.

Proofs (Heuser86), Th. 55.7, (Hirzebruch), Th. 25.4, (Luenberger), Th. 6.6.1, 6.6.3, (Wloka), Th. 2.4.12.

For the notations *Range* and *Ker* see Sect. 1.1**(e)**. A finite-dimensional subspace of a BANACH space is always closed. Of course, statement (2°) remains valid if $\mathcal{Y}$ is replaced by $\text{Range}\, L$ and $\text{Range}\, L$ is closed, as then $\text{Range}\, L$ is a BANACH space. (Properties (2°) and (3°) do not contradict themselves because they concern different sets.)

The next result of (Decker) is a generalization of Lemma 1.2 to operators in BANACH spaces.

Theorem 1.20. *(Bordering Lemma) Let the linear operator $E : \mathcal{X} \times \mathbb{R}^m \to \mathcal{Y} \times \mathbb{R}^m$ be of the form*

$$E = \begin{bmatrix} A & B \\ C & D \end{bmatrix}$$

where

$$A : \mathcal{X} \to \mathcal{Y}\,,\ B : \mathbb{R}^m \to \mathcal{Y}\,,\ C : \mathcal{X} \to \mathbb{R}^m\,,\ D : \mathbb{R}^m \to \mathbb{R}^m .$$

(1°) *Let A bijective. Then E is bijective if and only if $D - CA^{-1}B$ is bijective.*
(2°) *Let A be not bijective and* $\dim \text{Ker}\, A = \text{codim Range}\, L = m \geq 1$. *Then E is bijective if and only if*

$$\begin{aligned} &\dim \text{Range}\, B = m\,, \ \ \text{Range}\, D \cap \text{Range}\, A = \{0\}\,, \\ &\dim \text{Range}\, C = m\,, \ \ \text{Ker}\, A \cap \text{Ker}\, C = \{0\}\,. \end{aligned}$$

(3°) *Let A be not bijective and* $\dim \text{Ker}\, A > m$. *Then E is not bijective.*
(4°) *Let $\mathcal{X} = \mathcal{Y} = \mathbb{R}^n$ and A, $F = (D - CA^{-1}B)^{-1}$ are regular. Then*

$$E^{-1} = \begin{bmatrix} A^{-1}[I + BFCA^{-1}] & -A^{-1}BF \\ -FCA^{-1} & F \end{bmatrix} .$$

Proof in `SUPPLEMENT\chap01b`.

In a LAGRANGE-Matrix E we have $C = B^T$ and $D = 0$ but the matrix A is frequently singular; see Lemma 1.2. Suppose that $D = 0$ and A regular then the inverse of E exists if and only if the SCHUR *complement* $S := BA^{-1}B^T$ is regular; to this end the matrix B must have maximum rank. Suppose in particular that $m = 1$ and recall $\mathrm{Ker}(A) = [\mathrm{Range}(A_d)]^T$, then the regularity condition (2°) reduces to $B \notin \mathrm{Range}(A)$, $C \notin \mathrm{Range}(A_d)$ and this condition is equivalent to

$$\begin{array}{ll} \langle v, B\rangle \neq 0\,, & \text{for } A_d v = 0\,,\ v \neq 0 \\ Cu \neq 0\,, & \text{for } Au = 0\,,\ u \neq 0\,. \end{array} \tag{1.53}$$

(b) Projectors More generally as in Sect. 1.1**(f1)**, a not necessarily continuous mapping $P : \mathcal{X} \to \mathcal{X}$ is called *projector* if P is idempotent, i.e., if $P \circ P = P$. The projector is linear if and only if $\mathrm{Range}\, P$ is a linear subspace of $\mathcal{X}$ which is always supposed in the sequel; then $\mathrm{Ker}\, P$ is also a subspace. Let P be a projector in $\mathcal{X}$ and I be the identity then $I - P$ is also a projector. It then follows immediately from the identity $x = Px + (x - Px)$ that
(1°)

$$\boxed{\begin{array}{llll} \mathrm{Range}\, P & = \{x \in \mathcal{X},\ Px = x\}\,, & \mathrm{Ker}\, P & = \{x - Px,\ x \in \mathcal{X}\} \\ \mathrm{Range}(I - P) & = \mathrm{Ker}\, P\,, & \mathrm{Ker}(I - P) & = \mathrm{Range}\, P \end{array}}\ .$$

(2°) Every (linear) projector P yields a direct decomposition of its domain by

$$\boxed{\mathcal{X} = \mathrm{Range}\, P \oplus \mathrm{Ker}\, P}\ . \tag{1.54}$$

Theorem 1.21. (1°) *For every linear subspace $\mathcal{U} \subset \mathcal{X}$ there exists a linear projector P in $\mathcal{X}$ with* $\mathrm{Range}\, P = \mathcal{U}$.
(2°) *Let P be a continuous (linear) projector in $\mathcal{X}$. Then* $\mathrm{Range}\, P$ *is closed.*
(3°) *Let P is a (linear) projector in $\mathcal{X}$ and let* $\mathrm{Range}\, P$ *and* $\mathrm{Ker}\, P$ both *closed. Then P is continuous.*

Proof see (Taylor).

By (1.54), every linear subspace $\mathcal{U} \subset \mathcal{X}$ yields a direct decomposition

$$\mathcal{X} = \mathcal{U} \oplus \mathcal{V},\quad \dim \mathcal{V} = \mathrm{codim}\, \mathcal{U}\,. \tag{1.55}$$

The space $\mathcal{V}$ is uniquely determined in finite-dimensional spaces. However, if $\mathcal{X}$ is an infinite-dimensional BANACH space, there does not always exist a *continuous* projector P of $\mathcal{X}$ for every *closed* $\mathcal{U}$ so that $P(\mathcal{X}) = \mathcal{U}$. Consequently, there does not always exist a direct decomposition $\mathcal{X} = \mathcal{U} \oplus \mathcal{V}$ to every *closed* subspace $\mathcal{U}$ such that $\mathcal{V}$ is closed. This fact implies that $\mathcal{V}$ in (1.55) is not always determined uniquely; c.f. (Taylor), Sect. 4.8.

(c) Implicit Functions For every two open subsets $\mathcal{U} \subset \mathcal{X}$, $\mathcal{V} \subset \mathcal{Y}$, a mapping $f : \mathcal{U} \to \mathcal{V}$ is called *C^r-diffeomorphsm* ($r \geq 1$) if it is bijective and together with its inverse r-times continuously differentiable.

Theorem 1.22. *(Inverse Function Theorem) Let $\mathcal{U} \subset \mathcal{X}$ and $\mathcal{V} \subset \mathcal{Y}$ be open and let $a \in \mathcal{U}$ be fixed. Moreover, let $f \in \mathcal{C}^r(\mathcal{U}, \mathcal{V})$, $r \geq 1$, and let the F-derivative* $\operatorname{grad} f(a)$ *of f be bijective with bounded inverse in point a. Then there exist open subsets $a \in \mathcal{U}_0 \subset \mathcal{U}$ and $\emptyset \neq \mathcal{V}_0 \subset \mathcal{V}$ such that the restriction of f to $\mathcal{U}_0 \times \mathcal{V}_0$ is a $\mathcal{C}^r$-diffeomorphism.*

Corollary 1.6. *(Implicit Function Theorem) Let $\mathcal{X}$, $\mathcal{Y}$, $\mathcal{Z}$ be* BANACH *spaces, let $f \in \mathcal{C}^r(\mathcal{X} \times \mathcal{Y}; \mathcal{Z})$, $r \geq 1$, $c = f(a,b)$,* $\operatorname{grad}_b f(a,b)$ *bijective with bounded inverse. Then there exist open subsets $\mathcal{U}$, $\mathcal{W}$ with $a \in \mathcal{U} \subset \mathcal{X}$, $c \in \mathcal{W} \subset \mathcal{Z}$ and a unique function $\Phi \in \mathcal{C}^r(\mathcal{U} \times \mathcal{W}, \mathcal{Y})$ such that*

$$b = \Phi(a, c), \quad \forall\, x \in \mathcal{U} \;\forall\, z \in \mathcal{W} : z = f(x, \Phi(x, z)),$$

and Φ is as smooth as f.

For $c = 0$ we obtain the well-known form:

Corollary 1.7. *(Implicit Function Theorem) Let $\mathcal{X}$, $\mathcal{Y}$, $\mathcal{Z}$ be* BANACH *spaces, $f \in \mathcal{C}^r(\mathcal{X} \times \mathcal{Y}; \mathcal{Z})$, $r \geq 1$, $f(a,b) = 0$,* $\operatorname{grad}_b f(a,b)$ *bijective with bounded inverse. Then there exists an open subset $\mathcal{U}$ with $a \in \mathcal{U} \subset \mathcal{X}$ and a unique function $\Phi \in \mathcal{C}^r(\mathcal{U}, \mathcal{Y})$ such that*

$$\Phi(a) = b, \quad \forall\, x \in \mathcal{U} : f(x, \Phi(x)) = 0,$$

and Φ is as smooth as f.

Theorem 1.23. *(Generalized Inverse Function Theorem) Let $\mathcal{X}$, $\mathcal{Y}$ be* BANACH *spaces and $f : \mathcal{X} \to \mathcal{Y}$ continuously F-differentiable. For some $a \in \mathcal{X}$, let $b = f(a)$ and* $\operatorname{grad} f(a)$ *be surjective (but not necessarily invertible). Then there exists an open neighborhood $\mathcal{U} \ni b$ and a constant κ such that $f(x) = y$ has a solution x for every $y \in \mathcal{U}$ and $\|x - a\| \leq \kappa \|y - b\|$, in particular x depends continuously on y.*

Proofs of subsection **(c)** in `SUPPLEMENT\chap01b`, see also (Craven78); Proof of Theorem 1.23 (Luenberger).

1.10 Convex Sets and Functions

In dealing with finite- and infinite-dimensional extremal problems, a good knowledge of convex sets and functions is indispensable but also the handling of inequalities. Here, the *order cone* plays a crucial role because the concept of duality is founded on it and every numerical approach uses some duality aspects. Also, the Theorem of *Farkas* deserves the same attention in systems of linear inequalities as the Range Theorem 1.2 in systems of linear equations; nevertheless it leads a somewhat hidden life in literature. Because inequalities are subject to entire different rules in comparison with equalities, some proofs are presented below for a better understanding of the problems occuring in optimization, calculus of variations, and control theory.

(a) Convex Sets and Cones Let $\mathcal{X}$ be a real normed vector space and let $[\underline{x}, \underline{y}] := \{\underline{z} = \underline{x} + \lambda(\underline{y} - \underline{x}),\ 0 \le \lambda \le 1\}$ be the line segment joining two elements $x, y \in \mathcal{X}$.

Notations:

(1°) A set $\mathcal{C} \subset \mathcal{X}$ is *convex* if $\forall\, x, y \in \mathcal{C} : [x, y] \subset \mathcal{C}$.
(2°) A set $\mathcal{K} \subset \mathcal{X}$ is a *cone* ("with vertex in 0") if $\forall\, x \in \mathcal{K}\ \ \forall\, \alpha \ge 0 : \alpha x \in \mathcal{K}$.
(3°) A cone $\mathcal{K}$ is *pointed* if $\mathcal{K} \cap -\mathcal{K} = \{0\}$ $(\iff x \in \mathcal{K} \wedge -x \in \mathcal{K} \iff x = 0)$.

(4°) A *convex* cone $\mathcal{K}$ defines a *pre-order* by $y \ge x \iff x \le y :\iff y - x \in \mathcal{K}$ with the properties

$$\begin{aligned} \forall\, x \in \mathcal{X} \quad &\Longrightarrow x \le x \text{ (reflexivity)} \\ x \le y \wedge y \le z &\Longrightarrow y \le z \text{ (transitivity).} \end{aligned}$$

(5°) A pointed convex cone $\mathcal{K}$ defines a *partial order* with the additional property $x \le y \wedge y \le x \implies y = x$ (symmetry).
(6°) If a pre-order is defined by a convex cone $\mathcal{K} \ne \emptyset$, this cone is called *positive cone* or *order cone*.
(7°) If $\mathcal{M} \subset \mathcal{X}$ is an arbitrary subset, the set

$$\mathcal{M}_d := \{y \in \mathcal{X}_d,\ \forall\, x \in \mathcal{M} : y\,(x) \ge 0\}$$

is called *dual cone* or *adjoint cone* to $\mathcal{M}$. If $\mathcal{M}$ is a positive cone, $\mathcal{M}_d$ is called *dual positive cone*.

Many applications manage with a pre-order only in which the order cone is not pointed. A positive cone is never open but frequently closed, the dual cone, however, is always closed. After canonical imbedding $\mathcal{X} \subset (\mathcal{X}_d)_d$ we have $\mathcal{K} \subset (\mathcal{K}_d)_d$. If $\mathcal{X}$ is a reflexive BANACH space, i.e. $(\mathcal{X}_d)_d = \mathcal{X}$, and $\mathcal{K} \subset \mathcal{X}$ is a closed cone, then $\mathcal{K} = (\mathcal{K}_d)_d$. We write

$$\boxed{x \ge 0 \iff x \in \mathcal{K},\ x > 0 \iff x \in \operatorname{int}(\mathcal{K})}$$

where $\operatorname{int}(\mathcal{K})$ shall be the *interior of* $\mathcal{K}$.

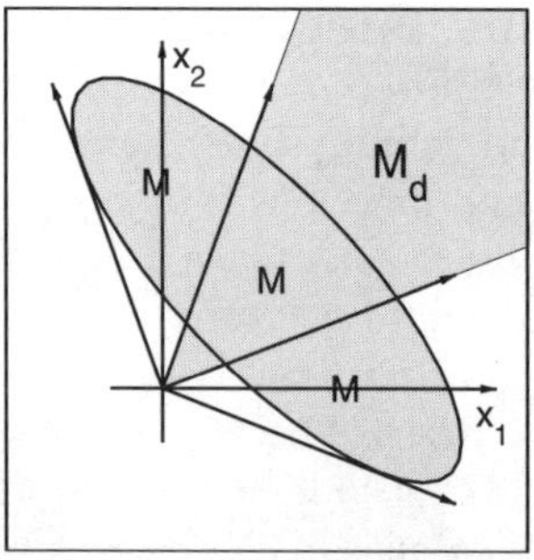

Figure 1.12. Dual cone of $\mathcal{M}$

Example 1.24.(1°) A cone $\mathcal{K} \subset \mathcal{X}$ is convex if and only if
$\mathcal{K} + \mathcal{K} := \{x + y\,,\ x, y \in \mathcal{K}\} \subset \mathcal{K}$.
(2°) Let $\mathcal{K} = \mathbb{R}^n_+ \subset \mathbb{R}^n$. Then $\mathcal{K}_d = \mathbb{R}^n_+$ after canonical identification.
(3°) The cone $\mathcal{K} = \{\underline{x} \in \mathbb{R}^2\,,\ x^1 \geq 0\,,\ x^2 < 0\} \cup \{(0,0)\}$ is not closed.
(4°) Let $\mathcal{K} \subset \mathbb{R}^n$ be a positive cone and $A \in \mathbb{R}^m{}_n$. Then $A(\mathcal{K}) := \{y \in \mathbb{R}^m,\ \exists\, \underline{x} \in \mathcal{K} : \underline{y} = A\underline{x}\}$ is the convex cone spanned by the columns of A relative to the cone $\mathcal{K}$.
(5°) Let $\mathcal{K} = \{f \in \mathcal{C}(a,b)\,,\ f(x) \geq 0\}$. Then $\mathcal{K}_d = \{y \in \mathcal{C}_d(a,b)\,,\ y(f) = \int_a^b f(x)\,dv(x)\,,\, v \in \mathrm{BV}(a,b)\,,$ weakly monotone increasing$\}$; cf. Sect. 12.5.
(6°) The natural cone in $\mathcal{L}_p(a,b)\,,\ 1 \leq p < \infty$, cf. Sect. 12.5, is $\mathcal{K} := \{x \in \mathcal{L}_p(a,b)\,,\ x(t) \geq 0$ a.e.$\}$ with the dual cone $\mathcal{K}_d := \{x \in \mathcal{L}_q(a,b)\,,\ x(t) \geq 0$ a.e.$\}$ where $p^{-1} + q^{-1} = 1$.

Example 1.25. Let $\underline{0} \neq \underline{a} = [a^1,\, a^2]^T \in \mathbb{R}^2$. Then $\underline{a}$ and $\underline{a}^\perp := [-a^2,\, a^1]^T$ constitute a right oriented system. Let now $\underline{a},\, \underline{b} \in \mathbb{R}^2$ be arbitrary and $\underline{c} = \underline{a}^\perp$, $\underline{d} = -\underline{b}^\perp$. Then

$$\begin{aligned} \mathcal{K} &= \{\underline{x} = \alpha\,\underline{a} + \beta\,\underline{b}\,,\ \alpha \geq 0\,,\ \beta \geq 0\} \text{ is a positive cone} \\ \mathcal{K}_d &= \{\underline{x} = \gamma\,\underline{c} + \delta\,\underline{d}\,,\ \gamma \geq 0\,,\ \delta \geq 0\} \ \text{ the dual cone,} \end{aligned}$$

and both coincide if $\underline{a}$ is perpendicular to $\underline{b}$.

Example 1.26. (Ben-Israel) Let the closed cone $\mathcal{K} \subset \mathbb{R}^3$ consist of all vectors $\underline{x} = [x,\, y,\, z]^T$ which have an angle $\leq \pi/4$ with the symmetry axis $\underline{a} = [1,\, 0,\, 1]^T$. Then $\underline{a}^T\underline{x} \geq \|\underline{a}\|\,\|\underline{x}\|/\sqrt{2}$, hence

$$\mathcal{K} = \{\underline{x} \in \mathbb{R}^3\,,\ x \geq 0\,,\ z \geq 0\,,\ 2xz \geq y^2\}\,.$$

Let now P be the orthogonal projection of $\mathbb{R}^3$ onto the (y,z)-plane then

$$P = \begin{bmatrix} 0 & 1 & 0 \\ 0 & 0 & 1 \end{bmatrix}\,,\quad P(\mathcal{K}) = \{[y, z]^T\,,\ x \geq 0\,,\ z \geq 0\,,\ 2xz \geq y^2\}\,.$$

But $2xz \geq y^2$ has only the solution $y = 0$ for $z = 0$ hence

$$P(\mathcal{K}) = \{(y,z)\,,\ z > 0\} \cup (0,0) = \{r\cos\varphi, r\sin\varphi)\,,\ r \geq 0\,,\ 0 < \varphi < \pi\}\,,$$

and the image of $\mathcal{K}$ under the continuous linear mapping P is *not closed.*

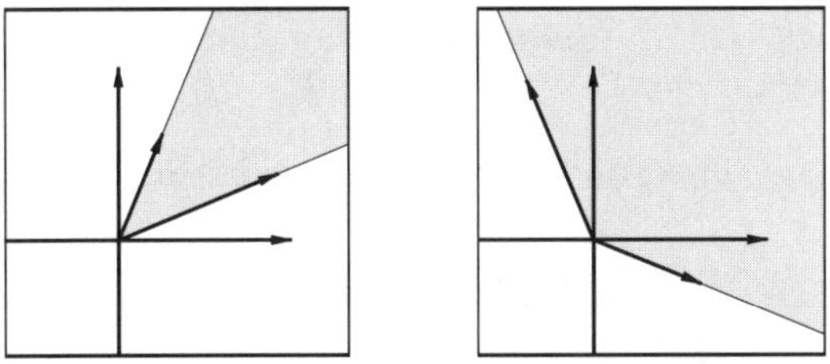

Figure 1.13. Cone and dual cone in $\mathbb{R}^2$

(b) Separation Theorems In many problems of optimization and control, the LAGRANGE multipliers describe hyperplanes which separate certain convex sets from each other. The subsequent *separation theorem* is of fundamental importance in the existence of such hyperplanes:

Theorem 1.24. *Let $\mathcal{X}$ be a normed vector space and let $\mathcal{C}$, $\mathcal{D} \subset \mathcal{X}$ be convex then there exists an $\alpha \in \mathbb{R}$ and a $0 \neq y \in \mathcal{X}_d$ so that*

$$\boxed{\sup_{x\in\mathcal{D}} y(x) \leq \alpha \leq \inf_{x\in\mathcal{C}} y(x)}\,,$$

if one of the following conditions is fulfilled:
(1°) $\mathcal{C}$ is open and $\mathcal{C} \cap \mathcal{D} = \emptyset$;
(2°) $\mathcal{C}$ is closed, $\mathcal{C} \cap \mathcal{D} = \emptyset$, and $\mathcal{D} = \{b\}$ is a point;
(3°) $\operatorname{int}(\mathcal{C}) \neq \emptyset$ and $\operatorname{int}(\mathcal{C}) \cap \mathcal{D} = \emptyset$ (EIDELHEIT's separation theorem).

Proof see (Schaeffer), (Luenberger), (Werner).

If $\mathcal{X}$ is an infinite-dimensional vector space then the crucial condition $\operatorname{int}(\mathcal{C}) \neq \emptyset$ in (3°) cannot be dropped even if $\mathcal{C}$ is compact; cf. (Marti). In finite-dimensional vector spaces however, the result of Theorem 1.24(2°) can be proved simply in a direct way:

Lemma 1.18. *Let $\emptyset \neq \mathcal{C} \subset \mathbb{R}^n$ be closed and convex and let $\underline{0} \notin \mathcal{C}$. Then there exists a $\underline{y} \in \mathbb{R}^n$ and $\alpha > 0$ so that $\underline{y}^T\underline{x} > \alpha > 0$ holds for all $\underline{x} \in \mathcal{C}$. In other words, the hyperplane $\mathcal{H} = \{\underline{x} \in \mathbb{R}^n,\ \underline{y}^T\underline{x} = \alpha\}$ separates $\mathcal{C}$ and $\{0\}$ in strong sense.*

Proof. The compact set $\mathcal{U} := \{\underline{x} \in \mathbb{R}^n,\ \|\underline{x}\| \leq \beta\} \cap \mathcal{C}$ is not empty for sufficiently large β hence the continuous function $f : \underline{x} \mapsto \|\underline{x}\|$ on $\mathcal{U}$ takes its minimum in a point $\underline{y} \neq 0$. Then

$$\|\underline{y} + \lambda(\underline{x} - \underline{y})\|^2 \geq \|\underline{y}\|^2\,,\quad \lambda \in [0,1]\,,$$

for all $\underline{x} \in \mathcal{C}$ because $\mathcal{C}$ convex or

$$\lambda^2(\underline{x} - \underline{y})^T(\underline{y} - \underline{x}) + 2\lambda\underline{y}^T(\underline{x} - \underline{y}) \geq 0\,,\quad \lambda \in [0,1].$$

It follows that $\underline{y}^T(\underline{x} - \underline{y}) \geq 0$ for $\lambda \to 0$, i.e., $\underline{y}^T\underline{x} \geq \underline{y}^T\underline{y} > \underline{y}^T\underline{y}/4 =: \alpha > 0$. □

Lemma 1.19. *Let $\mathcal{K} \subset \mathbb{R}^n$ be a positive, closed cone and let $\underline{b} \notin \mathcal{K}$ then there exists a $\underline{y} \in \mathcal{K}_d$ with $\underline{y}^T\underline{b} < 0$ (hence always $\mathcal{K}_d \neq \emptyset$).*

Proof. Because $\mathcal{K}$ is closed and convex, there exists a $\underline{y} \in \mathbb{R}^n$ that

$$\underline{y}^T\underline{b} < \alpha < \inf{}_{\underline{x}\in\mathcal{K}}\, \underline{y}^T\underline{x} \tag{1.56}$$

after a simple modification of Lemma 1.18. Because $\underline{0} \in \mathcal{K}$ we have $\alpha \leq 0$. Let $\underline{u} \in \mathcal{K}$ with $\underline{y}^T\underline{u} < 0$. Because of $\lambda\underline{u} \in \mathcal{K}$ for $\lambda \geq 0$, there exists a $\underline{w} = \lambda\underline{u}$

that $\underline{y}^T\underline{w} = \lambda\underline{y}^T\underline{u} < \alpha$. This is a contradiction of (1.56). Thus $\underline{y}^T\underline{u} \geq 0$ for all $\underline{u} \in \mathcal{K}$ and $\underline{y} \in \mathcal{K}_d$ by definition. □

The generalization of this result to arbitrary normed vector spaces can only be proved by using the Separation Theorem:

Lemma 1.20. *If $\mathcal{K} \subset \mathcal{X}$ is a convex closed cone and $b \notin \mathcal{K}$, there exists a $y \in \mathcal{K}_d$ with $y(b) < 0$.*

Proof. By Theorem 1.24 there exists a $y \in \mathcal{X}_d$ such that

$$y(b) < \alpha < \inf{}_{x\in\mathcal{K}}\, y(x)\,. \tag{1.57}$$

Because of $0 \in \mathcal{K}$ it follows that $\alpha \leq 0$. Let $u \in \mathcal{K}$ mit $y(u) < 0$. Because $\lambda u \in \mathcal{K}$, $\lambda \geq 0$ then there exists a $w = \lambda u$ such that $y(w) = \lambda y(u) < \alpha$ being a contradiction of inequality (1.57). Thus $y(u) \geq 0$ for all $u \in \mathcal{K}$ and $y \in \mathcal{K}_d$ by definition. □

(c) Cone Properties The so-called *cone corollary* is a simple inference of Lemma 1.20:

Lemma 1.21. *If $\mathcal{K} \subset \mathcal{X}$ is a convex closed cone then*

$$x \in \mathcal{K} \iff (\forall\, y \in \mathcal{K}_d \implies y(x) \geq 0)\ .$$

Proof. The left side implies the right side by definition. The negation of the left side implies the negation of the right side by Lemma 1.20. □

Lemma 1.22. *(Cone Inclusion Theorem) Let $\mathcal{K}$, $\mathcal{L} \subset \mathcal{X}$ be convex cones and $\mathcal{L}$ closed then*

$$\boxed{\mathcal{K} \subset \mathcal{L} \iff \mathcal{L}_d \subset \mathcal{K}_d}\ .$$

Proof. Obviously, the left side implies the right side. The negation of the left side implies the existence of an element $x \in \mathcal{K}$ with $x \notin \mathcal{L}$. By the cone corollary there exists a $y \in \mathcal{L}_d$ such that $y(x) < 0$ hence $y \notin \mathcal{K}_d$. This implies the negation of the right side. □

Lemma 1.23. (1°) *Let $\mathcal{K} \subset \mathcal{X}$ be a cone, let $y \in \mathcal{X}_d$, $\gamma \in \mathbb{R}$, and $y(x) \geq \gamma$ for all $x \in \mathcal{K}$. Then $y \in \mathcal{K}_d$ and $\gamma \leq 0$.*
(2°) *Let $\mathcal{M} \subset \mathcal{X}$ be arbitrary and $0 \neq y \in \mathcal{M}_d$, then $y(x) > 0$ for all $x \in \operatorname{int}(\mathcal{M})$.*

Proof. (1°) At first, $\gamma \leq 0$ follows from $y(0) = 0 \geq \gamma$. Suppose that there exists an $x \in \mathcal{K}$ with $y(x) < 0$, then

$$y(\lambda x) = \lambda y(x) \to -\infty,\ \lambda \to \infty\,,$$

which is a contradiction because $y(\lambda x) \geq \gamma$ for all $\lambda > 0$.
(2°) Suppose there exists an $x \in \operatorname{int}(\mathcal{M})$ with $y(x) = 0$. Then there exists for every $u \in \mathcal{X}$ a $\tau > 0$ so that $x \pm \tau u \in \mathcal{M}$. Then $\pm y(u) \geq 0$ for all $u \in \mathcal{X}$

hence $y(u) = 0$ for all $u \in \mathcal{X}$. This implies $y = 0$ in contradiction of the assumption. □

The next result leads to a necessary and sufficient condition for *sign bounded* solutions of linear systems of equations.

Lemma 1.24. *(*Farkas *1902) Let $\mathcal{X}$, $\mathcal{Y}$ be normed vector spaces, let $\mathcal{K} \subset \mathcal{X}$ be a closed convex cone, $A \in \mathcal{L}(\mathcal{X}, \mathcal{Y})$ and $b \in \mathcal{Y}$. Then*

$$\boxed{b \in A(\mathcal{K}) \iff (A_d u \in \mathcal{K}_d \implies u(b) \geq 0)}\,,$$

if $A(\mathcal{K})$ is closed.

Proof. "$\Longrightarrow$". Let $b = Ax$ for some $x \in \mathcal{K}$. If $A_d u \in \mathcal{K}_d$, then $u(Ax) = u(b) \geq 0$ because $A_d\, u = u\, A$, hence $u(Ax) = (A_d u)x$.
"$\Longleftarrow$". We have

$$\begin{aligned} A_d\, u \in \mathcal{K}_d &\iff \forall\, x \in \mathcal{K} : (A_d\, u)x = (u\, A)(x) \geq 0 \\ &\iff \forall\, x \in \mathcal{K} : u(Ax) \geq 0 \iff u \in [A(\mathcal{K})]_d\,. \end{aligned}$$

In the same way $u(b) \geq 0$ says that $u \in \mathcal{M}_d$ for the cone defined by $\mathcal{M} = \{\alpha\, b,\ \alpha \geq 0\}$. Thus the right side says that $[A(\mathcal{K})]_d \subset \mathcal{M}_d$. Because $A(\mathcal{K})$ is closed, the Cone Inclusion Theorem can be applied: $[A(\mathcal{K})]_d \subset \mathcal{M}_d \implies \mathcal{M} \subset A(\mathcal{K})$. But $\mathcal{M} \subset A(\mathcal{K})$ says that $b \in A(\mathcal{K})$. □

The following relations hold in the Lemma of Farkas:

$$\begin{aligned} \text{left side} &\iff \{x \in X,\ Ax = b,\ x \geq 0\} \neq \emptyset\,, \\ \text{right side} &\iff \{u \in Y,\ A_d u \geq 0\} \subset \{u \in Y,\ u(b) \geq 0\}\,, \\ \neg\,\text{right side} &\iff \exists\, u \in Y : A_d u \geq 0\ \wedge\ u(b) < 0 \\ &\iff \{u \in Y : A_d u \geq 0\ \wedge\ u(b) < 0\} \neq \emptyset\,. \end{aligned}$$

Therefore the Lemma of Farkas is equivalent to

$$\{x \in X : Ax = b,\ x \geq 0\} \neq \emptyset \iff \{u \in Y : A_d u \geq 0,\ u(b) < 0\} = \emptyset\,.$$

Let now especially $\mathcal{K} = \mathbb{R}^n_+$ hence $\underline{x} \geq \underline{0} \iff x^i \geq 0\ \forall\ i$ then $\mathcal{K} = \mathcal{K}_d$. The basic Theorem 1.2 for linear systems of equations has an analogue for sign-bounded solutions of linear systems by the Lemma of Farkas. In order to point out the similarity of both results, we write Theorem 1.2 below in the equivalent form (1°) and the result of Farkas in the form (2°):

Corollary 1.8.

$$\boxed{\begin{array}{lll} (1^\circ)\ A\underline{x} = \underline{y} & \iff & (A^T\underline{z} = 0 \implies \underline{y}^T\underline{z} = 0) \\ (2^\circ)\ A\underline{x} = \underline{y} \wedge \underline{x} \geq 0 & \iff & (A^T\underline{z} \geq 0 \implies \underline{y}^T\underline{z} \geq 0) \end{array}}\,.$$

(d) Convex Functions Let $\mathcal{C} \subset \mathcal{X}$ be *convex*. Then a function $f : \mathcal{C} \to \mathbb{R}$ is *convex* if

$$\forall\, x, y \in \mathcal{C} \quad \forall\, \lambda \in [0,1] : \; f(x + \lambda(y - x)) \leq (1 - \lambda) f(x) + \lambda f(y)\,.$$

$f : \mathcal{C} \to \mathbb{R}$ is called *concave*, if $-f$ is convex. If $\mathcal{X}, \mathcal{Y}$ are vector spaces, $\mathcal{C} \subset \mathcal{X}$ is convex, and $\mathcal{K} \subset \mathcal{Y}$ a positive cone, then $f : \mathcal{C} \to \mathcal{Y}$ is called $\mathcal{K}$-*convex* if

$$\forall\, x, y \in \mathcal{C}, \; \forall\, \lambda \in [0,1] : f(x + \lambda(y - x)) \leq (1 - \lambda) f(x) + \lambda f(y).$$

(Recall that $x \geq 0 \iff x \in \mathcal{K}$.) The convexity of f depends strongly on the current definition of the order cone. If this cone is fixed during computation, one writes "convex" instead of "$\mathcal{K}$-convex".

Example 1.27.(1°) A norm is always a convex function.
(2°) Precisely the affine linear functions are convex and concave at the same time.
(3°) Let f and g convex and $\alpha \in \mathbb{R}$. Then αf and $f + g$ are also convex.
(4°) For $\mathcal{X} = L^2[0,1]$, the function

$$f : \mathcal{X} \ni x \mapsto \int_0^1 (x^2 + |x|)dt \in \mathbb{R}$$

is convex in $\mathcal{X}$.
(5°) Let $f, g, h : \mathcal{C} \to \mathbb{R}$ are convex. Then the set $\{x \in \mathcal{C},\; f(x) \leq a,\; g(x) \leq b,\; h(x) \leq c\}$ is also convex.
(6°) Let $\mu = \inf_{x \in \mathcal{C}} f(x)$. Then $\{x \in \mathcal{C},\; f(x) = \mu\}$ is convex ($\emptyset$ convex).
(7°) Let $\mathcal{X}$, $\mathcal{Y}$ be two vector spaces , $\mathcal{C} \subset \mathcal{X}$ convex and $g : \mathcal{C} \to \mathcal{Y}$ convex. Then $\{x \in \mathcal{C},\; g(x) \leq y\}$ is a convex set for all $y \in \mathcal{Y}$.

Lemma 1.25. *Let $\mathcal{C} \subset \mathcal{X}$ convex and $f : \mathcal{C} \to \mathbb{R}$ convex.*

(1°) f is continuous in $\mathrm{int}(\mathcal{C})$ *(!) if* $\dim(\mathcal{X}) < \infty$.
(2°) A point x^* is a global minimum of f if x^* is a local minimum of f.
(3°) x^* is the unique global minimum of f in $\mathcal{C}$ if f is strongly convex and x^* is a local minimum of f.

Proof. (1°) (Luenberger), S. 194.
(2°) Let $f(x^*) \leq f(x)$ for all $x \in \mathcal{U}(x^*) \cap \mathcal{C}$ with $\mathcal{U}(x^*)$ being an open neighborhood of x^*, and let $y \in \mathcal{C}$ arbitrary, then there exists an $x \in \mathcal{U}(x^*)$ so that $x = \lambda x^* + (1 - \lambda) y$, $0 < \lambda < 1$. This implies that $f(x^*) \leq f(x) \leq \lambda f(x^*) + (1 - \lambda) f(y)$ or $(1 - \lambda) f(x^*) \leq (1 - \lambda) f(y)$ which is the assertion.
(3°) Two local minimums x^*, y^* are global minimums by (2°). Because $\lambda x^* + (1 - \lambda) y^* \in \mathcal{C}$, $0 < \lambda < 1$, it follows that

$$f(\lambda x^* + (1 - \lambda) y^*) < \lambda f(x^*) + (1 - \lambda) f(y^*) = f(y^*)\,.$$

This is a contradiction because x^* and y^* are both global minimum points.

□

Lemma 1.26. *Let $\mathcal{X}$, $\mathcal{Y}$ be normed vector spaces, $\mathcal{C} \subset \mathcal{X}$ convex, $\mathcal{K} \subset \mathcal{Y}$ positive cone, and $f : \mathcal{C} \to \mathcal{Y}$ F-differentiable in $\mathcal{D} \supset \mathcal{C}$ open.*
(1°) *f is $\mathcal{K}$-convex if and only if*

$$\forall\, x, y \in \mathcal{C} : f(y) - f(x) - \nabla f(x)(y - x) \geq 0\,, \quad \text{i.e.,} \ \in \mathcal{K}\,. \tag{1.58}$$

(2°) *Let $\mathcal{X} = \mathbb{R}^n$ and f two-times F-differentiable and $\mathcal{K}$-convex. Then*

$$\forall\, y \in \mathcal{X} : \nabla\nabla f(x)[yy] \geq 0, \quad \text{i.e.,} \ \in \mathcal{K}\,. \tag{1.59}$$

(3°) *Let $\mathcal{X} = \mathbb{R}^n$, $\mathcal{Y} = \mathbb{R}$, let f be two-times F-differentiable and let (1.59) hold. Then f is $\mathcal{K}$-convex.*

Proofs in `SUPPLEMENT\chap01b`.

Lemma 1.27. *Let $\mathcal{C} \subset \mathcal{X}$ be convex, let $f : \mathcal{C} \to \mathbb{R}$ convex and f F-differentiable in an open set $\mathcal{D} \supset \mathcal{C}$.*
(1°) $x^* = \arg\min_{x\in\mathcal{C}} f(x) \iff x^* \in \mathcal{C}\,,\ \forall\, x \in \mathcal{C} : \nabla f(x^*)(x - x^*) \geq 0\,.$
(2°) *If in addition $\mathcal{C}$ is a cone,*

$$x^* = \arg\min_{x\in\mathcal{C}} f(x) \iff x^* \in \mathcal{C}\,,\ \nabla f(x^*)x^* = 0\,,\ \forall\, x \in \mathcal{C} : \nabla f(x^*)x \geq 0\,.$$

Of course, $\nabla f(x^*)z \geq 0\ \forall\, z \in \mathcal{C}$ implies $\nabla f(x^*) = 0$ if $x^* \in \text{int}(\mathcal{C})$, but here x^* can be also a boundary point of $\mathcal{C}$.
Proof. (1°) "$\Longleftarrow$" by substituting $x = x^*$ in Lemma 1.26,

$$\forall\, y \in \mathcal{C} : f(y) \geq f(x^*) + \nabla f(x^*)(y - x^*) \geq f(x^*)\,.$$

"$\Longrightarrow$" We have for arbitrary $y = x^* + h \in \mathcal{U}$ and $\lambda \in [0, 1]$

$$f(x^* + \lambda h) - f(x^*) = \lambda[\nabla f(x^*)h + \varepsilon(\lambda)]\,,\ \lim_{\lambda\to 0} \varepsilon(\lambda) = 0\,.$$

Thus $\lambda\nabla f(x^*)h \geq 0$ because else $f(x^* + \lambda h) - g(x^*) < 0$ for sufficiently small $\lambda > 0$ in contradiction to the minimum property of x^*.
(2°) The left side follows from the right side by (1°). Conversely, let the left side be fulfilled. Then, again by (1°), $\nabla f(x^*)(x-x^*) \geq 0$. The points $x = x^*/2$ and $x = 2x^*$ are both contained in the cone $\mathcal{C}$. Substitution yields

$$x = x^*/2 \implies \nabla f(x^*)x^* \leq 0\,,\quad x = 2x^* \implies \nabla f(x^*)x^* \geq 0\,.$$

This implies that $\nabla f(x^*)x^* = 0$ and therefore $\nabla f(x^*)x \geq 0$. □

As a direct inference we obtain a *saddlepoint criterium* for F-differentiable functions.

Lemma 1.28. *Let $\mathcal{X}$, $\mathcal{Y}$ be normed vector spaces, let $\emptyset \neq \mathcal{C} \subset \mathcal{X}$, $\emptyset \neq \mathcal{D} \subset \mathcal{Y}$ be both convex and closed and let*

$f : \mathcal{C} \times \mathcal{D} \ni (x, y) \mapsto f(x, y) \in \mathbb{R}$
$\forall\, x \in \mathcal{C} : D \ni y \mapsto f(x, y)$ *concave, continuous, F-differentible*
$\forall\, y \in \mathcal{D} : C \ni x \mapsto f(x, y)$ *convex, continuous, F-differentiable.*

(1°) (x^*, y^*) is a local saddlepoint of f, i.e.,

$$\forall\,(x,y) \in \mathcal{C} \times \mathcal{D} : f(x^*, y) \le f(x^*, y^*) \le f(x, y^*)\,, \tag{1.60}$$

if and only if, with partial F-derivatives,

$$\forall\, x \in \mathcal{C} : \nabla_x f(x^*, y^*)(x - x^*) \ge 0\,,\;\; \forall\, y \in \mathcal{D} : \nabla_y f(x^*, y^*)(y - y^*) \le 0\,.$$

(2°) If, in addition, $\mathcal{C}$, $\mathcal{D}$ are convex cones with adjoint cones $\mathcal{C}_d$, $\mathcal{D}_d$, then (1.60) holds if and only if

$$\nabla_x f(x^*, y^*) \in \mathcal{C}_d\,,\;\; \nabla_x f(x^*, y^*)x^* = 0\,,$$
$$-\nabla_y f(x^*, y^*) \in \mathcal{D}_d\,,\;\; \nabla_y f(x^*, y^*)y^* = 0\,.$$

If $f : [a, b] \to \mathbb{R}$ is differentiable and attains a minimum in $x^* \in [a, b]$, then necessarily

$$\lim_{x \to x^*, x \in [a,b]} \frac{f(x) - f(x^*)}{|x - x^*|} \ge 0\,.$$

Therefore (Craven78) introduces the following generalization.

Definition 1.3. *Let $\mathcal{X}$ be a* BANACH *space and $\mathcal{U} \subset \mathcal{X}$ a subset. Then $f : \mathcal{U} \to \mathbb{R}$ has the* quasimin property *(QM) in $x \in \mathcal{U}$ if*

$$\liminf{}_{y \to x, y \in \mathcal{U}} \frac{f(y) - f(x)}{\|y - x\|} \ge 0\,.$$

If f is F-differentiable, the quasimin property in x is equivalent to

$$\forall\, y \in \mathcal{U} :\; \nabla f(x)(y - x) \ge 0\,.$$

If $f : \mathcal{U} \to \mathbb{R}$ has a minimum in x, it has obviously the (QM) in x. Conversely, we have

Lemma 1.29. *Let $\mathcal{U} \subset \mathcal{X}$ be convex and $f : \mathcal{U} \to \mathbb{R}$ convex. If f in x^* has the quasimin property, then x^* is a minimum point of f in $\mathcal{U}$.*

Proof. Let $x \in \mathcal{U}$ and $h = x - x^*$ then also $x^* + \lambda h \in \mathcal{U}$, $0 < \lambda < 1$, and we have by assumption

$$\liminf{}_{\lambda \to 0} \frac{f(x^* + \lambda h) - f(x^*)}{\lambda \|y - x\|} \ge 0\,.$$

But $f(x^* + \lambda h) \le (1 - \lambda) f(x^*) + \lambda f(x)$ hence

$$\liminf{}_{\lambda \to 0} \frac{(1 - \lambda) f(x^*) + \lambda f(x) - f(x^*)}{\lambda \|x - x^*\|} \ge 0$$

and therefore, for all $x \in \mathcal{U}$,

$$\frac{f(x) - f(x^*)}{\|x - x^*\|} \ge 0\,.$$

□

Note that no differentiability of f is required in this result but, of course, the convexity of f cannot be omitted.

References: (Craven95), (Ekeland), (Luenberger), (Marti), (Schaeffer).

1.11 Quadratic Functionals

The *principle of complementary energy* is studied in the context of LEGENDRE transformations, see Sect. 4.1**(e4)**.

Let $\mathcal{H}$ be a HILBERT space with canonical scalar product $(\diamond, \circ)$.

(a) The Energy Functional Let there be given

a symmetric bilinear form	$a : \mathcal{H} \times \mathcal{H} \to \mathbb{R}$
a $f \in \mathcal{H}_d$, i.e., a continuous linear functional	$f : \mathcal{H} \to \mathbb{R}$
a closed convex subset	$\emptyset \neq \mathcal{U} \subset \mathcal{H}$
a quadratic functional	$J(v) = a(v,v) - 2\,f(v)$

.

In many stationary problems of continuum mechanics, $a(v,v)$ is the (double) interior energy (strain energy), $-f(v)$ the outer potential energy, and $J(v)$ the (double) total energy of a system in state v, therefore the quadratic form $J(v)$ is frequently called *energy form* or *energy functional.* By the extremal principle of mechanics, it takes a stationary value in equilibrium being usually a minimum. Accordingly, the subject of the *energy method* consists of solving theoretically and numerically the extremal problem

$$J(v) = \min!\,, \quad v \in \mathcal{U}\,.$$

To this purpose the bilinear form a must define a *norm* $\|\circ\|_a = (\circ, \circ)^{1/2}$ on $\mathcal{U}$ and this *energy norm* has to be *equivalent* to the canonical norm $\|\circ\| = (\circ, \circ)^{1/2}$, i.e., the following assumption must be satisfied:

Definition 1.4. *The bilinear form a is a $\mathcal{U}$-elliptic functional with the following properties:*

(1°) $a(\circ, \circ)$ *is bounded in* $\mathcal{U}$,

$$\exists\, \beta > 0 \;\; \forall\, v \in \mathcal{U} : a(v,v) \leq \beta\, \|v\|^2.$$

(2°) a *is* uniformly *positive definite in* $\mathcal{U}$,

$$\exists\, \alpha > 0 \;\; \forall\, v \in \mathcal{U} : a(v,v) \geq \alpha\, \|v\|^2\,.$$

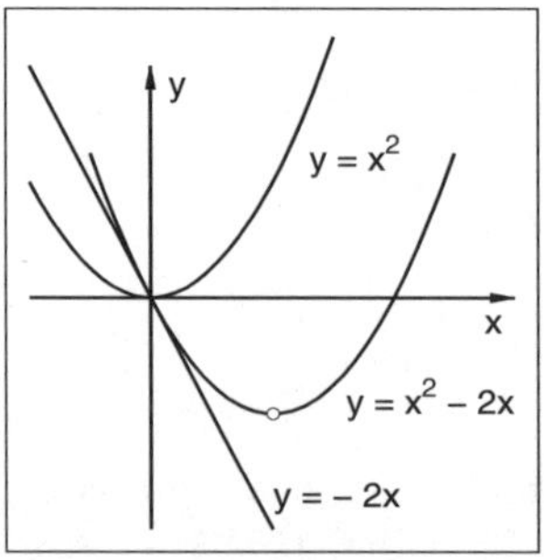

Figure 1.14. A simple energy functional

Then, by the CAUCHY-SCHWARZ inequality,

$$|a(u,v)| \le a(u,u)^{1/2} a(v,v)^{1/2} \le \beta \, \|u\| \|v\|$$

and thus the bilinear form a is bounded hence continuous in $\mathcal{H} \times \mathcal{H}$.

Theorem 1.25. *(Unique Existence) Let $\mathcal{U} \subset \mathcal{H}$ be convex and closed and let a be $\mathcal{U}$-elliptic then there exists a unique $u \in \mathcal{U}$ for every $f \in \mathcal{H}_d$ so that*

$$u = \arg\min_{v\in\mathcal{U}} J(v) \,.$$

Proof. (1°) $d := \mathrm{Inf}_{v\in\mathcal{U}} J(v)$ exists since $f \in \mathcal{H}_d$ is bounded, hence $J(v)$ is bounded from below:

$$J(v) \ge \alpha \, \|v\|^2 - 2\|f\| \, \|v\| = \alpha^{-1} \left(\alpha\|v\| - \|f\|\right)^2 - (\|f\|^2/\alpha) \ge -(\|f\|^2/\alpha) \,.$$

(2°) Let $\{v_m\}$ be a sequence with $\lim_{m\to\infty} J(v_m) = d$. Then it follows by the parallelogram identity for the norm $\|\cdot\|_a$ that

$$\begin{aligned}
\alpha \, \|v_m - v_n\|_a^2 &\le a(v_m - v_n \,,\, v_m - v_n) \\
&= 2a(v_m \,,\, v_m) + 2a(v_n \,,\, v_n) - a(v_m + v_n \,,\, v_m + v_n) \\
&\quad -4f(v_m) - 4f(v_n) + 4[f(v_m) + f(v_n)] \\
&= 2J(v_m) + 2J(v_n) - 4J((v_m + v_n)/2) \\
&\le 2J(v_m) + 2J(v_n) - 4d \to 0 \,,
\end{aligned}$$

because $(v_m + v_n)/2$ in $\mathcal{U}$. By definition of d, $J(v_m)$ and and $J(v_n)$ tend both to d hence $\{v_m\}$ is a CAUCHY sequence. Its limit u is contained in $\mathcal{U}$ because $\mathcal{U}$ is closed.
(3°) By the continuity of J we have $\lim_{n\to\infty} J(v_n) = J(u) = \inf_{v\in\mathcal{U}} J(v)$.
(4°) If u and u^* are two solutions then $u \,,\, u^* \,,\, u \,,\, u^* \,,\, \ldots$ is a minimum sequence, too. But every minimum sequence has a unique limit by (2°) hence $u = u^*$. □

In particular, let $\mathcal{U} \subset \mathcal{H}$ be a closed subspace and $u_0 \in \mathcal{H}$, then the affine subspace $u_0 + \mathcal{U}$ is closed and convex, and

$$u = \arg\min_{v\in u_0+\mathcal{U}} J(v) \tag{1.61}$$

does exist uniquely. For the *reduction* to a problem in vector space $\mathcal{U}$, let $v = u_0 + w \,,\ w \in \mathcal{U}$, then

$$J(v) = J(u_0 + w) = a(w,w) - 2g(w) + J(u_0)$$

with $g(w) = f(w) - a(u_0, w)$. Hence (1.61) is equivalent to the computation of

$$u = \arg\min_{v\in\mathcal{U}} (a(v,v) - 2g(v)) \,, \tag{1.62}$$

where g is bounded: $\forall\, v \in \mathcal{U} : |g\,(v)| \le (\|f\| + \beta \, \|u_0\|) \, \|v\| \,.$ □

Theorem 1.26. *(Characterization Theorem) Let $\mathcal{H}$ be a* HILBERT *space, then*

$$u = \arg\min_{v\in\mathcal{U}} J(v) \tag{1.63}$$

exists uniquely if and only if one of the following conditions is fulfilled:
(1°) $\emptyset \neq \mathcal{U} \subset \mathcal{H}$ *is a closed convex set and*

$$\exists\, u \in \mathcal{U}\ \ \forall\, v \in \mathcal{U} : a(u\,,\, v-u) \geq f(v-u)\,. \tag{1.64}$$

(2°) $\emptyset \neq \mathcal{U} \subset \mathcal{H}$ *is a closed subspace and*

$$\exists\, u \in \mathcal{U}\ \ \forall\, v \in \mathcal{U} : a(u,v) = f(v)\,. \tag{1.65}$$

(3°) $\emptyset \neq \mathcal{U} \subset \mathcal{H}$ *is a closed convex cone with vertex in origin and*

$$\exists\, u \in \mathcal{U}\ \ \forall\, v \in \mathcal{U} : a(u,v) \geq f(v)\,,\ \ a(u,u) = f(u)\,. \tag{1.66}$$

Proof. (1°). Let $v \in \mathcal{U}$ arbitrary. Then $w = (1-\lambda)\,u + \lambda v = u + \lambda\,(v-u) \in \mathcal{U}$ for $0 \leq \lambda \leq 1$ because $\mathcal{U}$ convex. Let $\varphi\,(\lambda) = J\,(w)$ with

$$J(w) = J(u) + 2\lambda\,[a(u\,,\, v-u) - f(v-u)] + \lambda^2 a(v-u, v-u). \tag{1.67}$$

By assumption, φ takes its minimum in $\lambda = 0$ hence $\varphi'(0) \geq 0$ which implies (1.64).
If conversely (1.64) is fulfilled then, by (1.67) for $\lambda = 1$,

$$\exists\, u \in \mathcal{U}\ \ \forall\, v \in \mathcal{U} : J(v) \geq J(u) + a(v-u, v-u) \geq J(u)\,,$$

and thus (1.63).
The proof of (2°) is carried out in the same way.
(3°) By (1.66) we obtain immediately (1.64) and thus (1.63) by (1°). Conversely,

$$\exists\, u\ \ \forall\, v \in \mathcal{U} : a(u,v) \geq f(v) \text{ and } a(u,u) \geq f(u)$$

follows by (1.64) for $v := u + v$ resp. $v := 2u$. If $\mathcal{U}$ is a cone with vertex in origin then $0 \in \mathcal{U}$ and thus

$$-a(u,u) \geq -f(u) \implies a(u,u) \leq f(u)\,.$$

for $v = 0$ by (1.64). □

The relations (1.64), (1.65) or (1.66) are called EULER equations of the variational problem $J\,(v) = \min!\,,\ v \in \mathcal{U}$. The Theorems 1.25 and 1.26 have some interesting inferences:

Lemma 1.30. (RIESZ' *Representation Theorem) Let $\mathcal{H}$ be a* HILBERT *space and $f \in \mathcal{H}_d$. Then there exists a unique $u_f \in \mathcal{H}$ such that*

$$\boxed{\forall\, v \in \mathcal{H} : f\,(v) = (u_f,\, v)}\,.$$

Proof. The result follows immediately from the Existence Theorem and the Charcterization Theorem for $\mathcal{U} = \mathcal{H}$ and the canonical scalar product $a(\cdot\,,\circ) = (\cdot\,,\circ)$. □

The bijective relation of $u_f \in \mathcal{H}$ and $f \in \mathcal{H}_d$ is called *canonical isomorphism* or RIESZ mapping. We do no longer make a difference between these both (different) elements but write f instead u_f hence $f(v) = (f, v)$ (one symbol with two meanings); by this way $\mathcal{H}_d$ and $\mathcal{H}$ are *canonically identified*, $\mathcal{H}_d = \mathcal{H}$.

(b) Operators in Hilbert Space Let $\mathcal{X}$, $\mathcal{Y}$ be HILBERT spaces and $L : \mathcal{X} \to \mathcal{Y}$ a bounded linear mapping, in short $L \in \mathcal{L}(\mathcal{X}, \mathcal{Y})$, then, by canonical identification, the *adjoint operator* $L_d : \mathcal{Y} \to \mathcal{X}$ is defined by

$$(x,\, L_d y)_{\mathcal{X}} = (y,\, Lx)_{\mathcal{Y}}\,.$$

If $\mathcal{X} = \mathcal{Y}$ and $L_d = L$, then

$$\forall\, x,\, y \in \mathcal{X} : (y,\, Lx) = (x,\, Ly) \tag{1.68}$$

and the operator L is called *selfadjoint* or *symmetric* in this case. Of course, every $L \in \mathcal{L}(\mathcal{X}, \mathcal{X})$ defines a bilinear form $a(u, v) := (u, Lv)$ but also the converse is true by the following result.

Theorem 1.27. (LAX-MILGRAM) *Let $a : \mathcal{H} \times \mathcal{H} \to \mathbb{R}$ be an elliptic but not necessarily symmetric bilinear form. Then there exists a unique $u \in \mathcal{H}$ for every $f \in \mathcal{H}_d$ $(= \mathcal{H})$ such that*

$$\forall\, v \in \mathcal{H} : a(u, v) = (f, v)\,. \tag{1.69}$$

By (1.69), a bijective and continuous operator $L^{-1} : \mathcal{H}_d \to \mathcal{H}$ is defined satisfying

$$\forall\, u,\, v \in \mathcal{H} : a(u, v) = (Lu, v)\,,$$

then the linear operator L is symmetric if and only if the bilinear form a is symmetric.

Example 1.28. Let $\mathcal{X} = \mathcal{Y} = L_2[0,\, 1]$ and let $L \in \mathcal{L}(\mathcal{X}, \mathcal{Y})$ be defined by

$$Lx = \int_0^t K(t, s)x(s)\, ds\,,\ t \in [0,\, 1]\,,$$

with continuous *kernel* $K(t, s)$ in unit square, and let $(f,\, g) = \int_0^1 f(x)g(x)\, dx$ be a scalar product. Interchange of integration yields

$$\begin{aligned}
(y,\, Lx) &= \int_0^1 y(t) \int_0^t K(t, s)x(s)\, dsdt &&= \int_0^1 \int_0^t y(t)K(t, s)x(s)\, dsdt \\
&= \int_0^1 \int_s^1 y(t)K(t, s)x(s)\, dtds &&= \int_0^1 x(s) \left(\int_s^1 K(t, s)y(t)\, dt \right) ds \\
&= \int_0^1 x(t) \left(\int_t^1 K(s, t)y(s)\, ds \right) dt = (x,\, L_d\, y)\,.
\end{aligned}$$

Accordingly, the adjoint operator L_d has the form

$$L_d\, y = \int_t^1 K(s,t)y(s)\, ds\,.$$

Example 1.29. Let $\mathcal{X} = \mathcal{Y} = H_0^2(\Omega)\,,\ \Omega \subset \mathbb{R}^2$ and let $Lu = \Delta u = u_{x_1x_1} + u_{x_2x_2}$ be the LAPLACE operator. Then an application of GREEN's Theorem (1.23) yields

$$\int_\Omega [\varphi\Delta\psi - \psi\Delta\varphi]\, dV = \oint_{\partial\Omega} \left[\varphi\frac{\partial\psi}{\partial n} + \psi\frac{\partial\varphi}{\partial n}\right] dO = 0\,.$$

Accordingly, this operator is symmetric on $H_0^2(\Omega)$.

(c) Projectors in Hilbert Space

Theorem 1.28. *(Projection Theorem) Let $\mathcal{H}$ be a* HILBERT *space and let $\emptyset \neq \mathcal{U} \subset \mathcal{H}$ be a closed convex subset. Then*

$$\boxed{\forall\, w \in \mathcal{H} \quad \exists!\; u \in \mathcal{U} : \|w - u\| = \inf{}_{v\in\mathcal{U}} \|w - v\|}\,. \tag{1.70}$$

Proof. Consider the quadratic functional

$$J(v) = \|w - v\|^2 - \|w\|^2 = (v,v) - 2(w,v)$$

for fixed $w \in \mathcal{H}$ and apply the Existence Theorem 1.25 . □

The element u of (1.70) is called *projection* of w onto $\mathcal{U}$ and one writes $u = Pw$ with the projection operator P. For $J(v) = (v,v) - 2(w,v)$ we obtain directly by the Characterization Theorem :

Lemma 1.31. *(Characterization of Projectors) Let $\mathcal{H}$ be a* HILBERT *space and $\emptyset \neq \mathcal{U} \subset \mathcal{H}$ a convex closed subset.*
(1°) *$u = Pw$ holds if and only if*

$$\forall\, v \in \mathcal{U} : (w - u\,,\, v - u) \le 0\,. \tag{1.71}$$

(2°) *If $\mathcal{U}$ is a closed linear subspace, $u = Pw$ holds if and only if*

$$\forall\, v \in \mathcal{U} : (w - u\,,\, v) = 0\,. \tag{1.72}$$

Defining

$$\cos\varphi := \frac{(u,v)}{\|u\|\,\|v\|}\,,\ 0 \le \varphi \le \pi\,,$$

for u, $v \in \mathcal{H}$, (1.71) says that the angle φ between $w - v$ and $v - u$ is not less than $\pi/2$.

Further properties of the projection operator:

Lemma 1.32. (1°) *(Contraction, Continuity) Under the assumption of Theorem 1.28*

$$\forall\, v,\, w \in \mathcal{H} : \|Pv - Pw\| \le \|v - w\|\,.$$

(2°) *(Linearity) The projection operator P is a linear mapping $P : \mathcal{H} \to \mathcal{U}$ if and only if $\mathcal{U}$ is a linear subspace.*

Proofs in `SUPPLEMENT\chap01b`.

A projector P of $\mathcal{H}$ is called *orthogonal* if $\mathrm{Ker}(P) = \mathrm{Range}(P)^{\perp}$. Then

$$\mathcal{H} = \mathrm{Range}(P) \oplus [\mathrm{Range}(P)]^{\perp} = [\mathrm{Ker}(P)]^{\perp} \oplus \mathrm{Ker}(P)$$

in the decomposition $\mathcal{H} = \mathrm{Range}(P) \oplus \mathrm{Ker}(P)$.

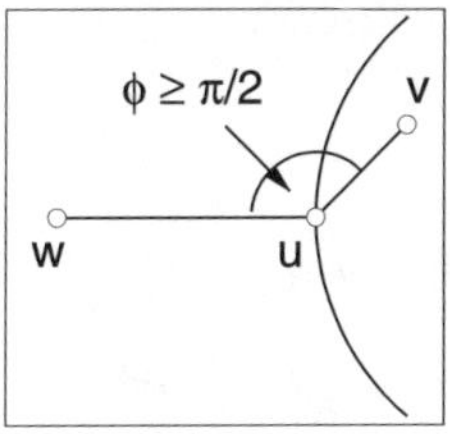

Figure 1.15. To Lemma 1.31(1°)

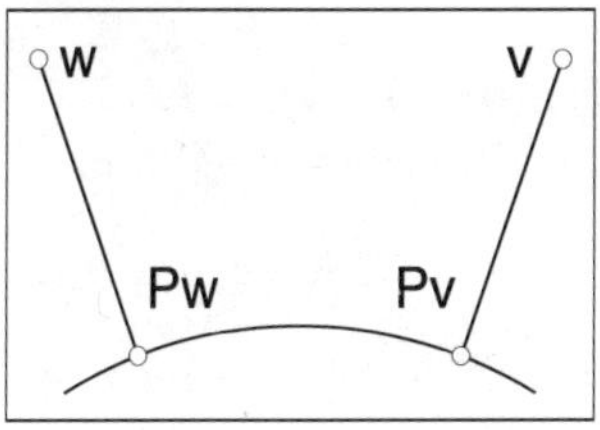

Figure 1.16. To Lemma 1.32(1°)

Theorem 1.29.(1°) *A projector P of a* HILBERT *space is orthogonal if and only if it is symmetric.*
(2°) *An orthogonal projector is continuous.*
(3°) *If $\mathcal{U} \subset \mathcal{H}$ is a subset, then $\mathcal{U}^{\perp}$ is a closed subspace.*
(4°) *If $\mathcal{U} \subset \mathcal{H}$ is a closed subspace, then $\mathcal{H} = \mathcal{U} \oplus \mathcal{U}^{\perp}$ and there exists an orthogonal projector P with* $\mathrm{Range}(P) = \mathcal{U}$.

Proof see (Taylor).

(d) Properties of the Energy Functional

Lemma 1.33. *Let $u = \arg\min_{v \in \mathcal{U}} J(v)$. Then*
(1°) $J(u) = -a(u, u)$ *("total energy = negative strain energy"),*
(2°) $\forall\, v \in \mathcal{U} : J(v) - J(u) = a(v - u, v - u)$ *("error of total energy = inner energy of error").*

Proof. (1°) The Characterization Theorem yields $a(u, v) = f(v)\ \forall\, v \in \mathcal{U}$. Then $a(u, u) = f(u)$ because $u \in \mathcal{U}$ hence $J(u) = a(u, u) - 2a(u, u) = -a(u, u)$.
(2°) Because $w = v - u \in \mathcal{U}$ and $a(u, w) = f(w)$ we have

$$\begin{aligned} J(v) &= J(u + w) = a(u, u) + 2a(u, w) + a(w, w) - 2f(u) - 2f(w) \\ &= J(u) + a(w, w)\,. \end{aligned}$$

□

(e) Ritz Approximation Let the assumptions of the Existence Theorem be satisfied further on and let $\mathcal{U} \subset \mathcal{H}$ be a linear subspace. The idea of RITZ consists in solving the extremum problem

$$J(v) = a(v,v) - 2(f,v) = \min!\,, \quad v \in \mathcal{U}\,,$$

resp. the equivalent variational problem on a *finite dimensional subspace* $\mathcal{R}$. The following result is frequently called *Main Theorem of* RITZ *Theory.*

Theorem 1.30. *Let $\mathcal{R} \subset \mathcal{U} \subset \mathcal{H}$ be closed linear subspaces and*

$$u = \arg\min\nolimits_{v\in\mathcal{U}} J(v)\,, \quad u_R = \arg\min\nolimits_{v\in\mathcal{R}} J(v)\,.$$

Then the following statements are equivalent:

(1°) *"u_R is* GALERKIN *approximation": $\forall\, v \in \mathcal{R} : a(u_R, v) = (f,v)$.*
(2°) *"$u - u_R$ is perpendicular to $\mathcal{R}$" with respect to energy:*
$\forall\, v \in \mathcal{R} : a(u - u_R, v) = 0$.
(3°) *"u_R minimizes energy of error":*

$$\forall\, w \in \mathcal{R} : a(u - u_R, u - u_R) \leq a(u - w, u - w)\,.$$

Proof. (1°) $\Longrightarrow$ (2°). We have $a(u,v) = (f,v)\ \forall\, v \in \mathcal{R} \subset \mathcal{U}$; hence

$$\forall\, v \in \mathcal{R} : a(u,v) - a(u_R,v) = (f,v) - (f,v) = 0\,.$$

(2°) $\Longrightarrow$ (1°). We have

$$\forall\, v \in \mathcal{R} \subset \mathcal{U} : a(u_R,v) = (f,v) = a(u,v)\,.$$

(2°) $\Longleftrightarrow$ (3°) By the Characterization Theorem we have for

$$J(w) = a(w-u, w-u) - a(u,u) = a(w,w) - 2a(u,w)\,, \quad w \in \mathcal{U} = \mathcal{R}\,,$$

the inequality

$$J(w) \geq J(u_R) \Longleftrightarrow \forall\, v \in \mathcal{R} : a(u_R, v) = a(u,v)\,.$$

□

Because (2°) and Lemma 1.31, u_R is the projection of u onto $\mathcal{R}$ w.r.t. the energy product $a(\diamond, \circ)$, therefore the RITZ method is also called *projection method.*

Lemma 1.34. *"Energy of error equals error of energy":*

$$a(u - u_R, u - u_R) = a(u,u) - a(u_R, u_R)\,,$$

*in particular, "*RITZ *approximation does never overestimate energy":*

$$a(u_R, u_R) \leq a(u,u)\,.$$

Proof. Using $\mathcal{R}$ instead of $\mathcal{U}$ we obtain $J(u_R) = -a(u_R, u_R)$ by Lemma 1.33(1°). Substituting $v = u_R$ in Lemma 1.33(2°), it follows

$$J(u_R) = -a(u_R, u_R) = a(u - u_R, u - u_R) - a(u, u).$$

□

Let now $\mathcal{R} = \text{span}\{\varphi_1, \ldots, \varphi_n\}$ be finite-dimensional then $u_R = u_0 + \sum_{j=1:m} y_j \varphi_j$ is called RITZ approximation of u. Substituting this basic approach and $v = \varphi_j$ in $a(v, u_r) = (v, f)$ successively we obtain for $\underline{y} = [y_1, \ldots, y_m]^T$ a linear system of equations,

$$\boxed{A\underline{y} = \underline{b}, \quad A = [a(\varphi_i, \varphi_j)]_{i,j=1}^m, \quad \underline{b} = [(\varphi_j, f) - a(\varphi_j, u_0)]_{j=1}^m}\,.$$

The leading matrix A is symmetric and positive semi-definite. If the functions φ_j are linear independent (form a basis of $\mathcal{R}$), A is positive definite and the system has a unique solution.

The finite element method (FEM) starts from the extremal problem or from the associated variational problem and uses only this formulation. The classical formulation as boundary value problem is abandoned here. For numerical approximation, the function space of (weak) solutions is replaced by a *finite-dimensional space* $\mathbb{R}$ (usually a subspace of the basic space $\mathcal{U}$ but not always). The finite difference method (FDM) starts from the classical formulation of a problem as boundary value problem disregarding the formulation as variational problem. For numerical approximation the domain on which the solution lives is replaced by a mesh with a finite number of knots.

References: (Ciarlet79), (Strang), (Velte). For an interesting mechanical interpretation of Figure 1.14 see (Hartmann).

2

Numerical Methods

Before the computer (*ordinateur* in French) changed the world, numerical mathematics — which mockers referred to as phenomenological — could hardly be counted as one of the supreme disciplines of the mathematical sciences. Whether that is the case today is beside the point, but combined with modeling and simulation it has risen in the hierarchy, and the latter even have to suffer to some extent to justify the existence of the other subjects. In the 1960s the integrimeter, integraph and harmonic oscillator were treated as *instrumental mathematics* in lectures. They have been long forgotten, as have all numerical methods for the hand calculator such as, e.g., extracting roots by subtracting odd numbers.

The turbulent evolution of the computer to the *laptop* allowed *numerical mathematics* to successfully keep pace at some distance. Logarithmic Tables, etc., have long been replaced by the pocket calculator, and linear systems — a central problem — are solved today with three inconspicuous glyphs `A\b`, without resulting in any inconsistencies in style. A multitude of monographs displays what has been achieved so far; only (Golub), (Hairer) and (Rheinboldt70) are stated as examples. The curve of the number of publications with purely numerical themes also seems to be getting somewhat flatter, while the number of problem-related applications is on the rise.

If numerical methods shall be described in a single chapter, it is necessary to concentrate on the essential aspects. The author assumes that the reader is interested primarily in applying existing codes, which is not possible without a minimum of understanding and intuition. This is the premise for the introduction to the mindsets of *numerics* provided here and the discussion of challenging developments as the multiple shooting method and differential-algebraic problems. The numerical part of this book is not limited to the topics treated here; for those possessing the necessary background, further issues are dealt with in later chapters.

2.1 Interpolation and Approximation

In many applications, functions are given only by discrete data sets. Or, a function cannot be integrated in closed form and must be replaced by a simpler one to that end. Then it is approximated *piecewise* by polynomials of *moderate* degree because polynomials of higher degree oscillate more or less strongly in larger intervals. But also approximations by rational functions, exponential functions, and, preferably in periodic problems, trigonometric interpolation are in common use. Let

$$\boxed{\Pi_n \text{ the set of real polynomials } p_n \text{ of degree } \leq n}\,.$$

With the usual addition and scalar multiplication, Π_n is a *vector space* of dimension $n+1$ whose basis $\{q_0(x), \ldots, q_n(x)\}$ is chosen according to the individual requirements.

(a) The General Interpolation Problem Let there be given

$$\begin{array}{ll} \text{a sequence of support abszissas} & \{x_i\}_{i=0}^{\infty}\,,\ x_i \in \mathbb{R}\,, \\ \text{a sequence of support ordinates} & \{f_i\}_{i=0}^{\infty}\,,\ f_i \in \mathbb{R}\,, \\ \text{a sequence of functions} & \{g_i\}_{i=0}^{\infty}\,,\ g_i \in C[a,b]\,. \end{array}$$

The support abszissas x_i shall be *mutually distinct*; for the other case we refer to (Hoellig) Sect. 3.1. Then, a sequence of functions

$$\{h_n\}_{n=1}^{\infty}\,,\ h_n(x) = \sum_{i=0}^{n-1} \alpha_i g_i(x)\,, \tag{2.1}$$

is to be found with the *interpolation property*

$$h_n(x_j) = f_j\,,\ j = 0 : n-1\,. \tag{2.2}$$

Writing $\underline{a} = [\alpha_0, \ldots, \alpha_{n-1}]^T$ and $\underline{f} = [f_0, \ldots, f_{n-1}]^T$, (2.1) is, for *fixed* n, equivalent to the linear system of equations

$$A\underline{a} = \underline{f}, \quad A = \left[g_i(x_j)\right]_{i,j=0}^{n-1}\,, \tag{2.3}$$

and the interpolation problem (2.1), (2.2) has a unique solution for a regular matrix A.

Theorem 2.1. *(Existence,* HAAR *Condition) Let all* n *support abscissas* x_j, $j = 0 : n-1$, *be mutually distinct and let every not identical disappearing linear combination of* n *functions* g_i, $i = 0 : n-1$, *have no more than* $n-1$ *zeros, then the matrix* A *is regular and the interpolation problem has a unique solution.*

Proof. If the matrix A is singular, there exists a row vector $\underline{c} \in \mathbb{R}_n$ with $\underline{c}\,A = 0 \in \mathbb{R}_n$. Then the linear combination $h(x) := \sum_{i=0:n-1} c_i g_i(x)$ has n different zeros $x_0, \ldots, x_{n-1}$ because

$$h(x_j) = \sum_{i=0}^{n-1} c_i g_i(x_j) = 0,\; j = 0 : n-1\,,$$

in contradiction to the assumption. □

In particular, the HAAR condition is fulfilled for a sequence $\{p_n\}_{n=0}^{\infty}$ of polynomials $p_n \in \Pi_n$ because every not identically disappearing linear combination $h(x) := \sum_{i=0:n-1} c_i p_i(x)$ is a polynomial of degree $\leq n-1$ having no more than $n-1$ zeros. However, the matrix A in (2.3) is ill-conditioned in general, hence this linear system of equations is not used commonly for numerical computation of the coefficients α_i.

(b) Interpolating Polynomials To find a *linear recursion formula* for *interpolating polynomials*, let $\{j_0, \ldots, j_m\} \subset \{0, \ldots, n\}$ be an index set with different elements. Then the interpolating polynomial $p_{j_0,\ldots,j_m}(x) \in \Pi_m$ is uniquely determined by

$$\begin{aligned} p_j(x) \quad &= f(x_j)\,,\; m = 0\,, \\ p_{j_0,\ldots,j_m}(x_i) &= f(x_i)\,,\; i = j_0, \ldots, j_m\,,\; m = 1 : n\,, \end{aligned} \tag{2.4}$$

following the Existence Theorem; thus, in particular, $p_{j_0,\ldots,j_m}(x)$ does not depend on a permutation of indices.

Lemma 2.1. (AITKEN) *For* $j = 0 : n-m\,,\; m = 1 : n$

$$p_{j,\ldots,j+m}(x) = \frac{1}{x_{j+m} - x_j}\left[(x - x_j)p_{j+1,\ldots,j+m}(x) - (x - x_{j+m})p_{j,\ldots,j+m-1}(x)\right].$$

This formula is used in various applications for computation of interpolating polynomials at a given point x.

Theorem 2.2. (CAUCHY*'s Error Representation) Let the function* f *be (n+1)-times differentiable in* $[a, b]$ *and let* $[u, v, \ldots, w]$ *be the smallest interval* $\mathcal{I} \subset \mathbb{R}$ *containing all* $u, v, \ldots, w \in \mathcal{I}$. *Then* $\forall\, x \in [a,\, b]\;\; \exists\, \xi_x \in [x_0, \ldots, x_n, x]$:

$$f(x) - p_n(x; f) = \frac{f^{(n+1)}(\xi_x)}{(n+1)!}\,\omega(x)\,,\;\; \omega(x) = (x - x_0)\cdots(x - x_n)\,. \tag{2.5}$$

Proof in SUPPLEMENT\chap02.

Note that the intermediate values ξ_x change with the value x.

(c) Interpolation after Lagrange We consider the approximation of a function f by an interpolating polynomial in *separated* form

$$\boxed{f(x) \approx p_n(x; f) = \sum_{i=0}^{n} f(x_i) q_i(x)}\,, \tag{2.6}$$

with the basis $\{q_0, \ldots, q_n\}$ of Π_n consisting of LAGRANGE polynomials

$$q_i(x) = \prod_{j=0, j\neq i}^{n} \frac{x - x_j}{x_i - x_j}, \quad i = 0 : n. \tag{2.7}$$

The interpolation property $p_n(x_i; f) = f(x_i)$ is guaranteed here by the specific property $q_i(x_j) = \delta^i{}_j$ (KRONECKER symbol).

Properties: (1°) Interpolation of LAGRANGE is of high theoretical but less practical use because all polynomials $q_i(x)$ have to be computed anew if the set of support nodes is changed or augmented.
(2°) By unique existence, the formula (2.6) is exact for all *monomials* $f(x) = x^k$, $k = 0 : n$, $\sum_{i=0:n}(x_i)^k q_i(x) = x^k$, $k = 0 : n$; in particular, we obtain a *partition of unity* $\sum_{i=0:n} q_i(x) = 1$ for $k = 0$.
(3°) For *equidistant abszissas*, $h = 1/n$, $x_i = x_0 + ih$, $x = x_0 + sh$, $s \in [0, n]$, we obtain

$$\frac{x - x_j}{x_i - x_j} = \frac{(x_0 + sh) - (x_0 + jh)}{(x_0 + ih) - (x_0 + jh)} = \frac{s - j}{i - j};$$

hence formula (2.6) is simplified considerably by this translation and scaling of the independent variable:

$$p_n(x(s); f) = \sum_{i=0}^{n} f(x_i) q_i(x(s)), \quad q_i(x(s)) = \prod_{j=0, j\neq i}^{n} \frac{s - j}{i - j}. \tag{2.8}$$

This representation applies in constructing numerical quadrature formulas as well as in constructing numerical devices for approximations of ordinary differential equations and systems.

(d) Interpolation after Newton Let $f[x_{j_0}, \ldots, x_{j_m}]$ be the highest term of $p_{j_0,\ldots,j_m}(x)$ then, by Lemma 2.1,

$$f[x_j, \ldots, x_{j+m}] = \frac{f[x_{j+1}, \ldots, x_{j+m}] - f[x_j, \ldots, x_{j+m-1}]}{x_{j+m} - x_j}. \tag{2.9}$$

These *divided differences* do not depend on the succession of indices because the associated polynomials have this property. On choosing the NEWTON basis for Π_n, $n_0(x) \equiv 1$, $n_j(x) = (x - x_0) \cdots (x - x_{j-1})$, $j = 1 : n$, we obtain

$$\begin{aligned} p_{0,\ldots,n}(x; f) &= \sum_{j=0}^{n} a_j n_j(x), \quad a_j = f[x_0, \ldots, x_j], \\ &= (\cdots(a_n(x - x_{n-1}) + a_{n-1})(x - x_{n-2}) + a_{n-2})\cdots)(x - x_0) + a_0 \end{aligned} \tag{2.10}$$

because of the interpolation property and the recursion formula

$$p_{0,\ldots,n}(x) = p_{0,\ldots,n-1}(x) + a_n \pi(x), \quad \pi(x) = (x - x_0)\cdots(x - x_{n-1}) \in \Pi_n$$

being typical for this form of the interpolating polynomial.

The well-known TAYLOR polynomial $p_n(x;f) = \sum_{i=0:n} \frac{f^{(i)}(x_0)}{i!}(x-x_0)^i$ may be said to stand on the opposite side of the scale of approximation by polynomials since it uses only one "interpolation point" x_0. By using (2.9), a natural relation may be found between TAYLOR coefficients and divided differences being the coefficients of NEWTON's polynomial:

Lemma 2.2.

$$f[x_i,\ldots,x_{i+k}] = \frac{f^{(k)}(\xi)}{k!}, \quad \xi \in [x_i,\ldots,x_{i+k}]\,.$$

Proof. Let $p_{i,\ldots,i+k}(x)$ be the NEWTON interpolating polynomial of degree $\leq k$ with nodes (x_j, f_j), $j = i : i+k-1$, and let $(x_{i+k}, f_{i+k}) = (\overline{x}, f(\overline{x}))$ where all abszissas x_j and $\overline{x}$ shall be mutually distinct. Then we have

$$f(\overline{x}) = p_{i,\ldots,i+k}(\overline{x}) = p_{i,\ldots,i+k-1}(\overline{x}) + f[x_i,\ldots,x_{i+k}](\overline{x}-x_i)\cdots(\overline{x}-x_{i+k-1})$$

at the point $x_{i+k} = \overline{x}$, and, on the other side, the error formula for $p_{i,\ldots,i+k}(x)$,

$$f(\overline{x}) = p_{i,\ldots,i+k-1}(\overline{x}) + \frac{f^{(k)}(\xi)}{k!}(\overline{x}-x_i)\cdots(\overline{x}-x_{i+k-1})\,.$$

□

(e) By additional **Interpolation of the Derivatives** of f at all abszissas x_i we obtain interpolating polynomials of HERMITE type:

$$f(x) \approx h_{2n+1}(x,f) = \sum_{i=0}^{n} [f(x_i)h_{0,i}(x) + f'(x_i)h_{1,i}(x)] \in \Pi_{2n+1}$$

with HERMITE polynomials

$$\begin{aligned} h_{0,i}(x) &= \big[1 - 2q_i'(x_i)(x-x_i)\big]q_i(x)^2 \Longrightarrow h_{0,i}(x_k) = \delta^i{}_k\,,\ h_{0,i}'(x_k) = 0\,,\\ h_{1,i}(x) &= (x-x_i)q_i(x)^2 \qquad\qquad\quad \Longrightarrow h_{1,i}(x_k) = 0\,,\ h_{1,i}'(x_k) = \delta^i{}_k\,, \end{aligned}$$

where $q_i(x)$ are the LAGRANGE polynomials again. The error has the same form as in (2.5):

$$f(x) - h_{2n+1}(x,f) = \frac{f^{(2n+1)}(\xi_x)}{(n+1)!}\,\omega(x)\,,\quad \omega(x) = (x-x_0)^2\cdots(x-x_n)^2.$$

Besides some few exceptions, enhancing the degree n of an interpolating polynomial does *not* improve the approximation, instead a *segmentwise* interpolation with simple polynomials is to be preferred. By requiring some global smoothness, the compound polynomials then lead to the *interpolating spline functions*; cf. **(g)**.

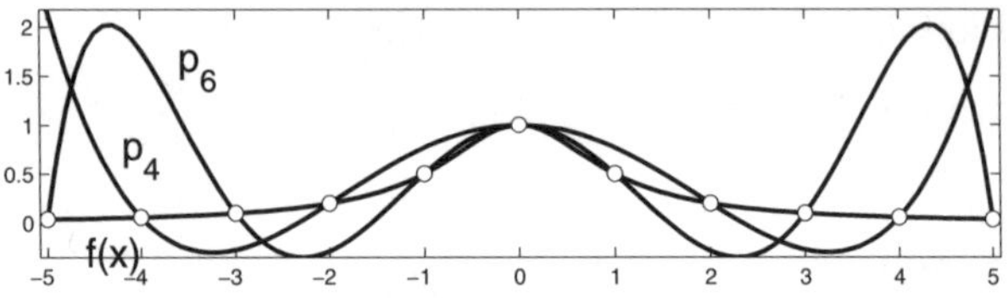

Figure 2.1. Interpolating polynomial of degree $n = 4, 6$ for $f(x) = 1/(1 + x^2)$

(f) Approximation by Beziér Polynomials In a fixed, unpartitioned interval, essential improvement of approximation is attained by abandoning the strong interpolating condition $p_n(x_i; f) = f(x_i)$ in the *interior* of the considered intervall. By the partition of unity,

$$1 = (x + (1 - x))^n = \sum_{i=0}^{n} \binom{n}{i} x^i (1 - x)^{n-i} =: \sum_{i=0}^{n} B_i^n(x)\,,$$

we obtain the basis of *Bernstein* polynomials $B_i^n(x)$ of Π_n and the general representation of $p_n \in \Pi_n$ as BEZIÉR *polynomial* with the BEZIÉR *points* b_i,

$$p_{n,\text{bez}}(x) = b_0 B_0^n(x) + \ldots + b_n B_n^n(x)\,. \tag{2.11}$$

The roots of all $B_i^n(x)$ are placed at the boundary of the interval $[0, 1]$, and precisely one extremal point exists in the interior (maximum point). Therefore BERNSTEIN polynomials possess no turning point in this interval. As a consequence, no *spurious* turning points are dragged in by approximating a function f in this way. However, the approximation is restricted to the unit interval $[0, 1]$ here, otherwise a rescaling becomes necessary. Besides, piecewise interpolation is to be preferred in the present case, too.

Properties: (1°) $\sum_{i=0}^{n} B_i^n(x) = 1\,,\ \sum_{i=0}^{n} \left(\frac{i}{n}\right) B_i^n(x) = x\,.$

(2°) By applying *forward differences* $\Delta b_i = b_{i+1} - b_i$, $\Delta^k b_i = \Delta(\Delta^{k-1} b_i)$, we obtain for the derivatives

$$p_{n,\text{bez}}^{(k)}(x) = \frac{n!}{(n-k)!} \sum_{i=0}^{n-k} (\Delta^k b_i) B_i^{n-k}(x)\,,$$

from which the above mentioned important property follows, namely

$$\boxed{\forall\, i : \Delta^k b_i \geq 0 \implies \forall\, x \in [0, 1] : p^{(k)}(x) \geq 0}\,.$$

Moreover, the k-th derivatives in $x = 0$ resp. $x = 1$ depend only on the BEZIÉR points $b_0, \ldots, b_k$ resp. $b_{n-k}, \ldots, b_n$.

(3°) Let there be given $s - r + 1$ successive BEZIÉR points $\{b_r, \ldots, b_s\}$ for the abszissas $x = (i - r)/(s - r)$, $i = r : s$, then

$$b_{r,\dots,s}(x) := \sum_{i=r}^{s} b_i B_{i-r}^{s-r}(x)$$

is the corresponding BEZIÉR polynomial of degree $\leq s-r$ (depending on the succession of points!). By means of the addition theorem for binomial coefficients, the *linear recursion formula* of DE CASTELJAU can be derived,

$$b_{r,\dots,s}(x) = (1-x)b_{r,\dots,s-1}(x) + xb_{r+1,\dots,s}(x)\,,$$

being applied for *pointwise* computation of $p_{n,\text{bez}}(x)$ in place of the algebraic representation (2.11).

(4°) The points $(x_i, b_i) = (i/n, f(i/n)) \in \mathbb{R}^2$, $i = 0 : n$, are also called BEZIÉR nodes or BEZIÉR points. The BEZIÉR polynomial is attached to the corresponding BEZIÉR polygon like a circus tent to its masts approaching it more and more closely by enhancing the degree resp. the node number. General curves in space are obtained by replacing the BEZIÉR points b_i in (2.11) by vectors,

$$\underline{p}_{n,\text{bez}}(x; \underline{f}) = \sum_{i=0}^{n} \underline{f}\left(\frac{i}{n}\right) B_i^n(x) \in \mathbb{R}^n\,, \quad \underline{f}(x) \in \mathbb{R}^n\,.$$

(5°) The *approximation by* BEZIÉR *polynomials* provides a basic result of *functional analysis*:

Theorem 2.3. *(*WEIERSTRASS*) Let $f \in \mathcal{C}[0,1]$ and*

$$B_n f : x \mapsto \sum_{i=0}^{n} f\left(\frac{i}{n}\right) B_i^n(x)$$

with BERNSTEIN *polynomials $B_i^n(x)$. Then*
$\lim_{n\to\infty} \max_{0\leq x\leq 1} |f(x) - B_n f(x)| = 0\,.$

The proof is a simple conclusion of a surprising result of KOROVKIN:

Theorem 2.4. *Let $L_n : \mathcal{C}[a,b] \mapsto \mathcal{C}[a,b]$ be a sequence of linear and positive operators L_n, i.e.,*

$$\forall\, f, g \in \mathcal{C}[a,b]\ \ \forall\, x \in [a,b] : f(x) \leq g(x) \Longrightarrow L_n(f) \leq L_n(g)\,,$$

and let $f_1(x) = 1$, $f_2(x) = x$, $f_3(x) = x^2$. If

$$\lim_{n\to\infty} \|L_n f_i - f_i\|_\infty = 0 \quad i = 1:3\,,$$

then
$\forall\, f \in \mathcal{C}[a,b] : \lim_{n\to\infty} \|L_n f - f\|_\infty = 0\,.$

Proof (Kosmol) Sect. 4.4.5.

Proof of Theorem 2.3. Apparently, the operators B_n are linear and positive, and we have

$$B_n(f_1, x) = 1\,, \quad B_n(f_2, x) = x\,, \quad B_n(f_3, x) = x^2 + \frac{x - x^2}{n}\,,$$

hence the assumptions of Theorem 2.2 are satisfied. □

A direct, likewise interesting proof of Theorem 2.3 is found in (Yosida) Sect. 0.2.

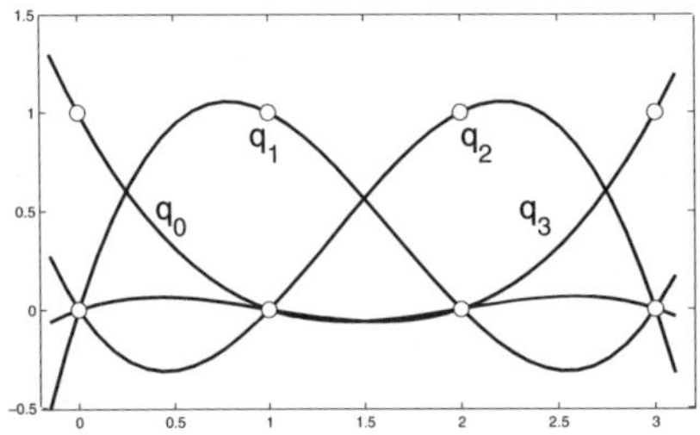

Figure 2.2. LAGRANGE polynomials, $n = 3$

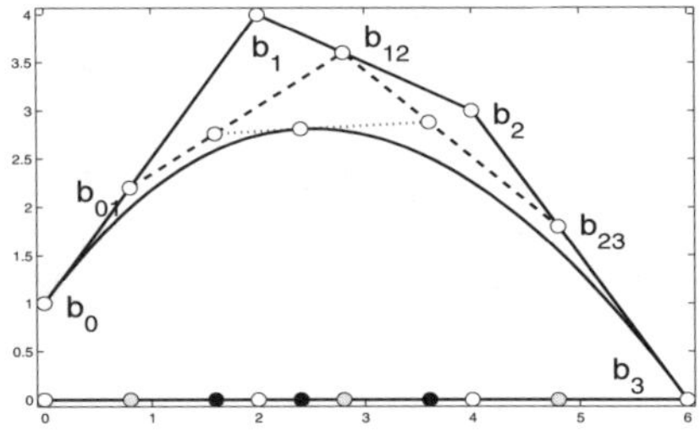

Figure 2.3. BEZIÉR polynomials, $n = 3$, with rescaling

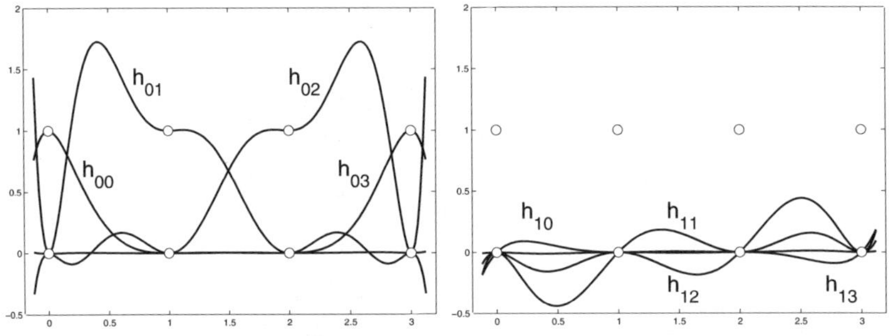

Figure 2.4. HERMITE polynomials, $n = 3$

(g) Interpolating Splines A (continuous) BEZIÉR *curve* consists piecewise of BEZIÉR polynomials having the interpolation property at the ends of their domain, respectively. We consider the special case of a BEZIÉR curve in interval $\mathcal{I} = [0, n \cdot m]$, $m \in \mathbb{N}$. The curve shall consist of BEZIÉR polynomials of degree n with BEZIÉR points $b_{nk}, \ldots, b_{n(k+1)}$ in the subintervals $\mathcal{I}_k = [n(k-1), nk]$, $k = 1 : m$, and it shall attain the values $f(nk)$ at the points nk (n fixed) (Fig. 2.5).

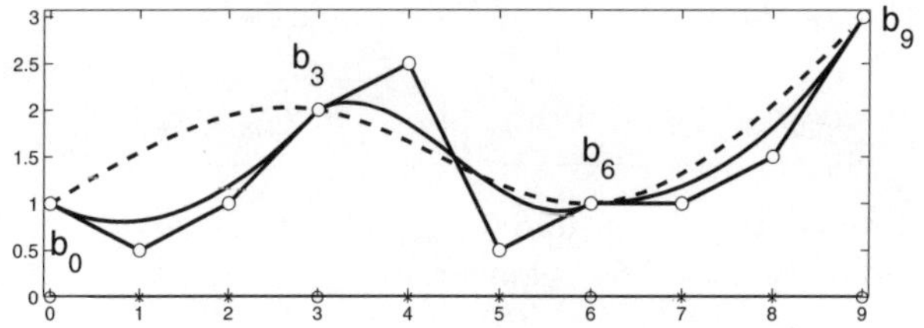

Figure 2.5. BEZIÉR curve and spline, m = n = 3

Definition 2.1. (1°) *A segmented continuous curve of polynomial segments of degree* $\leq n$ *is a* (polynomial) spline *if it is* $(n-1)$*-times continuously differentiable on the entire interval. For* $n = 3$, *the spline is called* cubic spline.
(2°) *Let* $\mathcal{I} = [a, b]$ *and let a partition of* Δ_m *of* $\mathcal{I}$ *be defined by* $a = x_0 < x_1 < \ldots < x_m = b$ *then*

$$S_3(\Delta_m) := \{s \in \mathcal{C}^2(\mathcal{I}),\ \forall\, x \in [x_{i-1}, x_i) : s^{(3)}(x) = \text{const},\ i = 1 : m\}$$

is the vector space of cubic splines.

The dimension of S_3 is $m + 3 = (m + 1) + 2$ hence there are two conditions free for further specification.

For $k \in \mathbb{N}_0$ let

$$\begin{aligned} p_k(x) &:= x^k, \\ q_k(t, x) &:= (t - x)_+^k := \max\{(t - x)^k, 0\} \quad \text{(FÖPPL symbol)}. \end{aligned}$$

The function $q_k(t, x)$ has $k - 1$ continuous derivatives in both arguments and the k-th derivative makes a jump of height $k!$ resp. $(-1)^k k!$.

Theorem 2.5. *The set* $S_n(\Delta_m)$ *is a linear space of dimension* $m + n$. *The elements*

$$p_0, \ldots, p_n, q_n(\,\cdot\,, x_1), \ldots, q_n(\,\cdot\,, x_{m-1})$$

constitute a basis of $S_d(\Delta_n)$.

Proof, e.g., (Haemmerlin), p. 246.

Now we consider the case $n = 3$ more exactly. Conceiving $s \in S_3$ as BEZIÉR curve, we have for the BEZIÉR points at distance $x_{i+1} - x_i = 1$:

$$\begin{aligned} s(x_k) &= b_{3k} && \text{because } s \in \mathcal{C}[a, b] \\ 2b_{3k} &= b_{3k-1} + b_{3k+1} && \text{because } s \in \mathcal{C}^1[a, b] \\ 2b_{3k-1} - b_{3k-2} &= d_k = 2b_{3k+1} - b_{3k+2} && \text{because } s \in \mathcal{C}^2[a, b]\,. \end{aligned} \tag{2.12}$$

By these relations we obtain

$$\begin{aligned} 4b_{3k-1} - 2b_{3k-2} &= 2d_k\,, & 4b_{3k+1} - 2b_{3k+2} &= 2d_k\,, \\ 2b_{3(k-1)+1} - b_{3(k-1)+2} &= d_{k-1}\,, & 2b_{3(k+1)-1} - b_{3(k+1)-2} &= d_{k+1}\,, \end{aligned}$$

and, by addition of the left and right sides separately,

$$\boxed{3b_{3k-1} = d_{k-1} + 2d_k,\ 3b_{3k+1} = 2d_k + d_{k+1}}\ . \tag{2.13}$$

Accordingly, the numbers b_{3k+1} und $b_{3(k+1)-1} = b_{3k+2}$ divide the line segment between d_k and d_{k+1} into *three* segments.

Furthermore, by (2.12) and (2.13),

$$6b_{3k} = 3b_{3k-1} + 3b_{3k+1} = d_{k-1} + 4d_k + d_{k+1}\,, \tag{2.14}$$

for all *interior* points b_{3k}, $k = 1 : m-1$. Together with (2.13) for b_1 and b_{3m-1}, i.e., for $k = m$ and $k = 0$,

$$3b_{3m-1} = d_{m-1} + 2d_m\,,\quad 3b_1 = 2d_0 + d_1\,,$$

we obtain the following linear system for the vector $[d_0, \ldots, d_m]^T$ of unknown coefficients (DEBOOR points) in case where all data on the right side are given:

$$\begin{bmatrix} 2 & 1 & 0 & 0 & 0 \\ 1 & 4 & 1 & \ddots & 0 \\ 0 & \ddots & \ddots & \ddots & 0 \\ 0 & \ddots & 1 & 4 & 1 \\ 0 & 0 & 0 & 1 & 2 \end{bmatrix} \begin{bmatrix} d_0 \\ d_1 \\ \vdots \\ d_{m-1} \\ d_m \end{bmatrix} = \begin{bmatrix} 3b_1 \\ 6b_3 \\ \vdots \\ 6b_{3m-3} \\ 3b_{3m-1} \end{bmatrix}. \tag{2.15}$$

The matrix is regular and well-conditioned; thus there exists precisely one spline $s \in S_3(\Delta)$ to the data set $\{d_0, \ldots, d_m, b_0, b_{3m}\}$ by the above construction. It is called *cubic interpolating spline* because $s(x_k) = b_{3k} = f_k$ for $k = 0 : m$ (Fig. 2.6).

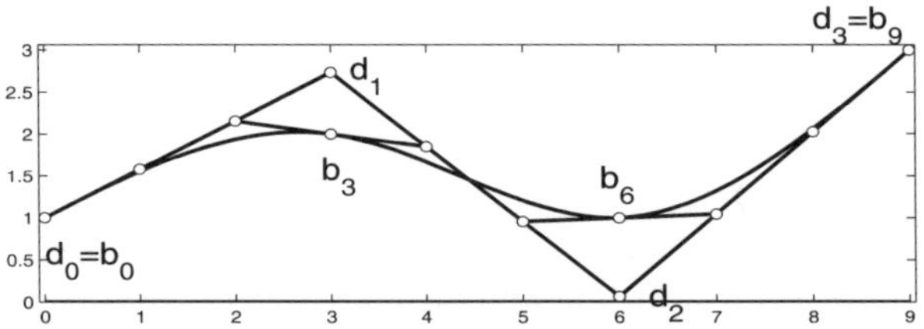

Figure 2.6. Interpolating spline, m = n = 3

Calculation Let $f_k = b_{3k}$ for $k = 0 : m$ be given and let the interpolating spline $s \in S_3(\Delta_m)$ in $[a, b] = [0, m]$, $x_{i+1} - x_i = 1$ to be found.
(1°) Let $f'(a)$, $f'(b)$ be specified. Find b_1, b_{3m-1} by solving

$$f'(a) = s'(0) = 3(b_1 - b_0),\quad f'(b) = s'(m) = 3(b_{3m} - b_{3m-1}),$$

compute $d_0, \ldots, d_m$ by (2.15), compute b_{3k+1}, b_{3k+2} by (2.13),
compute $s(x)$ in $[x_k, x_{k+1}]$ as BEZIÉR polynomial by DE CASTELJAU,

$$s(x_k + \xi) = \sum_{i=0}^{3} b_{3k+i} B_i^3(\xi)\,,$$

using local coordinates $\xi \in [0,\ 1]$.
(2°) Requiring $s''(a) = s''(b) = 0$ we obtain the *natural splines*, $s \in N_3(\Delta_m)$. For their computation, $d_0 = b_0$ and $d_m = b_{3m}$ are prescribed then, with $n = 3$,

$$\begin{aligned} s''(0) &= n(n-1)(b_2 - 2b_1 + b_0) = 6(-d_0 + b_0) = 0,\\ s''(m) &= n(n-1)(b_{3m} - 2b_{3m-1} + b_{3m-2}) = 6(-d_m + b_{3m}) = 0\,. \end{aligned}$$

Compute $(d_1, \ldots, d_{m-1})^T$ by (2.15) without first and last row (because d_0 and d_m fixed)

$$\begin{bmatrix} 4 & 1 & 0 & 0 & 0 \\ 1 & 4 & 1 & \ddots & 0 \\ 0 & \ddots & \ddots & \ddots & 0 \\ 0 & \ddots & 1 & 4 & 1 \\ 0 & 0 & 0 & 1 & 4 \end{bmatrix} \begin{bmatrix} d_1 \\ \vdots \\ \vdots \\ \vdots \\ d_{m-1} \end{bmatrix} = \begin{bmatrix} 6b_3 - b_0 \\ 6b_6 \\ \vdots \\ 6b_{3m-6} \\ 6b_{3m-3} - b_{3m} \end{bmatrix}. \tag{2.16}$$

Because $s''(x_k) = 6(b_{3k} - d_k)/h^2$, $k = 1 : m-1$, $h = x_{i+1} - x_i$ constant, the "moments" $s''(x_k)$ satisfy a similar linear system as the values d_k.

If the exact curvature $\kappa(x) = f''(x)/(1 + f'(x)^2)^{3/2}$ of $f : x \mapsto f(x)$ is replaced by $f''(x)$ approximatively, then the natural splines reveal to be *bending lines*:

Theorem 2.6. *Let Δ_m be an arbitrary partition and let $s \in N_3(\Delta_m)$, i.e., a natural spline, with $s(x_i) = f_i$, $i = 0 : m$. Then*

$$|s|_2^2 := \int_a^b (s''(x))^2 dx = \min\left\{|g|_2^2,\ g \in \mathcal{C}^2[a,b],\ g(x_i) = f_i\right\}.$$

Proof. Let g have the mentioned properties then, using $g''(x)^2 = (s''(x) + g''(x) - s''(x))^2$,

$$\int_a^b g''(x)^2 dx = \int_a^b (s'')^2 dx + 2\int_a^b s''(g'' - s'')\,dx + \int_a^b (g'' - s'')^2 dx\,. \tag{2.17}$$

By partial integration we obtain for the mixed term

$$\begin{aligned} \int_a^b s''(g'' - s'')\,dx &= s''(g' - s')|_a^b - \sum_{i=1}^m \int_{x_{i-1}}^{x_i} s'''(g' - s')\,dx,\\ \sum_{i=1}^m \int_{x_{i-1}}^{x_i} s'''(g' - s')\,dx &= \sum_{i=1}^m \int_{x_{i-1}}^{x_i} c_i(g' - s')\,dx = \sum_{i=1}^m c_i(g - s)\Big|_{x_{i-1}}^{x_i} = 0\,, \end{aligned}$$

because $g(x_i) = s(x_i) = f_i$ by assumption. Accordingly, the assertion follows if the boundary terms disappear, i.e., if

$$\boxed{s''(g'-s')\Big|_a^b = 0}\,.$$

This condition is fulfilled for instance in the following commonly used cases:

(1°) if $s''(a) = 0 = s''(b)$ (natural spline),
(2°) if $g'(a) = s'(a) = f'(a)$ fixed, $g'(b) = s'(b) = f'(b)$ fixed,
(3°) if s, g periodic with period $b - a$.

□

If calculation shall be performed in an interval $[a, b]$ instead $[0, m]$ then re-scaling becomes necessary; likewise, a change is necessary if the nodes are no longer equidistant. But we do not pursue this matter here; cf. however `SUPPLEMENT\chap02`.

2.2 Orthogonal Polynomials

> Let $\overline{\Pi}_n$ be the set of polynomial $p_n \in \Pi_n$ of *exact* degree n and with highest term *one*.

(a) Construction Let $-\infty \leq a < b \leq \infty$ and let $\omega : [a, b] \to \mathbb{R}_+$ be a non-negative *weight function* with the following properties:

Assumption 2.1. *The* moments $m_k := \int_a^b \omega(x)x^k\,dx$, $k \in \mathbb{N}_0$, *exist finitely (possibly being improper integrals), and* $m_0 > 0$.

Then two polynomials p, $q \in \Pi_n$ are called *orthogonal* (w.r.t. the considered interval of integration and the weight function ω) if

$$(p, q) := \int_a^b \omega(x)p(x)q(x)\,dx = 0\,.$$

Theorem 2.7. *(Existence and Construction) Adopt Assumption 2.1.*
(1°) $\forall\, i \in \mathbb{N}_0\ \ \exists!\ p_i \in \overline{\Pi}_i : i \neq k \implies (p_i, p_k) = 0$.
(2°) *The orthogonal polynomials are uniquely determined by the three-term recurrence relation (with $xp : x \mapsto xp(x)$)*

$$\begin{aligned} &p_{-1}(x) = 0\,,\ \ p_0(x) = 1\,,\ \ p_{i+1}(x) = (x - \delta_{i+1})p_i(x) - \gamma_{i+1}^2 p_{i-1}(x)\,,\ i \geq 0,\\ &\delta_{i+1} = (xp_i, p_i)/(p_i, p_i)\,,\ i \geq 0\,,\ \ \gamma_{i+1}^2 = \begin{cases} 0\,, & i = 0\,,\\ (p_i, p_i)/(p_{i-1}, p_{i-1})\,, & i \geq 1\,. \end{cases} \end{aligned} \tag{2.18}$$

Proof by GRAM-SCHMIDT *orthogonalization* (Stoer), see also `SUPPLEMENT\chap02`.

Obviously, for $p_n \in \overline{\Pi}_n$, it follows that $(p, p_n) = 0$ for all $p \in \Pi_{n-1}$ because orthogonal polynomials are linearly independent and thus form a basis of Π_n . In the remaining part of this section we consider othogonal polynomials p_n as introduced by Theorem 2.7.

Theorem 2.8. *The roots x_i of p_n are real and simple. They all lie in the open interval (a, b) .*

Proof. Let $x_1, \ldots, x_k$ be all roots of p_n of odd multiplicity contained in (a, b) then p_n changes sign precisely at these points. Let

$$a < x_1 < \ldots < x_k < b, \quad q(x) := \prod_{i=1}^{k}(x - x_i),\ k \leq n\,,$$

then $p_n(x)q(x)$ does *not* change sign in (a, b) hence $(p_n, q) \neq 0$. Therefore the degree of q must be $k = n$ otherwise we have a contradiction to the above inference to Theorem 2.7. □

(b) The Formulas of Rodriguez To compute orthogonal polynomials $p_n \in \Pi_n$ *explicitely*, we observe the general condition of orthogonality

$$\forall\, q_{n-1} \in \Pi_{n-1} : \int_a^b \omega(x)p_n(x)q_{n-1}(x)\,dx = 0\,, \quad n = 0, 1, \ldots \tag{2.19}$$

and choose for approach in skilful way

$$\omega(x)p_n(x) = \frac{d^n}{dx^n}u_n(x) \implies p_n(x) = \frac{1}{\omega(x)}\frac{d^n}{dx^n}u_n(x) \in \Pi_n\,.$$

Since p_n shall be a polynomial of degree not greater n, obviously

$$\frac{d^{n+1}}{dx^{n+1}}\left[\frac{1}{\omega(x)}\frac{d^n u_n(x)}{dx^n}\right] = \left[\frac{u_n^{(n)}(x)}{\omega(x)}\right]^{(n+1)} = 0\,. \tag{2.20}$$

On the other side, a n-fold partial integration of $\displaystyle\int_a^b u_n^{(n)}(x)q_{n-1}(x)\,dx = 0$ yields

$$\left[u_n^{(n-1)}q_{n-1} - u_n^{(n-2)}q_{n-1}' + - \ldots + (-1)^{n-1}u_n q^{(n-1)}\right]\Big|_a^b = 0\,.$$

This relation is certainly fulfilled for the boundary conditions

$$u_n^{(i)}(a) = 0,\ \ u_n^{(i)}(b) = 0,\ \ i = 0 : n-1\,. \tag{2.21}$$

The converse result does also hold and has been proved by (Szegoe):

Theorem 2.9. *Let Assumption 2.1 be fulfilled. Then the boundary value problem (2.20), (2.21) has always a solution* u_n *and* $p_n := u_n/\omega$ *is a polynomial of degree* n .

The above boundary value problem has $2n$ boundary conditions for a differential equation of order $2n+1$ hence one condition stands at disposition for normalization.

Example 2.1. LEGENDRE *polynomials:* Interval $(a,b) = (-1,1)$, weight function $\omega(x) \equiv 1$, $u_n^{(2n+1)} = 0$, $u_n^{(i)}(\pm 1) = 0$, $i = 0 : n-1$ (Fig. reffig0202.1).

$$p_n(x) = \gamma_n \frac{d^n}{dx^n}(x^2-1)^n .$$

The constants γ_n are specified in different ways.

Example 2.2. JACOBI *polynomials:* Interval (a,b) finite, weight function $\omega(x) = (x-a)^\alpha (b-x)^\beta$, $\alpha > -1$, $\beta > -1$.

$$p_n(x) = \gamma_n \frac{1}{(x-a)^\alpha (b-x)^\beta} \frac{d^n}{dx^n}[(x-a)^{n+\alpha}(b-x)^{n+\beta}] .$$

In particular, *shifted* LEGENDRE *polynomials* are obtained for $(a,b) = (0,1)$ and $(\alpha,\beta) = (0,0),\ (1,0),\ (0,1),\ (1,1)$ being applied later on in numerical integration:

$$\begin{aligned} p_{1,n}(x) &= \frac{d^n}{dx^n}\left(x^n(1-x)^n\right) , & p_{2,n}(x) &= \frac{1}{x}\frac{d^n}{dx^n}\left(x^{n+1}(1-x)^n\right) \\ p_{3,n}(x) &= \frac{1}{1-x}\frac{d^n}{dx^n}\left(x^n(1-x)^{n+1}\right) , & p_{4,n}(x) &= \frac{1}{x(1-x)}\frac{d^n}{dx^n}x^{n+1}(1-x)^{n+1} . \end{aligned} \tag{2.22}$$

Example 2.3. CHEBYSHEV *polynomials* $T_n(x)$ with $(a,b) = (-1,1)$, $\omega(x) = (1-x^2)^{-1/2}$ are special JACOBI polynomials as well as the above LEGENDRE polynomials. In expanding a function by these polynomials, the values at the boundaries of the interval are more strongly regarded because of the special weight function (Fig. 2.8). By the original condition of orthogonality,

$$\forall\, q_{n-1}(x) \in \Pi_{n-1} : \int_{-1}^{1} \frac{T_n(x) q_{n-1}(x)}{(1-x^2)^{1/2}}\, dx = 0 , \tag{2.23}$$

a substitution of $x = \cos\varphi$ yields the condition

$$\int_0^\pi T_n(\cos\varphi) q_{n-1}(\cos\varphi)\, d\varphi = 0 .$$

Because

$$\cos(n+1)\varphi + \cos(n-1)\varphi = 2\cos\varphi\cos n\varphi \tag{2.24}$$

for $n \in \mathbb{N}$, the function $\cos n\varphi$ is a polynomial in $\cos\varphi$, and $(\cos\varphi)^k$ is a linear combination of $\cos j\varphi$, $j = 0 : k$, hence

$$q_{n-1}(\cos\varphi) = \sum_{j=0}^{n-1} \gamma_j (\cos\varphi)^j = \sum_{k=0}^{n-1} \delta_k \cos(k\varphi)\,.$$

Thus (2.23) holds if and only if

$$\int_0^\pi T_n(\cos\varphi)\cos(j\varphi)\,d\varphi = 0,\ j = 0 : n-1,$$

$$\Longrightarrow\ T_n(\cos\varphi) = \cos(n\varphi) \Longrightarrow \boxed{T_n(x) = \cos(n\arccos x)}\,,\ n = 0, 1, \ldots .$$

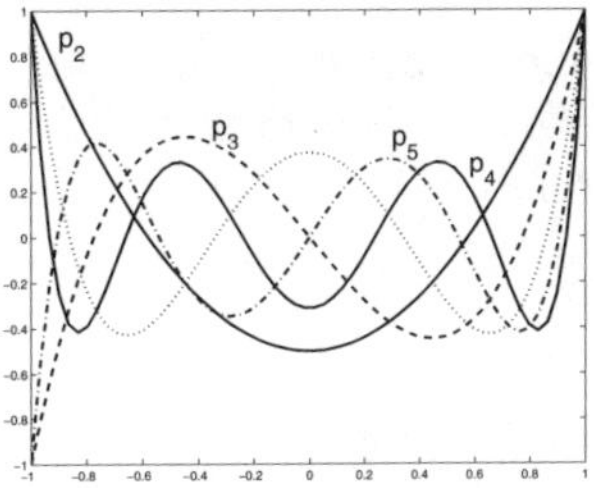

Figure 2.7. LEGENDRE polynomials, $n = 2 : 6$

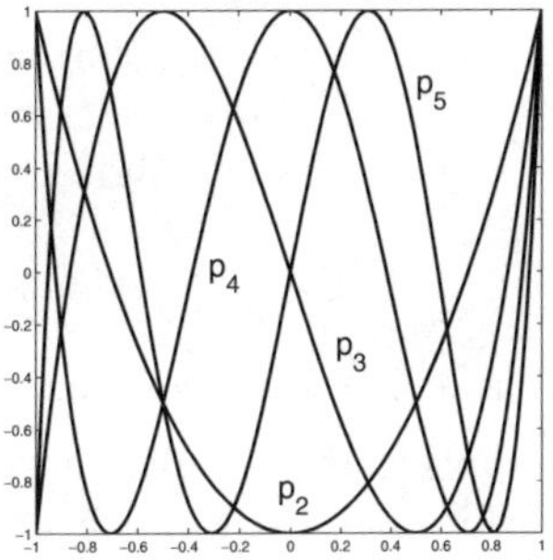

Figure 2.8. CHEBYSHEV polynomials, $n = 2 : 5$

(c) Minimum Property of Chebyshev Polynomials The recurrence formula (2.24) shows that $T_n(x) = \cos(n\arccos(x))$ has the highest term 2^{n-1}.

Theorem 2.10. *Let $p_n(x)$ be any polynomial of degree $\leq n$ with highest term 2^{n-1}. Then there exists at least one $x \in [-1, 1]$ such that $|p_n(x)| \geq 1$.*

Proof. Suppose that $|p_n(x)| < 1$ for all $x \in [-1, 1]$. $T_n(x)$ takes alternating the values ± 1 at its $n+1$ extremal points $x_i = \cos(i\pi/n)$, $i = 0 : n$, in $[-1, 1]$. Therefore, $T_n(x) - p_n(x)$ is alternating positive or negative at these extremal points and thus $T_n(x) - p_n(x)$ has at least n zero points in $(-1, 1)$. However, because of identical highest terms, $T_n(x) - p_n(x)$ is a polynomial of degree $\leq n-1$. Accordingly, $T_n(x) - p_n(x) \equiv 0$ in contradiction to the assumption. □

Corollary 2.1. *Let q_n be a polynomial of degree n with highest term a_n then there exists a value $x \in [-1, 1]$ such that $|q_n(x)| \geq a_n/2^{n-1}$.*

Proof. Let $a_n \neq 0$ and $q_n^*(x) = q_n(x)2^{n-1}/a_n$. The polynomial q_n^* has the highest term 2^{n-1} hence $|q_n^*(x)| \geq 1$ by Theorem 2.10 for at least one $x_0 \in [-1,1]$. Then $|q_n(x_0)| = |q_n^*(x_0)a_n/2^{n-1}| \geq a_n/2^{n-1}$. □

For an arbitrary polynomial $q_n(x)$ — especially also for a TAYLOR polynomial — there exists a unique expansion by CHEBYSHEV polynomials,

$$q_n(x) = \sum_{i=0}^{n} c_i T_i(x)\,, \quad x \in [-1,1]\,, \quad T_i(x) = \cos(i \arccos(x))\,,$$

because these polynomials are linearly independent by orthogonality.

Theorem 2.11. *If* $S_n(x) = \sum_{i=0}^{n} c_i T_i(x)$ *are the partial sums of an expansion by* CHEBYSHEV *polynomials then*

$$\max_{-1\leq x\leq 1} |S_{n+1}(x) - S_n(x)| = \inf_{p_n} \max_{-1\leq x\leq 1} |S_{n+1}(x) - p_n(x)|$$

where p_n *is an arbitrary polynomial of degree* $\leq n$.

Proof. We have $S_n(x) - S_{n-1}(x) = c_n T_n(x)$ hence $|S_n(x) - S_{n-1}(x)| \leq |c_n|\,,\ -1 \leq x \leq 1$. For any arbitrary polynomial p_{n-1} of degree $n-1$, the difference $S_n - p_{n-1}$ has the highest term $c_n 2^{n-1}$ hence

$$|S_n(x) - p_{n-1}(x)| \geq c_n 2^{n-1}/2^{n-1} = c_n\,.$$

at least for one $x \in [-1,1]$ by Corollary 2.1. □

Roughly spoken, the components of an expansion by CHEBYSHEV polynomials decrease in absolute value in the fastest way.

2.3 Numerical Integration

Integrating is an art and differentiating is a handicraft as everybody knows, but from numerical point of view fortunately the situation behaves conversely in some sense. Integration is a smoothing process which has advantageous consequences in numerical approximation whereas a differential quotient has to be replaced always by a difference quotient numerically. Then, in numerator and denominator, subtraction of nearly equal numbers does occur entailing the befeared extinction of leading numbers. However, it should be mentioned at this place that an asymptotic expansion in the sense of Sect. 2.4(**c**) may produce surprisingly exact results; cf. (Rutishauser).

As MATLAB does not know the index zero and also for applications later on, we work in this section throughout with n nodes instead of the usual $n+1$ nodes in interpolating problems.

(a) Integration Rules of Lagrange The computational effort of a numerical integration formula depends on the number of function evaluations. Note once more that we work with n nodes here to compare the individual rules with each other. Accordingly, we proceed from an interpolating polynomial $p_{n-1}(x)$ of degree $n-1$ of LAGRANGE type, i.e., by (2.6) in slightly modified form,

$$f(x) \approx p_{n-1}(x;f) = \sum_{i=1}^{n} f(x_i)q_i(x)\,,\quad q_i(x) = \prod_{j=1, j\neq i}^{n} \frac{x-x_j}{x_i-x_j} \in \Pi_{n-1}\,, \tag{2.25}$$

and obtain by integration over an interval $(a\,,\,b)$

$$I(f) := \int_a^b f(x)\,dx \approx \sum_{i=1}^{n} f(x_i)\int_a^b q_i(x)\,dx =: \sum_{i=1}^{n} f(x_i)\alpha_i =: I_n(f)\,. \tag{2.26}$$

The n support abszissas x_i shall be *mutually distinct* again, otherwise they can be chosen arbitrarily, in particular, they may lie also in the exterior of the interval of integration. But in this section we suppose always that

$$a \leq x_1 < \ldots < x_{n-1} < x_n \leq b\,.$$

An *integration* rule has *degree* N if (at least) all polynomials of degree $\leq N$ are integrated exactly. Apparently, a LAGRANGE formula (2.26) with n abszissas has degree $N = n-1$; the *maximum degree* N is not greater than $2n-1$, because, inserting the polynomial $f(x) = \Pi_{i=1}^{n}(x-x_i)^2 \in \Pi_{2n}$, we have $I_n(f) = 0$ for the rule and $I(f) > 0$ for the exact integral.

The integration formulas of NEWTON and COTES are of separated type (2.25), too, but with a uniform partition $x_i = a + (i-1)h$ and step length $h = (b-a)/(n-1)\,,\ n \geq 2$. Inserting the translation $x = a + sh$ we obtain by Sect. 2.1

$$\begin{aligned}
q_i(x) &= q_i(a+sh) =: \varphi_i(s) = \prod_{j=1,\, j\neq i}^{n} \frac{s-j}{i-j} \in \Pi_{n-1}(s)\,,\quad s \in [0,\, n-1]\,,\\
\alpha_i := \int_a^b q_i(x)dx &= \int_0^{n-1} q_i(a+sh)\frac{dx}{ds}ds = h\int_0^{n-1}\varphi_i(s)ds = h\,\beta_i\,,\\
I_n(f) = \sum_{i=1}^{n} f(x_i)\alpha_i &= h\sum_{i=1}^{n} f(x_i)\beta_i\,.
\end{aligned}$$

The new weights β_i are now *rational* numbers which depend only on the number n and no longer on the boundaries a, b of the integral, therefore they can be calculated once for all in tabular form; a substitution of $f(x) \equiv 1$ shows that $\sum_{i=1}^{n} \beta_i = n-1\,,\ n \geq 2\,.$

Example 2.4. Midpoint rule (1 node):

$$I(f) = (b-a)f\left(\frac{a+b}{2}\right) + \frac{1}{24}(b-a)^3 f''(\xi)\,,$$

Trapezoidal rule ($n = 2$ nodes, $h = b - a$):

$$I(f) = \frac{b-a}{2}[f(a) + f(b)] - \frac{1}{12}(b-a)^3 f''(\xi)\,,$$

SIMPSON*'s rule* ($n = 3$ nodes, $h = (b - a)/2$):

$$I(f) = \frac{b-a}{6}\left[f(a) + 4f\left(\frac{a+b}{2}\right) + f(b)\right] - \frac{(b-a)^5}{2^5 \cdot 90} f^{(4)}(\xi)\,.$$

Note that the intermediate values $\xi \in (a,\, b)$ differ in the individual integration rules.

One observes that the midpoint rule and SIMPSON'S also called KEPLER's rule have degree n instead of the expected lower degree $n-1$ of the underlying interpolating polynomial. It is however a general property of NEWTON-COTES rules that the degree is n instead $n - 1$ for n odd by reason of symmetry.

The general error term $R_n(f)$ in

$$I(f) = I_n(f) + R_n(f)\,, \tag{2.27}$$

is a linear operator satisfying $R_n(\alpha f + \beta g) = \alpha R_n(f) + \beta R_n(g)$. Introducing the FÖPPL-Symbol $(x-t)_+^N := \max\{(x-t)^N,\, 0\}$ again, the following classical result of PEANO displays $R_n(f)$ in elegant integral form; cf. e.g. (Stoer).

Theorem 2.12. *Let an integration rule (2.26) with n nodes have degree N then, for all $f \in \mathcal{C}^{N+1}[a,\, b]$,*

$$R_n(f) = \int_a^b f^{(N+1)}(t) K_n(t)\, dt\,, \quad K_n(t) = \frac{1}{N!} R_n(h_t), \quad h_t : x \mapsto (x-t)_+^N\,.$$

Proof see (Stoer) and `SUPPLEMENT\chap02a`.

$R_n(h_t)$ denotes here the error of the integration rule w.r.t the function $h_t : x \mapsto (x - t)_+^N$ instead of f. Frequently the *kernel* $K_n(t)$ does not change sign in $(a\,,\, b)$ as for instance in NEWTON-COTES rules. Then the mean value theorem of integration yields

$$R_n(f) = f^{(N+1)}(\xi) \int_a^b K_n(t) dt\,, \quad \xi \in (a,b)\,. \tag{2.28}$$

Inserting here the special function $\varphi : x \mapsto x^{N+1}$ for f we obtain

$$R_n(\varphi) = (N+1)! \int_a^b K_n(t) dt \Longrightarrow \int_a^b K_n(t) dt = R_n(\varphi)/(N+1)!\,. \tag{2.29}$$

Fazit: If the integration rule (2.26) has degree N and $K_n(t)$ does not change sign in interval of integration then (2.28) and (2.29) yields the error representation

$$\boxed{R_n(f) = \frac{f^{(N+1)}(\xi)}{(N+1)!}\,R_n(\varphi)\,,\quad \varphi : x \mapsto x^{N+1},\ \xi \in (a,b)} \tag{2.30}$$

(PEANO's error representation); but $R_n(\varphi)$ can be always calculated exactly!

As already noted above, it can be shown for NEWTON-COTES rules with n mutually distinct support abszissas that

$$R_n(f) = \begin{cases} \dfrac{f^{(n)}(\xi)}{(n)!}\,R_n(x^n) & n \text{ even} \\ \dfrac{f^{(n+1)}(\xi)}{(n+1)!}\,R_n(x^{n+1}) & n \text{ odd} \end{cases}$$

where $\xi \in (a,b)$.

(b) Composite Integration Rules As already mentioned in Sect. 2.1, the approximation of f by an interpolating polynomial $p_n \in \Pi_n$ is not improved by enhancing the degree n. Therefore one uses locally polynomials of low degree on a collection of subintervals. The full integral is then approximated by the sum of the approximations of the subintegrals. The resulting *composite rules* are arbitrarily exact in dependence of the node number even for continuous integrand f. For instance, writing $x_i = a + ih,\ i = 0 : m,\ h = (b-a)/m,\ m \in \mathbb{N}$, ($m+1$ support abszissas) we obtain the important *composite trapezoidal rule* from the simple trapezoidal rule

$$\boxed{T(h;f) = \frac{h}{2}\left[f(x_0) + 2\sum_{i=1}^{m-1} f(x_i) + f(x_m)\right] = I(f) - h^2(b-a)\frac{1}{12}f''(\xi)}\,, \tag{2.31}$$

and the simple SIMPSON rule leads to the corresponding composite rule

$$\begin{aligned} I(f) = {} & \frac{h}{6}\left[f(x_0) + 2\sum_{i=1}^{m-1} f(x_i) + 4\sum_{i=0}^{m-1} f\left(\frac{x_i + x_{i+1}}{2}\right) + f(x_m)\right] \\ & + h^4(b-a)\frac{1}{2880}f^{(4)}(\xi)\,. \end{aligned}$$

By applying EULER MCLAURIN's summation formula (Stoer) it can be shown that formula (2.31) has the following surprising property:

Lemma 2.3. *Let $f \in C^\infty(\mathbb{R})$ be an $(b-a)$-periodic function then*

$$T(h;f) = \int_a^b f(x)dx + \mathcal{O}(h^p) \quad \forall\, p \in \mathbb{N}\,.$$

In other words, the composite trapezoidal rule is faster convergent than every power of step length h for smooth periodic functions!

(c) Gauß Integration Using an interpolating polynomial of HERMITE form instead of LAGRANGE form for integration, cf. Sect. 2.1**(e)**, we obtain a further type of integration rules, namely,

$$I_n(f) := \sum_{i=1}^{n} \left[f(x_i) \int_a^b h_{0,i}(x)dx + f'(x_i) \int_a^b h_{1,i}(x)dx \right] \tag{2.32}$$

where

$$h_{0,i}(x) = \left[1 - 2q_i'(x_i)(x - x_i)\right] q_i(x)^2 \,, \quad h_{1,i}(x) = (x - x_i) q_i(x)^2 \,, \tag{2.33}$$

and $q_i(x) \in \Pi_{n-1}$ are the LAGRANGE polynomials. The formula has degree $N = 2n - 1$ for n evaluations of f and n evaluations of the derivative of f.

Choosing now for abszissas the roots x_i, $i = 1 : n$ of orthogonal polynomials $p_n(x) \in \Pi_n$ w.r.t. the *scalar product* $(f, g) = \int_a^b f(x)g(x)\,dx$ we obtain by Sect. 2.2

$$\int_a^b h_{1,i}(x)\,dx = \int_a^b (x - x_i) q_i(x)^2 \,dx = 0$$

because $(x - x_i)q_i(x) = p_n(x)$ and $q_i(x) \in \Pi_{n-1}$. As an inference we have also $\int_a^b h_{0,i}(x)\,dx = \int_a^b q_i(x)^2\,dx$ therefore, by (2.32) and (2.33), we obtain the following integration rules

$$I_n(f) := \sum_{i=1}^{n} f(x_i) \int_a^b q_i(x)^2\,dx \tag{2.34}$$

which have *maximum* degree $N = 2n - 1$ for n nodes.

We summarize the result for a general weight function $\omega(x)$ with the properties of Sect. 2.2 in the following theorem:

Theorem 2.13. *(GAUSS Integration) Let $p_n \in \overline{\Pi}_n$ be orthogonal polynomials w.r.t. the scalar product*

$$(f, g) := \int_a^b \omega(x) f(x) g(x)\,dx \,,$$

Let $x_1, \ldots, x_n$ be the roots of p_n, and let

$$A = [p_i(x_j)]_{i=0}^{n-1}{}_{j=1}^{n}, \quad \underline{c} = [(p_0, p_0), 0, \ldots, 0]^T .$$

(1°) *The matrix A is regular.*
(2°) *Let $\underline{b} = A^{-1}\underline{c}$ and $\underline{b} = [\beta_1, \ldots, \beta_n]^T$ then*

$$\forall\, p \in \Pi_{2n-1} : \int_a^b \omega(x) p(x)\,dx = \sum_{i=1}^{n} \beta_i p(x_i) \,, \tag{2.35}$$

i.e., the integration rule

$$\int_a^b \omega(x) f(x)\,dx = \sum_{i=1}^{n} \beta_i f(x_i) + R_{n,\omega}(x; f) \tag{2.36}$$

has maximum degree $N = 2n-1$.
(3°) *For error representation in (2.36) we have*

$$\forall\, f \in \mathcal{C}^{2n}[a,\, b] \;\exists\, \xi \in (a,\, b) : R_{n,\omega}(x; f) = \frac{f^{(2n)}(\xi)}{(2n)!}(p_n,\, p_n)\,.$$

(4°) *Conversely, if (2.35) holds then the abszissas* x_i *are the roots of the orthogonal polynomials* $p_n(x)$ *and* $A\underline{b} = \underline{c}$ *is fulfilled with* $\underline{b} = [\beta_1, \ldots, \beta_n]^T$.
(5°) *If a rule (2.36) has degree* $N \geq n-2$ *then the weights* β_i *are positive.*

Proof see (Stoer).

(d) Suboptimal Integration Rules are an important tool in constructing implicit RUNGE-KUTTA methods of maximum order in the next section. Let

$$\underline{b} = [\beta_1, \ldots, \beta_n]^T\,,\;\; \underline{x} = [x_1, \ldots, x_n]^T\,,\;\; F(\underline{x}) = [f(x_1), \ldots, f(x_n)]^T\,.$$

Theorem 2.14. *For* $\delta, \varepsilon \in \{0,\, 1\}$, *there exists a unique integration rule*

$$\int_0^1 f(x)\,dx \approx \delta\beta_0\, f(0) + \underline{b}^T F(\underline{x}) + \varepsilon\, \beta_{n+1} f(1) \tag{2.37}$$

of maximum degree $\widetilde{N} = 2n + \delta + \varepsilon - 1$.

Choose GAUSS weights and GAUSS nodes $\widetilde{\underline{b}}$, $\underline{x}$ by Theorem 2.13 w.r.t. the weight function $\omega^*(t) = t^\delta(1-t)^\varepsilon$ in $[0,\, 1]$ and insert $\underline{b} = [\widetilde{\beta}_i/\omega^*(x_i)]_{i=1}^n$. Then the rule is optimal for $(\delta,\varepsilon) = (0,0)$. For $(\delta,\varepsilon) = (1,0)$ or $(\delta,\varepsilon) = (0,1)$, the remaining weight is found by $1 = \delta\beta_0 f(0) + \underline{b}^T\underline{e} + \varepsilon\beta_{n+1} f(1)$. For $(\delta,\varepsilon) = (1,1)$, the remaining both weights are found by solving

$$\frac{1}{2} = 0 + \underline{b}^T\underline{x} + \beta_{n+1}\,,\;\; 1 = \beta_0 + \underline{b}^T\underline{e} + \beta_{n+1}\,.$$

A comparison with the shifted LEGENDRE polynomials in (2.22) shows that the node abszissas $x_1, \ldots, x_n$ of a rule (2.37) with together n nodes are the roots of the following polynomials:

$$\begin{aligned} &(\delta,\varepsilon) = (0,0) : p_{1,n}(x)\,, && (\delta,\varepsilon) = (1,0) : x p_{2,n-1}(x) \\ &(\delta,\varepsilon) = (0,1) : (1-x)p_{3,n-1}(x)\,, && (\delta,\varepsilon) = (1,1) : x(1-x)p_{4,n-2}(x)\,. \end{aligned} \tag{2.38}$$

Proof of Theorem 2.14 see `SUPPLEMENT\chap02a`. A program for the computation of nodes and weights in all four cases is found in `KAPITEL02\SECTION_1_2_3`. For integration over an interval (a, b) rescale nodes and weights by $\widehat{x}_i = a + (b-a)x_i$, $\widehat{b}_i = (b-a)b_i$, $i = 1:n$.

Example 2.5. GAUSS integration with LEGENDRE polynomials:

$$\int_{-1}^{1} f(x)\,dx \approx \sum_{i=1}^{n} \beta_i f(x_i)$$

Table 2.1. Gauss-Legendre formulas with n nodes

n	x_i	β_i
2	$\pm\sqrt{3}/3$	1
3	0	8/9
	$\pm\sqrt{15}/5$	5/9
4	$\pm\left[525-70\sqrt{30}\right]^{1/2}/35$	$(18+\sqrt{30})/36$
	$\pm\left[525+70\sqrt{3}\right]^{1/2}/35$	$(18-\sqrt{30})/36$
5	0	128/225
	$\pm\left[245-14\sqrt{70}\right]^{1/2}/25$	$(322+13\sqrt{70})/900$
	$\pm\left[245+14\sqrt{70}\right]^{1/2}/25$	$(322-13\sqrt{70})900$

These rules for integration over the interval $[a,\, b] = [-1,\, 1]$ are exact for polynomials of degree n. In a transformation to the interval $[a',\, b']$, the weights and nodes have to be transformed:

$$w_i' = \frac{b'-a'}{b-a}w_i\,,\quad x_i' = a' + \frac{b'-a'}{b-a}(x_i - a)\,.$$

For instance, in transformation to the unit interval $[0,\, 1]$, the weights must be divided by two and the nodes $x_i' = (1+x_i)/2$ are to be used.

(e) Barycentric Coordinates serve mainly to a lucid and concise representation of interpolating polynomials which *live* on arbitrary triangles T in the plane or, more general, on n-simplices in $\mathbb{R}^n$. Also integration rules for general polynomials on these domains can be simplified by this way. We restrict ourselves to the plane and consider an arbitrary triangle T in cartesian (x,y)-coordinates with vertices $P_i(x_i, y_i),\ i = 1,2,3$, being numerated counterclockwise. Then the *double surface area*

$$2|T| = (x_2 - x_1)(y_3 - y_1) - (x_3 - x_1)(y_2 - y_1) = x_{21}y_{31} - x_{31}y_{21}\,, \tag{2.39}$$

($x_{21} = x_2 - x_1$ etc.) is positive as long as T is non-degenerated. Using the notations of Figure 2.9, the (dimensionless) *barycentric* or *area* coordinates are defined for $0 \le \zeta_i \le 1$ by

$$\boxed{\zeta_i = \frac{\text{area of } T_i}{\text{area of } T}\,,\quad i = 1,2,3}\,.$$

Accordingly, we have $P_1 \simeq (1,0,0)$, $P_2 \simeq (0,1,0)$, $P_3 \simeq (0,0,1)$ and $\zeta_1 + \zeta_2 + \zeta_3 = 1$ whence the barycentric coordinates are *not* linearly independent.

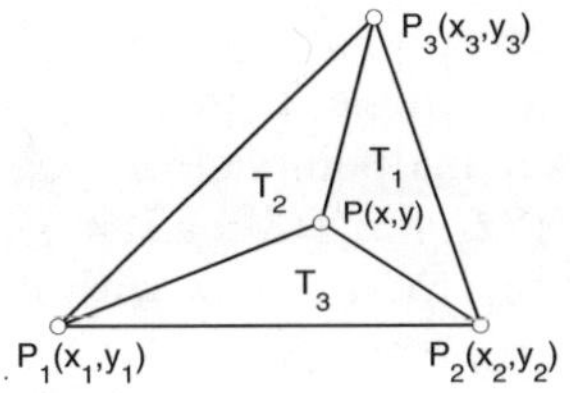

Figure 2.9. Barycentric coordinates

The connection of cartesian and barycentric coordinates is provided by the area rule:

$$2|T_1| = \begin{vmatrix} 1 & x & y \\ 1 & x_2 & y_2 \\ 1 & x_3 & y_3 \end{vmatrix}, \quad 2|T_2| = \begin{vmatrix} 1 & x & y \\ 1 & x_3 & y_3 \\ 1 & x_1 & y_1 \end{vmatrix}, \quad 2|T_3| = \begin{vmatrix} 1 & x & y \\ 1 & x_1 & y_1 \\ 1 & x_2 & y_2 \end{vmatrix}.$$

Expanding w.r.t. the first row and dividing by $2|T|$, we get the affin-linear relations

$$\begin{aligned} \zeta_1 &= [(x_2y_3 - x_3y_2) + y_{23}x + x_{32}y]\,/(2|T|) \\ \zeta_2 &= [(x_3y_1 - x_1y_3) + y_{31}x + x_{13}y]\,/(2|T|) \\ \zeta_3 &= [(x_1y_2 - x_2y_1) + y_{12}x + x_{21}y]\,/(2|T|) \end{aligned} \tag{2.40}$$

(note the cyclic permutation of indices modulo 3). These relations are valid for an arbitrary cartesian coordinate system not necessarily having origin in the center of the triangle. They are used in different applications, in particular for the calculation of partial derivatives, e.g. $\partial\zeta_1/\partial x = y_{23}/(2|T|)$ etc.. Resolution of two equations in (2.40) w.r.t. x and y yields the relation

$$\boxed{1 = \zeta_1 + \zeta_2 + \zeta_3\,,\ x = x_1\zeta_1 + x_2\zeta_2 + x_3\zeta_3\,,\ y = y_1\zeta_1 + y_2\zeta_2 + y_3\zeta_3} \tag{2.41}$$

between arbitrary cartesian and barycentric coordinates. In *unit triangle* $S(\xi,\eta)$ with vertices $Q_1(0,0)$, $Q_2(1,0)$, $Q_3(0,1)$, we have the relation

$$\boxed{\zeta_1 = 1 - \xi - \eta\,,\ \zeta_2 = \xi\,,\ \zeta_3 = \eta}\,, \tag{2.42}$$

and

$$\int_S \xi^p\eta^q\, d\xi d\eta = \int_0^1\int_0^{1-\eta} \xi^p\eta^q\, d\xi d\eta = \frac{p!q!}{(p+q+2)!}\,. \tag{2.43}$$

The formula of Holand and Bell (1969) for general triangles T,

$$\boxed{\int_T \zeta_1^m\zeta_2^n\zeta_3^p\, dxdy = 2|T|\,\frac{m!n!p!}{(m+n+p+2)!}} \tag{2.44}$$

then follows by substitution (Bell). Its straightforward generalization to tetrahedrons $T \subset \mathbb{R}^3$ with volume $|T|$ reads

$$\int_T \zeta_1^m\zeta_2^n\zeta_3^p\zeta_4^q\, dxdydz = 6|T|\,\frac{m!n!p!q!}{(m+n+p+q+3)!}\,. \tag{2.45}$$

Example 2.6. (Ciarlet79) Let $\underline{x}_i \in \mathbb{R}^n$, $i = 1 : n+1$, be the vertices of an n-simplex in $\mathbb{R}^n$, e.g., a triangle T in $\mathbb{R}^2$ or a tetrahedron in $\mathbb{R}^3$. Let $\underline{x}_{ij} = (\underline{x}_i + \underline{x}_j)/2$ for $i < j$ the midpoints of the edges, $\underline{x}_{ijk} = (\underline{x}_i + \underline{x}_j + \underline{x}_k)/3$ for $i < j < k$, and $\underline{x}_{iij} = (2\underline{x}_i + \underline{x}_j)/3$ for $i \neq j$. Denote again by Π_m the vector space of polynomials up to degree m with n variables in $\mathbb{R}^n$. Then the following *identities* are valid in $\mathbb{R}^n$:

$$
\begin{aligned}
\forall\, p \in \Pi_1 : p &= \sum\nolimits_{i=1:n+1} p(\underline{x}_i)\zeta_i \\
\forall\, p \in \Pi_2 : p &= \sum\nolimits_{i=1:n+1} p(\underline{x}_i)\zeta_i(2\zeta_i - 1) + \sum\nolimits_{i<j} p(\underline{x}_{ij})4\zeta_i\zeta_j \\
\forall\, p \in \Pi_3 : p &= 2^{-1}\sum\nolimits_{i=1:n+1} p(\underline{x}_i)\zeta_i(3\zeta_i - 1)(3\zeta_i - 2) \\
&\quad + 2^{-1}\sum\nolimits_{i<j} p(\underline{x}_{ij})9\zeta_i\zeta_j(3\zeta_i - 1) \\
&\quad + \sum\nolimits_{i<j<k} p(\underline{x}_{ijk})27\zeta_i\zeta_j\zeta_k \\
\forall\, p \in \Pi_3 : p &= \sum\nolimits_{i=1:n+1} p(\underline{x}_i)\left(-2\zeta_i^3 + 3\zeta_i^2 - 7\zeta_i \sum\nolimits_{j<k, j\neq i, k\neq i} \zeta_j\zeta_k\right) \\
&\quad + 27\sum\nolimits_{i<j<k} p(\underline{x}_{ijk})\zeta_i\zeta_j\zeta_k \\
&\quad + \sum\nolimits_{i\neq j} \nabla p(\underline{x}_i)(\underline{x}_j - \underline{x}_i)\zeta_i\zeta_j(2\zeta_i + \zeta_j - 1)\,.
\end{aligned}
$$

Up to the first both, these identities are not trivial and they are not unique w.r.t. the barycentric coordinates ζ_i because of the linear interdependence. A corresponding formula for MORLEY's second order polynomial and for ARGYRIS' fifth order polynomial is given in Sect. 12.5 (both in $\mathbb{R}^2$ and using normal derivatives in $\underline{x}_{ij}$). See also `SUPPLEMENT\chap09e\chap09f`.

Integration of interpolating polynomials over triangles and more general geometric configurations is a basic tool in the construction of finite elements; see Chap. 9. For triangles we may use (2.42) or, in case of (x, y)-coordinates, the affin-linear transformation

$$x = x_1 + x_{21}\xi + x_{31}\eta\,, \quad y = y_1 + y_{21}\xi + y_{31}\eta\,, \tag{2.46}$$

and then apply (2.43) to integrate over the unit triangle S. Or we integrate directly over area coordinates and use BELL's formula (2.44); for instance

$$
\begin{aligned}
\forall\, p \in \Pi_1 : \int_T p(x,y)\,dxdy &= \frac{|T|}{3}\sum\nolimits_{i=1:3} p(\underline{x}_i) \\
\forall\, p \in \Pi_2 : \int_T p(x,y)\,dxdy &= \frac{|T|}{3}\sum\nolimits_{1\leq i<j\leq 3} p(\underline{x}_{ij})\,.
\end{aligned}
$$

The use of area coordinates together with BELL's formula is the natural choice for triangular elements. It allows to obtain convenient expressions for various integrals in finite element approach without time consuming numerical procedures.

(f) Domain Integrals

(f1) GAUSSian rules apply also to integration over the unit square,

$$\int_{-1}^{1}\int_{-1}^{1} f(x,y)dx \approx \sum_{i=1}^{n}\sum_{j=1}^{n} \beta_i\,\beta_j\, f(x_i, x_j)\,;$$

other squares have to be rescaled properly. With the data of Table 2.1 these rules are exact for polynomials

$$p(x,y) = \sum_{i=0}^{N} \sum_{k=0}^{N} a_{ik}\, x^i y^k \text{ with } N \le 2n-1\,,\ n = 2:5\,.$$

(f2) Abszissas and weights of two commonly used GAUSSian rules in unit triangle $S(\xi,\,\eta)$ with vertices $Q(0,0)\,,\ Q(1,0)\,,\ Q(0,1)\,,$

$$\int_S f(\xi,\eta)\, d\xi d\eta \approx |S| \sum_{i=1}^{m} \gamma_i f(\xi_i,\eta_i)\,,\quad |S| = \frac{1}{2}\,,$$

are given in Table 2.2. These rules are exact for polynomials $p(\xi,\eta) = \sum_{0 \le i+k \le n} a_{ik}\, \xi^i \eta^k$ of total degree $n = 2, 5$. Integration rules for polynomials on an arbitrary triangle T,

$$\int_T f(x,y) dx dy \approx |T| \sum_{i=1}^{m} \gamma_i\, f(\widetilde{x}_i, \widetilde{y}_i)\,,$$

follow then easily by substitution with the mapping (2.46) (Fig. 2.10).

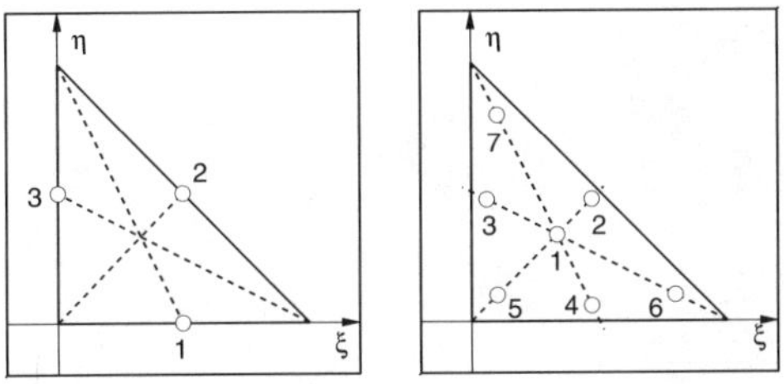

Figure 2.10. GAUSS abszissas in unit triangle, n = 2, 5

Table 2.2.

n	i	ξ_i	η_i	γ_i
2	1	1/2	0	1/3
	2	1/2	1/2	1/3
	3	0	1/2	1/3
5	1	1/3	1/3	0.225
	2	a	a	$(155+\sqrt{15})/1200$
	3	b	a	$(155+\sqrt{15})/1200$
	4	a	b	$(155+\sqrt{15}/1200$
	5	c	c	$(155-\sqrt{15})/1200$
	6	d	c	$(155-\sqrt{15})/1200$
	7	c	d	$(155-\sqrt{15})/1200$

a	$(6+\sqrt{15})/21$
b	$(9-2\sqrt{15})/21$
c	$(6-\sqrt{15})/21$
d	$(9+2\sqrt{15})/21$

(A rule for $n = 3$ using only four nodes is not recommended because of a negative weight γ and that with positive weights has seven nodes as the rule of order 5.)

(f3) In direct integration over triangle T w.r.t. global (x, y)-coordinates we lastly have to find integrals of monomials

$$\boxed{\begin{aligned} P_{rs} &= \int_T x^r y^s \, dxdy \\ &= 2|T| \int_S (x_1\zeta_1 + x_2\zeta_2 + x_3\zeta_3)^r (y_1\zeta_1 + y_2\zeta_2 + y_3\zeta_3)^s \, d\zeta_2 d\zeta_3 \\ &= 2|T| \int_S (x_1 + x_{21}\xi + x_{31}\eta)^r (x_1 + y_{21}\xi + y_{31}\eta)^s \, d\xi \, d\eta \end{aligned}} \quad . \tag{2.47}$$

By this way, the integrals P_{rs} are reduced to sums of integrals of the form (2.44) resp. (2.43). The last formula uses again the substitution rule (2.46) for the mapping $g : S \to T$ of (9.20).

Some results are assembled in Table 2.3 for polynomials up to degree $n = 5$ (Bell) where the origin of the cartesian KOS is the center of the triangle for simplicity. The concise representation in this table is however lost beyound degree 5 and for a coordinate system with different position but nowadays a *program* replaces large tables. `KAPITEL02\SECTION_1_2_3\bell1.m` supplies values of the integral (2.47) for arbitrary r, $s \in \mathbb{N}$ in an KOS with arbitrary origin by using SYMBOLIC MATHEMATICS.

Table 2.3.

Order $n = r + s$	$P_{rs}(x,y) = \int_T x^r y^s\, dxdy$
1	$P_{rs}(x,y) = 0$
2	$P_{rs}(x,y) = \lvert T\rvert \left(x_1^r y_1^s + x_2^r y_2^s + x_3^r y_3^s\right)/12$
3	$P_{rs}(x,y) = \lvert T\rvert \left(x_1^r y_1^s + x_2^r y_2^s + x_3^r y_3^s\right)/30$
4	$P_{rs}(x,y) = \lvert T\rvert \left(x_1^r y_1^s + x_2^r y_2^s + x_3^r y_3^s\right)/30$
5	$P_{rs}(x,y) = 2\lvert T\rvert \left(x_1^r y_1^s + x_2^r y_2^s + x_3^r y_3^s\right)/105$

References: (Kardestuncer), (Stoer).

2.4 Initial Value Problems

In this section, vectors are *not* underlined for simple representation. The letter x denotes always the exact solution and y the numerical approximation.

(a) Euler's Method We seek a solution $x : [0, T] \to \mathbb{R}^n$ of the initial value or CAUCHY problem

$$x'(t) = f(t, x(t)),\ 0 \le t \le T,\ x(0) = x_0\,. \tag{2.48}$$

The problem is said to be *autonomous* if f does not depend explicitely on the independent variable t, i.e. $x'(t) = f(x(t))$.

For solving (2.48) numerically either we can replace $x'(t)$ by a numerical differentiation formula or we can transform the differential equation into an integral equation

$$x(t+\tau) = x(t) + \int_t^{t+\tau} f(s, x(s))ds\,,\ \tau \text{ step length,}$$

and then replace the integral by a numerical integration rule. The most simple case $\int_t^{t+\tau} f(s, x(s))\,ds \doteq \tau f(t, x(t))$ leads immediately to the *explicit* EULER *method*,

$$y(t+\tau) = y(t) + \tau f(t, y(t)),\ t = j\tau\,,\ j = 0, 1, \dots\,,\ y(0) = x(0) = x_0\,. \tag{2.49}$$

A substitution of the (unknown) *exact* solution x into the approximation formula (2.49) yields the *defect* or, after dividing by the step length τ, the *discretization error* of this method:

$$d(t, x, \tau) = \frac{x(t+\tau) - x(t)}{\tau} - f(t, x(t))\,.$$

It measures the exactness with which the exact solution satisfies the *approximation formula* (2.49) and represents also the *local error* in explicit methods as in the present case. If namely $y(t) = x(t)$ is *exact* then we obtain for a single step

$$x(t+\tau) - y(t+\tau) = x(t+\tau) - x(t) + \tau f(t, x(t)) = \tau d(t, x, \tau)\,.$$

The discretization error is always calculated by using a TAYLOR expansion of the solution, e.g., with integral error term,

$$x(t+\tau) = x(t) + \tau\, f(t, x(t)) + \tau^2 \int_0^1 (1-\sigma) x''(t+\sigma\,\tau)\, d\sigma\,,$$

and thus, for the method (2.49),

$$\|d(t, x, \tau)\| \le \tau^p \int_0^1 \|x''(t+\sigma\,\tau)\|\, d\sigma\,,\quad p = 1\,.$$

Accordingly we say that the method (2.49) has order $p = 1$.

Furthermore, we obtain for the *global* error $e(t) = x(t) - y(t)$ by subtraction and application of LIPSCHITZ boundedness, i.e., $\|f(t,u) - f(t,v)\| \le L\,\|u - v\|$ in a suitable domain:

$$\begin{aligned} e(t+\tau) &= e(t) + \tau\,[f(t, x(t)) - f(t, y(t))] + \tau\, d(t, x, \tau)\,,\\ \|e(t+\tau)\| &\le (1 + L\tau)\|e(t)\| + \tau\,\|d(t, x, \tau)\|\,. \end{aligned}$$

An induction then yields an estimation of the global error where the inequality $(1+x)^n \le e^{nx}$, $x \ge -1$, is applied for optical reasons:

Lemma 2.4. *(Error Estimation, Convergence) Let $x \in C^2[0,T]$ be a solution of (2.48) and let f be* LIPSCHITZ*-bounded then*

$$\|e(t)\| \le e^{Lt}\|e(0)\| + \frac{e^{Lt} - 1}{\tau L} \max\nolimits_{0 \le s \le t} \tau\,\|d(s, x, \tau)\|\,,\; t = n\tau\,,\; n = 1, 2, \ldots\,.$$

The step length τ is canceled out once. Basically, one power of the step length τ is lost by passing from the local to the global error. This *a-priori error bound* contains the unknown solution x on the right side of the inequality. Therefore it makes only sense in theoretical studies or in comparing different methods with each other but not in practical applications. *A-posteriori error bounds* estimating the error by calculated data are much more difficult to find, therefore one contents himself here usually with *assessed valuation.* The above error bound is sharp for $x' = Lx$ with $L > 0$, and the problem is badly conditioned for large $L \cdot T$. For $L < 0$ this estimation does not make any sense and thus further criteria are necessary to qualify numerical approximations.

(b) General One-step Methods

Example 2.7. The computational device

$$\boxed{y(t+\tau) = y(t) + \tau\big[\omega f(t+\tau, y(t+\tau)) + (1-\omega) f(t, y(t))\big]}\,, \tag{2.50}$$

$0 \le \omega \le 1$, yields the explicit EULER method for $\omega = 0$, the *implicit* EULER *method* for $\omega = 1$, and the *trapezoidal rule* for $\omega = 1/2$. The method has order $p = 2$ for $\omega = 1/2$ and order $p = 1$ else.

A general one-step method can be written in the form

$$y(t+\tau) = y(t) + \tau\,\Phi(t, y(t), \tau)\,,\ t = j\tau\,,\ j = 0, 1, \dots\,,\ y(0) = x_0\,, \tag{2.51}$$

or, if the step length τ is constant throughout iteration, as

$$y_{j+1} = y_j + \tau\,\Phi_j(y_j, \tau)\,,\ j = 0, 1, \dots\,,$$

where $y_j := y(j\tau)$. The device function Φ must satisfy some obvious conditions which are not enumerated here and are fulfilled in normal case; cf. (Hairer). The method is then said to be *explicit* if a *finite* number of evaluations of the right side f (and derivatives of f) suffices for an *exact* computation of Φ, otherwise the method is called *implicit*.

The discretization error is defined in the same way as in **(b)**, and the method is *consistent* (with the differential equation) if

$$\Gamma(x) := \sup_{0\le\tau\le\tau^*}\ \sup_{0\le t\le T-\tau} \frac{1}{\tau^p}\|d(t, x, \tau)\| < \infty \tag{2.52}$$

for some $p \ge 1$ and for *all* solutions $x \in C^{p+1}[0, T]$. The maximum possible number $p \in \mathbb{N}$ in (2.52) is the order of the method for the considered differential equation and called *order* in general if the method has order p for all sufficiently smooth right sides f of (2.48). The content of Lemma 2.4 remains unchanged by this convention.

(c) Asymptotic Expansion, Extrapolation

Lemma 2.5. *Let the method (2.51) have order $p \ge 1$, let*

$$\Phi(t, x(t), 0) = f(t, x(t))\,,\ \ \operatorname{grad}_x \Phi(t, x(t), \tau) = \operatorname{grad}_x f(t, x(t)) + \mathcal{O}(\tau)\,,$$

and let $\partial\Phi/\partial\tau$ be continuous in τ near $\tau = 0$. Then there exists an error function r *being* independent *of the step length τ such that*

$$\boxed{y(t) = x(t) + r(t)\tau^p + \mathcal{O}(\tau^{p+1})\,,\ \tau \to 0}\,.$$

Proof see (Hairer), vol. I, Sect. 2.8.

This asymptotic representation of the numerical approximation $y(t)$ has two important consequences whenever we apply the method (2.51) once with step length τ and then once more with the reduced step length $q\tau$, $0 < q < 1$,

$$\begin{aligned} y(t, \tau) &= x(t) + r(t)\tau^p + \mathcal{O}(\tau^{p+1})\,,\\ y(t, q\tau) &= x(t) + r(t)(q\tau)^p + \mathcal{O}(\tau^{p+1})\,. \end{aligned}$$

(1°) The *weighted* difference

$$z(t) := \frac{q^{-p}y(t,q\tau) - y(t,\tau)}{q^{-p}-1} = x(t) + \mathcal{O}(\tau^{p+1}) \tag{2.53}$$

supplies an improved method of order $p+1$ instead of p with comparable few computational effort.

Example 2.8. The *model problem* $x' = \lambda x$ has the solution $x(t) = \kappa\, e^{\lambda t}$. We choose $x(0) = 1$, $\lambda = 1$, and apply the trapezoidal rule, once with step length $\tau = 1$ and, for comparison, twice with step length $\tau = 1/2$:

$$\begin{aligned} h = 1: \quad y(1) &= \frac{1+0.5}{1-0.5} &&= 3 \\ h = 0.5: \widetilde{y}(1) &= \frac{1+0.25}{1-0.25}\cdot\frac{1+0.25}{1-0.25} = \frac{25}{9} &&= 2.\overline{7}\,. \end{aligned}$$

An application of the *averaging* (2.53) with $p = 2$ and $q = 1/2$ yields the improvement

$$z(1) = \frac{1}{3}\left(4\cdot\frac{25}{9} - 3\right) = \frac{100-27}{27} = 2.703703\ldots$$

with error $\varepsilon = 0.0145\ldots$; the additional amount of work is a neglecting quantity in larger problems.

(2°) The *simple* difference

$$\begin{aligned} y(t,\tau) - y(t,q\tau) &= r(t)(q\tau)^p(q^{-p}-1) + \mathcal{O}(\tau^{p+1})\,, \\ r(t)(q\tau)^p &\simeq \frac{y(t,\tau)-y(t,q\tau)}{q^{-p}-1}\,, \end{aligned}$$

supplies a good estimation of the error function $r(t)$. If, further, D is a diagonal matrix containing suitable weights and C contains tolerances and suitable security factors then, with additional safety bounds,

$$\tau_{\text{new}} = \tau_{\text{old}}\cdot C\cdot \|D[y(q\tau) - y(t)]\|^{-1/p}$$

provides an excellent step length control. Particular advantages are obtained in the case of *imbedded* methods of order p supplying an approximation $z(t)$ of order $p-1$ at the same time,

$$\begin{aligned} y(t) &= x(t) + r(t)\tau^p + \mathcal{O}(\tau^{p+1})\,, \\ z(t) &= x(t) + \widetilde{r}(t)\tau^{p-1} + \mathcal{O}(\tau^p)\,. \end{aligned}$$

The difference yields here *directly* an estimation of the error of $z(t)$:

$$z(t) - y(t) = \widetilde{r}(t)\tau^{p-1} + \mathcal{O}(\tau^p) \simeq z(t) - x(t)\,.$$

(d) Runge-Kutta Methods

Example 2.9. The explicit method of HEUN is obtained from the (implicit) trapezoidal rule (2.50) by inserting

$$f_{j+1}(y_{j+1}) \simeq f_{j+1}\left(y_j + \tau f_j(y_j)\right) .$$

$$\boxed{\begin{array}{ll} \text{Trapezoidal rule (p = 2) : } & y_{j+1} = y_j + \dfrac{\tau}{2}\Big(f_j(y_j) + f_{j+1}(y_{j+1})\Big) \\ \text{Method of HEUN (p = 2) : } & y_{j+1} = y_j + \dfrac{\tau}{2}\Big(f_j(y_j) + f_j(y_j + \tau f_j(y_j))\Big) \end{array}} ;$$

the numerical computation is carried out by the scheme

$$k_1 = f_j(y_j)\,, \quad k_2 = f_j(y_j + \tau\, k_1)\,, \quad y_{j+1} = y_j + \frac{\tau}{2}\,(k_1 + k_2)\,.$$

A generalization of this concept leads to *multistage* methods or RUNGE-KUTTA methods.

Example 2.10. The classical RUNGE-KUTTA method is a four-stage method of order $p = 4$ in which the function f is four-times evaluated at intermediate steps. Ensuing, a linear combination of these terms forms the forward step; this last step originates mostly from a numerical integration rule, in the present case being SIMPSON's rule:

$$\begin{aligned} k_1 &= f(y_j)\,, \quad k_2 = f_j\left(y_j + \frac{\tau}{2}\,k_1\right)\,, \quad k_3 = f_j\left(y_j + \frac{\tau}{2}\,k_2\right)\,, \\ k_4 &= f_j(y_j + \tau\, k_3)\,, \quad y_{j+1} = y_j + \tau\,\frac{1}{6}\,(k_1 + 2k_2 + 2k_3 + k_4)\,. \end{aligned}$$

The example shows that the order of a one-step method can be enhanced in a skilful way if several intermediate steps are properly introduced. Even methods of arbitrary high order can be constructed by this way (methods of GRAGG-BULIRSCH-STOER).

A general r-stage one-step method for $x' = f(t,x) \in \mathbb{R}^n$ is a computational devicc of the form

$$\boxed{\begin{array}{l} k_i(t) = f\left(t + \gamma_i\tau\,,\, y(t) + \tau\displaystyle\sum_{j=1}^{r}\alpha_{ij}k_j(t)\right)\,,\; i = 1:r \\ y(t+\tau) = y(t) + \tau\displaystyle\sum_{i=1}^{r}\beta_i k_i(t) \end{array}}\,. \tag{2.54}$$

with the *function values* $k_i(t) := f(t+\gamma_i\tau, u_i(t))$ being the unknown quantities. However the subsequent representation is to be preferred in the studies of this method. Let I be the unit matrix, $A = [\alpha^i{}_j] \in \mathbb{R}^r{}_r$, and $b = [\beta_i]$, $c = [\gamma_j]\,, e = [1]$ all together in $\mathbb{R}^r$, and, moreover,

$$\begin{aligned} A \times B \quad &= [\alpha_{ij}B]_{i,j=1}^r \qquad &&\text{KRONECKER product},\\ U(t) \quad &= [u_i(t)]_{i=1}^r \qquad &&\text{auxiliary vectors}, \ u_i(t) \in \mathbb{R}^n,\\ F(t,U(t)) &= [f(t+\gamma_i\tau, u_i(t))]_{i=1}^r \in \mathbb{R}^{r\cdot n}\,. \end{aligned}$$

The computational device (2.54) is then equivalent to the form with *intermediate values* u_i of the approximation

$$\begin{aligned} U(t) \quad &= e \times y(t) + \tau(A \times I)F(t,U(t)) \in \mathbb{R}^{r\cdot n}\\ y(t+\tau) &= y(t) + \tau(b \times I)^T F(t,U(t)) \quad\ \ \in \mathbb{R}^n\,. \end{aligned} \tag{2.55}$$

For instance, the method of HEUN may now be written as

$$u_1 = y_j\,,\quad u_2 = y_j + \tau f_j(u_1)\,,\quad y_{j+1} = y_j + \tau\left(f_j(u_1) + f_j(u_2)\right)\,.$$

The auxiliary quantities $u_i(t)$ may be interpreted as approximations of the exact values $x(t+\gamma_i\tau)$ which however is only of interest in derivation of order conditions.

Properties and Further Notations:

(1°) The method (2.55) is called *explicit* resp. *semi-implicit* if (possibly after renumeration) the matrix A is a strongly lower resp. a lower triangular matrix.

(2°) In normal case, the points $\gamma^i\tau$ are contained in interval $[0, \tau]$, but they are not always mutually distinct; cf. Example 2.10. The intermediate values $u_i(t)$ are approximations of the exact solution at the intermediate points $t+\gamma^i\tau$ as already mentioned. The system of intermediate stages is uniquely solvable if f is LIPSCHITZ bounded and if, but only in implicit methods, the step length τ is sufficiently small.

(3°) The method (2.55) is frequently described by using the BUTCHER matrix $[A|b|c]$ or a similar form. For instance, Example 2.10 can be displayed as

$$\left[\begin{array}{c|c} A & c\\ \hline b & \end{array}\right] = \left[\begin{array}{cccc|c} 0 & 0 & 0 & 0 & 0\\ 1/2 & 0 & 0 & 0 & 1/2\\ 0 & 1/2 & 0 & 0 & 1/2\\ 0 & 0 & 1 & 0 & 1\\ \hline 1/6 & 1/3 & 1/3 & 1/6 & \end{array}\right]\,.$$

(4°) If $W(t)$ is the solution of

$$W(t) = e \times x(t) + \tau(A \times I)F(t,W(t))\,,$$

then

$$d(t,x,\tau) = \frac{x(t+\tau) - x(t)}{\tau} - (b \times I)^T F(t,W(t))$$

is the *discretization error*.

(5°) The method (2.54) has at least order $p = 1$ if and only if $b^T e = \sum_{i=1}^r \beta_i = 1$.

(6°) Table of attainable order p^* of explicit RUNGE-KUTTA methods in dependence of the stage number r (Butcher):

Table 2.4.

r	1	2	3	4	5	6	7	8	9	$r \geq 10$
p^*	1	2	3	**4**	**4**	5	6	6	7	$\leq r-2$

therefore the RUNGE-KUTTA method of order $p = 4$ plays a particular role among all explicit methods.

(e) Multistep Methods A multiple evaluation of the right side f of the underlying differential system can be rather cumbersome. But the order of a method can be enhanced also if the formerly obtained values $y_j\,,\ y_{j-1}\,,\dots$ are regarded in the sense of an extrapolation beyound the interval known at the present state. For instance, the device

$$3y_{j+1} - 4y_j + y_{j-1} = 2\tau\, f_j(y_{j+1})\,,\ j = 1, 2, \dots\,,$$

is a well-known implicit method of order $p = 2$ with extraordinary stability properties.

General multistep methods have the form

$$\sum_{i=0}^{k} \alpha_i y_{j+i} = \tau \sum_{i=0}^{k} \beta_i f_{j+i}(y_{j+i})\,,\ j = 0, 1, \dots\,, \tag{2.56}$$

where $\alpha_k \neq 0$ and $|\alpha_0| + |\beta_0| \neq 0$; the method is *explicit* for $\beta_k = 0$ and *implicit* else. The function f has to be evaluated here only once in every t-step. But, on the other side, the starting values $y_1, \dots, y_{k-1}$ must be supplied by some other method.

Properties and Notations:

(1°) Using the polynomials

$$\varrho(\zeta) = \sum_{i=0}^{k} \alpha_i \zeta^i\,,\quad \sigma(\zeta) = \sum_{i=0}^{k} \beta_i \zeta^i$$

and the translation operator $E : y(t) \mapsto Ey(t) := y(t+\tau)$, the device (2.56) can be written more simply as

$$\boxed{\varrho(E)y_j = \tau\sigma(E)f_j\,,\ j = 0, 1, \dots}\ . \tag{2.57}$$

(2°) Application to the model equation $x' = \lambda x$ yields the very simple device

$$\pi(E, \eta)y_j := \varrho(E)y_j - \eta\sigma(E)y_j = 0\,,\ j = 0, 1, \dots$$

where $\eta = \tau\lambda$ with step length τ, and $\pi(E,\eta)$ is the *characteristic polynomial* of the multistep methods describing it completely as well. Furthermore, the discretization error d of the method has the very simple form

$$\tau d(t,x,\tau) = \varrho(E)x(t) - \tau\sigma(E)x'(t)\,.$$

(3°) Order and consistence are defined in the same way as in **(b)**. Whereas the derivation of order conditions and the construction of individual RUNGE-KUTTA-methods and their relatives is a difficult matter and must be leaved to experts, cf. (Hairer), it is much simpler in multistep methods. By reasons of linearity the coefficients are to be adjusted here in a way that, for a prescribed order p, all "differential equations" $x'(t) = t^k$, $k = 0 : p$, are solved exactly. Further possibilities for constructing very special methods arise by using other, especially chosen elementary functions as e^{it}, $\sin(jt)$, $\cos(kt)$ for the adjustment above.

Lemma 2.6. *A multistep method* $(\varrho,\,\sigma)$ *has order* $p \geq 1$ *if and only if*

$$\varrho(1) = 0 \ \ \textit{and} \ \sum_{i=0}^{k}\left(\alpha_i\frac{i^m}{m} - \beta_i\, i^{m-1}\right) = 0\,,\ m = 1 : p\,.$$

In particular, the method has at least order $p = 1$ if $\varrho(1) = 0$ and $\varrho'(1) - \sigma(1) = 0$.

(4°) The discretization error satisfies in simple way

$$\|\tau d(t,x,\tau)\| \leq \text{const}\,\tau^p \int_t^{t+k\tau} \|x^{(p+1)}(s)\|\,ds\,,$$

and, by induction, an error estimation is deduced in a similar way as in Lemma 2.4 with the same qualitative properties. But, in the present case, the polynomial ϱ must satisfy the following *root condition* or *stability criterium*:

Every root λ of ϱ satisfies $\lvert\lambda\rvert \leq 1$ and every root λ with $\lvert\lambda\rvert = 1$ is a simple root.

(5°) In order to prevent "spurious solutions", a multistep method should also be *strongly stable* which means that the polynomial ϱ has precisely one root of modulus one, namely the always appearing root $\zeta = 1$, and this root must be a simple root according to the root condition.

(f) Summary

multistage methods	multistep methods
self-starting	not self-starting
high computational effort	low computational effort
simple step length control	difficult step length control

(g) Stability

(g1) A differential equation $x'(t) = f(t, x(t))$ is *stable* if the difference of any two solutions remains bounded in absolute value for all $t > 0$, it is *asymptotically stable* if the difference tends to zero in absolute value for $t \to \infty$; cf. Sect. 1.5(c); analogeous notations hold for differential systems w.r.t. an arbitrary submultiplicative norm. The instability of a differential equation is inherited to the numerical approximation in any case, nevertheless, a good step control may supply very acceptable results. If a differential equation is stable then the exact solutions decrease in absolute value during a long t-interval or they remain bounded at least. Of course, this property should be inherited to the numerical approximations, too, but this is not always the case:

Example 2.11. $x' = Ax,\ x_0 = [1, 0, -1]^T,\ x(t) = [x(t), y(t), z(t)]^T$.

$$A = \begin{bmatrix} -21 & 19 & -20 \\ 19 & -21 & 20 \\ 40 & -40 & -40 \end{bmatrix},\ \text{eigenvalues}: \lambda_1 = -2,\ \lambda_{2,3} = -40 \pm 40\,i.$$

The solution

$$\begin{aligned} x(t) &= \frac{1}{2}e^{-2t} + \frac{1}{2}e^{-40t}(\cos 40t + \sin 40t), \\ y(t) &= \frac{1}{2}e^{-2t} - \frac{1}{2}e^{-40t}(\cos 40t + \sin 40t), \\ z(t) &= -e^{-40t}(\cos 40t - \sin 40t). \end{aligned}$$

behaves like the solution of $x' = Bx$ for $t > 0.1$ where

$$B = \begin{bmatrix} -2 & 0 & 0 \\ 0 & -2 & 0 \\ 0 & 0 & 0 \end{bmatrix}.$$

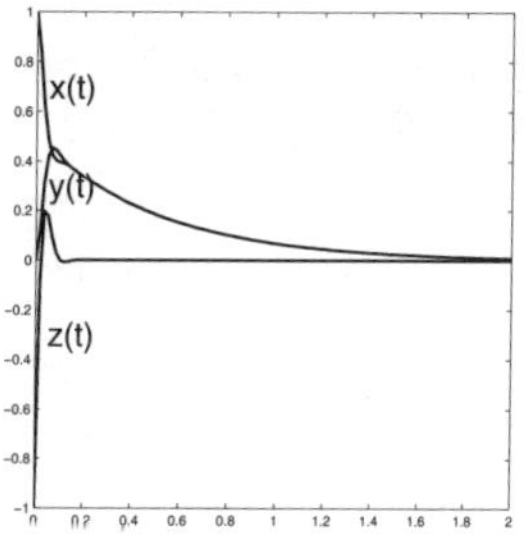

Figure 2.11. Ex. 2.11, solution

The trapezoidal rule provides good approximations whereas the explicit EULER method supplies entirely unacceptable results with the applied step length. In the explicit method, a small step length τ (with high computational effort) is unnecessary for large t whereas, on the other side, a large step length magnifies the high-frequent but fast decreasing parts of the solution in an explosive way.

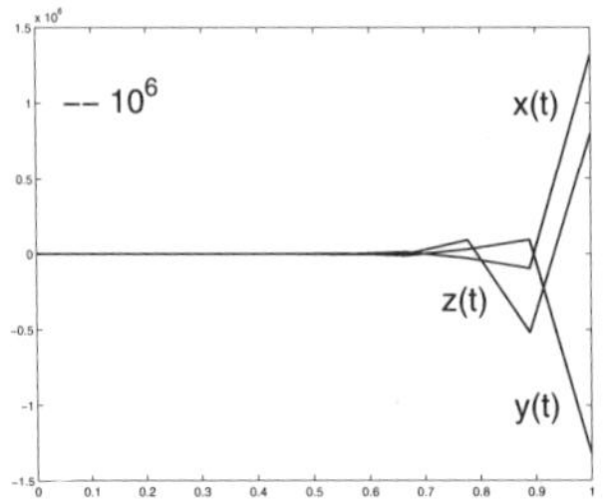

Figure 2.12. Ex. 2.11, EULER explicit, $\tau = 0.1$

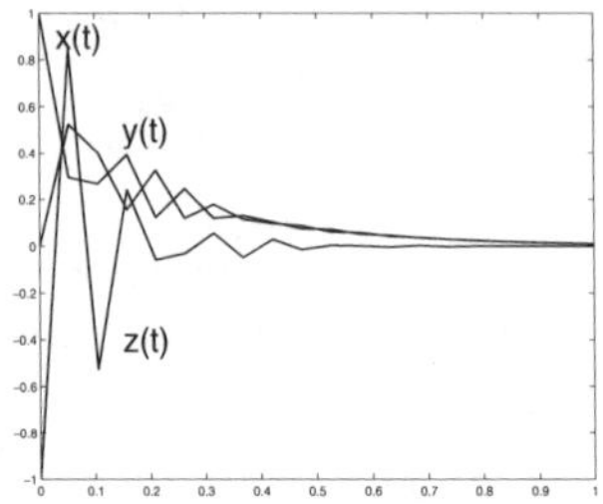

Figure 2.13. Ex. 2.11, Trapezoidal rule, $\tau = 0.1$

(g2) For a more thorough investigation of this phenomenon, we apply the multistage method (2.55) to the *model equation* $x'(t) = \lambda x(t)$ again and obtain with step length τ and $\eta = \tau\lambda$

$$\begin{aligned} U(t) &= ey(t) + \tau\lambda A U(t) \\ y(t+\tau) &= y(t) + \tau\lambda b^T U(t) \end{aligned} \Longrightarrow \begin{aligned} U(t) &= (I - \eta A)^{-1} e y(t) \\ y(t+\tau) &= \left[1 + \eta b^T (I - \eta A)^{-1} e\right] y(t)\,. \end{aligned}$$

Thereby the computational device

$$y_{j+1} = R(\eta) y_j\,, \quad R(\eta) = 1 + \eta b^T (I - \eta A)^{-1} e\,, \quad R(\infty) := 1 - b^T A^{-1} e \quad (2.58)$$

is derived for the model equation. By using CRAMER's rule, the *stability function* R can be written after brief computation as

$$R(\eta) = \frac{\det\left(I - \eta A + \eta e b^T\right)}{\det(I - \eta A)} =: \frac{P(\eta)}{Q(\eta)}\,, \quad (2.59)$$

with polynomials P and Q, and $Q(\eta) = 1$ in explicit methods. Now the closed set in complex η-plane

$$\boxed{\mathcal{S} := \{\eta \in \mathbb{C} \cup \{\infty\},\ |R(\eta)| \le 1\}}$$

is called *stability region* of the specific one-step method. If we consider more generally the system $x'(t) = Ax(t)$ with *diagonalizable* Matrix A, $A = U\Lambda U^{-1}$, (Λ diagonal matrix of eigenvalues λ_i of A), then the one-step methods obtain the form

$$y_{j+1} = UR(\tau\Lambda)U^{-1} y_j = UR(\tau\Lambda)^j U^{-1} y_0$$

with diagonal matrix

$$R(\tau\Lambda) = \operatorname{diag}(R(\tau\lambda_1), \ldots, R(\tau\lambda_n))\,.$$

As a consequence, all $\eta_i := \tau\lambda_i$ must be contained in the stability region $\mathcal{S}$ if every solution shall be at least bounded for all $t > 0$. This is the well-known COURANT-FRIEDRICHS-LEVY condition for the step length τ, and this condition must always be regarded in stable problems but also in various other cases; see e.g. Sect. 9.7 **(d)**.

Example 2.12. Consider the model equation $x' = \lambda x$ with $\eta = \tau\lambda$. Then

(A) EULER method explicit ($p = 1$):	$y_{j+1} = (1+\eta)y_j\,,$
(B) EULER method implizit ($p = 1$):	$y_{j+1} = (1-\eta)^{-1}y_j\,,$
(C) Trapezoidal rule ($p = 2$):	$y_{j+1} = \dfrac{2+\eta}{2-\eta}y_j\,,$
(D) Method of HEUN ($p = 2$):	$y_{j+1} = \left(1+\eta+\dfrac{1}{2}\eta^2\right)y_j\,.$

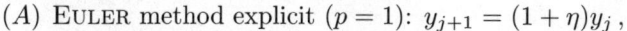

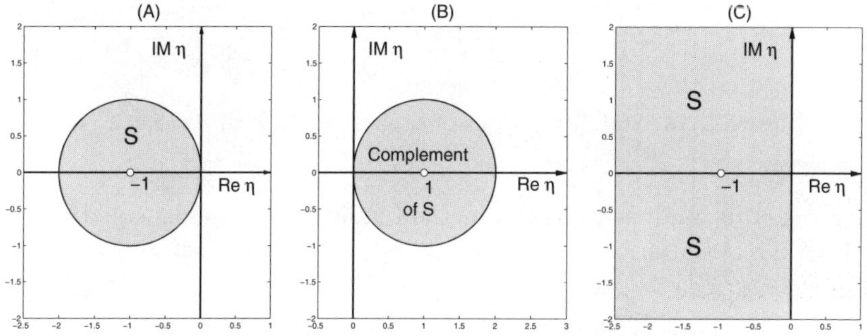

Figure 2.14. Stability regions for example 2.12 without (D)

The discretization error d of a method of type (2.55) of order $p \geq 1$ satisfies $\tau d(t,x,\tau) = \mathcal{O}(\tau^{p+1})$ and, on the other side, a substitution of the model equation with $\lambda = 1$ und $x(0) = 1$ yields the device

$$y(\tau) = R(\tau)y(0) = R(\tau)\,,\quad x(\tau) = e^\tau\,.$$

Subtraction yields

$$\tau d(\tau,x,\tau) = x(\tau) - y(\tau) = e^\tau - R(\tau) = \mathcal{O}(\tau^{p+1})\,.$$

Accordingly, we have

$$R(\eta) = 1 + \eta + \frac{\eta^2}{2} + \ldots + \frac{\eta^p}{p!} + \mathcal{O}(\eta^{p+1})\,,$$

in every RUNGE-KUTTA method of order p. On the other side $R(\eta) = 1 + \eta + \sum_{i=2}^{r} \kappa_i\eta^i$ in an explicit r-stage method by (2.59). As a consequence, all *explicit* RUNGE-KUTTA methods with same order $\boxed{p = r}$ have the same stability function $R(\eta) = \sum_{i=0}^{p} \eta^i/i!$ (where however $p^* = r \leq 4$ for the *optimum* order p^* by Table 2.4). By symmetry to the real axis, only the upper half of the implicit curve $|\sum_{i=0}^{p} \eta^i/i!| = 1$, $p = 1:6$, is plotted in Figure 2.15.

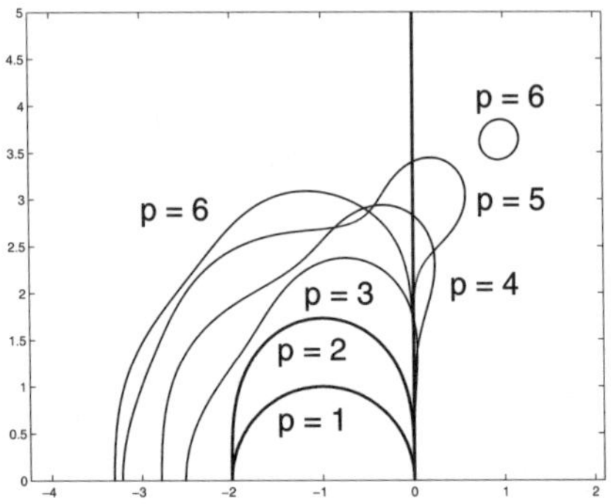

Figure 2.15. Stability regions of explicit RKM with $p = r = 1:4$

(g3) Let us apply a multistep method (2.56) to the model equation $x' = \lambda x$ then the result is

$$\pi(E,\eta)y_j = \sum_{i=0}^{k} \gamma_i(\eta)E^i y_j = \sum_{i=0}^{k} \gamma_i(\eta)y_{j+i} = 0\,,\; j = 0,1,\dots\,,$$

by (2.57) or, writing the device as one-step method by introducing the vector $Y_j = [y_j, y_{j+1}, \dots, y_{j+k-1}]^T \in \mathbb{R}^k$,

$$Y_{j+1} = F_\pi(\eta)Y_j\,,\quad F_\pi(\eta) = \begin{bmatrix} 0 & 1 & 0 & 0 & & 0 \\ 0 & 0 & 1 & \ddots & & 0 \\ 0 & & \ddots & \ddots & \ddots & 0 \\ 0 & & & \ddots & 0\;\;0 & 1 \\ -\gamma_0(\eta)0/\gamma_k(\eta) & \dots & \dots & \dots & \dots & -\gamma_{k-1}(\eta)/\gamma_k(\eta) \end{bmatrix}. \tag{2.60}$$

The FROBENIUS matrix $F_\pi(\eta)$ has the characteristic polynomial $\det(\lambda I - F_\pi(\eta)) = \pi(\lambda,\eta)$, hence it is called sometimes *accompanying matrix to the polynomial* π. Every eigenvalue of this matrix possesses precisely one eigenvector. On the other hand the matrix must be a M-matrix by Theorem 1.1; cf. Sect. 1.1 **(c4)**, if all iterations (2.60) shall remain bounded in absolute value (resp. in some norm). Therefore the concept of *stability regions* must be adapted to multistep methods as follows:

Definition 2.2. *Let* $(\varrho\,,\,\sigma)$ *be a multistep method with characteristic polynomial* $\pi(\zeta,\eta) = \varrho(\zeta) - \eta\sigma(\zeta)$*, and let* $\pi(\zeta,\infty) = \sigma(\zeta)$*. Then the stability region* $\mathcal{S} \in \mathbb{C} \cup \{\infty\}$ *is the set of all values* $\eta \in \mathbb{C}$ *with the following two properties:*

(1°) *All roots* $\zeta_i(\eta)$ *of* $\pi(\zeta,\eta)$ *satisfy* $|\zeta_i(\eta)| \leq 1$.
(2°) *All roots* $\zeta_i(\eta)$ *of* $\pi(\zeta,\eta)$ *satisfying* $|\zeta_i(\eta)| = 1$ *(unimodular roots) are simple roots of* $\pi(\zeta,\eta)$.

(h) Stiff Differential Systems A system $x'(t) = Ax(t)$ is said to be *stiff* if the eigenvalues λ_i of A have the property

$$\operatorname{Re}\lambda_i \leq 0 \text{ and } \max_i |\operatorname{Re}\lambda_i| \gg \min_i |\operatorname{Re}\lambda_i|\,.$$

Such systems occur e.g. in the motion of mass points being connected with each other by weak and stiff springs at the same time. Furthermore, they appear necessarily in discretization of differential equations as the following simple example shows impressively.

Example 2.13. The eigenvalue problem

$$y''(x) = \lambda^2\, y\,,\; y(0) = y(1) = 0\,, \tag{2.61}$$

has the characteristic pairs

$$(\lambda_j^2\,,\, y_j(x)) = (-j^2\pi^2\,,\, \sin(j\pi x))\,,\; j \in \mathbb{N}\,.$$

If the second derivative is approximated by the central divided difference

$$y''(jh) = h^{-2}\left[y((j+1)h) - 2y(jh) + y(t,(j-1)h)\right] + \mathcal{O}(h^2)\,,\quad h = 1/(n+1)\,,$$

then we obtain the discrete eigenvalue problem $AY = \widetilde{\lambda}^2 Y \in \mathbb{R}^n$,

$$AY = h^{-2}\begin{bmatrix} -2 & 1 & 0 & \cdots\cdots & 0 & 0 \\ 1 & -2 & 1 & \cdots\cdots\cdots & & 0 \\ 0 & \ddots & \ddots & \ddots & \ddots & \vdots \\ \vdots & \ddots & \ddots & \ddots & \ddots & \vdots \\ \vdots & \ddots & \ddots & \ddots & \ddots & 0 \\ 0 & \ddots & \ddots & 1 & -2 & 1 \\ 0 & 0 & \cdots\cdots & 0 & 1 & -2 \end{bmatrix}\begin{bmatrix} y(h) \\ \vdots \\ \vdots \\ \vdots \\ \vdots \\ \vdots \\ y(nh) \end{bmatrix} = \widetilde{\lambda}^2\begin{bmatrix} y(h) \\ \vdots \\ \vdots \\ \vdots \\ \vdots \\ \vdots \\ y(nh) \end{bmatrix}\,, \tag{2.62}$$

with characteristic pairs

$$(\widetilde{\lambda}_j^2, Y_j) = \left(-h^{-2}4\sin^2\left(\frac{jh\pi}{2}\right),\left[\sin\left(\frac{jk\pi}{n+1}\right)\right]_{k=1}^{n}\right)\,,\; j = 1:n\,.$$

Note here the very exceptional fact that the eigenvectors of the discretized problem have the same values as the eigenfunctions of the analytic problem (2.61) at corresponding points. The eigenvalues satisfy

$$\widetilde{\lambda}_j^2 = -h^{-2}4\sin^2\left(\frac{jh\pi}{2}\right) = -j^2\pi^2 + \mathcal{O}(j^4h^2) = \lambda_j^2 + \mathcal{O}(j^4h^2)\,,\; j = 1:n\,.$$

Accordingly, $\widetilde{\lambda}_j^2$ is a second-order approximation of the eigenvalue $-j^2\pi^2$ of the differential equation for every *fixed* j and, in particular, the eigenvalues of (2.62) increase beyond every bound in absolute value if the step length h tends to zero.

Let us now consider the *parabolic* initial boundary value problem

$$\begin{aligned} u_t(t,x) &= u_{xx}(t,x)\,,\ 0 \le x \le 1\,,\ 0 \le t\,, \\ u(t,0) &= a(t)\,,\ u(t,1) = b(t)\,, u(0,x) = u_0(x)\,,\ u_0(0) = a(0)\,,\ u_0(1) = b(0)\,, \end{aligned} \tag{2.63}$$

which can be solved exactly at least for $a = 0$ and $b = 0$. A discretization in the space variable x in the same way as in (2.61) leads to the initial value problem

$$\begin{aligned} U'(t) &= AU(t) + B(t)\,, \\ U(t) &= [u(t,h), \ldots, u(t,nh)]^T\,,\ B(t) = h^{-2}[a(t),\, 0, \ldots, 0,\, b(t)]^T\,, \end{aligned}$$

with the same matrix A as in (2.62). If this ordinary differential system shall be solved by one of the above considered methods then, by the COURANT-FRIEDRICHS-LEVY condition, the step length $\tau > 0$ must be chosen so small that the value $\widetilde{\lambda}_n^2 \simeq -4\tau/h^2$ still lies in the stability region $\mathcal{S}$. But this restriction of step length can be dropped in the special *implicit* methods (B) and (C) because here the entire negative semi-line belongs to $\mathcal{S}$. By this reason, further criteria on the *shape* of the stability region are introduced: A multistage/multistep method is called

A-stable	$\Longleftrightarrow$	$\{\eta \in \mathbb{C} \cup \{\infty\},\ \mathrm{Re}\,\eta < 0\} \subset \mathrm{int}\,\mathcal{S}$
A(α)*-stable*	$\Longleftrightarrow$	$\{\eta \in \mathbb{C} \cup \{\infty\},\ \eta \neq 0\,,\ \lvert\pi - \arg\eta\rvert < \alpha\} =: \mathcal{S}_\alpha \subset \mathrm{int}\,\mathcal{S}$
A(0)*-stable*	$\Longleftrightarrow$	$\exists\,\alpha > 0$ such that the method is $A(\alpha)$-stable
A_0*-stable*	$\Longleftrightarrow$	$(-\infty\,, 0) \subset \mathcal{S}$
L-stable	$\Longleftrightarrow$	method A-stable and $\infty \subset \mathrm{int}\,\mathcal{S}$

. (2.64)

The stability function (2.59) shows that the stability region of an *explicit* RUNGE-KUTTA method can never have one of these properties; the same can be verified easily for *explicit* multistep methods. One now recognizes the dilemma in the model Example 2.13: In explicit methods, the step length τ in t-direction must be chosen proportionally to the quadrat of the step length h in x-direction whereas in implicit methods the computational amount of work is considerably magnified.

By the way, not all implicit methods have one of the properties (2.64) which is shown best by plotting the individual stability regions.

Rule for application:

> Use only explicit methods with step control for solving *unstable* differential systems.

(i) Further Examples We consider briefly some methods of the MATLAB ODE suite.

(1°) MATLAB `ode45.m` Imbedded explicit RUNGE-KUTTA method due to DORMAND & PRINCE; cf. (Dormand):

$$\begin{bmatrix} A \\ b \end{bmatrix} =$$

$$\left[\begin{array}{ccccccc}
0 & 0 & 0 & 0 & 0 & 0 & 0 \\
1/5 & 0 & 0 & 0 & 0 & 0 & 0 \\
3/40 & 9/40 & 0 & 0 & 0 & 0 & 0 \\
44/45 & -56/15 & 32/9 & 0 & 0 & 0 & 0 \\
19372/6561 & -25360/2187 & 64448/6561 & -212/729 & 0 & 0 & 0 \\
9017/3168 & -355/33 & 46732/5247 & 49/176 & -5103/18656 & 0 & 0 \\
35/384 & 0 & 500/1113 & 125/129 & -2187/6784 & 11/84 & 0 \\
\hline
35/384 & 0 & 500/1113 & 125/129 & -2187/6784 & 11/84 & 0 \\
\hline
5179/57600 & 0 & 7571/16695 & 393/640 & -92097/339200 & 187/2100 & 1/40
\end{array}\right]$$

$$c = [0,\ 1/5,\ 3/10,\ 4/5,\ 8/9,\ 1,\ 1]$$

If one takes the pen-ultimate row for weight vector b in the forward step, then the result is a method of order $p = 5$ and with the last row instead a method of order $p = 4$. The six-stage method has order $p = 5$, the seventh stage being used only for error estimation in step length control. The stability region is given by the curve for $p = 6$ in Figure 2.15.

(2°) ROSENBROCK methods are perhaps not the *ultima ratio* but the result of a long investigation on the *efficiency* of methods for stiff systems. The contradicting requirements on high order, low computational effort and best stability properties as L-stability, cf. (2.64), have finally led to a compromise. Starting point of the deliberations is a RUNGE-KUTTA method of which the matrix A is a (weakly) lower triangular matrix (diagonal implicit methods). Confining ourselves to an autonomous system $x'(t) = f(x(t))$, the method has the form

$$\begin{aligned}
k_i(t) &= f\left(y(t) + \tau \sum_{j=1}^{i-1} a_{ij} k_j(t) + \tau a_{ii} k_i(t)\right), \quad i = 1 : r, \\
y(t+\tau) &= y(t) + \tau \sum_{i=1}^{r} \beta_i k_i(t).
\end{aligned}$$

These equations are now linearized, for instance the values $k_i(t)$ are replaced by

$$k_i(t) = f(g_i(t)) + \operatorname{grad} f(g_i(t)) a_{ii} k_i(t), \quad g_i(t) = y(t) + \tau \sum_{j=1}^{i-1} a_{ij} k_j(t).$$

Furthermore, the r matrices $\operatorname{grad} f(g_i(t))$ are replaced by a single matrix $J := \operatorname{grad} f(y(t))$ following a proposition of (Calahan) which once more reduces the computational effort considerably. Then the combination

$$
\begin{aligned}
k_i(t) &= f\left(y(t) + \sum_{j=1}^{i-1} a_{ij}k_j(t)\right) + J\sum_{j=1}^{i} d_{ij}k_j(t) \\
y(t+\tau) &= y(t) + \tau\sum_{i=1}^{r} \beta_i k_i(t)
\end{aligned}
$$

is chosen in ROSENBROCK methods to attain a higher degree of freedom in the choice of suitable coefficients. The MATLAB program `ode23s.m` of (Shampine82) is of this type where the last evaluation of f in the preceding step is used for first evaluation in the new step:

$$
\boxed{
\begin{aligned}
f_0 &= f(t, y(t)) \\
Wk_1 &= f_0 + \tau dT \\
f_1 &= f\left(t + \frac{1}{2}\tau, y(t) + \frac{1}{2}\tau k_1\right) \\
Wk_2 &= f_1 - k_1 + Wk_1 \\
y(t+\tau) &= y(t) + \tau k_2 \\
f_2 &= f(t+\tau, y(t+\tau)) \\
Wk_3 &= f_2 - e\,(k_2 - f_1) - 2\,(k_1 - f_0) \\
\widetilde{y}(t+\tau) &= y(t+\tau) + \frac{\tau}{6}\,(k_1 - 2k_2 + k_3)
\end{aligned}
}
$$

$$
d = 1/(2+\sqrt{2})\,,\quad e = 6+\sqrt{2}\,,
$$

$$
T = \frac{\partial}{\partial t}f(t, y(t))\,,\quad J = \operatorname{grad} f(t, y(t))\,,\quad W = I - hdJ\,.
$$

This two-stage method has order $p = 2$, the value $\widetilde{y}(t+\tau)$ is only used for error estimation. Substitution of the model equation $x'(t) = \lambda x(t)$ with $\eta = \tau\lambda$ again yields the device

$$
y_{n+1} = R(\eta)y_n\,,\quad R(\eta) = \frac{1 + (1-2d)\eta + (d^2 - 2d + 1/2)\eta^2}{1 - 2d\eta + d^2\eta^2}\,.
$$

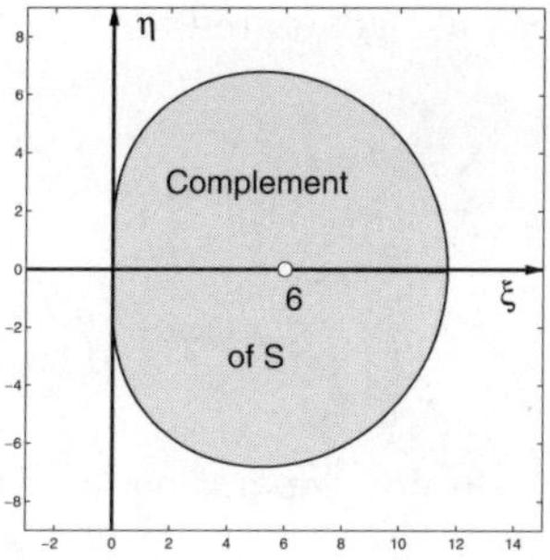

Figure 2.16. Stability region of the Rosenbrock method

(3°) MATLAB `ode113.m` predictor-corrector method
Let the *backward differences* be defined by

$$\nabla^0 f(t) = f(t)\,,\ \nabla f(t) = f(t) - f(t-\tau)\,,$$
$$\nabla^2 f(t) = \nabla(\nabla f(t)) = f(t) - 2f(t-\tau) + f(t-2\tau)\,,\ \text{etc.},$$

and let $p_k(x; f)$ be an interpolating polynomial of degree k with interpolation property

$$p_k((j+i)\tau) = f((j+i)\tau)\,,\ i = 0 : k\,,$$

where $j \in \mathbb{N}_0$ is fixed for the present. Then the polynomial may be written in NEWTON-GREGORY form:

$$p_k((j+k+s)\tau; f) = \sum_{i=0}^{k} \binom{s+i-1}{i} \nabla^i f((j+k)\tau)\,,\quad \binom{s-1}{0} = 1\,. \quad (2.65)$$

(3.1°) By integration of (2.65) over the s-interval $(-1, 0)$,

$$x_{j+k} - x_{j+k-1} = \tau \int_{-1}^{0} f((j+k+s)\tau, x(t))\, ds \simeq \tau \int_{-1}^{0} p_k((j+k+s)\tau; f)\, ds\,,$$

the implicit ADAMS methods are generated with k steps and the order $k+1$:

$$y_{j+k} - y_{j+k-1} = \tau \sum_{i=0}^{k} \gamma_i \nabla^i f_{j+k}(y_{j+k})\,,\ j = 0, 1, \ldots\,,$$

$$\gamma_i = \int_{-1}^{0} \binom{\xi+i-1}{i}\, d\xi\,.$$

(3.2°) By integration over the s-Intervall $(0, 1)$,

$$x_{j+k+1} - x_{j+k} = \tau \int_{0}^{1} f((j+k+s)\tau, x(t))\, ds \simeq \tau \int_{0}^{1} p_{k-1}((j+k+s)\tau; f)\, ds\,,$$

the explicit ADAMS methods are generated with k steps and order k:

$$y_{j+k} - y_{j+k-1} = \tau \sum_{i=0}^{k-1} \gamma_i^* \nabla^i f_{j+k-1}(y_{j+k-1}) \, , \; j = 0, 1, \ldots \, ,$$

$$\gamma_i^* = \int_0^1 \binom{\xi + i - 1}{i} \, d\xi \, .$$

The coefficients γ_i and γ_i^* do not depend on the step number k and can be computed in advance by recurrence.

(3.3°) Both methods together supply a *predictor-corrector method* for non-stiff differential equations:

In the predictor step, the explicit method (3.2°) of order k is applied once.

In the corrector step, the implicit method (3.1°) of order $k+1$ is repeatedly applied or only once since a single application suffices for order k in the combined method (Fig. 2.17).

Table 2.5. Implicit Adams methods

k	β_0	β_1	β_2	β_3	β_4	β_5
1	$\frac{1}{2}$	$\frac{1}{2}$				
2	$-\frac{1}{12}$	$\frac{8}{12}$	$\frac{5}{12}$			
3	$\frac{1}{24}$	$-\frac{5}{24}$	$\frac{19}{24}$	$\frac{9}{24}$		
4	$-\frac{19}{720}$	$\frac{106}{720}$	$-\frac{264}{720}$	$\frac{646}{720}$	$\frac{251}{720}$	
5	$\frac{27}{1440}$	$-\frac{173}{1440}$	$\frac{482}{1440}$	$-\frac{798}{1440}$	$\frac{1427}{1440}$	$\frac{475}{1440}$

Table 2.6. Explicit Adams methods

k	β_0	β_1	β_2	β_3	β_4	β_5
1	1					
2	$-\frac{1}{2}$	$\frac{3}{2}$				
3	$\frac{5}{12}$	$-\frac{16}{12}$	$\frac{23}{12}$			
4	$-\frac{9}{24}$	$\frac{37}{24}$	$-\frac{59}{24}$	$\frac{55}{24}$		
5	$\frac{251}{720}$	$-\frac{1274}{720}$	$\frac{2616}{720}$	$-\frac{2774}{720}$	$\frac{1901}{720}$	
6	$-\frac{475}{1440}$	$\frac{2877}{1440}$	$-\frac{7298}{1440}$	$\frac{9982}{1440}$	$-\frac{7923}{1440}$	$\frac{4277}{1440}$

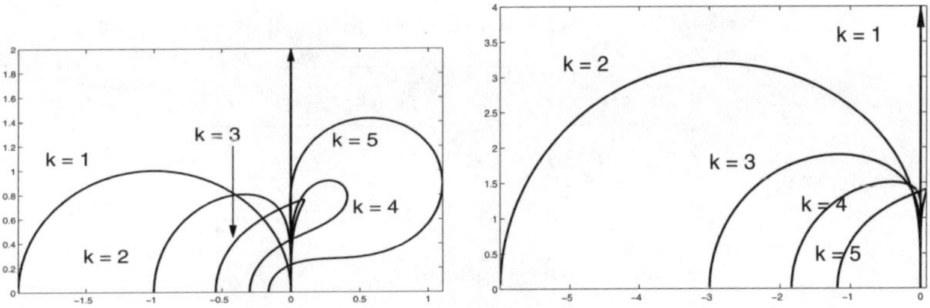

Figure 2.17. Stability regions of explicit and implicit ADAMS methods

(4°) Backward Differentiation Methods (similar methods are applied in `ode15s.m`). Because

$$\frac{d}{ds}\binom{s+i-1}{i}\bigg|_{s=0} = \begin{cases} 0 \ \text{ for } i=0 \\ \dfrac{1}{i} \ \text{ for } i \in \mathbb{N} \end{cases}$$

we obtain from (2.65) by differentiating w.r.t. the variable s

$$f'((j+k)\tau) \simeq \frac{d}{ds}p((j+k)\tau; f) = \frac{1}{\tau}\sum_{i=1}^{k}\frac{1}{i}\nabla^i f((j+k)\tau)\,.$$

In this equation the right side is known and the left side is unknown. We write $y(t) = F(t)$ instead $f(t)$ and $f(t) = F'(t)$ instead $f'(t)$ for the inversion and obtain

$$\tau f_{j+k}(y_{j+k}) = \sum_{i=1}^{k}\frac{1}{i}\nabla^i y_{j+k}\,.$$

Thereby the implicit backward differentiation methods are generated with k steps and order k:

$$\sum_{i=0}^{k}\alpha_i y_{j+i} = \tau\beta_k f_{j+k}(y_{j+k})\,,\ j = 0, 1, \ldots\,,$$

In Figure 2.18 the upper half of the stability regions consists of the *exterior domain* of the plotted curves and of the curves themselves, therefore the "point" ∞ is advantageously contained in the interior of $\mathcal{S}$. For $k > 6$ these methods do no longer fulfill the root criterium.

Table 2.7. Backward differentiation methods

k	β_k	α_0	α_1	α_2	α_3	α_4	α_5	α_6
1	1	- 1	1					
2	2	1	-4	3				
3	6	-2	9	-18	11			
4	12	3	-16	36	-48	25		
5	60	-12	75	-200	300	-300	137	
6	60	10	-72	225	-400	450	-360	147

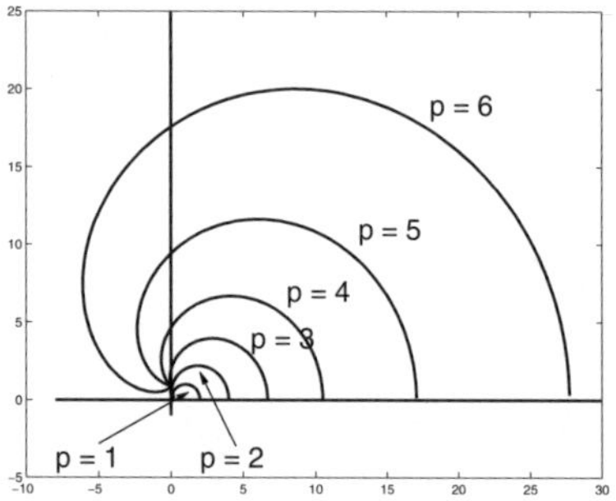

Figure 2.18. Stability regions of backward differentiation methods

(j) Full Implicit Runge-Kutta Methods have recently enjoyed new interest in connection with solving differential-algebraic equations which appear in many technical applications. The results of this subsection are already known since the pioneering work of BUTCHER, but the subsequent algebraized form of the order conditions is presumably due to (Crouzeix75) and (Crouzeix80). For detailled proofs see `SUPPLEMENT\chap02b`. Only for adaption to the notations in this subsection we make the following stipulation:

> A numerical integration rule has *order* p if it has *degree* $p-1$; cf. Sect. 2.3**(a)**, i.e., if it is exact for polynomials $p \in \Pi_{p-1}$.

Let a r-stage RUNGE-KUTTA method (RKM) of order ϱ with BUTCHER matrix $(A,\, b,\, c)$ apply to the trivial differential equation $x'(t) = f(t)$ then the exterior equation or forward step, namely

$$x(t+\tau) - x(t) = \int_t^{t+\tau} f(t)\,dt = \tau \sum_{i=1}^{r} \beta_i f(t+\gamma_i \tau) + \mathcal{O}(\tau^{\varrho+1})\,,$$

corresponds for $t = 0$, $\tau = 1$ to a *numerical integration rule*

$$\int_0^1 f(t)\,dt \sim \sum_{i=1}^{r} \beta_i f(\gamma_i)\,.$$

Insertion of the *monomials* $f(t) = t^{k-1}$ shows (by reasons of linearity) that it has order ϱ if and only if

$$\sum_{i=1}^{r} \beta_i \gamma_i^{k-1} = \frac{1}{k}\,,\ k = 1:\varrho\,. \tag{2.66}$$

In the same way, the interior equations may be considered as integration rules,

$$\int_0^{\gamma_i} f(t)\,dt \sim \sum_{i=1}^{r} \alpha_{ik} f(\gamma_k)\,,\ i = 1:r\,,$$

and they have order ϱ if and only if

$$\sum_{j=1}^{r} \alpha_{ij}\gamma_j^{k-1} = \frac{1}{k}\gamma_i^k\,,\ i = 1:r\,,\ k = 1:\varrho\,. \tag{2.67}$$

The exterior equation (integration rule) of a RKM of order ϱ has necessarily order ϱ. The *maximum common* order of the formulas (2.67) is called *interior order* of the underlying RKM. By this way, an explicit RKM has the interior order $\varrho = 1$ because the first equation has degree zero.

Following (Crouzeix75), the order conditions of implicit r-stage RKM may be described in a surprisingly simple algebraic form, but to this end we have to introduce a further condition: By partial integration we obtain the equation

$$\int_0^1 x^{k-1} \int_0^x f(s)\,ds\,dx = \frac{1}{k}\int_0^1 (1 - x^k) f(x)\,dx\,. \tag{2.68}$$

Approximating the exterior integral on *left side* by the exterior equation and the interior integrals by the interior equations of a r-stage RKM yields

$$\int_0^1 x^{k-1} \int_0^x f(s)\,ds\,dx \approx \sum_{i=1}^{r} \beta_i\gamma_i^{k-1} \int_0^{\gamma_k} f(s)\,ds \approx \sum_{i=1}^{r} \beta_i\gamma_i^{k-1} \sum_{j=1}^{r} \alpha_{kj} f(\gamma_j)\,.$$

On the other side, approximating the *right side* by the exterior equation yields

$$\frac{1}{k}\int_0^1 (1 - x^k) f(x)\,dx \approx \frac{1}{k}\sum_{i=1}^{r} \beta_i (1 - \gamma_i^k) f(\gamma_i)\,.$$

Equalizing both sides and substituting for f successively the LAGRANGE polynomials $q_i \in \Pi_{r-1}$ with $q_i(\gamma_j) = \delta^i{}_j$, $i = 1, \ldots, r$, yields finally the desired additional condition

$$\sum_{i=1}^{r} \beta_i \gamma_i^{k-1} a_{ij} = \frac{1}{k} \beta_j (1 - \gamma_j^k)\,,\ j = 1 : r\,,\ k = 1 : \varrho\,. \tag{2.69}$$

For simplicity we now introduce the following (nearly historical) abbreviations

$$\boxed{\begin{aligned} &\mathcal{A}(\varrho) :\Longleftrightarrow \text{ the RKM has (at least) order } \varrho \\ &\mathcal{B}(\varrho) :\Longleftrightarrow \text{ the exterior equation has (at least) order } \varrho \\ &\mathcal{C}(\varrho) :\Longleftrightarrow \text{ the RKM has (at least) interior order } \varrho \\ &\mathcal{D}(\varrho) :\Longleftrightarrow \text{ (2.69) holds for } k = 1, \ldots, \varrho \end{aligned}} \tag{2.70}$$

and the further notations

$$\begin{aligned} &b = [\beta_1, \beta_2, \ldots, \beta_r]^T \in \mathbb{R}^r\,,\ c = [\gamma_1, \gamma_2, \ldots, \gamma_r]^T \in \mathbb{R}^r\,,\ C = \text{diag}(c)\,, \\ &e = [1, \ldots, 1]^T \in \mathbb{R}^r\,, \qquad z_\varrho = [1,\, 1/2, \ldots,\, 1/\varrho]^T \in \mathbb{R}^\varrho\,. \end{aligned} \tag{2.71}$$

Then, by (2.66), (2.67) and (2.68), the stipulations (2.70) are equivalent to

$$\boxed{\begin{array}{lll} \mathcal{A}(\varrho) \Longleftrightarrow & \text{RKM has order } \varrho & \\ \mathcal{B}(\varrho) \Longleftrightarrow & b^T C^{k-1} e = \dfrac{1}{k}\,, & k = 1 : \varrho \\ \mathcal{C}(\varrho) \Longleftrightarrow & A C^{k-1} e = \dfrac{1}{k} C^k e\,, & k = 1 : \varrho \\ \mathcal{D}(\varrho) \Longleftrightarrow & b^T C^{k-1} A = \dfrac{1}{k}(b^T - b^T C^k)\,, & k = 1 : \varrho \end{array}}\,. \tag{2.72}$$

Theorem 2.15. (BUTCHER, CROUZEIX, EHLE) *Let a r-stage RKM be given and let all abszissas γ_i be mutually distinct. Then*
(1°) $\mathcal{B}(\varrho) \wedge \mathcal{C}(\xi) \wedge \mathcal{D}(\eta) \Longrightarrow \mathcal{A}(\min\{\varrho,\, 2\xi - 2,\, \xi + \eta + 1\})$.
(2°) $\mathcal{B}(\varrho) \wedge \mathcal{C}(r) \Longrightarrow \mathcal{D}(\varrho - r)$.
(3°) $\mathcal{B}(\varrho) \wedge \mathcal{D}(r) \Longrightarrow \mathcal{C}(\varrho - r)$ *if all weights* $\beta_i \neq 0$.

Consequently, if all abszissas γ_i are mutually distinct and the RKM has property $\mathcal{B}(\varrho)$, then $\mathcal{C}(r)$ or $\mathcal{D}(r)$ determine the crucial property $\mathcal{A}(p)$.

For a further *algebraization* of the order conditions let

$$V_\varrho = [\gamma_i^{j-1}]_{i=1}^{r}{}_{j=1}^{\varrho} = \begin{bmatrix} 1 & \gamma_1 & \gamma_1^2 & \ldots & \gamma_1^{\varrho-1} \\ \vdots & \vdots & \vdots & & \vdots \\ 1 & \gamma_r & \gamma_r^2 & \ldots & \gamma_r^{\varrho-1} \end{bmatrix} \in \mathbb{R}^r{}_\varrho$$

be the VANDERMONDE matrix. Then, using (2.71), we can write instead of (2.72) in matrix form

$$\boxed{\begin{array}{lcl} \mathcal{B}(\varrho) & \Longleftrightarrow & V_\varrho^T b = z_\varrho \\[2mm] \mathcal{C}(\varrho) & \Longleftrightarrow & AV_\varrho = \operatorname{diag}(c)V_\varrho \operatorname{diag}(z_\varrho) =: W_\varrho \in \mathbb{R}^r{}_\varrho \\[2mm] \mathcal{D}(\varrho) & \Longleftrightarrow & V_\varrho^T \operatorname{diag}(b)A = (z_\varrho e^T - W_\varrho^T) \operatorname{diag}(b) \end{array}} \; . \tag{2.73}$$

Therefore the matrix A of the RKM is uniquely determined by $\mathcal{C}(r)$ if all γ_i are different, and is determined uniquely by $\mathcal{D}(r)$ if in addition all weigths β_i are non-zero.

Corollary 2.2. *(GAUSS Methods) The maximum order of the exterior equation, i.e., of the integration rule (2.66), is $p = 2r$ for r stages by Sect. 2.3***(c)***. It is attained if one chooses the roots γ_i , $i = 1 : n$ of $p_{1,n}(x)$ in (2.38). However, $\mathcal{C}(r)$ and $\mathcal{D}(r)$ are equivalent for $\varrho = 2r$ by Theorem 2.15*(2°) *and* (3°)*. Under the above assumptions then, by* (1°),

$$\mathcal{B}(2r) \wedge \Big(\mathcal{C}(r) \vee \mathcal{D}(r)\Big) \Longrightarrow \mathcal{A}(2r).$$

Corollary 2.3. *(BUTCHER Methods) Let $\varrho \geq r$ and let all γ_i be different then, by Theorem 2.15*(1°) *and* (2°) ,

$$\mathcal{B}(\varrho) \wedge \mathcal{C}(r) \Longrightarrow \mathcal{A}(p)\,,\ p = \min\{\varrho, 2r+2, r+\varrho-r+1\} = \varrho\,.$$

For a fixed vector c of abszissas one obtains the BUTCHER *methods of order $\varrho \geq r$ by (2.71) choosing*

$$(A, b, c) = \big(W_r V_r^{-1},\, V_r^{-1} z_r\,,\, c\big)\,.$$

Corollary 2.4. *(EHLE Methods) Let $\varrho \geq 2r-2$, let all γ_i be different and all β_i non-zero, then by Theorem 2.15* (1°) *and* (3°)

$$\mathcal{B}(\varrho) \wedge \mathcal{D}(r) \Longrightarrow \mathcal{A}(p)\,,$$
$$p = \min\{\varrho, 2(\varrho - r)+2, \varrho - r + r + 1\} = \min\{\varrho, 2(\varrho - r)+2\} = \varrho\,.$$

For a fixed vector c of abszissas one obtains the EHLE *methods of order $\varrho \geq 2r-2$ by (2.71) choosing*

$$(A, b, c) = \big(\operatorname{diag}(b)^{-1} V_r^{-T}(z_r e^T - W_r^T)\operatorname{diag}(b)\,,\, V_r^{-T} z_r\,,\, c\big)\,.$$

The roots of the polynomials $p \in \Pi_r$ of (2.38) are chosen for components of the vector c:

$$\begin{aligned}
&\text{GAUSS methods:} && \varrho = 2r, && p_{1,r}(x) && = \left[x^r(1-x)^r\right]^{(r)}\\
&\text{methods of type I:} && \varrho = 2r-1, && x p_{2,r-1}(x) && = \left[x^r(1-x)^{r-1}\right]^{(r-1)}\\
&\text{methods of type II:} && \varrho = 2r-1, && (1-x)p_{3,r-1} && = \left[x^{r-1}(1-x)^r\right]^{(r-1)}\\
&\text{methods of type III:} && \varrho = 2r-2, && x(1-x)p_{4,r-2} && = \left[x^{r-1}(1-x)^{r-1}\right]^{(r-2)} .
\end{aligned} \tag{2.74}$$

The results are summarized in the following table:

Table 2.8.

Type	cond. for γ_i	order	BUTCHER		EHLE	CHIPMAN
GAUSS	(2.72)(1°)	$2r$	$\otimes^A$	$=$	$\otimes^A$	—
RADAU I B/A	(2.72)(2°)	$2r-1$	$\otimes$		$\otimes^{A,L}$	—
RADAU II A/B	(2.72)(3°)	$2r-1$	$\otimes^{A,L}$		$\otimes$	—
LOBATTO III A/B/C	(2.72)(4°)	$2r-2$	$\otimes^A$		$\otimes^A$	$\otimes^{A,L}$

The CHIPMAN methods have the properties $\mathcal{C}(r)$ and $Ae_1 = \beta_1 e_1$, $e_1 = [\delta^1{}_k]_{k=1}^r$ which determine uniquely the matrix A. A-stable methods are marked with $\otimes^A$, L-stable methods with an additional index L.

Example 2.14. The methods of type **Radau II A** have $\gamma_n = 1$ hence the exterior equation and the last row of A are identical. This property has advantages in application to *differential-algebraic problems.* The 3-stage method of order $p = 5$ is A-stable and L-stable, its data are given in the following BUTCHER matrix:

$$\left[\begin{array}{c|c} A & c \\ \hline b & \end{array}\right] = \left[\begin{array}{ccc|c}
\frac{88-7\sqrt{6}}{360} & \frac{296-169\sqrt{6}}{1800} & \frac{-2+3\sqrt{6}}{225} & \frac{4-\sqrt{6}}{10}\\
\frac{296+169\sqrt{6}}{1800} & \frac{88+7\sqrt{6}}{360} & \frac{-2-3\sqrt{6}}{225} & \frac{4+\sqrt{6}}{10}\\
\frac{16-\sqrt{6}}{36} & \frac{16+\sqrt{6}}{36} & \frac{1}{9} & 1\\ \hline
\frac{16-\sqrt{6}}{36} & \frac{16+\sqrt{6}}{36} & \frac{1}{9} &
\end{array}\right] .$$

References: (Hairer), (Shampine97).

2.5 Boundary Value Problems

We look for a solution $x : [0,\, 1] \to \mathbb{R}^n$ of the boundary value problem

$$x'(t) = f(t, x(t)) ,\ 0 \le t \le 1 ,\ g(x(0) ,\, x(1)) = 0 \in \mathbb{R}^n , \tag{2.75}$$

confining ourselves to the unit interval by optical reasons and for simple implementation later on. In the other case as, e.g., in periodic problems, a rescaling becomes necessary again: For a problem

$$u'(s) = h(s, u(s))\,,\ 0 \le s \le T\,,\ g(u(0)\,,\, u(T)) = 0 \in \mathbb{R}^n\,,$$

a substitution of $s = T\,t$ yields

$$x'(t) = T\,h(T\,t, x(t))\,,\ 0 \le t \le 1\,,\ g(x(0)\,,\, x(1)) = 0 \in \mathbb{R}^n$$

and the additional factor T has always to be regarded.

(a) The Linear Problem reads:

$$x'(t) = A(t)x(t) + c(t),\ 0 \le t \le 1\,,\ R_0 x(0) + R_1 x(1) = d \in \mathbb{R}^n\,. \tag{2.76}$$

We choose a uniform partition of the basic t-interval for simplicity,

$$0 = t_1 < t_2 < \ldots < t_m < t_{m+1} = 1\,,\ t_j = (j-1)\tau\,,\ \tau = 1/m\,,$$

beginning with index $j = 1$ w.r.t. the compatibility with MATLAB numeration. Suppose that a *numerical approach* to the differential system on a individual t-interval $[t_j,\, t_{j+1}]$ has the form

$$\boxed{P_j y_j + Q_j y_{j+1} = r_j\,,}$$

then we obtain altogether a large linear system of equations

$$\mathbf{L}(\tau)Y := \begin{bmatrix} P_1 & Q_1 & 0 & \ldots & 0 & 0 \\ 0 & P_2 & Q_2 & \ddots & \ddots & 0 \\ \vdots & \ddots & \ddots & \ddots & \ddots & \vdots \\ \vdots & \ddots & \ddots & \ddots & \ddots & 0 \\ 0 & \ddots & \ddots & \ddots & P_m & Q_m \\ R_0 & 0 & \cdots & \cdots & 0 & R_1 \end{bmatrix} \begin{bmatrix} y_1 \\ \vdots \\ \vdots \\ \vdots \\ \vdots \\ y_{m+1} \end{bmatrix} = \begin{bmatrix} r_1 \\ \vdots \\ \vdots \\ \vdots \\ r_m \\ d \end{bmatrix} =: R \tag{2.77}$$

for the unknown values y_j ; and the matrix $\mathbf{L}(\tau)$ must be regular.

Example 2.15. (1°) *Trapezoidal rule*

$$y_{j+1} - y_j - \frac{\tau}{2}\left[A_{j+1}y_{j+1} + A_j y_j\right] = \frac{\tau}{2}\left(c_{j+1} + c_j\right)\,,$$

$$P_j = -I - \frac{\tau}{2}A_j\,,\ Q_j = I - \frac{\tau}{2}A_{j+1}\,,\ r_j = \frac{\tau}{2}\left(c_j + c_{j+1}\right)\,.$$

(2°) *Box scheme*

$$y_{j+1} - y_j - \frac{\tau}{2}A_{j+1/2}\left(y_{j+1} + y_j\right) = \tau\, c_{j+1/2}\,,$$

$$P_j = -I - \frac{\tau}{2}A_{j+1/2}\,,\; Q_j = I - \frac{\tau}{2}A_{j+1/2}\,,\; r_j = \tau\, c_{j+1/2}\,, \quad \tau = 1/m\,.$$

(3°) *Multiple shooting method*
(3.1°) Solve, for $j = 1, \ldots, m$, the inhomogenous problem with homogenous initial condition

$$x'(t) = A(t)x(t) + c(t)\,,\; t_j \le t \le t_{j+1}\,,\; y(t_j) = 0\,;$$

and denote the solution at point t_{j+1} by r_j.
(3.2°) Solve, for $j = 1, \ldots, m$, the n homogenous initial value problems with inhomogenous initial condition

$$X'(t) = A(t)X(t)\,,\; t_j \le t \le t_{j+1}\,,\; X(t_j) = I \text{ (unit matrix)}\,;$$

and let the solution at point t_{j+1} be the *matrix* V_j. Then

$$y_{j+1} = r_j + V_j y_j \implies y_{j+1} - V_j y_j = r_j\,, \implies \boxed{P_j = -V_j\,,\; Q_j = I}\,.$$

(b) In **nonlinear case** a nonlinear system of equations is produced in the same way and is solved by NEWTON's method. We confine ourselves to the multiple shooting method and apply the flux integral $\Phi(t; t_0, x_0)$ of Sect. 1.6. The numerical solution is denoted by y again.

Multiple shooting method:

(1°) Choose a moderate number m of shooting points in intervall $[0,\, 1]$,

$$[(t_1, y_1), \ldots, (t_m, y_m)]\,,\; y_j \in \mathbb{R}^n\,,\; t_1 = 0\,,\; t_{m+1} = 1\,.$$

(2°) Compute

$$\Phi(t_{j+1}; t_j, y_j) := y_j + \int_{t_j}^{t_{j+1}} f(t, x(t))\, dt\,,\; j = 1 : m\,,$$

by solving the initial value problems

$$x'(t) = f(t, x(t))\;\; t_j \le t \le t_{j+1}\,,\; x(t_j) = y_j\,. \tag{2.78}$$

(3°) Solve the system

$$y_{j+1} - \Phi(t_{j+1}; t_j, y_j) = 0\,,\; j = 1 : m\,,\; g(y_1, y_{m+1}) = 0\,, \tag{2.79}$$

by NEWTON's method. The nonlinear system of equations (2.79) has the form

$$\mathbf{F}(Y) = 0 \in \mathbb{R}^{n(m+1)}\,,\text{ with node vector } Y = [y_1, \ldots, y_{m+1}]^T\,. \tag{2.80}$$

By solving this system with NEWTON's method, a computation of grad $\mathbf{F}(Y)$ becomes necessary which needs the gradients of Φ at the points (t_j, y_j),

$$\text{grad}_v\, \Phi(t_{j+1}; t_j, v) = I + \int_{t_j}^{t_{j+1}} \text{grad}\, f(t, x(t))\, \text{grad}_v\, \Phi(t; t_j, v)\, dt\,. \tag{2.81}$$

The vector field of this matrix-valued flux integral reads:

$$W'(t) = \mathrm{grad}\, f(t, x(t)W(t)\,,\; W(t)) \in \mathbb{R}^n{}_n\,,$$

and the initial condition for (2.81) is $W(t_j) = I$. Accordingly, in interval $[t_j, t_{j+1}]$, we have to solve again n initial value problems of the form

$$w_k'(t) = \mathrm{grad}_x\, f(t, x(t))w_k(t) \in \mathbb{R}^n\,,\; t_j \le t \le t_{j+1}\,,\; w_k(t_j) = e_k\,,\; k = 1:n\,, \tag{2.82}$$

where $x(t)$ plays the role of a parameter and $e_k \in \mathbb{R}^n$ is k-th unit vector.

The *simultaneous* solution of all $n + 1$ initial value problems (2.78) and (2.81) in every interval $[t_j\,,\, t_{j+1}]$ is essential for success of the method. Then the matrix $\mathrm{grad}\,\mathbf{F}(Y)$ has the same form as the matrix $\mathbf{L}(\tau)$ in (2.77) where

$$\boxed{P_j = -\,\mathrm{grad}_y\, \Phi(t_{j+1}; t_j, y_j)\,,\;\; Q_j = I}\,.$$

The method develops its full power only if the shooting points are chosen properly adapted to the individual problem. Moreover, the NEWTON method must be globalized by a suitable step control. Also, the starting values cannot be chosen arbitrarily but, in simple cases, a linear function respecting the boundary conditions may be sufficient for convergence.

Example 2.16. (Stoer)

$$x_1' = x_2\,,\;\; x_2' = 5\sinh(5\,x_1)\,,\;\; x_1(0) = 0\,,\; x_1(1) = 1\,.$$

For starting trajectory we choose the straight line connecting the boundary points $(0,\, x_1(0))$ and $(1, x_1(1))$. Observe however that $\lim_{t\to 1.0326...} x_1(t) = \infty$, therefore the initially chosen uniform partition of the interval [0, 1] must be adapted to the problem.

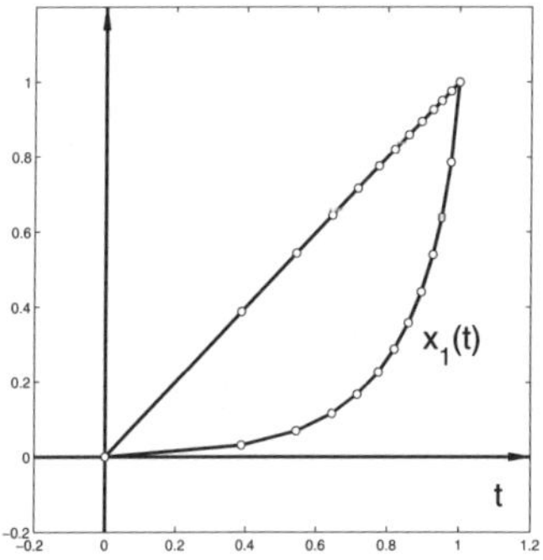

Figure 2.19. Ex. 2.16

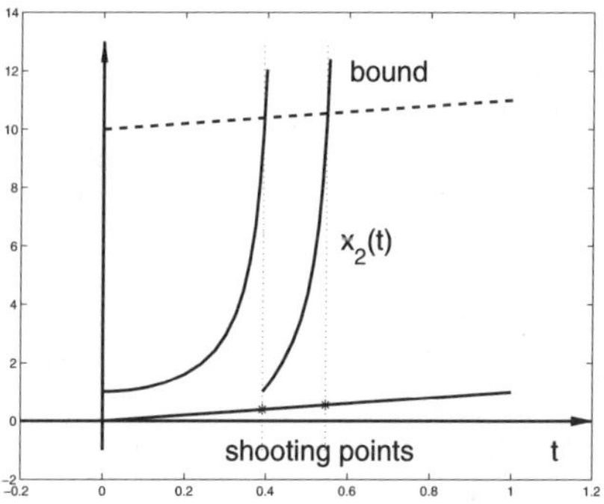

Figure 2.20. Ex. 2.16, adaption

In KAPITEL04\CONTROL02 some benchmark problems of control theory are solved by NEWTON's method and box scheme.

(c) Boundary Value Problems with Parameter We look for a solution $x(\,\cdot\,,\alpha) : [0\,,\,1] \to \mathbb{R}^n$ of the boundary value problem

$$x'(t;\alpha) = f(t, x(t;\alpha);\alpha)\,,\ 0 \le t \le 1\,,\ g(x(0;\alpha)\,,\,x(1;\alpha);\alpha) = 0 \in \mathbb{R}^n\,, \tag{2.83}$$

where the real parameter α may vary in some interval. Now, the problem has no longer a unique solution but an additional degree of freedom and therefore the numerical solution depends strongly on the chosen initial approximation of x and α which must be given rather accurately.

To apply the multiple shooting method again, let

$$\begin{aligned} \Phi(t;t_0,x_0,\alpha) &= x_0 + \int_{t_0}^{t} f(t,x(t;\alpha);\alpha)\,dt \\ \Phi(t_{j+1};t_j,y_j(\alpha),\alpha) &= y_j(\alpha) + \int_{t_j}^{t_{j+1}} f(t,x(t;\alpha);\alpha)\,dt \end{aligned}$$

be the flux integral belonging to (2.83). Solving the system (2.80) being now of the form

$$\mathbf{F}(V) = 0 \in \mathbb{R}^{n(m+1)+1}\,,\quad V = [y_1,\ldots,y_{m+1};\alpha]^T \text{ node vector}\,, \tag{2.84}$$

by NEWTON's method, one needs the additional derivative

$$\begin{aligned} H_j := \frac{\partial}{\partial\alpha}\Phi(t_{j+1};t_j,y_j(\alpha),\alpha) &= \frac{\partial}{\partial\alpha}y_j(\alpha) + \int_{t_j}^{t_{j+1}} \frac{\partial}{\partial\alpha}f(t,x(t;\alpha);\alpha)\,ds \\ &+ \int_{t_j}^{t_{j+1}} \operatorname{grad}_x f(t,x(t;\alpha);\alpha)\frac{\partial}{\partial\alpha}x(t;\alpha)\,dt\,. \end{aligned}$$

Therefore the additional initial value problem

$$\begin{aligned} v_j'(t) &= \frac{\partial}{\partial\alpha}f(t,x(t;\alpha);\alpha) + \operatorname{grad}_x f(t,x(t;\alpha);\alpha)v_j(s)\,, \\ v_j(t_j) &= \frac{\partial}{\partial\alpha}(y_j)(\alpha) \in \mathbb{R}^n\,,\ v_1(t_1) = 0\,, \end{aligned} \tag{2.85}$$

has to be solved in every interval $[t_j,\,t_{j+1}]$. Note that all $n+2$ initial value problems (2.78), (2.82) and (2.85) must be solved *simultaneously* again. Note also that during the entire iteration not only the numerical approximations V are to be calculated but als the partial derivatives $y_{\alpha,j}$, $j = 1 : m+1$ w.r.t. the parameter α else the method may fail.

The JACOBI matrix $\operatorname{grad}\mathbf{F}(V)$ is now a $(m\cdot n, m\cdot n+1)$-matrix in block form

$$\mathbf{L} := \begin{bmatrix} P_1 & Q_1 & 0 & \dots & 0 & 0 & H_1 \\ 0 & P_2 & Q_2 & \ddots & \ddots & 0 & H_2 \\ \vdots & \ddots & \ddots & \ddots & \ddots & \vdots & \vdots \\ \vdots & & \ddots & \ddots & \ddots & 0 & H_{m-1} \\ 0 & & \ddots & \ddots & P_m & Q_m & H_m \\ g_{x_0} & 0 & \cdots & \cdots & 0 & g_{x_1} & g_\alpha \end{bmatrix}, \tag{2.86}$$

therefore the MOORE-PENROSE inverse $[\operatorname{grad}\mathbf{F}(V)]^+$ must be used in the GAUSS-NEWTON method, cf. Sect. 1.1**(g)**. An underdetermined linear system of equations

$$[\operatorname{grad}\mathbf{F}(V_j)](V_{j+1} - V_j) = -\mathbf{F}(V_j)$$

is to be solved in every NEWTON step for which the algorithm in Sect. 1.1**(h3)** can be applied. Also, as already mentioned, a good initial approximation V_0 must be known to prevent a convergence to the trivial solution. In rather simple methods also the box scheme may be modified in a suitable way.

2.6 Periodic Problems

(a) Problems with Known Period We seek a T-periodic solution $x : \mathbb{R} \to \mathbb{R}^n$ of the boundary value problem

$$x'(t) = f(t, x(t)),\ 0 \le t \le T,\ x(0) = x(T) \in \mathbb{R}^n. \tag{2.87}$$

If $x(t)$ is a solution, also $x(t+\alpha)$ is a T-periodic solution here for arbitrary $\alpha \in \mathbb{R}$. Therefore an additional *phase condition* must be introduced to ensure uniqueness, but the problem remains nevertheless numerically unstable.

Some possible phase conditions are

(1°) $p(x(0)) := x_k - \eta = 0,\ \eta \neq 0;$

(2°) $p(x(0)) := f_k(x(0)) = 0 \implies x_k'(0) = 0.$

But then we have $n+1$ boundary conditions for n unknown functions. If a point on the unknown orbit is known then satisfying results may be obtained also by a good solver for initial value problems.

(b) In **problems with unknown period** T, a transformation to a parameter-dependent problem with known period suggests itself. For instance, the solution $\widetilde{x}$ of

$$\widetilde{x}'(s) = T\, f(Ts, \widetilde{x}(s)),\ \widetilde{x}(0) = \widetilde{x}(1) \tag{2.88}$$

has period one in s and $x(t) = \widetilde{x}(t/T)$ is a solution of (2.87) with period T. The further treatment of the problem is carried out as in Sect. 2.5**(c)** for parameter-dependent problems. The relatively simple box scheme however cannot be applied here because of the boundary condition (2.88) and the results of Sect. 2.5**(a)**. In the subsequent examples the *multiple shooting method* is used with *fixed* partition of the underlying t-interval and the solution of an initial value problem with *estimated* initial value for starting trajectory. In particular, this method may serve for final adjustment of periodic solutions.

Example 2.17. Nerve membrane model (Deuflhard84) (Fig. 2.21).

$$\begin{aligned} \dot{u}_1 &= 3(u_2 + u_1 - \frac{1}{3}u_1^3 + \lambda) \\ \dot{u}_2 &= -\frac{1}{3}(u_1 - 0.7 + 0.8u_2)\,. \end{aligned}$$

Transformation to a parameter-dependent problem with period one:

$$\begin{aligned} x_1' &= 3T(x_2 + x_1 - \frac{1}{3}x_1^3 + \lambda) \\ x_2' &= -\frac{T}{3}(x_1 - 0.7 + 0.8x_2)\,. \end{aligned}$$

The initial value problem (2.85) reads:

$$\begin{aligned} v_1' &= 3(x_2 + x_1 - \frac{1}{3}x_1^3 + \lambda) + 3T(1 - x_1^2)v_1 + 3Tv_2 \\ v_2' &= -\frac{1}{3}(x_1 - 0.7 + 0.8x_2) - \frac{T}{3}v_1 - \frac{0.8T}{3}v_2\,. \end{aligned}$$

Test problem : $\lambda = -1$, starting value $(x_1^0, x_2^0, T^0) = (3, 1.5, 12)$.

Example 2.18. Heated flow problem (Deuflhard84) (Fig. 2.22).

$$\dot{u}_1 = -\sigma(u_1 - u_2)\,, \quad \dot{u}_2 = u_1(r - u_3) - u_2\,, \quad \dot{u}_3 = u_1u_2 - bu_3\,.$$

Transformation to a parameter-dependent problem with period one:

$$x_1' = -\sigma T(x_1 - x_2)\,, \quad x_2' = T\,[x_1(r - x_3) - x_2]\,, \quad x_3' = T\,(x_1x_2 - b\,x_3)\,.$$

The initial value problem (2.85) reads:

$$\begin{aligned} v_1' &= -\sigma(x_1 - x_2) && - \sigma T(v_1 + v_2) \\ v_2' &= x_1(r - x_3) - x_2 && + T[(r - x_3)v_1 - v_2 - x_1v_3] \\ v_3' &= x_1x_2 - b\,x_3 + x_2v_1 && + T(x_1v_2 - b\,v_3)\,. \end{aligned}$$

Test problem : $\sigma = 16$, $b = 4$, $r = 153.083$, starting values $(x_1^0, x_2^0, x_3^0, T^0) = (0, -28, 140, 0.95)$.

Example 2.19. Arenstorf orbits (Arenstorf). In the degenerated three-body problem, three bodies (earth, moon, satellite) are given with masses m_1, m_2 und $m_3 = 0$, and with the following simplifications (Fig. 2.23):

(1°) Earth, moon, satellite move in a plane.
(2°) The distance earth-moon is constant and set to one.
(3°) The influence of the remaining celestial bodies is neglected.

The straight line between earth and moon is chosen for x-axis with common gravity center for origin. Furthermore, $\mu = m_2/(m_1+m_2) \sim 1/81.45$ is the relative moon mass, $\mu' = 1-\mu$. Then a system of *two* differential equations is obtained for the motion of the mass-free body in the rotating frame; see Sect. 6.5 **(b)**. For a transformation of the T-periodic problem into a problem with unit period, it is referred to the appertaining MATLAB program.

Example 2.20. Nonlinear oscillators occur in many technical applications. Forced DUFFING equations

$$\ddot{u} + \alpha\dot{u} + \beta u + \gamma u^3 = \delta\cos(\omega t)\,.$$

are a standard model problem in investigation of period doubling, transition to chaos, and bifurcation (in homogenous case); see e.g. (Seydel94). We look here for *harmonic* solutions which have the same period $T = 2\pi/\omega$ as the excitation (else we are led to *strange attractors*). Transformation to a parameter-dependent system with period one by substitution of $t = Ts\,,\ T = 2\pi/\omega\,,$ yields as above with $y_1(s) = u(t)$

$$\begin{aligned} y_1' &= Ty_2 \\ y_2' &= -T\big(\alpha y_2 + \beta y_1 + \gamma y_1^3 - \delta\cos(2\pi s)\big)\,. \end{aligned}$$

In Fig. 2.24 we have $\alpha = 0.2\,,\ \beta = 0\,,\ \gamma = 1\,,$ and at beginning $\delta = 5$. At first an initial value problem is solved to get a start trajectory, then a simple continuation is chosen up to $\delta = 7$. Initial guess of period = 12, final period = 10.2209.

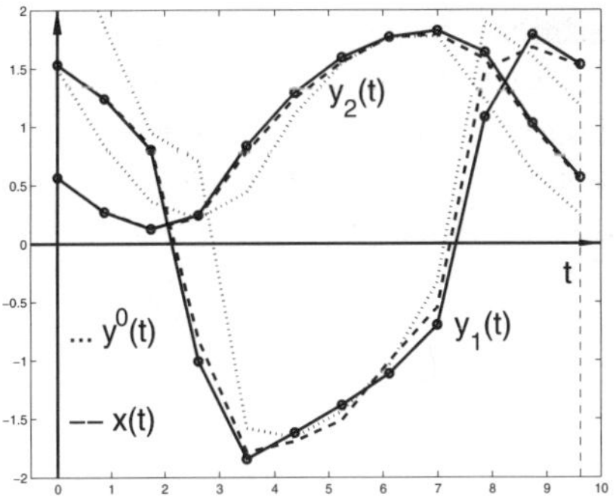

Figure 2.21. Example 2.17

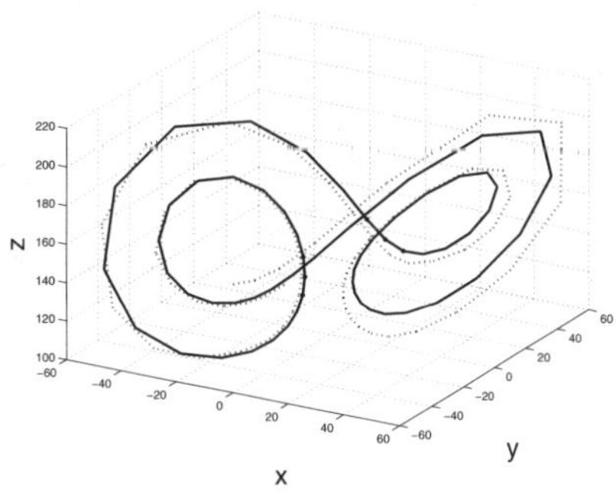

Figure 2.22. Example 2.18

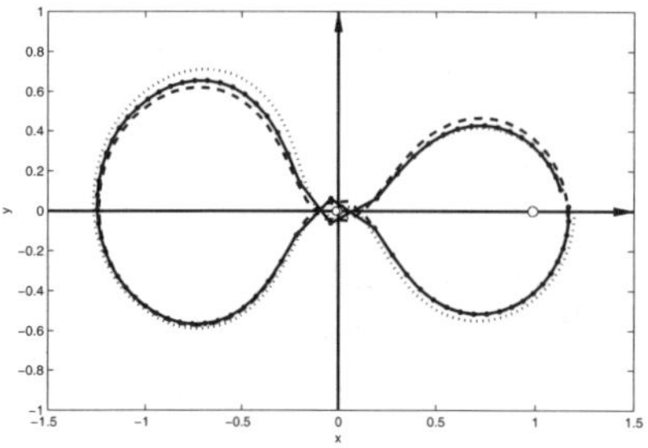

Figure 2.23. Example 2.19

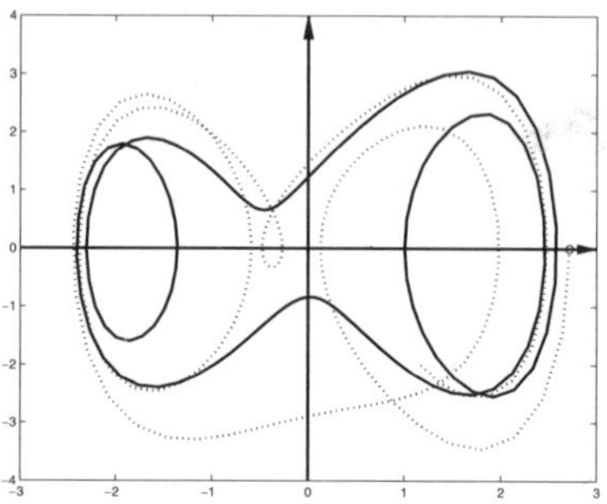

Figure 2.24. Example 2.20

2.7 Differential-Algebraic Problems

Extremal problems are solved in mechanics by their associated variational problem (Euler equations), and frequently a formulation of the extremal function or objective function itself is relinquished at all; cf. Sect. 4.1. Possible equality or inequality restrictions are taken into the objective function via Lagrange multiplicators as far as possible; indeed the entire Lagrange theory has originated in mechanics of mass points. But the side conditions have nevertheless to be regarded and thus a more or less (rather more) complicated system of differential equations, analytic equations (here apostrophized as "algebraic") and perhaps also inequalities is waiting for numerical approach. Ultimately one is faced with a highly nonlinear *boundary value problem* or a equally nonlinear *initial value problem* of which the solution must satisfy additional restrictions; cf. Chap. 3. Numerical devices for solving such families of problems need a *consistent* start trajectory of which the calculation is often more difficult than the remaining computation. If however the problem is transformed artificially into a control problem, then modern numerical methods can be applied working with (rather) *arbitrary* initial trajectory; cf. Sect. 4.4. If, on the other side, the differential-algebraic problem is a *pure initial value problem* then also special Runge-Kutta methods may be applied. Some of these methods, having been developed in more recent time, shall be considered in the present section. The problem of consistent initial values appears here, too, but is a nonlinear system of equations being solved by the usual methods in the generic case. For some practical applications of these methods we refer to Sect. 11.3 on *multibody problems.*

In this subsection let (x, y) be the theoretical solution and (u, v) its numerical approximation.

(a) Formulation of the Problem At first we consider a *singular* initial value problem in separated form

$$\begin{aligned} x'(t) &= f(t,x(t),y(t)) \in \mathbb{R}^n\,, \quad (x(0),y(0)) = (x_0,\, y_0)\,, \\ \varepsilon\, y'(t) &= g(t,x(t),y(t)) \in \mathbb{R}^m\,, \quad 0 \le \varepsilon \ll 1 \end{aligned} \tag{2.89}$$

depending on the parameter ε. Writing $z(t) = [x(t),\, y(t)]^T$, the system is equivalent to

$$Mz'(t) = F(t,z(t)) \in \mathbb{R}^{n+m}\,, \quad z(t) = [x(t),\, y(t)]^T \tag{2.90}$$

where the matrix $M \in \mathbb{R}^{n+m}{}_{n+m}$ is *singular* for $\varepsilon = 0$. Problems of the form $M(x(t))x'(t) = F(t,x(t))$ are transformed by preference in a system

$$x' = y\,, \quad M(x)y - F(x) = 0\,. \tag{2.91}$$

Assumption 2.2. (1°) *The problem (2.89) has a unique solution in* $[0,\, T]$, $0 < T$.
(2°) *The gradient* $\nabla_y g(x,y)$ *is* regular *near the solution* (x,y) *for* $\varepsilon = 0$.

The problem is said to be a *differential-algebraic problem* (DA problem) in the case where $\varepsilon = 0$. In this case the initial values (x_0, y_0) must be *consistent*, i.e., they must satisfy the side condition $g(x_0, y_0) = 0$. Also in the *numerical solution* such initial values must be known or calculated first, at least approximatively. The DA-problem is said to have *index 1* if assumption 2.2 (2°) does hold. Then the function g is invertible w.r.t. y near the solution, $y(t) = G(t,x(t))$, and one obtains by substitution at least *theoretically* an ordinary initial value problem with the differential system $x'(t) = f(t,x(t),G(t,x(t)))$.

Example 2.21. VAN DER POL's equation.

The *linear oscillator* $\ddot{x} + \alpha\dot{x} + x = 0$ is damped for $\alpha > 0$ and unstable for $\alpha < 0$. If the parameter α is replaced by $\mu(x^2 - 1)$, $\mu > 0$, then large $|x(t)|$-values lead to a damping and small $|x(t)|$ to an amplification. Transformation in a system of first order yields

$$\dot{x}_1 = x_2\,, \quad \dot{x}_2 = \mu\,(1 - x_1^2)x_2 - x_1\,;$$

If we now write $x_1 = y_1$, $y_2 = \mu\, x_2$, $s = t/\mu$ and ensuing $\mu^2 = 1/\varepsilon$ then we obtain after re-notation

$$\dot{x} = y\,, \; \varepsilon\,\dot{y} = (1 - x^2)y - x\,.$$

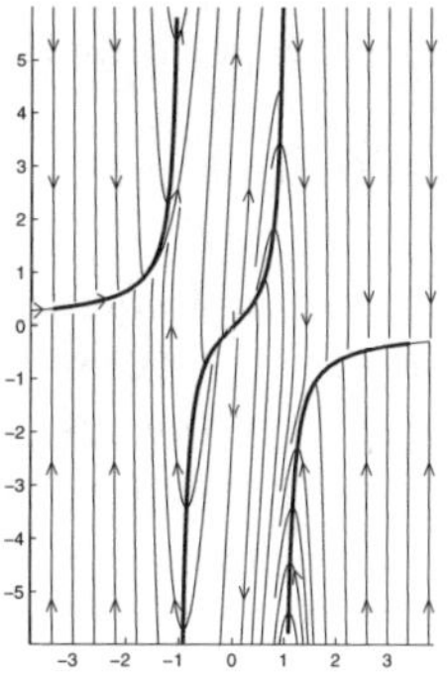

Figure 2.25. VAN DER POL's equation

In Figure 2.25 the phase portrait is plotted for $\varepsilon = 0.05$ and the curve $(1 - x^2)y - x = 0$ is marked boldface.

(b) DA-problems are mainly solved by special **Runge-Kutta methods** but also multistep methods may be applied, in particular backward differentiation methods as dealed with in Sect. 2.4**(i)**(4°). If $\varepsilon > 0$ for the present then, with the notations of Sect. 2.4**(d)**, we obtain as *common* method for the *separated* equations

$$\begin{aligned} U(t) &= e \times u(t) + \tau(A \times I)F(t, U(t), V(t)) &\in \mathbb{R}^{r \cdot n} \\ \varepsilon V(t) &= \varepsilon\, e \times v(t) + \tau(A \times I)G(t, U(t), V(t)) &\in \mathbb{R}^{r \cdot m} \\ u(t+\tau) &= u(t) + \tau(b \times I)^T F(t, U(t), V(t)) &\in \mathbb{R}^{n} \\ \varepsilon\, v(t+\tau) &= \varepsilon\, v(t) + \tau(b \times I)^T G(t, U(t), V(t)) &\in \mathbb{R}^{m}\,; \end{aligned} \tag{2.92}$$

where $U(t)$ and $V(t)$ are the vectors at the intermediate stages. If now the matrix A is *regular* then the second equation can be written as

$$\tau\, G(t, U(t), V(t)) = \varepsilon\,(A^{-1} \times I)[V(t)) - (e \times v(t))]\,,$$

and, by substitution of the last equation, the parameter ε may be *canceled.* By this way one obtains a *direct approximation* of the DA-problem by a RUNGE-KUTTA method:

$$\begin{aligned} U(t) &= e \times u(t) + \tau(A \times I)F(t, U(t), V(t)) &\in \mathbb{R}^{r \cdot n} \\ 0 &= G(t, U(t), V(t)) &\in \mathbb{R}^{r \cdot m} \\ u(t+\tau) &= u(t) + \tau(b \times I)^T F(t, U(t), V(t)) &\in \mathbb{R}^{n} \\ v(t+\tau) &= (1 - b^T A^{-1} e)v(t) + (b \times I)^T (A^{-1} \times I)V(t) &\in \mathbb{R}^{m}\,, \end{aligned} \tag{2.93}$$

and the stability function satisfies $R(\infty) = 1 - b^T A^{-1} e$; cf. (2.58). However, the algebraic side condition $g(u, v) = 0$ is fulfilled only approximatively in methods of this type (in normal case). This disadvantage is removed if the last equation in (2.93) is replaced by requiring $g(u_{n+1}, v_{n+1}) = 0$. The resulting *indirect* type of methods constitutes an approximation of $x' = f(x, G(x))$ in systems of index 1. If however, besides the regularity of A, also the last row of A is the same as the vector b of weights in the exterior equation (*stiffly accurate methods*) then $g(u_{n+1}, v_{n+1}) = 0$ is fulfilled automatically because of the second equation in (2.93), and the last equation can be dropped. For instance the RUNG-KUTTA methods of type RADAU II A described in Sect. 2.4**(j)** have the just mentioned additional property and thus are suited in a particular way for solving DA-problems.

Let us now apply a RUNGE-KUTTA method to a differential system $M\,x' = f(t, x)$ with *regular* matrix M, then we obtain the computational device

$$\begin{aligned} (I \times M)\big(U(t) - e \times u(t)\big) &= \tau(A \times I)F(t, U(t)) \in \mathbb{R}^{r \cdot n} \\ u(t+\tau) &= (1 - b^T A^{-1} e)u(t) + (b \times I)^T (A^{-1} \times I)U(t) \in \mathbb{R}^{m}\,, \end{aligned} \tag{2.94}$$

in the same way as in the transition of (2.92) to (2.93) and this device works also in a *singular* matrix M. But then the method depends on the condition of the matrix $I \times M - \tau(A \times I)$ hence in particular of the step length τ.

(c) Regular Matrix Pencils Let $(\lambda,\, u)$ be a characteristic pair of the *generalized eigenvalue problem*

$$(A + \lambda\, B)u = 0\,, \quad A, B \in \mathbb{R}^n{}_n\,.$$

Then $x(t) = e^{\lambda t} u$ is a solution of the differential system

$$Bx' + Ax = c(t) \in \mathbb{R}^n \tag{2.95}$$

for $c(t) \equiv 0$. If here, e.g., $A = B$ and $\det(A) = 0$ then $A + \lambda B$ is singular for *all* $\lambda \in \mathbb{R}$, therefore it is tacitly assumed in linear systems (2.95) that the *matrix pencil* $A + \lambda\, B$ is *regular* such that the associated generalized eigenvalue problem has a *finite* number of nonzero eigenvalues.

Theorem 2.16. *(*WEIERSTRASS, KRONECKER*) Let $A + \lambda\, B$ be a regular matrix pencil then there exist regular matrices P, Q such that*

$$PAQ = \begin{bmatrix} C & 0 \\ 0 & I \end{bmatrix}, \quad PBQ = \begin{bmatrix} I & 0 \\ 0 & T \end{bmatrix}. \tag{2.96}$$

The matrix $T = \operatorname{diag}(T_1, \ldots, T_k)$ is a block diagonal matrix with blocks $T_i \in \mathbb{R}^{n_i}{}_{n_i}$ of the form described in Sect. 1.1(**c3**)*, and $n_1 + \ldots + n_k = n$.*

Proof see e.g. (Hairer), vol. II, Sect. 6.5.

Let us now multiply (2.95) by P and use the partition

$$\begin{bmatrix} y \\ z \end{bmatrix} = Q^{-1}x\,, \begin{bmatrix} f \\ g \end{bmatrix} = Pc(t)\,,$$

then we obtain two separated systems for for y and z, namely

$$y' = Cy + f(t)\,, \quad Tz' + z = g(t)\,. \tag{2.97}$$

(d) Differential Index The second system in (2.97) has to be solved by recurrence. If for instance $k = 1$ in Theorem 2.16 and $T = T_1 \in \mathbb{R}^m{}_m$ then one starts out from the last row $z_m = g_m(t) \in \mathbb{R}$ and then has to calculate successively the components $z_i(t)$, $i = m - 1 : 1$, by $z_i(t) = g_i(t) - z_{i+1}^{(i+1)}(t)$ for which one needs the derivatives $g_m^{(m)}, \ldots, g_2^{(1)}$. Also, with these derivatives, the system $Tz' + z = g(t)$ can be written as *explicit* system,

$$z_{i+1}^{(i+1)} + z_i^{(i)} = g_i^{(i)}(t)\,, \quad i = 1, \ldots, m-1\,, \quad z_m^{(m)} = g_m^{(m)}(t)\,. \tag{2.98}$$

In general, the *differential index* is the number of derivatives being necessary to transform the implicit differential system $F(t, x'(t), x(t)) = 0$ *analytically* into an explicit system. The explicit system $x'(t) = f(t, x(t))$ has index zero by definition, and, e.g., the system (2.98) has index m because m derivatives of g are necessary.

System with index 1. Let the matrix $\nabla_y g(x,y)$ be regular near a solution (x,y) of

$$x'(t) = f(t,x(t),y(t))\,, \quad g(t,x(t),y(t)) \tag{2.99}$$

then we obtain by $0 = \nabla_x g(x,y)x' + \nabla_y g(x,y)y'$ together with (2.99)(1°) the explicit system

$$x' = f(x,y)\,, \quad y' = -\nabla_y g(x,y)^{-1}\nabla_x g(x,y)f(x,y)\,.$$

In this case the system (2.99) has *index 1*.

System with index 2. Let $\nabla_y g(x,y)$ be singular near a solution of (2.99). Then $h(x,y) := \nabla_x g(x,y)f(x,y) = 0$ follows from $g(x,y) = 0$ and

$$\begin{aligned}\nabla_x h(x,y) &= \nabla^2_{xx} g(x,y)f(x,y) + \nabla_x g(x,y)\nabla_x f(x,y)\\ \nabla_y h(x,y) &= \nabla_y \nabla_x g(x,y)f(x,y) + \nabla_x g(x,y)\nabla_y f(x,y)\,.\end{aligned}$$

If $\nabla_y h(x,y)$ is regular then $x' = f(x,y)\,,\ h(x,y) = 0$ is a system with index 1. The system (2.99) has *index 2* in this case. By solving $\nabla_x h(x,y)x' + \nabla_y h(x,y)y' = 0$ w.r.t. y' one obtains again an explicit system of first order,

$$x' = f(x,y)\,, \quad y' = -\nabla_y h(x,y)^{-1}\nabla_x h(x,y)f(x,y)\,.$$

System with index 3. If $\nabla_y g(x,y)$ and $\nabla_y h(x,y)$ are both singular near the solution of (2.99) then $k(x,y) := \nabla_x h(x,y)f(x,y) = 0$ follows from $h(x,y) = 0$. If now $\nabla_y k(x,y)$ is regular then the system $x' = f(x,y)\,,\ k(x,y) = 0$ has index 1. The system (2.99) has *index 3* in this case.

(e) In more recent time also **Semi-Explicit Runge-Kutta Methods**

$$\begin{aligned}U(t) &= e \times u(t) + \tau(A \times I)F(U(t),V(t)) && \in \mathbb{R}^{r\cdot n}\\ 0 &= G(U(t)) && \in \mathbb{R}^{r\cdot m}\\ u(t+\tau) &= u(t) + \tau(b \times I)^T F(U(t),V(t)) && \in \mathbb{R}^{n}\\ 0 &= g(u(t+\tau)) && \in \mathbb{R}^{m}\end{aligned} \tag{2.100}$$

have been proposed for solving DA-problems of the form

$$x'(t) = f(x(t),y(t))\,, \quad g(x(t)) = 0\,. \tag{2.101}$$

If the matrix A of coefficients is a *triangular matrix* with $\text{diag}(A) = 0$ then we obtain the following device for a single t-step:

(2.102)

Set $u_1 = u$ and compute v_1 with NEWTON's method by $g\big(u + \tau\alpha_{21}f(u_1,v_1)\big) = 0$ for $i = 2 : r$.
Set $u_i = u + \tau\sum_{j=1}^{i-1}\alpha_{ij}f(u_j,v_i)$ and compute v_i with NEWTON's method by $g(u_i) = 0\,,\ i = 2 : r$.
Set $u(t+\tau) = u + \tau\sum_{i=1}^{r}\beta_i f(u_i,v_i)$ and compute $v(t+\tau)$ with NEWTON's method by $g(u(t+\tau)) = 0$.

Theorem 2.17. (1°) *Let the problem (2.101) have a unique solution in* $[0, T]$, $0 < T$.
(2°) *Let the initial values satisfy* $g(x_0) = 0$, $\nabla g(x_0) f(x_0, y_0) = 0$.
(3°) *Let* $\nabla g(x)\nabla_y f(x, y)$ *be regular near the solution (system with index 2).*
(4°) *In the matrix* A *and vector* b *of the method (2.100), let*

$$\alpha_{i,i-1} \neq 0\,,\; i = 2 : r\,,\;\; \beta_i \neq 0\,,\; i = 1 : r\,.$$

Then the systems in (2.102) have a local unique solution for sufficiently small τ.

Proof see (Brasey92), (Brasey93).

Example 2.22. HEM4, 5-stage method of order $p = 4$ by (Brasey92).

$$\left[\begin{array}{c|c} A & c \\ \hline b & \end{array}\right] = \left[\begin{array}{ccccc|c} - & - & - & - & - & - \\ \frac{3}{10} & - & - & - & - & \frac{3}{10} \\ \frac{1+\sqrt{6}}{30} & \frac{11-4\sqrt{6}}{30} & - & - & - & \frac{4-\sqrt{6}}{10} \\ \frac{-79-31\sqrt{6}}{150} & \frac{-1-4\sqrt{6}}{30} & \frac{24+11\sqrt{6}}{25} & - & - & \frac{4+\sqrt{6}}{10} \\ \frac{14+5\sqrt{6}}{6} & \frac{-8+7\sqrt{6}}{6} & \frac{-9-7\sqrt{6}}{4} & \frac{9-\sqrt{6}}{4} & - & 1 \\ \hline 0 & 0 & \frac{16-\sqrt{6}}{36} & \frac{16+\sqrt{6}}{36} & \frac{1}{9} & - \end{array}\right].$$

2.8 Hints to the MATLAB programs

```
KAPITEL02/SECTION_1_2_3
Figures of Section 2.1 and 2.2
demo1.m       Test of four Gauss rules in interval
demo2.m       Test of Gauss and Bell rules in arbitrary triangle
bell.m        Exact integration of polynomial
              in arbitrary triangle
gauss_1.m:    Gauss-Legendre integration
gauss_2/3/4.m: Gauss integration, suboptimal, three cases
gauss_t5.m:   Gauss rule of order n = 5 in arbitrary triangle
divdif.m:     Generalized divided differences
KAPITEL02/SECTION_4: Initial Value Problems
Figures of Section 2.4 and stability regions
demo1.m       Arenstorf orbits by using dopri.m,
dopri.m       MATLAB version of FORTRAN version of HAIRER I
dreik_a.m     differential system of restricted
              three-body problem
stab_region.m Program for plots of the stability regions
              of one-step methods
```

```
KAPITEL02/SECTION_5: Boundary Value Problems
adapt01.m     Adaption of shooting points for example
box.m         Box scheme for Newton method
bsp01.m       Example Stoer-Bulirsch, Par. 7.3, Bsp. 1
demo1.m       Masterfile for multiple shooting method
mehrziel.m    Multiple shooting for Newton method
newton.m      Quasi-global Newton method
Kapitel02/SECTION_6: Periodic Problems
bsp01.m       Nerve membran model
bsp02.m       Heat flow problem
bsp03.m       Arenstorf orbit I
bsp04.m       Duffing's equation
demo1.m       Masterfile for multiple shooting method
demo2.m       Periodic solution of Duffing's equation
demo1.m       Some solutions of Duffing's equation
mehrziel_p.m Multiple shooting scheme for Newton's method
              and problems with unknown period
newton_p.m    Quasi-global Newton's method for periodic problems
```

3

Optimization

Leveraging a method by a suitable example applies in optimization more than in any other area; one must often settle for improvements compared to the *nominal solution* (start trajectory). The multitude of methods is hardly manageable, and new ones are constantly being added. E.g. (Himmelblau), (Spellucci) and (Polak) would have to be mentioned as standard works on *nonlinear optimization*; but also Monte-Carlo methods (based on random numbers) have made their entrance, see e.g. (Hajek), (VanLaarhoven). The basic idea of all methods is for the most part a simple geometric or physical principle, the execution of which becomes complicated (especially if, as in MATLAB, you have make do without `goto` commands), and saveguarding against operation errors knows no bound. This chapter should treat as examples the *projection method* and a *penalty method*, both of which are well established in the *community*. They basically differ in their nature and properties: In the projection method, the gradient of the objective function is projected in both the linear and nonlinear case onto the boundary of the feasible domain, thus determining the new search direction. In the penalty method, violating of the constraints are punished by an increasement of the objective function (in minimum problems). Here penalty parameters, which no longer have any physical importance, *formally* take the place of the LAGRANGE multipliers. For the method to deliver good results, the penalty parameters have to be *adapted* to the geometric situation in a suitable strategy over the iteration. Like elsewhere in optimization, the development is far from being completed here; see e.g. (Byrd). Nature took an infinitely long time to construct optimal systems. In contrast, we must carefully consider the relationship of the time required to the accuracy, and the evaluation of these contrary criteria perpetually changes with technical advances in hardware components.

Hint to this and the next chapter:

> 1. Vectors are *not* underlined!
> 2. *Row vectors* $a \in \mathbb{R}_n$ and *column vectors* $b \in \mathbb{R}^n$
> are used together such that, e.g., $a\,b \in \mathbb{R}$ and $b\,a \in \mathbb{R}^n{}_n$.
> Rule:
>
> Primal problem/primal space $\doteq$ column vectors
> Dual problem/dual space $\doteq$ row vectors.

3.1 Minimization of a Function

(a) Descend Methods On the search for a (possible local) minimum of a sufficiently smooth, scalar function f, a *stationary* or *critical* point x is determined commonly where $\nabla f(x) = 0$. Further investigations then have to verify whether that point is actually is a minimum point. Thereby one proceeds iteratively, in each step a *descend direction* d and a *step length* σ has to be chosen properly. This way of proceeding may be written as a preliminary computational device:

> Choose start vector x, tolerance tol; done = 0.
> WHILE NOT done
> Choose direction d and step length σ such that
> $\varphi(\sigma) := f(x - \sigma\, d) < f(x)$
> $x := x - \sigma\, d$
> done = $(\|\sigma\, d\| < \text{tol})$
> END

(3.1)

By lack of better possibilities, a simple stopping criterium is commonly applied here which is actually allowed only if convergence is warranted by some other way. Otherwise the criterium may *pretend* convergence as e.g. is illustrated by the well-known *harmonical* series. Therefore the results of iterations are always to be *examined critically.*

(b) Negative Examples Caution is also called for in choosing descend direction d and step length σ. Two simple examples due to (Dennis) demonstrate the reason for this impressively:

$$f(x) = x^2\,,\; x \in \mathbb{R}\,,\; x_0 = 2\,,\; x^* = 0\,.$$

(1°) Choose $d_i = (-1)^i$, $\sigma_i = 2 + 3 \cdot 2^{-i-1}$ for $i = 0, 1, \ldots$, then
$x_{i+1} = x_i - \sigma_i d_i = (-1)^{i+1}(1 + 2^{-i-1})$.
(2°) Choose $d_i = 1$, $\sigma_i = 2^{-i-1}$, for $i = 0, 1, \ldots$, then
$x_{i+1} = x_i - \sigma_i d_i = 1 + 2^{-i-1}$.

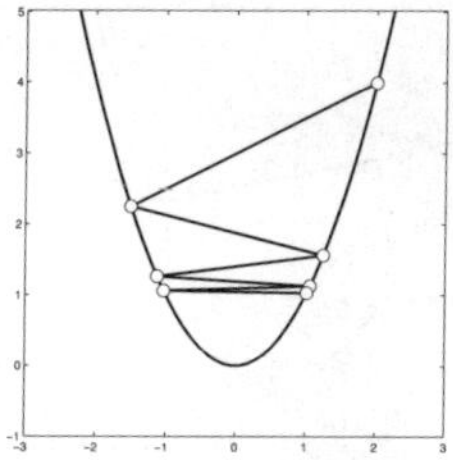

Figure 3.1. Example 1(a)

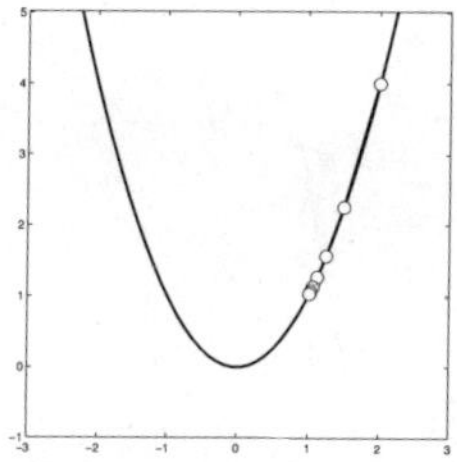

Figure 3.2. Example 1(b)

In both cases $f(x_i) \underset{i\to\infty}{\longrightarrow} 1 \neq 0$; decreasing of f relative to step length is too slow in the first case, and step length becomes too small in the second case.

(c) Convergence As well-known, the gradient $\nabla f(x)$ of f at point x shows into that direction where f increases most strongly, hence the vector $-\nabla f(x)$ shows into direction of the *falling line* at point x. The angle between ascend direction $d(x)$ and $\nabla f(x)$ must *always* be (uniformly) less than $\pi/2$ (Fig. 3.3). In other words, for all elements x of the below introduced *niveau set* $\mathcal{L}_f(f(x^0))$, we have to adopt the general rule

$$\boxed{\exists\, \varepsilon > 0 \;\; \forall\, x \;\; \forall\, d(x) : \; 0 \le \text{angle}(\nabla f(x), d(x)) \le \frac{\pi}{2} - \varepsilon} \, . \tag{3.2}$$

For *step length strategy*, the GOLDSTEIN-ARMIJO descend test (GA-test) is a good choice. Besides some technical details, it reads:

Let $\varphi(\sigma) = f(x - \sigma d)$ and $\varphi'(0) = -\nabla f(x) d < 0$, cf. (3.1). Choose $0 < \delta < 1$, e.g., $\delta = 1/100$, and in each step at first first (e.g.) $\sigma = 1$. Replace σ by $\sigma/2$ until (Fig. 3.4) $f(x - \sigma\, d) < f(x) + \sigma\, \delta \varphi'(0) =: g(\sigma)$	. (3.3)

There are some further step-length strategies of higher order but their higher computational effort yields seldom better results relative to total computational time. In Fig. 3.4 we have

$$\psi(\sigma) = \varphi(0) + \varphi'(0)(1 - \delta)\sigma + \|d\|^2 M \sigma^2/2$$

where $M = \max\{\|\nabla^2 f(\xi)\|\,,\ \xi \in \text{conv}\mathcal{L}_f(f(x^0))\}$; cf. the subsequent Theorem 3.1. This line remains always *above* of $\varphi(\sigma)$ and, by consequence, the step length after the GA-test satisfies always $\sigma \ge \varrho/2$ which says that, under the above assumptions, the step length cannot tend to zero!

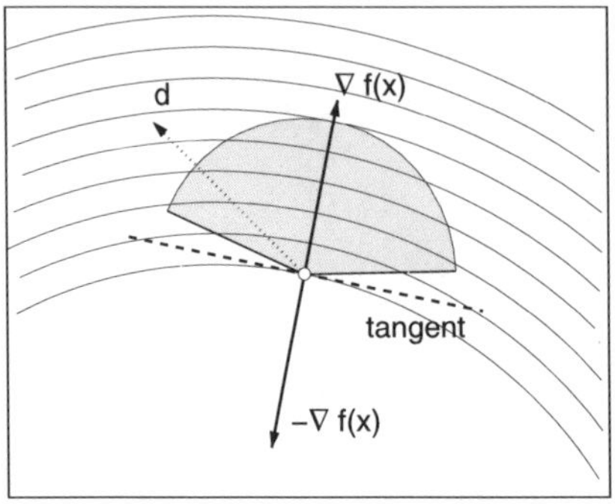

Figure 3.3. Feasible angular domain for direction d

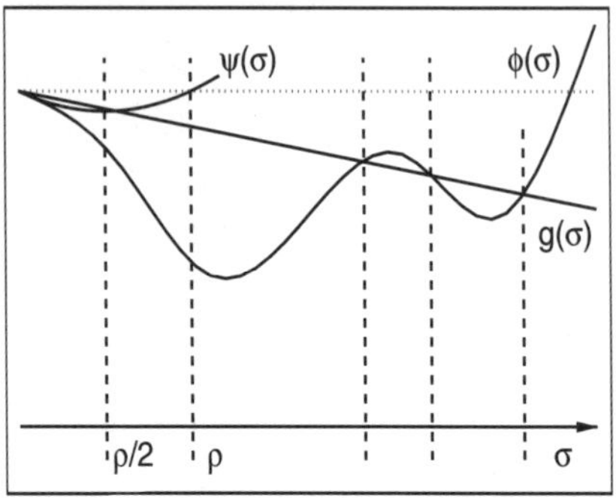

Figure 3.4. GOLDSTEIN-ARMIJO test

Theorem 3.1. (1°) *Let* $f \in \mathcal{C}^2(\mathbb{R}^n; \mathbb{R})$ *be bounded from below.*
(2°) *Let the* niveau set $\mathcal{L}_f(f(x_0)) := \{x \in \mathbb{R}^n\,,\ f(x) \leq f(x_0)\}$ *relative to the start value* x_0 *be* compact.
(3°) *Let the set* $\Omega = \{x^* \in \mathbb{R}^n\,,\ \nabla f(x^*) = 0\}$ *of critical points be finite and non-empty.*
(4°) *Let the directions* d *in* $\mathcal{L}_f(f(x_0))$ *be feasible in sense of (3.2).*
(5°) *Let the step lengths* σ *be chosen following (3.3).*
Then the sequence (3.1) with start value x_0 *converges to a critical point* x^*.

Proof see, e.g., (Spellucci), Sect. 3.1.2.2.

(d) Efficient Choice of Descend Direction Of course, the *speed* of the descend method depends in a crucial way on a suitable choice of the directions d. But here we have to regard the total computational amount of work and not only the number of iterations needed for the desired accuracy. In the method of *steepest descend* the self-suggesting direction $d = \nabla f(x)$ is chosen directly. This locally optimal choice leads however to the befeared "zigzagging" whereby the procedure is slowed down considerably in most cases. On the other side, "conjugate" directions may be applied which regard the local/global constellations more properly. Nevertheless, in these methods as well as in the below discussed NEWTON's method, the time-consuming Hessian $A = \nabla^2 f(x)$ as to be evaluated repeatedly. By these reasons the methods of *variable metric* have been introduced of which the name stems from the fact that every positive definite matrix defines a scalar product and thus a metric. These methods are described most simply by considering first a linear system of equations $Ax-b=0$ with regular matrix $A \in \mathbb{R}^n{}_n$. On choosing here the descend direction $-d = -B\nabla f(x)$ with some matrix B, the solution is obtained trivially in a single step if $B = A^{-1}$, namely $x_1 = x_0 - A^{-1}(Ax_0 - b) = A^{-1}b$. But acceptable results are obtained also by this way if B is only a more or less passable approximation of the inverse A^{-1}. The methods of variable metric choose generally

$$d_i = A_i^{-1}\nabla f(x_i)^T\,,\ \ i = 1, 2, \ldots\,,$$

for local descend direction implying that a *system* $A_i d_i = \nabla f(x_i)$ has to be solved in each step. The requirement now reads: Find a sequence of matrices $\{A_i\}$ by applying the ingredients x_j, $\nabla f(x_j)$, $j = 0 : i-1$, *known hitherto* such that $A_n = A$ does hold. After n iterations the procedure is re-started (of course by using the most recent values). Such a method has been found indeed by BROYDEN, FLETCHER, GOLDFARB and SHANNON after several preliminary investigations and is a mathematical icon today. The BFGS method leads off his trumps less in a convex quadratic objective function but in *general* uniformly convex function it beats even NEWTON's method relative to total computational time. (Note however that not necessarily $A_i \to \nabla^2 f(x^*)$ in BFGS methods.) For further discussion of this and related methods we have to refer to (Spellucci).

```
function [W,errorcode] = bfgs(name,X,TOL);
% BFGS Method, cf. Spellucci, S. 135
% f(x) = x'Ax/2 + b'x + c, A symm. positive definite
% errorcode = 1: Descend direction unfeasible
% errorcode = 2: GA_test fails
% errorcode = 3: Max. step number in iteration

MAXITER  = 10; errorcode = 0;
A = eye(length(X)); W = X; ITER = 0; DONE = 0;
GRAD = feval(name,X,2);
while ~DONE
   ITER = ITER + 1; D = A\GRAD';
   [Y,errorcode] = ga_test(name,GRAD,X,D,1);
   U = Y - X; V = A*U;
   GRAD1  = feval(name,Y,2);
   NORM   = norm(GRAD1);
   Z      = GRAD1 - GRAD;
   A      =  A   - V*V'/(V'*U) + Z'*Z/(Z*U);
   GRAD   = GRAD1;
   X = Y, W = [W,X];
   DONE   = norm(U) < TOL | ITER > MAXITER | NORM < TOL;
end
if ITER > MAXITER  errorcode = 3; end
```

(e) Newton's Method is the classical workhorse for solving nonlinear systems of equations $F(x) = 0 \in \mathbb{R}^n$. By introducing a *variable* step length σ, the originally local convergent method becomes a (nearly) global method, say, of the following form:

```
Choose a suitable start value x, choose tolerance tol
done = 0
WHILE NOT done
Choose step length σ e.g. by GA strategy,
(beginning with σ = 1 because of local quadratic
convergence)
```

$$\begin{aligned} y &= x - \sigma\, \nabla F(x)^{-1} F(x) \\ x &= y \end{aligned}$$

```
done = (‖x − y‖ < tol)
END
```

As already mentioned, the stopping criterium is to be scrutinized critically.

In order to apply the Convergence Theorem 3.1, we consider NEWTON's method in the form of *descending method* for

$$f(x) = \frac{1}{2} F(x)^T F(x) = \frac{1}{2} \|F(x)\|^2 . \tag{3.4}$$

The NEWTON descend direction $-d(x) = -\nabla F(x)^{-1} F(x)$ for the original system F then satisfies condition (3.2) for feasible descend relative to the scalar function f if the Jacobian ∇F remains regular in the compact niveau set $\mathcal{L}_f(f(x_0))$. NEWTON's method should always be the first choice in solving nonlinear systems of equations. It is rather insensitive w.r.t. inexact or approximative Jacobian $\nabla F(x)$ whence this matrix must not to be updated in every step of iteration. In case the method fails a systematical error is rather probably where the Jacobian is not evaluated correctly or becomes singular. The version after (Hoellig) presented in `KAPITEL03` attains $\sigma = 1$ automatically near the solution and thus warrants local quadratic convergence.

The minimization method of NELDER and MEAD is part of the OPTIMIZATION TOOLBOX of MATLAB. It is a very ingenious procedure which gets along *without* the costly derivatives but, on the other side, it is too slow (up to today) for most applications because of its complicated local optimum search. Equality and inequality constraints may be assigned, too, after (Himmelblau) enhancing operational time once more considerably. If gradients of objective function and side conditions are not accessible in large systems then this method is employed on mainframe computers or in compound of computers. For some examples see `KAPITEL03\FLEXIPLEX`.

Example 3.1.

$$\begin{aligned} f(x, y) &= 1.1x^2 + 1.2y^2 - 2xy + \sqrt{1 + x^2 + y^2} - 7x - 3y \\ (x^*, y^*) &= (15.376, 13.786) . \end{aligned}$$

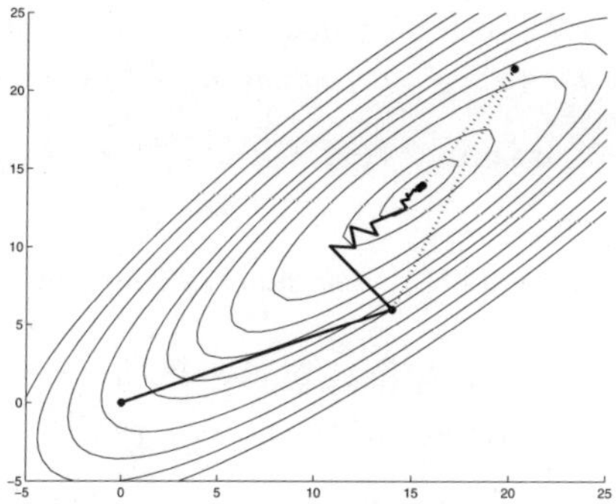

Figure 3.5. Method of steepest descend and BFGS method

Examples for NEWTON' method are considered in context of control theory.

3.2 Extrema with Constraints

Let $\mathbb{R}^n_+$ be the order cone in $\mathbb{R}^n$ hence $x \geq 0$ if $\forall\, i : x^i \geq 0$ and $x > 0$ if $\forall\, i : x^i > 0$, and remember that $x \geq y$ if $x - y \geq 0$ etc.; cf. Sect. 1.10.

(a) Formulation of the Problem Let $f : \mathbb{R}^n \to \mathbb{R}$, $g : \mathbb{R}^n \to \mathbb{R}^m$, $h : \mathbb{R}^n \to \mathbb{R}^p$ be three continuous functions with components g^i resp. h^j. We look for a solution x^* of the *minimum problem*

$$\boxed{\{f(x)\,;\ g(x) \leq 0\,,\ h(x) = 0\} = \min!} \tag{3.5}$$

where f is the *objective function.* The problem is called *differentiable* if f, g, h are (F-)differentiable. It is called *convex* if all three functions f, g, h are convex then the *set of feasible points*

$$\mathcal{S} := \{x \in \mathbb{R}^n,\ g(x) \leq 0\,,\ h(x) = 0\}$$

is convex; cf. Sect. 1.10. Letting $\mathcal{L}(f(x_0)) = \{x \in \mathbb{R}^n; f(x) \leq f(x_0)\}$ be again the *niveau set* there exists a not necessarily unique solution if the set $\mathcal{S} \cap \mathcal{L}(f(x_0))$ is compact and non-empty for some $x_0 \in \mathbb{R}^n$.

(b) Multiplier Rule By a likewise simple and ingenious idea of LAGRANGE (1736-1813), a function

$$L : \mathbb{R}^n \times \mathbb{R}_m \times \mathbb{R}_p \ni (x, y, z) \mapsto L(x, y, z) := f(x) + yg(x) + zh(x) \in \mathbb{R}$$

is associated to the problem (3.5) and the stationary points of this LAGRANGE function L are investigated instead of those of the actual objective function f. Originally this idea was developed for equations of motion where some artificial additional forces shall compel the mass points to regard the side conditions. The following result constitutes the basis of many approaches in optimization and control in spite of its apparent simpleness.

Theorem 3.2. *Let the following assumptions be fulfilled:*
(1°) *There exists a triple* (x^*, y^*, z^*) *where* $y^* \geq 0$ *such that*

$$x^* = \arg\min_x \{f(x) + y^* g(x) + z^* h(x)\}\,, \tag{3.6}$$

(2°) $x^* \in \mathcal{S}$, *i.e.* x^* *is feasible,*
(3°) $y^*\, g(x^*) = 0$ *(*complement(ary slackness) condition*).*
Then

$$x^* = \arg\min\{f(x)\,;\ g(x) \leq 0\,,\ h(x) = 0\},$$

i.e., x^* *is solution of the minimum problem (3.5).*

As $y^* \geq 0$ by (1°), assumptions (2°) and (3°) together can be replaced for g by
(4°) $y^* g(x^*) = 0$ *and* $y_i^* = 0 \iff g^i(x^*) < 0$.
Most numerical procedures, as far as they work with LAGRANGE multipliers at all, supply a multiplier with the last property automatically then $y^* \geq 0$ and $y^* g(x^*) = 0$ guarantee that $g(x^*) \leq 0$.
Proof. For all $x \in \mathcal{S}$

$$\begin{aligned} f(x^*) &\overset{(2),(3)}{=} f(x^*) + y^* g(x^*) + z^* h(x^*) \\ &\overset{(1)}{\leq} f(x) + y^* g(x) + z^* h(x) \overset{x \in \mathcal{S}}{\leq} f(x)\,. \end{aligned}$$

□

Equality constraints h may play a more or less trivial role here but the augmented formulation is justifed by the subsequent conclusion (3.7). Theorem 3.2 holds just the same way for a *maximum problem* if the positive sign of $y\,g(x)$ is replaced by the negative sign in the LAGRANGE function L. Note however that the existence of suitable LAGRANGE multipliers $0 \leq y \in \mathbb{R}_m$ and $z \in \mathbb{R}_p$ is *supposed* here. The result of Theorem 3.2 may not be very convincing at first glance since the original objective function f of the problem is replaced by the more complicated LAGRANGE function L which contains some unknown parameters namely the multipliers. But the assertion says that an *extremal* of L is solution of the constrained problem under the enumerated additional conditions therefore (3.6) is to be understood as *preliminary device* for further orientation. In a naive way one may compute the extremals of the LAGRANGE function and then, in a second step, select the multipliers by computation or by "intuition" such that conditions (2°) and (3°) are satisfied.
A scalar inequality constraint $g^i(x) \leq 0$ is called

$$\boxed{\begin{array}{ll} \text{active in } x & \text{if } g^i(x) = 0 \\ \text{inactive in } x & \text{if } g^i(x) < 0 \end{array}}\,.$$

Let now the problem (3.5) be differentiable then, by assumption (1°), there results a *necessary condition* for a minimum point (or more generally for a stationary point)

$$\nabla f(x^*) + y^* \nabla g(x^*) + z^* \nabla h(x^*) = 0 \in \mathbb{R}_n\,, \quad G(x^*) := \begin{bmatrix} \nabla g(x^*) \\ \nabla h(x^*) \end{bmatrix}. \tag{3.7}$$

It says that the gradient of the objective function can be written in x^* as a linear combination of the gradients of all constraints or, in other words, that $\nabla f(x^*)$ is an element of the (row-wise) range of the matrix $G(x^*)$.

> A point x^* obeys the *multiplier rule* (MR) or also, x^* is a KUHN-Tucker point, if condition (3.7) as well as the assumptions (2°) and (3°) of Theorem 3.2 do hold in x^* for some pair $(y^*, z^*) \in \mathbb{R}_m \times \mathbb{R}_p$ and $y^* > 0$.

(3.8)

If the minimum problem contains only equality constraints, the complement condition (3°) is dropped of course. One the other side, if the problem contains only inequality constraints, the system

$$\nabla_x L(x, y) = 0\,, \quad y \ge 0\,, \quad y\, g(x) = 0\,, \quad g(x) \le 0\,,$$

is to be solved for $(x, y) \in \mathbb{R}^n \times \mathbb{R}_m$. By attempting to find a solution directly, all 2^m possibilities $y_i > 0$, $y_i = 0, i = 1 : m$, have to be tested for feasible x therefore this way of procedure is rejected in general. The minimum problem (3.5) carries frequently the additional constraint that the state variable x is not allowed to vary in the entire space $\mathbb{R}^n$ but is restricted to, say, a bounded closed set $\mathcal{C}$. Then the necessary condition (3.7) must be replaced by a weaker inequality form because extrema may appear also at the boundary of $\mathcal{C}$; cf. Sect. 3.6**(f)**.

The multiplier rule reveals to be the basic criterium in solving the problem (3.5) which also is manifested in the following Theorems 3.3 and 3.5.

Theorem 3.3. *(MR sufficient in convex problems) Let the minimum problem (3.5) be convex (hence h affine linear) and differentiable, then every* KUHN-TUCKER *point is a global minimum point of f in $\mathcal{S}$ and thus a solution of the minimum problem.*

Proof see `SUPPLEMENT\chap03a`.

Following the proof, the assertion remains also true if MR is replaced by

$$[\nabla f(x^*) + y^* \nabla g(x^*)](x - x^*) \ge 0$$

on convex subsets $\mathcal{C} \subset \mathbb{R}^n$ and $h = 0$.

Assume that the gradients of the active inequalities and the gradients of all equalities together span the entire row space $\mathbb{R}_n$, then there exist surely LAGRANGE multipliers y and z such that (3.7) does hold where however y is not necessarily sign-bounded. Feasible points are called *singular* if there do *not* exist LAGRANGE multipliers $y \ge 0$ and z such that (3.7) i.e. the multiplier rule holds. All these special points have to be studied separately. As concerns the general *existence* of LAGRANGE multipliers there are about twenty different *constraint qualifications* (regularity conditions) which unfortunately may only partially be ordered in hierarchical way; cf. (Peterson). Let us specify one of the most popular qualifications more exactly (being admittedly a rather strong one). To this end let

$$\mathcal{A}(x) = \{i \in \{1, \ldots, m\},\ g^i(x) = 0\}$$

be the index set of all active inequalities at point $x \in \mathcal{S}$ and let $\nabla g^{\mathcal{A}}(x)$ be the matrix of gradients with rows $g^i(x)$, $i \in \mathcal{A}(x)$.

The feasible point $x \in \mathcal{S}$ satisfies the *rank condition* if the matrix of gradients of *all* active constraints in x,

$$\begin{bmatrix} \nabla h(x) \\ \nabla g^{\mathcal{A}}(x) \end{bmatrix} \tag{3.9}$$

has maximum row rank $p + |\mathcal{A}(x)|$, i.e., all rows of this matrix are linearly independent; then necessarily $p + |\mathcal{A}(x)| \leq n$.

Let now $\mathcal{I}$ be a closed interval with non-empty interior $\mathrm{int}(\mathcal{I})$ and let $x : \mathcal{I} \ni t \to x(t) \in \mathbb{R}^n$ be a sufficiently smooth curve. Then by Lemma 1.27

$$\begin{aligned}
&t^* \in \mathrm{int}(\mathcal{I}) \text{ and } f(x(t^*)) = \min_{t\in\mathcal{I}} f(x(t)) \Longrightarrow \nabla f(x(t^*))x'(t^*) = 0\,,\\
&t^* \in \mathcal{I} \qquad \text{ and } f(x(t^*)) = \min_{t\in\mathcal{I}} f(x(t)) \Longrightarrow \nabla f(x(t^*))x'(t^*) \geq 0\,,\\
&\qquad\qquad\qquad h(x(t^*)) = \qquad 0 \qquad\quad \Longrightarrow \nabla h(x(t^*))x'(t^*) = 0\,.
\end{aligned}$$

For the inversion of these conclusions we define an illustrative property of both functions g and h:

Definition 3.1. (1°) *Let $h(x^*) = 0$, then $h(x) = 0$ is* locally solvable *in x^* if*

$$\begin{aligned}
&\forall\, v \in \mathbb{R}^n\ \ \exists\, \varepsilon > 0\ \ \exists\, \varphi : \mathbb{R} \to \mathbb{R}^n,\ \varphi(\alpha) = o(|\alpha|) :\\
&\nabla h(x^*)v = 0\,,\ 0 < \alpha \leq \varepsilon \implies h(x^* + \alpha v + \varphi(\alpha)) = 0\,.
\end{aligned}$$

(2°) *Let $g(x^*) \leq 0$, then $g(x) \leq 0$ is locally solvable in x^* if*

$$\begin{aligned}
&\forall\, v \in \mathbb{R}^n\ \ \exists\, \varepsilon > 0\ \ \exists\, \varphi : \mathbb{R} \to \mathbb{R}^n,\ \varphi(\alpha) = o(|\alpha|) :\\
&\nabla g^{\mathcal{A}}(x^*)v \leq 0\,,\ 0 < \alpha \leq \varepsilon \Longrightarrow g^{\mathcal{A}}(x^* + \alpha v + \varphi(\alpha)) \leq 0\,.
\end{aligned}$$

The inactive inequalities at x^* do not play any role in this local property.

Theorem 3.4. (LJUSTERNIK, *Regularity Condition)*
(1°) *Let $h(x^*) = 0$ and let $\nabla h(x^*)$ have maximum row rank p then $h(x) = 0$ is locally solvable in x^*.*
(2°) *Let $g(x^*) \leq 0$ and let $\nabla g^{\mathcal{A}}(x^*)$ have maximum row rank $|\mathcal{A}|$ then $g(x) \leq 0$ is locally solvable in x^*.*

Proof see e.g. (Craven95), Sect. 3.7.

Theorem 3.5. *(MR local necessary) Let the minimum problem (3.5) be differentiable, let x^* be a local solution and let x^* satisfy the rank condition (3.9), then the multiplier rule does hold in x^*.*

Proof. We carry the proof out separately for equalities h and inequalities g.
(1°) For every v satisfying $\nabla h(x^*)v = 0$ there exists locally a curve $x(t) = x^* + tv + \varphi(t)$ by the local solvability assumption such that

$$h(x(t)) = h(x^* + tv + \varphi(t)) = 0\,, \quad 0 \le t\,.$$

But then also $-\nabla h(x^*)v = 0$ and the same result does hold if v is replaced by $-v$. Consequently, x^* is an interior point of a curve being entirely contained in the feasible domain. Therefore $\nabla f(x^*)x'(0) = \nabla f(x^*)v = 0$ hence

$$\begin{aligned}&\Big\{\forall\, v : \nabla h(x^*)v = 0 \implies \nabla f(x^*)v = 0\Big\}\\ &\implies \nabla f(x^*) \in \mathrm{Ker}(\nabla h(x^*))^{\perp} = \mathrm{Range}(\nabla h(x^*)^T)\,.\end{aligned}$$

Accordingly, there exists a $z^* \in \mathbb{R}^p$ satisfying $\nabla f(x^*)^T = -\nabla h(x^*)^T z^*$ by the Range Theorem 1.2.
(2°) For every v satisfying $\nabla g^{\mathcal{A}}(x^*)v \le 0$ there exists local a curve $x(t) = x^* + tv + \varphi(t)$ such that

$$g(x(t)) = g(x^* + tv + \varphi(t)) \le 0\,, \quad 0 \le t\,.$$

Now x^* is a local minimum by assumption therefore $\nabla f(x^*)x'(0) = \nabla f(x^*)v \ge 0$. Consequently

$$-\nabla g^{\mathcal{A}}(x^*)v \ge 0 \implies \nabla f(x^*)v \ge 0\,.$$

By the Lemma 1.24 of Farkas there exists a $0 \le y^{*\mathcal{A}} \in \mathbb{R}^{|\mathcal{A}|}$ such that $-\nabla g^{\mathcal{A}}(x^*)^T y^{*\mathcal{A}} = \nabla f(x^*)^T$. This proves the assertion by setting $y^{*i} = 0$ for $i \notin \mathcal{A}$. □

Example 3.2. $n = 2$, $m = 3$, $p = 0$ (Fig. 3.6).

$$\begin{aligned}&f(x^1, x^2) = x^1 = \min!\\ &g^1(x^1, x^2) = -x^1\,, \quad g^2(x^1, x^2) = -x^2\,,\\ &g^3(x^1, x^2) = -(1 - x^1)^3 + x^2\,.\end{aligned}$$

The point $x^* = (1\,,\,0)$ is a minimum point where g^1 is inactive, and

$$\nabla f(x^*) = (1\,,\,0)\,, \quad \nabla g^2(x^*) = (0\,,\,-1)\,, \quad \nabla g^3(x^*) = (0\,,\,1)\,.$$

We have $y^{*1} = 0$ but y^{*2} and y^{*3} do *not* exist.

The Theorems 3.3 and 3.5 display the connection between KUHN-TUCKER points and the solution of extremal problems. The multiplier rule however is only a local property whereas Theorem 3.2 makes a global statement without any assumption of smoothness. It says that the extremum points of the LAGRANGE-function L coincide with those of the objective function f in the feasible domain *in case where* the LAGRANGE multiplier are chosen properly. The inversion of this question namely the question for the existence of *these* LAGRANGE multipliers requires considerably more auxiliaries and can be answered only under relatively restrictive assumptions. But this latter question is not of that importance in practice as the question for the existence of KUHN-TUCKER points. Therefore we refer to Sect. 3.6 for further discussion.

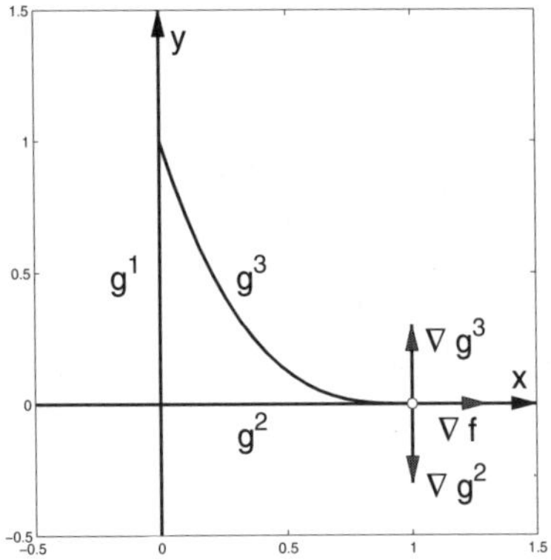

Figure 3.6. Example 3.2

3.3 Linear Programming

The famous *simplex algorithm* has been developed by DANTZIG in the fifties of the last century and since that time, where only mechanical calculating machines have been in common use, linear optimization in finite-dimensional spaces is handled under the name "linear programming".

(a) Examples All problems of linear programming follow the same pattern to some extent:

(a1) Somebody has n activities $j = 1 : n$ (e.g. the production of n articles) with profit α_j per unit, ξ^j units of activity j yielding $\alpha_j\xi^j$ units of profit. To this end there are m resources $i = 1 : m$ at disposal (e.g. appliances and manpower). One unit of activity j requires ${\beta^i}_j$ units of resource i (e.g. hours per month). But there are *at most* γ^i units at disposal, ${\beta^i}_1\xi^1 + \ldots + {\beta^i}_n\xi^n \leq \gamma^i$. How are the resources applied optimally? Let

$$a := [\alpha_i] \in \mathbb{R}_n\,, \quad B = [{\beta^i}_j] \in {\mathbb{R}^m}_n\,, \quad c = [\gamma^i] \in \mathbb{R}^m\,, \quad x = [\xi^j] \in \mathbb{R}^n\,,$$

then there results a maximum problem

$$\max\{ax\,;\; Bx \le c\,,\; 0 \le x\}$$

where the inequalities are to be understood componentwise.

(a2) Somebody has n resources $j = 1 : n$ at disposal (e.g. manpower, time, money) with costs α_j per unit, ξ^j units of resource j hence cost $\alpha_j\xi^j$ units. With these resources he has to satisfy m requirements $i = 1 : m$. One unit of requirement i needs ${\beta^i}_j$ units of resource j, altogether *at least* γ^i units of requirement i must be met, ${\beta^i}_1\xi^1 + \ldots {\beta^i}_n\xi^n \ge \gamma^i$. How are the resources applied optimally such that the costs are minimized? This problem may be considered in some way as *dual* to **(a1)** and leads to the minimum problem

$$\min\{ax\,;\; Bx \ge c\,,\; x \ge 0\}\,.$$

(b) Formulation of the Problem We consider the linear optimization problem

$$\boxed{\min\{a\,x\,;\; Bx \le c\}\,,\quad a \in \mathbb{R}_n\,,\quad c \in \mathbb{R}^m} \tag{3.10}$$

where $m \gg n$ in normal case. The feasible domain $\mathcal{S} = \{x \in \mathbb{R}^n;\; Bx \le c\}$ of (3.10) is a *convex polytope.*

In general, a point z of a convex set $\mathcal{C}$ is called *extreme point* if

$$\forall\, x,\, y \in C : z \in [x, y] \implies z = x \;\vee\; z = y\,.$$

Geometrically an extreme point is a "corner" of the polytope $\mathcal{S}$ and sometimes also called so. The extreme points of the feasible domain $\mathcal{S}$ allow a simple characterization but, to this end, we have to modify slightly the index set $\mathcal{A}(x)$ introduced in the preceding section. Note also that the gradient of $g^k(x) = b^kx - \gamma^k$ is now the k-th row b^k of the matrix B in the present situation of linear inequality constraints. Let

$$\begin{aligned}
\mathcal{A}^*(x) &:= \{k \in \{1, \ldots, m\}\,,\; b^kx = \gamma^k\}\,,\\
\mathcal{A}(x) &:= \{k \in \mathcal{A}^*(x)\,,\; b^k \text{ linearly independent}\},\\
\mathcal{N}(x) &:= \{1, \ldots, m\}\backslash\mathcal{A}(x)\,.
\end{aligned}$$

$\mathcal{A}^*(x)$ denotes the *(index-)set of active side conditions* in x. The set $\mathcal{N}(x)$ contains the indices of all inactive side conditions in x if and only if $\mathcal{A}(x)$ coincides with $\mathcal{A}^*(x)$. In the other case, neither $\mathcal{A}(x)$ nor $\mathcal{N}(x)$ are determined uniquely.

Lemma 3.1. *$x \in \mathcal{S}$ is extreme point if and only if there exists an index set $\mathcal{A}(x)$ with $|\mathcal{A}(x)| = n$.*

Proof see `SUPPLEMENT\chap03a`.

By consequence, the feasible domain $\mathcal{S}$ has extreme points only if rank$(B) = n$ maximum otherwise it consists of an unbounded "strip" if non-empty. In order to exclude this pathological case we introduce a *modified* rank condition:

$$\boxed{\text{The matrix } B \in \mathbb{R}^m{}_n \text{ in (3.10) has rank } n}\,. \tag{3.11}$$

As an inference of Lemma 3.1, at least n side conditions $b^k x = \gamma^k$ are active in an extreme point $x \in \mathcal{S}$ and exactly n gradients b^k of these side conditions are linearly independent. An extreme point is *degenerated* if more than n side conditions are active, i.e., if $|\mathcal{A}^*(x)| > n$. Let x be extreme point and $k \in \mathcal{A}(x)$, then we say briefly that "b^k *is in the basis of* x" since the rows b^k of B with $k \in \mathcal{A}(x)$ form a basis of the row-space $\mathbb{R}_n$. But note that this basis of rows of B is *not* determined uniquely in a *degenerated* corner.

The following geometrically obvious result is also called *fundamental theorem of linear programming* or sometimes corner theorem.

Theorem 3.6. *(Existence) Let the feasible domain $\mathcal{S}$ be non-empty and let the objective function $x \mapsto ax$ be bounded on $\mathcal{S}$. Then the problem (3.10) has a solution being extreme point of $\mathcal{S}$.*

The proof in `SUPPLEMENT\chap03a` supplies also a method to find an initial extreme point.

(c) The **Projection Method** starts in an extreme point of the feasible domain $\mathcal{S}$ which has to be determined first (phase 1). Let x be any extreme point then the gradient a of the objective function is projected onto the hyperplanes of active side conditions and it is investigated which of these projections leads to the optimal improvement (descend in a mimium problem) without leaving the feasible domain. Subsequently, it is proceeded along the selected edge until the next extreme point is attained. The procedure is then repeated until an optimal extreme point is found. Accordingly, the iteration moves permanently on the boundary of the feasible domain (polytope) which may be rather expensive. Therefore "interior-point methods" have been developed in more recent time which move to optimum from the interior of the feasible domain (but need an *interior* point for start value by consequence).

Now we need a proper criterium for the optimal search direction at a given extreme point. It is provided by the following *Duality Theorem* or *Equivalence Theorem* of FARKAS being displayed here in a more general form for later applications.

Theorem 3.7. *x^* is solution of*

$$\min\{ax\,;\; Bx \le c\,,\; Cx = d\}\,,\quad a \in \mathbb{R}_n\,,\quad c \in \mathbb{R}^m\,,\quad d \in \mathbb{R}^p\,, \tag{3.12}$$

if and only if there exists a triple $(x^*, y^*, z^*) \in \mathbb{R}^n \times \mathbb{R}_m \times \mathbb{R}_p$ with the following three properties:

$$\begin{array}{lll}(1^\circ)\; Bx^* \le c\,,\;\; Cx^* = d\,, & & \textit{primal feasibility,}\\ (2^\circ)\; y^*B + z^*C & = -a\,,\;\; y^* \ge 0\,, & \textit{dual feasibility,}\\ (3^\circ)\; y^*(Bx^* - c) & = 0\,, & \textit{complement(ary slackness) condition}.\end{array}$$

Proof. The theorem is of crucial importance in linear optimization therefore the proof is displayed in full length:

Let (1°) – (3°) hold and let $x \in \mathcal{S}$ be arbitrary then

$$\begin{aligned} ax - ax^* &\overset{(2)}{=} -(y^*B + z^*C)x + (y^*B + z^*C)x^* \\ &\overset{(1),(3)}{=} -y^*Bx - z^*Cx + y^*c + z^*d \overset{(1)}{=} y^*(c - Bx) \overset{(1),(2)}{\geq} 0 \,. \end{aligned}$$

Conversely, let (3.12) be fulfilled then x^* is feasible hence (1°). Let

$$\widetilde{B} = \begin{bmatrix} B \\ -C \\ C \end{bmatrix}, \quad \widetilde{c} = \begin{bmatrix} c \\ -d \\ d \end{bmatrix}, \quad p = \widetilde{c} - \widetilde{B}x^*$$

then $p \geq 0$. Let (x, ϱ) be given such that $q := \varrho p - \widetilde{B}x \geq 0$, then $Cx = 0$, and, for sufficiently small $\sigma \geq 0$,

$$0 \leq (1-\sigma\varrho)p + \sigma q = p + \sigma(q - \varrho p) = p + \sigma\varrho p - \sigma\widetilde{B}x - \sigma\varrho p = c - \widetilde{B}(x^* + \sigma x) \geq 0 \,.$$

Accordingly,

$$\begin{bmatrix} c - B(x^* + \sigma x) \\ -d + C(x^* + \sigma x) \\ d - C(x^* + \sigma x) \end{bmatrix} \geq 0 \,.$$

By this way it is shown that $x^* + \sigma x$ is feasible for sufficiently small $\sigma > 0$, therefore

$$a(x^* + \sigma x) \geq ax^* \implies ax \geq 0 \,,$$

since x^* is optimal by assumption. Together

$$\Big\{q \geq 0 \implies ax \geq 0\Big\} \iff \Big\{[-\widetilde{B},\, p]\begin{bmatrix} x \\ \varrho \end{bmatrix} \geq 0 \implies [a,\, 0]\begin{bmatrix} x \\ \varrho \end{bmatrix} \geq 0\Big\} \,.$$

But this inference constitutes the right side of the Lemma 1.24 of Farkas therefore the left side holds also and guarantees the existence of a vector $w^{*T} \geq 0$ such that

$$\begin{bmatrix} a^T \\ 0 \end{bmatrix} = \begin{bmatrix} -\widetilde{B}^T \\ p^T \end{bmatrix} w^{*T} \,. \tag{3.13}$$

Writing $w^* = (y^*, z_1^*, z_2^*)$ und $z^* = z_2^* - z_1^*$, we obtain first $a = -w^*\widetilde{B}$ or

$$a = -y^*B - z^*C \implies y^*B + z^*C = -a \,, \ \ y^* \geq 0 \,,$$

hence condition (2°). Furthermore, $w^*p = 0$ follows from (3.13). Substitution of $p = (c - Bx^*,\, -d + Cx^*,\, d - Cx^*)$ yields

$$y^*(c - Bx^*) + z_1^{*T}(Cx^* - d) - z_2^{*T}(Cx^* - d) = 0 \,,$$

which implies, finally, the complement condition (3°) since $Cx^* = d$. □

The vectors $y^* \in \mathbb{R}_m$ and $z^* \in \mathbb{R}_p$ represent the LAGRANGE multipliers of the problems (3.12) by Theorem 3.2.

Let now $x \in \mathbb{R}^n$ be any initial extreme point then without loss of generality

$$\mathcal{A} := \mathcal{A}(x) = \{\varrho_1, \ldots, \varrho_n\}\,, \quad \mathcal{N} := \mathcal{N}(x) = \{\sigma_1, \ldots, \sigma_{m-n}\}\,,$$

$$B^{\mathcal{A}} := \begin{bmatrix} b^{\varrho_1} \\ \vdots \\ b^{\varrho_n} \end{bmatrix}\,, \quad A := [B^{\mathcal{A}}]^{-1} =: [a_1, \ldots, a_n] \in \mathbb{R}^n{}_n\,.$$

Then $B^{\mathcal{A}}$ is regular by assumption and $B^{\mathcal{A}}A = I$, i.e., $b^j a_k = \delta^j{}_k$ (KRONECKER symbol). By this way, every column a_k of A reveals to be parallel to some edge of the feasible domain $\mathcal{S}$ or, in other words, a_k is contained in the intersection of $n-1$ hyperplanes $b^j x = 0\,,\ j = 1:n\,,\ j \neq k$ which are parallel to the corresponding boundary surfaces $b^j x = \gamma^j$ of $\mathcal{S}$:

$B^{\mathcal{A}}$	is the basis or *gradient matrix* of the extreme point x where the rows are gradients of active side conditions in x.
A	is the *edge matrix* of the extreme point x where the columns are edges of $\mathcal{S}$ pointing *into* direction of x.

(c1) Optimality Condition Let us return to the original problem (3.10). By Theorem 3.7

$$\boxed{x \text{ optimal} \iff -a = wB^{\mathcal{A}}\,,\ w = -aA \geq 0}\,.$$

If the extreme point x is not optimal, there exist some $w_i < 0$ and we have to proceed from x in direction of an edge $-a_j$ to a (hopefully) better extreme point $\widetilde{x} = x - \tau a_j\,,\ \tau > 0$, such that the objective function decreases most strongly. Because $a\widetilde{x} \equiv a(x - \tau a_j) = ax - \tau a a_j = ax + \tau w_j$, the optimal edge j is given by

$$\boxed{\begin{aligned} w_j &= \min\{w_k\} \ \ (< 0, \text{ else } x \textit{ optimal}) \\ j &= \min \arg_k \min\{w_k\} \in \mathbb{N} \end{aligned}} \tag{3.14}$$

(double "min" because of possible ambiguity).

(c2) The **Optimal Step Length** τ results form the requirement that we have to proceed along the chosen edge as far as possible without leaving the feasible domain, namely until the next extreme point is reached or, in other words, until one of side conditions being inactive hitherto becomes active: Choose τ *maximal* such that

$$B^{\mathcal{N}}(x - \tau a_j) \leq c^{\mathcal{N}} \text{ or } r^{\mathcal{N}} := B^{\mathcal{N}}x - c^{\mathcal{N}} \leq \tau B^{\mathcal{N}} a_j\,.$$

If $B^{\mathcal{N}} a_j \geq 0$ then $\tau > 0$ may be chosen arbitrarily and the problem has *no solution.* Otherwise the index i of the new active side condition is found by

$$\boxed{\begin{array}{l} \tau = \min_{k\in\mathcal{N}}\{\varphi(j,k)\,,\ \varphi(j,k)\geq 0\}\,,\ \varphi(j,k) := \dfrac{b^k x - \gamma^k}{b^k a_j} \\ i = \min\arg_k \min_{k\in\mathcal{N}}\{\varphi(j,k)\,,\ \varphi(j,k)\geq 0\} \in \mathbb{N} \end{array}}\ . \tag{3.15}$$

(c3) Change of Basis (GAUSS-JORDAN step of exchange) Now a basis of the new corner $\widetilde{x}$ has to be found whereby the pair (i,j) of (3.15) and (3.14) is called *pivot point.* Recall

$$\begin{array}{ll} \text{Index sets for } x\text{:} & \mathcal{A} := \mathcal{A}(x) = \{\varrho_1,\ldots,\varrho_n\}\,, \\ & \mathcal{N} := \mathcal{N}(x) = \{\sigma_1,\ldots,\sigma_{m-n}\}\,, \\ \text{Requirement:} & \varrho_j \text{ resp. } b^{\varrho_j} \text{ out of basis, } \ \sigma_i \text{ resp. } b^{\sigma_i} \text{ into basis,} \\ \text{Index set for } \widetilde{x}\text{:} & \widetilde{\mathcal{A}} = \{\varrho_1,\ldots,\varrho_{j-1},\sigma_i,\varrho_{j+1},\ldots,\varrho_n\}, \\ & \widetilde{\mathcal{N}}\{\sigma_1,\ldots,\sigma_{i-1},\varrho_j,\sigma_{i+1},\ldots,\sigma_{m-n}\}\,. \end{array} \tag{3.16}$$

We consider the following situation without loss of generality: Let $B = [b^i]_{i=1}^n$ be arbitrary regular and the *row* b^i of B shall be replaced by the row vector $d \in \mathbb{R}_n$. The new matrix $\widetilde{B}$ reads simply

$$\widetilde{B} = B + e_j(d - b^j) = (I + e_j(d - b^j)B^{-1})B =: TB\,,$$

where e_j denotes the j-th column unit vector. But $b^j B^{-1} = e^j$ (row vector) hence the transformation matrix T reads:

$$T = I + e_j(dB^{-1}) - e_j e^j\,.$$

In this matrix the j-th row of the unit matrix I is replaced by $b = [\beta_1,\ldots,\beta_n]$ $:= dB^{-1}$. The actual job however is to compute $\widetilde{A} = \widetilde{B}^{-1}$ by means of the matrix $A = B^{-1}$. To this end observe that $\widetilde{B}^{-1} = B^{-1}T^{-1} = AT^{-1}$ and that the inverse of T may be obtained in a *simple way*,

$$T^{-1} = I - \frac{1}{\beta_j}e_j b + \frac{1}{\beta_j}e_j e^j$$

where the *pivot element* β_j is fortunately non-zero by the above choice of pivot point (i,j) . Because

$$\left[A - \frac{1}{\beta_j}a_j b\right]_j = 0\,,\quad \text{(column } j\text{)}$$

the exchange method (GAUSS-JORDAN step) reads in matrix form:

$$\boxed{\begin{array}{l} d = b^{\sigma_i}\,,\quad b = dA\,, \\ G = A - \dfrac{1}{\beta_j}a_j b\,,\ [G]_j = \dfrac{1}{\beta_j}a_j\,,\ \widetilde{A} = G \end{array}}\ . \tag{3.17}$$

(d) The frequently appearing problem $\min\{ax\,;\ \widetilde{B}x \le \widetilde{c},\ x \ge 0\}$ is equivalent to the problem

$$\min\{ax\,;\ Bx \le c\},\ \ B = \begin{bmatrix} -I \\ \widetilde{B} \end{bmatrix},\ \ c = \begin{bmatrix} 0 \\ \widetilde{c} \end{bmatrix}. \tag{3.18}$$

The point $x = 0$ is here already an extreme point and can always be chosen for initial point, then $\mathcal{A}(0) = \{1, \ldots, n\}$, $\mathcal{N}(0) = \{n+1, \ldots, m\}$. Therefore $B^{\mathcal{A}} = -I$, $A = -I$ at the beginning and the initial matrix inversion is dropped. The complete algorithm for the problem (3.18) may be written as follows (to give an example):

START: $x = 0$, $A = -I$, $w = -aA = -a$, $r = -c$.
WHILE NOT $w \ge 0$
(1°) Find pivot point (i, j):

$$\begin{aligned} j &= \min\arg_k \min\{w_k\},\ \ (w \ge 0 \Longrightarrow x \text{ optimal}) \\ i &= \min\arg_k \min\{\varphi(j,k),\ \varphi(j,k) \ge 0\} \\ &\quad (i = [\,] \Longrightarrow \text{solution does not exist}) \end{aligned}$$

(2°) Exchange step: $b^{\varrho_j} \longleftrightarrow b^{\sigma_i}$ by (3.17)
(3°) Update: $\mathcal{A}$, $\mathcal{N}$ and matrix A, compute

$$x = Ac^{\mathcal{A}},\ \ r = B^{\mathcal{N}}x - c^{\mathcal{N}},\ \ w = -aA$$

END

The third step is handled automatically by the exchange method. The Lagrange multipliers y are found by $y_{\mathcal{A}} = w$, $y_{\mathcal{N}} = 0$.

(e) Degenerated Corners (having more than n active side consitions) appear frequently. Then *cycles* may arise theoretically where bases of the same corner are mutually exchanged whereas the step length τ in (3.15) remains zero. Such cycles occur very seldom in practice and may also be prevented by "Bland's rule": Choose (i, j) by the rule

$$\begin{aligned} j &= \arg_k \min\{\varrho_k \in \mathcal{A}(x)\,;\ w^{\varrho_k} < 0\} \\ i &= \arg_k \min\{\sigma_k \in \mathcal{N}(x)\,;\ \varphi(j,k) = \min_l\{\varphi(j,l)\,;\ \varphi(j,l) \ge 0\}\}. \end{aligned}$$

The projection method is however slowed down by this device and this criterium should be only applied if necessary.

(f) Multiple Solutions arise in case of a non-degenerated solution x^* if and only if the non-trivial component $y^*_{\mathcal{A}}$ of the Lagrange multiplier contains zeros in optimum. Without loss of generality let $y^*_{\mathcal{A}} = (u^*,\ v^*)$ where $v^* = 0$, and let

$$B^{\mathcal{A}} = \begin{bmatrix} C \\ D \end{bmatrix},\ \ A = [B^{\mathcal{A}}]^{-1} = [P,\ Q]$$

be decomposed correspondingly. Then $CP = I$, $DQ = I$, $CQ = 0$, $DP = 0$, and $u^*C + 0 \cdot D = -a$. Because $CQ = 0$ in optimum,

$$-ax^* = y_{\mathcal{A}}^* B^{\mathcal{A}} x^* = u^* C x^* = u^* C(x^* - Qd)$$

where d denotes an arbitrary vector at present.
(1°) Insert $x^* - Qd$ into the active side conditions:

$$\begin{aligned} C(x^* - Qd) &= Cx^* = c^{\mathcal{C}} && \text{conditions not violated,} \\ D(x^* - Qd) &= Dx^* - d \le c^{\mathcal{D}} \Longrightarrow && 0 \le d \text{ but elsewhere free.} \end{aligned}$$

(2°) Insert $x^* - Qd$ into the inactive side conditions:

$$B^{\mathcal{N}}(x^* - Qd) \le c^{\mathcal{N}} \Longrightarrow r^{\mathcal{N}} := B^{\mathcal{N}} x^* - c^{\mathcal{N}} \le B^{\mathcal{N}} Qd \,,$$

where $r^{\mathcal{N}} \le 0$ is the residuum. Consequently, the general solution set reads:

$$\boxed{\{x^* - Qd\,,\ d \ge 0\,,\ r^{\mathcal{N}} \le B^{\mathcal{N}} Qd\} \subset \mathbb{R}^n}\,.$$

(g) Equality Constraints are assigned easily to the algorithm. They are made active once at the beginning and then handled as inequalities which *never* are inactivated again and of which the LAGRANGE multipliers obey no sign restriction. But the row numbers of the equality restrictions (linear equations) must always be part of the set $\mathcal{A}$ of active indices throughout the iteration.

(h) Sensitivity In optimum $a = -w^* B^{\mathcal{A}}$, $w^* = y_{\mathcal{A}}^*$ hence also

$$a\,x^* = -w^* B^{\mathcal{A}} x^* = -w^* B^{\mathcal{A}} [B^{\mathcal{A}}]^{-1} c^{\mathcal{A}} = -w^* c^{\mathcal{A}} = -y_{\mathcal{A}}^* c^{\mathcal{A}} \,. \tag{3.19}$$

This relation does hold for varying c as long as the index set $\mathcal{N}$ of inactive conditions does not change. Let

$$x^*(c) = \arg\min\{ax\,;\ Bx \le c\}\,,$$

be a unique non-degenerated solution depending on the right side c of side conditions then, by (3.19),

$$\frac{\partial}{\partial \gamma^k} a\,x^* \Big|_{c=c^*} = \pm\, y_k^*(c^*)\,, \quad k = 1 : m$$

where the positive sign holds in *maximum problems*. Thus the LAGRANGE multiplier y_k^* is the *rate of change* of the maximum yield in optimum relative to the employed resource k. By this result, the LAGRANGE multipliers y_k^* reveal to be the well-known *shadow prices* of operational research: The profit $a\,x^*$ increases by y_k^* units if the k-th resource is modified by one unit. Accordingly, buying of some additional units of resource k makes sense if the prize per unit is *less than* y_k^*.

(i) The Dual Problem Recalling the notations of (3.12) we introduce a *row-wise* problem

$$\max\{-yc - zd\,;\; yB + zC = -a\,,\;\; y \geq 0\}\,,\;\; y \in \mathbb{R}_m\,,\;\; z \in \mathbb{R}_p\,, \tag{3.20}$$

and write it in the same column form as (3.12),

$$\min\{\widetilde{a}\widetilde{x}\,;\; \widetilde{B}\widetilde{x} \leq \widetilde{c}\,,\; \widetilde{C}\widetilde{x} = \widetilde{d}\}\,, \tag{3.21}$$

then

$$\begin{aligned} &\widetilde{x} = (y\,,\,z)^T \in \mathbb{R}^{m+p}\,, \quad \widetilde{a} = (c\,,\,d)^T \in \mathbb{R}_{m+p}\,, \qquad \widetilde{B} = [-I,\,O] \in \mathbb{R}^m{}_{m+p}\,,\\ &\widetilde{c} = 0 \in \mathbb{R}^m\,, \qquad\qquad \widetilde{C} = [B^T\,,\,C^T] \in \mathbb{R}^n{}_{m+p}\,,\;\; \widetilde{d} = -a^T \in \mathbb{R}^n\,. \end{aligned}$$

Applying the Equivalence Theorem 3.7 to (3.21), conditions (2°) and (3°) read:

$$\widetilde{y}^*[-I,\,O] + \widetilde{z}^*[B^T\,,\,C^T] = -[c^T\,,\,d^T] \in \mathbb{R}_{m+p}\,,\;\; \widetilde{y}^* \geq 0\,,\;\; \widetilde{y}^* y^{*T} = 0\,,$$

where $\widetilde{y}^* \in \mathbb{R}_m$ and $\widetilde{z}^* \in \mathbb{R}_n$ are now the associated LAGRANGE multipliers. Then, by separation,

$$\begin{aligned} -\widetilde{y}^* + \widetilde{z}^* B^T &= -c^T &&\Longleftrightarrow\;\; \widetilde{y}^{*T} &&= B\widetilde{z}^{*T} + c \geq 0\,,\\ \widetilde{z}^* C^T &= -d^T &&\Longleftrightarrow\;\; C\widetilde{z}^{*T} &&= -d\,,\\ y^*\widetilde{y}^{*T} &= 0 &&\Longleftrightarrow\;\; y^*(B\widetilde{z}^{*T} + c) = 0\,. \end{aligned} \tag{3.22}$$

Insert now $\widetilde{z}^* = -x^*$, then the equations on the right side constitute the conditions (1°) and (3°) for the primal problem (3.12), and, besides,

$$ax^* = -y^*Bx^* - z^*Cx^* = -y^*c - z^*d\,.$$

Thus the both problems (3.12) and (3.20) reveal to be *dual to each other*; the latter being the classical *simplex problem.* The primal problem (P), (3.12), has a solution x^* if and only if the dual problem (D), (3.20), has a solution (y^*, z^*), and the values of the respective objective functions *coincide* in optimum. Moreover, the solution of (D) is composed by the LAGRANGE multipliers of (P) whereas the solution of (P) is the negative second LAGRANGE multiplier of (D). Noting that $Bx^* + \widetilde{y}^* = c$ the first multiplier in (D) consists of the *slack variables* of the inequalities in (P).

(j) The Tableau At the times of manual computation all relevant data of the method have been assembled into an appropriate *tableau* and then updated in each step of iteration. This tableau reads for the minimum problem $\min\{ax\,;\; Bx \leq c\}$:

$$\mathbf{P}(x) = [p^k{}_l] := \begin{bmatrix} A & x \\ B^{\mathcal{N}}A & r^{\mathcal{N}} \\ y_{\mathcal{A}} & f \end{bmatrix} \tag{3.23}$$

where

$$x = Ac^{\mathcal{A}}\,, \quad r^{\mathcal{N}} = B^{\mathcal{N}}x - c^{\mathcal{N}}\,, \quad y_{\mathcal{A}} = -aA\,, \quad f = aAc^{\mathcal{A}} = -y_{\mathcal{A}}c^{\mathcal{A}}\,.$$

The first block row may be dropped if the index set $\mathcal{A}$ is always updated since the solution x has to be computed only once at the end of the procedure. After having found the pivot elements, algorithm (3.17) (exchange after GAUSS-JORDAN) may be applied to the *entire* tableau at once (!) whereby it is updated in a simple way. The tableau $\mathbf{Q}(\widetilde{x}) = [q^k{}_l]$ of the new corner $\widetilde{x}$ is derived from the old tableau $\mathbf{P}(x)$ by the well-known GAUSS-JORDAN step:

$$\begin{array}{llll} q^i{}_j = 1/p^i{}_j & \text{(pivot element)}\,, & q^k{}_j = p^k{}_j/p^i{}_j\,,\ k \neq i & \text{(pivot column)}\,, \\ q^i{}_l = -p^i{}_l/p^i{}_j\,,\ l \neq j & \text{(pivot row)}\,, & q^k{}_l = p^k{}_l - p^k{}_j p^i{}_l/p^i{}_j & \text{(others)}. \end{array}$$

If x^* denotes the unique non-degenerate solution of the primal problem $\min\{a\,x\,;\ Bx \leq c\}$ then y^* is the unique non-degenerate solution of the dual problem

$$\max\{-yc\,;\ yB = a\,,\ y \geq 0\}\,, \quad y \in \mathbb{R}_m\,,$$

and conversely. The tableau $\widetilde{\mathbf{P}}(y)$ of the dual method may be written in a way that in optimum $\mathbf{P}(x^*) = \widetilde{\mathbf{P}}(y^*)$. In particular, the edge matrix A has the same dimension in both problems if it is written appropriately. Therefore one may say that the computational amount of work is by and large the *same* in both the primal and the associated dual problem.

If x is a non-optimal corner of (P) then y in tableau $\mathbf{P}(x)$ is non-feasible w.r.t. (D) and conversely, if y is a non-optimal corner of (D) then the point x associated by (3.22) is non-feasible for (P). Consequently the solution x^* of (P) is approximated from outside the feasible domain if the problem (D) is solved instead of (P). Likewise the solution y^* of (D) is approximated outside the feasible domain if the primal problem (P) is solved instead of (D). Similar results hold also for the more general problems dealed with in Theorem 3.7.

A more detailed discussion of the primal-dual interdependencies, described here briefly, is found in directory `SUPPLEMENT\chap03b`.

Example 3.3. $f(x, y) := x + y = \max!\,,\ \ 0 \leq x \leq 6\,,\ \ 0 \leq y \leq 5\,,$
$3x + 2y \leq 20\,,\ \ x + 2y \leq 12$ (Figs. 3.7, 3.8).

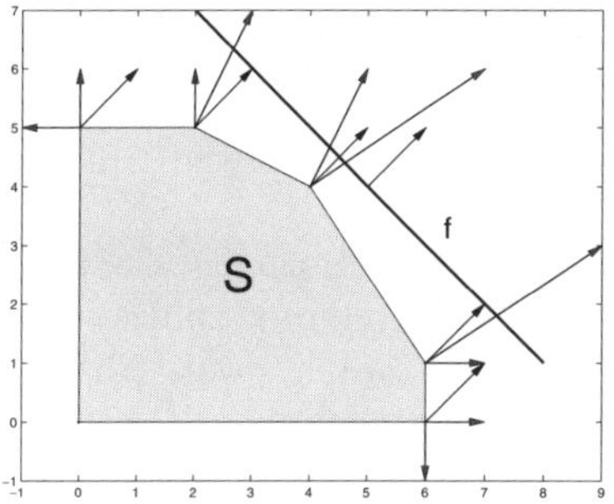

Figure 3.7. Basis of corners

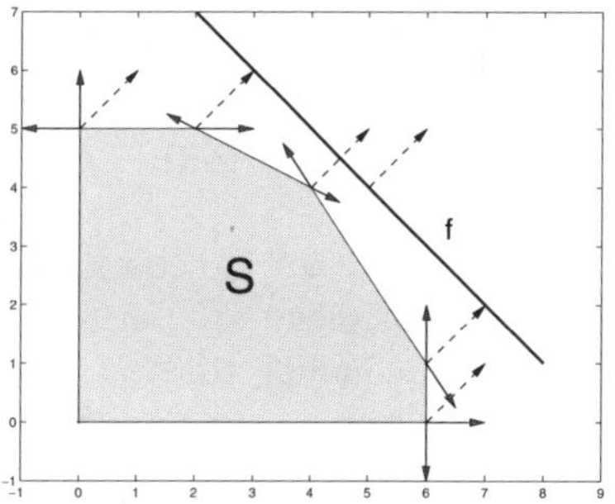

Figure 3.8. Reziprocal basis of corners

The present approach of linear optimization goes back to RITTER & BEST (Best). An extensive documentation with associated MATLAB suite is found in KAPITEL03\LOP.

3.4 Linear-Quadratic Problems

(a) Primal Projection Method Let $A \in \mathbb{R}^n{}_n$ be a symmetric positive definite matrix and let $B \in \mathbb{R}^m{}_n$ rank-maximal with $m \le n$. Then the linear-quadratic optimization problem (LQP) with equality constraints reads:

$$\boxed{\begin{aligned} f(x) &= \frac{1}{2}x^T Ax - a^T x = \min! \\ h(x) &= Bx + b = 0 \end{aligned}} \;. \tag{3.24}$$

The problem is convex and the multiplier rule of Sect. 3.2 provides therefore a necessary and sufficient condition for a *regular* solution x:

$$\boxed{x \text{ feasible and } \exists\, z : \nabla f(x) + z\nabla h(x) = 0 \in \mathbb{R}_n} \;.$$

Theorem 3.8. *(LQP primal) Let the assumption of Lemma 1.2 be fulfilled and let*

$$\begin{bmatrix} A & B^T \\ B & 0 \end{bmatrix}\begin{bmatrix} x_0 - x^* \\ z^* \end{bmatrix} = \begin{bmatrix} \nabla f(x_0)^T \\ h(x_0) \end{bmatrix} \tag{3.25}$$

for some $x_0 \in \mathbb{R}$ then x^* is solution of (3.24).

Proof. The first row of (3.25) is equivalent to

$$A(x_0 - x^*) + B^T z^* = Ax_0 - a \equiv \nabla f(x_0)^T$$

or

$$0 = (-Ax^* + a) + B^T z^* \equiv -\nabla f(x^*)^T + \nabla h(x^*)^T z^*$$

which is the multipier rule for $-z^*$ instead of z^*. The second row yields $h(x^*) = 0$ because h linear. Therefore x^* is feasible and optimal by Theorem 3.3 since the problem is convex. □

The optimal solution is found here in a single descend step $x^* = x_0 - \sigma d$ with *optimal* step length $\sigma = 1$. In more general linear-quadratic problems (3.25) is repeatedly applied to the current *active restrictions* in the same way as in linear problems.

Equalities are again active constraints being *never* inactivated and of which the LAGRANGE multipliers are not sign-restricted. Therefore it suffices again to consider only inequality constraints in the sequel. Note however that the primal method needs a *feasible* point x_0 at the beginning.

Consider now the linear-quadratic problem with inequality restrictions,

$$\boxed{\begin{aligned} f(x) &= \frac{1}{2}x^T Ax - a^T x = \min!\,, \quad x \in \mathbb{R}^n \\ g(x) &= Bx + b \qquad\quad \geq 0 \in \mathbb{R}^m \end{aligned}} \tag{3.26}$$

and let again $\mathcal{S} = \{x \in \mathbb{R}^n,\ g(x) \geq 0\}$ be the set of feasible points. The slight modifications of the inequality constraints has only optical reasons and we write correspondingly

$$L(x, y) = f(x) - yg(x)\,, \quad g(x) \geq 0$$

for the LAGRANGE function in order that the LAGRANGE multipliers y become non-negative again. The set of KUHN-TUCKER points, i.e., the set of all points satisfying the multiplier rule, then reads:

$$\Omega = \{x \in \mathcal{S}\,,\ \exists\, y \geq 0 : \nabla_x L(x, y) = 0\,,\ \ y\,g(x) = 0\}\,.$$

As customary in this context, degenerated extreme points shall be excluded in advance and, by consequence, the rank condition in Sect. 3.3 has to be modified slightly:

In each point x of the iteration, let $\mathcal{A}(x)$ be the index set of *all* active constraints in x and let the matrix $B^{\mathcal{A}} := B^{\mathcal{A}(x)} := [b^i]_{i \in \mathcal{A}(x)} \equiv [\nabla g^i(x)]_{i \in \mathcal{A}(x)} \in \mathbb{R}^n{}_n$ be regular.	(3.27)

Theorem 3.9. *Let the* rank condition *hold in a point $x \in \mathbb{R}^n$, let $H \in \mathbb{R}^n{}_n$ be an arbitrary symmetric matrix such that $\forall\, 0 \neq u \in \mathrm{Ker}(B^{\mathcal{A}}) : u^T Hu > 0$, and let $(d\,, y^{\mathcal{A}})$ be a solution of the linear system ($\mathcal{A} := \mathcal{A}(x)$)*

$$\begin{bmatrix} H & [B^{\mathcal{A}}]^T \\ B^{\mathcal{A}} & 0 \end{bmatrix} \begin{bmatrix} d \\ y^{\mathcal{A}} \end{bmatrix} = \begin{bmatrix} \nabla f(x)^T \\ 0 \end{bmatrix}. \tag{3.28}$$

Then x is KUHN-TUCKER *point if and only if x feasible and $d = 0$ as well as $y^{\mathcal{A}} \geq 0$.*

Proof. The system (3.28) has a unique solution by Lemma 1.2. If x is a solution of (3.26) hence KUHN-TUCKER point then the multiplier rule does hold and (3.28) has a solution w with the named properties. On the other side, let x be feasible and let the system (3.28) have a solution where $d = 0$ and $y^{\mathcal{A}} \geq 0$ then the multiplier rule is fulfilled if the remaining components of y, namely $y^{\mathcal{N}}$, are set equal to zero. Thus x is a KUHN-TUCKER point. □

Theorem 3.9 represents the kernel of the algorithms `plgp.m` and `dlqp.m` of quadratic programming which on their side constitute the kernel of the nonlinear gradient projection method `gp.m` resp. of the penalty method `sqp.m`.

All these methods work by using the gradient of *active* constraints which have to be selected in each step by means of the LAGRANGE multipliers in a way that a descend to the minimum takes place altogether. The speed of the method is heavily influenced by a proper choice of the matrix H but here we have to refer to the respective literature. In most simple case $H = I$ is chosen which corresponds to the simple method of steepest descend with all its drawbacks. But, on the other side, different choices of H destroy normally a possible *sparsity* of the system (3.28).

(b) The Algorithm `plqp.m` for (3.24):

START: Find a *feasible* x such that $B^{\mathcal{A}}$ regular, let $y^{\mathcal{A}} \not\geq 0$ arbitrary, $y^{\mathcal{N}} = 0$.
WHILE NOT $y^{\mathcal{A}} \geq 0$
(1°) Get $(d,\, y^{\mathcal{A}})$ by solving the linear system (3.28).
By Theorem 3.8, $\widetilde{x} = x - d$ is global minimum of f on the set

$$\mathcal{M} := \{u \in \mathbb{R}^n,\ \ g^{\mathcal{A}(x)}(u) = 0\}\,.$$

(2°) If $d = 0$ and $y^i := \min\{y^k\,,\ k \in \mathcal{A}\} < 0$, inactivate g^i:
set $\mathcal{A}(x) := \mathcal{A}(x)\backslash\{i\}$. Repeat step 1.
(3°) If $d \neq 0$ then find optimal step length for descend direction $-d$:

$$\sigma := \min\left\{\varphi(k) > 0\,,\ \ \varphi(k) = \frac{g^k(x)}{b^k d} > 0\,,\ \ k \notin \mathcal{A}(x)\right\}\,,$$
$$j := \min \arg_k \min\{\varphi(k) > 0\,,\ \ k \notin \mathcal{A}(x)\}$$

σ is the maximum feasible step length; if σ does not exist, set $\sigma = 1$.
Set $x := x - \sigma d$ and $\mathcal{A}(x) := \mathcal{A}(x) \cup \{j\}$.
END

Lemma 3.2. *The direction d in (3.28) is non-zero after inactivation of an active constraint hence actually $f(x - \sigma d) < f(x)$.*

Proof in `SUPPLEMENT\chap03c`.

Cycles cannot arise in this algorithm under the regularity condition that $B^{\mathcal{A}}$ is always rank-maximal. At least in every second step $d \neq 0$ by Lemma 3.2 and a genuine descend takes place. Therefore also the index set $\mathcal{A}$ does change in every step without repeating. Since there is only a finite number of subsets of the total index set, the algorithm breaks off sometime. A feasible initial point x_0 may be found by the procedure described in the proof of Theorem 3.6.

(c) Dual Projection Method In the same way as above, equality constraints are handled here as active inequality constraints never being inactivated if once activated, and the corresponding LAGRANGE multipliers are not sign-restricted again. If the algorithm breaks off, all equality constraints must be active otherwise the problem has no solution. The ingenious method of (Goldfarb) presented here plays a crucial role in *sequential quadratic programming* discussed in next section. Mainly it allows to start from an arbitrary, not necessarily feasible point. But, on the other side, the solution is approximated from outside the feasible domain in normal case such that the obtained numerical result does *not* satisfy the constraints exactly.

We use the same notations as in **(a)**. By Sect. 3.6**(d)** the minimum problem is equivalent to the LAGRANGE *problem*

$$\boxed{\begin{array}{l}\text{Find } (x^*,y^*)\in\mathbb{R}^n\times\mathbb{R}^m \text{ such that}\\ L(x^*,y^*)=\max_{y\geq 0}\inf_x\left\{\dfrac{1}{2}x^TAx-a^Tx-y^T(Bx+b)\right\}\end{array}} \qquad (3.29)$$

where we have only a simple sign restriction as inequality constraint. Of course the problem is convex as before hence the multiplier rule is sufficient for a solution. Starting from the absolute minimum $x = A^{-1}a$ of the objective function in (3.24), $f(x)$ is magnified until the multiplier rule is satisfied or, in other words, until the iterated point x reaches the feasible domain (as far as possible).

Notations and Conventions:

(1°) $g^i(x) > 0 \Longrightarrow y^i = 0$.
(2°) $\mathcal{K} := \{1,\ldots,m\}$ und $\mathcal{J} \subset \mathcal{K}$ index sets; at beginning mostly $\mathcal{J} = \emptyset$.
(3°) The problem

$$\{f(x)\,;\ g^j(x) := b^jx + \beta^j \geq 0\,,\ j\in\mathcal{J}\} = \min!$$

is briefly denoted by LQ($\mathcal{J}$).
(4°) $(x,\mathcal{B})$ is called solution pair of LQ($\mathcal{J}$) if x is solution of LQ($\mathcal{J}$) and $g^j(x) = 0$ for all $j \in \mathcal{B} \subset \mathcal{J}$ as well as $B^{\mathcal{B}} = [\nabla g^j(x)]_{j\in\mathcal{B}}$ has maximum row-rank.

Then always $\mathcal{B} \subset \mathcal{A}(x)$ but not necessarily $\mathcal{A}(x) \subset \mathcal{J}$. Conversely, if (4°) holds for some $(x,\mathcal{B})$ then there exists an index set $\mathcal{J} \supset \mathcal{B}$ such that $(x,\mathcal{B})$ is solution pair of LQ($\mathcal{J}$); the index set $\mathcal{J}$ however plays a minor role below. Now we obtain at once the following intermediate result by (4°) and Theorem 3.8:

Let $(d, y^{\mathcal{B}})$ be solution of

$$\begin{bmatrix} H & [B^{\mathcal{B}}]^T \\ B^{\mathcal{B}} & O \end{bmatrix}\begin{bmatrix} d \\ y^{\mathcal{B}} \end{bmatrix} = \begin{bmatrix} \nabla f(x)^T \\ 0\ (\equiv g^{\mathcal{B}}(x)) \end{bmatrix},$$

let $y^{\mathcal{B}} \geq 0$ and $g^{\mathcal{B}}(x) = 0$. Then $(x+d,\mathcal{B})$ is a solution pair of LQ($\mathcal{B}$) and solution of LQ($\mathcal{J}$) for all $\mathcal{J}$ such that $\mathcal{B} \subset \mathcal{J}$ and $g^j(x) \geq 0\,,\ j \in \mathcal{J}$.

($H = I$ in the MATLAB suite.) Therefore $S(\mathcal{J}) := \{x \in \mathbb{R}^n,\ g^j(x) \geq 0,\ j \in \mathcal{J}\}$ is the set of feasible points for LQ($\mathcal{B}$).

After these preliminaries, the algorithm `dlqp.m` for (3.24) reads roughly:

START: $x = A^{-1}a = \arg\min f(x)$. Let a solution pair $(x, \mathcal{B})$ (possibly $\mathcal{B} = \emptyset$) be specified $\Longrightarrow\ \exists\, \mathcal{J}$ where $\mathcal{B} \subset \mathcal{J}$.
WHILE NOT $g(x) \geq 0$
(1°) Choose p such that $g^p(x) < 0$, i.e. $p \notin \mathcal{J}$.
(2°) If LQ($\mathcal{B} \cup \{p\}$) unsolvable then STOP, ((3.26) unsolvable, $\mathcal{S} = \emptyset$) .
(3°) Else: Find new solution pair $(\widetilde{x}\,,\ \widetilde{\mathcal{B}} \cup \{p\})$ such that $\widetilde{\mathcal{B}} \subset \mathcal{B}$ and $f(\widetilde{x}) > f(x)$.
Set $(x\,,\ \mathcal{B}) := (\widetilde{x}\,,\ \widetilde{\mathcal{B}} \cup \{p\})$ (also $\mathcal{B} = \widetilde{\mathcal{B}}$ (!)).
END

However, in activation of a further restriction with index $p \notin \mathcal{B}$, some or even all conditions of $\mathcal{B}$ may become infeasible! The latter case implies a complete re-start of the procedure in practice with different initial value.

A more detailed description of the method `dlqp.m` is found in `SUPPLEMENT\chap03g`.

Example 3.4.

$$f(x) = 6x_1 + 2(x_1^2 - x_1x_2 + x_2^2) = \min!$$
$$x_1 \geq 0\,,\ \ x_2 \geq 0\,,\ \ x_1 + x_2 \geq 2\,.$$

For solving with the dual method, the absolute minimum $\underline{x}_0 = [-2, -1]$ is taken for initial vector where all three side conditions are inactive. In the first step, the most strongly violated condition is activated being the condition with index

$$i = \min \arg_k \min\{g^k(\underline{x}_0)\}\,.$$

In Figure 3.9 however all conditions g^1 , g^2 , g^3 are taken for first condition one after the other. The different iterational sequences then are

Path 1: $\underline{x}_0 \longrightarrow \underline{x}_1 \longrightarrow \underline{x}_2 \longrightarrow \underline{x}_3$
Path 2: $\underline{x}_0 \longrightarrow \underline{x}_4 \longrightarrow \underline{x}_5 \longrightarrow \underline{x}_3$
Path 3: $\underline{x}_0 \longrightarrow \underline{x}_3$

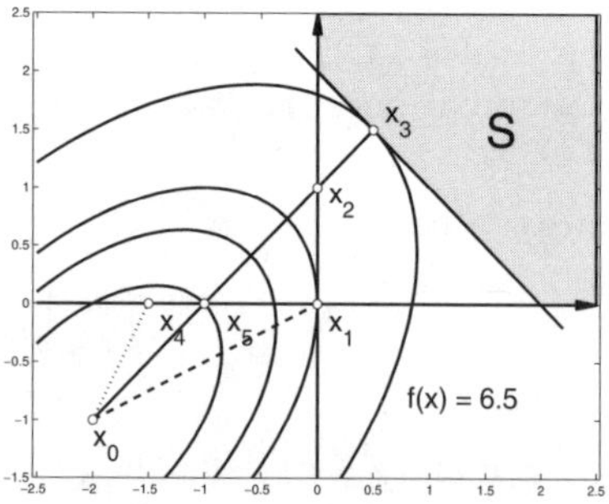

Figure 3.9. Example 3.4

Example 3.5. Solve by the dual method $\{f(x)\,,\ g(x) \geq 0\} = \min!$ where

$$
\begin{aligned}
f(x) &= \frac{1}{2}(x_1^2 + x_2^2) + 10x_1 + 2x_2 \\
g_1(x) &= x_1 + x_2 \\
g_2(x) &= 3 - x_2 \\
g_3(x) &= -x_1 - x_2 + 5 \\
g_4(x) &= x_1 - x_2 + 2 \\
g_5(x) &= 5 - x_1 \\
g_6(x) &= x_2 + 1 \\
g_7(x) &= x_1 + 2x_2 \\
g_8(x) &= x_1 - 2x_2 + 4
\end{aligned}
$$

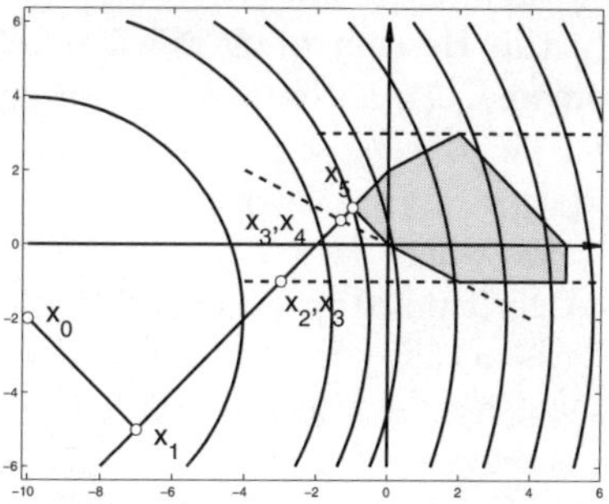

Figure 3.10. Example 3.5

References: (Goldfarb), (Spellucci).

3.5 Nonlinear Optimization

We consider the general nonlinear optimization problem

$$\{f(x)\,;\ x \in \mathbb{R}^n\,,\ g(x) \geq 0 \in \mathbb{R}^m\,,\ h(x) = 0 \in \mathbb{R}^p\} = \min! \tag{3.30}$$

and use the same notations and conventions as in the preceding section. In particular, $L(x,y,z) = f(x) - y\,g(x) + z\,h(x)$ is the LAGRANGE function, $\mathcal{A}(x)$ is the index set of *all* active constraints, and the rank condition (3.27) is supposed to hold in each point x during iteration.

(a) The **Gradient Projection Method** for (3.30) runs by and large as in the linear-quadratic case but the gradients of objective function and side conditions now depend on x which makes of course some modifications necessary. *Linear equations* $h(x) = 0$ are handled without problems if one proceeds in the same way as before. But *nonlinear* equations may lead to some difficulties since the method works only with points of the feasible domain. Above all, a *feasible initial vector* has to be found and afterwards equalities $h(x) = 0 \in \mathbb{R}^p$ must always remain active during iteration. In the sequel, equalities are handled again as inequalities never being inactivated and, by consequence, only inequality restrictions are studied as in the preceding section.

(b) In a **Typical Iteration Step** the linear system (3.28),

$$\begin{bmatrix} H & [B^{\mathcal{A}}]^T \\ B^{\mathcal{A}} & 0 \end{bmatrix} \begin{bmatrix} d \\ y^{\mathcal{A}} \end{bmatrix} = \begin{bmatrix} \nabla f(x)^T \\ 0 \end{bmatrix}, \tag{3.31}$$

has to be solved. The computational speed of the method is again controlled by the choice of the symmetric, positive definite matrix H where the unit matrix is chosen for H in the most simple case. In convex problems, the choice $H = \nabla_x^2 L(x, y, z)$ leads to asymptotic quadratic convergence since the method then resembles NEWTON's method. But the evaluation of this matrix, namely the Hessian of L relative to the variable x in each step is rather cumbersome. An acceptable compromise between the both extremals $H = I$ and $H = \nabla_x^2 L(x, y, z)$ may be an application of the BFGS-method of Sect. 2.1 for updating of H. Observe however that a possible *sparsity* of H is destroyed in all cases where $H \neq I$.

In Algorithm `plqp.m`, a condition $g^i(x),\ i \in \mathcal{A}(x)$, selected for inactivation in $(2°)$, is actually inactivated if $d = 0$. This rule does no longer hold in the nonlinear case. A selected condition is also inactivated in case $d \neq 0$ if this action is *successful*. This means that the *angle* between the descend direction d and the local optimal direction $\nabla f(x)$ is not too large or, more precisely,

$$\nabla f(x)\, d \geq \gamma\, |y_i| \tag{3.32}$$

must be fulfilled with a suitable constant $\gamma > 0$, say $\gamma = 1/2$. The involved LAGRANGE multipliers are taken from (3.31) numerically. In case condition (3.32) is violated, the iteration is repeated with same d but without inactivation. Also a second modification is necessary to prevent permanent *jamming* or *zigzagging* between different active side conditions. The inactivation of an active side condition is prohibited as long as $\mathcal{A}(\widetilde{x}) \not\subset \mathcal{A}(x)$ until the first time $\mathcal{A}(\widetilde{x}) \subset \mathcal{A}(x)$ or $d = 0$. Earlier "proofs" of convergence did not regard this necessary condition. See also algorithm `gp.m` in `KAPITEL03\SECTION_5`.

(c) Restoration Suppose that the (possibly local) *absolute* minimum of the objective function f resp. the stationary point lies outside or on the boundary of the feasible domain. If the initial point lies in interior then iteration moves first to the boundary. In linear-quadratic problems, iteration remains on the boundary of the polytope if once arrived there. In nonlinear problems the situation is entirely different. Iteration may return into the interior by the above two modifications. On the other side, the straight search direction d leads to the exterior in curvilinear boundaries (in normal case). But all points of iteration must be feasible in this method therefore the new point $\widehat{x}$ must be moved back onto the boundary of the feasible domain by a suitable projection. Under unfavorable geometric constellations this pull back does not succeed in a single step and the direction d has perhaps to shortened several times until the projection meets the boundary really. This expensive operation called "restoration" may slow down the iteration considerably and may even deteriorate exactness in some cases.

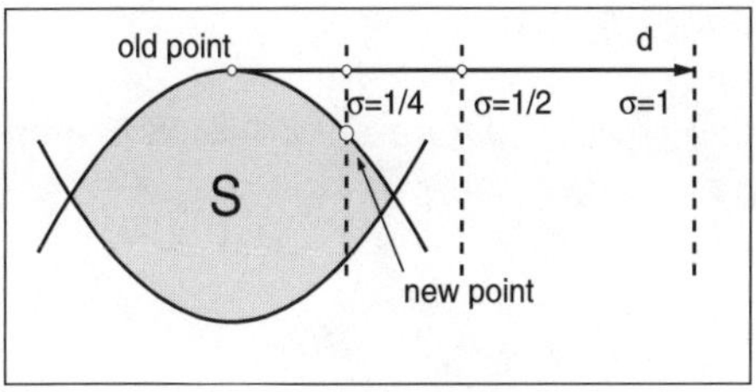

Figure 3.11. Restoration

As mentioned repeatedly, the gradient method needs a feasible point at beginning which is frequently difficult to find. Then *restoration* ensures in each step that iteration moves back to the boundary of the feasible domain. This is but the shortcoming of the method but warrants on the other side that the approximative solution regards all restrictions. Penalty methods are able to start with an arbitrary point (at least theoretically) but in normal case the solution is approached from outside the feasible domain. Consequently such methods are not applicable in a problem, say, with the objective function

$$f(x) = \sum_{i=1}^{10} x^i \left(c^i + \ln \left[x^i / \sum_{i=1}^{10} x^i \right] \right)$$

and side condition $x \geq 0$, cf. (Himmelblau), p. 395.

(d) Penalty Methods In this iterative methods, a stationary point of the LAGRANGE function is no longer the aim of desire but a *penalty function* is introduced which prosecutes violations of feasibility. From a formal point of view the penalty function resembles the LAGRANGE-function to some degree but the penalty costs play an entire different role (geometrically and physically) by contrast with the former LAGRANGE multipliers. Among the various possibilities we choose here the penalty function after (Zangwill),

$$\boxed{\begin{aligned} P(x,y,z) = f(x) - yg(x)^- + z|h(x)|\,, \quad [g^i(x)]^- = \min\{g^i(x)\,,\, 0\} \\ g(x)^- = \left[[g^i(x)]^-\right]_{i=1}^m, \quad |h(x)| = \left[|h^j(x)|\right]_{j=1}^p \end{aligned}}\,.$$

This scalar function is but only one-sided differentiable at some points but it is *exact*: $P(x,y,z) = f(x)$ does hold for all *feasible* x. The first, negative property may be got under control whereas the second, positive property entails crucial advantages numerically because of sharper modelling near an optimum point. Also it is essential in applying this method that the penalty costs y and z, once specified, remain *not* constant during iteration but vary in adaption to the local geometric situation in the individual step of iteration.

In the following two illustrations, ZANGWILL' function is compared with the likewise exact classical (differentiable) penalty function

$$Q(x,y,z) = f(x) + \sum_{i=1}^{m} y_i \max\{0, -g_i(x)\}^2 + \sum_{i=1}^{p} z_i h_i(x)^2\,.$$

Example 3.6. (Spellucci), p. 456.

$$f(x) = x(x-1)(x-3)(x-5)/8 = \min!$$
$$g(x) = x(4-x) \geq 0,$$

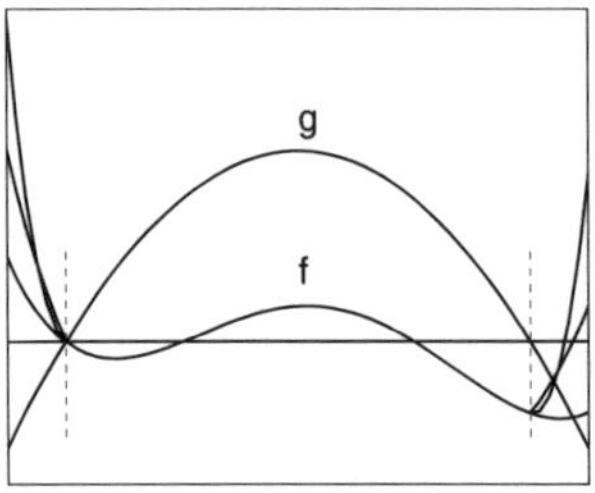

Figure 3.12. Example 3.6

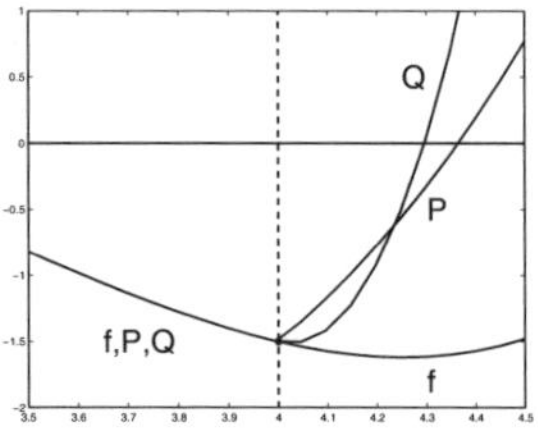

Figure 3.13. Example 3.6, scaled

Example 3.7. Same as Example 3.9 below.

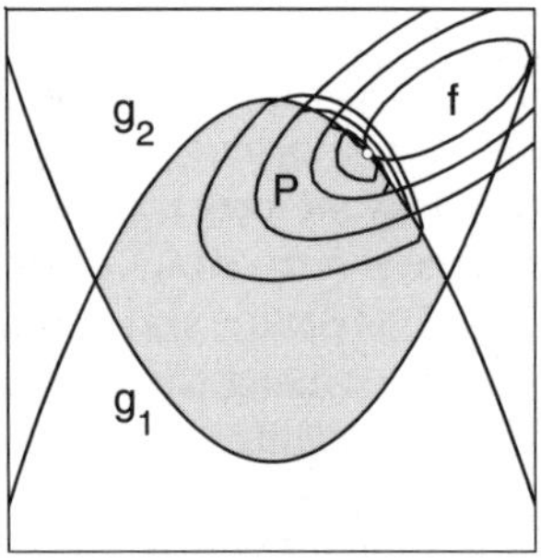

Figure 3.14. Example 3.7, $\beta = 5$, Zangwill penalty function

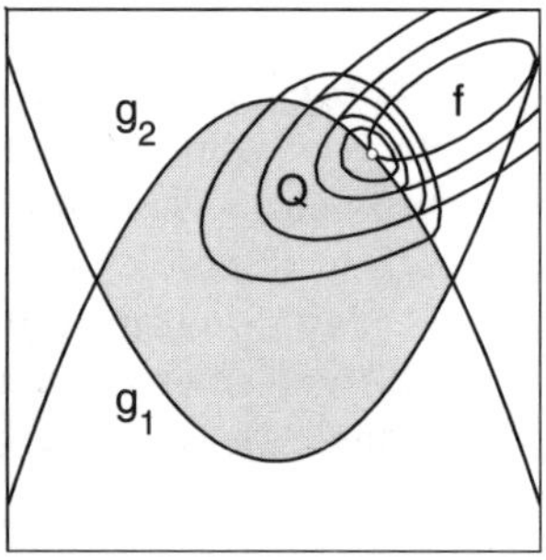

Figure 3.15. Example 3.7, $\beta = 5$, classic penalty function

Notations: (1°)

$$\begin{aligned}
e &= [1, \dots, 1]^T \in \mathbb{R}^m \\
\mathcal{A}(x) &= \{i \in \{1, \dots, m\}\,,\ g^i(x) = 0\} \\
\mathcal{B}(x) &= \{i \in \{1, \dots, m\}\,,\ g^i(x) > 0\} \\
\mathcal{V}(x) &= \{i \in \{1, \dots, m\}\,,\ g^i(x) < 0\} \quad \text{(violated constraints)} \\
\mathcal{S}(\tau) &= \{x \in \mathbb{R}^n\,,\ g(x) \geq -\tau\, e\,,\ |h(x)| \leq \tau\, e\}\,,\ \tau \geq 0 \\
&\quad \text{hence } \mathcal{S}(0) = \mathcal{S}\,,\ \ \mathcal{S}(\infty) = \mathbb{R}^n.
\end{aligned}$$

(2°) The "augmented MANGASARIAN-FROMOWITZ (constraint) qualification" (AMF) is used in this method for regularity condition:

$$\exists\, \tau_0 > 0 \;\; \forall\, x \in \mathcal{S}(\tau_0) \;\; \exists\, v \in \mathbb{R}^n :$$
$$\nabla h(x)v = 0\,, \;\; \forall\, i \in \mathcal{A}(x) \cup \mathcal{V}(x) : \nabla g(x)_i v > 0\,, \;\; \nabla h(x) \text{ is row-regular.} \tag{3.33}$$

(3°) Let x be *fixed* and $H \in \mathbb{R}^n{}_n$ a symmetric, positive definite matrix. The following linear-quadratic problem for $u \in \mathbb{R}^n$ is applied below and called briefly QP(x,H):

$$\begin{aligned}
\mathit{f}(u;x) &:= f(x) + \nabla f(x)(u-x) + \frac{1}{2}(u-x)^T H(u-x)\\
g(u;x) &:= g(x) + \nabla g(x)(u-x) \geq 0\\
h(u;x) &:= h(x) + \nabla h(x)(u-x)\,.
\end{aligned}$$

(4°) SLATER Condition for QP(x,H):

$$\exists\, u \in \mathbb{R}^n : g(x) + \nabla g(x)(u-x) > 0\,,\; h(x) + \nabla h(x)(u-x) = 0\,.$$

The essential properties of the penalty function are summarized in the following theorem:

Theorem 3.10. (1°) *Let AMF hold for some $\tau_0 > 0$ and let $\mathcal{S}(\tau_0)$ be compact. Then the penalty function $P(x,y,z)$ has no local minimum point in the complementary set $\mathcal{S}(\tau_0)\backslash\mathcal{S}$.*
(2°) *Let AMF hold for some $\tau_0 > 0$ and let x^* be a strong local minimum of the objective function f then x^* is a local minimum of $P(x,y,z)$ if y and z are sufficiently large.*
(3°) *Let the* SLATER *condition hold for QP(x,H) then there are penalty vectors $y > 0$ and $z > 0$ such that*

$$\delta^+ P((x;d),y,z) < 0 \tag{3.34}$$

where u^* is the unique solution of QP(x,H), $d = u^* - x$ and $\delta^+ P((x;d),y,z)$ denotes the one-sided directional derivative of P w.r.t. x in direction of d.
(4°) *Let u^* be solution of QP(x,H) and $d = u^* - x$ then, for $0 \leq \sigma \leq 1$,*

$$\begin{aligned}
\|h(x+\sigma\, d)\|_1 &= (1-\sigma)\|h(x)\|_1 + \mathcal{O}(\sigma^2)\\
\|g(x+\sigma\, d)^-\|_1 &= (1-\sigma)\|g(x)^-\|_1 + \mathcal{O}(\sigma^2)\,.
\end{aligned}$$

Proof see e.g. (Spellucci).

The last assertion says that the infeasibility of the equality as well as the inequality restrictions *decreases* for sufficiently small $\sigma > 0$.

(e) The Algorithm `sqp.m` for (3.30):

PARAMETER: $\varepsilon > 0$, δ, $\alpha \in (0,1)$, e.g. $\delta = 0.1$, $\alpha = 0.5$, $\tau_0 > 0$, $d = e$, tol.
START: Choose start vector $x_0 \in \mathcal{S}(\tau_0)$, $y = 0$, $z = 0$.
WHILE NOT $|\sigma\, d| <$ tol
(1°) Choose H symmetric positive definite, e.g. $H = I$;
Solve QP(x, H) completely with solution, say, $(u, \widetilde{y}, \widetilde{z})$, set $d = u - x$, $\widetilde{y}_{\mathcal{N}} = 0$
(2°) Set for all i

$$\begin{aligned} y_i &:= \widetilde{y}_i + 2\varepsilon \;\; \text{if } \widetilde{y}_i + \varepsilon \geq y_i \;\; \text{else unchanged} \\ z_i &:= |\widetilde{z}_i| + 2\varepsilon \text{ if } |\widetilde{z}_i| + \varepsilon \geq z_i \text{ else unchanged.} \end{aligned}$$

(3°) Find step length $\sigma > 0$ by repeated halving such that σ maximal and

$$P(x,y,z) - P(x+\sigma\, d, y, z) \geq \sigma\delta\left[d^T H d + \varepsilon\|g(x)^-\|_1 + \varepsilon\|h(x)\|_1\right] =: \eta$$

as well as — if $x \in \mathcal{S}(\tau_0)\backslash\mathcal{S}(\tau_0/2)$ —

$$\begin{aligned} &\|g(x)^-\|_1 - \|g(x+\sigma d)^-\|_1 + \|h(x)\|_1 - \|h(x+\sigma d)\|_1 \\ &\geq \sigma\delta\left[\|g(x)^-\|_1 + \|h(x)\|_1\right] \end{aligned}$$

(4°) Set $x := x + \sigma\, d$
END

(f) Supplements

(f1) In the algorithm of **(e)**, the penalty parameters become unnecessarily large near the solution with bad effect on numerical stability. Therefore (Spellucci) has proposed a modification of (2°) for calculation of these weights where however the initial vector x_0 is involved; see `SUPPLEMENT\chap03c` and the algorithm `sqp.m` in `KAPITEL03\SECTION_5`. By this way the penalties y and z may be reduced again during iteratin and altogether they are more properly adapted to geometric constellation of the individual step of iteration.

(f2) If the problem is non-convex but well-behaved, the feasible set of the quadratic subproblem QP(x, H) may be empty. In this case the *modified* problem QP(x, H, ξ)

$$\begin{aligned} &\nabla f(x)d + \frac{1}{2}d^T H d && = \min! \\ &\xi\, g(x)^{\mathcal{A}(x)\cup\mathcal{V}(x)} + \nabla g(x)^{\mathcal{A}(x)\cup\mathcal{V}(x)} d && \geq 0 \\ &g(x)^{\mathcal{B}(x)} + \nabla g(x)^{\mathcal{B}(x)} d && \geq 0 \\ &\xi\, h(x) + \nabla h(x) d && = 0 \end{aligned} \tag{3.35}$$

has to be solved instead. Hereby the parameter $\xi \in (0,1]$ has to be diminuished repeatedly until the resulting problem (3.35) enjoys a non-empty feasible set and thus a solution. Such a ξ exists always under assumption AMF. In less well-behaved non-convex problems, this parameter ξ may become too small during the repeated search for a feasible domain then the entire procedure is overcharged and breaks off.

(f3) The speed of convergence may be improved by a proper choice of the matrix H also in this method and, with some restrictions, also the attainable exactness. The optimal choice of $H = \nabla_x^2 L(x,y,z)$ is however too time-consuming hence the BFGS-method is preferred for updating. In non-convex case some modification of the BFGS-method may be lead to a certain improvement, too. Note however once more that both modifications destroy the sparsity of the matrix H which entails again an increasement of computational effort especially in large systems.

(f4) If convergence becomes too slow near the solution, correctures of second order $\widetilde{x} = x + \sigma d + \sigma^2 q$ may provide some improvement (Maratos effect).

(g) Examples for gradient method (GPM) and penalty method (SQP).

Example 3.8. (Spellucci), p. 397, p. 457. $a(x,y) = (x+y-3.5)^2$, $b(x,y) = 4(y-x+0.5)^2$.

$$\begin{aligned} f_1(x,y) &= \frac{100}{a(x,y)+b(x,y)} = \min!\,, && (x^*,y^*) = (-1,0)\,,\\ f_2(x,y) &= (x-y)^2 + (y-1)^2 = \min!\,, && (x^*,y^*) = (0.546, 0.702)\,,\\ & 1-x^2+y \geq 0\,,\ \ 1-x^2-y \geq 0\,. \end{aligned}$$

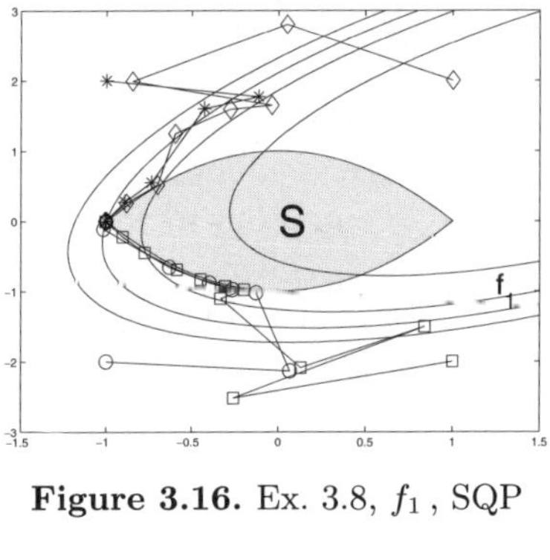

Figure 3.16. Ex. 3.8, f_1, SQP

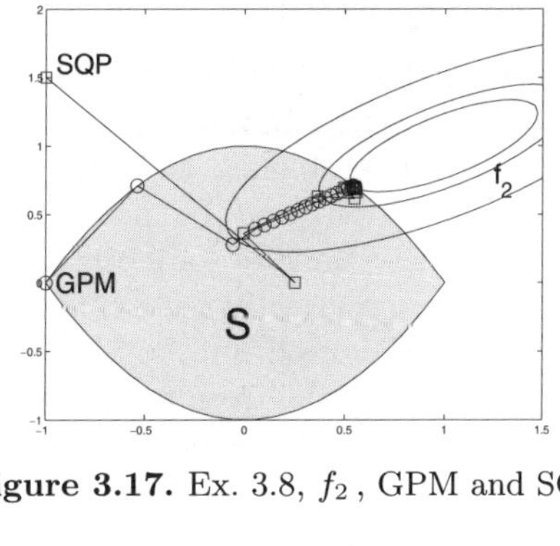

Figure 3.17. Ex. 3.8, f_2, GPM and SQP

Example 3.9.

$$\begin{aligned} & f_1(x,y) = (x-8)^2/4 + (y-1)^2 = \min!\,, \ \ (x^*,y^*) = (3.45, 0.19)\,.\\ & f_2(x,y) = (x-8)^2/4 + (y-3)^2 = \min!\,, \ \ (x^*,y^*) = (2.43, 1.27)\,.\\ & 25-(x-4)^2-y^2 \geq 0\,,\ \ 30-(x+1)^2-(y+3)^2 \geq 0\\ & 30-(x+1)^2-(y-3)^2 \geq 0 \end{aligned}$$

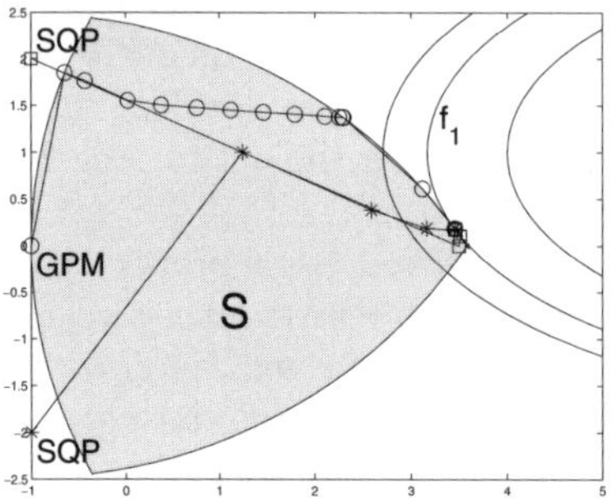

Figure 3.18. Ex. 3.9, f_1 , GPM and SQP

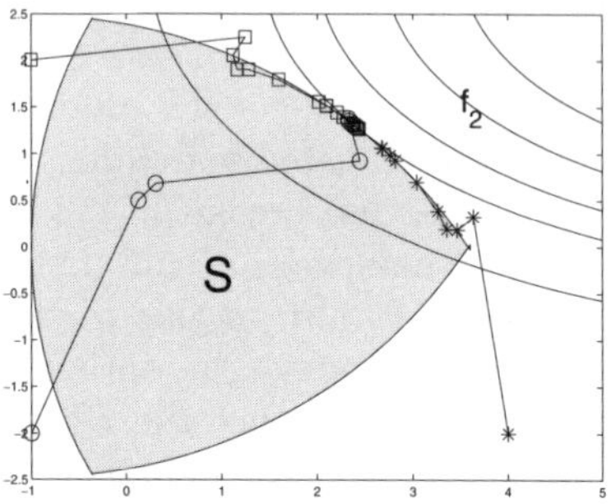

Figure 3.19. Ex. 3.9, f_2 , SQP

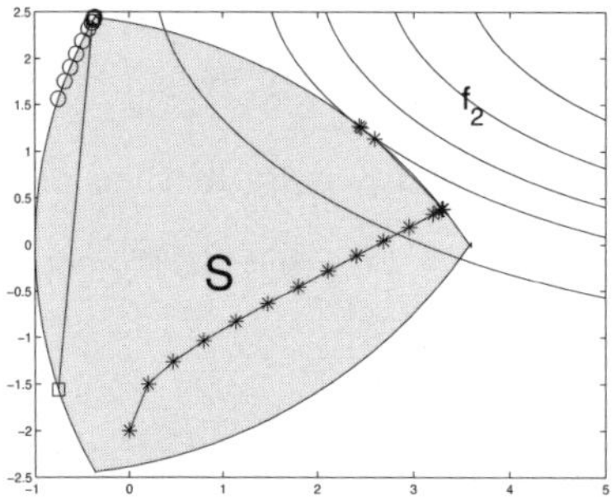

Figure 3.20. Ex. 3.9, f_2 , GPM, $\gamma = 0.5$

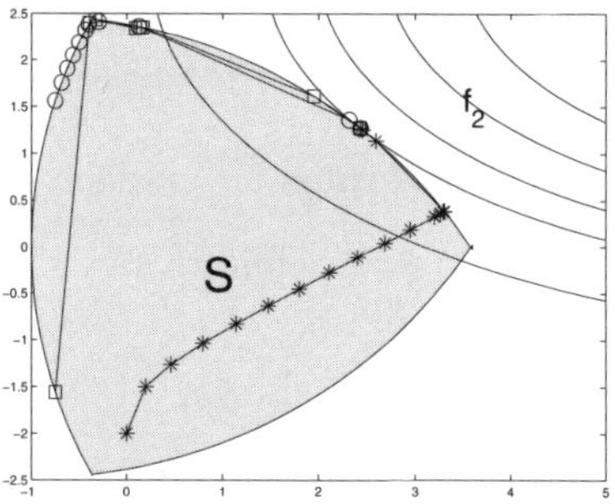

Figure 3.21. Ex. 3.9, f_2 , GPM, $\gamma = 0.1$

γ denotes the parameter of inactivation in (3.32)

Up today the numerical approach of nonlinear optimization problems is not so elaborated as e.g. that of linear-quadratic problems. In any case there remains some imponderableness since the systems may be very large and the geometric constellations may be arbitrary complicated. An adaption to the specific problem may provide some improvement or also the gradients are no longer computed analytically but approximated by discretization (MATLAB program available). The SQP-method needs a careful adjusting of the various parameters for acceptable results and convergence is only verified for a minor class of problems. In others one has to be content with a "good" result improving the nominal solution whereas the optimal solution remains unreached. This remark holds in particular for non-convex problems which appear frequently in technical applications and may have several local solutions.

The algorithms `gp.m` and `sqp.m` presented in `KAPITEL03` have admittedly some model character. Especially the second procedure may still be improved by applying some propositions dealed with in (Spellucci).

3.6 A Brief on Lagrange Theory

FARKAS theorem and the efficient dual method of quadratic programming — being more genuinely a primal-dual method — show already that some dual approach enters automatically into studies of optimization and control problems. *Saddlepoint problems* appear not only in this context but also in continuum theory. For instance the (weak) stationary NAVIER-STOKES equations have such a form, cf. Theorem 9.1, and in the treatment of solid bodies they appear as so-called *mixed methods* which allow a higher flexibility in numerical approach (Brezzi). Therefore we re-consider here once more the features of Sect. 3.2 from a more general point of view.

(a) Formulation of the Problem Let $\mathcal{X}$, $\mathcal{Y}$, $\mathcal{Z}$ be real normed vector spaces, let $\emptyset \neq \mathcal{C} \subset \mathcal{X}$ be an arbitrary set being not necessarily open or a subspace, and let $\emptyset \neq \mathcal{K} \subset \mathcal{Y}$ be an order cone with dual cone $\mathcal{K}_d$ in dual space $\mathcal{Y}_d$ of $\mathcal{Y}$, cf. Sect. 1.10. Further, let

$$f : \mathcal{C} \to \mathbb{R}\,, \quad g : \mathcal{C} \to \mathcal{Y}\,, \quad h : \mathcal{C} \to \mathcal{Z}\,,$$

be three mappings and consider the general *minimum problem* (MP)

$$\{f(x)\,;\; x \in \mathcal{C}\,,\; -g(x) \in \mathcal{K}\,,\; h(x) = 0\} = \min! \tag{3.36}$$

with feasible set $\mathcal{S} = \{x \in \mathcal{C}\,,\; -g(x) \in \mathcal{K}\,,\; h(x) = 0\}$ and LAGRANGE function

$$L : \mathcal{C} \times \mathcal{Y}_d \times \mathcal{Z}_d \ni (x, y, z) \mapsto L(x, y, z) = f(x) \pm y \circ g(x) + z \circ h(x) \in \mathbb{R}\,.$$

The sign of y is positive in the minimum problem and negative in the corresponding maximum problem; cf. Sect. 3.2**(b)**. The problem (3.36) is called *convex* again if $\mathcal{C}$, f convex, g $\mathcal{K}$-convex, and h *affine linear*. For $-g(x) \in \mathcal{K}$ we write briefly $g(x) \leq 0$ and observe that, in the present situation, the LAGRANGE multipliers y and z are elements of the dual spaces $\mathcal{Y}_d$ and $\mathcal{Z}_d$, resp. may be canonically identified with elements of these spaces. Altogether we are faced with the following constellation:

mapping:	f	g	h
range:	$\mathbb{R}$	$\mathcal{Y}$	$\mathcal{Z}$
order cone:	$\mathbb{R}_{\geq 0}$	$\mathcal{K}$	$\mathcal{L} = \{0\}$
dual elements:	$\varrho \in \mathbb{R}$	$y \in \mathcal{Y}_d$	$z \in \mathcal{Z}_d$

.

The equality restrictions $h(x) = 0$ may *not* be replaced by double inequalities because we have to suppose sometimes that the interior $\mathrm{int}(\mathcal{K})$ of $\mathcal{K}$ is not empty. In slight generalization, the restrictions differ from each other by the two order cones $\mathcal{K}$ and $\mathcal{L}$ according to

$$g(x) \overset{\mathcal{K}}{\leq} 0\,,\; \mathrm{int}(\mathcal{K}) \neq \emptyset\,, \quad h(x) \overset{\mathcal{L}}{\leq} 0\,,\; \mathrm{int}(\mathcal{L}) = \emptyset\,.$$

The fundamental Theorem 3.2 now reads:

Theorem 3.11. *Let* $(x^*, y^*, z^*) \in \mathcal{C} \times \mathcal{K}_d \times \mathcal{Z}_d$ *be a triple such that*

$$x^* = \arg \begin{matrix}\min\\ \max\end{matrix} \{f(x^*) \pm y^* \circ g(x^*) + z^* \circ h(x^*)\,,\; x \in \mathcal{C}\}\,, \tag{3.37}$$

and let $x^* \in \mathcal{S}$ *as well as* $y^* \circ g(x^*) = 0$ *. Then*

$$x^* = \arg \begin{matrix}\min\\ \max\end{matrix} \{f(x)\,,\; x \in \mathcal{C}\,,\; g(x) \leq 0\,,\; h(x) = 0\}\,.$$

(b) Lagrange Problem Originally, LAGRANGE multipliers have been introduced as proportionality constants of artificial constraint forces in the equation of motion (NEWTON's axiom) and nobody has had some reason to doubt their existence. Later on problems became more difficult and applicants were interested whether a solution exists at all when the computer supplies only some nonsense. So the *proof of existence* of LAGRANGE multipliers became more and more important and takes today a large place in mathematical theory of optimization and control. The alternative notation "costate" (variables) demonstrates already their importance here. Different attempts have been made to create a *unified theory* for both types of constraints, inequalities and equalities, together but the latter cause some difficulties here because the associated trival order cone $\mathcal{L}$ has no interior. The general impression remains somewhat inlucid up today certainly also since regularity conditions can be ordered just as little hierarchically in infinite-dimensional problems as in finite-dimensional problems; cf. Sect. 3.2**(b)**. For problems with equality constraints only, (Luenberger) has proved the existence of LAGRANGE multipliers z in a different way by a simple application of the generalized Range Theorem; see Theorem 1.19. Separation theorems are not used in LUENBERGER's proof but the likewise deep result of Theorem 1.23 on generalized inverse functions.

Let us first summarize some properties of the LAGRANGE function:

Lemma 3.3. (1°) $x \in \mathcal{S} \implies f(x) = L(x, 0, 0) = \max_{(y,z) \in \mathcal{K}_d \times \mathcal{Z}_d} L(x, y, z)\,.$
(2°) *Let the order cone* $\mathcal{K}$ *in* $\mathcal{Y}$ *be closed, let* $x \in \mathcal{C}$ *and suppose*

$$\exists\, (y^*, z^*) \in \mathcal{K}_d \times \mathcal{Z}_d : L(x, y^*, z^*) = \max_{(y,z) \in \mathcal{K}_d \times \mathcal{Z}_d} L(x, y, z)\,,$$

then $x \in \mathcal{S}$ *and* $y^* \circ g(x) = 0$ *.*
(3°) *Let the cone* $\mathcal{K}$ *be closed and* $x \in \mathcal{C}$ *then*
$\mathcal{S} = \emptyset \iff \sup_{(y,z) \in \mathcal{K}_d \times \mathcal{Z}_d} L(x, y, z) = \infty\,.$

By these three results we obtain directly

Theorem 3.12. *(*LAGRANGE *problem sufficient) Let the order cone* $\mathcal{K} \subset \mathcal{Y}$ *be closed and let*
$(x^*, y^*, z^*) \in \mathcal{C} \times \mathcal{K}_d \times \mathcal{Z}_d$ *be a triple such that*

$$\boxed{(x^*, y^*, z^*) = \arg\min_{x \in \mathcal{C}} \sup_{(y,z) \in \mathcal{K}_d \times \mathcal{Z}_d} L(x, y, z)}\,, \tag{3.38}$$

then x^* *is solution of the minimum problem (3.36).*

$\sup_{(y,z)\in\mathcal{K}_d\times Z_d} L(x,y,z)$ is not necessarily attained for all $x \in \mathcal{C}$ but for $x^* \in \mathcal{C}$ by assumption.

Corollary 3.1. *The minimum problem (MP) (3.36) and the* LAGRANGE *problem (LP) (3.38) are equivalent, i.e., if x^* is a solution of (MP) and $(\widetilde{x}^*, y^*, z^*)$ is a solution of (LP) then $x^* = \widetilde{x}^*$.*

The inversion of Theorem 3.12, namely the question for existence of LAGRANGE multipliers, is more difficult to answer as in all proofs of existence.

Definition 3.2. (SLATER *Condition) Let $\mathcal{X}$, $\mathcal{Y}$ be real normed vector spaces, let $\mathcal{C} \subset \mathcal{X}$ convex,*
$\mathcal{K} \subset \mathcal{Y}$ a positive cone with non-empty interior, and $g : \mathcal{C} \to \mathcal{Y}$ $\mathcal{K}$-convex. Then g satisfies the SLATER *condition if*

$$\boxed{\exists\, x \in \mathcal{C} : g(x) < 0\,,\ h(x) = 0}\,,$$

i.e. $\{x \in \mathcal{C}\,,\ g(x) < 0\} \neq \emptyset$. in the case where $h = 0$.

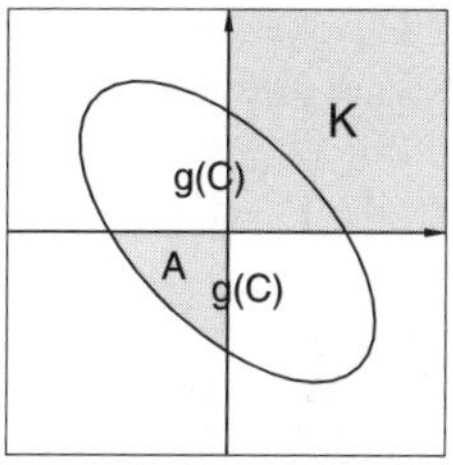

Figure 3.22. SLATER condition: $A \neq \emptyset$

Theorem 3.13. (F. JOHN, *Existence)* (1°) *Let $\mathcal{X}$, $\mathcal{Y}$ be normed vector spaces and let $\mathcal{K} \subset \mathcal{Y}$ be an order cone with non-empty interior,* $\mathrm{int}(\mathcal{K}) \neq \emptyset$.
(2°) *Let x^* be a solution of the convex minimum problem $\{f(x)\,;\ x \in \mathcal{C}\,,\ g(x) \leq 0\}$, i.e. $\mathcal{C} \subset \mathcal{X}$ convex, $f : \mathcal{C} \to \mathbb{R}$ convex and $g : \mathcal{C} \to \mathcal{Y}$ $\mathcal{K}$-convex.*
(3°) $\exists\, x \in \mathcal{C}\ \forall\, 0 \neq y \in \mathcal{K}_d : y \circ g(x) < 0$.
Then there exists a $0 \neq y^* \in \mathcal{K}_d$ such that

$$x^* = \arg\min\{f(x) + y^* \circ g(x)\,,\ x \in \mathcal{C}\}\,.$$

Assumption (3°) is a slightly weakened SLATER condition. The proof uses the Separation Theorem 1.24 as essential tool; see `SUPPLEMENT\chap03d`. Note also that no assumptions at all occur concerning smoothness of the data.

F. JOHN's theorem can be generalized to mixed problem with additional equality restrictions but observe that $h(x) = 0$ is convex if and only h is affine linear. Also some additional regularity conditions are necessary if we maintain the requirement that the LAGRANGE multipliers are elements of the *topological* dual spaces and not only elements of the *algebraic* dual spaces (Kirsch). We present here two results of (Kosmol) which display the general situation rather completely.

Definition 3.3. *Let $\mathcal{X}$ be a normed vector space and $\mathcal{C}, \mathcal{D} \subset \mathcal{X}$.*
(1°) $\mathrm{aff}(\mathcal{C})$ *is the smallest affine subspace of $\mathcal{X}$ which contains $\mathcal{C}$, $\mathcal{C} \subset \mathrm{aff}(\mathcal{C})$.*
(2°) *Let $x \in \mathcal{C} \cap \mathcal{D}$ then x is interior point of $\mathcal{C}$ relative to $\mathcal{D}$ if there exists a neighborhood of x in $\mathcal{D}$ which is entirely contained in $\mathcal{C}$:*

$$\exists\, \varepsilon > 0,\ \forall\, u \in \mathcal{D} : \|u - x\| \le \varepsilon \Longrightarrow u \in \mathcal{C}.$$

(3°) $\mathrm{relint}(\mathcal{C})$ *is the set of interior points of $\mathcal{C}$ relative to* $\mathrm{aff}(\mathcal{C})$.

Let e.g. $h : \mathcal{C} \to \mathcal{Z}$ affine linear then $\mathrm{relint}(h(\mathcal{C})) \ne \emptyset$, if $\mathcal{Z}$ finite-dimensional or $\mathrm{relint}(\mathcal{C}) \ne \emptyset$.

Theorem 3.14. *(Existence) Let x^* be solution of the convex minimum problem*

$$\{f(x)\,;\ x \in \mathcal{C},\ g(x) \le 0,\ h(x) = 0\} = \mathrm{min!}, \tag{3.39}$$

(h affine linear), let $\mathcal{G} = \{x \in \mathcal{C},\ g(x) \le 0\}$. Suppose that
(1°) $\mathrm{int}(\mathcal{K}) \ne \emptyset$, (2°) $h : \mathcal{C} \to Z = \mathbb{R}^m$ finite-dimensional,
(3°)*(i)* $\exists\, x \in \mathcal{C} : g(x) < 0$ *and* $0 \in \mathrm{relint}(h(\mathcal{G}))$ **or**
(3°)*(ii)* $0 \in \mathrm{relint}(h(\mathcal{C}))$ *and* $\exists\, x \in \mathcal{C} : g(x) \le 0,\ h(x) = 0$.
Then there exists a pair $(y^*, z^*) \in \mathcal{K}_d \times \mathcal{Z}_d$ with $y^* \ne 0$ such that

$$x^* = \min\{f(x) + y^* \circ g(x) + z^* \circ h(x),\ x \in \mathcal{C}\}$$

and $y^* \circ g(x^*) = 0$.

This result together with Lemma 3.3 provides an inversion of Theorem 3.12:

Corollary 3.2. *(LAGRANGE problem necessary) Adopt the assumptions of Theorem 3.14, and let x^* be a solution of the minimum problem (3.39). Then there exists*

$$(x^*, y^*, z^*) = \arg\min_{x \in \mathcal{C}} \sup_{(y,z) \in \mathcal{K}_d \times \mathcal{Z}_d} L(x, y, z)$$

and $y^* \circ g(x^*) = 0$.

The restriction that the image space $\mathcal{Z}$ of h is finite-dimensional can be cancelled by involving some fundamental results of functional analysis but then continuity comes into the play and $\mathcal{C} = \mathcal{X}$ must be the full vector space.

Theorem 3.15. *(Existence) Let $\mathcal{X}, \mathcal{Z}$ be BANACH spaces and $\mathcal{Y}$ is normed with order cone $\mathcal{K} \subset \mathcal{Y}$. Let x^* be solution of the convex minimum problem*

$$\{f(x)\,;\ x \in \mathcal{C},\ g(x) \le 0,\ h(x) = 0,\ r(x) = 0\} = \mathrm{min!},$$

where $f : \mathcal{X} \to \mathbb{R}$ *is convex,* $g : \mathcal{X} \to \mathcal{Y}$ *is* $\mathcal{K}$*-convex,* $h : \mathcal{X} \to \mathbb{R}^m$ *is affine linear,* $r : \mathcal{X} \to \mathcal{Z}$ *is affine linear and continuous, and* $r(\mathcal{X})$ *closed. Suppose also that* $\mathrm{int}(\mathcal{K}) \neq \emptyset$ *and* $\exists\, x \in \mathcal{X} : g(x) < 0\,,\ h(x) = 0\,,\ r(x) = 0\,.$

Then there exists a triple $(y^*, z^*, w^*) \in \mathcal{K}_d \times \mathbb{R}_m \times \mathcal{Z}_d$ *with* $y^* \neq 0$ *such that*

$$x^* = \arg\min\{f(x) + y^* \circ g(x) + z^* \circ h(x) + w^* \circ r(x),\ x \in \mathcal{X}\}$$

and $y^* \circ g(x^*) = 0\,.$

The proofs of both theorems 3.14 and 3.15 use F. John's theorem and Eidelheit's theorem again, see (Kosmol) and SUPPLEMENT\chap03d. Of course an analogous result to Corollary 3.2 does holds also w.r.t. Theorem 3.15.

(c) Saddlepoint Problems In the following both results, $\mathcal{A}$, $\mathcal{B}$ denote arbitrary sets and $\Phi : \mathcal{A} \times \mathcal{B} \ni (x, y) \mapsto \Phi(x, y) \in \mathbb{R}$ is an arbitrary function.

Lemma 3.4.

$$\sup{}_{y\in\mathcal{B}} \inf{}_{x\in\mathcal{A}} \Phi(x, y) \leq \inf{}_{x\in\mathcal{A}} \sup{}_{y\in\mathcal{B}} \Phi(x, y)\,.$$

A pair $(x^*, y^*) \in \mathcal{A} \times \mathcal{B}$ is called *saddlepoint* of Φ if

$$\forall\, x \in \mathcal{A}\ \ \forall\, y \in \mathcal{B} : \Phi(x^*, y) \leq \Phi(x^*, y^*) \leq \Phi(x, y^*)\,. \tag{3.40}$$

Theorem 3.16. *The function* Φ *has a saddlepoint in* $\mathcal{A} \times \mathcal{B}$ *if and only if*

$$\boxed{\max{}_{y\in\mathcal{B}} \inf{}_{x\in\mathcal{A}} \Phi(x, y) = \min{}_{x\in\mathcal{A}} \sup{}_{y\in\mathcal{B}} \Phi(x, y)}\,. \tag{3.41}$$

Note that e.g. "max" instead "sup" says that the supremum is attained actually. Note also that a saddlepoint (x^*, y^*) satisfies

$$\max{}_{y\in\mathcal{B}} \inf{}_{x\in\mathcal{A}} \Phi(x, y) = \Phi(x^*, y^*) = \min{}_{x\in\mathcal{A}} \sup{}_{y\in\mathcal{B}} \Phi(x, y)\,,$$

but not every point with this property must be a saddlepoint as the simple example $\Phi : \mathbb{R}^2 \ni (x, y) \mapsto \Phi(x, y) = x \cdot y$ shows.

Now a *saddlepoint problem* (SPP) is associated to the minimum problem by means of the Lagrange function L:

Find a *saddlepoint* $(x^*, y^*, z^*) \in \mathcal{C} \times \mathcal{K}_d \times \mathcal{Z}_d$ such that

$$\boxed{\forall\, (x, y, z) \in \mathcal{C} \times \mathcal{K}_d \times \mathcal{Z}_d :\ L(x^*, y, z) \leq L(x^*, y^*, z^*) \leq L(x, y^*, z^*)}\,. \tag{3.42}$$

Theorem 3.17. *((SPP) sufficient) Let the order cone* $\mathcal{K}$ *be closed. If* (x^*, y^*, z^*) *is a saddlepoint hence solution of (3.42) then* x^* *is minimum point hence solution of (3.36).*

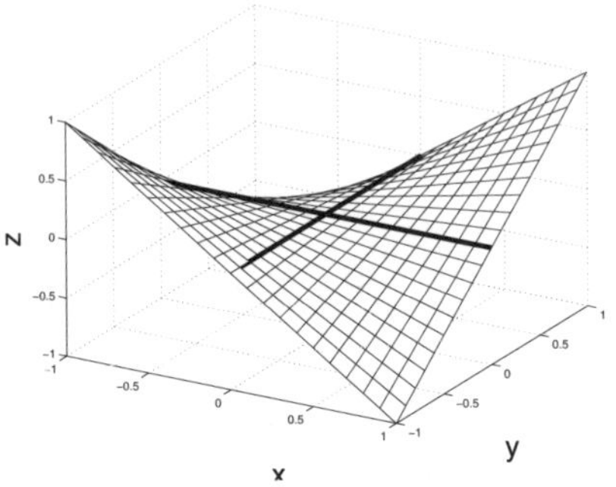

Figure 3.23. Image of $\Phi(x, y) = x \cdot y$

Proof. (1°) The left inequality of (3.42) shows for $z = z^*$

$$y \circ g(x^*) \leq y^* \circ g(x^*) \,. \tag{3.43}$$

For all $y \in \mathcal{K}_d$ we have $y + y^* \in \mathcal{K}_d$ because $\mathcal{K}_d$ convex. Substituting $y + y^*$ for y in (3.43) we obtain $\forall\, y \in \mathcal{K}_d : y \circ g(x^*) \leq 0$. By the Cone Corollary 1.21 then $g(x^*) \leq 0$ and also $y^* \circ g(x^*) \leq 0$. Setting $y = 0$ in (3.43), $y^* \circ g(x^*) \geq 0$ hence together $y^* \circ g(x^*) = 0$.
(2°) Setting $y = y^*$ in the left inequality of (3.42), $(z - z^*) \circ h(x^*) \leq 0$. Then $h(x^*) = 0$ since $z - z^* \in Z_d$ arbitrary. Therefore x^* is feasible.
(3°) Using (1°) and (2°) we obtain by the right inequality of (3.42) for feasible x^*

$$f(x^*) + 0 \leq f(x) + y^* \circ g(x) \leq f(x)$$

since $y^* \in \mathcal{K}_d$ hence x^* is a minimum point. □

Theorem 3.18. *((SPP) necessary in convex problems) Adopt the assumptions of Theorem 3.14, and let x^* be a solution of the minimum problem (3.39). Then there exists a pair $(0,0) \neq (y^*, z^*) \in \mathcal{K}_d \times \mathcal{Z}_d$ such that the* Lagrange *function $L(x, y, z) = f(x) + y \circ g(x) + z \circ h(x)$ has a saddlepoint (x^*, y^*, z^*),*

$$\forall\, (x, y, z) \in \mathcal{C} \times \mathcal{K}_d \times \mathcal{Z}_d : L(x^*, y, z) \leq L(x^*, y^*, z^*) \leq L(x, y^*, z^*)$$

and $y^* \circ g(x^*) = 0$.

Proof. Because of Theorem 3.14, only the left inequality has to be verified. But it is equivalent to $y \circ g(x^*) \leq y^* \circ g(x^*) = 0$ and is thus true because $y \geq 0$ and $g(x^*) \leq 0$. □

An analogous result does hold also w.r.t Theorem 3.15. A direct consequence of Theorem 3.16 is now

Corollary 3.3.

$$\boxed{\max{}_{y \in \mathcal{K}_d} \inf{}_{x \in \mathcal{C}} L(x, y) = \min{}_{x \in \mathcal{C}} \sup{}_{y \in \mathcal{K}_d} L(x, y)}$$

does hold if and only if the minimum problem (MP) has a saddlepoint.

(d) Primal and Dual Problems By the results shown hitherto in this section, three problems stand in mutually relation to each other:

(1°) *Minimum problem* (MP): Find $x^* \in \mathcal{X}$ such that

$$x^* = \arg\min\{f(x)\,,\ x \in \mathcal{C} \subset \mathcal{X}\,,\ g(x) \le 0\,,\ h(x) = 0\}\,. \tag{3.44}$$

(2°) *Primal* LAGRANGE *problem* (LP): Find a triple $(x^*, y^*, z^*) \in \mathcal{X} \times \mathcal{K}_d \times \mathcal{Z}_d$ such that

$$L(x^*, y^*, z^*) = \min{}_{x\in\mathcal{C}} \sup{}_{(y,z)\in\mathcal{K}_d\times\mathcal{Z}_d} L(x, y, z)\,. \tag{3.45}$$

(3°) *Dual* LAGRANGE *problem* (DLP): Find a triple $(x^*, y^*, z^*) \in \mathcal{X} \times \mathcal{K}_d \times \mathcal{Z}_d$ such that

$$L(x^*, y^*, z^*) = \max{}_{(y,z)\in\mathcal{K}_d\times\mathcal{Z}_d} \inf{}_{x\in\mathcal{C}} L(x, y, z)\,. \tag{3.46}$$

We introduce the *primal functional* φ and the *dual functional* ψ:

$$\begin{aligned}
&\varphi : \mathcal{X} && \ni x \mapsto \varphi(x) && := \sup{}_{(y,z)\in\mathcal{K}_d\times\mathcal{Z}_d} L(x, y, z) \in \mathbb{R} \cup \{\infty\}\,,\\
&\psi : \mathcal{Y}_d \times \mathcal{Z}_d && \ni (y, z) \mapsto \psi(y, z) && := \inf{}_{x\in\mathcal{C}} L(x, y, z) \in \mathbb{R} \cup \{\infty\}\,.
\end{aligned}$$

φ is convex and ψ is concave if the minimum prroblem is convex. Then by (3.45) and (3.46) directly

$$\begin{aligned}
&(\text{LP}): && \{\varphi(x)\,,\ x \in \mathcal{S}\} && = \min!\,,\ \ \mathcal{S} = \{x \in \mathcal{C}\,,\ \varphi(x) < \infty\}\,,\\
&(\text{DLP}): && \{\psi(y, z)\,,\ (y, z) \in \mathcal{T}\} && = \max!\,,\\
&&&&& \mathcal{T} = \{(y, z) \in \mathcal{K}_d \times \mathcal{Z}_d\,,\ \psi(y, z) > -\infty\}\,,
\end{aligned} \tag{3.47}$$

Theorem 3.19. *(Weak Duality)*
(1°) *Let $x \in \mathcal{S}$ and $(y, z) \in \mathcal{T}$ then $\psi(y, z) \le \varphi(x)$.*
(2°) *Let x^*, y^* be feasible and $\psi(y^*, z^*) = \varphi(x^*)$ then x^* is solution of (LP) and (y^*, z^*) is solution of (DLP).*

Proof. Because $x \subset \mathcal{S}$ and $y \in \mathcal{K}_d$ we have

$$\psi(y, z) \le f(x) + y \circ g(x) + z \circ h(x) \le f(x) \le \varphi(x)\,.$$

The rest is clear. □

By this way, the solution of the dual problem is a lower bound of the solution of the primal problem and the solution of the primal problem is an upper bound of the solution of the dual problem. If

$$\psi(y^*, z^*) < f(x^*) = \varphi(x^*)\,,$$

for the respective solutions of (MP) and (DLP) then one speaks of a *duality gap*. Additional assumptions become necessary to avoid duality gaps. For instance such gaps do *not* occur if (MP) is linear or if a saddlepoint does exist.

Summary (LP := LAGRANGE problem)
Problem 1:

$$
\begin{aligned}
x^* &= \arg\min_{x\in\mathcal{C}}\{f(x)\,;\ g(x)\le 0\,,\ h(x)=0\} && \text{(minimum problem)}\\
L(x,y,z) &= f(x)+y\circ g(x)+z\circ h(x) && \text{(LAGRANGE function)}\\
(x^*,y^*,z^*) &= \arg\min_{x\in\mathcal{C}}\sup_{y\ge 0,z} L(x,y,z) && \text{(primal LP)}\\
(x^*,y^*,z^*) &= \arg\max_{y\ge 0,z}\inf_{x\in\mathcal{C}} L(x,y,z) && \text{(dual LP)}
\end{aligned}
$$

Problem 2:

$$
\begin{aligned}
x^* &= \arg\max_{x\in\mathcal{C}}\{f(x)\,;\ g(x)\le 0\,,\ h(x)=0\} && \text{(maximum problem)}\\
L(x,y,z) &= f(x)-y\circ g(x)+z\circ h(x) && \text{(LAGRANGE function)}\\
(x^*,y^*,z^*) &= \arg\max_{x\in\mathcal{C}}\inf_{y\ge 0,z} L(x,y,z) && \text{(primal LP)}\\
(x^*,y^*,z^*) &= \arg\min_{y\ge 0,z}\sup_{x\in\mathcal{C}} L(x,y,z) && \text{(dual LP)}
\end{aligned}
$$

When the transformation of (LP) into (MP) resp. of (LP) in the dual problem (DLP) is not sufficiently transparent, the following result may supply some additional information.

Lemma 3.5. *Let the order cone $\mathcal{K}\subset\mathcal{Y}$ be closed.*
(1°) *For (LP)*

$$
\begin{aligned}
&\textit{Problem 1:}\ \mathcal{S}=\{x\in\mathcal{C}\,,\ \sup_{y\ge 0,z} L(x,y,z)<\infty\}\,,\\
&\textit{Problem 2:}\ \mathcal{S}=\{x\in\mathcal{C}\,,\ \inf_{y\ge 0,z} L(x,y,z)>-\infty\}\,.
\end{aligned}
$$

(2°) *For (DLP)*

$$
\begin{aligned}
&\textit{Problem 1:}\ \mathcal{T}=\{(y,z)\in\mathcal{K}_d\times\mathcal{Z}_d\,,\ \inf_{x\in\mathcal{C}} L(x,y,z)>-\infty\}\,,\\
&\textit{Problem 2:}\ \mathcal{T}=\{(y,z)\in\mathcal{K}_d\times\mathcal{Z}_d\,,\ \sup_{x\in\mathcal{C}} L(x,y,z)<\infty\}\,.
\end{aligned}
$$

(e) Geometric Interpretation Beside the analytic interpretation of the primal-dual interdependencies by means of saddlepoints there is also a geometric version due to (F.John) where we confine ourselves to inequalities for simplicity. Let $\varGamma=\{u\in\mathcal{Y}\,,\ \exists\,x\in\mathcal{C}:g(x)\le u\}$,

$$
\boxed{\omega:\varGamma\ni u\mapsto\omega(u)=\inf\{f(x)\,;\ x\in\mathcal{C}\,,\ g(x)\le u\}\in\mathbb{R}\cup\{\infty\}}
$$

and

$$
\begin{aligned}
\mathcal{A}_x &= \{(u\,,\ \varrho)\in\mathcal{Y}\times\mathbb{R}\,,\ g(x)\le u\,,\ f(x)\le\varrho\}\\
\mathcal{A} &= \{(u,\,\varrho)\in\mathcal{Y}\times\mathbb{R}\,,\ \exists\,x\in\mathcal{C}:g(x)\le u\,,\ f(x)\le\varrho\}\\
&= \{(g(x)+k,f(x)+\sigma)\,,\ x\in\mathcal{C}\,,\ k\in\mathcal{K}\,,\ \sigma\ge 0\}\,,\\
\mathcal{B} &= \{(u,\,\sigma)\in\mathcal{Y}\times\mathbb{R}\,,\ u\le 0\,,\ \sigma\le\omega(0)\}\,,
\end{aligned}
$$

Then in particular $(g(x)\,,\ f(x))\in\mathcal{A}_x$ and moreover

$$
\mathcal{A}=\bigcup_{x\in\mathcal{C}}\mathcal{A}_x=\{(u\,,\ \varrho)\in\varGamma\times\mathbb{R}\,,\ \omega(u)\le\varrho\}\,,
$$

hence the *perturbation function* ω describes the boundary surface of the set $\mathcal{A}$. The set $\mathcal{S}$ of feasible points is non-empty if and only if there exists a $\varrho \in \mathbb{R}$ such that $(0, \varrho) \in \mathcal{A}$, and then

$$\inf\{f(x)\,;\; x \in \mathcal{S}\} = \inf\{\varrho;\; (0, \varrho) \in \mathcal{A}\}\,.$$

We summarize some properties of ω and $\mathcal{A}$ in the following auxiliary result:

Lemma 3.6. (1°) *ω is weakly monotone decreasing.*
(2°) *The dual functional satisfies* $\psi(y) = \inf_{u\in\Gamma}\{\omega(u) + y \circ u\}\,,\; y \in \mathcal{K}_d\,.$
Let the minimum problem be convex then
(3°) *Γ is convex,*
(4°) *$\mathcal{A}$ is convex,*
(5°) $\mathrm{int}(\mathcal{A}) \cap \mathcal{B} = \emptyset\,,$
(6°) *ω is convex.*

Note also that $\psi(-y)$ is the LEGENDRE transformation of $\omega(u)$ because $\psi(-y) = \sup_{u\in\mathcal{Y}}\{y \circ u - \omega(u)\}\,.$

The computation of a solution of a convex minimum problem is now equivalent to the computation of the scalar value $\omega(0)$ which is the smallest point of intersection of $\mathcal{A}$ and the $\mathbb{R}$-axis. The *dual problem* consists then in the computation of a supporting hyperplane $\mathcal{H}$ to the set $\mathcal{A}$ which contains $\mathcal{A}$ in the positive half-space on the one side hand has a maximum point of intersection $(0, \sigma)$ with the $\mathbb{R}$-axis on the other side.

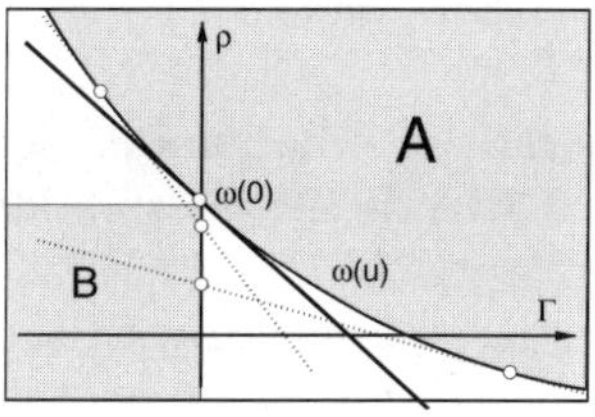

Figure 3.24. Duality, convex minimum problem

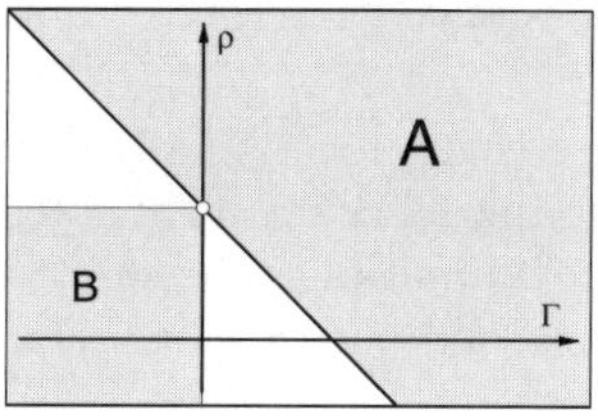

Figure 3.25. Linear problem

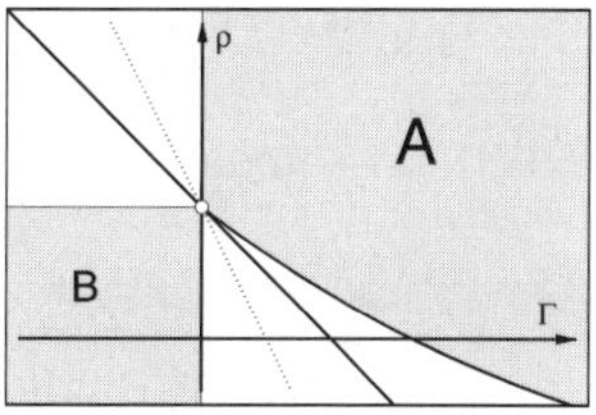

Figure 3.26. SLATER condition violated

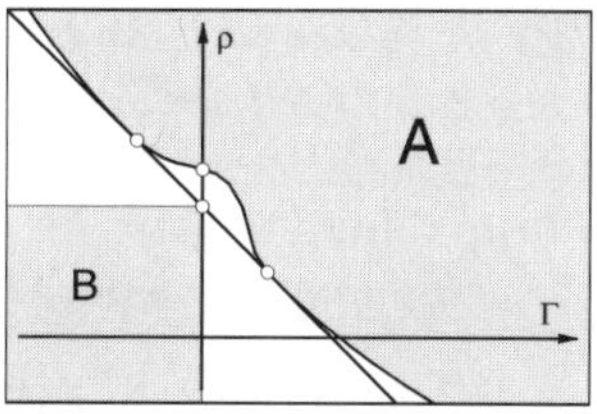

Figure 3.27. Duality gap, nonconvex problem

Let $\mathcal{H} \subset \mathcal{Y} \times \mathbb{R}$ denote a hyperplane not standing "perpendicular" on $\mathcal{Y}$ which means here that it does *not* contain the set $\{(0, \varrho) \in \mathcal{Y} \times \mathbb{R}\}$. Then in general, by Lemma 1.12

$$\mathcal{H}(y, \sigma) = \{(u, \varrho) \in \mathcal{Y} \times \mathbb{R},\ 1 \cdot \varrho + y \circ u = \sigma\},$$
$$\mathcal{H}(y, \sigma)_+ = \{(u, \varrho) \in \mathcal{Y} \times \mathbb{R},\ 1 \cdot \varrho + y \circ u \geq \sigma\} \quad \text{(positive half-space)}$$

where $(y, \sigma) \in \mathcal{Y}_d \times \mathbb{R}$. Consequently the dual problem reads:

$$\text{(DP):} \quad \{\sigma \in \mathbb{R} \cup \{\infty\},\ \mathcal{A} \subset \mathcal{H}(y, \sigma)_+\} = \max!$$

Set $\sigma = -\infty$ if none of these hyperplanes does exist. The following result characterizes the hyperplanes $\mathcal{H}$ in a somewhat different way:

Theorem 3.20. (1°)

$$\mathcal{A} \subset \mathcal{H}(y, \sigma)_+ \iff \forall\, x \in \mathcal{C}\ \ \forall\, k \in \mathcal{K} : f(x) + y \circ (g(x) + k) \geq \sigma\,,$$

(2°)

$$y \in \mathcal{K}_d \textit{ and } \sigma \leq \psi(y) \iff \forall\, x \in \mathcal{C}\ \ \forall\, k \in \mathcal{K} : f(x) + y \circ (g(x) + k) \geq \sigma\,.$$

By this result we now obtain the geometric form of the dual LAGRANGE problem: Find $(y^*, \sigma^*) \in \mathcal{K}_d \times \mathbb{R}$ such that

$$\boxed{\sigma^* = \max\{\sigma\,;\ x \in \mathcal{C},\ k \in \mathcal{K},\ y \in \mathcal{K}_d : f(x) + y \circ (g(x) + k) \geq \sigma\}}\,.$$

Note also that $y^* = -\nabla\omega(0)$ in the case where the perturbation function ω is differentiable at the point $u = 0$; cf. also Sect. 3.3**(h)** on the interpretation of LAGRANGE multipliers as shadow prices.

The second part of Theorem 3.20 provides the equivalence with the dual LAGRANGE problem $\sigma^* = \max_{y \geq 0} \psi(y)$ here; cf. (3.46). One sees that the geometric form (DP) of the dual problem is much less transparent than the analytic form (DLP) but it is undispensable for theoretical investigations.

The above equivalence relations can also be expressed by *geometric* properties of the set $\mathcal{A}$:

Theorem 3.21. *(Werner), Th. 4.3.2. Let the minimum problem (MP) be convex, let $\mathcal{Y}$ be normed, and let* $\text{int}(\mathcal{A}) \cap \{0\} \times \mathbb{R} \neq \emptyset$. *Then the feasible set $\mathcal{S}$ is not empty. Further*
(1°) *If* $\inf_{x \in \mathcal{S}} f(x) > -\infty$, *the dual problem (DLP) has a solution y^* and* $\max_{y \in \mathcal{T}} \psi(y) = \inf_{x \in \mathcal{S}} f(x)$.
(2°) *If the minimum problem (MP) has a solution* x^*, $y^* \circ g(x^*) = 0$.

Condition $\text{int}(\mathcal{A}) \cap \{0\} \times \mathbb{R} \neq \emptyset$ is fulfilled e.g. if the SLATER condition is fulfilled, i.e., if $\text{int}(K) \neq \emptyset$ and $g(x_0) < 0$ for some $x_0 \in \mathcal{C}$. Namely, in this case,

$$\{(g(x)+k, f(x)+\varrho) \in \mathcal{Y}\times\mathbb{R}\,,\ x\in\mathcal{C}\,,\ k\in\mathrm{int}(\mathcal{K})\,,\ \varrho>0\}\subset\mathrm{int}(\mathcal{A})$$

and thus in particular

$$(g(x_0)+(-g(x_0)), f(x_0)+\varrho) = (0, f(x_0)+\varrho)\in\mathrm{int}(\mathcal{A})\cap\{0\}\cap\mathbb{R}$$

for all $\varrho > 0$.

Finally, the following theorem supplies information on the equivalence of (LP) and (DLP); cf. Lemma 3.5:

Theorem 3.22. *(Werner), Th. 4.3.1. Let the minimum problem (MP) be convex, $\mathcal{Y}$ normed, and $\mathcal{A}$ closed. Then*
(1°) $\mathcal{S}\neq\emptyset$ and $\inf_{x\in\mathcal{S}} f(x) > -\infty \iff \mathcal{T}\neq\emptyset$ and $\sup_{y\in\mathcal{T}}\psi(y) < +\infty$.
(MP) has a solution in both cases, moreover
$-\infty < \sup_{y\in\mathcal{T}}\psi(y) = \min_{x\in\mathcal{S}} f(x) < \infty$.
(2°) $\mathcal{S}=\emptyset$ and $\mathcal{T}\neq\emptyset \Longrightarrow \sup_{y\in\mathcal{T}}\psi(y) = +\infty$.
(3°) $\mathcal{T}=\emptyset$ and $\mathcal{S}\neq\emptyset \Longrightarrow \inf_{x\in\mathcal{S}} f(x) = -\infty$.

(f) Local Lagrange Theory The problem (3.36),

$$\{f(x)\,,\ x\in\mathcal{C}\,,\ g(x)\le 0\,,\ h(x)=0\} = \min!$$

is again called *differentiable* if f, g, h are FRÉCHET-differentiable. Then (3.37) provides *necessary* conditions for an optimum x^*:

$$\begin{aligned}
&\forall\, x\in\mathcal{C}:\ \nabla_x L(x^*,y^*,z^*)(x-x^*)\ge 0\ \ \text{(minimum problem)},\\
&\forall\, x\in\mathcal{C}:\ \nabla_x L(x^*,y^*,z^*)(x-x^*)\le 0\ \ \text{(maximum problem)},\\
&\forall\, x\in\mathcal{C}:\ \nabla_x L(x^*,y^*,z^*)(x-x^*)= 0\ \ \text{(if } \mathcal{C} \text{ in } \mathcal{X} \text{ open or subspace)}.
\end{aligned}\tag{3.48}$$

Theorem 3.23. *(Multiplier rule (MR) sufficient in convex problems) Let the minimum problem (MP) (3.36) be convex (h affine linear), differentiable, $x^*\in\mathcal{S}$, and let the MR be fulfilled:*

$$\exists\,(y^*,z^*)\in\mathcal{K}_d\times\mathcal{Z}_d:\nabla_x L(x^*,y^*,z^*)(x-x^*)\ge 0\ \textit{and}\ y^*\circ g(x^*)=0\,.$$

Then x^* is solution of (MP).

Proof. Since f, g convex,

$$\begin{aligned}
f(x)&\ge f(x^*)+\nabla f(x^*)(x-x^*)\,,\\
g(x)&\ge g(x^*)+\nabla g(x^*)(x-x^*)\,,\\
h(x)&= h(x^*)+\nabla h(x^*)(x-x^*)\,.
\end{aligned}$$

Therefore by (MR) for $x\in\mathcal{C}$

$$\begin{aligned}
f(x)&\ge f(x)+y^*\circ g(x)+z^*\circ h(x)\\
&\ge f(x^*)+\nabla f(x^*)(x-x^*)+y^*\circ[g(x^*)+\nabla g(x^*)(x-x^*)]\\
&\quad +z^*\circ[h(x^*)+\nabla h(x^*)(x-x^*)]\\
&= f(x^*)+y^*\circ g(x^*)+z^*\circ h(x^*)+\nabla_x L(x^*,y^*,z^*)(x-x^*)\ge f(x^*)\,.
\end{aligned}$$

□

In place of Theorem 3.5 we now have

Theorem 3.24. *(MR local necessary) Let the minimum problem (3.36) be differentiable, and suppose further that:*
(1°)
$$x^* = \arg\min\{f(x),\ x \in \mathcal{C},\ g(x) \le 0,\ h(x) = 0\}\,,$$
(2°) $\mathrm{int}(\mathcal{K}) \ne \emptyset$,
(3°) $\mathrm{relint}(\nabla h(x^*)(\mathcal{C})) \ne \emptyset$,
(4°) *h is locally solvable in x^* w.r.t. $\mathcal{C}$.*
Then there exists a triple $(0,0,0) \ne (\varrho^*, y^*, z^*) \in \mathbb{R}_{\ge 0} \times \mathcal{K}_d \times \mathcal{Z}_d$ such that
(1°)
$$\forall\, x \in \mathcal{C} :\ [\varrho^* \nabla f(x^*) + y^* \circ \nabla g(x^*) + z^* \circ \nabla h(x^*)](x - x^*) \ge 0\,,$$
(2°) $y^* \circ g(x^*) = 0$.
(3°) *If there exists a $x \in \mathcal{C}$ such that*
$$g(x^*) + \nabla g(x^*)(x - x^*) < 0,\ \nabla h(x^*)(x - x^*) = 0\,,$$
and if $x^* \in \mathrm{int}(\nabla h(x^*)(\mathcal{C}))$ then $\varrho^* = 1$ may be chosen, and $y^* \ne 0$.

Condition (3°) is a modified SLATER condition. The somewhat intransparent assumption (4°) may be concretized by using a more general result of (Robinson) on local solvability.

Theorem 3.25. *Let $\mathcal{X}$ be a BANACH space, $\mathcal{Y}$ a normed space, $\mathcal{K} \subset \mathcal{Y}$ closed and $g : \mathcal{X} \to \mathcal{Y}$ continuously differentiable. Then $g(x) \le 0$ is locally solvable in $x^* \in \mathcal{X}$ if*
$$0 \in \mathrm{int}[g(x^*) + \nabla g(x^*)(\mathcal{X}) + \mathcal{K}]\,. \tag{3.49}$$

Let now $\mathcal{Z} = \nabla h(x^*)(\mathcal{X})$ then $\mathrm{relint}(\nabla h(x^*)(\mathcal{X})) = \mathrm{int}(\mathcal{Z}) \ne \emptyset$, $0 \in \mathrm{int}(\nabla h(x^*)(\mathcal{X})$ and h in x^* locally solvable by ROBINSON's theorem if $h(x^*) = 0$. Then Theorem 3.24 may be modified as follows:

Corollary 3.4. *Let the minimum problem (3.36) be continuously differentiable. Suppose that:*
(1°) $x^* = \arg\min\{f(x)\,;\ x \in \mathcal{X}\,,\ g(x) \le 0\,,\ h(x) = 0\}$,
(2°) $\mathrm{int}(\mathcal{K}) \ne \emptyset$,
(3°) *$\nabla h(x^*) : \mathcal{X} \to \mathcal{Z}$ surjective,*
(4°) $\exists\, x \in \mathcal{X} :\ g(x^*) + \nabla g(x^*)x < 0\,,\ \nabla h(x^*)x = 0$.
Then there exists a pair $(y^*, z^*) \in \mathcal{K}_d \times \mathcal{Z}_d$ with $y^* \ne 0$ such that
$\nabla f(x^*) + y^* \circ \nabla g(x^*) + z^* \circ \nabla h(x^*) = 0$ and $y^* \circ g(x^*) = 0$.

The rank condition (3.9) implies the condition of (Robinson) in finite-dimensional problems; cf. however (Craven78), Ex. 2.6.2.

(g) Everybody, concerned with optimization, knows the classic book of (Luenberger) where this discipline is joined with functional analysis in a felicitous way. This section would not be complete without refering a surprisingly simple result on extremal problems with equality constraints in BANACH spaces. The result uses however the Generalized Inverse Function Theorem 1.23 being not at all trivial and also due to (Luenberger).

Assumption 3.1. *Let $\mathcal{X}$, $\mathcal{Z}$ be* BANACH *spaces, let $\mathcal{C} \subset \mathcal{X}$ be open and $f : \mathcal{C} \to \mathbb{R}$, $h : \mathcal{C} \to \mathcal{Z}$ F-differentiable. Further suppose that $x^* = \arg\min_{x \in \mathcal{C}}\{f(x)\,,\ h(x) = 0\}$ exists and is a regular point such that* Range $\nabla h(x^*) = \mathcal{Z}$.

Lemma 3.7. *Adopt Assumption 3.1. Then*

$$\forall\, v \in \mathcal{X} : \nabla h(x^*)v = 0 \implies \nabla f(x^*)v = 0\,, \tag{3.50}$$

i.e., $\nabla f(x^*) \in [\mathrm{Ker}(\nabla h(x^*))]^\perp \in \mathcal{X}_d$.

Proof. Consider the composed mapping $T : \mathcal{X} \ni x \mapsto (f(x), h(x)) \in \mathbb{R} \times \mathcal{Z}$ and suppose that there exists a $v \in \mathcal{X}$ such that $\nabla h(x^*)v = 0$ and $\nabla f(x^*)v \neq 0$. Then necessarily $\nabla f(x^*) \neq 0$ and $\nabla T(x^*) = (\nabla f(x^*), \nabla h(x^*)) : \mathcal{X} \to \mathbb{R} \times \mathcal{Z}$ is surjective. Therefore x^* is a regular point of T and, by Theorem 1.23, for any $\varepsilon > 0$ there exists a x and $\delta > 0$ such that $T(x) = (f(x^*) - \delta, h(x^*)) = (f(x^*) - \delta, 0)$ with $\|x - x^*\| < \varepsilon$. This is a contradiction to the assumption that x^* is a local minimum. □

Theorem 3.26. *(Multiplier rule necessary) Adopt Assumption 3.1. Then there exists a $z^* \in \mathcal{Z}_d$ such that $L(x) = f(x) + z^* \circ h(x)$ is stationary in x^*, $\nabla_x L(x^*) = \nabla f(x^*) + z^* \circ \nabla h(x^*) = 0$.*

Proof. Let first $\mathcal{X} = \mathbb{R}^n$ and $\mathcal{Z} = \mathbb{R}^p$ be finite-dimensional and write briefly $A = \nabla h(x^*) \in \mathbb{R}^p{}_m$. Then the Range Theorem 1.2 says that Range$(A^T) = [\mathrm{Ker}\, A]^\perp$. Therefore, by Lemma 3.7, there is a $z \in \mathbb{R}^p$ such that $[\nabla f(x^*)]^T = -A^T z$ thus $\nabla f(x^*) + z^T A = 0,\ z^T \in \mathbb{R}_m = \mathcal{Z}_d$. The general proof follows in a similar way by using the Closed Range Theorem 1.19 (6°) and the fact that Range$(\nabla h(x^*)) = \mathcal{Z}$ is closed. □

(h) Examples

Example 3.10. Consider first the linear problem

$$\{ax\,;\ Bx \le c\,,\ Cx = d\} = \max!$$

with LAGRANGE-Function

$$L(x, y, z) = ax - y(Bx - c) + z(Cx - d)\,.$$

The primal and dual LAGRANGE problems then read:

$$\begin{aligned}(x^*, y^*, z^*) &= \max_x \inf_{y \ge 0, z}[ax - y(Bx - c) + z(Cx - d)]\,,\\ (x^*, y^*, z^*) &= \min_{y \ge 0, z} \sup_x [x(a - yB + zC) - yc - zd]\,.\end{aligned}$$

$a - yB + zC$ must be equal to zero that $\sup \ldots < \infty$ hence the dual problem reads:

$$\{-yc - zd\,;\ yB - zC = a\,,\ y \ge 0\} = \min!\,.$$

So the dual problem contains only the simple sign constraint $y \ge 0$.

Example 3.11. Consider the linear-quadratic problem (3.26)

$$\left\{\frac{1}{2}x^T Ax - a\,x\,;\ -(b+Bx) \le 0\right\} = \min!$$

where $A \in \mathbb{R}^n{}_n$ is a symmetric, positive definite matrix and $B \in \mathbb{R}^m{}_n$. Then the dual functional reads

$$\psi(y) = \inf_x \left\{\frac{1}{2}x^T Ax - a\,x - y(b+Bx)\right\}).$$

Note that the argument is convex in x therefore we have by (3.48) the necessary condition for x^*

$$(x^*)^T A - yB - a = 0 \in \mathbb{R}_n \implies x^* = A^{-1}\left(B^T y^T + a^T\right) \in \mathbb{R}^n .$$

Substitution into $y^* = \arg\max_{y\ge 0}\varphi(y)$ yields

$$y^* = \arg\max_{y\ge 0}\left[-\frac{1}{2}\,yBA^{-1}B^T y^T - y\,b + \frac{1}{2}\,aA^{-1}a^T\right].$$

The dual problem has only a simple sign restriction for constraint but, on the other side, needs the inversion of the matrix A which can be a serious drawback in large, sparse problems. Moreover the dimension is reduced in the (less frequent) case $m < n$.

Example 3.12. Consider the control problem

$$\begin{aligned} J(u) &= \frac{1}{2}\int_0^T u(t)^2 dt = \min! \\ \dot{x} &= A(t)x + b(t)u(t) \in \mathbb{R}^n\,,\ x(0) = a\,,\ x(T) \ge c\,,\ u(t) \in \mathbb{R}\,, \end{aligned}$$

and let $X(t)$ be a fundamental matrix of the differential system such that $X(0) = I$, cf. Sect. 1.5**(d)**. Then by (1.31)

$$x(t) = X(t)a + \int_0^t X(t)X(s)^{-1}b(s)u(s)\,ds =: X(t)a + k(t,u) \in \mathbb{R}^n .$$

Writing $d = c - X(T)a$, the control problem has now the form

$$\min\{J(u)\,,\ k(T,u) \ge d\}\,,\ u \in C[0,T].$$

This problem is convex and the SLATER condition applies. The minimum problem and the dual LAGRANGE-Problem

$$(u^*, y^*) = \arg\max_{y\ge 0}\inf_u\left(J(u) + y(d - k(T,u))\right)\,, \tag{3.51}$$

$y \in \mathbb{R}_n$, are therefore equivalent. The control is unrestricted therefore differentiation of $J(u) + y(d - k(T,u))$ w.r.t. u is allowed. Setting the result equal to zero, we obtain the optimal control depending from y,

$$u^*(t) = y\,X(T)X(t)^{-1}b(t)\,.$$

A substitution into (3.51) yields the *finite-dimensional* maximum problem

$$y^* = \arg\max_{y\geq 0}\left(-yQy^T + y\,d\right)\,, \tag{3.52}$$

where Q is a symmetric positive semi-definite matrix,

$$Q = \frac{1}{2}\int_0^T X(T)X(s)^{-1}b(s)b(s)^T X(s)^{-T}X(T)^T\,ds\,.$$

The problem (3.52) has at least one solution $y^* \in \mathbb{R}_n$ and then an optimal control reads: $u^*(t) = y^* X(T)X(t)^{-1}b(t)\,,\ \ 0 < t < T\,.$

More detailed documentation in `SUPPLEMENT\chap03d,chap03e,chap03f`.

References: (Craven78), (Craven95), (Ekeland), (Gelfand), (Kirsch), (Kosmol), (Krabs), (Luenberger), (Petrov), (Schaeffer), (Teo89), (Teo91), (Werner).

3.7 Hints to the MATLAB Programs

```
KAPITEL03/SECTION_1_4, Linear-Quadratic Programming
bfgs.m       BFGS method
demo1.m      Example, bfgs.m and desc.m
demo2.m      Test of dlqp.m
demo3.m      Test of dlqp.m  with random variables
desc.m       Steepest descend
dlqp.m       Linear-quadratic programming after Goldfarb-Idnani
dlqp_g.m     as dlqp.m, but only inequalities
ga_test.m    Goldstein-Armijo descend test
```

```
KAPITEL03/SECTION_5, Nonlinear Programming
bsp01.m--bsp16.m  Examples
demo1.m     Masterfile for gradient projection
demo2.m     Masterfile for sequential quadratic programming
demo3.m     Masterfile for flexiplex method after Himmelblau
flexiplex.m Flexible tolerance method after Himmelblau
gp.m        Gradient projection method general
gp_g.m      Gradient projection method, only inequalities
restor.m    Restoration in gp.m
sigini.m    Start vector for step length sigma in gp.m
sqp.m       Sequential quadratic programming general
sqp_g.m     Sequential quadratic programming, only inequalities
```

```
KAPITEL03/FEXIPLEX, Method of Nelder and Mead
demo1.m    Minimization of a function (3 Ex.)
demo2.m    Minimization with constraints (4 Ex.)
simplex.m Minimization after Nelder and Mead
flexiplex Constraint Minimization after Himmelblau
```

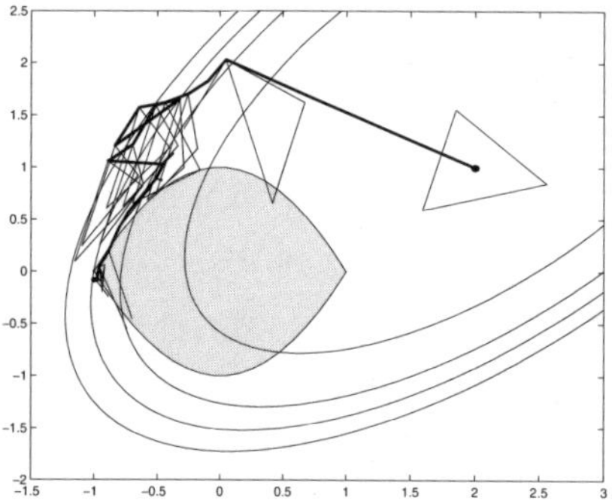

Figure 3.28. Example 3.8, f_1

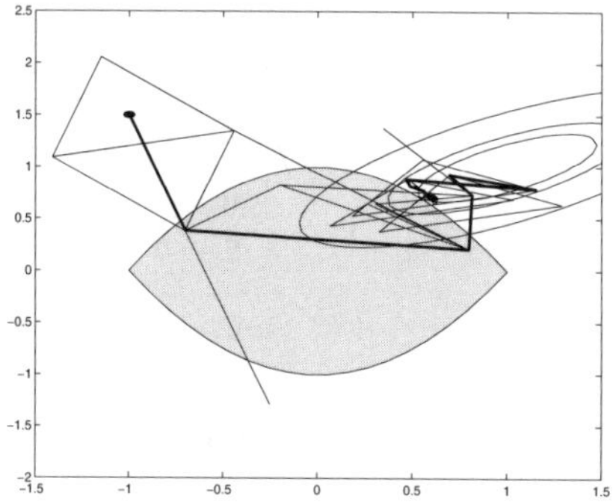

Figure 3.29. Example 3.8, f_2

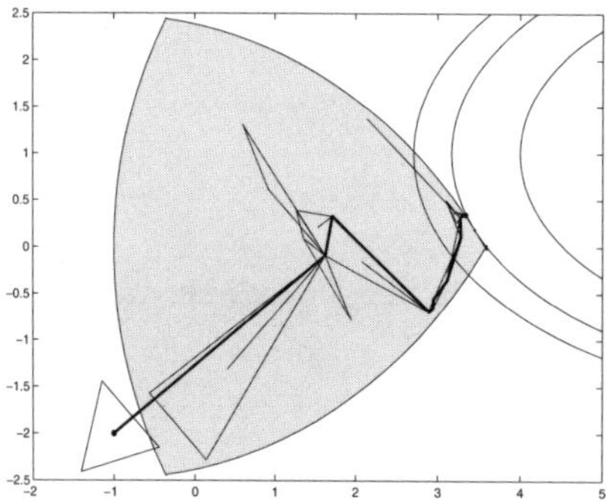

Figure 3.30. Example 3.9, f_1

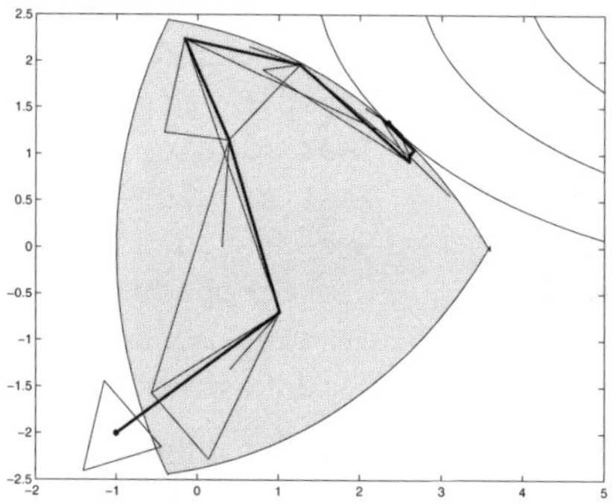

Figure 3.31. Example 3.9, f_2

Flexiplex with three total steps

4

Variation and Control

Calculus of variations is primarily concerned with functionals living on *infinite-dimensional* spaces as the action integral of HAMILTON's principle and to a much less degree with functionals over finite-dimensional spaces. (Note that critical or stationary points are sometimes called "extreme points" without being in fact extremal points.) To find a stationary point of a functional $f : \mathcal{V} \to \mathbb{R}$ as a candidate for a maximum or minimum point, the first or GÂTEAUX variation $\partial(f; h)$ is set equal to zero being simply the directional derivative, and this is, by the way, the most simple form of a derivative at all; see Sect. 1.8. For functionals on finite-dimensional spaces this process is the same as setting the first derivative (gradient) equal to zero.

The variational increment $h \in \mathcal{V}$ is frequently called *test function* in particular in the context of finite element methods. In engineering sciences it is called *virtual displacement* but mathematicians prefer other attributes perhaps since their discipline is anyhow a virtual one. The result of this process, namely the variational equations are obtained uniquely whereas the converse, namely finding the assigned extremal function (frequently the total energy of a mechanical system) cannot be performed in unique way as all integrals. Also, it plays a minor role from the practical point of view so that one works often without that appealing functional in applications and formulates directly the variational problem with its boundary conditions.

Although DIDO's problem has been known long before, calculus of variations emerged with the well-known *brachystochrone* problem posed by JOHANN BERNOULLI in 1669, which is dealt with below. This fascinating new type of problem, namely the matter of an unknown *curve*, was first fully perceived and acted upon by his brother JAKOB. Since that time, calculus of variations has evolved into an indispensable part of applied mathematics. It plays an important role in numerical realization in both continuum mechanics and control theory; that is why the latter is treated as an *application* in this chapter. But it supplies also the fundamentals of analytic mechanics and their basic principles which have been developed more or less at the same time (which has caused some discrepancies in the mutual "estime" of the main actors, D. BERNOULLI and D'ALEMBERT).

Control problems are extremal problems with at least one equality constraint in form of a dynamical system. For the computation of stationary values, they are transformed into an differential-algebraic problem by variation and application of LAGRANGE's multiplier method. In contrast with D'ALEMBERT's principle, this method dispenses with the elimination of surplus variables and takes the constraints into account without any modification. Simple control problems without further constraints can also be transformed into a pure boundary value problem — at least theoretically.

The boundary value problems of control and the EULER-LAGRANGE equations of variation can be highly nonlinear and defy sometimes simply numerical realization. The main problem is to find a suitable start trajectory for further treatment by multiple shooting methods or specially adapted numerical devices as the MATLAB code `bvp4c`. However, control problems can be discretized directly to yield a discrete optimization problems for application of the methods proposed in the previous chapter. In this way, one can often obtain, if not the solution itself, suitable "initial guesses" for subsequent treatment as a boundary value problem.

If, in a variational problem, the derivative $\dot{x}$ of the desired solution x appears as an unknown variable which is more likely to be the rule than the exception, then one can fall back to the underlying extremal problem and introduces a control by virtue of $u := \dot{x}$. The result is then also a control problem which can be treated in the same way by procedures for nonlinear optimization.

Numerical examples are considered in the context of Control Theory; see Sect. 4.4.

4.1 Variation

The conventions at the beginning of Chap. 3 apply also in the present chapter.

(a) Extremal Problem, Variational Problem and Boundary Value Problem We seek a solution (curve) $x : [0, T] \ni t \mapsto x(t) \in \mathbb{R}^n$ of the extremal problem

$$\boxed{J(x) = \int_0^T q\big(t, x(t), \dot{x}(t)\big)\, dt = \text{ extr!}\,, \quad x \in \mathcal{C}^1([0, T]; \mathbb{R}^n)} \tag{4.1}$$

where q is a sufficiently smooth scalar function. This extremal problem relates directly to HAMILTON's principle which plays a crucial role in classical physics as well as in quantum mechanics. In a similar way as in Sect. 3.2, conditions for a stationary point of (4.1) are derived in this chapter which then form a *necessary* condition for a solution again. Thereafter more general problems are considered with various types of equality and inequality constraints.

The null vector is the only vector standing perpendicular on *all* vectors in $\mathbb{R}^n$ w.r.t. any scalar product. The generalization of this simple truth to vector spaces of functions is the crucial auxiliary result in passing from an extremal problem via *variational problem* to a boundary value problem for a system of ordinary or partial differential equations.

Lemma 4.1. *(*LAGRANGE*, Fundamentallemma of Calculus of Variations) Let $f,\, g \in \mathcal{C}([0,T];\mathbb{R}^n)$ and $h \in \mathcal{C}_0^1([0,T];\mathbb{R}^n)$ then*

$$(1^\circ)\ \forall\, h: \int_0^T f(t)^T h(t)\,dt = 0 \qquad\qquad \Longrightarrow f \equiv 0\,,$$

$$(2^\circ)\ \forall\, h: \int_0^T f(t)^T \dot h(t)\,dt = 0 \qquad\qquad \Longrightarrow f = \text{ constant},$$

$$(3^\circ)\ \forall\, h: \int_0^T [f(t)^T h(t) + g(t)^T \dot h(t)]\,dt = 0 \Longrightarrow g \in \mathcal{C}^1([0\,,\,T];\mathbb{R}^n) \text{ and } f = \dot g\,.$$

Proof (Amann), (Kosmol), `SUPPLEMENT\chap04a`.

The proof of (1°) is of purely technical nature, (2°) is also known as Lemma of DUBOIS-REYMOND and (3°) is an inference from (2°).

Theorem 4.1. *Let the "terminal time" T be fixed, $0 < T < \infty$. Every solution $x \in \mathcal{C}^2([0,\,T];\mathbb{R}^n)$ of the extremal problem (4.1) satisfies the differential system*

$$\boxed{\operatorname{grad}_x q(t,x,\dot x) - \frac{d}{dt}\operatorname{grad}_{\dot x} q(t,x,\dot x) = 0 \in \mathbb{R}_n\,,\ 0 < t < T} \tag{4.2}$$

named after EULER *and* LAGRANGE.

Proof. Following 1.14 a necessary condition for an extremal value of J is obtained by setting the first variation equal zero: $\forall\, v \in \mathcal{C}^1([0,\,T];\mathbb{R}^n)$:

$$\partial J(x;v) = \int_0^T [\operatorname{grad}_x q(x,\dot x)v + \operatorname{grad}_{\dot x} q(x,\dot x)\dot v]\,dt = 0\,. \tag{4.3}$$

Partial integration of the second term,

$$\int_0^T \operatorname{grad}_{\dot x} q\,\dot v\,dt = \operatorname{grad}_{\dot x} q\,v\Big|_{t=0}^{t=T} - \int_0^T \left[\frac{d}{dt}\operatorname{grad}_{\dot x} q\right] v\,dt\,,$$

yields the *variational problem* $\forall\, v \in \mathcal{C}^1([0,\,T];\mathbb{R}^n)$:

$$\operatorname{grad}_{\dot x} qv\Big|_{t=0}^{t=T} + \int_0^T \left[\operatorname{grad}_x q(x,\dot x) - \frac{d}{dt}\operatorname{grad}_{\dot x} q(x,\dot x)\right] v\,dt = 0\,. \tag{4.4}$$

On choosing for v a test function where $v(0) = v(T) = 0$, the boundary term disappears and an application of Lemma 4.1(1°) yields the assertion. □

The above partial integration requires continuous differentiability of the argument $t \mapsto \nabla_{\dot{x}} q(t, x(t), \dot{x}(t))$ but an application of Lemma 4.1(3°) cancels this requirement again, thus continuity of q suffices for the proof.

The second term on the left side of (4.2) is the *total* derivative w.r.t. t,

$$\frac{d}{dt}[\nabla_{\dot{x}} q]^T = \frac{\partial}{\partial t}[\nabla_{\dot{x}} q]^T + [\nabla_{\dot{x}} \nabla_x q]\dot{x} + [\nabla_{\dot{x}} \nabla_{\dot{x}} q]\ddot{x} \in \mathbb{R}^n .$$

The (columnwise written) EULER-LAGRANGE equations — or briefly EULER equations — constitute a system of ordinary differential equations of the form

$$\boxed{A(t, x, \dot{x})\ddot{x} + b(t, x, \dot{x}) = 0 \in \mathbb{R}^n} \tag{4.5}$$

which is affine linear in the second derivative of x (semi-linear system).

For solving (4.5) uniquely, the mappings A and b have to be sufficiently smooth and $2n$ *appropriate boundary conditions* have to be introduced. The boundary term in (4.4) disappears exactly in *four* cases which leads to a diversification of boundary conditions into *two* different classes being typically for *all* variational problems.

$$\begin{array}{lll}
x(0) = x_0 , & x(T) = x_T , & \text{both essential (geometrical)} \\
x(0) = x_0 , & \text{grad}_{\dot{x}}\, q\Big|_{t=T} = 0 , & \text{essential/natural} \\
\text{grad}_{\dot{x}}\, q\Big|_{t=0} = 0 , & x(T) = x_T , & \text{natural/essential} \\
\text{grad}_{\dot{x}}\, q\Big|_{t=0} = 0 , & \text{grad}_{\dot{x}}\, q\Big|_{t=T} = 0 , & \text{both natural (dynamical)} .
\end{array} \tag{4.6}$$

All *functions of variation* $x + \varepsilon v$, $\varepsilon \in \mathbb{R}$, in the variation (4.3) have to regard the essential boundary conditions, therefore the *test functions* v have to satisfy the corresponding homogeneous essential conditions but *not* the natural boundary conditions in (4.6). In consequence, the following conditions must hold:

$$v(0) = v(T) = 0 \text{ in } (4.6)(1°) , \quad v(0) = 0 \text{ in } (4.6)(2°) , \quad v(T) = 0 \text{ in } (4.6)(3°) .$$

By the above partial integration, the solution of the boundary value problem (BVP) has to be by one degree *smoother* than the solution of (4.3); hence the problem (4.3) is also called *weak problem* (weaker requirements on smoothness). The way displayed here, namely

$$\boxed{\text{extremal problem} \overset{\text{variation}}{\longrightarrow} \text{weak problem} \underset{\text{fundamentallemma}}{\overset{\text{partial integration}}{\longrightarrow}} \text{BVP}}$$

is typically for variational problems. Naturally it may be gone also the opposite way but then the *possible* boundary conditions appear in a less evident form and are not determined uniquely. Contrary to the boundary value problem (4.5) and (4.6), natural boundary conditions do not appear explicitely in the weak problem (4.3) and this fact remains also true in numerical approximations: The numerical ansatz functions in the weak problem do not "know" anything from the natural boundary conditions and satisfy them as well as the differential equation (4.5) only in transition to the limit of infinitisimal refinement. The numerical approximation becomes bad or even wrong if one chooses natural instead of essential conditions or vice versa in the numerical ansatz for the weak problem; cf. the instructive examples in (Collatz60), p. 241, and (Strang), Sect. 1.3.

Special Cases:

Type	EULER Equation	Remark
$q(t,x,\dot{x}) = q_1(t,x)$ $+ q_2(t,x)\dot{x}$	$[q_1]_x - [q_2]_t = 0$	no differential equation
$q(t,x,\dot{x}) = q(t,\dot{x})$	$\dfrac{d}{dt}\operatorname{grad}_{\dot{x}} q = 0$	$\operatorname{grad}_{\dot{x}} q(t,\dot{x}) =$ constant
$q(t,x,\dot{x}) = q(t,x)$	$\operatorname{grad}_x q = 0$	implicit representation of x
$q(t,x,\dot{x}) = q(x,\dot{x})$	$\dfrac{d}{dt}(q - \operatorname{grad}_{\dot{x}} q\,\dot{x}) = 0$	after multiplication by $\dot{x}$
	$q - \operatorname{grad}_{\dot{x}} q\,\dot{x} = \text{const.}$	DUBOIS-REYMOND-condition

See **(f)** for the interesting special case that a variable x_i does not appear in q whereas $\dot{x}_i$ is present (*cyclic* variable).

(b) Modified Problems Instead of (4.1) we consider the augmented problem

$$\boxed{\begin{aligned} J(x) = p(x(0),x(T)) + \int_0^T q(t,x(t),\dot{x}(t))\,dt = \text{ extr!} \\ 0 = r(x(0),x(T)) \in \mathbb{R}^{|r|} \end{aligned}} \tag{4.7}$$

where again all data shall be sufficiently smooth. The function p is called *terminal payoff*, and the boundary conditions are assembled in a possibly nonlinear function r. For simplicity we also write

$$\nabla_2 q(t,x,\dot{x}) = \operatorname{grad}_x q(t,x,\dot{x}),\ \ \nabla_3 q(t,x,\dot{x}) = \operatorname{grad}_{\dot{x}} q(t,x,\dot{x})$$

for the gradients of q, etc.. The generalized boundary condition plays the role of an equality constraint; therefore LAGRANGE multipliers $z \in \mathbb{R}_r$ (row vector!) have to be introduced by Sect. 3.6 and a regularity condition: Let x^* be the unique solution of the problem and let

$$\forall\, c \in \mathbb{R}^r\ \ \exists\, u,\, v \in \mathbb{R}^n : \nabla_1 r(x^*(0),x^*(T))u + \nabla_2 r(x^*(0),x^*(T))v = c\,. \tag{4.8}$$

Following Sect. 3.6, instead of the objective function J, now the first variation of the LAGRANGE function $L = J + z\,r$ is set equal to zero under application of (4.4):

$$\begin{aligned}
0 &= \partial L(x; v) \\
&= \big[\nabla_3 q(T, x(T), \dot{x}(T)) + \nabla_2 p(x(0), x(T)) + z\nabla_2 r(x(0), x(T))\big]v(T) \\
&+ \big[-\nabla_3 q(0, x(0), \dot{x}(0)) + \nabla_1 p(x(0), x(T)) + z\nabla_1 r(x(0), x(T))\big]v(0) \\
&+ \int_0^T \left[\nabla_2 q(t, x, \dot{x}) - \frac{d}{dt}\nabla_3 q(t, x, \dot{x})\right] v\,dt\,.
\end{aligned}$$

Let v be an arbitrary test function without any restrictions then we obtain the system

$$\begin{aligned}
&\nabla_2 q(t, x, \dot{x}) - \frac{d}{dt}\nabla_3 q(t, x, \dot{x}) &&= 0 \text{ EULER eq.} \\
&r(x(0), x(T)) &&= 0 \text{ ess. BC} \\
&-\nabla_3 q(0, x(0), \dot{x}(0)) + \nabla_1 p(x(0), x(T)) + z\nabla_1 r(x(0), x(T)) &&= 0 \text{ nat. BC} \\
&\nabla_3 q(T, x(T), \dot{x}(T)) + \nabla_2 p(x(0), x(T)) + z\nabla_2 r(x(0), x(T)) &&= 0 \text{ nat. BC}
\end{aligned} \tag{4.9}$$

by applying the Fundamentallemma among others again. Up to now the parameters z are not subjected to any conditions. They rule the interplay between essential and natural boundary conditions:

(1°) If no essential boundary conditions appear at all then $r = 0$ and thus the gradients of r disappear. There remain only natural boundary conditions

$$\begin{aligned}
-\nabla_3 q(0, x(0), \dot{x}(0)) + \nabla_1 p(x(0), x(T)) &= 0\,, \\
\nabla_3 q(T, x(T), \dot{x}(T)) + \nabla_2 p(x(0), x(T)) &= 0\,.
\end{aligned}$$

(2°) If, e.g., $x(0) = x_0$ is required being an essential condition then $z\nabla_1 r(x(0), x(T)) = z$ and the natural boundary condition (4.9)(3°) are free ($\rightarrow$ drop it).

(3°) If, e.g., $n = 2$ and $x^1(0) = \alpha \in \mathbb{R}$ is required (single essential boundary condition) then $z\nabla_1 r(x(0), x(T)) = (z_1,\, 0)$, and the first component of (4.9)(3°) is free whereas the second yields the natural boundary condition

$$-D_5 q(0, x(0), \dot{x}(0)) + D_4 p(x(0), x(T)) = 0$$

D_i denoting the partial derivativion w.r.t. the i-th scalar argument.

(c) Variable Terminal Point As an instructive example for the efficiency of variational calculus we consider the problem

$$\begin{aligned}
J(T, x) = p(T, x(T)) + \int_0^T q(t, x, \dot{x})\,dt = \min! \\
x \in \mathcal{C}^1[0,\, T_0]\,,\;\; x(0) = a \in \mathbb{R}^n\,,\;\; g(T, x(T)) = 0 \in \mathbb{R}
\end{aligned} \tag{4.10}$$

where the terminal time T, $0 < T < T_0$, is an independent variable, the upper bound T_0 is sufficiently large and the data are again sufficiently smooth. As simple example may serve the following function where $n = 1$

$$\begin{aligned} &g(T, x(T)) = x(T) - h(T), \\ &h(0) < a, \ \ h'(t) > 0, \ 0 < t < T_0 . \end{aligned} \tag{4.11}$$

Here it is intuitively clear that we may *vary* in optimum (T^*, x^*) only if the trajectory $x^*(t)$ and the constraint curve $h(t)$ do *intersect* in optimum point $t = T^*$ and are *not* tangential to each other; hence necessarily $\dot{x}^*(T^*) - \dot{h}(T^*) \neq 0$ in optimum.

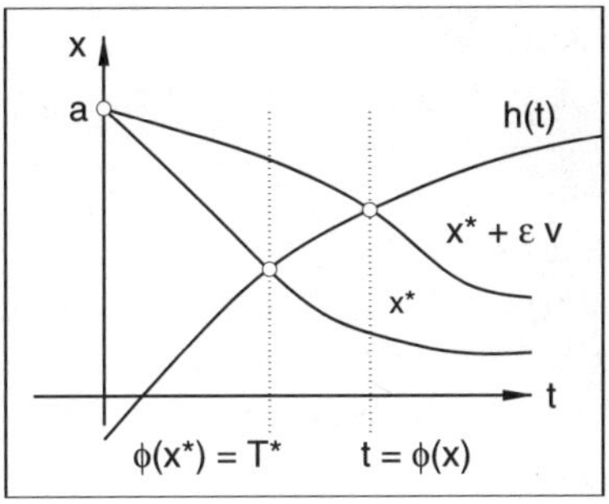

Figure 4.1. Transversality condition

The side condition g in problem (4.10) describes a hyper-surface which depends also on t. In optimum the tangential vector

$$\frac{d}{dt}(t, x^*(t))\Big|_{t=T^*} = [1\,,\, \dot{x}^*(T^*)]^T \in \mathbb{R}^{n+1}$$

is *not* allowed to be tangent to this hyper-surface i.e. it is not allowed to stand perpendicular on the normal vector at this point.

Lemma 4.2. *(Tranversality Condition) Let* $(T^*, x^*) \in \mathbb{R}_+ \times C^2([0, T]; \mathbb{R}^n)$ *of be a solution of the extremal problem (4.10) with free terminal time* T *and suppose that*

$$\nabla_x g(t, x)\dot{x} + g_t(t, x) \neq 0 \tag{4.12}$$

at $(T^*, x^*(T^*))$. *Then necessarily*

$$(\nabla_x g\dot{x} + g_t)\,[\nabla_x p + \nabla_{\dot{x}} q] + (\nabla_x p\dot{x} + p_t + q)\,\nabla_x g = 0 \in \mathbb{R}_n \tag{4.13}$$

at the point $(t, x(t)) = (T^*, x^*(T^*))$ *in addition to the* EULER *equation.*

Proof in `SUPPLEMENT\chap04a`. For instance, let $n = 1$ and $g(t, x(t)) = x(t) - h(t) \in \mathbb{R}$ then (4.12) yields the above condition $\dot{x}(T) - \dot{h}(T) \neq 0$ in optimum.

(d) Legendre Transformation

(d1) (Analytic Interpretation) Let $f \in C^2(\mathbb{R};\mathbb{R})$ be strictly convex then $f''(x) > 0$ and the *derivative* f' of f is invertible with inverse function h: $p = f'(x)$, $x = h(p)$. By this way the derivative (slope) p of f may be introduced for *new variable.* How does now the *antiderivative* g of h look like? By

$$g(p) := p\,h(p) - f(h(p)) = xf'(x) - f(x)\,, \quad x = h(p)$$

we obtain at once

$$g'(p) = p\,h'(p) + h(p) - f'(h(p))h'(p) = p\,h'(p) + h(p) - p\,h'(p) = h(p)\,.$$

The LEGENDRE transformation g of f is a *new* function with a *new* variable p, e.g., now

$$Q(x, xy' - y, y') = 0 \iff Q(g'(p), g(p), p) = 0\,,$$

but the dash means derivation w.r.t. x in the left equation and differentiatin w.r.t. p in the right equation.

(d2) (Geometric Interpretation). Let f again be strictly convex. Choose a "slope" p and form $y = p\,x$. Then choose the point $x(p)$ on the x-axis such that the distance between the straight line $y = p\,x$ and the curve $y = f(x)$ becomes *maximum,*

$$x(p) = \arg\max_x \{p\,x - f(x)\}\,.$$

Then the function $F(p, x) := px - f(x)$ has a unique maximum at point $x(p)$. Now the LEGENDRE transformation g of f is geometrically defined by

$$g(p) = \max_x F(p, x) = F(p, x(p)) = px(p) - f(x(p))\,.$$

Then naturally $\dfrac{\partial F}{\partial x}(p, x) = 0$ at point $x(p)$ hence $p - f'(x) = 0$, and $p = f'(x)$ is the new variable by this way.

(d3) (Involution) Let f be strictly convex and let g be the LEGENDRE transformation of f then f is the LEGENDRE transformation of g. The LEGENDRE transformation is thus involutoric (inverse to itself); this property does hold also geometrically

$$g(p) = \max_x\{px - f(x)\}\,, \quad f(x) = \max_p\{xp - g(p)\}\,.$$

The *proof* follows immediately by the analytic definition:

$$\begin{aligned} &g(z) = xf'(x) - f(x)\,, \quad z = f'(x)\,, \quad x = g'(z) \\ &\Longrightarrow\ f(x) = xf'(x) - g(z) = zg'(z) - g(z)\,. \end{aligned}$$

Example 4.1. (Fig. 4.2).

$$
\begin{aligned}
f(x) &= x^2 \\
F(p,x) &= px - x^2 \\
x(p) &= p/2\,,\ \ g(p) = p^2/4 \\
f(x) &= \frac{mx^2}{2} \longrightarrow g(p) = \frac{p^2}{2m}\,.
\end{aligned}
$$

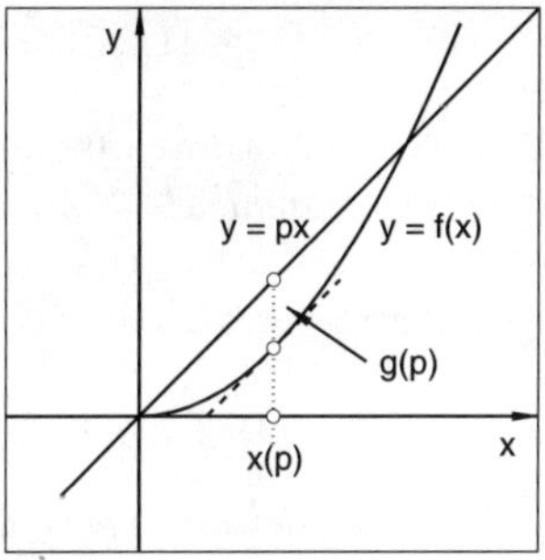

Figure 4.2. LEGENDRE transformation

Example 4.2. Let $f(x) = x^\alpha/\alpha$ then $g(p) = p^\beta/\beta$ where $(1/\alpha) + (1/\beta) = 1\,,\ \alpha > 1\,,\ \beta > 1\,.$

Example 4.3. CLAIRAUT's (implicit) differential equation, $y = xy' - g(y')\,,\ g \in \mathcal{C}^1(\mathcal{I})\,,$ has obviously all straight lines $y = cx - g(c)\,,\ c \in \mathbb{R}\,,$ for solutions. Let $y = f(x)$ be a further solution such that $p := y' = f'(x)$ is invertible, $x = [f']^{-1}(p)$ then $\widetilde{y}(p) = y(x(p))$ hence $\widetilde{y}'(p) = px'(p)\,.$ But on the other side, differentiating $\widetilde{y}(p) = xp - g(p)$ w.r.t. p yields

$$\widetilde{y}'(p) = px'(p) + x(p) - g'(p) \stackrel{!}{=} px'(p).$$

Therefore the *envelope* of the above family of straight lines is a further (nonlinear) solution with parameter representation

$$\widetilde{y}(p) = pg'(p) - g(p)\,,\ x(p) = g'(p)\,.$$

(d4) To generalize the concept of LEGENDRE transformations, let $\mathcal{X}$ be a normed vector space with dual space $\mathcal{X}_d$ and write again $\langle y, x\rangle$ for $x \in \mathcal{X}$ and $y \in \mathcal{X}_d$ to emphasize the dual pairing.

Definition 4.1. *Let f be a convex functional on a convex set $\mathcal{C} \subset \mathcal{X}$.*
(1°) $\mathcal{C}^* := \{y \in \mathcal{X}_d\,,\ \sup_{x\in\mathcal{C}}[\langle y,x\rangle - f(x)] < \infty\} \subset \mathcal{X}_d$ *is the set conjugate to* $\mathcal{C}$.
(2°) $f^* : \mathcal{C}^* \ni y \mapsto \sup_{x\in\mathcal{C}}[\langle y,x\rangle - f(x)] = f^*(y) \in \mathbb{R}$ *is the functional conjugate to* f.

The conjugate set $\mathcal{C}^*$ and the conjugate functional f^* are convex. The conjugate functional f^* is *involutoric*

$$f(x) = \sup_{y\in\mathcal{C}^*}[\langle y,x\rangle - f^*(y)] \tag{4.14}$$

when the bidual $[\mathcal{X}_d]_d$ can be identified with $\mathcal{X}$ as, e.g., in HILBERT spaces. If $\mathcal{C}$ and f convex, the *epigraph* of f,

$$[\mathcal{C}, f] := \{(x, \xi) \in \mathcal{C} \times \mathbb{R}\,,\; \xi \geq f(x)\} \subset \mathcal{X} \times \mathbb{R}\,,$$

is also convex; it contains obviously all points "above" f on $\mathcal{C}$. Recall now that, given an element $x^* \in \mathcal{X}_d$, the affine linear equation $\langle x^*, x\rangle = d \in \mathbb{R}$ represents a *hyperplane* in the vector space $\mathcal{X}$, and $\langle x^*, x\rangle \leq d$ is a family of half-spaces increasing with d. If $\mathcal{C} \subset \mathcal{X}$ convex and $h(x^*) := \sup_{x\in\mathcal{C}}\langle x^*, x\rangle \in \mathbb{R}$ then

$$\mathcal{H} := \{x \in \mathcal{X}\,,\; \langle x^*, x\rangle = h(x^*)\} \subset \mathcal{X}$$

is obviously a *supporting hyperplane* to $\mathcal{C}$. But we are faced here with spaces $\mathcal{X} \times \mathbb{R}$ where

$$\widetilde{\mathcal{H}} := \{(x, \xi) \in \mathcal{X} \times \mathbb{R}\,,\; \langle x^*, x\rangle - \eta \cdot \xi = d\} \subset \mathcal{X} \times \mathbb{R}$$

is a hyperplane. $\widetilde{H}$ is *non-vertical* if $\eta \neq 0$ and then $\eta = 1$ without loss of generality. In that case $\widetilde{\mathcal{H}}$ intersect the vertical axis $\mathbb{R}$ at the point $\xi = -d$ (negative sign only chosen for compatibility with the functional f^*). This hyperplane is a *supporting* hyperplane of the convex epigraph $[\mathcal{C}, f]$ of f if and only if $d = f^*(x^*)$, by definition. By this way, conjugate functionals and supporting hyperplanes are strongly connected geometrically.

(d5) Conjugate functionals (resp. their inverses) are well-suited for transforming minimum problems into saddlepoint problems. Consider for instance

$$x^* = \arg\inf{}_{x\in\mathcal{X}}\{f(x) + g(x)\}$$

with convex f and some scalar function g. Inserting the conjugate functional, we get

$$x^* = \arg\inf{}_{x\in\mathcal{X}} \sup{}_{y\in\mathcal{X}_d}\{\langle y, x\rangle - f^*(y) + g(x)\}$$

which leads to the primal and the dual LAGRANGE problem

$$\begin{aligned}(x^*, y^*) &= \arg\min{}_{x\in\mathcal{X}} \sup{}_{y\in\mathcal{X}_d}\{\langle y, x\rangle - f^*(y) + g(x)\}\\ &= \arg\max{}_{y\in\mathcal{X}_d} \inf{}_{x\in\mathcal{X}}\{\langle y, x\rangle + g(x) - f^*(y)\}\,.\end{aligned}$$

Apparently we have to know the conjugate function explicitely in practical applications. This is most properly explained by giving two simple examples:

Example 4.4. (Principle of complementary energy) Reconsider the quadratic minimum problem of Sect. 1.11 but with some linear operator L such that $a(u, u) = (Lu, Lu) = |Lu|^2$, i.e.,

$$u^* = \arg\min{}_u\big\{(Lu, Lu) - 2f(u)\big\}\,,\;\; u \in \mathcal{U}\,. \tag{4.15}$$

By using the conjugate function (or by solving the associated variational problem) we get immediately

$$|Lu|^2 = \sup_{v\in\mathcal{U}}\{2(v, Lu) - |v|^2\}\,,$$

and an insertion in (4.15) yields the associated primal and dual LAGRANGE problem

$$\begin{aligned}
u^* \quad &= \arg\min_u \Big\{ \sup_v\{2(v, Lu) - |v|^2\} - 2f(u)\Big\}\,, \quad u, v \in \mathcal{U}\\
(u^*, v^*) &= \arg\min_u \sup_v\{2(v, Lu) - 2f(u) - |v|^2\}\\
(u^*, v^*) &= \arg\max_v \inf_u\{2(v, Lu) - 2f(u) - |v|^2\}\,.
\end{aligned}$$

Let $(L_d v, u) = (v, Lu)$ and observe that the infimum over u must exist finitely that a saddlepoint exists then the *dual maximum problem* reads:

$$v^* = \arg\max_v\{-|v|^2\,;\ L_d v - f = 0\}\,.$$

One observes that the objective function has become more simple but its domain more complicated (for numerical implementation).

Example 4.5. (Application of the complementary energy principle.) Consider the simple DIRICHLET problem: Find $u^* \in H_0^1(\Omega)$ such that

$$u^* = \arg\min\left\{\frac{1}{2}\int_\Omega \left[|\operatorname{grad} u|^2\, dx - f\,u\right] dx\,;\ u \in H_0^1(\Omega)\right\} \tag{4.16}$$

where $u = 0$ on the entire boundary of the domain Ω; see Sects. 1.7, 9.1 for notations. We insert the relation

$$\frac{1}{2}\int_\Omega |\operatorname{grad} u|^2\, dx = \sup_{\underline{v}\in(\mathcal{L}^2(\Omega))^2}\int_\Omega\left[\underline{v}\cdot\operatorname{grad} u - \frac{1}{2}|\underline{v}|^2\right] dx$$

and recall that

$$\int_\Omega \operatorname{grad} u\cdot \underline{v}\, dx = -\int_\Omega u\ \operatorname{div}\underline{v}\, dx \tag{4.17}$$

in the present case of homogeneous DIRICHLET boundary conditions. Then we get again the saddlepoint problem corresponding to (4.16) in two alternative forms: Find $(u^*, \underline{v}^*) \in \mathcal{U}\times\mathcal{V}$ such that

$$\begin{aligned}
(u^*, \underline{v}^*) &= \arg\min_u \sup_{\underline{v}}\left\{\int_\Omega\left[-\frac{1}{2}|\underline{v}|^2 - f\,u + \operatorname{grad} u\cdot\underline{v}\right] dx\,,\ u\in\mathcal{U}\,,\ \underline{v}\in\mathcal{V}\right\}\\
&= \arg\max_{\underline{v}} \inf_u\left\{\int_\Omega\left[-\frac{1}{2}|\underline{v}|^2 - f\,u - u\operatorname{div}\underline{v}\right] dx\,,\ u\in\mathcal{U}\,,\ \underline{v}\in\mathcal{V}\right\}\,.
\end{aligned} \tag{4.18}$$

where $\mathcal{U} = \mathcal{H}_0^1(\Omega)$ and $\mathcal{V} = \mathcal{L}^2(\Omega)\times\mathcal{L}^2(\Omega)$. Both representations are equivalent in the case where a saddlepoint exists. Again the infimum over u must exist finitely and the *dual maximum problem* is readily obtained from the second form (written as minimum problem): Find $\underline{v}^* \in \mathcal{W}$ such that

$$\underline{v}^* = \arg\inf_{\underline{v}\in\mathcal{W}}\int_\Omega |\underline{v}|^2\, dx\,,\quad \mathcal{W} = \{\underline{v}\in\mathcal{V}\,,\ \operatorname{div}\underline{v} + f = 0\}\,. \tag{4.19}$$

The saddlepoint $(u^*, \underline{v}^*)$ is also characterized by the two variational equations associated, e.g., to the first form in (4.18) where we use (4.17) once more

$$\begin{aligned}\int_\Omega \left[\underline{v} \cdot \underline{w} - \operatorname{grad} u \cdot \underline{w}\right] dx &= 0\,, \forall\, \underline{w} \in \mathcal{V}\\ \int_\Omega \left[u \operatorname{div} \underline{v}\, dx + f\, u\right] dx &= 0\,, \forall\, u \in \mathcal{U}\,.\end{aligned}$$

Passing to the associated boundary value problem of this *dual* form of the DIRICHLET problem we get two *separated* first order equations

$$\boxed{\underline{v} = \operatorname{grad} u\,, \quad \operatorname{div} \underline{v} + f = 0\,, \quad u \in H_0^1(\Omega)} \tag{4.20}$$

which of course are also found directly by decomposing $-\Delta u = -\operatorname{div}\operatorname{grad} u = f$. The dual form is well-suited in cases where the *gradient* of the unknown function u is more important than u as maybe in stationary heat distribution of a disc etc., see for instance Example 9.2.)

(e) Lagrange Function and Hamilton Function For a brief introduction of *generalized coordinates* let us quote from (Lanczos):
Analytical problems of motion require a generalization of the original coordinate concept of DESCARTES. *Any set of parameters which can characterize the position of a mechanical system may be chosen as a suitable set of coordinates. They are called "generalized coordinates" of the mechanical system.*

The generalized coordinates q can be considered as the rectangular coordinates of a *single point* in a *configuration space. This space is no longer the ordinary physical space. It is an abstract space with as many dimensions as the nature of the problem requires.* (Lanczos).

The integrand of the objective function J in (4.1) is called LAGRANGE function in this context and denoted by $L(t, q, \dot{q})$. Letting $L = T - U$ where $T(\dot{q})$ and $U(q)$ are the kinetic and potential energy of, say, a system of n mass points, HAMILTON's principle ("principle of least action") says that motions $q(t) \in \mathbb{R}^N$, $N = 3n$, of this system coincide with the stationary points (functions) of the *action integral* $\int_{t_0}^{t_1} L(q, \dot{q})\, dt$ and are even *minimum points.* (Unfortunately the word "extremal" is often used with the meaning "stationary" in this context.) The EULER-LAGRANGE equations $\partial[\nabla_{\dot{q}} L]/\partial t - \nabla_q L = 0 \in \mathbb{R}^N$ then are obviously the associated equations of motion.

For a general lagrangian $L(t, q, \dot{q})$, it suggests itself to introduce *new* time-dependent coordinates $p \in \mathbb{R}_N$ by defining $p := \nabla_{\dot{q}} L$ since the EULER equations (4.2) then read simply $\dot{p} = \nabla_q L(t, q, \dot{q})$ (generalized forces). In a LEGENDRE transformation of L w.r.t. $\dot{q}$, we replace that variable by p under the assumption that L is strictly convex in $\dot{q}$. As the kinetic energy T is in general a positive definite quadratic form in $\dot{q}$, this assumption is not an undue requirement. Then the EULER equations $p - \nabla_{\dot{q}} L(t, q, \dot{q}) = 0$ can be solved for $\dot{q}$, say, $\dot{q} = K(t,p,q)$. The result of this transformation is the (scalar) HAMILTON *function* or simply *hamiltonian*

$$H(t,p,q) := p\dot{q} - L(t,q,\dot{q}) = pK(p,q) - L(t,q,K(p,q)) \in \mathbb{R}.$$

The reason for the introduction of this new function H is supplied by the following result:

Theorem 4.2. *Let $p = \nabla_{\dot{q}} L$ and let $L(t,q,\dot{q})$ be strictly convex in $\dot{q}$. Then the differential system of N EULER equations*

$$\frac{d}{dt}\nabla_{\dot{q}} L(t,q,\dot{q}) = \nabla_q L(t,q,\dot{q})$$

is equivalent to the differential system of $2N$ HAMILTON equations

$$\dot{p} = -\nabla_q H(p,q)\,, \quad \dot{q} = \nabla_p H(p,q)\,, \tag{4.21}$$

i.e., the sets of solutions q are the same in both cases.

The *skew-symmetric* system of first order HAMILTON equations has various advantages in theoretical studies compared with the system of second order EULER equations and leads to an entire individual geometry (symplectic geometry). (Actually, both equations are written in row form since the gradient of a scalar function shall always be a row vector. After proof however they may be written in column form of course.)

Proof. Using $p - \nabla_{\dot{q}} L = 0$ and $\dot{p} - \nabla_q L = 0$ we find immediately by partial differentiation

$$\begin{aligned} \nabla_q H &= p\nabla_q K - \nabla_q L - \nabla_{\dot{q}} L \nabla_q K &&= -\nabla_q L = -\dot{p} \\ \nabla_p H &= K(p,q) + p\nabla_p K - \nabla_{\dot{q}} L \nabla_p K = K &&\equiv \dot{q}\,. \end{aligned}$$

On the other side, solving the second system in (4.21) for p yields $p = \nabla_{\dot{q}} L$ again and a substitution into the first system leads to the original EULER equations. □

Corollary 4.1. *Let $t \mapsto (p(t), q(t))$ be a solution of the* HAMILTON *system (4.21) then*

$$\frac{dH}{dt}(t,p(t),q(t)) = \frac{\partial H}{\partial t}(t,p(t),q(t))$$

Proof by substituting (4.21) into $\dot{H} = \nabla_p H\dot{p} + \nabla_q H\dot{q} + H_t$. □

In particular, $\partial H/\partial t = 0$ if H does not depend explicitely on t then $H(p(t), q(t)) = \text{const}$ (law of conservation of the HAMILTON function).

A coordinate q_i is called *cyclic, ignorable* or *kinostenic* if it does not enter into the lagrangian L although $\dot{q}_i$ is present. Then $\dot{p}_i = \partial L/\partial q_i = 0$ and the associated generalized moment p_i reveals to be an invariant of the system, $p_i(t) = c_i = \text{constant}$. Let, e.g., $i = n$ and suppose that $\partial L/\partial \dot{q}_n = c_n$ can be solved analytically with respect to $\dot{q}_n$ to yield

$$\dot{q}_n = f(t, q_1, \ldots, q_{n-1}; \dot{q}_1, \ldots, \dot{q}_{n-1}, c_n)\,. \tag{4.22}$$

Then we can replace $\dot{q}_n$ in the EULER-LAGRANGE equations by the right side of (4.22) and the problem of integrating these equations is reduced by one dependent variable because q_n does no longer occur. After integration, the q_i and $\dot{q}_i$, now being functions of t, are inserted into (4.22) and the cyclic variable q_n is found by simple quadrature. The entire process is however much simpler to manage in passing to HAMILTON's equations (Lanczos). For an application see, e.g., the theory of top in Sect. 6.7.

Example 4.6. Cf. Sect. 6.2. By NEWTON's law, the motion $x(t) \in \mathbb{R}^3$ of a point of mass m in a potential field satisfies

$$\begin{aligned} m\ddot{x} &= f(x) = -\operatorname{grad} U(x) && U \text{ potential energy} \\ T &= \frac{m}{2}|\dot{x}|^2 && \text{kinetic energy} \\ E &= T + U = \frac{m}{2}|\dot{x}|^2 + U(x) && \text{total energy}\,. \end{aligned} \tag{4.23}$$

By DuBOIS-REYMOND's condition,

$$-m\ddot{x} - \operatorname{grad} U(x) = -\Big(m\ddot{x} + \operatorname{grad} U(x)\Big) = 0$$

is obtained as EULER's equation of the variational problem

$$J(x) = \int_{t_1}^{t_2} L(x, \dot{x})dt = \int_{t_1}^{t_2} \left[\frac{m}{2}|\dot{x}|^2 - U(x)\right] dt = \text{ extr!}$$

where $L = T - U$ is the LAGRANGE function. This result is again the above mentioned HAMILTON's principle of (least) action (dimension of J = energy $\cdot$ time) and does hold in great generality. Introduction of $y = m\dot{x} = L_{\dot{x}} \in \mathbb{R}^3$ as new time-dependent variable (the mass m being constant) leads to the differential system

$$\begin{aligned} \dot{x} &= \operatorname{grad}_y H \;\; := y/m && \text{(definition)}, \\ \dot{y} &= -\operatorname{grad}_x H := -\operatorname{grad} U(x) && \text{NEWTON's law.} \end{aligned}$$

It follows immediately that $H = E$ is an invariant of the system:

$$\begin{aligned} \text{const} = E &= \frac{1}{2}y^T\dot{x} + U(x) = \frac{1}{2m}y^Ty + U(x) = H(x, y) \\ &= T + U = 2T - (T - U) = y^T\dot{x} - L\,. \end{aligned}$$

(f) A Classic Example We consider more exactly the following variational problem where x is the independent space variable (instead of t) and y denotes the dependent variable; cf. e.g. (Clegg):

$$J(y) = \int_\alpha^\beta y(x)^n(1 + y'(x)^2)^{1/2}dx = \text{ min!}\,, \;\; y \in \mathcal{C}^1([\alpha, \beta]; \mathbb{R})\,.$$

The DuBois-Reymond condition then reads

$$y^n(x) = c[1 + y'(x)^2]^{1/2} \implies y^n(x)[1 + y'(x)^2]^{-1/2} = c\,. \tag{4.24}$$

Here we have to insert $y'(x) = \tan\varphi$ in order that $[1 + y'(x)^2]^{-1/2} = \cos\varphi$ which is of obvious advantage. Therefore the reparametrization $\varphi \mapsto x(\varphi)$, $\varphi \mapsto y(x(\varphi)) = \widetilde{y}(\varphi)$ is to be recommended by

$$\frac{dy}{dx} = \frac{d\widetilde{y}}{d\varphi}\frac{d\varphi}{dx} = \tan\varphi(x) \Longrightarrow \frac{d\widetilde{y}}{d\varphi} \equiv \widetilde{y}'(\varphi) = \tan(\varphi)\frac{dx}{d\varphi} = \tan(\varphi)x'(\varphi)\,. \tag{4.25}$$

Ensuing we write again y instead $\widetilde{y}$ and obtain by (4.24)

$$\boxed{y(\varphi)^n \cos\varphi = c}\,. \tag{4.26}$$

Derivation yields

$$ny(\varphi)^{n-1}y'(\varphi)\cos\varphi - y(\varphi)^n \sin\varphi = 0 \implies n\frac{y'(\varphi)}{y(\varphi)} - \tan\varphi = 0\,.$$

Using the relation (4.26), a representation of the solution is found by (4.25) in parameter form

$$y'(\varphi) = \tan(\varphi)x'(\varphi),\ x'(\varphi) = y(\varphi)/n\,. \tag{4.27}$$

Here the cases $n = 1,\ 1/2,\ 0,\ -1/2,\ -1$, are of particular interest.

Case 1: $n = 0$. The *shortest curve* $x \mapsto (x, y(x)) \in \mathbb{R}^2$ from $(0,0)$ to $(x, h(x))$ is solution of the variational problem

$$J(y) = \int_0^a (1 + y'(x)^2)^{1/2}\,dx = \min!\,,\ y(0) = 0\,,\ y(a) = h(a)\,.$$

By (4.24) we obtain $y'(x) = \alpha$ constant hence $y(x) = \alpha\, x + \beta$ where $\beta = 0$ because of the initial condition. The transversality condition is here an orthogonality condition for the tangents of $(x, y(x))$ and $(x, h(x))$:

$$\begin{aligned} 0 &= q + q_{y'(x)}(h'(x) - y'(x)) \\ &= \left[(1 + y'(x)^2)^{1/2} + \frac{1}{2}\frac{2y'(x)}{(1 + y'(x)^2)^{1/2}}(h'(x) - y'(x))\right]_{x=a} \\ &\doteq (1 + y'(x)^2) + y'(x)h'(x) - y'(x)^2\Big|_{x=a}\,. \end{aligned}$$

Accordingly, $y'(a^*) = -1/h'(a^*)$ hence

$$\left[1, -\frac{1}{h'(a^*)}\right]\left[\frac{1}{h'(a^*)}\right] = 0$$

for the both tangents.

Case 2: $n = 1$. *Minimal surface* of revolution between two coaxial annuli of radius a resp. b and distance $\beta - \alpha$:

$$J(y) = 2\pi \int_\alpha^\beta y(1+(y')^2)^{1/2} dx = \min!\,, \quad y(\alpha) = a\,, \ y(\beta) = b\,.$$

EULER's equation (4.2) becomes $1+y'(x)^2 = y\,y''$ and the DUBOIS-REYMOND condition (4.26) yields by derivation the differential equation of the *catenary curve*

$$y(x) = c[1+y'(x)^2]^{1/2} \Longrightarrow y''(x) = c^{-1}[1+y'(x)^2]^{1/2}\,.$$

Also, for $n = 1$, we obtain directly from (4.24)

$$y' = \left(\frac{y^2-c^2}{c^2}\right)^{1/2} \Longrightarrow dx = \frac{c\,dy}{(y^2-c^2)^{1/2}}$$

and then, by consulting some *formula collection*,

$$f(x) := \frac{x+d}{c} = \ln\left[\frac{y+(y^2-c^2)^{1/2}}{c}\right] \Longrightarrow (c\exp f(x) - y)^2 = y^2 - c^2$$

or

$$c^2 e^{2f(x)} - 2cye^{f(x)} + y^2 = y^2 - c^2 \Longrightarrow ce^{f(x)} - 2y + ce^{-f(x)} = 0$$

hence also the *catenary* $y(x) = c\cosh\big((x+d)/c\big)$. If especially $\alpha = -L$, $\beta = L$ and $a = b$ then $d = 0$ by symmetry and furthermore the right boundary condition

$$\frac{a}{c} = \frac{a}{L}\cdot\frac{L}{c} = \cosh\left(\frac{L}{c}\right) \Longrightarrow \frac{a}{L}\cdot\xi = \cosh(\xi)\,.$$

is obtained. The straight line to the left and the curve to the right may have *zero, one, or two* intersection points and the slope a/L must be sufficiently small in order that they have at least one intersection point.

Case 3: $n = -1/2$. *Brachistochrone problem (Greek: shortest time).* A bead shall move on a wire from $(0,0)$ to (a,b) as fast as possible under gravity g without friction. Let the y-axis point to "below" for simplicity. The law of conservation of energy supplies

$$\frac{1}{2}m\,v^2 = m\,g\,y(x) \Longrightarrow v(x) = \sqrt{2gy(x)}\,,$$

for the velocity $v(x)$ at point $(x, y(x))$. The length $s(x)$ of the unknown trajectory satisfies

$$\frac{ds(x)}{dx} = \sqrt{1+y'(x)^2} \Longrightarrow v(x(t)) = \frac{ds(x(t))}{dt} = \frac{ds}{dx}\cdot\frac{dx(t)}{dt}$$

where the parameter t denotes time. Together

$$\frac{dx(t)}{dt} = \sqrt{\frac{2gy(x(t))}{1+y'(x(t))^2}} \Longrightarrow \frac{dt(x)}{dx} = \frac{\sqrt{1+y'(x)^2}}{\sqrt{2gy(x)}},$$

since $x(t)$ increases strictly monotone. Now we obtain the extremal problem

$$J(y) = \int_0^a \left[\frac{1+y'(x)^2}{2gy(x)}\right]^{1/2} dx = \min!$$

for operational time. However this integral has a singularity at the point $x = 0$ whence the existence of a solution can be proved in full accuracy only by using results of LEBESGUE Theory; cf. (Kosmol). Now, by (4.24) for $n = -1/2$

$$y = 2\kappa\cos^2\varphi = \kappa(1+\cos(2\varphi)) = \kappa(1+\cos(\psi))$$

where $2\varphi = \psi$ and by (4.27)

$$\begin{aligned} x'(\varphi) &= -2\widetilde{y}(\varphi) = -2\kappa(1+\cos(2\varphi)) \\ x(\varphi) &= \gamma - \kappa(2\varphi + \sin(2\varphi)) = \gamma - \kappa(\psi + \sin(\psi)). \end{aligned}$$

Altogether, we obtain the solution in parameter form as

$$\boxed{(x(\psi), y(\psi)) = \Big(\gamma - \kappa(\psi + \sin(\psi)),\ \kappa(1+\cos(\psi))\Big)}\ .$$

For $\kappa = -\varrho < 0$ and $\psi = \varphi + \pi$ this is the equation of an *orthocycloid* which has a vertex in $(x, y) = (0, 0)$ for $\varphi = 0$ and $\gamma = -\varrho\pi$, and passes moreover below the x-axis. Keeping ϱ fixed and changing γ implies that the cycloid is moved on the x-axis whereas a modification of ϱ leads to an enlargement resp. diminution of the original cycloid. In other words, a unique extremal exists here for solution; cf. (Clegg), p. 49; (Kosmol), Sect. 4.2 (Fig. 4.3).

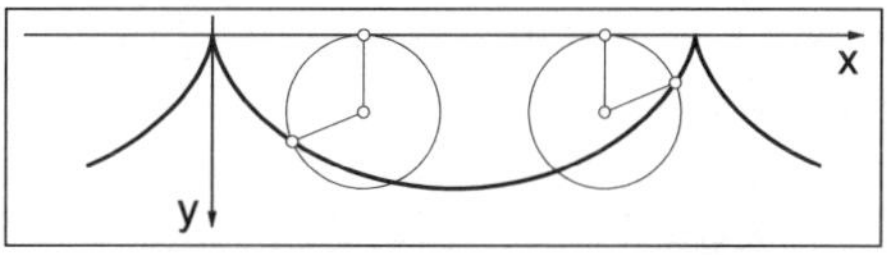

Figure 4.3. Orthocycloid in brachistochrone

Case 4: Motion in a homogeneous field. In an (x, y)-coordinate system let a particle of unity mass have kinetic energy T, potential energy $U = gy$ and total energy $E = T + U$, hence $T = E - U$. By the principle of least action in JACOBI's form, see Sect. 6.9**(f)**, the trajectory between two points is a stationary value of the modified integral of action:

$$\int_{s_0}^{s_1} \sqrt{(E-U)}\, ds = \int_{s_0}^{s_1} \sqrt{(E-gy)}\, ds .$$

But $E = \text{const}$ by Example 4.6, and the maximum reachable point on the y-axis is obviuosly $y_0 = g/E$. After a simple translation and re-notation we may consider the problem

$$\int_{s_0}^{s_1} \sqrt{y}\, ds = \int_a^b \sqrt{y(1+y'^2)}\, dx = \text{min!}$$

By (4.26) and (4.27) we obtain for $n = -1/2$ immediately

$$y(\varphi) = \frac{c}{\cos^2\varphi}\,,\ x'(\varphi) = \frac{2c}{\cos^2\varphi}\,.$$

Integration of the second equation and elimination of φ yields $(x-d)^2 = 4c\,(y-c)$ which are the well-known trajectory parabolas. Substitution of the points $P_1 = (x_1, y_1)$ and $P_2 = (x_2, y_2)$, $x_1 \neq x_2$, and ensuing substraction yields

$$\begin{matrix} (x_1-d)^2 = 4c\,(y_1-c) \\ (x_2-d)^2 = 4c\,(y_2-c) \end{matrix} \implies d = \frac{1}{2}(x_1+x_2) - \frac{2c(y_1-y_2)}{x_1-x_2}$$

Substitution in, say, $(x_1-d)^2 = 4c\,(y_1-c)$ yields a quadratic equation for the parameter c which has two, one or zero solutions in dependence of the points P_1 and P_2. Therefore there exist in this case two, one or zero extremals through two different points. In the *fountain* of Figure 4.4 however we solve NEWTON's equations $\ddot{x} = 0$, $\ddot{y} = -g$ directly with initial conditions $x(0) = 0$, $y(0) = 0$, $\dot{x}(0) = \cos\varphi$, $\dot{y}(0) = \sin\varphi$ that the initial velocity vector has unity length and, by this way, the total energy is constant as required above. After elimination of time we obtain the family of curves as

$$f(x, y; \varphi) := y + \frac{g\,x^2}{2\cos^2\varphi} - x\tan\varphi = 0\,.$$

The *envelope* of the family is obtained up to an additive constant by elimination of the parameter φ from the equations

$$f(x, y; \varphi) = 0\,,\quad \frac{\partial}{\partial\varphi} f(x, y; \varphi) = 0\,.$$

Regarding the limited size of the present volume, we refer to (Clegg) for the last case $n = -1$.

Figure 4.4. Fountain with envelope

4.2 Control Problems without Constraints

(a) Let $0 < T < \infty$ be a fixed terminal time. We seek a *state* $x : [0\,,\,T] \ni t \mapsto x(t) \in \mathbb{R}^n$ and a *control* $u : [0\,,\,T] \ni t \mapsto u(t) \in \mathbb{R}^m$ such that the pair $(x,\,u)$ is a solution of the control problem

$$\boxed{\begin{aligned} J(x,u) &= p(x(0),x(T)) + \int_0^T q(t,x(t),u(t))\,dt = \max!\\ x(t) &= x(0) + \int_0^t f(s,x(s),u(s))\,ds\,,\quad t \in [0\,,\,T]\\ 0 &= r(x(0),x(T)) \in \mathbb{R}^{|r|} \end{aligned}} \tag{4.28}$$

where all data shall be continuously F-differentiable in an open neighborhood of the solution. The differential equation is transformed here into an integral equation (4.28)(2°) because it is treated in this form later on. It plays the role of an equality constraint as well as the boundary conditions (being possibly nonlinear). Therefore LAGRANGE multipliers have to be introduced. As concerns the integral equation they are in fact RIEMANN-STIELTJES integrals with an associated function $y : [0,\,T] \ni t \mapsto y(t) \in \mathbb{R}_n$ (row vector!), cf. Sect. 12.5, whereas the multipliers belonging to the boundary condition are simple row vectors $z \in \mathbb{R}_r$. In the same way as in the preceding section we obtain a boundary value problem for the HAMILTON function

$$H(t,x(t),u(t),y(t)) = q(t,x(t),u(t)) + y(t)f(t,x(t),u(t))\,,\ t \in [0\,,\,T]\,.$$

As already displayed in Section 3.6, the constraints have to be feasible in some way. There, it was required for equality constraints that the gradient is a surjective mapping in optimum. Let $\mathcal{X}$ and $\mathcal{U}$ be spaces of functions to be specified later which describe the smoothness of the solution. Then the above regularity condition corresponds in the present situation to the following *regularity condition*:

Definition 4.2. *A solution $(x^*,u^*) \in \mathcal{X} \times \mathcal{U}$ of the control problem (4.28) is regular if*

$$\begin{aligned} &\forall\,(w,c) \in \mathcal{X} \times \mathbb{R}^r\ \ \exists\,(x,v) \in \mathcal{X} \times \mathcal{U} :\\ &x(t) - x(0) - \int_0^t \left[\nabla_x f(x^*,u^*)x + \nabla_u f(x^*,u^*)v\right] ds = w\,,\ \ t \in (0,T)\,,\\ &\nabla_1 r(x^*(0),x^*(T))x(0) + \nabla_2 r(x^*(0),x^*(T))x(T) = c\,. \end{aligned}$$

The second condition is equivalent to

$$\operatorname{rank}\left[\nabla_1 r(x^*(0),x^*(T))\,,\ \nabla_2 r(x^*(0),x^*(T))\right] = |r|\,.$$

Consequently a solution (x^*,u^*) is regular if the *linearized* boundary problem has a solution (x,v) in optimum for all right sides (w,c) or, in other words, the solution of the linearized problem is *controllable*. However we may choose $w = 0$ without restriction; cf. (Luenberger), p. 256.

Theorem 4.3. *Let $\mathcal{X} = \mathcal{C}^1([0, T]; \mathbb{R}^n)$, $\mathcal{U} = \mathcal{C}^1([0, T]; \mathbb{R}^m)$, and let $(x^*, u^*) \in \mathcal{X} \times \mathcal{U}$ be a regular solution of the problem (4.28). Then there exists a pair $(y^*, z^*) \in \mathcal{X} \times \mathbb{R}_{|r|}$ such that the quadruple (x^*, u^*, y^*, z^*) is solution of the differential-algebraic boundary problem*

$$\boxed{\begin{array}{rll} \dot{x}(t) = & [\nabla_y H]^T(t, x, u, y) & \in \mathbb{R}^n \\ \dot{y}(t) = - & \nabla_x H(t, x, u, y) & \in \mathbb{R}_n \\ 0 = & \nabla_u H(t, x, u, y) & \in \mathbb{R}_m \\ 0 = & r(x(0), x(T)) & \in \mathbb{R}^{|r|} \\ y(0) = - & \nabla_1 (p + z\, r)(x(0), x(T)) & \in \mathbb{R}_n \\ y(T-) = & \nabla_2 (p + z\, r)(x(0), x(T)) & \in \mathbb{R}_n \end{array}} \; . \tag{4.29}$$

Obviously, one may also write the *costate* $y(t)$ as column vector but the present form is advantageous in later computation and in implementation. If $\nabla_u \nabla_u H \in \mathbb{R}^m{}_m$ is regular near the solution, the control $u(t) \in \mathbb{R}^m$ can be eliminated by $\nabla_u H = 0$ at least theoretically and also in many practical cases (also in the re-entry problem of Sect. 4.5). Then the resulting system is a pure boundary value problem, and one may try to solve it by a suitable device as, e.g., the multiple shooting method. The difficulty in this way of procedure however consists in the computation of an appropriate initial trajectory for both the state x and the costate y.

Hint to the proof. The proof of this theorem is developed completely in `SUPPLEMENT\chap04b` because of its model character. In particular it is shown that the costate y enjoys the same smoothness as the state x. The Lagrange function of the problem with equality constraints reads by Sect. 3.6 and Sect. 12.5

$$\begin{aligned} L((x, u), y, z) &= [p + z\, r](x(0), x(T)) \\ &+ \int_0^T q(x, u)\, dt + \int_0^T dy(t) \left[x(t) - x(0) - \int_0^t f(x, u)\, ds \right] \end{aligned} \tag{4.30}$$

where $z \in \mathbb{R}_{|r|}$ and $y \in \mathrm{NBV}([0, T]; \mathbb{R}_n)$. The second integral is a Riemann-Stieltjes integral counting jumps of y; cf. Sect. 12.5. The F-derivative of L w.r.t. (x, u) and increment (ξ, η) has the form

$$\begin{aligned} &\nabla_{(x,u)} L((x, u), y, z)(\xi, \eta) \\ &= \nabla_1 [p + z\, r](x(0), x(T))\xi(0) + \nabla_2 [p + z\, r](x(0), x(T))\xi(T) \\ &+ \int_0^T [\nabla_x q\, \xi + \nabla_u q\, \eta] dt \\ &+ \int_0^T dy(t) \left[\xi(t) - \xi(0) - \int_0^t (\nabla_x f\, \xi + \nabla_u f\, \eta)\, ds \right]. \end{aligned}$$

By Corollary 3.4 there exists a pair $(y^*, z^*) \in \mathrm{NBV}(0,T) \times \mathbb{R}_{|r|}$ such that in optimum

$$\forall\, (\xi, \zeta) \in \mathcal{X} \times \mathcal{U} : \nabla_{(x,u)} L((x^*, u^*), y^*, z^*)(\xi, \eta) = 0\,.$$

The remaining proof is carried out in four steps by applying partial integration and the Fundamentallemma 4.1:
1. Step: Choose $\eta = 0$; *2. Step:* Choose $\eta = 0$ and $\xi \in C^1$ arbitrary such that $\xi(0) = \xi(T) = 0$; *3. Step:* Choose $\eta = 0$ and ξ arbitrary such that either $\xi(T) = 0$ or $\xi(0) = 0$; *4. Step:* Choose $\xi = 0$ and $\eta \in U$ arbitrary.

Corollary 4.2. *Suppose that both functions q and f in (4.28) do not depend explicitely on t then the* HAMILTON *function H is an invariant (first integral) of the system (4.29), i.e., $H(x^*(t), u^*(t), y^*(t)) = \text{constant}$.*

Proof. Cf. Corollary 4.1. By assumption

$$\begin{aligned}
H(x,u,y) &= q(x(t),u(t)) + y(t) f(x(t),u(t)) \\
\frac{d}{dt} H(x,u,y) &= H_t + \nabla_x H \dot{x} + \nabla_u H \dot{u} + \nabla_y H \dot{y} = H_t + \nabla_u H \dot{u} + (\nabla_x H + \dot{y}) f\,.
\end{aligned}$$

By assumption also $H_t = 0$ and, by Theorem 4.3, $\nabla_u H = 0$ and $\nabla_x H + \dot{y} = 0$ at the solution. Therefore $\dot{H} = 0$ and accordingly $H = \text{constant}$ in (x^*, u^*, y^*). □

(b) Free Terminal Time Let the terminal time $0 < T < \infty$ be *free*, i.e., an independent variable, then a triple (T, x, u) is sought for solution of the control problem (4.28) where the payoff function p and the boundary condition r may now depend on T.

Theorem 4.4. *Let (T^*, x^*, u^*) be a solution of the control problem (4.28) with free terminal time T, and let the assumptions of Theorem 4.3 be fulfilled. Then there exists a pair $(y^*, z^*) \in \mathcal{X} \times \mathbb{R}_{|r|}$ such that the quintuple $(T^*, x^*, u^*, y^*, z^*)$ is solution of the differential-algebraic boundary problem (4.29) with additional* transversality condition *for T^**

$$\boxed{0 = \frac{\partial}{\partial T}[p + z\,r](T, x(0), x(T)) + H(T, x(T), u(T), y(T)) \in \mathbb{R}}\,. \tag{4.31}$$

Proof. The F-derivative of the LAGRANGE function L w.r.t. (T, x, u) and increment (Θ, ξ, ζ) has now the form

$$\begin{aligned}
&\nabla_{(T,x,u)} L((T,x,u),y,z)(\Theta,\xi,\zeta) \\
&= \nabla_{(x,u)} L((T,x,u),y,z)(\xi,\zeta) + L_T((T,x,u),y,z)\Theta \\
&= \nabla_{(x,u)} L((T,x,u),y,z)(\xi,\zeta) \\
&+ \Big[\nabla_2(p + z\,r)(x(0),x(T),T)\dot{x}(T) + \frac{\partial}{\partial T}(p + z\,r)(x(0),x(T),T)\Big]\Theta \\
&+ \left[q(T,x,u) + \big(y(T) - y(T-)\big)\left[x(T) - x(0) - \int_0^T f(x,u)dt\right]\right]\Theta\,.
\end{aligned}$$

This term must disappear in optimum for all (Θ, ξ, ζ) as necessary condition for the solution. Because $x^*(T) - x^*(0) - \int_0^{T^*} f(x,u)dt = 0$ we obtain by this way the following additional necessary condition where $\dot{x} = f(x,u)$

$$\begin{aligned} 0 &= \nabla_2[p + z\,r](x(0), x(T), T) f(T, x(T), u(T)) \\ &+ \frac{\partial}{\partial T}[p + z\,r](x(0), x(T), T) + q(T, x(T), u(T)) \,. \end{aligned}$$

Substitution of $\nabla_2 \ldots = y(T-)$ yields the assertion. □

An independent terminal time T may also be introduced as a new *dependent* variable $T = \widetilde{x}_{n+1}(s)$, $\dot{\widetilde{x}}_{n+1}(s) = 0$, or as an *independent control parameter*. In the latter case e.g. the substitution $t = T\,s$ yields

$$x(t) = \widetilde{x}(s)\,, \quad \frac{d}{dt} x(t) = \frac{1}{T}\frac{d}{ds}\widetilde{x}(s)\,.$$

If we now write x instead $\widetilde{x}$ and u instead $\widetilde{u}$ again, we obtain by (4.28) a modified problem with fixed terminal time 1 and independent variable s,

$$\boxed{\begin{aligned} J(T,x,u) &= p(T, x(0), x(1)) + T \int_0^1 q(Ts, x(s), u(s))\,ds = \max! \\ x(s) &= x(0) + T \int_0^s f(T\sigma, x(\sigma), u(\sigma))\,d\sigma\,, \quad s \in [0\,,\,1] \\ 0 &= r(T, x(0), x(1)) \end{aligned}} \,. \qquad (4.32)$$

In this modified problem, T is a *control parameter* which underlies no longer any restriction.

(c) The Free Lagrange Multipliers $z \in \mathbb{R}_{|r|}$ rule the distribution of the boundary conditions onto the both combined boundary problems, namely for the state x

$$\begin{aligned} \dot{x}(t) &= [\nabla_y H]^T(t, x, u, y) \in \mathbb{R}^n \\ 0 &= r(x(0), x(T)) \qquad \in \mathbb{R}^{|r|}\,, \end{aligned} \qquad (4.33)$$

and for the *costate* y

$$\begin{aligned} \dot{y}(t) &= -\ \nabla_x H(t, x, u, y)] &&\in \mathbb{R}_n \\ y(0) &= -\ \nabla_1(p + z\,r)(x(0), x(T)) &&\in \mathbb{R}_n \\ y(T) &= \ \ \nabla_2(p + z\,r)(x(0), x(T)) &&\in \mathbb{R}_n\,. \end{aligned} \qquad (4.34)$$

Let (x^*, u^*) be a unique regular solution of the problem (4.28), then the matrix

$$\Big[\nabla_1 r(x(0), x(T))\,,\ \nabla_2 r(x(0), x(T))\Big]$$

must have full rank $|r|$ by the regularity condition in Definition 4.2. We consider the following simple special cases where $n = 2$ and $p = 0$:

Case 1: $x_1(0) = a_1$, $x_2(0) = a_2$, i.e., $x(0)$ fixed, $x(T)$ free, then, where $z = [z_1, z_2] \in \mathbb{R}_2$ are free,

$$r(x(0), x(T)) := \begin{bmatrix} x_1(0) - a_1 \\ x_2(0) - a_2 \end{bmatrix} = \begin{bmatrix} 0 \\ 0 \end{bmatrix},$$

$$\nabla_1 r(x(0), x(T)) = \begin{bmatrix} 1 & 0 \\ 0 & 1 \end{bmatrix}, \quad \nabla_2 r(x(0), x(T)) = \begin{bmatrix} 0 & 0 \\ 0 & 0 \end{bmatrix},$$

i.e., $y(0) = z\,\nabla_1 r = [z_1,\ z_2]$ free, $y(T) = z\,\nabla_2 r = [0,\ 0]$ fixed.

Case 2: $x_1(0) = a_1$, $x_2(T) = b_2$, $x_2(0)$ and $x_1(T)$ free, then in the same way

$$r(x(0), x(T)) := \begin{bmatrix} x_1(0) - a_1 \\ x_2(T) - b_2 \end{bmatrix} = \begin{bmatrix} 0 \\ 0 \end{bmatrix},$$

$$\nabla_1 r(x(0), x(T)) = \begin{bmatrix} 1 & 0 \\ 0 & 0 \end{bmatrix}, \quad \nabla_2 r(x(0), x(T)) = \begin{bmatrix} 0 & 0 \\ 0 & 1 \end{bmatrix},$$

$$y(0) = z\,\nabla_1 r = [z_1,\ 0], \quad y(T) = z\,\nabla_2 r = [0,\ z_2].$$

i.e., $y_1(0)$, $y_2(T)$ free, $y_2(0) = y_1(T) = 0$ fixed.

Case 3: $x_1(T) = b_1$, $x_2(T) = b_2$, i.e., $x(0)$ free.

$$r(x(0), x(T)) := \begin{bmatrix} x_1(T) - b_1 \\ x_2(T) - b_2 \end{bmatrix} = \begin{bmatrix} 0 \\ 0 \end{bmatrix},$$

$$\nabla_1 r(x(0), x(T)) = \begin{bmatrix} 0 & 0 \\ 0 & 0 \end{bmatrix}, \quad \nabla_2 r(x(0), x(T)) = \begin{bmatrix} 1 & 0 \\ 0 & 1 \end{bmatrix},$$

$$y(0) = z\,\nabla_1 r = [0,\ 0], \quad y(T) = z\,\nabla_2 r = [z_1\,,z_2]$$

i.e., $y(0) = 0$ fixed, $y(T)$ free.

In each case we so obtain four boundary conditions for four differential equations as a necessary condition for a unique solution. The boundary conditions have to be filled up properly by zeros in the case where $|r| < n$; thereafter one may proceed in the same way as above.

(d) The Costate $y \in \mathbb{R}_n$ is also called *shadow price*; cf. Sect. 3.3**(h)**. We consider a simple problem with fixed terminal time T to study its role in a control problem more properly:

$$\begin{aligned} J(x,u) &= p(x(T)) + \int_0^T q(t,x,u)\,dt = \max! \\ \dot{x} &= f(t,x,u)\,, \quad x(0) = a \in \mathbb{R}\,, \end{aligned} \tag{4.35}$$

and make the following assumption:

Assumption 4.1. (1°) *Let* (x^*, u^*) *be a unique regular solution.*
(2°) *For all* $\tau \in (0\,, T)$, *let (4.35) be uniquely solvable in a neighborhood* $\mathcal{U}$ *of* $x^*(\tau)$ *for* $t \in [\tau, T]$.
(3°) *For all solutions of* (2°) *let the optimal control be a function of the optimal state (feedback control).*

Let (x, u) be solution of the partial problem in the segment $[\tau, T]$,

$$\begin{aligned} J(x,u) = p(x(T)) + \int_\tau^T q(t,x,u)\,dt = \max! \\ \dot{x} = f(t,x,u)\,, \quad \tau \le t \le T\,, \quad x(\tau) = x^*(\tau)\,. \end{aligned} \tag{4.36}$$

Then (x, u) coincides with the optimal solution (x^*, u^*) on the interval $[\tau, T]$ (*principle of optimality*), and the *profit function*

$$V(\tau, x(\tau)) = p(x(T)) + \int_\tau^T q(t,x,u(x))\,dt$$

is a function of τ and $x(\tau)$ in $\mathcal{U}$. Applying Leibniz' rule

$$\begin{aligned} \frac{dV}{d\tau}(\tau,x(\tau)) &= V_\tau(\tau,x(\tau)) + \nabla_x V(\tau,x(\tau))\, f(\tau,x(\tau),u(\tau,x(\tau))) \\ &= -q(\tau,x(\tau),u(\tau,x(\tau)))\,; \end{aligned}$$

hence — writing again t instead τ

$$V_t(t,x(t)) + \nabla_x V(t,x(t))\, f(t,x(t),u(t,x(t))) + q(t,x(t),u(t,x(t))) = 0\,. \tag{4.37}$$

Now we set

$$\boxed{y(t) = \nabla_x V(t,x(t))}\,, \tag{4.38}$$

then $y(T) = \nabla p(x(T))$ and by (4.37) and permutation of the derivatives w.r.t. t and x

$$\begin{aligned} \dot{y} &= f^T \nabla_x \nabla_x V + \frac{\partial}{\partial t} \nabla_x V(t,x(t)) = f^T \nabla_x \nabla_x V + \nabla_x(-\nabla_x V\, f - q) \\ &= f^T \nabla_x \nabla_x V - f^T \nabla_x \nabla_x V - \nabla_x V \nabla_x f - \nabla_x q = -y\, \nabla_x f - \nabla_x q\,. \end{aligned}$$

Then, applying the Hamilton function $H = q + y\,f$, y is solution of the terminal value problem

$$\dot{y} = -\nabla_x H(t,x(t),u(t),y(t))\,, \quad y(T) = \nabla_x p(T,x(T))\,,$$

and the notation (4.38) is justified.

(e) Maximum Principle The profit function V of the problem (4.35) satisfies

$$V_t(t, x^*(t)) + \nabla_x V(x^*(t))\, f(t, x^*(t), u^*(t)) + q(t, x^*(t), u^*(t)) = 0 \qquad (4.39)$$

under Assumption 4.1 by the above result. Moreover, the *principle of optimality* does hold, namely that an optimal solution is also optimal on subintervals,

$$\begin{aligned} & p(x^*(T)) + \int_t^T q(s, x^*(s), u(s))\, ds \\ & \le p(x^*(T)) + \int_t^T q(s, x^*(s), u^*(s))\, ds = V(t, x^*(t)), \end{aligned} \qquad (4.40)$$

$0 \le t \le T$. The relation

$$V_t(t, x^*(t)) + \nabla_x V(x^*(t))\, f(t, x^*(t), u(t)) + q(t, x^*(t), u(t)) \le 0 \qquad (4.41)$$

can be derived from this inequality by passing to the limit. Equation (4.39) and (4.41) supply together — without argument t

$$\begin{aligned} & V_t(x^*) + \nabla_x V(x^*)\, f(x^*, u) + q(t, x^*, u) \\ & \le V_t(x^*) + \nabla_x V(x^*)\, f(x^*, u^*) + q(x^*, u^*). \end{aligned}$$

As an inference to this relation we obtain on the one side the functional equation

$$\boxed{\begin{aligned} 0 &= V_t^*(t, x(t)) + \max\nolimits_u \Big\{ q(t, x(t), u) + \nabla_x V^*(t, x(t))\, f(t, x(t), u) \Big\} \\ u^*(t) &= \arg\max\nolimits_u \Big\{ q(t, x(t), u) + \nabla_x V^*(t, x(t))\, f(t, x(t), u) \Big\} \end{aligned}}$$

named after HAMILTON-JACOBI-BELLMAN and, on the other side the *maximum principle of* PONTRJAGIN by applying (4.38) and the HAMILTONian H,

$$\boxed{H(t, x^*(t), u^*(t), y^*(t)) = \max\nolimits_u H(t, x^*(t), u, y^*(t)),\ 0 \le t \le T}\,. \qquad (4.42)$$

Example 4.7. To give an econometrical interpretation of the problem (4.35), let

$x(t) \in \mathbb{R}$	capital of a company
$u(t) \in \mathbb{R}$	corporate policy
$J(x, u)$	planned earnings, in period $[0, T]$
$q(t, x, u)$	growth rate of expected earnings
$f(t, x, u)$	capital growth rate
$y(t)$	shadow price
$H = q + y\, f$	*total growth rate*, i.e., growth rate of total assets (total assets = accumulated dividends + capital assets)
$\dot{y}(t)$	inflation rate.

The optimal political solution maximizes both earnings and total growth according to PONTRJAGIN's maximum principle. But for this occurrence the shadow price must be known.

Example 4.8. The *linear-quadratic* control problem, also called *state regulator problem* is one of the most popular examples, in a similar way as the corresponding problem in finite-dimensional optimization, see Sect. 3.4. Here, it can even be shown by solving the well-known RICATTI *equation* that the entire problem is equivalent to a single linear differential system for the state x :

$$\begin{aligned} J(x,u) &= x(T)^T P x(T) + \frac{1}{2}\int_0^T \left[x^T Q(t)x + u^T R(t)u\right] dt = \min! \\ \dot{x} &= A(t)x + B(t)u \in \mathbb{R}^n\,,\ x(0) = a\,. \end{aligned} \tag{4.43}$$

The matrices $Q(t)$, $R(t)$ shall be symmetric and positive definite, and the control $u(t) \in \mathbb{R}^m$ is commonly assumed to be unconstrained hence smooth. Introducing time-dependent "LAGRANGE multiplier" $y(t) \in \mathbb{R}^n$, two alternative problems may be formulated by means of the LAGRANGE function

$$L(x,u,y) = J(x,u) + \int_0^T y(t)^T \big[-\dot{x} + A(t)x + B(t)u\big]\, dt \tag{4.44}$$

and the associated Hamiltonian

$$H(x,u,y) = \frac{1}{2}\left[x^T Q(t)x + u^T R(t)u\right] + y^T\left[A(t)x + B(t)u\right].$$

(1°) Find a triple (x^*, u^*, y^*) such that $x(0) = a$ and

$$L(x^*, u^*, y^*) = \max\nolimits_y \min\nolimits_{x,u} L(x,u,y)\,. \tag{4.45}$$

(2°) Find (x^*, u^*, y^*) such that $x(0) = a$ and

$$\begin{aligned} L(x^*, u^*, y^*) &= \max\nolimits_y L(x,u,y) \\ \nabla_x L(x^*, u^*, y^*) = 0\,,\ &\ \nabla_u L(x^*, u^*, y^*) = 0\,. \end{aligned} \tag{4.46}$$

For the solution, three different accesses are possible:
Case 1: Differential-algebraic problem using the costate equations,

$$\begin{aligned} \dot{x} = A(t)x + B(t)u\,,\ &\ \dot{y} = -Q(t)x - A^T(t)y\,, \\ x(0) = a\,,\ y(T) = Px(T)\,,\ &\ R(t)u + B^T(t)y = 0\,. \end{aligned} \tag{4.47}$$

The boundary value problem follows by substituting $u = -R(t)^{-1}B(t)^T y$ (no feed-back control).
Case 2: Dynamic optimization approach. By using (4.38), i.e., $y(t) = \nabla_x V(t, x(t))$, we obtain $u = -R^{-1}B^T V_x$, and the HAMILTON-JACOBI-BELLMAN equation (4.37) reads

$$V_t + V_x(Ax + Bu) + \frac{1}{2}(x^TQx + u^TRu) = 0\,.$$

Substitution for u yields

$$\begin{aligned}0 &= V_t + V_xAx - V_xBR^{-1}B^TV_x + \frac{1}{2}\left[x^TQx + V_xBR^{-1}RR^{-1}B^TV_x\right]\\ &= V_t + V_xAx + \frac{1}{2}x^TQx - \frac{1}{2}V_xBR^{-1}B^TV_x\,.\end{aligned} \tag{4.48}$$

This equation happens to have a product solution of the form $V(t) = x^T(t)S(t)x(t)/2$ where $V(T) = x(T)^TS(T)x(T)/2$. The matrix $S(t) \in \mathbb{R}^n{}_n$ is symmetric without loss of generality since always $x^TSx = x^T(S + S^T)x/2$. Substitution into (4.48) yields

$$0 = \frac{1}{2}x^T[\dot{S} + 2SA - SBR^{-1}B^TS + Q]x$$

and by consequence the *matrix* RICCATI *equation*

$$\dot{S} + 2SA - SBR^{-1}B^TS + Q = 0\,,\ S(T) = P$$

(the simple RICCATI equation reads $\dot{y} + ay + by^2 = c$). Let S be a solution of this equation and insert $V_x = Sx$ into u once more then $u = -R^{-1}B^TSx$ is a *feedback* control and the state x is a solution of the linear initial value problem $\dot{x} = [A - BR^{-1}B^TS]x\,,\ x(0) = a\,.$

Case 3: Dual problem. By partial integration we obtain from (4.44)

$$\begin{aligned}L(x,u,y) &= x(T)^TPx(T) + \frac{1}{2}\int_0^T \left[x^TQx + u^TRu + 2y^T(Ax + Bu) + 2\dot{y}^Tx\right]\\ &\quad - y(t)x(t)\Big|_0^T\,,\end{aligned}$$

$y(T) = Px(T)$. Regarding (4.45), a substitution of (4.47) into (4.44) yields the dual problem $L(x,u;y) = \max!$ where

$$\begin{aligned}&L(x,u,y)\\ &= y(T)^TP^{-1}y(T) + \frac{1}{2}\int_0^T \left[(\dot{y} + Ay)^TQ^{-1}(\dot{y} + Ay) + y^TBR^{-1}B^Ty\right]dt\\ &+ \int_0^T [y(t)^T\dot{x}(t) + \dot{y}^Tx(t)]\,dt - y(t)x(t)\Big|_0^T\\ &= y(T)^TP^{-1}y(T) + \frac{1}{2}\int_0^T \left[(\dot{y} + Ay)^TQ^{-1}(\dot{y} + Ay) + y^TBR^{-1}B^Ty\right]\,dt\end{aligned}$$

if P regular ($Y(T) = 0$ if $P = 0$). This problem is solved in the best way by introducing a new control $u = \dot{y}$ without constraints.

4.3 Control Problems with Constraints

(a) The control u is often a bounded resource in technical applications, i.e., it is allowed to vary only in a bounded domain. Let for instance a vehicle drive on a straight line from a point a to a point b as fast as possible. Then it moves at first with *maximum* positive acceleration and then with *maximum* negative acceleration, and only the calculation of the switching point is of interest. This is a typical example of the frequently appearing *bang-bang control* where the optimum u takes its values always on the *boundary* of the feasible domain. This implies however normally that u has jumping points, i.e., becomes *discontinuous.*

Let us first consider a control problem where, besides the boundary value problem, no constraints on the state variables x appear at all, but the restrictions on the control u are rather general formulated. $\mathcal{X}$ and $\mathcal{U}$ are again BANACH spaces that are specified instantly as well as the set of possible controls Ω.

$$\boxed{\begin{aligned} J(x,u) = p(x(T)) + \int_0^T q(t,x(t),u(t),t)dt &= \max!\\ x(t) = x(0) + \int_0^t f(t,x(s),u(s),s)ds,\ t \in [0,T]&,\\ 0 = r(x(T)),\ r:\mathbb{R}^n \to \mathbb{R}^{|r|},\ x \in \mathcal{X},\ u \in \mathcal{U}&,\\ \forall t \in [0,T]:\ u(t) \in \Omega \subset \mathbb{R}^m&. \end{aligned}} \tag{4.49}$$

Theorem 4.5. *(Werner) For the problem (4.49) suppose that*
(1°) $\mathcal{X} = \mathcal{W}^1_\infty([0,T];\mathbb{R}^m)$, $\mathcal{U} = \mathcal{L}_\infty([0,T];\mathbb{R}^n)$, *see Sect. 12.5,*
(2°) $f:[0,\,T]\times\mathbb{R}^n\times\mathbb{R}^m \to \mathbb{R}^m$, $p:\mathbb{R}^n\to\mathbb{R}$, $q:[0,\,T]\times\mathbb{R}^n\times\mathbb{R}^m\to\mathbb{R}$, $r:\mathbb{R}^n\to\mathbb{R}^r$ *are continuously* FRÉCHET-*differentiable,*
(3°) (x^*,u^*) *is a local solution,*
(4°) Ω *nonempty, convex and closed,*
(5°) $\operatorname{rank}\nabla r(x^*(T)) = |r|$.
Then there exists a $\lambda \geq 0$, *a* $y \in W^{1,\infty}_n[0,T]$ *and* $z \in \mathbb{R}_k$ *with* $(\lambda,\,y) \neq 0$ *and*

$$\begin{aligned} \dot y(t)\ &= \lambda\nabla_x q(t,x^*(t),u^*(t)) - y(t)\nabla_x(t,x^*(t),u^*(t))\ \text{a.e. on}\ [0,\,T]\\ y(T) &= -\lambda\nabla p(x^*(T)) - z^*\nabla r(x^*(T))\,. \end{aligned}$$

For almost all $t \in [0,\,T]$ *one has the local* PONTRJAGIN *maximum principle*

$$\forall\, u \in \Omega: (y(t)\nabla_u f(t,x^*(t),u^*(t)) - \lambda\nabla_u q(t,x^*(t),u^*(t))(u-u^*(t)) \geq 0\,.$$

For practical applications the parameter λ must be prevented to become zero by a suitable (further) *regularity* or *controllability* condition.

We now introduce two notions concerning the smoothness of dependent variables which suffice in most applications to represent the underlying function spaces.

Definition 4.3. *Let* $[a, b] \subset \mathbb{R}$ *a finite interval and* $f : [a, b] \to \mathbb{R}^n$.
(1°) *A function* f *is* piecewise continuous, $f \in \mathcal{C}_{pc}([a, b]; \mathbb{R}^n)$ *or briefly* $f \in \mathcal{C}_{pc,n}[a, b]$, *if* f *is continuous in* $[a, b]$ *up to a finite number of points, and the one-sided limit values exist at these points of exception; in addition* f *shall everywhere be continuous from right.*
(2°) *A piecewise continuous function* f *is* piecewise continuously differentiable, $f \in \mathcal{C}^1_{pc,n}[a, b]$, *if the derivative is also piecewise continuous.*

Accordingly, a function $f \in \mathcal{C}^1_{pc,n}[a, b]$ is not necessarily continuous but the derivative may be defined everywhere if the one-sided derivatives from right are taken at the jumping points. Moreover, f may have a jumping point at the right end of the interval of definition. Hence the space $\mathcal{C}^1_{pc,n}[a, b]$ corresponds to the space NBV$[a, b]$ defined in Sect. 12.5 with the restriction that only a *finite* number of jumping points is allowed. On the other side, *chattering controls*, having an *infinite* number of jumping points with a cluster point, do not belong to one of the just introduced function spaces.

Let now $0 < T < \infty$ be a fixed terminal time and let

$$\begin{aligned} &\mathcal{X} = \mathcal{C}^1_{pc,n}[0, T] \cap C[0, T] && \text{the function space of states } x\,, \\ &\mathcal{U} = \mathcal{C}_{pc,m}[0, T] && \text{the function space of controls } u\,. \end{aligned}$$

We seek for a pair $(x, u) \in \mathcal{X} \times \mathcal{U}$ of state x and control u being a solution of the control problem

$$\boxed{\begin{aligned} J(x,u) &= p(x(0), x(T)) + \int_0^T q(t, x(t), u(t))\, dt = \max! \\ x(t) &= x(0) + \int_0^t f(s, x(s), u(s))\, ds\,, \quad t \in [0\,, T] \\ 0 &= r(x(0), x(T)) \in \mathbb{R}^{|r|} \\ 0 &\le g(t, x(t), u(t)) \in \mathbb{R}^{|g|}\,, \quad t \in [0, T] \\ 0 &\le h(t, x(t)) \in \mathbb{R}^{|h|}\,, \quad t \in [0, T] \end{aligned}} \tag{4.50}$$

where all data are continuously F-differentiable. Besides the differential equation (transformed into an integral equation) and the boundary conditions (being independent of u), no further *equality constraints* are commonly assigned to the problem. The *inequality constraints* are separated into two different types, one depending explicitely on the control and the other not but just the latter causes some severe difficulties in proving the *existence* of solutions (Hartl).

When the terminal time T is free then $t = Ts$ is substituted again and the modified problem

$$\boxed{\begin{aligned} &J(x,u) = p(T,x(0),x(T)) + T\int_0^1 q(Ts,x(s),u(s))\,ds = \max!\\ &x(s) = x(0) + T\int_0^s f(T\sigma,x(\sigma),u(\sigma))\,d\sigma\,,\quad s\in[0\,,\,1]\\ &0 = r(T,x(0),x(T)) \in \mathbb{R}^{|r|}\\ &0 \le g(Ts,x(s),u(s)) \in \mathbb{R}^{|g|},\quad s\in[0\,,\,1]\\ &0 \le h(Ts,x(s)) \in \mathbb{R}^{|h|},\quad s\in[0\,,\,1]\end{aligned}} \tag{4.51}$$

is considered where the operation horizon T has now become an independent variable.

(b) Necessary Conditions Reconsider once more the unconstrained problem with regard to Definition 4.3.

Theorem 4.6. *Let $g = 0$ and $h = 0$, and let $(x^*, u^*) \in \mathcal{X}\times\mathcal{U}$ be a regular solution of (4.50) then Theorem 4.3 does hold at all points where u is continuous.*

Proof. Let $0 < t_1 < \ldots < t_{N-1}$ be the jumping points of u^* and let $t_N = T$. By the proof of Theorem 4.3, the costate y^* may have jumping points at t_i, $i = 1 : N$ whereas $t = 0$ is not a jumping point. Consider for the moment a general function Φ, then partial integration yields

$$\begin{aligned}
&\int_0^T dy^*(t)\Phi(t)\,dt = \sum_{i=1}^N \Big[\int_{t_{i-1}}^{t_i} dy^*(t)\Phi(t)\,dt + [y^*(t_i) - y^*(t_i-)]\Phi(t_i)\Big]\,dt\\
&= \sum_{i=1}^N \Big[y^*(t_i-)\Phi(t_i) - y^*(t_{i-1})\Phi(t_{i-1}) - \int_{t_{i-1}}^{t_i} y^*(t)\dot\Phi(t)\,dt\\
&+[y^*(t_i) - y^*(t_i-)]\Phi(t_i)\Big]\\
&= \sum_{i=1}^N \left[-y^*(t_{i-1})\Phi(t_{i-1}) - \int_{t_{i-1}}^{t_i-} y^*(t)\dot\Phi(t)\,dt + y^*(t_i)\Phi(t_i)\right]\\
&= y^*(T)\Phi(T)0 \le y(t) \in \mathbb{R}_n\,, - \sum_{i=1}^N \int_{t_{i-1}}^{t_i} y^*(t)\dot\Phi(t)\,dt\,.
\end{aligned}$$

As a consequence, the values of the costate y^* at jumping points are canceled out in the Riemann-Stieltjes integral if partial integration is applied as in the proof of Theorem 4.3. The differential equations of this Theorem hold in those partial intervals where u^* is continuous. □

Let $H(t,x,u,y) = q(t,x,u) + y f(t,x,y)$ be again the Hamiltonian of the problem (4.50) again. Let $\mathcal{K} = \{f \in \mathcal{C}[a,b]\,,\ f(x) \ge 0\}$ be the order cone in $\mathcal{C}[a,b]$, then

$$\mathcal{K}_d = \{y \in \mathcal{C}_d(a,b)\,,\ y(f) = \int_a^b f(x)\,dv(x)\,,\\ v \in \mathrm{BV}(a,b) \text{ weakly monotone increasing}\}$$

is the dual cone; cf. Example 1.24(3°). Regarding Example 12.3(2°) we introduce *two* LAGRANGE functions $\widetilde{L}$ and L to the problem (4.50):

$$\begin{aligned}
\widetilde{L}((x,u),y,z,v,w) &= [p + z\,r](x(0),x(T)) \\
&\quad + \int_0^T q(x,u)\,dt + \int_0^T [dv\,g(x,u) + dw\,h(x)] \\
&\quad + \int_0^T dy(t)\left[x(t) - x(0) - \int_0^t f(x,u)\,ds\right] \\
L(t,x,u,y,\dot{v},\dot{w}) &= H(t,x,u,y) + \dot{v}\,g(t,x,u) + \dot{w}\,h(t,x)
\end{aligned} \tag{4.52}$$

where $y(t) \in \mathbb{R}_n$, $0 \le v(t) \in \mathbb{R}_{|g|}$, $0 \le w(t) \in \mathbb{R}_{|h|}$ are row vectors and all integrals over the operational interval $[0,T]$ are again RIEMANN-STIELTJES integrals by Sect. 12.5. Then we obtain a simple analogue to Theorem 3.2:

Theorem 4.7. *Let the following assumptions be fulfilled:*

(1°) *There exists*

$$(x^*,u^*,y^*,v^*,w^*,z^*) \in \mathcal{X}\times\mathcal{U}\times\mathcal{C}^1_{pc,n}[0,\,T]\times\mathcal{C}_{pc,|g|}[0,\,T]\times\mathcal{C}_{pc,|h|}[0,\,T]\times\mathbb{R}_{|r|}$$

where v^* and w^* are wakly monotone increasing and

$$(x^*,u^*) = \arg\max_{(x,u)\in X\times U} \widetilde{L}((x,u),y^*,z^*,v^*,w^*)\,. \tag{4.53}$$

(2°) *The pair (x^*,u^*) satisfies all constraints.*

(3°) $\displaystyle\int_0^T v^*(t)\,g(t,x^*,u^*)\,dt = 0$ *(*complementarity condition*).*

(4°) $\displaystyle\int_0^T w^*(t)\,h(t,x^*)\,dt = 0$ *(*complementarity condition*).*

Then (x^*,u^*) is a solution of (4.50).

This result is to be understood as *basic approach* for further studies. The proof of *existence* for the occuring LAGRANGE multipliers takes a large space in control theory. Frequently the requirements on smoothnes have to weakened considerably here; and, moreover, the system has to satisfy rather complicated criteria of regularity. We will not pursue the theory of existence further, the less so since many results are not yet proved in full detail; cf. (Hartl).

Lemma 4.3. *Adopt the assumptions of Theorem 4.7 and suppose in addition that $v \in \mathcal{C}^1_{pc,|g|}[0,\,T]$, $w \in \mathcal{C}^1_{pc,|h|}[0,\,T]$. Then $(x^*,u^*,y^*,\dot{v}^*,\dot{w}^*,z^*)$ is solution of the following system at all points t where none of the components of this six-tuple has a jump,*

$$
\boxed{\begin{aligned}
\dot{x}(t) &= \quad [\nabla_y H]^T(t,x,u,y)\\
\dot{y}(t) &= -\ \nabla_x L(t,x,u,y,\dot{v},\dot{w})\\
0 &= \quad \nabla_u L(t,x,u,y,\dot{v},\dot{w})\\
0 &= \quad r(x(0),x(T))\\
y(0) &= -\ \nabla_1(p+z\,r)(x(0),x(T))\\
y(T-) &= \quad \nabla_2(p+z\,r)(x(0),x(T))\\
0 &= \quad \dot{v}(t)\,g(t,x,u)\\
0 &= \quad \dot{w}(t)\,h(t,x)\\
0 &\le \quad \dot{v}(t)\,,\ 0 \le \dot{w}(t)\,.
\end{aligned}} \tag{4.54}
$$

Proof see `SUPPLEMENT\chap04b`.

The proof is carried out in the same way as the proof of Theorem 4.3 by a two-fold partial integration. Observe that a solution of the system (4.54) is a stationary point of which the feasibility as well the extremal property has to be verified. However, many numerical procedures set LAGRANGE multipliers automatically equal to zero if the corresponding inequality restriction becomes inactive, for instance

$$g_i(t,x(t),u(t)) > 0 \implies v_i(t) = 0\,.$$

In this case, a solution (x,u) of the system (4.54) is feasible w.r.t. the inequality constraints (at least at those discrete points of the interval $[0,\,T]$ which are used by the algorithm); cf. also Assumption (4°) of Theorem 3.2.

In the sequel, we write sometimes for brevity $f^*[t] = f(t,x^*(t),u^*(t))$ etc.. An interval $\emptyset \neq [\tau_1,\,\tau_2] \subset [0,\,T]$ is called *boundary interval* of the constraint h if at least one component h_i of h in $[\tau_1,\,\tau_2]$ is active in optimum:

$$\exists\, 1 \le i \le |h| \quad \forall\, t \in [\tau_1,\,\tau_2] : h_i^*[t] = h_i(t,x^*(t)) = 0\,.$$

A boundary interval may consist also of a single point only (contact time). The costate y^* has jumpings at the same points as the control u^*, in normal case where one condition $g^i(t,x,u) \ge 0$ becomes active or inactive. But, in normal case, the costate y^* has additional jumpings at those points where a condition $h_i(t,x(t)) \ge 0$ becomes active or inactive:

Theorem 4.8. *(Hartl), Theorem 4.2. For all $t_0 < t_1$ in $[0,\,T]$*

$$
\begin{aligned}
&y^*(t_1^+) - y^*(t_0^+)\\
&= -\int_{t_0}^{t_1} \Big[\nabla_x H^*[t]dt + v^*(t)\nabla_x g^*[t]\Big]\,dt + \int_{(t_0,t_1]} dw^*(t)\nabla_x h^*[t],\\
&H^*[t_1^+] - H^*[t_0^+] = \int_{t_0}^{t_1} \Big[H_t^*[t] + v^*(t)g_t^*[t]\Big]dt - \int_{(t_0,t_1]} dw^*(t)h_t^*[t]\,.
\end{aligned}
$$

In consequence, the costate y^* may have a point of discontinuity at an arbitrary point τ of a boundary interval which then satisfies the following *jump condition*: There exists a vector $c(\tau)$ such that

$$y^*(\tau-) = y^*(\tau+) + c(\tau)\nabla_x h(\tau, x^*(\tau))$$
$$H^*[\tau-] = H^*[\tau+] - c(\tau)\frac{\partial}{\partial t}h^*[\tau];$$

cf. (Hartl), Theorem 4.1. For the sake of completeness we give finally also a sufficient condition for an optimum:

Theorem 4.9. *(Hartl), Theorem 8.1. Let (x^*, u^*) be a feasible pair of (4.50). Let there exist further a $y \in C^1_{pc,n}[0, T]$ such that for all feasible pairs (x, u):*
(1°) $H(t, x^*(t), u^*(t), y(t)) - H(t, x(t), u(t), y(t)) \geq \dot{y}(t)[x(t) - x^*(t)]$ a.e. in $[0, T]$,
(2°) $[y(t-) - y(t+)][x(t) - x^*(t)] \geq 0$ at every jumping point of y,
(3°) the transversality condition does hold,

$$y(0)[x(0) - x^*(0)] \geq p(x(0), x(T)) - p(x^*(0), x^*(T))$$
$$y(T)[x(T) - x^*(T)] \geq p(x(0), x(T)) - p(x^*(0), x^*(T)).$$

Then (x^*, u^*) is optimal.

Generically, the above problems are not considered with additional equality restrictions $h(t, x, u) = 0$.

(c) The **Maximum Principle** of PONTRJAGIN

$$u^*(t) = \arg\max_{u \in \Omega(t, x^*(t))} H(t, x^*(t), u, y^*(t)), \text{ a.e. in } [0, T],$$

where $\Omega(t, x^*(t)) = \{u \in \mathbb{R}^m;\ g(t, x^*(t), u) \geq 0\}$, is also not yet proved in full strength for the above constellation but is generally adopted as correct; cf. (Hartl). The following comment shall be understood only as *consideration of plausibility*:

By the principle of optimality and under assumptions to be specified more exactly, a general function Φ and $J(u) = \int_0^T \Phi(t, u(t))\, dt$ may satisfy

$$\begin{aligned} J(u^*) &= \max\left\{\int_0^T \Phi(t, u(t))\, dt,\ u(t) \in \Omega(t)\right\} \\ &= \int_0^T \max_{u(t)\in\Omega(t)} \Phi(t, u(t))\, dt, \end{aligned} \tag{4.55}$$

which constitutes in same way an inversion of the well-known majorant theorem for definite integrals. If now all components besides u are already in optimum then (4.53) is equivalent to

$$\begin{aligned} &L(x^*, u^*, y^*, z^*, v^*, w^*) \\ &= \max\Big\{\int_0^T q(x^*, u)\, dt + \int_0^T dy^*(t) \left[x^*(t) - x^*(0) - \int_0^t f(x^*, u)\, ds\right], \\ &\qquad u(t) \in \Omega(t, x^*(t))\Big\} \\ &= \max\left\{\int_0^T q(x^*, u)\, dt - \int_0^T dy^*(t) \int_0^t f(x^*, u)\, ds,\ u(t) \in \Omega(t, x^*(t))\right\}, \end{aligned}$$

since the constant terms may be dropped. Partial integration yields in the same way as above by applying (4.55)

$$\begin{aligned}
&L(x^*, u^*, y^*, z^*, v^*, w^*) \\
&= \max\left\{\int_0^T q(x^*, u)\,dt + \int_0^T y^*(t)\,f(x^*, u)\,dt\,,\ u(t) \in \Omega(t, x^*(t))\right\} \\
&= \max\left\{\int_0^T H(x^*, u, y^*)\,dt\,,\ u(t) \in \Omega(t, x^*(t))\right\} \\
&= \int_0^T \max_{u\in\Omega(t,x^*(t))} H(x^*, u, y^*)\,dt = \int_0^T H(x^*, u^*, y^*)\,dt\,.
\end{aligned}$$

References: (Hartl) and the references given therein.

4.4 Examples

(a) Numerical Approach The differential-algebraic boundary value problems (4.29) and (4.54) are in general difficult to solve.

Hint:

> The simple control problem (4.29) contains no constraints of the control u which then is normally smooth. If the control $u(t) \in \mathbb{R}^m$ is unconstrained and smooth, it (or m other dependent variables) can (at least theoretically) be eliminated by means of the equation $\nabla_u H = 0$. This equation then disappears in the system (4.29) and the result is a pure nonlinear boundary value problem. After having found a suitable start trajectory, this problem can be solved by NEWTON's method and, e.g., the box scheme, cf. Sect. 2.5**(a)**.

Experiments with the gradient method after (Dyer) are less encouraging. In more general problems of the form (4.50), successes are perhaps reached in the simplest way if the costate y is dropped entirely and the problem is discretized in its original form:

Numerical Solution

(1°) An introduction of the costate y is dispensed with entirely.

(2°) A pair (x, u) of solution is computed where state x and control u are equally ranked.

(3°) The intervall $[0, T]$ is partitioned uniformly, x and u are replaced by step functions; the integrals are replaced by the *composite trapezoidal rule.*

(4°) The resulting finite-dimensional optimization problem is solved by the method `sqp.m` of *sequential quadratic programming*; cf. Sect. 3.5. In this method several parameter and weights appear which must adapted *carefully* to the individual problem.

In the sequel all differential equations are transformed into integral equations. The interval $[0, T]$ is uniformly partitioned into n subintervals and the composed trapezoidal rule is applied to discretize each integral. For instance, the scalar differential equation $x' = f(t,x,u)$ is transposed into an integral,

$$x(t) = x(0) + \int_0^t f(t,x,u)\,dt\,,$$

$$X - x(0)\,\underline{e} - \frac{1}{T}\,\mathbf{E}\,F(X,U) = 0\,,\ \ \underline{e} = [1] \in \mathbb{R}^{n+1}\,,\ \ X = [x_1,\ldots,x_{n+1}]^T \text{ etc..}$$

If, e.g., $n = 5$, the $(6,6)$-matrix $\mathbf{E}$ has the form

$$\mathbf{E} = \frac{1}{2}\begin{bmatrix} 0 & 0 & 0 & 0 & 0 & 0 \\ 1 & 1 & 0 & 0 & 0 & 0 \\ 1 & 2 & 1 & 0 & 0 & 0 \\ 1 & 2 & 2 & 1 & 0 & 0 \\ 1 & 2 & 2 & 2 & 1 & 0 \\ 1 & 2 & 2 & 2 & 2 & 1 \end{bmatrix}.$$

The first row of $\mathbf{E}$ must be canceled in the case where $x(0)$ is not specified. This simple approximation provides acceptable results in many cases and may serve for computation of suitable starting values in others. For demonstration, some *well-known* examples of the literature are solved below. If the terminal time T is free then we proceed as proposed in (4.32) and transform the problem into a problem with fixed terminal time in the interval $[0, 1]$ and additional free control parameter T.

(b) Examples Let H be always the HAMILTON function.

Example 4.9. (Thrust problem) (Bryson), p. 59. A spacecraft of mass m with thrust force $m\,a(t)$ in direction of its body axis is accelerated in an inertial (x_1, x_2)-coordinate system (Fig. 4.5).
Notations: $(x_1(t), x_2(t))$ position of the ship, $x_3(t)$ velocity in x_1-direction, $x_4(t)$ velocity in x_2-direction, $u(t)$ angle between the ships axis and x_1-direction (control).

The ship shall be transferred to a path parallel to the x_1-axis at height h in fixed time T such that the horizontal velocity $x_3(T)$ becomes maximum. The final x_1-coordinate does *not* play any role. Obviously the attainable height depends on the given operational time T.
Problem: (normalized)

$$J(x,u) = x_3(T) = \max!,\ \ x(0) = 0\,,\ \ x_2(T) = h\,,\ \ x_4(T) = 0\,,$$
$$\dot{x}_1 = x_3(t)\,,\ \ \dot{x}_2 = x_4(t)\,,\ \ \dot{x}_3 = a(t)\cos(u(t))\,,\ \ \dot{x}_4 = a(t)\sin(u(t))\,.$$

Hamilton function, costate equations and associated boundary conditions:

$$\begin{aligned} H(x,u,y) &= y\,f = y_1x_3 + y_2x_4 + y_3a(t)\cos(u) + y_4a(t)\sin(u)\,,\\ H_u(x,u,y) &= -\,y_3a\sin(u) + y_4a\cos u\,,\\ \dot{y}_1 &= 0\,,\ \ \dot{y}_2 = 0\,,\ \ \dot{y}_3 = -y_1\,,\ \ \dot{y}_4 = -y_2\,,\\ y_1(T) &= 0\,,\ \ y_3(T) = 1\,. \end{aligned}$$

Special features: Using $H_u = 0$, $\sin(u)$ and $\cos(u)$ may expressed by y_4. The results reads, writing $x_{4+i} := y_i$,

$$\begin{aligned}
&\dot{x_1} = x_3\,,\ \dot{x_2} = x_4\,,\ \dot{x_3} = a(t)/(1+x_8^2)^{1/2}\,,\ \dot{x_4} = a(t)x_8/(1+x_8^2)^{1/2}\,,\\
&\dot{x_5} = 0\,,\quad \dot{x_6} = 0\,,\quad \dot{x_7} = -x_5\,,\qquad\qquad \dot{x_8} = -x_6\,.
\end{aligned}$$

The problem is solved in `KAPITEL04\CONTROL01` by the above recommended method for $T = 2$ and $h = 0.8, 0.6, 0.4, 0.2$. In `KAPITEL04\CONTROL02` the control u is eliminated and the problem is solved as pure boundary value problem with the box-scheme and NEWTON's method; in `KAPITEL04\CONTROL03` the gradient method is used for solution.

Example 4.10. (Orbit problem) (Bryson), p. 66. A spaceship shall be transferred from a given initial circular orbit with radius r_0 to the largest possible circular orbit where the operational time T is fixed and the thrust is constant. The influence of other celestial bodies is neglected and the ship moves in mathematical positive angular direction (Fig. 4.6).
Notations: By using polar coordinates (r, φ) with (point-shaped) central body for center, let $r(t)$ radial distance of the ship, $u = \dot{r}$ radial component of velocity, $v = r\dot{\varphi}$ tangential component of velocity, $m(t)$ mass of the ship, $|\dot{m}(t)|$ *constant* rate of fuel consumption, $S = |\dot{m}(t)|\sigma$ thrust, σ a machine constant [m/s], $\alpha(t)$ thrust angle, γ gravitational constant, M mass of central body, $G = \gamma M$, $\varrho = |\dot{m}(t)|/m(0)$ [1/s] constant, $S = \varrho\,\sigma\, m(0)$. The control $\alpha(t)$ is the angle between the tangent to the current circular orbit at time t and the ships axis into flight-direction (local coordinate system). Hence r, $\dot{r}$ and the velocity $v = r\dot{\varphi}$ are chosen for dependent variables; cf. Sect. 6.2**(f)**.
Problem:

$$\begin{aligned}
&r(T) = \max!\,,\ r(0) = r_0\,,\ u(0) = 0\,,\ v(0) = \sqrt{G/r_0}\,,\\
&u(T) = 0\,,\ v(T) = \sqrt{G/r(T)}\,.
\end{aligned}$$

$$\begin{aligned}
\dot{r} &= u(t)\,,\\
\dot{u} &= \frac{v(t)^2}{r(t)} - \frac{G}{r(t)^2} + \frac{S(t)\sin(\alpha(t))}{m(0)(1-\varrho t)} = \frac{v(t)^2}{r(t)} - \frac{G}{r(t)^2} + \frac{\varrho\sigma\,\sin(\alpha(t))}{1-\varrho t}\,,\\
\dot{v} &= -\frac{u(t)\,v(t)}{r(t)} + \frac{S(t)\cos(\alpha(t))}{m(0)(1-\varrho t)} \quad = -\frac{u(t)\,v(t)}{r(t)} + \frac{\varrho\sigma\,\cos(\alpha(t))}{1-\varrho t}\,.
\end{aligned}$$

Reduction to a dimensionless system: The radius r and time t are replaced by quantities without dimension,

$$s = \frac{G^{1/2}}{r_0^{3/2}}\,t\,,\quad R(s) = \frac{1}{r_0}\,r(t)\,,$$

and the second and third equation of motion is multiplied by r_0^2/G. Let $\kappa = \varrho\sigma\cdot r_0^2/G$ be a constant without dimension then, for $U(s) = R'(s)$ and $V(s) = R(s)\varphi'(s)$,

$$R' = U\,,\ U' = \frac{V^2}{R} - \frac{1}{R^2} + \kappa\frac{\sin(\alpha)}{1-\varrho t}\,,\ V' = -\frac{UV}{R} + \kappa\frac{\cos(\alpha)}{1-\varrho t}\,.$$

we write again t instead s, T instead $s_f = TG^{1/2}/r_0^{3/2}$, u instead α and $x = [x_1, x_2, x_3]^T = [R, U, V]^T$.
Transformed problem:

$$J(x,u) = x_1(T) = \max!\,,\ \dot{x}_1 = x_2(t)\,,$$

$$\dot{x}_2 = \frac{x_3^2(t)}{x_1(t)} - \frac{1}{x_1^2(t)} + \kappa\frac{\sin(u(t))}{1-\varrho t}\,,\ \dot{x}_3 = -\frac{x_2(t)x_3(t)}{x_1(t)} + \kappa\frac{\cos(u(t))}{1-\varrho t}\,,$$
$$x_1(0) = 1\,,\ x_2(0) = 0\,,\ x_3(0) = 1\,,\ x_2(T) = 0\,,\ x_3(T)^2x_1(T) - 1 = 0\,.$$

Hamilton function, costate equations and their boundary conditions:

$$H(x,y,u,t) = y_1x_2 + y_2\left[\frac{x_3^2}{x_1} - \frac{1}{x_1^2} + \kappa\frac{\sin(u)}{1-\varrho t}\right] + y_3\left[\kappa\frac{\cos(u)}{1-\varrho t} - \frac{x_2x_3}{x_1}\right]$$
$$H_u = \kappa\left[y_2\cos(u) - y_3\sin(u)\right]\frac{1}{1-\varrho t}$$
$$\dot{y}_1 = \frac{y_2x_3^2}{x_1^2} - \frac{2y_2}{x_1^3} - \frac{y_3x_2x_3}{x_1^2}\,,\ \dot{y}_2 = \frac{y_3x_3}{2x_1} - y_1\,,\ \dot{y}_3 = \frac{y_3x_2}{x_1} - \frac{2y_2x_3}{x_1}$$
$$y_1(T) = 1 + 0.5\,y_3(T)x_1(T)^{-3/2}\,.$$

Special features: Using $H_u = 0$, $\sin(u)$ and $\cos(u)$ may be expressed by y_2 and y_3. The result is, writing $x_{3+i} = y_i$,

$$\dot{x}_1 = x_2\,,\ \dot{x}_2 = \frac{x_3^2}{x_1} - \frac{1}{x_1^2} + \kappa\frac{x_5}{(x_5^2+x_6^2)^{1/2}(1-\varrho t)}$$
$$\dot{x}_3 = \kappa\frac{x_6}{(x_5^2+x_6^2)^{1/2}(1-\varrho t)} - \frac{x_2x_3}{x_1}\,,\ \dot{x}_4 = \frac{x_3^2x_5}{x_1^2} - \frac{2x_5}{x_1^3} - \frac{x_2x_3x_6}{x_1^2}$$
$$\dot{x}_5 = \frac{x_3x_6}{x_1} - x_4\,,\ \dot{x}_6 = \frac{x_2x_6}{x_1} - \frac{2x_3x_5}{x_1}\,.$$

Physical and technical data for a small spacecraft with ion-propulsion engine after (Bryson) in `KAPITEL04\Beispiele\bsp02`.

The problem is solved in `KAPITEL04\CONTROL01` by the method proposed in **(a)**; in `KAPITEL04\CONTROL02` the control u is eliminated again and the problem is solved as boundary problem; in `KAPITEL04\CONTROL03` the gradient method is applied.

Example 4.11. (ZERMELO's problem) (Bryson), p. 77. A ship shall travel in minimum time from a point $a = (a_1, a_2)$ to a point $b = (b_1, b_2)$ in an inertial (x_1, x_2)-coordinate system ($x = (x_1, x_2)$) where the velocity S relative to water is constant; however there is a strong current flow (Fig. 4.7).
Notations: $(x_1(t), x_2(t))$ position of the ship, $v_1(x)$ velocity of the current flow in x_1-direction, $v_2(x)$ velocity of the current flow in x_2-direction, u angle of

the ship's axis relative to x_1-direction (control).
Problem:

$$J(T) = T = \min!\,,\quad x(0) = a\,,\quad x(T) = b\,,$$
$$\dot{x}_1 = S\cos(u(t)) + v_1(x(t))\,,\quad \dot{x}_2 = S\sin(u(t)) + v_2(x(t))\,.$$

Hamilton function, costate equation (no boundary conditions for y):

$$\begin{aligned} H(x,u,y) &= y_1[S\cos(u) + v_1(x)] + y_2[S\sin(u) + v_2(x)] + 1 \\ H_u(x,u,y) &= -\,y_1\sin(u) + y_2\cos(u) \\ \dot{y}_1 &= -y_1(v_1)_{x_1} - y_2(v_2)_{x_1}\,,\quad \dot{y}_2 = -y_1(v_1)_{x_2} - y_2(v_2)_{x_2}\,. \end{aligned}$$

Special features I: H does not depend on t explicitely; hence it is an invariant of the system (first integral), H = constant on the optimal trajectory, cf. Corollary 4.2. But $H(x(T),u(T),y(T)) = 0$ follows from the transversality condition (4.35) for the free terminal time T and thus $H(x,u,y) = 0$. Now by means of $H = 0$ or $H_u = 0$ the costate y may be expressed as function of u. Substitution into one of the costate equations yields then

$$\dot{u} = \sin^2(u)(v_2)_{x_1} + \sin(u)\cos(u)[(v_1)_{x_1} - (v_2)_{x_2}] - \cos^2(u)(v_1)_{x_2}\,.$$

Special features II: On choosing the substitution $t = Ts$ the unknown terminal time T is introduced as a new *dependent* variable;

$$\begin{aligned} X_3(s) &= T\,, & X_1(s) &= x_1(sT) = x_1(sX_3(s))\,, \\ X_2(s) &= x_2(sT) = x_2(sX_3(s))\,, & U(s) &= u(sX_3(s))\,, \end{aligned}$$

and we obtain a boundary value problem for four unknowns

$$\begin{aligned} X_3'(s) &= 0\,, & X_1'(s) &= \dot{x}_1(sX_3)\cdot X_3(s)\,, \\ X_2'(s) &= \dot{x}_2(sX_3)\cdot X_3(s)\,, & U'(s) &= \dot{u}(sX_3)\cdot X_3(s)\,. \end{aligned}$$

Writing for simplicity $y(s) = [X_1, X_2, X_3, U]$, this boundary problem reads

$$\begin{aligned} & y_1' = [S\cos(y_4) + v_1(y_1,y_2)]y_3\,,\quad y_2' = [S\sin(y_4) + v_2(y_1,y_2)]y_3\,,\quad y_3' = 0\,, \\ & y_4' = [\sin^2(y_4)(v_2)_{y_1} + \sin(y_4)\cos(y_4)((v_1)_{y_1} - (v_2)_{y_2}) - \cos^2(y_4)(v_1)_{y_2}]y_3\,, \\ & y_1(0) = a_1\,,\quad y_2(0) = a_2\,,\quad y_1(1) = b_1\,,\quad y_2(1) = b_1\,. \end{aligned}$$

Special features III: The unknown terminal time T can also be introduced as *free* variable by the above transformation. Then the transformed problem reads

$$J(T) = T = \min!\,,\quad x(0) = a\,,\quad x(1) = b\,,$$
$$x_1'(s) = [S\cos(u(s)) + v_1(x(s))]T\,,\quad x_2'(s) = [S\sin(u(s)) + v_2(x(s))]T\,,$$

in which form it is solved in `KAPITEL04\CONTROL01` by using the program `sqp.m`. In `KAPITEL04\CONTROL02` the costate variable y is eliminated and

NEWTON's method is applied again. In the considered example, $S = 1$ and $v = [-S^* x_2, 0]$ is chosen for current velocity where $S^* = S/16$, $S/8$, $S/4$, $S/2$, S, $2S$; the straight line-segment connecting the points a and b is chosen for start trajectory.

Example 4.12. Servo-problem (Burges), p. 281.

$$T = \min!\,, \quad x(0) = a_1\,, \quad x'(0) = a_2\,, \quad x(T) = x'(T) = 0\,, \quad |u(t)| \le 1\,,$$
$$\ddot{x} + a\dot{x} + \omega^2 x = u\,, \quad 0 < t < T, \quad a \ge 0\,.$$

For simplicity we consider the case $\omega^2 = 1$ and $a = 0$ (no damping). The system is transformed into a system of first order where the unknown terminal time T is a *free* control parameter. The resulting system reads

$$\begin{gathered} J(T) = T = \min! \\ X_1'(s) = T \cdot X_2(s)\,, \quad X_2'(s) = T \cdot (U(s) - X_1(s))\,, \\ X_1(0) = a_1\,, \quad X_2(0) = a_2\,, \quad X_1(1) = 0\,, \quad X_2(1) = 0\,, \\ 1 - U(s) \ge 0\,, \quad U(s) - 1 \ge 0\,, \end{gathered}$$

and is solved in this form.
Special features I: The optimal control is a bang-bang control where $u^*(t) = \pm 1$. A switching curve may be computed (shifted sinus functions) above of which $u^*(t) = -1$ and below $u^*(t) = 1$. By consequence, the trajectory consists in this special case of segments of circles with center $(1, 0)$ resp. $(-1, 0)$, the change happens at the points of intersection with the switching curve.
Special features II: In Sect. 3.4**(a)** and 3.5**(b)**, several modifications have been mentioned to accelerate the methods `plqp.m`, `dlqp.m` and `sqp.m` by suitable adaption of the matrix H. However, these modifications do not yield here any advantages in comparison with the choice $H = I$ and, by the way, destroy the sparsity of the fundamental linear system of equations. On the other side, the MARATOS effect is usually applied *near* the solution, i.e. if the difference between old and new result of the iteration becomes sufficiently small. Also this device does not work here, instead we took for Figure 4.8 twenty steps without and then twenty steps with MARATOS effect; cf. Sect. 3.5**(f4)**.

Example 4.13. (Hartl), p. 204.

$$J(x, u) = \int_0^3 x\,dt = \min! \;\; \dot{x} = u\,, \;\; x(0) = 1,\; x(3) = 1\,, \;\; -1 \le u \le 1\,, \;\; 0 \le x\,.$$

$$H = x + y\,u\,, \quad L = H + v_1(1 + u) + v_2(1 - u) + wx\,.$$

Special features: Using the necessary conditions

$$\begin{gathered} L_u = y + v_1 - v_2 = 0\,, \\ \dot{y} = -L_x = -1 - w\,, \quad y(3) = z\,, \quad z \in \mathbb{R}\,, \\ v_1 \ge 0\,, \quad v_2 \ge 0\,, \quad v_1(1 + u) = v_2(1 - u) = 0\,, \quad w \ge 0\,, \quad w\,x = 0\,, \end{gathered}$$

the solution may be computed explicitely:

$$x^* = \begin{cases} 1-t \\ 0 \\ t-2 \end{cases}, \quad u^* = \begin{cases} -1 \\ 0 \\ 1 \end{cases} \text{ for } t \in \begin{cases} [0,\,1) \\ [1,\,2] \\ (2,\,3] \end{cases}.$$

Beginning with the interval $(1, 2)$ we obtain the table of mulipliers:

Interval	y	v_1	v_2	w
[0, 1)	$t-1$	$1-t$	0	0
[1, 2]	0	0	0	1
(2, 3]	$t-2$	0	$t-2$	0

In Figure 4.9 it is averaged over the state x. The optimal control u^* is a bang-bang control; the jumps become continually sharper during iteration but remain always smoothed in some sense which is a natural property of discretization.

Example 4.14. (Hartl), p. 207.

$$J(x,u) = \int_0^3 e^{-rt} u\, dt = \min!\,, \quad r \geq 0\,,$$
$$\dot{x} = u\,, \quad x(0) = 0\,, \quad 0 \leq u \leq 3\,, \quad x - 1 + (t-2)^2 \geq 0\,.$$

$$H = -e^{-rt}u + yu\,, \quad L = H + v_1 u + v_2(3-u) + w\left[x - 1 + (t-2)^2\right].$$

Special features: Using the necessary conditions

$$\dot{y} = -L_x = -w\,, \quad L_u = -e^{-rt} + y + v_1 - v_2 = 0\,,$$
$$v_1 \geq 0\,, \quad v_2 \geq 0\,, \quad v_1 u = v_2(3-u) = 0\,, \quad w \geq 0\,, \quad w\left[x - 1 + (t-2)^2\right] = 0\,,$$
$$y(3) = 0\,, \quad y(2-) = y(2+) - c\,, \quad c \geq 0\,,$$

the solution may be computed explicitely again:

$$x^* = \begin{cases} 0 \\ 1-(t-2)^2 \\ 1 \end{cases}, \quad u^* = \begin{cases} 0 \\ 2(2-t) \\ 0 \end{cases}, \text{ for } t \in \begin{cases} [0,\,1) \\ [1,\,2] \\ (2,\,3] \end{cases}.$$

Beginning with the interval (2, 3) we obtain the table of multipliers:

Interval	y	v_1	v_2	w
[0, 1)	e^{-r}	$e^{-rt} - e^{-r}$	0	0
[1, 2]	e^{-rt}	0	0	re^{-rt}
(2, 3]	0	e^{-rt}	0	0

The optimal control makes a sharp jump at $t = 1$, hence 200 steps of iteration are taken for Figure 4.10.

Example 4.15. (Hartl), p. 208. $x = (x_1, x_2)$.

$$J(x,u) = \int_0^3 2x_1\,dt = \min!$$

$$\dot{x} = \begin{bmatrix} x_2 \\ u \end{bmatrix}, \quad x(0) = \begin{bmatrix} 2 \\ 0 \end{bmatrix}, \quad -2 \le u \le 3, \quad x_1 \ge \alpha, \quad \alpha \in \mathbb{R}, \quad \alpha \le 0.$$

Special features:
Case 1: $\alpha \le -7$. Solution:

$$x^* = \begin{cases} 2 - t^2 \\ -2t \end{cases}, \quad u^* = -2, \quad \text{for } 0 \le t \le 3.$$

Case 2: $-7 < \alpha \le 2.5$. In this case there exists a switching time

$$\sigma = 3 - \frac{1}{4}(56 + 8\alpha)^{1/2}.$$

Solution:

$$x_1^* = \begin{cases} 2 - t^2 \\ 2 + t^2 + 2\sigma^2 - 4\sigma t \end{cases}, \quad x_2^* = \begin{cases} -2t \\ 2(t - 2\sigma) \end{cases}, \quad u^* = \begin{cases} -2 \\ 2 \end{cases} \text{for } t \in \begin{cases} [0, \sigma) \\ [\sigma, 3] \end{cases}.$$

Case 3: $-2.5 < \alpha \le 0$. There exists a switching time σ and a junction time ϱ where $0 < \sigma < \varrho < 3$. This point ϱ is also an entry point into the boundary curve with the relation

$$\varrho = 2\sigma = (4 - 2\alpha)^{1/2}.$$

Solution:

$$x_1^* = \begin{cases} 2 - t^2 \\ 2 + t^2 + 2\sigma^2 - 4\sigma t \end{cases}, \quad x_2^* = \begin{cases} -2t \\ 2(t - 2\sigma) \end{cases}, \quad u^* = \begin{cases} -2 \\ 2 \end{cases} \text{for } t \in \begin{cases} [0, \sigma) \\ [\sigma, \varrho] \end{cases}.$$

In interval $(\varrho, 3)$ we have $x_1 = \alpha, \quad x_2 = 0, \quad u = 0$.

Example 4.16. (Hartl), p. 210.

$$\int_0^1 \left[10x^2 - u^2\right] dt = \max!,$$
$$\dot{x} = x^2 - u, \quad x(0) = x(1) = 1, \quad x(t) \le 1.5.$$

The state x increases monotonically at first, meets the boundary $x = 1.5$ at time $t_1 = 0.345037$ and leaves it again at time $t_2 = 1 - t_1$. The control u is continuous and "tangential" at the points t_1 and t_2 because the problem is regular.

Example 4.17. Brachistochrone problem, cf. Sect. 4.1**(g)**. Let the x_2-axis point to below. On setting

$$\dot{x}_1 = (2gx_2(t))^{1/2}\cos u(t)\,,\quad \dot{x}_2 = (2gx_2(t))^{1/2}\sin u(t)\,, \tag{4.56}$$

the conservation law of energy $mv(t)^2 = 2mgx_2(t)$ is regarded and the problem reads together with (4.56)

$$\begin{aligned} &T = \min!\,,\quad x(0) = (0,0)\,,\quad x_1(T) = a\,,\quad 0 < a\,,\\ &0 \le g(x_1, x_2)\,. \end{aligned}$$

The control u is the angle between the tangent of the trajectory and the x_1-axis. Substitution of $t = Ts$ yields the problem

$$\begin{aligned} &T = \min!\,,\quad x(0) = (0,0)\,,\quad x_1(1) = a\,,\\ &x_1' = T(2gx_2(s))^{1/2}\cos u(s)\,,\quad x_2' = T(2gx_2(s))^{1/2}\sin u(s)\,,\\ &0 \le g(x_1, x_2)\,. \end{aligned}$$

The exact solution of the unconstrained problem is computed in (Bryson) and solved in `KAPITEL04\CONTROL03` by the gradient method. In `KAPITEL04\CONTROL01` the constraints

$$g(x_1, x_2) = H + 0.5x_1 - x_2 \ge 0\,,\quad H = 0.1\,,\ 0.15\,,\ 0.2\,,$$

are chosen as example.

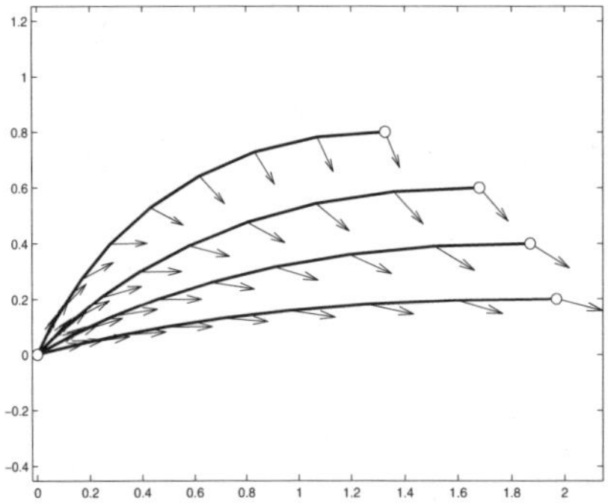

Figure 4.5. Example 4.9

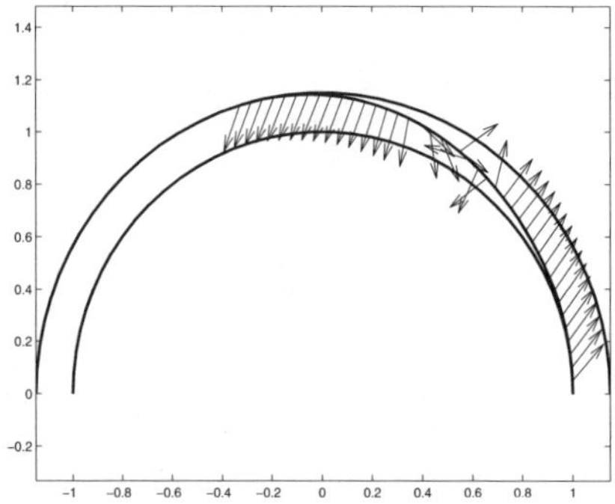

Figure 4.6. Example 4.10

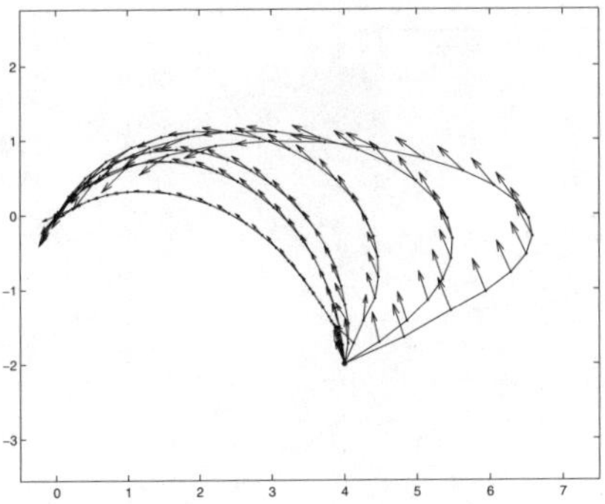

Figure 4.7. Example 4.11

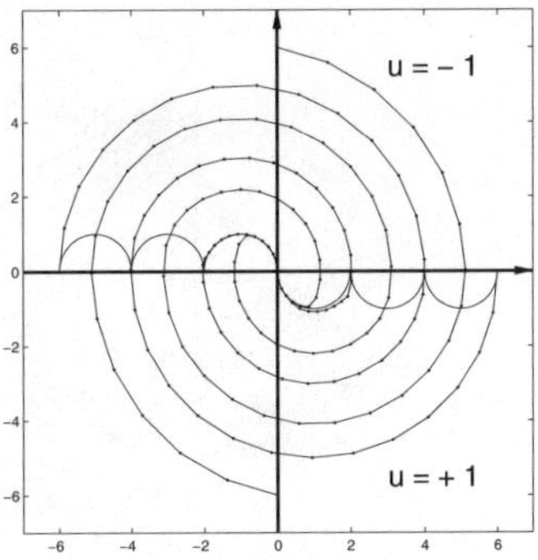

Figure 4.8. Example 4.12

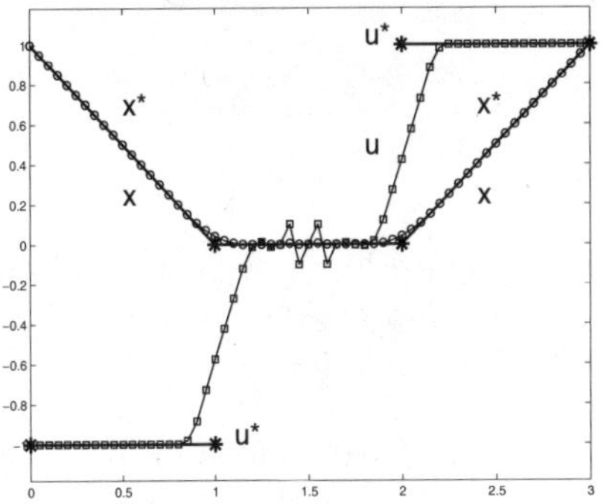

Figure 4.9. Example 4.13

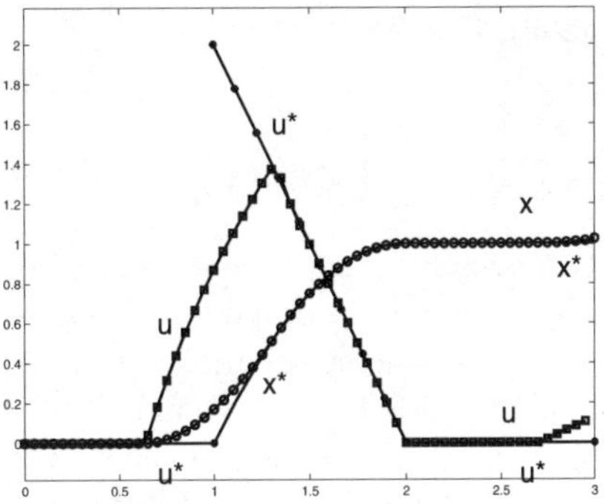

Figure 4.10. Example 4.14

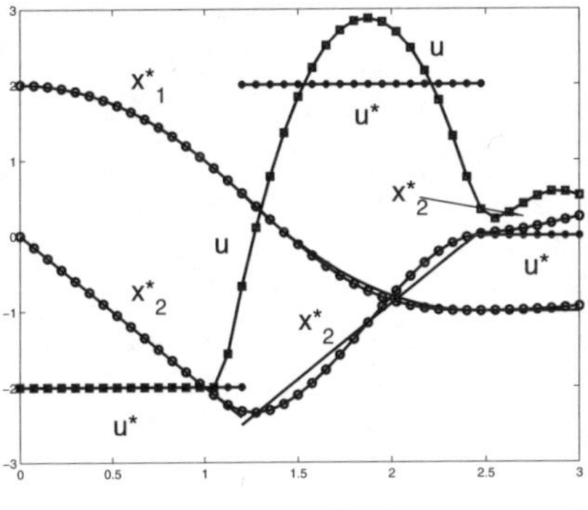

Figure 4.11. Example 4.15, $\alpha = -1$

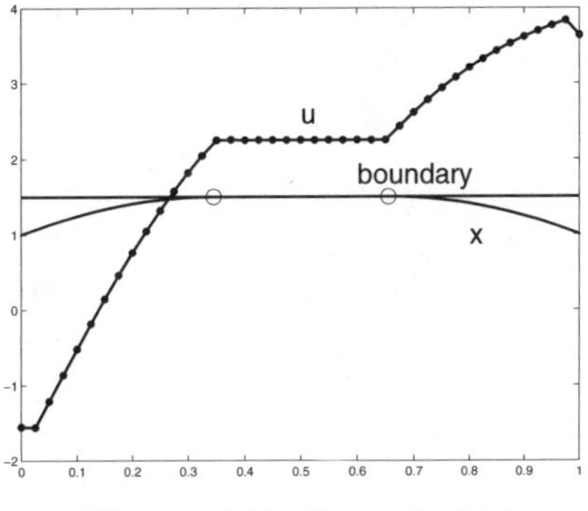

Figure 4.12. Example 4.16

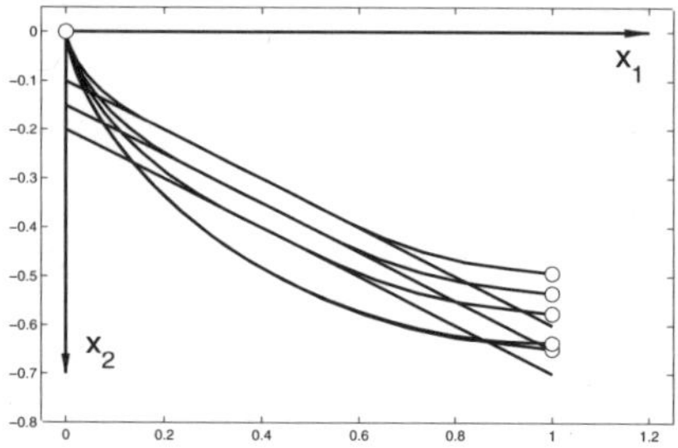

Figure 4.13. Example 4.17

References: (Berkowitz), (Bryson), (Craven78), (Craven95), (Dyer), (Ekeland), (Gelfand), (Gregory), (Hartl), (Kosmol), (Luenberger), (Petrov), (Teo89), (Teo91).

4.5 On the Reentry Problem

When reentrying into the atmosphere of earth, a spaceship shall bear a thermal charge as small as possible. We consider a space-ship of APOLLO type which flies from outer space into a circular earth-orbit and the spaceglider X-38 which descends from a circular orbit to a height of 25 km. Both optimization problems are discretized in the same way as the other examples and solved by the method `sqp_h.m` resp. by the method `sqp.m` in case of the sign restriction on the flight-angle in X-38.

Simplifications : Ball-shaped earth, point mass model (no attitude dynamics), symmetric flight (velocity vector in symmetry plane of the craft), no wind.

Notations : Γ gravitational constant $[m^3/kg]$, $R_{\rm E}$ radius of earth $[m]$, $M_{\rm E}$ mass of earth $[kg]$, $g_E = \Gamma M_E/R_E^2$ gravitational acceleration $[m/s^2]$, ω_E rotation velocity of earth $[rad/s]$, ϱ_0 air density at sea level $[kg/m^3]$, v velocity $[m/s]$, γ flight path angle $[rad]$, $r = R_{\rm E} + h$ distance to earth's center $[m]$, h height above surface of earth $[m]$, A lift $[N]$ (Newton), W drag $[N]$, c_A lift coefficient $[\,]$, c_W drag coefficient $[\,]$, S wing reference area $[m^2]$, M mass of the craft $[kg]$, $g(h) = \gamma M_E^2/(R_E + h)^2$ height-dependent gravitational acceleration (earth) $[m/s^2]$, $G(h) = M\,g(h)$ weight $[N]$, $\varrho(h)$ height-dependent air density $[kg/m^3]$, α angle of attack $[rad]$, μ_A bank angle $[rad]$, λ geographic latitude $[rad]$, χ heading angle $[rad]$, τ geographic longitude $[rad]$.

Transformations : Length: 1 $[ft]$ = 0.3048 $[m]$, weight: 1 $[sl]$ = 14.59 $[kg]$ (slugs), force: 1$[lbs]$ (pound) = 0.45359$[kp]$.

Constants : $\Gamma = 6.672 \cdot 10^{-11}\ [m^3/kg\,s^2]$, $R_E = 6370320\ [m]$, $\Gamma \cdot M_E = 3.98603 \cdot 10^{14}\ [m^3/s^2]$, $g_E = 9.806\ [m/s^2]$, $\omega_E = 7.292115 \cdot 10^{-5}[rad/s]$, $\varrho_0 = 1.225\ [kg/m^3]$. For approximation at height beyond 50 km it is but chosen

$$\varrho(h) = \varrho_1\, e^{-\beta h}\,, \quad \varrho_1 = 1.3932\ [kg/m^3]\,, \quad \beta = 1.3976 \cdot 10^{-4}\ [1/m]\,.$$

The ratio S/m is chosen as follows:

$$\begin{array}{ll} \text{APOLLO:} & S/M = 3.3876 \cdot 10^{-3}\ [m^2/kg]\,, \\ \text{X-38:} & S = 21.31\ [m^2]\,,\ M = 9060\ [kg]\,. \end{array}$$

The following values are chosen for lift A and drag W:

Typ	A	W
APOLLO	$S\varrho(h)v^2c_A(\alpha)/2$	$S\varrho(h)v^2c_W(\alpha)/2$
X-38	$S\varrho(h)v^2c_A(v)/2$	$S\varrho(h)v^2c_W(v)/2$

For X-38, the values c_A and c_W are given in tabular form and then are interpolated quadratically; for APOLLO these values are chosen by

$$c_A(\alpha) = 0.6\ \sin(\alpha)\,,\quad c_W(\alpha) = 1.174 - 0.9\ \cos(\alpha)\,.$$

The angles γ and χ describe the direction of the velocity vector, i.e., the tangent of the trajectory; γ denotes the angle relative to the local horizontal plane being positive if the velocity vector points to above. χ describes the direction of the local horizontal plane of east to north. In APOLLO the control $u = \alpha$ represents the angle between the axis of the cone (pointing to the vertex (!)) and the tangent of the trajectory. In X-38 the *control* $u = \mu_A$ is the angle between lift vector and the vertical plane through the current tangent of the trajectory.

In APOLLO the total heating is minimized and in X-38 the total thermal flux both at the stagnation point and per unit of area:

$$\begin{array}{ll} \text{APOLLO:} & J(v,h) = 10 \displaystyle\int_0^T [\varrho(h)]^{1/2} v^3\, dt \\ \text{X-38:} & J(v,h) = 10^{-4} \displaystyle\int_0^T [\varrho(h)]^{1/2} v^{3.15}\, dt\,. \end{array}$$

Equations of motion for APOLLO:

$$\begin{aligned} \dot v &= -\frac{W}{M} - g\sin\gamma \\ \dot\gamma &= \frac{A}{M\,v} + \frac{v\cos\gamma}{R} - \frac{g\cos\gamma}{v} \\ \dot r &= v\sin\gamma \end{aligned}$$

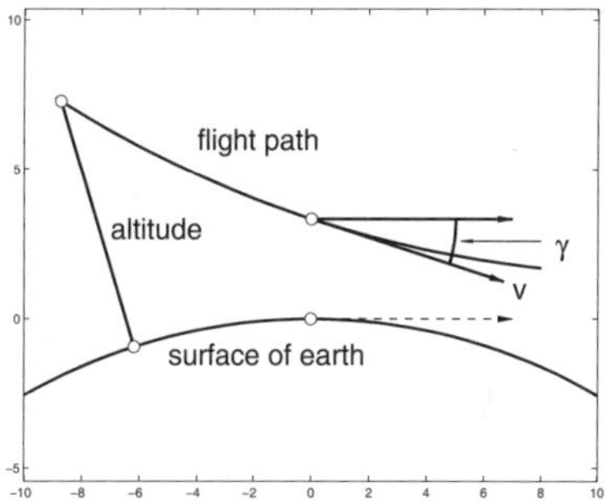

Figure 4.14. Reentry problem

Equations of motion for X-38:

$$\begin{aligned}
\dot{v} &= -\frac{W}{M} - g\sin\gamma \\
\dot{\gamma} &= \frac{A\ \cos\mu}{M\,v} + \frac{v\cos\gamma}{R} - \frac{g\cos\gamma}{v} + 2\omega_E \sin\chi\cos\lambda \\
\dot{r} &= v\sin\gamma \\
\dot{\chi} &= \frac{A\sin\mu}{Mv\cos\gamma} + \frac{v\ \cos\gamma\sin\chi\tan\lambda}{r} - 2\omega_E\tan\gamma\cos\chi\cos\lambda + 2\omega_E\sin\lambda \\
\dot{\lambda} &= \frac{v\cos\gamma\cos\chi}{r} \\
\dot{\tau} &= \frac{v\cos\gamma\sin\chi}{r\cos\lambda}
\end{aligned}$$

Operational time: APOLLO: T = 225 [s], X-38: T = 1150 [s].
Boundary conditions: ($\gamma(0)$ is especially critical, length in [km], angle in [rad] (!))
APOLLO (Fig. 4.17):

$$v(0) = 11\,,\ \ \gamma(0) = -0.14137167\,,\ \ h(0) = 121.92\,,\ \ v(T) = 8\,,\ \ h(T) = 76.2\,.$$

X-38 (Figs. 4.15, 4.16):

$$\begin{aligned}
&v(0) = 7.669\,,\ \ \gamma(0) = -0.0025656\,,\ \ h(0) = 80\,,\\
&\chi(0) = 1.9199\,,\ \ \lambda(0) = 1.22171\,,\ \ \tau(0) = -0.41888\,,\\
&v(T) = 1\,,\ \ h(T) = 25\,,\ \ \lambda(T) = 2.35584\,,\ \ \tau(T) = -0.49292\,.
\end{aligned}$$

Rescaling: v and r are to be rescaled for numerical solution and adapted by this way to the comparatively small quantities of the appearing angles values: $v = 10^5\widetilde{v}$, $r = 10^5\widetilde{r}_E(1+\widetilde{h})$, $R_E = 10^5\widetilde{R}_E$.

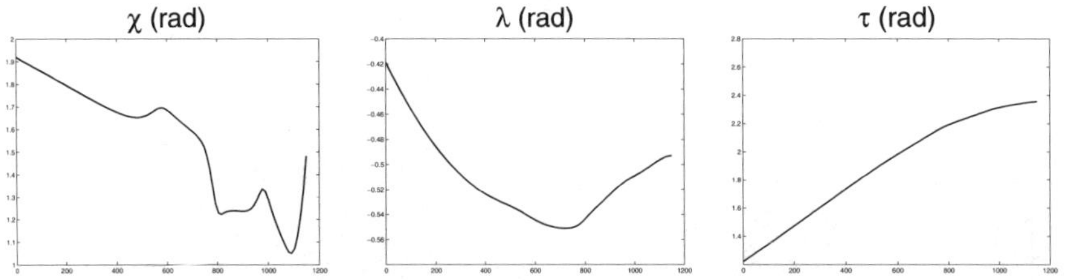

Figure 4.15. Diagr. 1 for X-38 without constraints

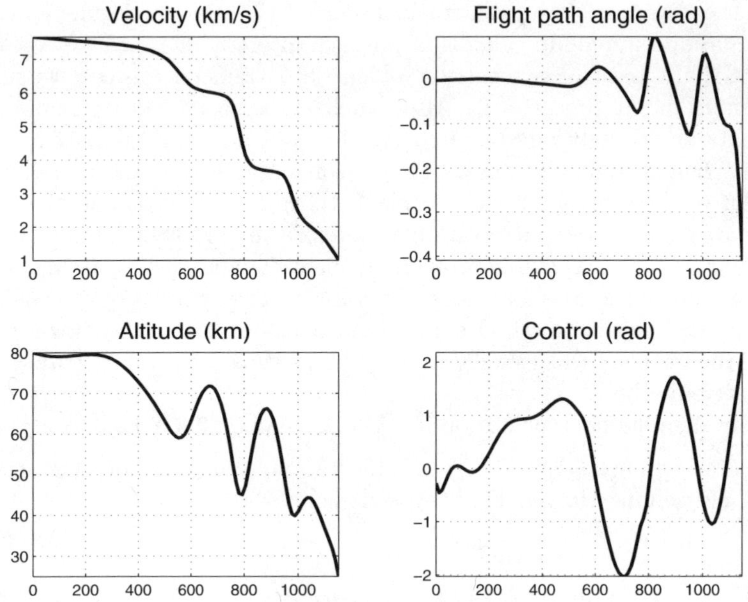

Figure 4.16. Diagr. 2 for X-38 without constraints

The characteristic swinging flight disappears under the sign restriction $\gamma \leq 0$ and the altitude decreases monotonically.

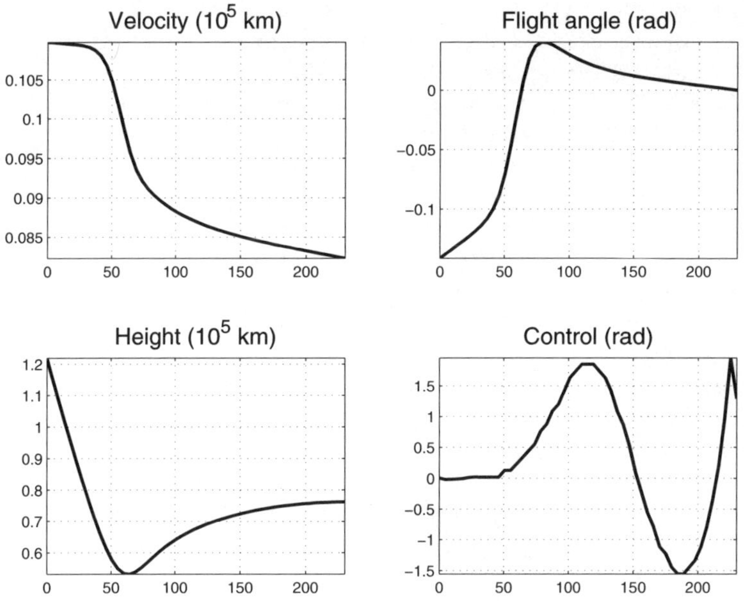

Figure 4.17. Diagrams for Apollo, SI units

In the diagram of the control in Figure 4.17 only the segment between maximum and minimum value is of physical interest.

The flight path of a reentry problem is extremely sensitive versus initial conditions at time $t = 0$. Also, small modifications of parameters and constants do strongly affect the control but less strongly the path variables velocity, height and path angle. The reason may be that the extremals are very flat in the control for such problems (Grimm). Nevertheless, the control is here a smooth unbounded variable therefore the problem can also be handled as pure boundary value problem as mentioned at the beginning of Sect. 4.2. The difficult search for a suitable start trajectory is however avoided by applying the penalty method `sqp.m` (and a rather coarse discretization); of course the result of this calculation may be used as such a start trajectory in alternative methods.

More detailed documentation in `KAPITEL04\EXAMPLES\bsp10\bsp11`.

References: (Stoer); GESOP Software User Manual. Institut für Flugmechanik und Flugregelung, Universität Stuttgart, Febr. 2004.

4.6 Hints to the MATLAB programs

```
KAPITEL04/CONTROL01, Control Problems
Solution by the method sqp.m of KAPITEL03/SECTION_5
demo1.m  Masterfile with sqp.m, examples 1--9
demo3.m  Reentry problem, Stoer, p. 491, US units, SI units
demo4.m  Space craft X-38 without constraints
demo5.m  Space craft X-38 with constraints of sign of
         attacking angle GAMMA
KAPITEL04/CONTROL02, Control Problem transformed into
          Boundary value problem
box.m     Box scheme for NEWTON's method
bsp01.m   Thrust problem, control eliminated
bsp02.m   Orbit problem, control eliminated
bsp03.m   Zermelo's problem, costate eliminated
demo.m    Masterfile for NEWTON's method
newton.m  Globalised NEWTON's method
KAPITEL04/CONTROL03, Control Problem and Gradient Method
demo1.m   Simple example after Dyer-McReynolds, p. 127
demo2.m   Brachistochrone,  Dyer-McReynolds, p. 128
demo3.m   Orbit problem, Bryson-Ho, p.66, Dyer-McReynolds, p.73
demo4a.m  Thrust problem Bryson-Ho, sect. 2.4, Start trajectory
demo4b.m  Thrust problem Bryson-Ho, dect. 2.4, solution
grad01.m -- grad04.m  Gradient method
```

5

The Road as Goal

Everybody knows that the implicit function $F(x, y) = y - x^2 = 0$ may be written as a single explicit function if x is the path parameter but not if y is the independent variable. Apparently there is a *turning point* in the second case which disappears by re-parametrization. Numerical *path following* shall not only pursue an (unknown) path correctly under moderate smoothness but also pass through possible turning points without difficulties. To this end, turning points have to be detected and ruled out by change of parameter. Entirely different is the situation in an implicit curve, say $F(x, y) = x\,y = 0$, where two solution branches $x = 0$ and $y = 0$ *intersect* in origin (here). In pursuing one of the branches one should hunt up the branching point and verify its topological character. As occasion demands, path following should take an other direction if "nature" does so, i.e., if *stability* passes, e.g., from the trivial branch to an non-trivial branch. So there are various problems to be studied before a numerical approach may be formulated with hope for success.

More precisely, let $A \in \mathbb{R}^n{}_n$ then the well-known system $F(\lambda, x) := Ax - \lambda x = 0$ has the trivial solution $x = 0 \in \mathbb{R}^n$ everywhere, i.e., for all $\lambda \in \mathbb{R}$. However, at certain isolated points on the λ-axis, namely at the eigenvalues of A, there exist one or more non-trivial solutions in addition to the trivial one, namely the eigenvectors appertaining to the eigenvalue λ. So not every phenomenon appearing in nature is continuous as, e.g., the brakes of a bicycle may demonstrate audibly. The investigation of corresponding situations in *nonlinear* systems $F(\lambda, x) = 0$ is subject of *branching* or *bifurcation theory.*

Besides *path following* in a smaller second part, this chapter is dedicated to the theoretical investigation of branching points and branching solutions as well as their numerical approach. We confine ourselves to some exemplary cases which are relevant in practice and at which the arising problems may be studied in sufficient way. *Branchings* are considered first from a more geometrical point of view and *necessary conditions* for the existence of *singular points* are derived which are checked for sufficiency in the simplest way by *numerical experiments.* Then a *scaling* is introduced being essentially due to (Keller72) which represents the key for numerical approach and allows to transform the problems in a suitable form for application of the Implicit Function Theorem.

For instance, HOPF bifurcation in periodic problems demonstrates impressively how bifurcation theory supplies new solutions which remain "terra incognita" under the usual assumptions on existence and uniqueness; cf., e.g., (Kirchgaessner), (Gekeler89). In some other applications, the results are less spectacular in comparison with the linearized problem, for instance, if the solutions are forced to obey strongly restrictive boundary conditions.

The numerical approach departs likewise from the branching point and must be persuaded frequently to leave the "attractive" trivial solution or not to return to it during iteration. It shall supply a *solution germ* from which a suitable *path-following* procedure then leads to solutions with acceptable orders of magnitude.

As in all mathematical problems to be solved numerically, a thorough investigation of the analytic problem is recommended before discretization since numerical approximation may alter *singularities* here or even destroy them entirely if one does not exercise proper care. To keep the application open for differential systems, the studies below are made generally in BANACH spaces. Accordingly, in the entire chapter, $\mathcal{X}$, $\mathcal{Y}$ and $\widetilde{\mathcal{X}}$ are BANACH spaces and $\mathcal{X}_d$, $\mathcal{Y}_d$ resp. $\widetilde{\mathcal{X}}_d$ are the associated dual spaces; also, in normal case, $\mathcal{X} \subset \mathcal{Y}$ hence $\mathcal{Y}_d \subset \mathcal{X}_d$.

Not only for representation as well as for checking of formulas in $\mathbb{R}^n$ but also for implementation (e.g., with MATLAB) it reveals to be of advantage to conceive all elements of any primal space consistently as *formal column vectors* and all elements of a dual space as *formal row vectors*. Dual pairing $\langle v, u\rangle$, $v \in \mathcal{X}_d$, $u \in \mathcal{X}$, then may be written simply as $v\,u$ in *formal* way. If $\mathcal{X}$ denotes a HILBERT space, $\langle u, v\rangle$ is the canonical scalar product, and if $\mathcal{X}$ is the coordinate space $\mathbb{R}^n$ of *column vectors* then the dual space $\mathcal{X}_d$ may be adopted to be the coordinate space $\mathbb{R}_n$ of *row vectors*, and dual pairing becomes a MATLAB-correct scalar product namely a special matrix product. Also, skew brackets are used here to some degree in order to emphasize the application of a linear or multilinear operator to its arguments, then, e.g., $A\langle x\rangle = Ax$ and $A\langle x, y\rangle = x^T Ay$ for $A \in \mathbb{R}^n{}_n$.

5.1 Bifurcation Problems

Notations not explained here are found in Sect. 1.9; see also §1.1(**e**).

(a) Fredholm Operators are the suitable generalizations of linear operators $L : \mathbb{R}^n \to \mathbb{R}^m$ to BANACH spaces (mostly HILBERT spaces) because the crucial eigenspaces remain finite-dimensional.

Let $\mathcal{X}$, $\mathcal{Y}$ be BANACH spaces with their dual spaces $\mathcal{X}_d$, $\mathcal{Y}_d$. The dual pairing is denoted by skew brackets: $x \in \mathcal{X}$, $x_d \in \mathcal{X}_d : \langle x_d, x\rangle := x_d(x)$ as x_d is a functional (mapping). In a HILBERT space $\mathcal{H}$, $\langle y, x\rangle$ denotes the canonical scalar product after identification of $\mathcal{H}$ with $\mathcal{H}_d$ by RIESZ' representation theorem. Let $L : \mathcal{X} \to \mathcal{Y}$ be a linear bounded operator, briefly written $L \in \mathcal{L}(\mathcal{X}, \mathcal{Y})$, and recall that the *adjoint operator* $L_d : \mathcal{Y}_d \to \mathcal{X}_d$ is defined by $\forall\, u \in \mathcal{X}\;\; \forall\, v \in \mathcal{Y}_d : \langle L_d v, u\rangle_{\mathcal{X}} = \langle v, Lu\rangle_{\mathcal{Y}}$.

Definition 5.1. (1°) *A linear bounded operator* $L : \mathcal{X} \to \mathcal{Y}$ *is a* FREDHOLM *operator if*

$$\boxed{\operatorname{Range} L \textit{ closed},\ \dim \operatorname{Ker} L < \infty\,,\ \dim \operatorname{Ker} L_d < \infty}\,.$$

(2°) *The number* $\operatorname{ind} L := \dim \operatorname{Ker} L - \dim \operatorname{Ker} L_d \in \mathbb{Z}$ *is called* index *of* L.

Additional Remarks (1°) Consider a *matrix* $L \in \mathbb{R}^m{}_n$ of rank p and recall that row rank equals column rank, then $\dim \operatorname{Ker} L = n - p$, $\dim \operatorname{Range} L = p$, $\dim \operatorname{Ker} L_d = m - p$, and $\operatorname{ind} L = n - m$.

(2°) In case L is a FREDHOLM operator, the adjoint operator $L_d : \mathcal{Y}_d \to \mathcal{X}_d$ of Sect. 1.9**(a)** is also a FREDHOLM operator, and $\operatorname{ind} L_d = -\operatorname{ind} L$ by Theorem 1.19.

(3°) A symmetric FREDHOLM operator — cf. Sect. 1.11 — has always index zero.

(4°) A FREDHOLM operator obeys by definition FREDHOLM's alternative $\operatorname{Range} L = [\operatorname{Ker} L_d]^\perp$ where the sign "$\perp$" is to be understood here in the sense of dual pairing in BANACH spaces (1.52).

(5°) The assumption that $\operatorname{Range} L$ is closed can be dropped in the above definition because it follows from the other properties (Hirzebruch), Lemma 25.7.

Let now $\operatorname{span}\{u_1, \ldots, u_\alpha\}$ denote the vector space spanned by some independent vectors $u_1, \ldots, u_\alpha$. In this chapter we use throughout the following notations concerning FREDHOLM operators:

$$\boxed{\begin{array}{l} \dim \operatorname{Ker} L \;\; = \alpha\,,\ \operatorname{Ker} L \;\; = \operatorname{span}\{u_1, \ldots, u_\alpha\} \subset \mathcal{X} \\ \dim \operatorname{Ker} L_d = \beta\,,\ \operatorname{Ker} L_d = \operatorname{span}\{v^1, \ldots, v^\beta\} \subset \mathcal{Y}_d \end{array}}\,. \tag{5.1}$$

Of course u_i are the right- and v^j the left eigenvectors to the eigenvalue zero in case $L \in \mathbb{R}^n{}_n$ is a square matrix but a right eigenvector u is here a column vector, $u \in \mathbb{R}^n$, and a left eigenvector is a row vector, $v \in \mathbb{R}_n$. In appropriate generalization we write all elements of a dual space as *formal left vectors* in the sequel, e.g., vL_d instead $L_d v$. For instance, let $L := A \in \mathbb{R}^m{}_n$ be a matrix then, using the scalar product $(u, v) = u^T v$,

$$L_d = A^T\,,\ (A^T v, u) = (v, Au)\,,\ Au : v \mapsto (v, Au)\,,\ vA^T : u \mapsto (A^T v, u)$$

but *without scalar product*, L_d is the same matrix A with *left* multiplication, $L_d : v \mapsto vA \in \mathbb{R}_n$. Regarding (5.1), we introduce also some matrices with formal rows and formal columns:

$$\boxed{\begin{array}{lll} U \;\; := [u_1, \ldots, u_\alpha]\,, & u_i \in \mathcal{X} & \text{matrix of } \textit{formal} \text{ column vectors} \\ U^d := [u^i]_{i=1}^\alpha\,, & u^i \in \mathcal{X}_d & \text{a dual basis of } U \text{ where } U^dU = I \in \mathbb{R}^\alpha{}_\alpha \\ & & \text{matrix of } \textit{formal} \text{ row vectors} \\ V^d := [v^k]_{k=1}^\beta\,, & v^k \in \mathcal{Y}_d & \text{matrix of } \textit{formal} \text{ row vectors} \\ V \;\; := [v_1, \ldots, v_\beta]\,, & v_k \in \mathcal{Y} & \text{a dual basis of } V^d \text{ with } V^dV = I \in \mathbb{R}^\beta{}_\beta \\ & & \text{matrix of } \textit{formal} \text{ column vectors} \end{array}} \tag{5.2}$$

(I unit matrix or identity operator). Note that $vL : u \mapsto \langle v, Lu\rangle$ and $L_d u : v \mapsto \langle vL_d, u\rangle$ are to be understood as mappings here and

$$\forall\, u \in \mathcal{X} \;\; \forall\, v \in \operatorname{Ker} L_d : 0 = \langle vL_d, u\rangle = \langle v, Lu\rangle \implies \forall\, v \in \operatorname{Ker} L_d : vL = 0$$
$$\forall\, v \in \mathcal{Y}_d \;\; \forall\, u \in \operatorname{Ker} L : 0 = \langle v, Lu\rangle = \langle vL_d, u\rangle \implies \forall\, u \in \operatorname{Ker} L : L_d u = 0$$

$$\boxed{V^dL_d = 0 \Longrightarrow V^dL = 0\,, \quad LU = 0 \Longrightarrow L_dU = 0}\,.$$

Further *formal* computational rules (*realiter* in coordinate space $\mathbb{R}^n$):

$$\begin{array}{llll} \xi \in \mathbb{R}^\alpha & \Longrightarrow U\xi & = u_1\xi^1 + \ldots + u_\alpha\xi^\alpha & \in \mathcal{X} \\ \zeta \in \mathbb{R}_\beta & \Longrightarrow \zeta V^d & = \zeta_1 v^1 + \ldots + \zeta_\beta v^\beta & \in \mathcal{Y}_d \\ x \in \mathcal{X}_d & \Longrightarrow xU & = [\langle x, u^1\rangle, \ldots, \langle x, u^\alpha\rangle] & \in \mathbb{R}_\alpha \\ y \in \mathcal{Y} & \Longrightarrow V^dy & = [\langle v^1, y\rangle, \ldots, \langle v^\beta, y\rangle]^T & \in \mathbb{R}^\beta \\ x \in \mathcal{X} & \Longrightarrow UU^dx & = u_1\langle u^1, x\rangle + \ldots + u_\alpha\langle u^\alpha, x\rangle & \in \operatorname{Ker} L \subset \mathcal{X} \\ y \in \mathcal{Y} & \Longrightarrow VV^dy & = v_1\langle v^1, y\rangle + \ldots + v_\beta\langle x^\beta, y\rangle & \in \mathcal{Y}\,. \end{array}$$

The most important properties of a FREDHOLM operator are briefly summarized in the following theorem:

Theorem 5.1. *Let $L \in \mathcal{L}(\mathcal{X}, \mathcal{Y})$ be a* FREDHOLM *operator.* (1°)

$$\operatorname{codim} \operatorname{Range} L = \dim \operatorname{Ker} L_d\,, \quad \operatorname{codim} \operatorname{Range} L_d = \dim \operatorname{Ker} L$$

($\operatorname{codim} \operatorname{Range} L := \dim(\mathcal{Y} \backslash \operatorname{Range} L)$*).*
(2°) The projectors $P := UU^d \in \mathcal{L}(\mathcal{X}, \mathcal{X})$ and $Q =: VV^d \in \mathcal{L}(\mathcal{Y}, \mathcal{Y})$ are (linear,) continuous and

$$\operatorname{Ker} L = \operatorname{Range} P\,, \quad \operatorname{Range} L = \operatorname{Ker} Q\,. \tag{5.3}$$

Proof. (1°) follows directly from the Range Theorem 1.19(5°).
(2°) It is obvious that P and Q are continuous operators because their ranges are finite-dimensional, more exactly

$$Pw = \sum_{i=1}^{\alpha} u_i \langle u^i, w\rangle_{\mathcal{X}} \in \mathcal{X}\,, \quad Qw = \sum_{i=1}^{\beta} v_i \langle v^i, w\rangle_{\mathcal{Y}} \in \mathcal{Y}\,.$$

The first equation in (5.3) follows directly from the definition of P and the second follows from the Range Theorem again. □

According to (5.3) and (1.54) we now introduce the fundamental decompositions

$$\boxed{\begin{array}{l} \mathcal{X} = \operatorname{Range} P \oplus \operatorname{Ker} P = \operatorname{Ker} L \quad\;\; \oplus \operatorname{Ker} P \\ \mathcal{Y} = \operatorname{Range} Q \oplus \operatorname{Ker} Q = \operatorname{Range} Q \oplus \operatorname{Range} L \end{array}} \,. \tag{5.4}$$

Both are *unique* because all subspaces are *closed*; the subspace $\operatorname{Range} Q$ of finite dimension β is sometimes also called *corange of* L.

For $L \in \mathcal{L}(\mathcal{X}, \mathcal{Y})$ with eigenvalue μ, each subspace $\operatorname{Ker}(L - \mu I)^m$ is trivially contained in the next for $m = 1, 2, \dots$, up to an order $m = p$ called Riesz *index* of μ. The *dimension* of $\operatorname{Ker}(L - \mu I)^p$ is the *multiplicity* of the eigenvalue. Obviously it is identical with the *algebraic* multiplicity whenever $L \in \mathbb{R}^n{}_n$ is a matrix, and, in this case, p is the maximum length of all associated chains of principal vectors. If $p = 1$, as for instance in symmetric operators and diagonalizable matrices, the eigenvalue is called *semi-simple.* Then the algebraic multiplicity of μ, namely $\dim \operatorname{Ker}(L - \mu I)$, coincides with the geometric multiplicity, and no principal vectors exist in finite-dimensional case.

Let us now return to Fredholm operators and recall the above introduced notations.

Lemma 5.1. *Let* $\mathcal{X} \subseteq \mathcal{Y}$ *and let* $L \in \mathcal{L}(\mathcal{X}, \mathcal{Y})$ *be a* Fredholm *operator with index zero and* $\dim \operatorname{Ker} L > 0$.
(1°). *The following statements are equivalent:*
(i) Zero is a semi-simple eigenvalue of L.
(ii) $V^d U \in \mathbb{R}^\alpha{}_\alpha$ *is a regular matrix.*
(iii) $\operatorname{Range} L \cap \operatorname{Ker} L = \{0\}$.
(2°) *If zero is a semi-simple eigenvalue of* L *then* $P = Q$ *in (5.4) without loss of generality and* $\mathcal{Y} = \operatorname{Range} L \oplus \operatorname{Ker} L$.

Proof. (1°)(a) Let $L \in \mathbb{R}^n{}_n$ be a quadratic *matrix*, let $Lu = 0$ and $Lv = u$ for some vectors u, v then the eigenvalue zero has Riesz index $p > 1$ because $L^2 v = 0$ and $Lv \neq 0$. But $V^d u = 0$ by the Range Theorem hence $V^d U$ singular. Conversely, let $V^d U$ regular then there is no such situation possible. Therefore (i) and (ii) are equivalent.
(1°)(b) $\operatorname{Range} L \cap \operatorname{Ker} L \neq \{0\} \iff \exists\, u \in \operatorname{Ker} L \quad \exists\, w \in \mathcal{X} : Lw = u \iff V^d u = 0$.
(2°) Let $V^d U = C$ regular then $C^{-1} V^d U = I$ and the transition from V^d to $C^{-1} V^d$ means only a change of the basis of $\operatorname{Ker} L_d$. Therefore we can assume without loss of generality that $V^d U = I$. Then V^d satisfies the assumptions of U^d and U the assumptions of V by definition, therefore $P = UV^d$ and $Q = UV^d$ without loss of generality. Then $\mathcal{Y} = \operatorname{Range} L \oplus \operatorname{Ker} L$ by (5.4). □

Note that the formal matrices U^d and V are not uniquely determined relative to U and V^d respectively. V^d can play the *role* of U^d but U^d can play the role of V^d only in self-adjount operators L (also left eigenvectors are transposed right eigenvectors only in symmetric matrices).

(b) Formulation of the Problem The Implicit Function Theorem says that a mapping $F : \mathbb{R}^n \to \mathbb{R}^n$ is locally invertible at a point x_0 whenever $\det \operatorname{grad} F(x_0) \neq 0$. This may be considered as the normal case since an equally distributed random variable takes *almost never* the value zero. Therefore a point x_0 with $\det \operatorname{grad} F(x_0) = 0$ is correctly called singular in this context. On the other side, most exciting things happen exactly at those points following Sect. 1.6. On leaving the realm of unique existence, the situation alters drastically, but a smaller problem can be distilled out from the original problem by an ingenious method, commonly called LJAPUNOV-SCHMIDT reduction. The associated sub-problem reflects completely the behavior of the solution near the singular point and is finite-dimensional.

Let $F : \mathbb{R} \times \mathcal{X} \ni (\mu, x) \mapsto F(\mu, x) \in \mathcal{Y}$ be a smooth mapping. Up to Theorem 5.2, we consider in this section only the case where F has always the trivial solution, i.e., $F(\mu, 0) = 0$, and, without loss of generality, the *origin* $(\mu, 0) = (0, 0)$ is the possible bifurcation point. We write henceforth briefly $F_x^0 = \nabla_x F(0,0)$, $F_\mu^0 = F_\mu(0,0)$ etc., then the system to be solved reads near the origin:

$$F(\mu, x) = Lx + f(\mu, x) = 0\,, \quad f(\mu, x) = o(\|x\|)\,, \quad \boxed{L := F_x^0}\,. \tag{5.5}$$

The trivial solution is the unique solution by the Implicit Function Theorem whenever $L : x \mapsto \operatorname{Range} L$ is invertible. Let now L be singular, i.e., $\dim \operatorname{Ker} L > 0$, and suppose that $\mu \mapsto x(\mu)$ is a *nontrivial* solution passing through the origin, $F(\mu, x(\mu)) = 0$, then the origin is a bifurcation point where this non-trivial solution branches off from $x = 0$. Also we have $F_\mu^0 + Lx'(0) = 0$ by differentiation but $(\partial^k F/\partial \mu^k)(\mu, 0) = 0$ for all $k \in \mathbb{N}$ hence $x'(0)$ must be an eigenvector to the eigenvalue zero of the operator L. Of course $0 = F_\mu^0 \in \operatorname{Range} L$ but $0 \neq F_\mu^0 \in \operatorname{Range} L$ remains also a necessary condition in branching off from a *non-trivial* solution with general bifurcation point (μ_0, x_0).

(c) Ljapunov-Schmidt Reduction We suppose always that L is a FREDHOLM operator. Then, regarding (5.4), the mapping F and its argument, x can be decomposed into *two* partial functions using the projectors of Theorem 5.1, $x = Px + (I - P)x$ and $F = QF + (I - Q)F$:

$$\boxed{\begin{array}{llll} u := Px & \in \operatorname{Ker} L\,, & w := (I-P)x & \in \operatorname{Ker} P \\ QF(\,\cdot\,, u+w) \in \operatorname{Range} Q\,, & & (I-Q)F(\,\cdot\,, u+w) \in \operatorname{Range} L & \end{array}} \tag{5.6}$$

and of course $F = 0$ if and only if $QF = 0$ and $(I - Q)F = 0$. Now the *operator equation*

$$(I - Q)F(\mu, u + w) = Lw + (I - Q)f(\mu, u + w) = 0 \tag{5.7}$$

has always a unique solution $w^* : (\mu, u) \mapsto w^*(\mu, u)$ in $\operatorname{Ker} P$ for sufficiently small $|\mu|$ because $(I-Q)f(\mu, u+w) \in \operatorname{Range} L$ and the restriction of L to $L_0 \in$

$\mathcal{L}(\operatorname{Ker} P, \operatorname{Range} L)$ is trivially injective and trivially surjective. Furthermore, $w \mapsto L_0{}^{-1}(I - Q)f(\mu, u + w)$ satisfies the assumptions of the Contraction Theorem near $(\mu, u) = (0, 0)$, e.g., by (Stakgold), p. 310, and w^* is as smooth as F. A substitution of w^* into the equation $QF(\mu, u + w) = 0$ yields the local *branching equation* near the origin,

$$\boxed{G(\mu, u) := QF(\mu, u + w^*(\mu, u)) = Qf(\mu, u + w^*(\mu, u)) = 0}\,. \qquad (5.8)$$

This equation for the remaining part $u \in \operatorname{Ker} L$ of the argument x does not necessarily have a non-trivial solution but every solution of $F(\mu, x) = 0$ corresponds uniquely to a solution (μ, u) of (5.8) whether it is isolated or not. This equivalence constitutes the basis of many results in bifurcation theory as well as concerns *existence* of branching solutions as *numerical approximations* in Sects. 5.4 – 5.6.

Note that $Q = VV^d$, $V^d L = 0$, and $u = U\zeta$, $\zeta \in \mathbb{R}^\alpha$, for $u \in \operatorname{Ker} L$. The matrices U and V^d of (5.2) are supposed to be calculated in advance. From a more practical aspect we then solve *simultaneously*

$$V^d f(\mu, U\zeta + w) = 0 \in \mathbb{R}^\beta\,, \quad Lw + f(\mu, U\zeta + w) = 0 \in \mathcal{Y}\,,\ U^d w = 0$$

for both ζ and w depending on μ. By the Range Theorem, the first equation is the *consistency condition* for solving the second system. The slightly modified branching equation is now finite-dimensional over the finite coordinate space $\mathbb{R}^{\alpha+1}$!

(d) The Branching Equation

(d1) Unfortunately, not much can be said about the function w^*. But consider once more the operator equation

$$H(\mu, u) := (I - Q)F(\mu, u + w^*(\mu, u)) = 0\,,\ u \in \operatorname{Ker} L\,,\ w^* \in \operatorname{Ker} P$$

at a general bifurcation point (μ_0, x_0). Then, for increment $h \in \operatorname{Ker} L$, we have also $w_u^{*0} h \in \operatorname{Ker} P$ and, as $L = F_x^0$,

$$\partial H((\mu_0, x_0); (0, h)) = (I - Q)L(h + w_u^{*0} h) = 0\,.$$

But $Lh = 0$ and the operator $(I - Q)L$ is bijective on $\operatorname{Ker} P$ hence $\boxed{w_u^{*0} = 0}$. Derivation for μ yields

$$H_\mu^0 = (I - Q)(F_\mu^0 + F_x^0 w_\mu^{*0}) = (I - Q)F_\mu^0 + Lw_\mu^{*0}$$

with the unique solution $w_\mu^{*0} \in \operatorname{Ker} P$. In a general branching point we have $0 \neq F_\mu^0 \in \operatorname{Range} L$ and thus w_μ^{*0} is not necessarily zero but it is zero if $F(\mu, 0) = 0$ hence $F_\mu^0 = 0$ and also, e.g., if F odd, $-F(\mu, x) = F(\mu, -x)$.

(d2) Returning to the case $F(\mu, 0) = 0$ with branching point $(0, 0)$, we find that

$$G_u^0 = Qf_x^0 + Qf_x^0 w_u^{*0} = 0\,,\ G_\mu^0 = Qf_\mu^0 + Qf_x^0 w_\mu^{*0} = 0 \qquad (5.9)$$

because $f_x^0 = 0$ by (5.5) and the definition of L. Therefore the Implicit Function Theorem cannot be applied *directly* to solve the branching equation neither for μ as a function of ζ nor for ζ as function of μ. Thus frequently a new path parameter ε is introduced to obtain non-trivial solution paths in the form $(\mu(\varepsilon), \zeta(\varepsilon))$; see Sect. 5.2.

Because of (5.9), a TAYLOR expansion of G has the form

$$G^i(\mu,\zeta) = \frac{1}{2}\begin{bmatrix}\mu\\ \zeta\end{bmatrix}^T \left[\nabla^2_{[\mu,\zeta]} G^i(0,0)\right]\begin{bmatrix}\mu\\ \zeta\end{bmatrix} + \mathcal{O}(\|[\mu,\zeta]\|^3)\,, \quad i = 1:\beta\,. \tag{5.10}$$

Using the results of **(d1)**, the second partial derivatives of the bifurcation equation are

$$\begin{aligned}
G^{i\;\;0}_{\zeta\,\zeta} &= v^i F^0_{xx}\langle U + w^{*0}_u\langle U\rangle, U + w^{*0}_u\langle U\rangle\rangle + v^i F^0_x\langle w^{*0}_{uu}\langle U,U\rangle\rangle\\
&= v^i F^0_{xx}\langle U,U\rangle + v^i F^0_x\langle w^{*0}_{uu}\langle U,U\rangle\rangle\\
G^0_{\mu\,\mu} &= v^i[F^0_{\mu\mu} + 2F^0_{x\mu}w^{*0}_\mu + F^0_{xx}w^{*0}_\mu w^{*0}_\mu] + v^i F^0_x w^{*0}_{\mu\mu}\\
G^0_{\mu\,\zeta} &= v^i[F^0_{\mu x}\langle U\rangle + F^0_{\mu x}\langle w^{*0}_u\langle U\rangle\rangle]\\
&\quad + v^i[F^0_{xx}\langle U, w^{*0}_\mu\rangle + F^0_{xx}\langle w^{*0}_u\langle U\rangle, w^{*0}_\mu\rangle + F^0_x\langle w^{*0}_{\mu u}\langle U\rangle\rangle]\,,
\end{aligned}$$

where $G^{i,0}_{\zeta\,\zeta}\langle\zeta,\zeta\rangle = v^i F^0_{xx}\langle U\zeta, U\zeta\rangle \in \mathbb{R}$ and $G^{i,0}_{\mu\,\zeta}\langle\mu,\zeta\rangle = v^i F^0_{\mu x}\langle\mu, U\zeta\rangle \in \mathbb{R}$. Under the assumption $w^{*0}_\mu = 0$ we thus obtain

$$\nabla^2_{[\mu,\zeta]} G^{i,0} = \begin{bmatrix} G^{i,0}_{\mu\mu} & G^{i,0}_{\mu\zeta}\\ G^{i,0}_{\zeta\mu} & G^{i,0}_{\zeta\zeta}\end{bmatrix} = \begin{bmatrix} O & v^i F^0_{\mu x} U\\ [v^i F^0_{\mu x} U]^T & v^i F^0_{xx}\langle U,U\rangle\end{bmatrix}, \quad i = 1:\beta\,. \tag{5.11}$$

The representation (5.11) remains literally the same if μ is a higher-dimensional parameter vector.

(d3) Let $\mathcal{X} \subseteq \mathcal{Y}$, let L be a FREDHOLM operator with eigenvalue zero and *index zero*. We consider the system

$$F(\mu,x) = Lx + \mu Bx + C\langle x,x\rangle + \text{ h.o.t.} = 0 \tag{5.12}$$

where $B = F^0_{\mu x}$ and $C = F^0_{xx}$.

Case 1: $\dim \operatorname{Ker} L = 1$, i.e. $\alpha = \beta = 1$. Choose $\zeta = \xi\mu$. After division by μ^2, (5.10) has the form

$$\begin{aligned}
\mu^{-2}G^1(\mu,\zeta) &= \left[a\,\xi + b\,\xi^2 + \text{ h.o.t. in } \mu \text{ and } \xi\right] = 0\,,\\
a = v^1 F^0_{\mu x} u_1 = v^1 B u_1\,,\quad & b = v^1 F^0_{xx}\langle u_1,u_1\rangle/2 = v^1 C\langle u_1,u_1\rangle\,.
\end{aligned}$$

For a and b both nonzero there exists a nontrivial isolated solution $\xi_0 = -a/b$ of $a\xi + b\xi^2 = 0$ hence a general solution $\zeta(\mu) = \mu\xi(\mu)$ with path parameter μ, and $\zeta(0) = 0$.

Case 2: $\dim \operatorname{Ker} L = 2$, i.e. $\alpha = \beta = 2$.
Choose $\zeta = \xi\mu \in \mathbb{R}^2$. After division by μ^2, we have now *two* equations

$$\mu^{-2}G(\mu,\zeta) = \big[Q(\xi) + \text{ h.o.t. in } \mu \text{ and } \xi\big] = 0 \in \mathbb{R}^2$$
$$Q(\xi) = V^dB\langle U\xi\rangle + V^dC\langle U\xi, U\xi\rangle\,,\; V^d = \begin{bmatrix} v^1 \\ v^2 \end{bmatrix}\,. \tag{5.13}$$

Lemma 5.2. *(McLeod). Let the matrix $V^dBU \in \mathbb{R}^2{}_2$ be regular and let the two quadratics given by $V^dC\langle U\xi, U\xi\rangle$ have no (real) common factor (in particular neither vanishes identically). Then every solution ξ^0 of $Q(\xi) = 0 \in \mathbb{R}^2$ with regular* $\operatorname{grad}_\xi Q(\xi^0)$ *gives a non-trivial solution branch*

$$x(\mu) = \mu(u_1\xi_1(\mu) + u_2\xi_2(\mu)) + \mu^2 w(\mu)\,,\; \xi_1(0) = \xi_1^0\,,\; \xi_2(0) = \xi_2^0\,,\; U^d w(\mu) = 0\,.$$

Furthermore, every solutions of (5.12) near $(\mu, u) = (0,0)$ lies on one of these branches.

For solutions of the quadratic form Q we refer to Example 5.1 below.

Case 3: To find higher-dimensional solution manifolds is a difficult matter (Keener74). Suppose that $\mu = (\mu_1, \mu_2)$ is a two-dimensional parameter vector. To find at least one of the many possible solution *paths* choose, e.g., $\zeta_0 \in \mathbb{R}^2$ arbitrary with $|\zeta_0| = 1$ and $\zeta = \zeta_0\varepsilon$. After division by ε we have

$$\varepsilon^{-1}G(\mu,\zeta) = \big[A\mu + \varepsilon\, c + \text{ h.o.t. in } \mu \text{ and } \varepsilon\big] = 0\,, \tag{5.14}$$

$$A = \begin{bmatrix} v^1F^0_{\mu_1 x}\langle U\zeta_0\rangle & v^1F^0_{\mu_2 x}\langle U\zeta_0\rangle \\ v^2F^0_{\mu_1 x}\langle U\zeta_0\rangle & v^2F^0_{\mu_2 x}\langle U\zeta_0\rangle \end{bmatrix}\,,\quad c = \begin{bmatrix} 2^{-1}v^1F^0_{xx}\langle U\zeta_0, U\zeta_0\rangle \\ 2^{-1}v^2F^0_{xx}\langle U\zeta_0, U\zeta_0\rangle \end{bmatrix}\,.$$

Let A be regular then there exists a non-trivial and locally unique solution $[\mu_1, \mu_2, \zeta]$ depending on the *new path parameter* ε.

Example 5.1. (1°) Solve the quadratic bifurcation problem

$$F(\mu,x) = \mu x + C\langle x,x\rangle = 0 \in \mathbb{R}^2\,,\; C\langle x,x\rangle = \begin{bmatrix} x^TPx \\ x^TQx \end{bmatrix}$$

at the origin $(0,0)$. The non-zero matrices $P, Q \in \mathbb{R}^2{}_2$ are symmetric without loss of generality and shall have no common factor: $\forall\, \alpha \in \mathbb{R} : P \neq \alpha Q$. The branching equation is here also $F(\mu,x) = 0$ since $F^0_x = 0 \in \mathbb{R}^2{}_2$. If $F(\mu,x) = \mu Bx + C\langle x,x\rangle$ and $B \in \mathbb{R}^2{}_2$ *regular*, set $x = B^{-1}y$ or consider $\mu x + B^{-1}C\langle x,x\rangle = 0$. Setting $x = \mu\zeta$ and dividing by μ^2 we have to find the intersections of the two quadratic forms $F(1,\zeta) = 0$. After simple manipulations we obtain a scalar equation $P\langle\zeta,\zeta\rangle\zeta_2 - Q\langle\zeta,\zeta\rangle\zeta_1 = 0$ as *necessary* condition. Suppose for instance that $\zeta_1 \neq 0$ then

$$p_{22}y_2^3 + (2p_{12} - q_{22})y^2 + (p_{11} - 2q_{12})y - q_{11} = 0\,.$$

for $y = \zeta_2/\zeta_1$. For $p_{22} \neq 0$ this polynomial of degree 3 has one or three real roots α_i and $\alpha_1 = 0$ for $q_{11} = 0$; further

$$\zeta_1 = -\big(p_{11} + 2p_{12}\alpha + p_{22}\alpha^2\big)^{-1}\,.$$

Similar result for $q_{22} \neq 0$ and $x_2 \neq 0$. Allowing some degeneracy, the problem $F(1,\zeta) = 0$ can have $N \in \{0,\dots,3\}$ real non-zero solutions $\zeta \in \mathbb{R}^2$ (Fig. 5.1). For instance, in MATLAB notation,

$$\begin{array}{lll} P = [0,1;1,0]\,, & Q = [-1,1;1,0] & \Longrightarrow N = 0 \\ P = [1,0;0,1]\,, & Q = [-1,1;1,0] & \Longrightarrow N = 1 \\ P = [0,-1;-1,0]\,, & Q = [-1,1;1,0] & \Longrightarrow N = 2 \\ P = [3,0;0,1]\,, & Q = [1/2,0;0,4] & \Longrightarrow N = 3\,. \end{array}$$

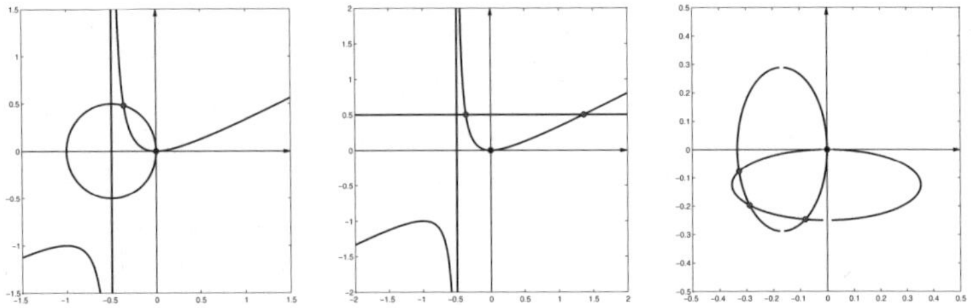

Figure 5.1. Example 5.1, N = 1,2,3

All solutions of $F(\mu, x) = 0$ reveal to be straight lines $\mu \mapsto (\mu\zeta_1, \mu\zeta_2)$; see also `KAPITEL05\BIFURKATION\demo_quad.m`.

(2°) Let $\widetilde{B}\langle x_1,\dots,x_q\rangle$ be a q-linear mapping and write $B_q\langle x\rangle = \widetilde{B}\langle x,\dots,x\rangle$ then $B_q\langle \gamma x\rangle = \gamma^q B_q\langle x\rangle$. Consider the problem

$$F(\mu, x) = \mu x + C_r\langle x\rangle = 0 \in \mathbb{R}^\alpha\,, \quad |x| = 1 \tag{5.15}$$

for α *odd* and $r \geq 2$. Suppose that $C_r\langle x\rangle \neq 0$ for $|x| = 1$. Then $F(\mu, x) = 0$ has at least one solution with $\mu \neq 0$ by a famous result of topological degree theory that "*every continuously combed hedgehog has at least one bald point*"; see (Birkhoff). In the other case there is of course a solution with $|x| = 1$ and $\mu = 0$. Also there is at least one further solution $(\mu, -x)$ for r odd and $(-\mu, -x)$ for r even.

(e) Some Further Results The following result does not use the LJAPUNOV-SCHMIDT reduction but applies the theory of topological degree in an ingenious way.

Theorem 5.2. (LERAY-SCHAUDER). *Let $L \in \mathcal{L}(\mathcal{X}, \mathcal{Y})$ be a* FREDHOLM *operator with index zero and let $\mu_0 \neq 0$ be a real eigenvalue of L with odd multiplicity then the system*

$$F(\mu, x) = Lx - \mu x + f(\mu, x)\,, \quad f(\mu, x) = o(\|x\|)\,, \quad \|x\| \to 0$$

has at least one non-trivial branch emanates from $(\mu_0, 0)$.

Proof (Krasnoselki). See also (Rabinowitz73) where it is shown that, in case of an algebraic simple eigenvalue μ_0 of L, the local non-trivial solution is unique and has the form

$$(\mu(\varepsilon), x(\varepsilon)) = (\mu_0 + \varepsilon\nu(\varepsilon), \varepsilon(u_1 + w(\varepsilon))\,, \quad |\varepsilon| \ll 1\,,$$

where $u_1^d w = 0$, $\nu(0) = 0$ and $w(0) = 0$.

The next result of (Crandall) & Rabinowitz concerns bifurcation from a simple eigenvalue but supplies a very simple consistency condition (3°) instead of solving the branching equation.

Theorem 5.3. *Let $\mathcal{X}, \mathcal{Y}$ be* BANACH *spaces, let $0 \in \mathcal{I} \subset \mathbb{R}$ be an open interval and let $F \in \mathcal{C}^2(\mathcal{I} \times \mathcal{X}, \mathcal{Y})$. Assumption:*
(1°) $F(\mu, 0) = 0$.
(2°) $F_x^0 := \nabla_x F(0,0)$, $\dim \operatorname{Ker} F_x^0 = 1$, $\operatorname{codim} \operatorname{Range} F_x^0 = 1$.
(3°) $\operatorname{Ker} F_x^0 = \operatorname{span}\{u_1\}$ *and* $F_{\mu x}^0 u_1 \notin \operatorname{Range} F_x^0$.
Then there exists a solution

$$(\mu(\varepsilon), x(\varepsilon)) = \varepsilon u_1 + \varepsilon w(\varepsilon)\,, \; u^1 w(\varepsilon) = 0\,, \; \mu(0) = 0\,, \; x(0) = 0 \qquad (5.16)$$

of $F(\mu, x) = 0$ for sufficiently small $|\varepsilon|$ depending continuously on ε.

Suppose that we have *any function* $v \in \mathcal{Y}_d$ such that $\operatorname{Range} F_x^0 = \{y \in \mathcal{Y} : v\,y = 0\}$ then assumption (3°) reads simply $vF_{\mu x}^0 u_1 \neq 0$. It is therefore not necessary by this result to calculate the adjoint operator $[F_x^0]_d$ explicitely and its kernel. If $\mathcal{X} \subset \mathcal{Y}$ and $F_{\mu x}^0 = I$ (identity in $\mathcal{Y}$) then $u_1 \notin \operatorname{Range} F_x^0$ means simply that zero is an *algebraic simple* eigenvalue of $\mathcal{F}_x^0$.

For further existence theorems we refer to the literature relevant to this subject.

(f) Preliminary **Examples of Bifurcation Problems**

Example 5.2. Of course the simplest system is a scalar equation with $L = 0 \in \mathbb{R}$. Let, e.g.,

$$F(\mu, x) := \mu x - cx^2 = 0\,, \quad 0 \neq c \in \mathbb{R}\,,$$

then $\mu = 0$ is the branching point in which the non-trivial solution $x(\mu) = \mu/c$ (straight line) branches off from the trivial solution $x = 0$. The same holds for

$$F(\mu, x) := \mu x - cx^3 = 0\,, \quad c > 0\,,$$

but now the branching solution $x(\mu) = \pm\sqrt{\mu/c)}\,, \mu \geq 0$, is a parabola ("pitch-fork" bifurcation). Fig. 5.2 shows also the flow of the associated Hamiltonian vector field.

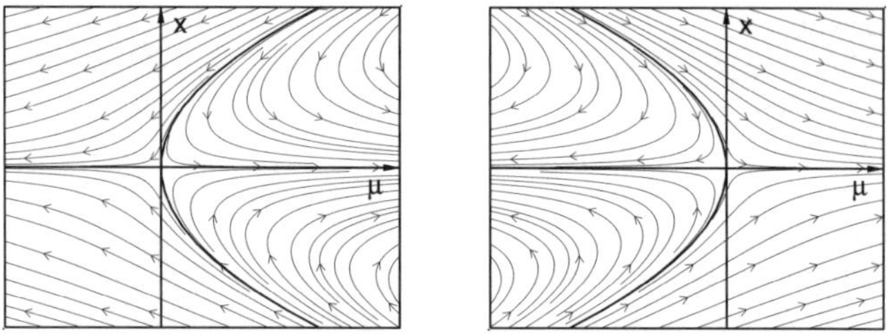

Figure 5.2. Pitchfork bifurcation, $c = 1$, $c = -1$

In this example we have for $c = 1$

$$F^0 = F_x^0 = F_{xx}^0 = F_\mu^0 = 0\,, \; F_{xxx}^0 F_{\mu x}^0 < 0\,. \tag{5.17}$$

A well-known result of *Singularity Theory* says that *every* mapping F with properties (5.17) has a pitchfork bifurcation at the origin. If $F_{xxx}^0 F_{\mu x}^0 > 0$, the number of solutions jumps from three to one instead from one to three.

Bifurcation equations may be *imbedded* into augmented equations to study their geometric and topological properties in a higher-dimensional frame. This process is called *(universal) unfolding.* As concerns pitchfork bifurcation $F(\mu, x) = x^3 + \mu x = 0 \in \mathbb{R}$, one obtains the "cusp" catastrophe by this way. Note in Figures 5.3, 5.4 that the derivative of the parameter κ of unfolding for x must vanish along the "fold". This condition yields $\mu = -3x^2$; substitution yields $\kappa = 2x^3$, thereafter x can be eliminated.

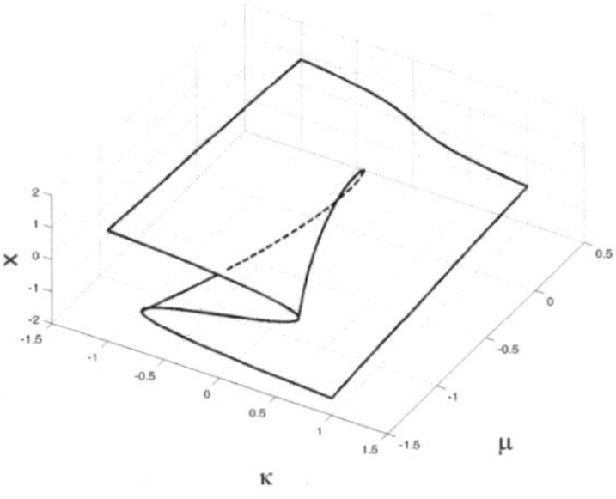

Figure 5.3. $x^3 + \mu x + \kappa = 0$

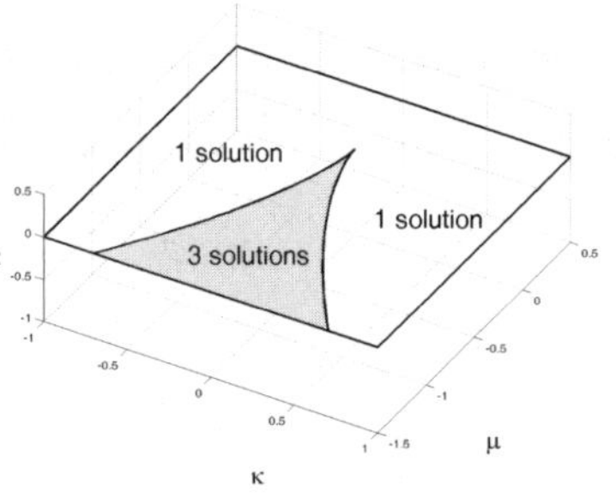

Figure 5.4. $[-\mu/3]^{1/2} = [\kappa/2]^{1/3}$

Example 5.3. ((Golubitsky), vol. I, p. 30 mod.) Consider the problem

$$F(\mu, x) := \begin{bmatrix} 0 & 0 \\ 0 & 1 \end{bmatrix} \begin{bmatrix} x_1 \\ x_2 \end{bmatrix} + \begin{bmatrix} x_1(\mu - x_1^2)/2 \\ -x_1^3/5 \end{bmatrix} = 0\,, \implies L = \begin{bmatrix} 0 & 0 \\ 0 & 1 \end{bmatrix}. \tag{5.18}$$

Then $\operatorname{Ker} L = \operatorname{span}\{\underline{e}_1\}$ where $\underline{e}_1 = [1,\, 0]^T$ is the first unit vector therefore

$$P = Q = \underline{e}_1 \underline{e}_1^T = \begin{bmatrix} 1 & 0 \\ 0 & 0 \end{bmatrix}, \quad I - Q = \begin{bmatrix} 0 & 0 \\ 0 & 1 \end{bmatrix} \ (= L)$$

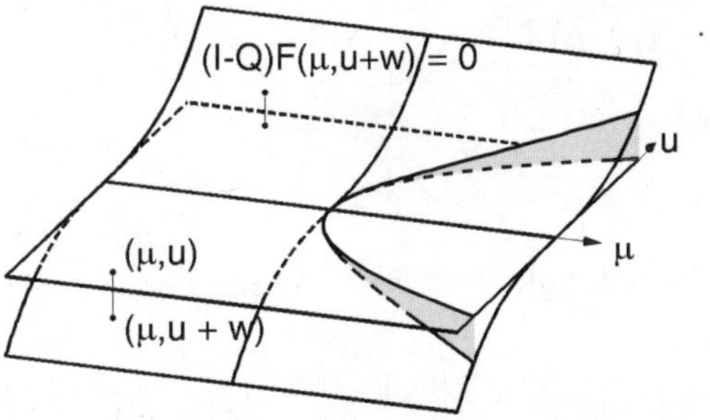

Figure 5.5. Example 5.3

Accordingly, bifurcation equation $\underline{e}_1^T F(\mu, x) = 0$ and operator equation $(I - Q)F(\mu, x) = 0$ are the first resp. second row of (5.18). The situation is completely illustrated in Figure 5.5 and shows that we have a pitchfork bifurcation at the origin, also the picture shows that the operator equation is rather insensitive against alterations. This example opens the way for designing various bifurcation problems with scalar branching equation. Note however that the linear operator L has the most simple form where one element of the unit matrix is replaced by zero.

Example 5.4. A less trivial example is due to (Crandall) & Rabinowitz and reads (Fig. 5.6):

$$F(\mu, x) = \begin{bmatrix} x_1 + \mu x_1(x_1^2 - 1 + x_2^2) \\ 10x_2 - \mu x_2(1 + 2x_1^2 + x_2^2) \end{bmatrix} = 0\,.$$

Branching points:

μ	x_1	x_2
1	0	0
5.5	0	$\pm 3/\sqrt{11}$
10	0	0
4	$\pm\sqrt{3}/2$	0

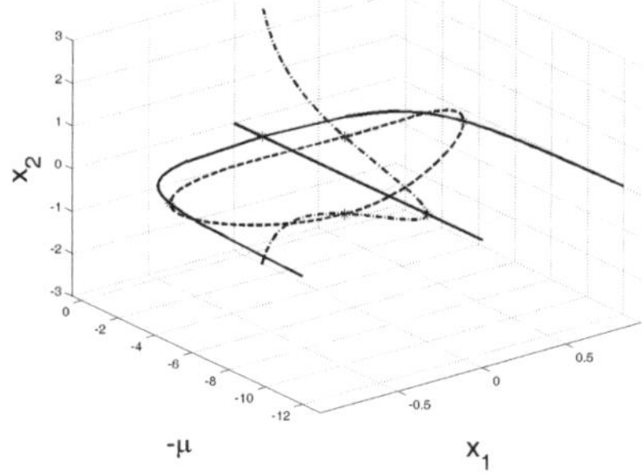

Figure 5.6. Example of (Crandall)

Solution:

$$\begin{aligned}
&x_1 = \pm((\mu-1)/\mu)^{1/2}\,, && x_2 = 0\,, && \text{for } 1 \le \mu\,,\\
&x_1 = 0\,, && x_2 = \pm((10-\mu)/\mu)^{1/2}\,, && \text{for } 0 \le \mu \le 10\,,\\
&x_1 = \pm((11-2\mu)/\mu)^{1/2}\,, && x_2 = \pm((3\mu-12)/\mu)^{1/2}\,, && \text{for } 4 \le \mu \le 5.5\,.
\end{aligned}$$

Example 5.5. (Stakgold) Let $L \in \mathbb{R}^2{}_2$ and let

$$F(\mu, x) = Lx - \mu x + f(x) = 0 \in \mathbb{R}^2\,,\quad f(x) = o(\|x\|)$$

then $F(\mu, 0) = 0$ but not every eigenvalue of L is necessarily a branching point!

(1°) Let $x - \mu x + f(x) = 0 \in \mathbb{R}^2\,,\ \ f(x) = [x_2^3,\ -x_1^3]^T$, then $L = I$ is the unit operator with *double eigenvalue* one. Multiply the first row by x_2, the second by x_1 and subtract one from the other then $(\mu, x_1, x_2) = (1, 0, 0)$ reveals to be no branching point (the branching equation has only the trivial solution).

(2°) Let $x - \mu x + f(x) = 0 \in \mathbb{R}^2\,,\ \ f(x) = [2x_1x_2,\ x_1^2 + 2x_2^2]^T$ where $L = I$ has has again the *double eigenvalue* one. Besides the trivial solution, the system has the solution $x = [0, (\mu-1)/2]$; by consequence *one* solution branches off from the trivial one.

(3°) Let

$$Lx - \mu x + f(x) = 0 \in \mathbb{R}^2\,,\quad L = \begin{bmatrix} 0 & 1\\ 0 & 0\end{bmatrix}\,,\quad f(x) = [x_1^3,\ x_1^2x_2 - x_1^3]^T$$

then L has the *double eigenvalue* zero with only one eigenvector, but no further solution exists besides the trivial one.

Example 5.6. Two examples with undiagonalizable matrix L, $F(\mu, 0) = 0$, and $(\mu_0, x_0) = (0, 0)$ (Landman).

$$(1°)\quad F(\mu, x) = \begin{bmatrix} x_2 + \mu a x_1 + \text{ h.o.t.}\\ \mu b x_2 + \mu^2 c x_1 + d x_1^2 + \text{ h.o.t.}\end{bmatrix} = 0 \tag{5.19}$$

for $x \in \mathbb{R}^2$ and real a, b, c, d. We have

$$\begin{aligned}
&L = \begin{bmatrix} 0 & 1\\ 0 & 0\end{bmatrix}\,,\quad B = \begin{bmatrix} a & 0\\ 0 & b\end{bmatrix}\,,\quad C_2(x) = \begin{bmatrix} 0\\ d\,x_1^2\end{bmatrix}\,,\quad F_\mu^0 = 0\,,\\
&v^d = [0,\ 1]\,,\quad u_1 = [1,\ 0]^T,\quad v^d B u_1 = 0
\end{aligned}$$

and thus $w_2 = x_2\,,\ \zeta = x_1$ in the above notation. The first row of F is the operator equation and its solution for w_2 into the second equation yields the bifurcation equation $-ab\mu^2x_1 + c\mu^2x_1 + dx_1^2 = 0$ of the reduced system. The general non-trivial solution of (5.19) is therefore

$$x_1 = \mu^2(ab-c)/d + \mathcal{O}(\mu^3)\,,\ x_2 = -\mu^3 a(ab-c)/d + \mathcal{O}(\mu^4)$$

for $ab - c \neq 0$ and $d \neq 0$.

$$(2^\circ)\quad F(\mu,x)=\begin{bmatrix} x_2+\mu a x_1+g x_1^2+\text{ h.o.t.}\\ \mu f x_1+d x_1^2+\text{ h.o.t.}\end{bmatrix}=0\,.$$

We have

$$L=\begin{bmatrix}0&1\\0&0\end{bmatrix},\ B=\begin{bmatrix}a&0\\f&0\end{bmatrix},\ C_2(x)=\begin{bmatrix}gx_1^2\\dx_1^2\end{bmatrix},\ F_\mu^0=0$$
$$v^d=[0,\,1]\,,\quad u_1=[1,\,0]\,,\quad v^dBu_1=f\,,$$

Again the first row of F is the operator equation and the second the bifurcation equation. The reduced equations

$$f\mu x_1+dx_1^2=0\,,\quad x_2+a\mu x_1+gx_1^2=0$$

have the unique non-trivial solution $x_1=-\mu f/d\,,\ x_2=\mu^2 f(ad-gf)/d^2$ and the general solution is

$$x_1=-\mu f/d+\mathcal{O}(\mu^2)\,,\ x_2=\mu^2 f(ad-gf)/d^2+\mathcal{O}(\mu^3)\,.$$

(g) Symmetry properties yield always additional information about a set of solutions and their behavior. We ask for some linear operations $C\in\mathcal{L}(\mathcal{X},\mathcal{X})$ and $\widetilde{C}\in\mathcal{L}(\mathcal{Y},\mathcal{Y})$ which do not or only few "affect" the system (5.5). These operations shall be invertible and composable. In other words, they shall form a *group* $\varGamma\subset\mathcal{L}(\mathcal{X},\mathcal{X})$ resp. $\widetilde{\varGamma}\subset\mathcal{L}(\mathcal{Y},\mathcal{Y})$, e.g. for $\varGamma$: $\forall\, C,D\in\varGamma: CD^{-1}\in\varGamma$. Both groups have to be compact in the infinite-dimensional case which means that $\varGamma$ is a compact (bounded and closed) subset of $\mathcal{L}(\mathcal{X},\mathcal{X})$ and $\widetilde{\varGamma}$ is a compact subset of $\mathcal{L}(\mathcal{Y},\mathcal{Y})$. Both groups however may (but must not) consist of entirely different *actions*. Consider for instance the Hopf bifurcation then $\varGamma$ consists of amplitude translations and $\widetilde{\varGamma}$ of rotations; cf. also Sect. 10.6.

Definition 5.2. (1°) *A subset $\mathcal{S}\subset\mathcal{Y}$ is* $\varGamma$-invariant, *if it is closed w.r.t. $\varGamma$: $\forall\, C\in\varGamma: C(\mathcal{S})\subset\mathcal{S}$.*

(2°) *The subset $\mathcal{U}\subset\mathcal{Y}$ is a* fixed-point subspace *w.r.t. $\varGamma$ or briefly a $\varGamma$-symmetric subspace if* $\mathcal{U}=\text{Fix}(\varGamma):=\{x\in\mathcal{Y},\ \forall\, C\in\varGamma: Cx=x\}$ *(this set is indeed a subspace).*

(3°) *Let $\mathcal{S}\subset\mathcal{Y}$ be a $\varGamma$-invariant subspace and let f be a mapping living on $\mathcal{S}$ then f is* $\varGamma$-invariant *if $\ \forall\, x\in\mathcal{S}\ \forall\, C\in\varGamma: f(Cy)=f(y)$.*

(4°) *Let $\mathcal{S}\subset\mathcal{Y}$ be a $\varGamma$-invariant subset then $f:\mathcal{X}\to\mathcal{Y}$ is* $\varGamma$-equivariant *(oder $\varGamma$-symmetric or "commutes with $\varGamma$") if there exists a symmetry group $\widetilde{\varGamma}\subset\mathcal{L}(\mathcal{Y},\mathcal{Y})$ such that*

$$\forall\, x\in\mathcal{S}\ \ \forall\, C\in\varGamma\ \ \exists!\,\widetilde{C}\in\widetilde{\varGamma}\ \text{ and }\ \forall\,\widetilde{C}\in\widetilde{\varGamma}\ \ \exists!\, C\in\varGamma:\widetilde{C}f(x)=f(Cx)\,.$$

(5°) *Let $\mu_0\in\mathcal{M}\subset\mathbb{R}^k$ and let $x_0\in\mathcal{U}\in\mathcal{X}$ be open then the mapping F of (5.5) is* $\varGamma$-equivarent *in $\mathcal{U}$ if $F(\mu,\circ)$ is $\varGamma$-equivariant in $\mathcal{U}$ for all $\mu\in\mathcal{M}$.*

In the next theorem we assemble some results for the system (5.5), $F(\mu,x)=Lx+f(\mu,x)$, following directly from the definition:

Theorem 5.4. *Let F be $\mathit{\Gamma}$-equivariant in $\mathcal{U} \subset \mathcal{X}$ and let $F(\mu_0, x_0) = 0$.*

(1°) *If (μ_0, x_0) is a solution of $F(\mu, x) = 0$, also (μ_0, Cx_0) is solution for all $C \in \mathit{\Gamma}$.*
(2°) $\widetilde{C} F_x^0 = F_x^0 C$ *holds for all $C \in \mathit{\Gamma}$, i.e. F_x^0 is $\mathit{\Gamma}$-equivariant.*
(3°) $\forall\, C \in \mathit{\Gamma} : C(\mathrm{Ker}\, F_x^0) = \mathrm{Ker}\, F_x^0$, *i.e.* $\mathrm{Ker}\, F_x^0$ *is $\mathit{\Gamma}$-invariant.*
(4°) $\forall\, \widetilde{C} \in \widetilde{\mathit{\Gamma}} : \widetilde{C}(\mathrm{Range}\, F_x^0) = \mathrm{Range}\, F_x^0$, *i.e.* $\mathrm{Range}\, F_x^0$ *is $\widetilde{\mathit{\Gamma}}$-invariant.*
(5°) $\forall\, (\mu, x) \in \mathcal{M} \times (\mathcal{U} \cap \mathrm{Fix}(\mathit{\Gamma})) : F(\mu, x) \in \mathrm{Fix}(\widetilde{\mathit{\Gamma}})$.
(6°) *Let F be $\mathit{\Gamma}$-equivariant and let F_x^0 be a* FREDHOLM *operator. Then the projectors P and Q of Theorem 5.1 can be chosen such that they commute with $\mathit{\Gamma}$ resp. with $\widetilde{\mathit{\Gamma}}$, i.e.,*

$$\forall\, C \in \mathit{\Gamma} : CP = PC\,, \;\; \forall\, \widetilde{C} \in \widetilde{\mathit{\Gamma}} : \widetilde{C}Q = Q\widetilde{C}\,.$$

Now the following results can be established by using the above auxiliaries:

Lemma 5.3. *Let F be $\mathit{\Gamma}$-equivariant and let F_x^0 be a* FREDHOLM *operator.*

(1°) *The function w^* of* **(c)** *is $\mathit{\Gamma}$-equivariant:*
$\forall\, C \in \mathit{\Gamma} : Cw^*(\mu, u) = w^*(\mu, Cu)$.
(2°) *The mapping $x^* : (\mu, u) \mapsto u + w^*(\mu, u)$ is $\mathit{\Gamma}$-equivariant:*
$\forall\, C \in \mathit{\Gamma} : Cx^*(\mu, u) = x^*(\mu, Cu)$.
(3°) *The branching equation $G : (\mu, u) \mapsto G(\mu, u)$ is $\mathit{\Gamma}$-equivariant:*
$\forall\, C \in \mathit{\Gamma} : \widetilde{C}G(\mu, u) = G(\mu, Cu)$.

As concerns the permutation of matrices, we know that with some few exceptions only diagonal matrices are commutative. Somewhat more generally, the following result is found in (Householder), p. 30:

Lemma 5.4. *Let A, $B \in \mathbb{R}^n{}_n$ be both normalizable, i.e. similar to a normal matrix then $AB = BA$ if and only if there is a regular matrix X such that $X^{-1}AX$ and $X^{-1}BX$ are both diagonal matrices.*

Apparently the columns of X are common eigenvectors of A and B.

(h) Examples with Symmetry cf. also Sect. 10.6 and in particular Sect. 5.6**(c)**.

(h1) $\mathbb{Z}_2$-Symmetry: $\mathit{\Gamma} = \{-\boldsymbol{\delta}, \boldsymbol{\delta}\}$ with unit operator $\boldsymbol{\delta}$ in $\mathcal{X}$ and $\mathcal{Y}$. If F odd, $-F(\mu, x) = F(\mu, -x)$ then, by Lemma 5.3, also the branching equation is odd, $-G(\mu, u) = G(\mu, -u)$.

(h2) The LORENTZ equation — cf. Example 5.16 — is likewise $\mathbb{Z}_2$-symmetric in $\mathbb{R}^3$ but the *action* of the group is defined in a different way:

$$\mathit{\Gamma} = \{\text{identity}\,,\; C\}\,, \;\; C[x^1, x^2, x^3]^T := [-x^1, -x^2, x^3]\,.$$

Accordingly, with one periodic solution x always Cx is a solution, too.

(h3) 2π-periodic functions with group $\mathit{\Gamma}$ of phase translations:

$$x(\circ) \in \mathcal{C}_{2\pi}\,, \;\; \widetilde{C}_\varphi \in \mathit{\Gamma}\,, \;\; \widetilde{C}_\varphi x(\circ) = x(\circ + \varphi) \;\; \text{phase translation}\,,$$
$$F(\mu, x)(t) = D_t x(t) + Ax(t) + f(\mu, x(t))\,;$$

cf. Sect. 5.5. The Γ-equivariance of F does hold in trivial way if F does not depend explicitely on t:

$$\begin{aligned}\widetilde{C}_\varphi F(\mu,x)(t) &= F(\mu,x)(t+\varphi) = F(\mu,\widetilde{C}_\varphi x)(t)\\ &= D_t x(t+\varphi) + Ax(t+\varphi) + f(\mu, x(t+\varphi))\,.\end{aligned}$$

Let now $A \in \mathbb{R}^2{}_2$ and $D_t + A = L : \mathcal{C}^1_{2\pi} \to \mathcal{C}_{2\pi}$ be a FREDHOLM operator with index zero and $\dim \operatorname{Ker} L = 2$ then A must have the eigenvalues $\pm i$. Let $\underline{c}$ and $\overline{\underline{c}}$ be the associated eigenvectors and let $u_1(t) = \operatorname{Re}(\underline{c}e^{it})$, $u_2(t) = \operatorname{Im}(\underline{c}e^{it})$, $U = [u_1, u_2]$. Then $\operatorname{Ker} L = \{U(t)\zeta\,,\ \zeta \in \mathbb{R}^2\}$. Furthermore, let

$$C_\varphi = \begin{bmatrix}\cos\varphi & -\sin\varphi\\ \sin\varphi & \cos\varphi\end{bmatrix}$$

be the rotation matrix in $\mathbb{R}^2$ then $U(t+\varphi)\zeta = U(t)C^T\zeta$ holds for the phase translation. Consider now the normal form of HOPF bifurcation

$$F(\mu,x) = x_t + Ax + f(\mu,|x|)x = 0 \in \mathbb{R}^2\,,\ A = \begin{bmatrix}0 & -1\\ 1 & 0\end{bmatrix}$$

with a scalar function f and $x \in \mathbb{R}^2$. Then A has the eigenvalues $\pm i$ again and $C_\varphi A = AC_\varphi$ therefore

$$C_\varphi F(\mu,x) = C_\varphi x_t + AC_\varphi x + f(\mu,|C_\varphi x|)C_\varphi x = F(\mu, C_\varphi x)$$

because $|C_\varphi x| = |x|$. Hence F is equivariant w.r.t. the group Γ of rotations in $\mathbb{R}^2$.

References: (Chow), (Golubitsky), (Kuznetsov), (Seydel94), (VanderBauwhede).

5.2 Scaling

(a) Modified Ljapunov-Schmidt Reduction Frequently the situation near a branching point (μ_0, x_0) can be displayed more properly if a new local parameter, say ε, is introduced into the LJAPUNOV-SCHMIDT reduction which does work like a magnifying glass, and this re-parametrization is also commonly applied in constructing numerical methods. The variables μ and x have to be written first as a suitable expansion w.r.t. that parameter ε. Then branching equation and operator equation are divided by a *properly chosen power* of ε; therafter a TAYLOR expansion is applied and a transition to the limit $\varepsilon \to 0$ is carried out. By this way the essential terms of the branching equation are filtered out and the rest is called *algebraic branching equation. Besides preparation for numerical approach, it is the aim of the scaling procedure to eliminate as many higher order terms as possible in that equation.* Eventually, some of the free variables have still to be shut down in order that the Implicit Function Theorem succeeds in the proof of *one* selected solution branch.

We consider again the nonlinear problem

$$F : \mathbb{R} \times \mathcal{X} \ni (\mu, x) \mapsto F(\mu, x) := Lx + f(\mu, x) = 0 \in \mathcal{Y} \tag{5.20}$$

with sufficiently smooth function f. We study the system (5.20) at the origin $(\mu, x) = (0, 0)$ under the assumption that $f(\mu, 0) = 0$ and suppose throughout this section that $L = F_x^0$ is a FREDHOLM operator with index zero and $\alpha := \dim \operatorname{Ker} L > 0$ (mostly $\alpha = 1$ or $\alpha = 2$). According to the decomposition in the LJAPUNOV-SCHMIDT reduction, we choose among the many possibilities at first the following scaling

$$\begin{aligned} \mu(\varepsilon) &= \varepsilon^r \xi(\varepsilon) && \in \mathbb{R}\,, && 0 < r \in \mathbb{Q}\,, \\ x(\varepsilon) &= \varepsilon(U\zeta(\varepsilon) + \varepsilon^r w(\varepsilon)) && \in \mathcal{X}\,, && \zeta(\varepsilon) \in \mathbb{R}^\alpha\,,\ U^d w(\varepsilon) = 0 \end{aligned} \tag{5.21}$$

The scalar parameter ε shall be always the *path parameter* in the sequel. The system of *branching equation* and *operator equation* reads now in scaled form

$$\boxed{\begin{aligned} \Phi(\xi, \zeta, w, \varepsilon) &:= \varepsilon^{-s} V^d F(\mu, x) && = 0 \in \mathbb{R}^\alpha \\ \Psi(\xi, \zeta, w, \varepsilon) &:= \varepsilon^{-(r+1)}(I - Q)F(\mu, x) && = 0 \in \operatorname{Range} L\,,\ U^d w(\varepsilon) = 0 \end{aligned}}\,. \tag{5.22}$$

Our goal is to select *one* solution branch and to calculate it numerically later. Recall that $QL = 0$ for the *fixed* projection operator $Q = VV^d$ satisfying $\operatorname{Ker} Q = \operatorname{Range} F_y^0$. Therefore f instead F can be written in the branching equation and the operator equation reads also

$$\varepsilon^{-(r+1)}(I - Q)F(\mu, x) = Lw(\varepsilon) + (I - Q)\psi(\xi(\varepsilon), \zeta(\varepsilon), w(\varepsilon), \varepsilon) = 0\,.$$

Assumption 5.1. *Let* $\xi_0 = \xi(0)$, $\zeta_0 = \zeta(0)$, $w_0 = w(0)$. Φ *and* Ψ *exist smoothly in some neighborhood of* $\varepsilon = 0$ *with finite limit values*

$$\begin{aligned} \Phi(\xi_0, \zeta_0, w_0, 0) &= && \lim_{\varepsilon \to 0} \Phi(\xi(\varepsilon), \zeta(\varepsilon), w(\varepsilon), \varepsilon) \\ \Psi(\xi_0, \zeta_0, w_0, 0) &= Lw_0 + (I - Q)\psi(\xi_0, \zeta_0, 0) &&= \lim_{\varepsilon \to 0} \Psi(\xi(\varepsilon), \zeta(\varepsilon), w(\varepsilon), \varepsilon)\,. \end{aligned}$$

The function ψ does not depend on w for $\varepsilon = 0$ by definition of L and because $F(\mu, 0) = 0$ hence this is mainly an assumption on the exponent s of the branching equation.

The system

$$\Phi(\xi_0, \zeta_0, w_0, 0) = 0\,, \quad \Psi(\xi_0, \zeta_0, w_0, 0) = 0 \tag{5.23}$$

consists of $\alpha+1$ equations for $\alpha+2$ variables ξ_0, $\zeta_0 \in \mathbb{R}^\alpha$ and $w(0) \in \operatorname{Ker} UU^d$ which must be regarded in the solution. Recall that L_0 is the *invertible* restriction of the operator L to $\operatorname{Ker} P$ and *onto* $\operatorname{Range} L$ and that e.g. ${\Phi_w}^0 = \operatorname{grad}_w \Phi(\xi_0, \zeta_0, w_0, 0)$. An application of the Implicit Function Theorem then yields:

Lemma 5.5. *Let the system (5.23) have a solution (ξ_0, ζ_0, w_0) with ξ_0 and ζ_0 both non-zero such that the linear operator*

$$G := \begin{bmatrix} \Phi_\zeta{}^0 & \Phi_w{}^0 \\ \psi_\zeta{}^0 & L_0 \end{bmatrix} \tag{5.24}$$

has a bounded inverse. Then the system (5.20) has a non-trivial local solution of the form (5.21) near $(\varepsilon, \xi) = (0, \xi_0)$ and $\mu = \varepsilon^r \xi$.

This simple result allows many modifications and generalizations. In particular we can consider one component of ζ as free variable and ξ as dependent variable instead.

Supplements: (1°) The solution (5.21) depends in unique way on (ξ_0, ζ_0) since w_0 solves uniquely $Lw = -(I - Q)\psi(\xi_0, \zeta_0)$, $U^d w = 0$.
(2°) The assumption $\zeta_0 \neq 0$ is necessary to avoid the trivial solution $(\mu(\varepsilon), x(\varepsilon)) = (\varepsilon^r \xi_0, 0)$ which is also obvious from the geometrical point of view.
(3°) Frequently it is of advantage to augment the branching equation by the additional normalizing condition $\zeta(\varepsilon)^T \zeta(\varepsilon) = 1$ instead to require the separated condition $\zeta_0 \neq 0$. In this case we have as many equations as variables (ξ_0, ζ_0, w_0) and the operator G must be augmented by one row and one column.
(4°) Suppose for instance that $\Phi_w{}^0 = 0$ then G is invertible if $\Phi_\zeta{}^0$ enjoys this property because $\psi_\zeta{}^0$ is bounded.

(b) Let $\widetilde{B}\langle x_1, \ldots, x_q\rangle$ be a q-linear mapping and write $B_q\langle x\rangle = \widetilde{B}\langle x, \ldots, x\rangle$ then $B_q\langle \gamma x\rangle = \gamma^q B_q\langle x\rangle$. More specifically, we consider the problem

$$F(\mu, x) = \big[Lx + \mu^p B_q\langle x\rangle + C_r\langle x\rangle + \text{ h.o.t. in } \mu \text{ and } x\big] = 0\,, \quad p, q, r \in \mathbb{N}\,,\ r > q\,. \tag{5.25}$$

L is again a Fredholm operator with index zero and $\dim \operatorname{Ker} L > 0$ and both mappings B and C are not identically zero. In the scaled ansatz

$$\mu(\varepsilon) = \varepsilon^s \xi\,, \quad x(\varepsilon) = \varepsilon U\zeta(\varepsilon) + \varepsilon^{s+1} w(\varepsilon)\,, \tag{5.26}$$

let $s = (r - q)/p$ be the solution of $ps + q = r$ then $0 < s \leq r - 1$. The scaled bifurcation equation is then

$$\begin{aligned} \Phi(\xi, \zeta, w, \varepsilon) &\equiv V^d\big[\xi^p B_q\langle U\zeta\rangle + C_r\langle U\zeta\rangle + \varepsilon^\gamma r(\xi, U\zeta, w, \varepsilon)\big] = 0 \\ r(\xi, \zeta, w, \varepsilon) &= \mathcal{O}(1)\,,\ \varepsilon \to 0\,,\ \ \gamma \geq \min\{s, 1\}\,, \end{aligned} \tag{5.27}$$

where ζ and w depend on ε.

Corollary 5.1. *Let the set*

$$Z = \{\zeta \in \mathbb{R}^\alpha : V^d B_q\langle U\zeta\rangle \neq 0 \text{ and } V^d C_r\langle U\zeta\rangle \neq 0\}$$

be not empty. For all $\zeta_0 \in Z$ the algebraic branching equation of the problem (5.25) with ansatz (5.26) and $s = (r - q)/p$ does not depend on w_0 and has the form

$$\boxed{\Phi(\xi_0,\zeta_0) := \Phi(\xi_0,\zeta_0,w_0,0) = \xi_0^p V^d B_q\langle U\zeta_0\rangle + V^d C_r\langle U\zeta_0\rangle = 0} \,. \tag{5.28}$$

For $p = q = 1$ and $r = 2$ we can refer to Example 5.1; see also Example 5.18. Note also that $V^dBU \in \mathbb{R}^\alpha{}_\alpha$ regular for a regular linear operator B if V^dU regular since then $V^d = CU^d$ with a regular matrix C. General results concerning the existence of isolated solutions of (5.28) are found in (Keller72).

Example 5.7.

$$F(\mu,x) = \begin{bmatrix} \mu x_2 + x_1^3 + \text{ h.o.t.} \\ x_2 + \mu x_2 + x_1^2 + \text{ h.o.t.} \end{bmatrix} = 0$$

$$L = \begin{bmatrix} 0 & 0 \\ 0 & 1 \end{bmatrix},\; B = \begin{bmatrix} 0 & 1 \\ 0 & 1 \end{bmatrix},\; C_2\langle x\rangle = \begin{bmatrix} 0 \\ x_1^2 \end{bmatrix},\; C_3\langle x\rangle = \begin{bmatrix} x_1^3 \\ 0 \end{bmatrix},$$
$$v^d = [1,\,0],\quad u_1 = [1,\,0]^T,\quad v^dBu_1 = 0\,.$$

The first equation is the bifurcation equation and the second equation the operator equation. The reduced equations have the solution

$$\widetilde{x}_1(\mu) = \mu/(1+\mu) = \mu + \mathcal{O}(\mu^2)\,,\quad \widetilde{x}_2(\mu) = -\mu^2/(1+\mu)^3 = -\mu^2 + \mathcal{O}(\mu^3)$$

and the general solution has the form $x_1(\mu) = \widetilde{x}_1(\mu) + \mathcal{O}(\mu^2)$, $x_2(\mu) = \widetilde{x}_2(\mu) + \mathcal{O}(\mu^3)$. Using the ansatz $\mu = \varepsilon\xi$, $x = \varepsilon U\zeta + \varepsilon^2 w$, the scaled reduced equations $\xi w_2 + \zeta^3 = 0$, $w_2 + \zeta^2 = 0$ depend here on $w^0 \neq 0$ and have the solution $\zeta = \xi$ and $w_2^0 = -\xi^2$. The matrix G of (5.50),

$$G = \begin{bmatrix} 3\zeta^2 & \xi \\ 2\zeta & 1 \end{bmatrix} = \begin{bmatrix} 3\xi^2 & \xi \\ 2\xi & 1 \end{bmatrix}$$

is regular for $\xi \neq 0$ and the solution is isolated.

See also Example 5.6 for a system with non-diagonalizable matrix L.

The next result of (Keener74) concerns the system (5.25) in the more commonly used form with $p = q = 1$ such that $s = r - 1$ in the ansatz (5.26). The point $(\mu_0, 0)$, $\mu_0 \in \mathbb{R}$, is the bifurcation point. The proof is managed by applying a suitable modification of the Contraction Theorem to a sequence of iterations and provides by this way a simple device for practical applications.

Let L again be FREDHOLM operator with index zero and $\alpha = \dim \operatorname{Ker} L > 0$, zero not necessarily being a semi-simple eigenvalue. We consider the problem

$$\begin{aligned} F(\mu,x) &= Lx + f(\mu,x) \\ &= \big[Lx + (\mu-\mu_0)Bx + C_r\langle x\rangle + \text{ h.o.t. in } \mu-\mu_0 \text{ and } x\big] = 0\,,\ r \geq 2\,, \end{aligned} \tag{5.29}$$

$L = \operatorname{grad}_x F(\mu_0,0)$, $B = F_{\mu x}^0 := \operatorname{grad}_x F_\mu(\mu_0,0)$ etc..

Lemma 5.6. *Suppose that there exists a pair (ξ_0,ζ_0) with $|\zeta_0| = 1$ solving*

$$\xi V^d B\langle U\zeta\rangle + V^d C_r\langle U\zeta\rangle = 0$$

such that the matrix $\xi_0 V^d BU + \mathrm{grad}_\zeta V^d C_r \langle U\zeta_0\rangle \in \mathbb{R}^\alpha{}_\alpha$ *is regular.*
(1°) *There exists locally a unique solution* $x(\varepsilon) = \varepsilon U\zeta(\varepsilon) + \varepsilon^r w(\varepsilon)$, $\mu(\varepsilon) = \mu_0 + \varepsilon^{r-1}\xi_0$ *of (5.29) with* $\zeta(0) = \zeta_0$ *depending continuously on* ε *near* $\varepsilon = 0$.
(2°) *This solution is the limit of sequence* $\{x^k(\varepsilon)\}$ *defined by*

$$x^k(\varepsilon) = \varepsilon U\zeta^k(\varepsilon) + \varepsilon^r w^k(\varepsilon) \quad \zeta^k := \zeta(w^k(\varepsilon), \varepsilon),$$

$w^0(\varepsilon) = 0$, $\zeta^k(0,0) = \zeta_0$, *where* ${\zeta^k}_j$, $j = 1 : \nu$, *are the unique functions which satisfy* $V^d f(\mu_0 + \varepsilon^{r-1}\xi_0, \varepsilon U\zeta^k + \varepsilon^r w^k) = 0$ *and* w^{k+1} *is the unique solution of*

$$Lw^{k+1} = -\varepsilon^{-r} f(\mu_0 + \varepsilon^{r-1}\xi_0, x^k(\varepsilon)), \quad U^d w^{k+1} = 0.$$

(3°) *The errors in the iteration satisfy* $|x^k(\varepsilon) - x(\varepsilon)| = \mathcal{O}(|\varepsilon|^{(k+1)(r-1)+1})$.

To avoid solving a system of nonlinear equations for the parameter vector $\zeta \in \mathbb{R}^\alpha$ see a proposition of (Demoulin) in Sect. 5.6.

(c) The **Nonlinear Eigenvalue Problem** has the basic equation

$$F(\lambda, x) = Ax - \lambda f(x) = 0, \quad f(0) = 0, \quad f_x^0 = \nabla f(0), \tag{5.30}$$

and $A \in \mathcal{L}(\mathcal{X}, \mathcal{X})$ is a Fredholm-Operator which is frequently symmetric and positive definite. The scaling ansatz (5.26) fails whenever $\lambda = 0$ is an eigenvalue of A. Let $\lambda_0 \neq 0$ be an eigenvalue of the generalized eigenvalue problem $Lx = 0$ with $L := A - \lambda_0 f_x^0$ being a Fredholm operator with index zero. Then

$$F(\lambda, x) = Lx - \lambda f(x) + \lambda_0 f_x^0 x = 0. \tag{5.31}$$

Let now for instance

$$f(x) = Bx + C_r(x) + \mathcal{O}(\|x\|^{r+1}), \tag{5.32}$$

so that $B = f_x^0$ then operator equation and bifurcation equation are

$$\begin{aligned} Lw &= (\lambda - \lambda_0)Bx + \lambda C_r(x) + \lambda(\text{h.o.t. in } x) \\ 0 &= \lambda\big[(\lambda - \lambda_0)\lambda^{-1} V^d Bx + V^d C_r(x) + \text{ h.o.t. in } x\big]. \end{aligned}$$

Using the scaling ansatz

$$\lambda(\varepsilon) = \lambda_0(1 + \varepsilon^{r-1}\xi(\varepsilon)), \; x(\varepsilon) = \varepsilon U\zeta(\varepsilon) + \varepsilon^r w(\varepsilon), \tag{5.33}$$

the scaled bifurcation equation after division by $\varepsilon^r\lambda$ and the operator equation after division by ε^r read:

$$\begin{aligned} \Phi(\xi, \zeta, w, \varepsilon) &= \kappa V^d BU\zeta + V^d C_r\langle U\zeta\rangle + \varepsilon r_1(\xi, \zeta, w, \varepsilon) = 0, \; \kappa = \xi/(1 + \varepsilon^{r-1}\xi) \\ \Psi(\xi, \zeta, w, \varepsilon) &= Lw - (I - Q)\big[\lambda_0 \xi BU\zeta + \lambda_0 C_r\langle U\zeta\rangle + \varepsilon r_2(\xi, \zeta, w, \varepsilon)\big] = 0, \\ U^d w = 0, &\qquad r_i(\xi, \zeta, w, \varepsilon) = \mathcal{O}(1), \; \varepsilon \to 0, \; i = 1, 2. \end{aligned}$$

As an application of Lemma 5.5 we then have:

Corollary 5.2. (1°) *Let* $\lambda_0 \neq 0$, *let* $L = A - \lambda_0 B$ *be a* FREDHOLM *operator with index zero and* $\alpha = \dim \operatorname{Ker} L > 0$.
(2°) *Let the set* Z *defined in Corollary 5.1 be not empty.*
(3°) *Let the algebraic bifurcation equation* $\kappa V^d BU\zeta + V^d C_r\langle U\zeta\rangle = 0$, $\zeta \in Z$, *have a solution* (κ_0, ζ_0) *with* $\kappa_0 \neq 0$ *and regular matrix* $\kappa_0 V^d BU + V^d \nabla_x [C_r\langle U\zeta_0\rangle]\langle U\rangle \in \mathbb{R}^\alpha{}_\alpha$.
Then the problem (5.31), (5.32) has a unique solution (5.33) with $\xi(0) = \kappa_0$ *and* $\zeta(0) = \zeta_0$ *depending continuously on* ε *near* $\varepsilon = 0$.

(d) Perturbated Eigenvalue Problem Let $\mathcal{X} \subset \mathcal{Y}$ be *complex* BANACH spaces, let $0 \in \mathcal{I} \subset \mathbb{R}$ be an open interval and let $L(\cdot) : \mathcal{I} \ni \mu \mapsto L(\mu) \in \mathcal{L}(\mathcal{X}, \mathcal{Y})$ an operator-valued function; in particular $\mathcal{X} = \mathcal{Y} = \mathbb{C}^n$ may be the complex coordinate space. We consider the mapping

$$\boxed{F : \mathcal{I} \times (\mathbb{C} \times \mathcal{X}) \ni (\mu, (\lambda, x)) \to L(\mu)x - \lambda x = 0 \in \mathcal{Y}} \,, \qquad (5.34)$$

where now (λ, x) plays the role of x in **(c)**, i.e. (λ, x) is the dependent variable. For a general overview on the perturbed eigenvalue problem it is referred to [Chow], Chap. 14, and the references there. Here we suppose that $L(\mu)$ is a non-selfadjoint FREDHOLM operator with index zero and derive a necessary and sufficient condition for the smoothness of characteristic pairs emanating from a semi-simple eigenvalue of $L(0)$ thus examining the point $\mu = 0$ for possible branching. Concerning the finite-dimensional case, some hints are also found in [Golub], p. 204, and [Wilkinson], chap. II.

Assumption 5.2. (1°) $L := L(0)$ *is a* FREDHOLM *operator with index zero and* $\operatorname{Ker} L = \operatorname{span}\{u_1, \ldots, u_\alpha\}$, $U = [u_1, \ldots, u_\alpha]$, $\alpha \geq 1$.
(2°) $\mathcal{X} \subset \mathcal{Y}$ *and* $\operatorname{Ker} L \cap \operatorname{Range} L = \{0\}$.
(3°) L *is continuously differentiable in* $\mathcal{I}$ *such that*

$$L(\mu) = L + \mu B + o(|\mu|) \,, \quad |\mu| \to 0 \,,$$

where $B = L_\mu^0 := L_\mu(0) \in \mathcal{L}(\mathcal{X}, \mathcal{Y})$.

Assumption (2°) excludes the existence of *principal vectors* w of L such that $Lw = u_i$; see Theorem 5.1, 3°. It is necessary since principle vectors always depend discontinuously on the data.

Theorem 5.5. *Adopt Assumption 5.2 and* $V^d U = I$. *There exist* α *different, near* $\mu = 0$ *continuously differentiable branches of characteristic pairs* $(\lambda_i(\mu), x_i(\mu))$ *such that* $(\lambda_i(0), x_i(0)) = (0, u_i)$, $i = 1 : \alpha$, *if and only if the matrix* $V^d BU$ *is diagonalizable.*

For instance, let $\mathcal{X}$ be a HILBERT space and $L(x)$ symmetric then B and thus $V^d BU$ are likewise symmetric hence diagonalizable.

Hint to the proof. Supposing existence, differentiation of $Lx + \mu Bx = \lambda x$ w.r.t. μ at $\mu = 0$ yields for $k = 1 : \alpha$

$$Lx_k'(0) + Bu_k = \lambda_k'(0)u_k + \lambda_k(0)x_k'(0) = \lambda_k'(0)u_k$$

because $\lambda_k(0) = 0$. Then $v^i Bu_k = \lambda_k'(0)v^i u_k = \lambda_k'(0)\delta^i{}_k$ by multiplication from left with v^i, and $V^d BU$ is even diagonal. The sufficient part of the proof is managed by application of the above scaling methods to the matrix-valued eigenvalue problem $L(\mu)U(\mu) = \lambda(\mu)U(\mu)$ (Gekeler95).

(e) Consider finally a **General Branching Point** $(\mu_0, x_0) \neq (0,0)$ where $F(\mu_0, x_0) = 0$. Scaling finds here only limited application. We proceed from a TAYLOR expansion

$$F(\mu_0+\mu, x_0+x) = \big[Lx+K_1\mu+K_2\mu^2+\mu Bx+C_2\langle x\rangle + \text{h.o.t. in } \mu \text{ and } x\big] = 0 \tag{5.35}$$

where $L = F_x^0 = \nabla_x F(\mu_0, x_0)$, $K_1 = F_\mu^0$, etc., and $\alpha = \dim \operatorname{Ker} L \geq 1$ as necessary assumption for a genuine branching point. We use the scaled ansatz

$$\mu = \varepsilon\xi\,, \quad x = \varepsilon U\zeta(\varepsilon) + \varepsilon w(\varepsilon)\,, \; U^d w(\varepsilon) = 0\,, \tag{5.36}$$

Suppose that there exists a local solution branch $y(\lambda)$ with $F(\lambda, y(\lambda)) = 0$ and $y(\mu_0) = x_0$. Then $F_\lambda{}^0 + F_x{}^0 y'(x_0) = 0$ therefore $K_1 = F_\mu^0$ is contained in $\operatorname{Range} L$ which means that $V^d K_1 = 0$. Conversely, let $V^d K_1 = 0$ and ξ arbitrary fixed then there exists *always* a solution (5.36) with $\zeta(\varepsilon) = 0$ by the Implicit Function Theorem since L can be replaced by the above introduced operator L_0 with bounded inverse. For further solutions, we consider again the system of scaled bifurcation and operator equation:

$$\begin{aligned}
\Phi(\xi,\zeta,w,\varepsilon) &= \varepsilon^{-2}V^d F(\mu_0+\mu, x_0+x)\\
&= V^d K_2\xi^2 + \xi V^d B\langle U\zeta + w\rangle + V^d C_2\langle U\zeta + w\rangle + \varepsilon r_1(\xi,\zeta,w,\varepsilon)\\
\Psi(\xi,\zeta,w,\varepsilon) &= \varepsilon^{-1}(I-Q)F(\mu_0+\mu, x_0+x)\\
&= Lw + (I-Q)K_1\xi + \varepsilon r_2(\xi,\zeta,w,\varepsilon) = 0\,, \; U^d w = 0\,,\\
r_i(\xi,\zeta,w,\varepsilon) &= \mathcal{O}(1)\,, \; \varepsilon \to 0\,, \; i = 1,2\,.
\end{aligned}$$

For an isolated branch emanating from (μ_0, x_0) we can apply Lemma 5.5 again. Then we have to look for a solution (ξ_0, ζ_0, w_0) of the reduced scaled equations

$$\begin{aligned}
\Phi(\xi,\zeta,w,0) &= V^d K_2\xi^2 + \xi V^d B\langle U\zeta + w\rangle + V^d C_2\langle U\zeta + w\rangle = 0 \in \mathbb{R}^\alpha\\
\Psi(\xi,\zeta,w,0) &= L_0 w + (I-Q)K_1\xi =: L_0 w + \psi(\xi,\zeta) = 0 \in \operatorname{Range} L
\end{aligned} \tag{5.37}$$

with $\xi_0 \neq 0$ and $\zeta_0 \neq 0$.

Case 1: Fix ξ_0, say $\xi_0 = 1$. Then the linear operator

$$G := \begin{bmatrix} \Phi_\zeta{}^0 & \Phi_w{}^0 \\ \psi_\zeta{}^0 & L_0 \end{bmatrix}$$

must have a bounded inverse by Lemma 5.5. But $\psi_\zeta = 0$ thus only

$$\Phi_\zeta{}^0 := \Phi_\zeta(\xi_0, \zeta_0, w_0, 0) = \xi_0 V^d B\langle U\rangle + V^d \nabla_x [C_2\langle U\zeta_0 + w_0\rangle]\langle U\rangle \in \mathbb{R}^\alpha{}_\alpha$$

must be a regular matrix. The algebraic branching equation $\Phi = 0$ in (5.37) is a system of second order for $\zeta \in \mathbb{R}^\alpha$ after substitution of w_0 and inhomogeneous in case $V^d K_2 \neq 0$.
Case 2: Let $\alpha = \dim \operatorname{Ker} L = 2$. To find a branching solution emanating from, say the first eigenvector u_1 of L, set $\zeta_2 = 0$. Then the algebraic branching equation $\Phi = 0 \in \mathbb{R}^2$ has the independent variable w and $z = (\xi, \zeta_1)$ of which both components must be nonzero. Also the linear operator

$$G := \begin{bmatrix} \Phi_z^0 & \Phi_w{}^0 \\ \psi_z^0 & L_0 \end{bmatrix}$$

must have a bounded inverse where $\psi_z^0 = [(I - Q)K_1, 0]$ is no longer zero and

$$\Phi_z{}^0 := V^d [2K_2\xi_0 + B\langle v\rangle,\ \xi_0 B\langle u_1\rangle + \nabla_x [C_2\langle v\rangle]\langle u_1\rangle] \in \mathbb{R}^2{}_2\,,\ v = u_1\zeta_{1,0} + w_0\,.$$

Because L_0 has a bounded inverse, G has a bounded inverse by the bordering lemma, Theorem 1.20, if and only if $\Phi_z{}^0 - \Phi_w{}^0 L_0{}^{-1}\psi_z{}^0 \in \mathbb{R}^2{}_2$ is a regular matrix.

By a simple change of variables in Theorem 5.3, (Crandall) & Rabinowitz have also proved the following result in which no component of the argument of F is declared as specific bifurcation parameter such that $F(y) = 0$ instead $F(\mu, x) = 0$ is the basic system.

Lemma 5.7. *Let $\mathcal{X}, \mathcal{Y}$ be* BANACH *spaces, let $\Omega \subset \mathcal{X}$ open and $F \in C^2(\Omega, \mathcal{Y})$. Further, let $\mathcal{I} \subset \mathbb{R}$ be an open interval such that $0 \in \mathcal{I}$, and let $y \in C^1(\mathcal{I}, \Omega)$ denote a curve such that $F(y(t)) = 0$.*
Assumption: (1°) $\dot y(0) \neq 0$.
(2°) $\dim \operatorname{Ker} F_y^0 = 2$, $\operatorname{codim} \operatorname{Range} F_y^0 = 1$.
(3°) $\operatorname{Ker} F_y^0 = \operatorname{span}\{\dot y(0), v\}$ *and*
(4°) $F_{yy}^0\langle \dot y(0), v\rangle \notin \operatorname{Range} F_y^0$.
Then $y(0)$ is a branching point of $F(y) = 0$ w.r.t. the curve $\mathcal{I} \ni t \mapsto y(t)$, and, in some neighborhood of $y(0)$, there exist two different continuous curves intersecting in $y(0)$.

Of course, in case $y(0) = 0$, one of the both curves may be the trivial curve $y(0) \equiv 0 \in \mathcal{X}$.

5.3 Calculation of Singular Points

(a) A Classification Suppose that the nonlinear system $F(\mu, x) = 0$ with scalar parameter μ has a sufficiently smooth solution $\mathbb{R} \supset \mathcal{I} \ni s \to (\mu(s), x(s))$ passing through the point (μ_0, x_0) for $s = 0$. Then a two-fold differentiation for the path parameter s yields

$$\begin{aligned} &F_\mu^0\mu'(0) + F_x^0 x'(0) &&= 0\,,\\ &F_\mu^0\mu''(0) + F_x^0 x''(0) + F_{\mu\mu}^0\mu'(0)^2 + 2F_{\mu x}\mu'(0)x'(0) + F_{xx}^0\langle x'(0), x'(0)\rangle &&= 0\,. \end{aligned} \tag{5.38}$$

Let again $F_x^0 = \nabla_x F(\mu_0, x_0)$ be a FREDHOLM operator with index zero then we have four different cases:

$$\begin{aligned} &(1^\circ)\ \dim\operatorname{Ker}[F_\mu^0, F_x^0] = 1\ \dim\operatorname{Ker} F_x^0 = 0 \text{ and } F_\mu^0 \in \operatorname{Range} F_x^0\\ &(2^\circ)\ \dim\operatorname{Ker}[F_\mu^0, F_x^0] = 1\ \dim\operatorname{Ker} F_x^0 = 1 \text{ and } F_\mu^0 \notin \operatorname{Range} F_x^0\\ &(3^\circ)\ \dim\operatorname{Ker}[F_\mu^0, F_x^0] > 1\ \dim\operatorname{Ker} F_x^0 \geq 1 \text{ and } F_\mu^0 \in \operatorname{Range} F_x^0\\ &(4^\circ)\ \dim\operatorname{Ker}[F_\mu^0, F_x^0] > 1\ \dim\operatorname{Ker} F_x^0 \geq 2 \text{ and } F_\mu^0 \notin \operatorname{Range} F_x^0\,. \end{aligned} \tag{5.39}$$

For instance, let $u \in \operatorname{Ker} F_x^0$ then $(0, u) \in \operatorname{Ker}[F_\mu^0, F_x^0]$ and further elements in $\operatorname{Ker}[F_\mu^0, F_x^0]$ exist only if $F_\mu^0 \in \operatorname{Range} F_x^0$. In case (1°) the point (μ_0, x_0) is *regular* and there exists a local unique solution by the Implicit Function Theorem. In case (2°) the point (μ_0, x_0) is a *turning point* or *limit point* with respect to the path parameter μ in which a reparametrization is necessary. In case (3°) we have a (possible) bifurcation point. In the last case we have a turning point which may at the same time be a bifurcation point; this case is not further investigated here.

(b) In a **Turning Point** let $\operatorname{Ker} F_x^0 = \operatorname{span}\{u_1\}$ and $F_\mu^0 \neq 0$. Let $P = u_1 u^1$ be the projector with $\operatorname{Range} P = \operatorname{Ker} F_x^0$ and let

$$\mathcal{X} = \operatorname{Ker} L \oplus \operatorname{Ker} P \implies \mathbb{R} \times \mathcal{X} = \mathcal{W} \oplus \operatorname{Range} P\,,\ \ \mathcal{W} = \mathbb{R} \times \operatorname{Ker} P$$

be the decomposition (5.4). Moreover, let $\widetilde{F_x^0}$ be the restriction of F_x^0 to $\operatorname{Ker} P$, then $[F_\mu^0, \widetilde{F_x^0}] : \mathcal{W} \to \operatorname{Range} F_\mu^0 \times \operatorname{Range} \widetilde{F_x^0}$ is bijective with bounded inverse. The Implicit Function Theorem then says that $F(\mu, x) = 0$ is local solvable for $v \in \mathcal{W}$ with argument $s\,u_1 \in \operatorname{Ker} F_x^0 = \operatorname{Range} P$. Therefore we have a parameter representation $\mathbb{R} \ni \mathcal{I} \ni s \to (\mu(s\,u_1), x(s\,u_1))$ where $F(\mu(s\,u_1), x(s\,u_1)) = 0$ and $(\mu(0), x(0)) = (\mu_0, x_0)$. Together with the result for regular points, there follows:

Lemma 5.8. *For* $\dim\operatorname{Ker}[F_\mu^0, F_x^0] = 1$ *there exists exactly one solution path of* $F(\mu, x) = 0$ *passing through* (μ_0, x_0) *which is as smooth as* F.

(b) Characterization of Turning Points Let us return to the original path representation $\mathcal{I} \ni s \to (\mu(s), x(s))$ considered in **(a)**. Then the following two results on the *geometrical form* at a turning point $s = 0$ are a simple inference of (5.38) (recall that $\operatorname{Ker}[F_x^0]_d = \operatorname{span}\{v^1\}$).

Lemma 5.9. *Let* (μ_0, x_0) *be a turning point with* $x'(0) \neq 0$. *Then* $\mu'(0) = 0$ *and*

(1°) (μ_0, x_0) *is a quadratic turning point with* $\mu''(0) \neq 0$ *if and only if* $v^1 F_{xx}^0\langle u_1, u_1\rangle \neq 0$.

(3°) (μ_0, x_0) *is a cubic turning point with* $\mu''(0) = 0$ *and* $\mu'''(0) \neq 0$ *if and only if* $v^1 F_{xx}^0\langle u_1, u_1\rangle = 0$ *and* $v^1 F_{xxx}^0\langle u_1, u_1, u_1\rangle \neq 0$.

Proof. Note that $v^1F_x^0 = 0$ and $v^1F_\mu^0 \neq 0$ hence $\mu'(0) = 0$, by multiplying (5.38),(1°) by v^1. We write $x'(0) = u_1$ without loss of generality and obtain $v^1F_\mu^0\mu''(0) + v^1F_{xx}^0\langle u_1, u_1\rangle = 0$ by (5.40,(2°)). Therefore $\mu''(0) = -v^1F_{xx}^0\langle u_1, u_1\rangle / v^1F_\mu^0 \neq 0$. The proof of (2°) is carried out in much the same way as (1°) by three-fold differentiation of $F(\mu(s), x(s)) = 0$. □

(c) For the **Calculation of Turning Points**, (Keener73) and (MooreB) have proposed *augmented accompanying systems*. The first one reads:

$$\Phi_1(z) := \Phi_1(\mu, x, u) := \begin{bmatrix} F(\mu, x) \\ F_x(\mu, x)u \\ a\,u - 1 \end{bmatrix} = 0\,, \quad z = (\mu, x, u)\,, \ a \in \mathcal{X}_d \text{ fixed}\,. \tag{5.40}$$

Lemma 5.10. *Let (μ_0, x_0) be a turning point and $a\,u_1 = 1$. Then $\Phi_1(z_0) = \Phi_1(\mu_0, x_0, u_1) = 0$ and*

$$\operatorname{grad} \Phi_1(z_0) : \mathbb{R} \times \mathcal{X} \times \mathcal{X} \to \mathcal{Y} \times \mathcal{Y} \times \mathbb{R}$$

has a bounded inverse if and only if $v^1{F^0}_{xx}\langle u_1, u_1\rangle \neq 0$.

Conversely, a regular solution of (5.40) is a quadratic turning point if in addition $\dim \operatorname{Ker} F_x^0 = 1$ and $v^1F_\mu^0 \neq 0$ which is fulfilled in normal case. Besides, the inequality $v^1F_\mu^0 \neq 0$ is much more often fulfilled as the equality $v^1F_\mu^0 = 0$. Accordingly, for a test on possible turning points, NEWTON's method may be applied to solve $\Phi_1(z) = 0$. Note however that this result and the following yields only a *necessary* condition and the defining properties of a turning point have still to be verified after the calculation.

The second accompanying system is dual to the first system (5.40) in some way:

$$\Phi_2(z) := \Phi_2(\mu, x, v) := \begin{bmatrix} F(\mu, x) \\ v\,F_x(\mu, x) \\ v\,F_\mu(\mu, x) - 1 \end{bmatrix} = 0\,, \ z = (\mu, x, v)\,, \ v \in \mathcal{Y}_d\,. \tag{5.41}$$

Lemma 5.11. *Let (μ_0, x_0) be a turning point. Then $\Phi_2(z_0) = \Phi_2(\mu_0, x_0, v^1) = 0$ (after some scaling of v^1) and*

$$\operatorname{grad} \Phi_2(z_0) : \mathbb{R} \times \mathcal{X} \times \mathcal{Y}_d \to \mathcal{Y} \times \mathcal{X}_d \times \mathbb{R}$$

has a bounded inverse if and only if $v^1{F^0}_{xx}\langle u_1, u_1\rangle \neq 0$.

Proofs of Lemmas 5.9 and 5.10 see (MooreB) and `SUPPLEMENT\chap05b`.

(d) Calculation of Simple Branching Points Here we have $F_\mu^0 \in \operatorname{Range} L$. Let $\operatorname{Ker} F_x^0 = \operatorname{span}\{u_1\}$ and F_x^0 be again a FREDHOLM operator with index zero then $\operatorname{Ker}[F_x^0]_d = \operatorname{span}\{v^1\}$ and $v^1F_\mu^0 = 0$ by the Range

Theorem. Let w be the unique solution of $F^0_\mu + F^0_x w = 0$ with $u^1 w = 0$. Every tangent $x'(0)$ in (5.38) has the representation $x'(0) = \alpha u_1 + \beta w$ where $\beta = \mu'(0)$. Insertion of $(\beta, x'(0))$ in (5.38) yields after multiplication by the vector v^1

$$v^1 F^0_{\mu\mu}\beta^2 + 2v^1 F^0_{\mu x}\beta(\alpha u_1 + \beta w) + v^1 F^0_{xx}\langle \alpha u_1 + \beta w, \alpha u_1 + \beta w\rangle = 0\,.$$

or $a^T Q_1(u_1, w, v^1)a = 0$, $a^T = [\alpha,\ \beta]$ where $Q_1 \in \mathbb{R}^2{}_2$,

$$Q_1(u_1, w, v^1) = \begin{bmatrix} v^1 F^0_{xx}\langle u_1, u_1\rangle & v^1\left[F^0_{\mu x}u_1 + F^0_{xx}\langle u_1, w\rangle\right] \\ v^1\left[F^0_{\mu x}u_1 + F^0_{xx}\langle u_1, w\rangle\right] & v^1\left[F^0_{\mu\mu} + 2F^0_{\mu x}w + F^0_{xx}\langle w, w\rangle\right] \end{bmatrix}. \tag{5.42}$$

This quadratic form has two different real solutions in case $\det(Q_1(u_1, w, v^1)) < 0$. By this way we obtain a *necessary* condition for (μ_0, x_0) to be a branching point. For $F^0_\mu = 0$ we have $w = 0$. The case $F^0_\mu = 0$ and $F^0_{\mu\mu} = 0$ mirrors branching off from the trivial solution $(\mu, 0)$ and yields the necessary condition

$$\alpha^2 v^1 F^0_{xx}\langle u_1, u_1\rangle + 2\alpha\beta v^1 F^0_{\mu,x}u_1 = 0$$

for the tangent of the non-trivial branch; see Sect. 5.1**(d3)**, *Case 1* and (5.28) for $U = u_1$.

For the *calculation* of a branching point with the above constellation, we need here a result on the accompanying system (5.41), in which *by exception* F^0_x is a FREDHOM operator with index *one*:

Lemma 5.12. *Let $F^0_x \in \mathcal{L}(\mathcal{X}, \mathcal{Y})$ be a* FREDHOM *operator with index one and*

$$\operatorname{Ker} F^0_x = \operatorname{span}\{u_1, u_2\}\,,\quad \operatorname{Ker}[F^0_x]_d = \operatorname{span}\{v^1\}\,,$$

and let $\Phi_2(z_0) = 0$ for $z_0 = (\mu_0, x_0, v^1)$ in (5.41). Then $\operatorname{grad}\Phi_2(z_0)$ *has a bounded inverse if and only if the matrix $Q_2(u_1, u_2, v^1) = [v^1 F^0_{xx}\langle u_i, u_k\rangle]^2_{i,k=1}$ is regular.*

Proof in `SUPPLEMENT\chap05b`.

By a proposition of (MooreA), branching points are computed as *regular* points of an *augmented* accompanying system in a similar way as in turning points. We consider the perturbated system

$$\Phi_3(\lambda, (\mu, x)) := \Phi_2(\mu, x) + \lambda r = 0 \tag{5.43}$$

where λ now plays the role of the former parameter μ and (μ, x) the role of the former x. In Lemma 5.11 it has been supposed that $v^1 F^0_\mu \neq 0$ since (μ_0, x_0) shall be a turning point. To apply this lemma we therefore require that

$$\frac{\partial}{\partial\lambda}\Phi_3(0, (\mu_0, x_0)) = r \notin \operatorname{Range}(\widetilde{F}_{(\mu,x)}(0, (\mu_0, x_0))) = \operatorname{Range}([F^0_\mu,\ F^0_x])\,.$$

The accompanying system with *specified* r and $v \in \mathcal{Y}_d$ then reads:

$$\Phi_3(z) := \Phi_3(\lambda, (\mu, x), v) = \begin{bmatrix} F(\mu, x) + \lambda r \\ v\,[F_\mu(\mu, x),\, F_x(\mu, x)] \\ v\,r - 1 \end{bmatrix} = \begin{bmatrix} F(\mu, x) + \lambda r \\ v\,F_x(\mu, x) \\ v\,F_\mu(\mu, x) \\ v\,r - 1 \end{bmatrix} = 0\,,\ z = (\lambda, (\mu, x), v)\,. \tag{5.44}$$

Lemma 5.13. *Let* (μ_0, x_0) *be a simple branching point with* $\dim \operatorname{Ker} {F_x}^0 = 1$ *(and* $F_\mu^0 \in \operatorname{Range} F_x^0$ *). Let* $z_0 = (0, \mu_0, x_0, v^1)$ *then* $\Phi_3(z_0) = 0$ *. Further, let* w *denote the unique solution of* $F_\mu^0 + F_x^0 w = 0\,,\ v^1\, w = 0\,,$ *then*

$$\operatorname{grad} \Phi_3(z_0) : \mathbb{R}^2 \times \mathcal{X} \times \mathcal{Y}_d \to \mathcal{Y} \times \mathcal{Y}_d \times \mathbb{R}^2$$

has a bounded inverse if and only if the matrix Q_1 *in (5.42) is regular.*

A similar result can be derived for branching points with $\dim \operatorname{Ker} F_x^0 = 2$. Note however that this result yields again a test on possible branching points and is by no means sufficient.

Hint to the proof. Here $\operatorname{Ker}([F_\mu^0, F_x^0]) = \operatorname{span}\{(0, u_1), (1, w)\}$, therefore the matrix $Q_2(u_1, u_2, v^1)$ in Lemma 5.12 relative to the augmented system has now the form of the matrix $Q_1(u_1, w, v^1)$ relative to the system (5.44). □

Consider now the finite-dimensional case $\mathcal{X} = \mathcal{Y} = \mathbb{R}^n$ then we may choose $r = v^T \in \mathbb{R}^n$ (because $v \in \mathbb{R}_n$) and there follows the accompanying system

$$\begin{bmatrix} F(\mu, x) + \lambda v^T \\ v\,F_x(\mu, x) \\ v\,F_\mu(\mu, x) \\ v\,v^T - 1 \end{bmatrix} = 0 \tag{5.45}$$

which is employed by (Deuflhard87) as accompanying system in continuation methods for detecting branching points.

5.4 Ordinary Differential Systems

It may be surprising or not but ordinary differential systems are more difficult to handle than the usual partial differential equations of second order. The reason may be that the former do not enjoy similar symmetry properties. In normal case, the system is transformed ad first into a system of first order (MATLAB knows only them). By the change of sign due to partial integration there arises some *skewness* which entails negative effects on the primal-dual way of consideration as is shown immediately.

(a) Linear Boundary Value Problem; cf. Sect. 1.5. Let $D_t = d/dt$ denote the simple differential operator. We consider the homogeneous boundary problem (1.33),

$$D_t\, x(t) + A(t)x(t) = 0 \in \mathbb{R}^n\,,\;\; R_a x(a) + R_b x(b) = 0 \in \mathbb{R}^m\,,\; m \leq n\,, \quad (5.46)$$

in interval $\mathcal{I} = [a,\, b]$ where $D_t = \partial/\partial t$. Let the matrix $A(t)$ be continuous and let rank $[R_a, R_b] = m$.

Notations:

$\mathcal{Y}$	$= \{x \in \mathcal{C}(\mathcal{I};\mathbb{R}^n)\,,\; \Vert x\Vert = \max \vert x(t)\vert\}$	BANACH space
$\mathcal{X}$	$= \{x \in \mathcal{C}^1(\mathcal{I};\mathbb{R}^n\}\,,\; \Vert x\Vert_1 = \Vert x\Vert + \Vert D_t\, x\Vert\}$	BANACH space
$\mathcal{X}_b$	$= \{x \in \mathcal{X}\,,\; R_a x(a) + R_b x(b) = 0\} \subset \mathcal{X}$	BANACH space
L	$= D_t + A(\,\cdot\,)$	operator $L \in \mathcal{L}(\mathcal{X}_b, \mathcal{Y})$
L^*	$= -D_t + A(\,\cdot\,)^T$	*formal* adjoint operator to L
$(x,\, y) =$	$\dfrac{1}{b-a}\displaystyle\int_a^b x(t)^T y(t)\, dt$	scalar product

More exactly

$$Lx(t) = \dot{x}(t) + A(t)x(t)\,,\;\; L^*x(t) = -\dot{x}(t) + A(t)^T x(t)\,.$$

Lemma 5.14. *Let A and f be continuous. Then the* CAUCHY *problem*

$$Lx = \dot{x} + A(\cdot)x = f(\cdot)\,,\;\; x(t_0) = x_0\,,$$

has a unique solution $x \in \mathcal{C}^1(\mathcal{I}, \mathbb{R}^n)$ for every initial point $(t_0, x_0) \in \mathcal{I} \times \mathbb{R}^n$.

Proof see Sect. 1.5.

(b) Adjoint Boundary Value Problem The *formal* adjoint operator L^* is defined by the relation

$$\begin{aligned}(y,\, Lx) - (L^*y, x) &= \frac{1}{b-a}\int_a^b \Big[y(t)^T\dot{x}(t) + \dot{y}(t)^T x(t]\Big]\, dt \\ &= \frac{y(b)^T x(b) - y(a)^T x(a)}{b-a}\,.\end{aligned} \quad (5.47)$$

In order that L^* becomes the adjoint operator L_d, the domain of definition has to be specified such that the right side in (5.47) is zero:

$$y(a)^T x(a) - y(b)^T x(b) = [y(a)^T, -y(b)^T]\begin{bmatrix} x(a) \\ x(b)\end{bmatrix} = 0\,.$$

By this condition together with the boundary conditions for the primal problem, namely

$$[R_a,\, R_b]\begin{bmatrix} x(a) \\ x(b) \end{bmatrix} = 0 \in \mathbb{R}^m ,$$

we obtain the following condition

$$[y(a)^T,\, -y(b)^T] = q[R_a,\, R_b] \tag{5.48}$$

for some row vector $q \in \mathbb{R}_m$. Since rank $[R_a,\, R_b] = m \le n$ let S_a and S_b be $(n-m, n)$-matrices such that the $(n-m, 2n)$-matrix $[S_a,\, S_b]$ has maximum rank $n-m$ and that

$$R_a\, S_a^T - R_b\, S_b^T = 0 \in \mathbb{R}^m{}_{n-m}\,, \quad \text{rank}\,[S_a,\, S_b] = n-m\,. \tag{5.49}$$

Then the rows of $[S_a, -S_b]$ form a basis of Ker $[R_a,\, R_b]$, and we have

$$[y(a)^T,\, -y(b)^T]\begin{bmatrix} S_a^T \\ -S_b^T \end{bmatrix} = q^T[R_a,\, R_b]\begin{bmatrix} S_a^T \\ -S_b^T \end{bmatrix} = 0.$$

Now the adjoint problem has the form

$$\boxed{L^*y = 0\,,\ \ S_a y(a) + S_b y(b) = 0}\,, \tag{5.50}$$

where the $(n-m, 2n)$-matrix $[S_a,\, S_b]$ is determined uniquely up to multiplication from right by a regular matrix. Conveniently we introduce the BANACH space

$$\mathcal{X}_b^* := \{y \in \mathcal{C}^1(\mathcal{I}, \mathbb{R}^n)\,,\ \ S_a y(a) + S_b y(b) = 0\} \subset \mathcal{X}\,,$$

then the operator L and its adjoint operator L^* satisfy

$$\boxed{L \in \mathcal{L}(\mathcal{X}_b, \mathcal{Y})\,,\ \ L^* \in \mathcal{L}(\mathcal{X}_b^*, \mathcal{Y})}\,, \tag{5.51}$$

which has always to be regarded in the sequel.

Summary and Recapitulation for L and L^* ($I \in \mathbb{R}^n{}_n$ unit matrix) the *(principal) fundamental matrices* $X(t)$ *and* $Y(t)$ satisfy:

$$\dot X + A(t)X(t) = 0\,,\ \ X(a) = I\,,\ \ -\dot Y + A(t)^T Y(t) = 0\,,\ \ Y(b) = I\,.$$

WRONSKI *matrices*:

$$L:\ W(s,t) = X(s)X(t)^{-1}\,,\ \ L^*:\ W^*(s,t) = W(t,s)^T\,.$$

Note that $Y(a)^T = Y(t)^T X(t) = X(b)$ and thus, for the *characteristic matrices* C and D,

$$C = R_a + R_b X(b) = R_a + R_b Y(a)^T\,,\ \ D = S_a Y(a) + S_b = S_a X(b)^T + S_b \tag{5.52}$$

and, for the solutions,

$$\begin{aligned}x(t) &= X(t)\left\{x(a) + \int_a^t X(s)^{-1} f(s)\,ds\right\}\\ &= Y^{-T}(t)\left\{Y^T(a)x(a) + \int_a^t Y(s)^T f(s)\,ds\right\}\\ y(t) &= Y(t)\left\{y(a) + \int_a^t Y(s)^{-1} f(s)\,ds\right\}\\ &= X^{-T}(t)\left\{X^T(a)y(a) + \int_a^t X(s)^T f(s)\,ds\right\}\end{aligned}$$

or more general

$$\begin{aligned}x(t;t_0,x_0) &= W(t,t_0)x_0 + \int_{t_0}^t W(t,s)f(s)\,ds\\ y(t;t_0,y_0) &= W(t_0,t)^T y_0 - \int_{t_0}^t W(s,t)^T f(s)\,ds\,.\end{aligned}$$

Lemma 5.15. *Let* $m = \operatorname{rank}[R_a,\ R_b]$ *and* $r = \operatorname{rank} C$. *Then*
(1°) $\dim \operatorname{Ker} L = \dim \operatorname{Ker} C = n - r$.
(2°) *Let the columns of* $\widetilde{C} \in \mathbb{R}^n{}_{n-r}$ *form a basis of* $\operatorname{Ker} C$ *then the columns of* $U(\,\cdot\,) = X(\,\cdot\,)\widetilde{C}$ *form a basis of* $\operatorname{Ker} L$.
(3°) $\dim \operatorname{Ker} L^* = \dim \operatorname{Ker} C^T = \dim \operatorname{Ker} D = m - r$, $\operatorname{rank} D = n - m - r$.
(4°) *Let the columns of* $\widehat{C} \in \mathbb{R}^m{}_{m-r}$ *form a basis of* $\operatorname{Ker} C^T$ *then the columns of* $\widetilde{D} := -R_b^T\widehat{C} = Y^{-1}(a)R_a^T\widehat{C}$ *form a basis of* $\operatorname{Ker} D$ *and the columns of* $V^d(\,\cdot\,) = Y(\,\cdot\,)\widetilde{D}$ *form a basis of* $\operatorname{Ker} L^*$.

Lemma 5.16. (1°) $\operatorname{Range} L$ *and* $\operatorname{Range} L^*$ *are closed.*
Relative to the above scalar product
(2°) $\operatorname{Range} L = [\operatorname{Ker} L^*]^\perp$, $[\operatorname{Range} L]^\perp = \operatorname{Ker} L^*$.
(3°) $\operatorname{Range} L^* = [\operatorname{Ker} L]^\perp$, $[\operatorname{Range} L^*]^\perp = \operatorname{Ker} L$.

Corollary 5.3. (1°) L *is a* FREDHOLM *operator with index* $n - m$.
(2°) L^* *is a* FREDHOLM *operator with index* $m - n$.

Accordingly, $n = m$ must hold in order that L is a FREDHOLM operator with index zero; the next-important case would be $m = n - 1$.

Example 5.8. (Linear Periodic Systems) Let $\mathcal{Y} = \mathcal{C}_{2\pi} = \{x \in \mathcal{C}(\mathbb{R};\mathbb{R}^n)\,,\ x(t+2\pi) = x(t)\}$ and $\mathcal{X} = \mathcal{C}^1_{2\pi} = \{x \in \mathcal{C}^1(\mathbb{R};\mathbb{R}^n\}\,,\ x(t+2\pi) = x(t)\}$. Then $\mathcal{X} \subset \mathcal{Y}$ and both are (non-reflexive) BANACH spaces with norms $\|x\| = \max|x(t)|$ and $\|x\|_1 = \|x\| + \|D_t\, x\|$ respectively. The operators $L,\ L^* \in \mathcal{L}(\mathcal{C}^1_{2\pi};\mathcal{C}_{2\pi})$ now satisfy

$$(y, Lx) - (L^*y, x) = 0$$

hence the formal adjoint operator L^* is already the adjoint operator and there is no troublesome determination of adjoint boundary conditions in periodic problems. The boundary conditions $x(0) - x(2\pi) = 0$ yield $R_a = -R_b = I$ therefore $\text{rank}[R_a, R_b] = n$. The system $Lx = \dot{x} + Ax = 0$ has a 2π-periodic solution $x \in \mathcal{C}^1_{2\pi}$ if and only if there exists a $0 \neq x_0 \in \mathbb{R}^n$ such that $x_0 = W(2\pi, 0)x_0$. Let the characteristic matrices C and D of (5.52) be written by the regular monodromy matrix $W(2\pi, 0)$,

$$C = (I - W(2\pi, 0))X(0), \quad D = (I - W(0, 2\pi)^T)Y(0)$$

where $W(0, 2\pi)^T = W(2\pi, 0)^{-T}$. Then we obtain directly

$$\begin{aligned} \text{Ker}(L) &= \{x = W(\cdot, 0)x_0,\ x_0 \in \text{Ker}(I - W(2\pi, 0))\}, \\ \text{Ker}(L^*) &= \{y = W(0, \cdot)^T y_0,\ y_0 \in \text{Ker}(I - W(2\pi, 0)^{-T})\}. \end{aligned}$$

Because $\dim \text{Ker}(I - M) = \dim \text{Ker}(I - M^{-T})$ for every regular matrix M, it follows that

$$\dim \text{Ker}(L) = \dim \text{Ker}(L^*) = \alpha < n$$

and thus L is now a FREDHOLM operator with *index zero*; again $\text{Range}(L) = [\text{Ker}(L^*)]^\perp$ as above.

(c) We consider the **Nonlinear Boundary Value Problem**

$$\begin{aligned} &D_t\, x(t) + A(t)x + f(t, \mu, x(t)) = 0 \in \mathbb{R}^n, \quad f(t, \mu, 0) = 0, \quad \mu \in \mathbb{R}, \\ &R_a x(a) + R_b x(b) = 0 \in \mathbb{R}^m, \end{aligned} \tag{5.53}$$

where all data shall be sufficiently smooth in interval $\mathcal{I} = [a,\, b]$.

Assumption 5.3. (1°) $\text{rank}\,[R_a,\, R_b] = m \geq n - 1$.
(2°) *Let the operators* $L \in \mathcal{L}(\mathcal{X}_b, \mathcal{Y})$ *and* $L^* \in \mathcal{L}(\mathcal{X}_b^*, \mathcal{Y})$ *have the form*

$$Lx := D_t x + A(t)x + f_x(t, \mu_0, 0)x, \quad L^*y := -D_t y + (A(t) + f_x(t, \mu_0, 0)]^T y$$

and

$$\begin{aligned} &\dim \text{Ker}\, L = \alpha \geq 1, \quad \text{Ker}\, L = \text{span}\{u_1, \ldots, u_\alpha\}, \quad U = [u_1, \ldots, u_\alpha] \\ &\dim \text{Ker}\, L^* = \beta, \qquad \text{Ker}\, L^* = \text{span}\{v^1, \ldots, v^\beta\}, \quad V^d = [v^1, \ldots, v^\beta]. \quad \square \end{aligned}$$

The asymptotic expansion

$$\begin{aligned} f(t, \mu, x) &= (\mu - \mu_0)B(t)\langle x\rangle + C(t)\langle x, x\rangle + g(t, \mu, x) \\ B(t) &= f_{\mu,x}(t, \mu_0, 0), \quad C(t) = f_{xx}(t, \mu_0, 0) \\ g(t, \mu, x) &= \mathcal{O}(|\mu|^2), \ |\mu| \to \mu_0, \quad g(t, \mu, x) = \mathcal{O}(\|x\|)^3), \ \|x\| \to 0 \end{aligned}$$

holds near a possible branching point $(\mu_0, x_0) = (\mu_0, 0)$. We use the scaled ansatz

$$\mu(\varepsilon) = \mu_0 + \varepsilon\xi(\varepsilon), \quad x(t, \varepsilon) = \varepsilon U(t)\zeta(\varepsilon) + \varepsilon^2 w(t, \varepsilon), \quad U^d w = 0 \tag{5.54}$$

and adopt the assumptions of Corollary 5.1. Then the pair of scaled reduced branching equation and operator equation reads after division by ε^2:

$$\begin{aligned}
\Phi(\xi_0,\zeta_0) &= V^d\Big[\xi_0 B\langle U\zeta_0\rangle + C_2\langle U\zeta_0, U\zeta_0\rangle\Big] = 0\\
\Psi(\xi_0,\zeta_0,w_0) &= Lw_0 + (I-Q)\big[\xi_0 B\langle U\zeta_0\rangle + C_2\langle U\zeta_0, U\zeta_0\rangle\big] = 0\\
U^d w_0 &= 0\,,\quad w_0 \in \mathcal{L}(\mathcal{X}_b,\mathcal{Y})\ (!).
\end{aligned} \tag{5.55}$$

Let e.g. $m = n$, and $\alpha = \dim\operatorname{Ker} L = 1$ then $\beta = \dim\operatorname{Ker} L^* = 1$, and L as well as L^* are FREDHOLM operators with index zero. Setting $\xi_0 := 1$, one then obtains a unique solution ζ_0 and w_0 under the classical assumption $v^1 B u_1 \neq 0$, cf. Sect. 5.1**(d3)**. (But also $\zeta_0 = 1$ may be set here without loss of generality.)

Supposing further that $m = n$ and $\alpha = 1$, a substitution of the expansion (5.54) into the differential system (5.53) yields directly an *overdetermined boundary problem* for w,

$$\begin{aligned}
Lw := D_t w + A(t)w + f_x(t,\mu_0,0)w &= -\xi B\langle u_1 + \varepsilon w\rangle\\
- \frac{1}{\varepsilon^2}\Big[C(t)\langle(\varepsilon u_1 + \varepsilon^2 w, \varepsilon u_1 + \varepsilon^2 w\rangle &+ g(t,\mu_0+\varepsilon\xi,\varepsilon u_1 + \varepsilon^2 w)\Big]
\end{aligned} \tag{5.56}$$

$$R_a w(a) + R_b w_b = 0\,,\quad (u_1, w) = 0\,.$$

The right side of (5.56) must be contained in Range L that a solution exists (since the projector $I-Q$ does not apply). By the Range Theorem 1.2, a *consistency condition* must therefore be fulfilled which is the branching equation of course,

$$\xi v^1 B(u_1 + \varepsilon w) + \varepsilon^{-2} v^1[C + R] = 0\,. \tag{5.57}$$

(If $v^1 B u_1 \neq 0$, also $v^1 B(u_1 + \varepsilon w) = 0$ for sufficiently small $|\varepsilon|$.) Instead of $v_1 \in \operatorname{Ker} L_d$ we can however choose any function $v \in \mathcal{Y}_d$ such that $y \in \operatorname{Range} L$ if and only if $(v, y) = 0$.

In a simple iterational device, the parameter ξ is calculated successively by (5.57) and inserted into (5.56) thereafter w is found by solving this system. The condition $(u_1, w) = 0$ leads to an overdetermined system as already mentioned but the complete system can nevertheless be solved by the *box scheme* or a similar device; cf. Sect. 2.5**(a)**. (Weber79) has proposed an interesting modification to solve this overdetermined system by applying a result of (Reid); see (Reid), problem III.10(3).

Example 5.9. A nonlinear BERNOULLI beam; (Golubitsky), vol. I, p. 296. s arc length, φ angle w.r.t. x-axis (neutral fibre in equilibrium), μ axial load, π length of the beam (normalized). The terminal points of the beam are constrained to lie on a line and to be fixed otherwise as in Sect. 7.2, Figure 7.3, *Case 4*. The displacement $y(s)$ of a point of the neutral fibre of the beam then reads:

$$y_1(s) = \int_0^s \cos\varphi(\sigma)\,d\sigma\,,\ \ y_2(s) = \int_0^s \sin\varphi(\sigma)\,d\sigma\,.$$

In a similar way as in mathematical pendulum, the differential form of the law of conservation of energy leads to a boundary value problem

$$\varphi''(s) + \mu\,\sin\varphi(s) = 0\,,\ \ \varphi'(0) = \varphi'(\pi) = 0\,.$$

Notations in context: $x(s) := \varphi(s)\,,\ \mathcal{X} = \{f \in \mathcal{C}^2[0,\pi]\,, f(0) = f(\pi) = 0\}\,,$ $\mathcal{Y} = \mathcal{C}[0,\pi]\,.$

$F(\mu,x)$	$=$	$x'' + \mu\,\sin x\,,\ x'(0) = 0\,,\ x'(\pi) = 0$	
$Lv = F_x^0 v$	$:=$	$v'' + \mu\,v$	linearization
μ_k	$=$	$(1+k)^2\,,\ k = 0,\ldots$	eigenvalues of F_x^0
$u_k(s)$	$=$	$\cos((1+k^2)^{1/2}s)\,,\ k = 0,\ldots$	eigenfunctions of F_x^0
(u,v)	$=$	$\int_0^\pi u(s)v(s)\,ds$	scalar product.

The operator L is self-adjoint (symmetric) hence $P = Q$ does hold for the both involved projection operators. We consider the situation at the first branching point (eigenvalue) $\mu^0 = 1$ and list up the specified data of Sect. 5.1:

$$\begin{aligned} u_1(s)\ &:\ [0,\pi] \ni s \mapsto \cos(s)\ \text{ single eigenvector of } \mu^0 \\ \operatorname{Ker} L = \mathbb{R}\cos\ &:\ [0,\,\pi] \ni s \mapsto r\cos(s)\,,\ r \in \mathbb{R} \text{ arbitrary} \\ P = Q\ &:\ \mathcal{E} \ni f \to u_1(u_1,f) = (f,\cos)\cos(\circ) \\ \operatorname{Ker} P &= \{f \in \mathcal{X}\,,\ (u_1,f) = 0\}\,. \end{aligned}$$

Of course, the problem has the trivial solution $(\mu,0) \in \mathbb{R}\times\mathcal{X}$ for all μ, and it is apparent that the branching is a simple pitchfork bifurcation.

Example 5.10. ((Crandall) & Rabinowitz). Consider the sufficiently smooth system

$$F(\mu,x) = Ax + f(x) + \mu(Bx + g(x)) = 0\,,\ f(x), g(x) = o(\|x\|)\,,\ \|x\| \to 0\,,$$

and let $L := A - \mu_0 B$ be a FREDHOLM operator with index zero and $\operatorname{Ker} L = \operatorname{span}\{u_1\}$. Let $v \in \mathcal{Y}_d$ be any function such that $y \in \operatorname{Range} L$ if and only if $v\,y = 0$ then, by Theorem 5.3, there exists a non-trivial solution of the form (5.16) whenever $vBu_1 \neq 0$.

Let $\mathcal{X} = \mathcal{Y} = \mathcal{C}([0,\pi])$ under maximum norm and $Lx = -(px')' + qx$ where p is continuously differentiable, positive and q is continuous on $[0,\,\pi]$. The dash denotes differentiation for the *space variable* s. As domain $\mathcal{D}$ of L we take e.g. $\mathcal{D}(L) = \{x \in \mathcal{C}^2([0,\,\pi])\,:\,x(0) = x(\pi) = 0\}$. Consider now the differential equation

$$F(\mu,x)(s) = Lx + f(s,x,x',x'') - \mu(u + g(s,x,x',x''))\,,\ F(\mu,0) = 0$$

and let μ_0 be an eigenvalue of L. Then L satisfies the assumptions of Theorem 5.3, as is well known. In particular, μ_0 is an algebraic simple eigenvalue and that Theorem is applicable.

Even more specially, consider the scalar boundary value problem

$$-x'' + h(x^2 + x'^2)x - \mu(x + k(x^2 + x'^2)x) = 0\,,\; x(0) = x(\pi) = 0 \qquad (5.58)$$

where h and k are smooth with $h(0) = 0$ and $k(0) = 0$. Then $L = -x''$ has the eigenvalue $\mu_0 = 1$ and $\mathrm{Ker}(L - \mu_0 I) = \mathrm{span}\{u_1\}$. To find the unique curve of solutions bifurcating from $(\mu_0, 0)$ we try $x = c\sin\, s$, c a constant. This gives $1 + h(c^2) - \mu(1 + k(c^2)) = 0$ and therefore $\mu(c) = (1 + h(c^2))/(1 + k(c^2))$. For testing numerical devices, we may choose $k = 0$, $h(c^2) = \exp(-1/c^2)\sin(1/c^2)$, $c \neq 0$, and $h(0) = 0$ following (Crandall).

It may be interesting to look for the above mentioned function $v \in \mathcal{Y}_d$. Because $\mathrm{Range}\, L = \{-x'' - x\,,\, x \in \mathcal{D}(L)\}$ we find by partial integration

$$\int_0^\pi \sin(s)\big(x''(s) + x(s)\big)\, ds = 0 \implies v(s) = \sin\, s\,.$$

References: (Langford77a), (Langford77b), (Langford78), (Weber79).

5.5 Hopf Bifurcation

Consider an *autonomous* differential system

$$\dot{x}(t) + g(\mu, x(t)) = 0 \in \mathbb{R}^n \qquad (5.59)$$

with real parameter μ and smooth vector field g. We look for periodic solutions branching off from a steady state solution or critical point $(\mu_0, x_0) \in \mathbb{R}^{n+1}$ with $g(\mu_0, x_0) = 0$. This phenomenon is called Hopf bifurcation after (Hopf) (1942) and is in the meanwhile one of the best investigated cases in bifurcation theory.

Roughly spoken, Hopf bifurcation arises at a critical point whenever two conjugate complex conjugate eigenvalues of $\nabla_x g(\mu, x)$ pass through the imaginary axis in dependence of the parameter μ. More exactly we have the following celebrated result which is, e.g., proved completely in (Golubitsky) vol. I, chap. VIII; but see also (Marsden76) and many others.

Theorem 5.6. *Let $\mathcal{I} = (\mu_0 - \varepsilon, \mu_0 + \varepsilon)$ be a non-empty parameter interval and suppose for the system (5.59) that:*
(1°) *There* exists *a smooth branch of steady state solutions $x_0(\mu)$ for $\mu \in \mathcal{I}$, i.e., $g(\mu, x_0(\mu)) = 0$,* and *$x_0(\mu_0))$ is an* isolated *solution of $g(\mu_0, x) = 0$.*
(2°) *The* Fréchet *derivative* $\mathrm{grad}_x\, g(\mu, x_0(\mu))$ *with $\mu \in \mathcal{I}$ has a pair of simple complex conjugate eigenvalues $\lambda_{1,2}(\mu) = \sigma(\mu) \pm i\omega(\mu)$ with $\omega(\mu_0) = \omega_0 \neq 0$, $\sigma(\mu_0) = 0$ and* $\mathrm{grad}_x\, g(\mu_0, x_0)$ *has no other eigenvalues lying on the imaginary*

axis also not the eigenvalue zero (simple eigenvalue condition).
(3°) $\sigma'(\mu_0) \neq 0$ *(eigenvalue crossing condition).*
Then there is a one-parametric family of periodic orbits bifurcating from the critical point (μ_0, x_0) .

Note also that (1°) follows from (3°) by the Implicit Function Theorem since $\mathrm{grad}_x\, g(\mu_0, x_0)$ is supposed to be regular.

Without loss of generality we do now shift the critical point into the origin $(0,0)$ and suppose also that $g(\mu, 0) = 0$. Then the system (5.59) can be written as

$$\dot{x}(t) + A(\mu)x(t) + f(\mu, x(t)) = 0\,, \quad f(\mu, 0) = 0\,,$$

where $f(\mu, x)$ contains the higher order terms in μ and x .

(a) Simple Examples

Example A The system

$$\dot{x} = A(\mu)x\,, \quad A(\mu) = \begin{bmatrix} \mu & -1 \\ 1 & \mu \end{bmatrix} \tag{5.60}$$

has the eigenvalues $\lambda_{1,2} = \mu \pm i$ and the phase portraits as in Fig. 5.7.

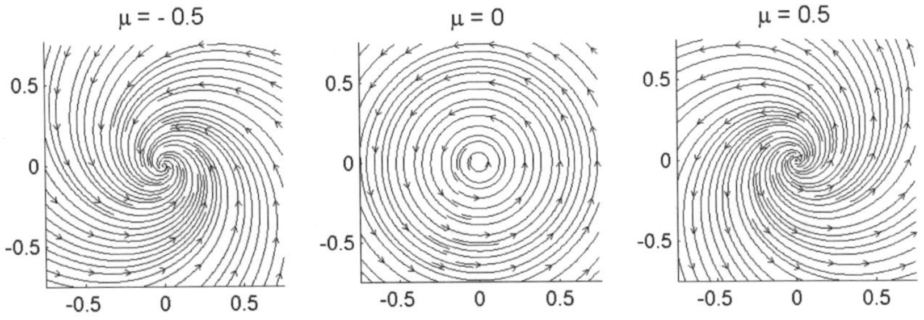

Figure 5.7. Example A

Exactly for $\mu \neq 0$ further solutions appear besides the trivial solution $x = 0$ and all have the same time-period 2π . □

The famous result of (Hopf) now says in essential that these fragile periodic solutions do *not* all disappear again if a nonlinear term is added in equation (5.60) but generically one solution is left over which however can (but must not) change its period in dependence of the parameter μ (if not re-scaled).

In *supercritical* Hopf *bifurcation* the nontrivial periodic solution is stable and the trivial solution unstable:

Example B The nonlinear system

$$\dot{x} - A(\mu)x + |x|x = 0 \in \mathbb{R}^2\,, \quad A(\mu) = \begin{bmatrix} \mu & -1 \\ 1 & \mu \end{bmatrix},$$

has the phase portraits as in Fig. 5.8.

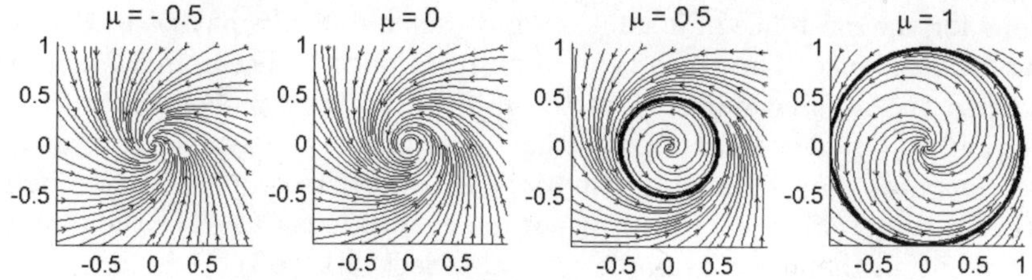

Figure 5.8. Example B

The figures show that a non-trivial periodic solution exists for $\mu > 0$ and all solutions besides the trivial one tend to that solution for $t \to \infty$, therefore the periodic solution is stable. □

In *subcritical* HOPF *bifurcation* the nontrivial periodic solution is unstable and the trivial solution stable:

Example C The nonlinear system

$$\dot{x} - A(\mu)x - |x|x = 0 \in \mathbb{R}^2\,, \quad A(\mu) = \begin{bmatrix} \mu & -1 \\ 1 & \mu \end{bmatrix},$$

has the phase portraits as in Fig. 5.9.

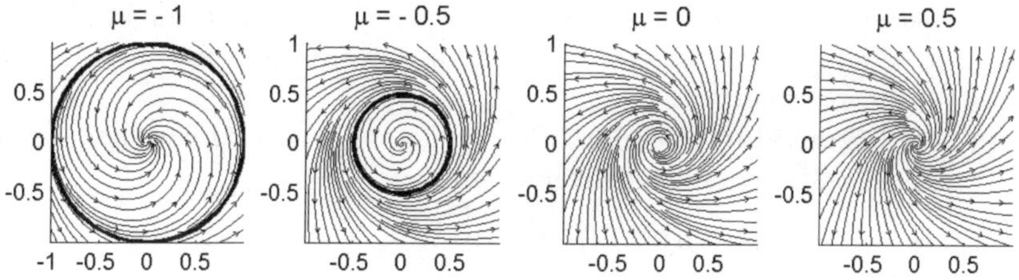

Figure 5.9. Example C

These figures show that a non-trivial periodic solution exists for $\mu < 0$ and all solutions besides the trivial one tend to that solution for $t \to -\infty$, therefore the periodic solution is unstable. □

Both examples have the exact solution $|\mu|(\cos t, \sin t)$ and the period does not change. The following theorem says that all *two-dimensional* HOPF bifurcations satisfying the assumption of Theorem 5.6 have more or less the same shape.

Theorem 5.7. *Adopt the assumptions of Theorem 5.6 and a second genericity condition. Then all systems $\dot{x} + A(\mu)x + f(\mu, x) = 0 \in \mathbb{R}^2$ are topologically equivalent near the origin to one of the systems*

$$\dot{x} - A(\mu)x \pm |x|^2 x = 0 \in \mathbb{R}^2\,, \quad A(\mu) = \begin{bmatrix} \mu & -1 \\ 1 & \mu \end{bmatrix}.$$

Proof see (Kuznetsov), Sect. 3.5; in particular Theorem 3.3. The word "generic" is frequently encountered in bifurcation theory and means that some *non-specific* cases shall be excluded. Genericity conditions are sufficient conditions for the existence of a solution and thus a difficult matter as all statements on existence. The *second* condition mentioned in Theorem 5.7 concerns of course the nonlinear part f of the differential system. For an exact formulation we however must replace that two-dimensional system equivalently by a *single complex* differential equation and study its properties thoroughly. Note also that $f(x) = |x|^2 x$ is invariant under rotations as well as the function $f(x) = |x|x$ in the above examples; see also Sect. 5.1 **(h3)**.

(b) Transformation to Uniform Period Functions with different periods do not form a vector space but the theory of FREDHOLM operators shall apply in the sequel. Therefore, and for numerical implementation, the system (5.59) must be transformed to a system with, say, period 2π. Let $v(s)$ be a T-periodic solution of the *autonomous, homogenous* differential system

$$\dot{v}(s) + g(\mu, v(s)) = 0 \in \mathbb{R}^n , \quad g(\mu, 0) = 0 .$$

Let $\omega = 2\pi/T$ be the *frequence*, then the substitution $s = t/\omega$ yields

$$\dot{x}(t) = \dot{v}(t/\omega)/\omega , \quad G(\mu, \omega, x)(t) := \omega\dot{x}(t) + g(\mu, x(t)) = 0 \tag{5.61}$$

for the transformed function $x(t) = v(t/\omega)$ and x is now a 2π-periodic solution. In the sequel we study therefore the re-scaled problem

$$F(\mu, \omega, x) := G(\mu_0 + \mu, \omega_0 + \omega, x) = 0$$

with *two parameters* for possible branching of 2π*-periodic* solution at the point $(\mu_0, \omega_0, 0)$. Note that the trivial solution $x_0 = 0$ having every period is here always the steady solution.

(c) An Eigenvalue Problem Recall Example 5.8 and consider the linearly perturbed 2π-periodic eigenvalue problem

$$(\omega_0 D_t + A + \mu B)x(t, \mu) = \lambda(\mu)x(t, \mu), \ \lambda(0) = 0 , \ x(\cdot, \mu) \in C^1_{2\pi} \tag{5.62}$$

($D_t = d/dt$) under the following assumptions and notations:

Assumption 5.4. (1°) *$i\omega_0 \neq 0$ is an α-fold semi-simple eigenvalue of the matrix $A \in \mathbb{R}^n{}_n$, $\alpha \in \mathbb{N}$.*
(2°) *Besides $\pm i\omega_0$, the matrix A has no further purely imaginary eigenvalues and also not the eigenvalue zero (A regular).*

The operators $L,\ L^* \in \mathcal{L}(C^1_{2\pi}; C_{2\pi})$, $L = \omega_0 D_t + A$, $L^* = -\omega_0 D_t + A^T$, have the same properties as in Example 5.8. In particular, they are FREDHOLM operators with index zero. Let

$$\begin{aligned} \mathrm{Ker}(L) &= \{u_1, \dots, u_\kappa\},\ U = [u_1, \dots, u_\kappa],\ u_i \ \text{ formal columns} \\ \mathrm{Ker}(L^*) &= \{v^1, \dots, v^\kappa\},\ V^d = [v^k]_{k=1}^\kappa \quad v^k \ \text{ formal rows}. \end{aligned}$$

Then $\kappa = 2\alpha$ by the following Lemma 5.17 and Range $L \cap \mathrm{Ker}\, L = \{0\}$ because $i\omega_0$ is a semi-simple eigenvalue, cf. Lemma 5.1.

Lemma 5.17. *Adopt Assumption 5.4.*
(1°) *There exist linear independent vectors* $\underline{c}_1, \ldots, \underline{c}_\alpha \in \mathbb{C}^n$ *and* $\underline{d}^1, \ldots \underline{d}^\alpha \in \mathbb{C}_n$ *such that*

$$\boxed{\begin{array}{l} A\underline{c}_k = -i\omega_0\underline{c}_k\,,\ \ \underline{c}_j^H\underline{c}_k = 2\delta^j{}_k\,, \\ \underline{d}^k A = i\omega_0\underline{d}^k\,,\ \ \overline{\underline{d}}^j\underline{c}_k = 2\delta^j{}_k\,,\ \ \underline{d}^j\underline{c}_k = 0\,,\ \ j,k = 1:\alpha \end{array}}\,. \tag{5.63}$$

(2°) *The eigensolutions of* L *resp. of* L^* *for the eigenvalue zero have the* real *form*

$$\begin{aligned} &u_k : t \mapsto \mathrm{Re}(\underline{c}_k e^{it})\,,\ k = 1:\alpha\,,\ u_k : t \mapsto \mathrm{Im}(\underline{c}_{k-\nu}e^{it})\,,\ k = \alpha+1:2\alpha\,, \\ &v^k : t \mapsto \mathrm{Re}(\underline{d}_k e^{it})\,,\ k = 1:\alpha\,,\ v^k : t \mapsto \mathrm{Im}(\underline{d}_{k-\nu}e^{it})\,,\ k = \alpha+1:2\alpha\,. \end{aligned}$$

See also (Golubitsky), Sect. 8.2 for $\alpha = 1$ and (Kielhoefer). (1°) is a simple conclusion from the fact that the left eigen- (and principal) vectors of a square matrix are a reciprocal basis to the right eigen- (and principal) vectors also in complex case. The second part follows by insertion. We assemble some properties where $\langle u, v\rangle = \dfrac{1}{2\pi}\displaystyle\int_0^{2\pi} \overline{u}(t)^T v(t)\,dt\,,\ u, v \in \mathcal{C}_{2\pi}\,,$ is the corresponding scalar product.

Lemma 5.18. *Adopt Assumption 5.4.* (1°)

$$\begin{aligned} \dot{u}_k &= -u_{\alpha+k}\,,\ \ \dot{u}_{\alpha+k} = u_k\,,\ \ k = 1:\alpha\,, \\ \dot{v}^k &= -v^{\alpha+k}\,,\ \ \dot{v}^{\alpha+k} = v^k\,,\ \ k = 1:\alpha\,. \end{aligned}$$

(2°) $u_1, \ldots, u_{2\alpha}$ *form a real orthonormal basis of* $\mathrm{Ker}\,L$.
(3°) $v^1, \ldots, v^{2\alpha}$ *form a basis of* $\mathrm{Ker}\,L^*$.
(4°) $v^1, \ldots, v^{2\alpha}$ *form the dual basis of* $u_1, \ldots, u_{2\alpha}$, $\langle v^j, u_k\rangle = \delta^j{}_k\,,\ j,k = 1:2\alpha\,.$
(5°) $B_1 := V^d\dot{U} = [\langle v^j,\, \dot{u}_k\rangle]_{j,k=1}^{2\alpha} = \begin{bmatrix} 0 & I \\ -I & 0 \end{bmatrix}$ *(symplectic normal form)*.

(6°) $B_2 := V^dBU = [\langle v^j,\, Bu_k\rangle]_{j,k=1}^{2\alpha} = \begin{bmatrix} \mathrm{Re}[\Lambda'(0)] & \mathrm{Im}[\Lambda'(0)] \\ -\mathrm{Im}[\Lambda'(0)] & \mathrm{Re}[\Lambda'(0)] \end{bmatrix}$

where B *is the matrix of (5.62) and* $\Lambda'(0) = \mathrm{diag}[\lambda_1'(0), \ldots, \lambda_\nu'(0)]$ *is a complex diagonal matrix.*

All assertions except the last one follow straightforward by insertion; for (6°) see the remark after Theorem 5.5. Now we can apply this theorem directly observing that B_2 is a normal matrix hence diagonalizable.

Corollary 5.4. *Let the system (5.62) satisfy Assumption 5.4. Then there exist* α *distinct smooth eigensolutions for sufficiently small* $|\mu|$.

(d) Scaled Problem Consider a two-parametric *autonomous* and *homogenous* differential system

$$F(\omega,\mu,x)(t) := (\omega_0+\omega)\dot{x}(t) + g(\mu,x(t)) = 0\,,\quad F(\omega,\mu,0) = 0 \in \mathbb{R}^n$$

where $F : \mathbb{R}^2 \times C^1_{2\pi} \to C_{2\pi}$ is sufficiently smooth. The origin $(0,0,0)$ is a possible branching point if the Fréchet derivative $F^0_x = \nabla_x F(0,0,0)$ has a nontrivial kernel. We suppose that the system has the form

$$F(\omega,\mu,x)(t) = Lx + \omega\dot{x}(t) + Ax(t) + \mu Bx(t) + C_r(x) + \text{ h.o.t.}\,,\quad Lx = \omega_0\dot{x} + Ax\,, \tag{5.64}$$

where $A = g^0_x$, $B = g^0_{\mu x}$, $C_r(\alpha x) = \alpha^r C_r(x) \neq 0$. We use the notations of **(c)** and adopt Assumption 5.4, then zero is a semi-simple eigenvalue of L therefore $\mathcal{Y} = \text{Range}(L) \oplus \text{Ker}(L)$ and we can assume without loss of generality that $V^d U = [< v^i, u_k >]^{\nu}_{i,k=1} = I$. The projector $Q : \mathcal{Y} \to \text{Range}(L)$ needed in the operator equation is given by $Q : u \mapsto Qu := UV^d u$.

For scaled ansatz we choose as above

$$\begin{aligned} (\omega(\varepsilon),\mu(\varepsilon)) &= \varepsilon^{r-1}(\kappa(\varepsilon),\eta(\varepsilon)),\\ x(t,\varepsilon) &= \varepsilon U(t)\zeta(\varepsilon) + \varepsilon^r w(t,\varepsilon),\ V(t)w(t,\varepsilon) = 0\,. \end{aligned} \tag{5.65}$$

Then Lemma 5.5 and Corollary 5.1 apply and yield:

Corollary 5.5. *For all ζ with non-zero $V^d BU\zeta$ the algebraic bifurcation equation of the system (5.64) is*

$$\Phi(\kappa,\eta,\zeta) := \kappa V^d\dot{U}\zeta + \eta V^d BU\zeta + V^d C_r(U\zeta) = 0\,. \tag{5.66}$$

Let it have a solution $(\kappa_0,\eta_0,\zeta_0) \in \mathbb{R}^{2\nu+2}$ and let the Jacobian $\Phi_{\kappa,\eta,\zeta}(\kappa_0,\eta_0,\zeta_0)$ have maximum rank 2ν. Then the problem (5.64) has a nontrivial solution (5.65) emanating from the bifurcation point $(\omega,\mu,x) = (0,0,0)$ for sufficiently small $|\varepsilon| > 0$.

Example 5.11. Adopt the assumptions of Theorem 5.6 for the system (5.64) in $\mathbb{R}^2$ then by Lemma 5.18

$$B_1 := V^d\dot{U} = \begin{bmatrix} 0 & 1\\ -1 & 0\end{bmatrix}\,,\quad B_2 := V^d BU = \begin{bmatrix} \sigma'(0) & \omega'(0)\\ -\omega'(0) & \sigma'(0)\end{bmatrix}\,.$$

Now $B_1\zeta_0$ and $B_2\zeta_0$ are linear independent for each $0 \neq \zeta_0 \in \mathbb{R}^2$ if and only if $\sigma'(0) \neq 0$ (eigenvalue crossing condition). In this case

$$\begin{bmatrix}\kappa_0\\ \eta_0\end{bmatrix} = [B_1\zeta_0, B_2\zeta_0]^{-1} V C_r(U\zeta_0) \tag{5.67}$$

yields a desired solution of (5.66). For $\alpha > 1$ we can augment the bifurcation equation (5.66) by the condition $|\zeta| = 1$ and consider μ as independent variable. Then the augmented system consists of $\alpha + 1$ equations for $\alpha + 1$ dependent variables (ω,ζ); see also (Kielhoefer).

(e) Discretization Let $(\omega, \mu, x) = (\omega_0, 0, 0) \in \mathbb{R} \times \mathbb{R} \times C^1_{2\pi}$ be a simple HOPF branching point of the homogeneous system

$$\boxed{F(\omega, \mu, x) := \omega\dot{x} + Ax + g(\mu, x) = 0 \in \mathbb{R}^n\,, \quad g(\mu, x) = \mu Bx + f(\mu, x)}\,, \tag{5.68}$$

and let $x(\circ\,, \mu) : t \mapsto x(t, \mu) \in \mathbb{R}^n$ be a 2π-periodic solution for sufficiently small $\mu > 0$.

(e1) Approximation by Differences Let

$$\boxed{t_k = k\tau\,, \quad 1 = 1 : 2m\,, \quad \tau = 2\pi/2m}\,, \tag{5.69}$$

be an uniform partition of the t-interval $[\tau, 2\pi]$ and let y_k, $k = 1 : 2m$, be a *numerical approximation* of $x(t_k) \in \mathbb{R}^n$. (MATLAB does not know the index zero.) Then we write

$$\begin{aligned} Y \quad &:= [y_1, \ldots, y_{2m}]^T \in \mathbb{R}^{2m\cdot n} \text{ (global node vector)}\,,\\ G(\mu, Y) &:= [g(\mu, y_1), \ldots, g(\mu, y_{2m})]^T \in \mathbb{R}^{2m\cdot n}. \end{aligned}$$

In an approximation of (5.68) by differences, the derivative $\dot{y}$ of y is replaced by a numerical differentiation rule. For instance, choose the backward differentiation rule of order $p = 2$ then

$$\dot{y}(t) = \frac{1}{2\tau}\,[3y(t) - 4y(t-\tau) + y(t - 2\tau)] + \mathcal{O}(|\tau|^2)\,, \quad \tau \to 0\,. \tag{5.70}$$

Regarding the partition (5.69) and the required periodicity, a discrete system for the node vector Y is found:

$$\boxed{\omega(C \times I_n)Y + (I_{2m} \times A)Y + G(\mu, Y) = 0}\,, \tag{5.71}$$

where $P \times Q = [p^i{}_k Q]$ denotes the usual KRONECKER product. Further, $I_n \in \mathbb{R}^n{}_n$ denotes the unit matrix and $C \in \mathbb{R}^{2m}{}_{2m}$ is a *circulant* matrix of the form

$$C = \begin{bmatrix} c_1 & c_2 & c_3 & \cdots & c_{2m}\\ c_{2m} & c_1 & c_2 & \cdots & c_{2m-1}\\ c_{2m-1} & c_{2m-2} & c_1 & \cdots & c_{2m-2}\\ \vdots & \vdots & \vdots & \ddots & \vdots\\ c_2 & c_3 & c_4 & \cdots & c_1 \end{bmatrix}. \tag{5.72}$$

For instance in the backward differentiation rule (5.70)

$$[c_1, c_2, c_3, \ldots, c_{2m}] = \frac{1}{2\tau}\,[3, 0, \ldots, 0, 1, -4]\,, \quad \tau = 2\pi/2m\,. \tag{5.73}$$

Let $\varrho = e^{2\pi i/2m}$ be a primitive unit root then the JORDAN normal form $C = P\Lambda P^H$ of a circulant matrix C may be written explicitely:

$$\begin{aligned}
P &= [p^i{}_j], \quad p^i{}_j = (2m)^{-1/2}\varrho^{(i-1)(j-1)} && \text{FOURIER matrix}\\
\Lambda &= \operatorname{diag}(\lambda_1, \lambda_2, \ldots, \lambda_{2m}) && \text{matrix of eigenvalues of } C\\
\lambda_k &= \textstyle\sum_{l=1}^{2m} c_l[\varrho^{k-1}]^{l-1}, \quad k = 1:2m && \text{eigenvalues;}
\end{aligned}$$

cf. (M.Chen). If C regular, the inverse $C^{-1} = P\Lambda^{-1}P^H$ is likewise a circulant matrix and, by consequence, the system $Cx = b$ has the solution $x = P\Lambda^{-1}P^H b$ which can be found by the *fast* FOURIER *transformation* in particular time-saving way. (In example (5.73) and in **(e2)** the matrix C is however singular.)

The eigenvectors u and v with the property $Cu = cu$ and $Av = av$ satisfy

$$a(C \times I_n)(u \times v) \pm c(I_{2m} \times A)(u \times v) = \begin{cases} 2ac(u \times v) & \text{for “+”} \\ 0 & \text{for “-”} \end{cases} .$$

Let now $a = i\omega_0$ be a pure imaginary eigenvalue of A and $c = i\gamma$ a pure imaginary eigenvalue of C then ac is real and $\operatorname{Im}(u \times v)$ is a real eigenvector of the real matrix $\omega_0(C \times I) - \gamma(I \times A)$ to the eigenvalue zero. Altogether, the eigenvectors of the leading matrix in (5.71) can be found without difficulty whenever a characteristic pair of A with imaginary eigenvalue is known; this is however exactly the situation in HOPF bifurcation. Unfortunately the matrix C of example (5.73) does not have an eigenvalue $i\gamma$ with real $\gamma \neq 0$ therefore we use the discretization of the exact eigenvectors of Lemma 5.18(2°) for numerical implementation.

(e2) Periodic Approach Suppose that the solution x of (5.68) has a absolutely continuous derivative, then the (unknown) solution x and the function g in (5.68) can be expanded into a real-valued FOURIER series with complex components,

$$\begin{aligned}
x(t) &= \textstyle\sum_{j=-\infty}^{\infty} x_j^*\, e^{ijt}, \quad x_{-j}^* = \overline{x}_j^*,\\
g(\mu, x(t)) &= \textstyle\sum_{j=-\infty}^{\infty} g(\mu, x)_j^*\, e^{ijt}, \quad g(\mu, x)_j^* = \frac{1}{2\pi}\int_0^{2\pi} g(\mu, x(t))\, e^{-ijt} dt\,.
\end{aligned}$$

We insert the FOURIER series of x and g into the system (5.68) and obtain an equivalent system for the FOURIER coefficients

$$\begin{aligned}
F(\omega, \mu, x)(t) &= \sum_{j=-\infty}^{\infty} \left[i\omega j x_j^* + A x_j^* + g(\mu, x)_j^*\right] e^{ijt} = 0\\
0 &= i\omega j x_j^* + A x_j^* + g(\mu, x)_j^*\,,\ j = -\infty, \ldots, \infty\,;
\end{aligned} \tag{5.74}$$

cf. also (Dellnitz).

Of course, it is summed up over a *finite* index set for numerical approach, preferably over $j = (1-m) : m$ in the present case. The resulting procedure is then the well-known RITZ method w.r.t. an expansion relative to eigenvectors of the differential operator d/dt in the present case of 2π-periodic functions. We refer to Sect. 12.4 in particular to (12.8) and write

$$y(t) = \sum_{j=1-m}^{m-1} y_j^* \, e^{ijt} + \frac{1}{2} y_m^* (e^{imt} + e^{-imt}) \in \mathbb{R}^n$$

for the numerical equivalent of the exact solution x. Then, by (5.74), the finite-dimensional system for the $2m$ unknown FOURIER coefficients reads:

$$\begin{aligned} i\omega j y_j^* + A y_j^* + g(\mu, y)_j^* &= 0\,, \quad j = (1-m) : (m-1) \\ 0 + A y_m^* + g(\mu, y)_m^* &= 0\,. \end{aligned} \tag{5.75}$$

Remember that, by Lemma 2.1, the composed trapezoidal rule is first choice in numerical integration of (smooth) periodic functions. Application yields

$$g(\mu, y)_j^* \sim \frac{1}{2m} \sum_{k=0}^{2m-1} g(\mu, y_k)\, e^{-ijk\tau}\,, \quad j = (1-m) : m\,,$$

by regarding $y(0) = y(t_0) = y(t_{2m}) = y(2\pi)$. Let conveniently

$$\begin{aligned} Y^* &:= [y_{1-m}^*, \ldots, y_m^*]^T \in \mathbb{R}^{2m \cdot n} \\ G(\mu, Y)^* &:= [g(\mu, y)_{1-m}^*, \ldots, g(\mu, y)_m^*]^T \\ D &:= i \operatorname{diag}[1-m, \ldots, m-1, 0]\,, \end{aligned}$$

then the rules (12.4) of discrete FOURIER tranformation apply and yield

$$Y = (Q \times I_n) Y^*\,, \quad Y^* = \frac{1}{2m} (Q^H \times I_n) Y\,. \tag{5.76}$$

By this way, the finite-dimensional approach of (5.67) has eventually the form

$$\begin{aligned} 0 &= \omega (D \times I) Y^* + (I \times A) Y^* + G(\mu, Y)^* \\ G(\mu, Y)^* &= \frac{1}{2m} (Q^H \times I) G(\mu, Y)\,. \end{aligned} \tag{5.77}$$

In order to find a system for the global node vector Y itself, we use (5.76) and multiply the first equation of (5.77) by $Q \times I$:

$$\boxed{\omega(\widetilde{C} \times I) Y + (I \times A) Y + G(\mu, Y) = 0}\,. \tag{5.78}$$

Recall that $(A \times B)(C \times D) = AC \times BD$ for compatible matrices, therefore the matrix $\widetilde{C}$ reads more detailed

$$\widetilde{C} = [c^j{}_k]_{j,k=0}^{2m-1} := \frac{1}{2m} Q\, D\, Q^H\,. \tag{5.79}$$

This matrix stands now in the discrete system for the differential operator d/dt. It is real, skew-symmetric as well as circulant, and it has $2m$ distinct entries

$$c^j{}_k = -\frac{1}{m}\sum_{l=1}^{m-1} l \ \sin(l(j-k)\tau)\,, \ \ j,k = 0:(2m-1)\,.$$

Note that the elements of D are the eigenvalues of $\widetilde{C}$ by (5.79) hence $\widetilde{C}$ has the double eigenvalue zero and purely imaginary eigenvalues else. The corresponding complex eigenvectors are the columns of Q. Accordingly this circulant matrix satisfies now the assumptions of **(e1)**.

(f) Numerical Solution We consider again a simple HOPF bifurcation of the system (5.68)

$$F(\omega,\mu,x) := Lx + \omega\dot{x} + \mu Bx + f(\mu,x) = 0 \in \mathbb{R}^n\,, \ \ Lx = \omega_0\dot{x} + Ax\,,$$

at the point $(\omega,\mu,x) = (0,0,0)$ where the *regular* matrix A has *exactly two* non-zero imaginary eigenvalues $\pm\, i\omega_0$ and the operator L has the double semi-simple eigenvalue zero. The following algorithm supplies a non-trivial solution germ by direct iteration. Updating of the parameters ω and μ follows in a similar way as in (5.67) whereas the component vector ζ remains fixed. Convergence follows for sufficiently small $|\varepsilon|$ from the Contraction Theorem in a similar way as in Lemma 5.6 by (Keener74). Ensuing a simple continuation method is applied again w.r.t. the parameter μ where (ω,x) are dependent variables with the aim that the solution leaves the solution germ and is transferred to some larger geometric orders of magnitude. Different initial vectors ζ with same absolute value produce only a *phase translation* and one of the both branching equations suffices for investigation of a possible stability transfer from the trivial solution (Golubitsky).

START: Choose tolerance *tol*, set $\widetilde{\mu} = (\omega,\mu) = 0\,,\ w = 0 \in \mathbb{R}^n\,.$
(1°) Find the associated matrices $U = [u_1,u_2]$ and $V^d = [v^1,v^2]$ of eigensolutions.
(2°) Choose a vector $\zeta \in \mathbb{R}^2$ with $|\zeta| = 1$ and $\varepsilon > 0$ sufficiently small, set $u_0 = \varepsilon U\zeta$ and $x = u_0\,.$
WHILE NOT $|F(\widetilde{\mu},x)| \leq tol$
(1°) Set $H = [V^d\dot{x}, V^dBx] \in \mathbb{R}^2{}_2$. If H regular, solve

$$H\widetilde{\mu}^* = -V^d f(\mu,x)\,,\ \omega = \widetilde{\mu}_1^*\,,\ \mu = \widetilde{\mu}_2^*\,.$$

If $[V^d\dot{u}_0, V^dBu_0]$ singular: STOP (algorithm fails), try a new ζ.
(2°) Solve

$$L\langle w\rangle = -\omega\dot{x} - \mu Bx - f(\mu,x)\,,\ V^d\langle w\rangle = 0\,.$$

and set $x = u_0 + w\,.$
END
In case of divergence try a smaller ε.

Of course, the algorithm is applied to one of the above *discretizations.* Then, for instance,

$$x = Y\,, \quad L = \omega_0(C \times I_n) + (I_{2m} \times A) \in \mathbb{R}^{2m+n}{}_{2m+n}$$

and the operator D_t of derivation for t is replaced by the operator $C \times I_n$; see the procedures `hopf_bdf.m` and `hopf_trig.m` in the MATLAB suite of this chapter.

(g) Examples We present same well-known examples with simple HOPF bifurcation at the point $(\omega, \mu, x) = (\omega_0, 0, 0)$ having the homogenous form

$$\omega\dot{x} + (A + \mu B)x + f(\mu, x) = 0\,, \quad f(\mu, 0) = 0\,, \tag{5.80}$$

where $f(\mu, x) = o\,(|\mu|)\,,\ f(\mu, x) = o\,(|x|)\,.$

Example 5.12. VAN DER POL's equation (Hairer), vol. I, p. 107 (Fig. 5.10).

$$\boxed{\begin{aligned} &\omega\dot{x}_1 - x_2 &&= 0 \\ &\omega\dot{x}_2 + x_1 + \mu(x_1^2 - 1)x_2 &&= 0\,, \quad \mu > 0 \end{aligned}}\;.$$

We multiply both equations by $\sqrt{\mu}$ and write $x := \sqrt{\mu}x$ then

$$A = \begin{bmatrix} 0 & -1 \\ 1 & 0 \end{bmatrix}, \quad B = \begin{bmatrix} 0 & 0 \\ 0 & -1 \end{bmatrix}, \quad f(\mu, x) = \begin{bmatrix} 0 \\ x_1^2 x_2 \end{bmatrix},$$
$$\omega_0 = 1\,, \quad \underline{c}_1 = \underline{d}_1 = [1,\, i]^T\,.$$

Example 5.13. Feedback inhibition model (Glass), (Langford77a) (Fig. 5.11).

$$\omega\dot{x} + Ax + h(\mu, x) = 0\,,$$

$$A = \begin{bmatrix} 1 & 0 & 2 \\ -2 & 1 & 0 \\ 0 & -2 & 1 \end{bmatrix}, \quad h(\mu, x) = \begin{bmatrix} g(\mu, x_3) \\ -g(\mu, x_1) \\ -g(\mu, x_2) \end{bmatrix}$$

$$g(\mu, x) = \frac{1}{2}\frac{(1 + 2x)^{\mu+4} - 1}{(1 + 2x)^{\mu+4} + 1} - 2x = \frac{1}{2}\mu x - 2x^2 - 8x^3 + \cdots\,.$$

Consequently

$$B = \frac{1}{2}\begin{bmatrix} 0 & 0 & 1 \\ -1 & 0 & 0 \\ 0 & -1 & 0 \end{bmatrix}, \quad f(\mu, x) = \begin{bmatrix} g(\mu, x_3) - \mu x_3/2 \\ -g(\mu, x_1) + \mu x_1/2 \\ -g(\mu, x_2) + \mu x_2/2 \end{bmatrix}.$$

The eigenvalues of A are $3,\ \pm i\sqrt{3}$. Recalling the system (5.63), by consequence, $\omega_0 = \sqrt{3}\,,\ \underline{c}_1 = \underline{d}_1 = [i\sqrt{3} - 1,\ \sqrt{3} + 1,\ 2]^T/\sqrt{6}\,.$

Example 5.14. Small brusselator (Hairer) vol. I, p. 112 (Fig. 5.12).

$$\boxed{\begin{array}{ll} \omega\dot{x}_1 + (b+1)x_1 - x_1^2 x_2 - a = 0\,, & a \neq 0\,, \\ \omega\dot{x}_2 - bx_1 + x_1^2 x_2 \qquad\qquad = 0 & \end{array}}\,.$$

Singular points are $(x_1, x_2) = (a, b/a)$ and HOPF bifurcation arises for $b = a^2 + 1$. We set $b = a^2 + 1 + \mu$ and then obtain

$$\begin{aligned} \omega\dot{x}_1 + (a^2 + 2 + \mu)x_1 - x_1^2 x_2 - a &= 0 \\ \omega\dot{x}_2 - (a^2 + 1 + \mu)x_1 + x_1^2 x_2 \quad &= 0\,. \end{aligned}$$

The value $\omega = 0$ leads to $x_1 = a$ and $x_2 = (a^2 + 1 + \mu)/a$. A substitution of $x_1 = a + u$ and $x_2 = v + (a^2 + 1 + \mu)/a$ then yields a problem with agreeable branching point $(\mu, x) = (0, 0)$:

$$\boxed{\begin{array}{l} \omega\dot{u} - (a^2 + \mu)u - a^2 v - (a + \mu a^{-1} + a^{-1})u^2 - 2auv - u^2 v \qquad = 0 \\ \omega\dot{v} + (a^2 + 1 + \mu)u + a^2 v + (a + \mu a^{-1} + a^{-1})u^2 + 2auv + u^2 v = 0 \end{array}}\,.$$

The basic equation (5.80) thus has the entries:

$$A = \begin{bmatrix} -a^2 & -a^2 \\ a^2 + 1 & a^2 \end{bmatrix}, \quad B = \begin{bmatrix} -1 & 0 \\ 1 & 0 \end{bmatrix},$$

$$f(\mu, x) = \begin{bmatrix} -(a + \mu a^{-1} + a^{-1})x_1^2 - 2ax_1x_2 - x_1^2 x_2 \\ (a + \mu a^{-1} + a^{-1})x_1^2 + 2ax_1x_2 + x_1^2 x_2 \end{bmatrix}.$$

One finds for the respective relations (5.63) that $\omega_0 = a$, $\underline{c}_1 = \gamma\,[a/(i-a),\ 1]^T$, $\underline{d}_1 = \delta\,[(a^2+1)/(a^2+ia),\ 1]^T$ where $\gamma^2 = 2(a^2+1)/(2a^2+1)$ und $\delta = (1-ia)/\gamma$.

Example 5.15. Full brusselator (Hairer) vol. I, p. 114 (Fig. 5.13).

$$\boxed{\begin{aligned} \omega\dot{x}_1 - x_1^2 x_2 + (x_3 + 1)x_1 - 1 &= 0 \\ \omega\dot{x}_2 - x_1 x_3 + x_1^2 x_2 \qquad &= 0 \\ \omega\dot{x}_3 + x_1 x_3 - a \qquad\qquad &= 0 \end{aligned}}\,.$$

The value $\omega = 0$ yields $x = [1,\ a,\ a]^T$ for singular points. The substitution $x_1 = 1 + u$, $x_2 = a + v$, $x_3 = a + w$ then leads to a system with trivial solution $x = 0$,

$$\omega\dot{x} + A(a)x + h(a, x) = 0\,,$$

where

$$A(a) = \begin{bmatrix} 1 - a & -1 & 1 \\ a & 1 & -1 \\ a & 0 & 1 \end{bmatrix}, \quad h(a, x) = \begin{bmatrix} -ax_1^2 - 2x_1x_2 - x_1^2 x_2 + x_1 x_3 \\ ax_1^2 + 2x_1x_2 + x_1^2 x_2 - x_1 x_3 \\ x_1 x_3 \end{bmatrix}.$$

HOPF bifurcation occurs at $a_0 = (9 - \sqrt{17})/4 = 1.21922\ldots$. Therefore one has to insert the translation $a = a_0 + \mu$ in (5.61).

Example 5.16. A LORENTZ equation (Seydel94) (Fig. 5.14).
Notations:

$$\begin{array}{lll} P_r & = \text{PRANDTL number} & (P_r = 16) \\ R_a & = \text{relative RAYLEIGH number} & (R_{a,0} = 368/11) \\ b & = \text{constant} & (b = 4) \\ S & = (bR_a - b)^{1/2} & (S = (bR_{a,0} - b)^{1/2} \,. \end{array}$$

The differential system $\omega\dot{x} + h(R_a, x) = 0$,

$$\boxed{\begin{array}{ll} \omega\dot{x}_1 + P_r(x_1 - x_2) & = 0 \\ \omega\dot{x}_2 + x_1x_3 - R_a x_1 + x_2 = 0 \\ \omega\dot{x}_3 - x_1x_2 + bx_3 & = 0 \end{array}}$$

has the singular points $x_1 = [0,\, 0,\, 0]$, $x_2 = [S,\, S,\, R_a - 1]$, $x_3 = [-S,\, -S,\, R_a - 1]$. For instance, for the above specified data, the gradient $\text{grad}_x\, h(R_a, x_2)$ has a pair of imaginary eigenvalues $\pm i\omega_0$, and HOPF bifurcation with parameter $R_a = R_{a,0} + \mu$ arises by consequence. The translation $x = x_2 + u$ yields the system $\omega\dot{u} + A(R_a)u + h(u) + c = 0$ where the constants are chosen such that $c = 0$:

$$A(\mu) = \begin{bmatrix} P_r & -P_r & 0 \\ -1-\mu & 1 & S \\ -S & -S & b \end{bmatrix}, \quad h(u) = \begin{bmatrix} 0 \\ u_1u_3 \\ -u_1u_2 \end{bmatrix}, \quad c = \begin{bmatrix} 0 \\ S(R_a - 1) - R_a S + S \\ -S^2 + b(R_a - 1) \end{bmatrix}$$

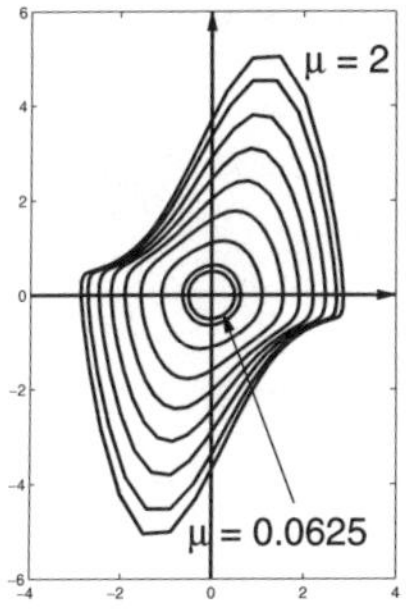

Figure 5.10. Example 5.12

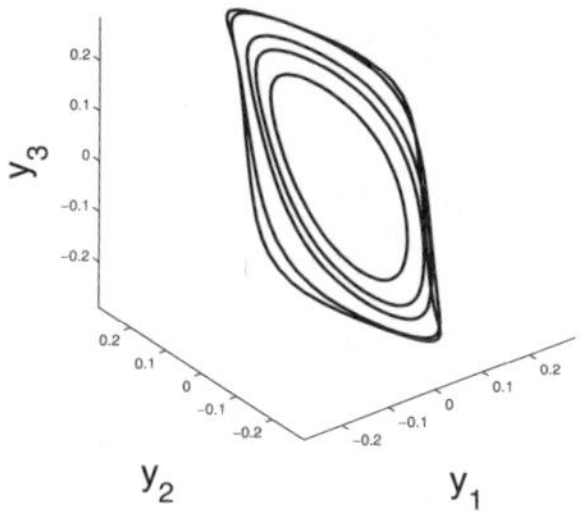

Figure 5.11. Example 5.13

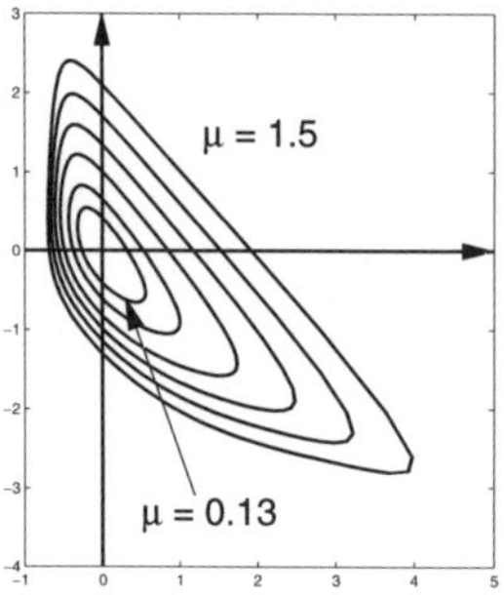

Figure 5.12. Example 5.14

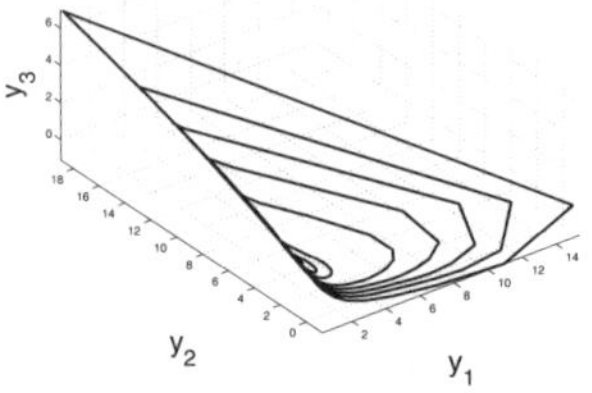

Figure 5.13. Example 5.15

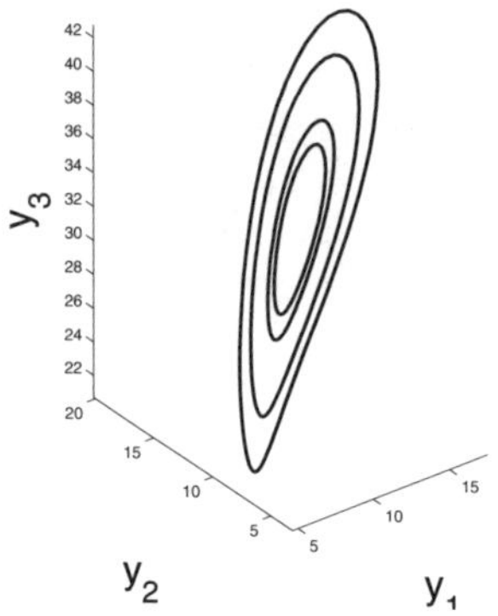

Figure 5.14. Example 5.16

References: (Golubitsky), (Kuznetsov), (Seydel94).

5.6 Numerical Bifurcation

In this section bifurcating solutions are calculated near singular points in *coordinate space* $\mathbb{R}^n$ therefore $\mathcal{X} = \mathcal{Y} = \mathbb{R}^n$, $\mathcal{X}_d = \mathcal{Y}_d = \mathbb{R}_n$, $U^d = U^T$, $V^d = V^T$, and $\langle a, b\rangle = a\, b$ where $a \in \mathbb{R}_n$ und $b \in \mathbb{R}^n$.

(a) We look for solutions of a system

$$F(\mu, x) = Ax + f(\mu, x) = 0 \in \mathbb{R}^n \tag{5.81}$$

with trivial solution, $F(\mu, 0) = 0$, at a point $(\mu_0, 0)$ where the linearized problem $(A + \nabla_x f(\mu_0, 0))x = 0$ has non-trivial solutions. Write briefly $B(\mu) = \nabla_x f(\mu, 0)$ then the *matrix* $L = A + B(\mu_0)$ is a FREDHOLM operator with index zero and $\alpha := \dim \operatorname{Ker} L \geq 1$. Of course $B(\mu) = \mu B$ whenever $f(\mu, x) = \mu B x$ + h.o.t. in μ and x. The nonlinear eigenvalue problem $Ax + \mu g(x) = 0$ is prepared in Sect. 5.2 **(c)**. For $\alpha = 1$ it can be solved by direct iteration as in the preceding section. For $\alpha > 1$ KEENER's procedure of Lemma 5.6 can be applied where however the nonlinear bifurcation equation must be solved in every step of iteration.

The following algorithm of DEMOULIN & CHEN (Demoulin) concerns systems (5.81) with $f(\mu, x) \neq \mu g(x)$ but involves only *linear* equations in each step of iteration. As above, let $U = [u_1, \ldots, u_\alpha] \in \mathbb{R}^n{}_\alpha$ and $V^T = [v^1; \ldots; v^\alpha] \in \mathbb{R}^\alpha{}_n$ be the matrices of of right and left eigenvectors such that $LU = 0 \in \mathbb{R}^n{}_\alpha$ and $V^T L = 0 \in \mathbb{R}^\alpha{}_n$; of course $V^T = U^T$ whenever L is symmetric.

(a1) $\alpha = 1$. Following the pattern of the LJAPUNOV-SCHMIDT reduction we consider the system $Lx + f(\mu, x) - B(\mu_0)x = 0$ and $v^1(f(\mu, x) - B(\mu_0)x)) = 0$ but, as proposed by (Demoulin), we look for solutions $x = \varepsilon u_1 + w$, $u^1 w = 0$ of the *perturbed* problem

$$Lw + f(\mu, x) - B(\mu_0)Bx + \varepsilon B_\mu(\mu_0)u_1(\Delta\mu) = 0\,.$$

Then $\varepsilon v^1 B_\mu(\mu_0)\underline{u}_1(\Delta\mu) = v^1(B(\mu_0)x - f(\mu, x))$. Let $v^1 B_\mu(\mu_0)u_1 =: a \neq 0$ and $\mu^0 = \mu_0$, $w^0 = 0$, $x^0 = \varepsilon u_1 + w^0$. Choose $|\varepsilon|$ not too small and not too large. Repeat for $n = 0, 1, \ldots$

$$\begin{aligned} \Delta\mu &= (\varepsilon a)^{-1} v^1 (B(\mu_0)x^n - f(\mu^n, x^n)) \\ Lw^{n+1} &= -f(\mu^n, x^n) + B(\mu_0)x^n - \varepsilon B_\mu(\mu_0)u_1\Delta\mu\,, \quad u^1 w^{n+1} = 0 \\ \mu^{n+1} &= \mu^n + \Delta\mu\,, \ x^{n+1} = \varepsilon u_1 + w^{n+1}\,. \end{aligned}$$

(a2) For $\alpha > 1$ let $f(\mu, x) = B(\mu)x + C_r(\mu)\langle x\rangle$+h.o.t. in x where $C_r(\mu)$ is a r-linear mapping in x as previously introduced in this chapter. Let (μ^*, ζ^*) be an (isolated) solution of the algebraic bifurcation equation

$$\mu V^d B_\mu(\mu_0)\langle U\zeta\rangle + V^d C_r(\mu_0)\langle U\zeta\rangle = 0\,, \ \langle U\zeta, U\zeta\rangle = 1$$

such that the matrix

$$G := \begin{bmatrix} V^d B_\mu(\mu_0)\langle U\zeta\rangle & V^d[\mu B_\mu(\mu_0) + \nabla_x[C_r(\mu_0)\langle U\zeta\rangle]]U \\ 0 & 2\zeta^T U^T U \end{bmatrix} \in \mathbb{R}^{\alpha+1}{}_{\alpha+1}$$

is regular. Let w_0 be a solution of $Lw = -\mu^* B_\mu(\mu_0)\langle U\zeta^*\rangle - C_r(\mu_0)\langle U\zeta^*\rangle$, $U^T w = 0$, and let $\zeta^0 = \zeta^*$, $\mu^0 = \mu_0 + \varepsilon^{r-1}\mu^*$, $w^0 = \varepsilon^r w_0$, $x^0 = \varepsilon U\zeta^0 + w^0$. Choose $|\varepsilon|$ not too small and not too large. Repeat for $n = 0, 1, \ldots$

$$\begin{bmatrix} \varepsilon V^d B_\mu(\mu_0)\langle U\zeta^*\rangle & \varepsilon V^d\big[(\mu^* - \mu_0)B_\mu(\mu_0) + \nabla_x[C_r(\mu_0)\langle \varepsilon U\zeta^*\rangle]\big]U \\ 0 & \zeta^{*T} U^T U \end{bmatrix} \begin{bmatrix} \Delta\mu \\ \Delta\zeta \end{bmatrix}$$

$$= \begin{bmatrix} \mu_0 V^d B(\mu_0)x^n - V^d f(\mu^n, x^n) \\ 2^{-1}(1 - \zeta^{nT} U^T U\zeta^n) \end{bmatrix} \in \mathbb{R}^{\alpha+1}$$

$$\begin{aligned} Lw^{n+1} = &-f(\mu^n, x^n) + B(\mu_0)x^n \\ &- \varepsilon(\Delta\mu)B_\mu(\mu_0)\langle U\zeta^*\rangle - \varepsilon(\mu^* - \mu_0)B\langle U(\Delta\zeta)\rangle \\ &- \varepsilon\nabla_x[C_r(\mu_0)\langle \varepsilon U\zeta^*\rangle]\langle U(\Delta\zeta)\rangle\,, \ U^T w^{n+1} = 0 \end{aligned}$$

$$\mu^{n+1} = \mu^n + \Delta\mu\,, \ \zeta^{n+1} = \zeta^n + \Delta\zeta\,, \ x^{n+1} = \varepsilon U\zeta^{n+1} + w^{n+1}\,.$$

(b) The following simple algorithm for general branching point (μ_0, x_0) of $F(\mu, x) = 0$ does not obey fully the agreements of the LJAPUNOV-SCHMIDT reduction but supplies non-trivial solution germs in many cases:

(1°) Find the matrices U and V^T of right and left eigenvectors to the eigenvalue zero of the matrix $L = \nabla_x F(\mu_0, x_0)$.
(2°) Solve the algebraic branching equation, and let ξ_0, ζ_0 be an isolated solution with non-zero ζ_0.
(2°) Choose $\varepsilon > 0$ properly (neither too small nor too large) and $(\lambda, w) = (0, 0)$ for start value.
(3°) Solve the system

$$H(\lambda, w) := \begin{bmatrix} F(\mu_0 + \lambda, x_0 + \varepsilon U\zeta_0 + w) \\ U^T w \end{bmatrix} = 0$$

for (λ, w) by the damped NEWTON method, set $\mu = \mu_0 + \lambda$, $x = x_0 + \varepsilon U\zeta_0 + w$.

The system $H(\lambda, w) = 0$ is overdetermined in case $\alpha = \dim \operatorname{Ker} L > 1$ entailing, however, no difficulties in MATLAB implementation. Instead of solving the bifurcation equation we may also *try* a specific eigenvector u_k for start tangent directly such that $x = x_0 + \varepsilon u_k + w$. The procedure can be checked by solving e.g. Example 5.4 of (Crandall) in Sect. 5.1 where also secondary bifurcation occurs.

(c) A Classic Example Let x be the unknown exact solution further on to preserve the context to the above introduced notations. Let $\Omega = \{(\xi, \eta) \in \mathbb{R}^2,\ 0 < \xi < a,\ 0 < \eta < b\}$ be a *rectangle* in (ξ, η)-plane. We consider the simple boundary value problem for POISSON's equation

$$F(\mu, x) = -\Delta x + f(\mu, x) = 0\,,\ (\xi, \eta) \in \Omega\,;\ x = 0\,,\ (\xi, \eta) \in \partial\Omega\,, \tag{5.82}$$

(c1) At first the solution of the linear eigenvalue problem

$$-\Delta x = \lambda x\,,\ (\xi, \eta) \in \Omega\,;\ x = 0\,,\ (\xi, \eta) \in \partial\Omega\,, \tag{5.83}$$

is determined. The LAPLACE operator Δ is a self-adjoint operator, $-\Delta$ is positive definite under the above boundary conditins. A *separation* $x(\xi, \eta) = u(\xi)v(\eta)$ leads to the both ordinary boundary problems

$$-u_{\xi\xi} = \lambda_1 u\,,\ u(0) = u(a) = 0\,,\ \ -v_{\eta\eta} = \lambda_2 v\,,\ v(0) = v(b) = 0\,,\ \lambda = \lambda_1 + \lambda_2\,,$$

of which the solutions have already been found in Sect. 2.4**(h)**. Together there result the characteristic pairs for (5.83)

$$\boxed{[\lambda_{pq}, u_{pq}] = \left[\pi^2\left(\frac{p^2}{a^2} + \frac{q^2}{b^2}\right),\ \sin\left(p\pi\frac{\xi}{a}\right)\cdot\sin\left(q\pi\frac{\eta}{b}\right)\right]\,,\ p, q \in \mathbb{N}}\,, \tag{5.84}$$

and it can be shown that *all* characteristic pairs of the problem are obtained by this way.

(c2) Discretization of the two-dimensional problem (5.82) by the same way as in the one-dimensional problem (2.61) yields the discrete problem $CY = \Lambda Y$ with matrix

$$C = \frac{1}{\Delta\xi^2}[I_m \times A_n] + \frac{1}{\Delta\eta^2}[A_m \times I_n]$$

for the *node vector*

$$\begin{aligned} Y &= ({y^1}_1, \ldots, {y^1}_n, \ldots, {y^m}_1, \ldots, {y^m}_n)^T\,, \\ {y^j}_k &= y(k\Delta\xi, j\Delta\eta)\,,\ \Delta\xi = a/(n+1)\,,\ \Delta\eta = b/(m+1)\,,\ m, n \in \mathbb{N}\,. \end{aligned}$$

$I_m \in {\mathbb{R}^m}_m$ denotes the unit matrix again and $A_m \in {\mathbb{R}^m}_m$ the matrix of (2.62) *without* prefix $1/h^2$. (It would have been somewhat more problem related if the vector Y were be written as *matrix* with rows $[{y^i}_1, \ldots, {y^i}_n]$, $i = m, m-1, \ldots 1$, since then the position ${y^i}_k$ in the matrix would have corresponded to the position in the domain Ω.) Also the zero-boundary conditions are still assigned to complete the shape. Now, by Sect. 2.4**(h)**, we obtain directly the characteristic pairs of C:

$$\begin{aligned} C\,U_{pq} = C(\underline{v}_q \times \underline{u}_p) &= \Delta\xi^{-2}(\underline{v}_q \times A_n\underline{u}_p) + \Delta\eta^{-2}(A_m\underline{v}_q \times \underline{u}_p) \\ &= \sigma_p(\underline{v}_q \times \underline{u}_p) + \tau_q(\underline{v}_q \times \underline{u}_p) = \Lambda_{pq}(\underline{v}_q \times \underline{u}_p) \end{aligned}$$

where

$$\begin{aligned} \sigma_p &= \frac{4}{\Delta\xi^2}\sin^2\left(\frac{p\pi a}{2(n+1)a}\right) = \frac{4}{\Delta\xi^2}\sin^2\left(\frac{p\pi}{2a}\Delta\xi\right) \\ \tau_q &= \frac{4}{\Delta\eta^2}\sin^2\left(\frac{q\pi b}{2(m+1)b}\right) = \frac{4}{\Delta\eta^2}\sin^2\left(\frac{q\pi}{2b}\Delta\eta\right) \\ \underline{u}_p &= \left[\sin\left(\frac{kp\pi a}{(n+1)a}\right)\right]_{k=1}^n = \left[\sin\left(\frac{p\pi}{a}k\Delta\xi\right)\right]_{k=1}^n \\ \underline{v}_q &= \left[\sin\left(\frac{jq\pi b}{(m+1)b}\right)\right]_{j=1}^m = \left[\sin\left(\frac{q\pi}{b}j\Delta\eta\right)\right]_{j=1}^m\,. \end{aligned}$$

Accordingly, in this particular model problem, the characteristic pairs of the discretization can be found *analytically.* Moreover, the eigenvectors of the discrete problem coincide by exception with the exact eigensolutions at the node points — as already shown in Sect. 2.4. Multiple eigenvalues appear, e.g., if $a = b$ and und $p^2 + q^2 = r^2$, $r \in \mathbb{N}$.

(c3) Consider further on the problem $-\Delta x - \mu x = 0$ in unit square with homogenous zero Dirichlet boundary conditions. Then $\mu_0 = \lambda_{11} = 2\pi^2$ is the smallest eigenvalue of $A = -\Delta$ as well as $u_1(\xi, \eta) = \sin(\pi\,\xi)\sin(\pi\,\eta)$ and $v^1 = u_1$ are the right and left eigenvector. In multiple eigenvalues the start tangent depends normally on the nonlinear part f by Sect. 5.2 which then has to be specified; this is however not the case here by reasons of symmetry. We choose

$$-\Delta x + f(\mu, x) = 0\,, \quad f(\mu, x) = -\mu x + x^3 + o(\|x\|^3) \tag{5.85}$$

Then $\mu_0 = \lambda_{12} = \lambda_{21} = 5\pi^2$ is a double eigenvalue. This problem enjoys two symmetries in case higher order terms $o(\|x\|^3)$ neglected: The arguments ξ and η may be permutated and on the other side $F(\mu, -x) = -F(\mu, x)$. Letting $\xi_0 = \xi$ and $\zeta_0 = \zeta = (\zeta_1, \zeta_2)$, the algebraic branching equation of the analytic problem reads:

$$\Phi(\xi, \zeta) = \begin{bmatrix} \big(u_1, -\xi\zeta_1 u_1 + (\zeta_1 u_1 + \zeta_2 u_2)^3\big) \\ \big(u_2, -\xi\zeta_2 u_2 + (\zeta_2 u_1 + \zeta_2 u_2)^3\big) \end{bmatrix} = 0 \tag{5.86}$$

where $(u, v) = \int_\Omega uv\, dV$ denotes the usual scalar product. However, in discrete case

$$\Phi(\xi, \zeta) = \begin{bmatrix} -\xi\zeta_1 u_1^T u_1 + u_1^T(\zeta_1 u_1 + \zeta_2 u_2)^{[3]} \\ -\xi\zeta_2 u_2^T u_2 + u_2^T(\zeta_1 u_1 + \zeta_2 u_2)^{[3]} \end{bmatrix} = 0 \tag{5.87}$$

where $v^{[2]}$ shall denote the *pointwise* multiplication of a vector v by itself. The second system has to be multiplied still by $(\Delta\xi\Delta\eta)^{-1}$ in order that the results are comparable to the first system. In any case we obtain the system

$$-{a^i}_1\xi\zeta_i + {a^i}_2{\zeta_1}^3 + {a^i}_3{\zeta_1}^2\zeta_2 + {a^i}_4\zeta_1{\zeta_2}^2 + {a^i}_5{\zeta_2}^3 = 0\,, \quad i = 1, 2,$$

of which the components may be found *explicitely* in the present problem. Multiplying by 64, the result reads:

$$\begin{aligned} -16\xi\zeta_1 + 12\zeta_1{\zeta_2}^2 + 9{\zeta_1}^3 &= 0 \\ -16\xi\zeta_2 + 12{\zeta_1}^2\zeta_2 + 9{\zeta_2}^3 &= 0\,; \end{aligned} \tag{5.88}$$

cf. (Budden). It becomes apparent that the second equation follows from the first equation by permutation of ζ_1 and ζ_2 which also can be predicted by the underlying symmetry. Now, one finds for $\xi = 1$

$$\zeta_1 = \pm 4/3\,, \quad \zeta_2 = 0\,, \quad bzw. \;\; \zeta_2 = \pm 4/3\,, \quad \zeta_1 = 0\,. \tag{5.89}$$

whereby the possible start tangents needed in Algoritmus I are supplied.

On choosing $f(\mu, x) = -\mu x - x^3$ then $\xi = -1$ has to be inserted and then $\mu < \mu_0$.

(c4) Let for instance $\mu_0 = \lambda_{55} = \lambda_{17} = \lambda_{71} = 50\pi^2$ be a triple eigenvalue then the algebraic branching equation for (5.85) reads:

$$-\xi u_i^T u_i \zeta_i + u_i^T(\zeta_1 u_1 + \zeta_2 u_2 + \zeta_3 u_3)^3 = 0\,, \; i = 1, 2, 3\,,$$

resp. for $i = 1, 2, 3$

$$\begin{aligned} &-{a^i}_1\xi\zeta_i + {a^i}_2{\zeta_1}^3 + {a^i}_3{\zeta_1}^2\zeta_2 + {a^i}_4\zeta_1{\zeta_2}^2 + {a^i}_5{\zeta_2}^3 + {a^i}_6{\zeta_1}^2\zeta_3 \\ &+{a^i}_7{\zeta_2}^2\zeta_3 + {a^i}_8\zeta_1{\zeta_3}^2 + {a^i}_9\zeta_2{\zeta_3}^2 + {a^i}_{10}\zeta_1\zeta_2\zeta_3 + {a^i}_{11}{\zeta_3}^3 = 0\,. \end{aligned}$$

Then

$$\begin{aligned}
-16\xi\zeta_1 + 9{\zeta_1}^3 + 27\zeta_1{\zeta_2}^2 + 27\zeta_1{\zeta_3}^2 &= 0 \\
-16\xi\zeta_2 + 9{\zeta_2}^3 + 27\zeta_2{\zeta_3}^2 + 27\zeta_2{\zeta_1}^2 &= 0 \\
-16\xi\zeta_3 + 9{\zeta_3}^3 + 27\zeta_3{\zeta_1}^2 + 27\zeta_3{\zeta_2}^2 &= 0
\end{aligned} \; .$$

Again it becomes apparent that the solution of the second and third equation arises by permutation of the components ζ_i; the three different solutions are

$$\zeta_1 = \pm 4/3\,,\ \zeta_2 = \zeta_3 = 0 \quad \zeta_2 = \pm 4/3\,,\ \zeta_3 = \zeta_1 = 0 \quad \zeta_3 = \pm 4/3\,,\ \zeta_1 = \zeta_2 = 0\,.$$

In more general cases also some more components ζ_i may be different from zero.

The example fascinates by the possibility to solve the algebraic branching equations analytically. From numerical point of view it remains however less impressive since the solution is hardly impressed by modifications of the "right side" f; in essential the image remains always the same (in the tested problems). The algorithm of **(b)** has been applied to the three functions

$$f(\mu, x) = -\mu(x \pm x^3)\,,\ \ f(\mu, x) = -\mu \sin x\,,\ \ f(\mu, x) = -\mu x \pm x^3\,,$$

and thereafter a simple continuation w.r.t. the parameter μ has been carried out. In all three examples however the graphical output has been more or less the same. The problem behaves also similarily by applying mixed boundary conditions, e.g.,

$$x(0,\eta) = x(a,\eta) = 0\,,\ \ x_\eta(\xi,0) = x_\eta(\xi,b) = 0\,.$$

Example 5.17. With the specification of **(c)** in unit square, let $f(\mu, x) = \mu(-x + x^3)$ in system (5.82), further, let $m = n = 24$ and let $u_0 = \varepsilon u_1$ be the start tangent. $\widetilde{u}_1$ is the unnormed eigenvector, the normed eigenvector u_1 is illustrated in the respective figure "below" the solution.
Case 1: $u = 0$ on $\partial\Omega$.

	(a)	(b)	(c)
μ_0	$2\pi^2 \sim 20$	$5\pi^2 \sim 49$	$50\pi^2 \sim 493$
$\widetilde{u}_1$	$\sin(\pi\xi)\sin(\pi\eta)$	$\sin(\pi\xi)\sin(2\pi\eta)$	$\sin(\pi\xi)\sin(7\pi\eta)$
μ_{end}	~ 59	~ 142	~ 657
ε	15	15	10

Case 2: $u = 0$ for $x = 0$ and $x = 1$.

	(a)	(b)	(c)
μ_0	$2\pi^2 \sim 20$	$5\pi^2 \sim 49$	$50\pi^2 \sim 493$
$\widetilde{u}_1$	$\sin(\pi\xi)\cos(\pi\eta)$	$\sin(\pi\xi)\cos(2\pi\eta)$	$\sin(\pi\xi)\cos(7\pi\eta)$
μ_{end}	~ 84	~ 67	~ 538
ε	17	10	8

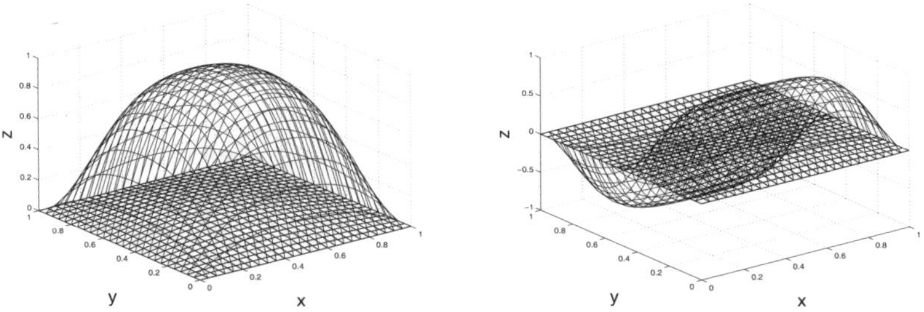

Figure 5.15. Case 1, (a), (b)

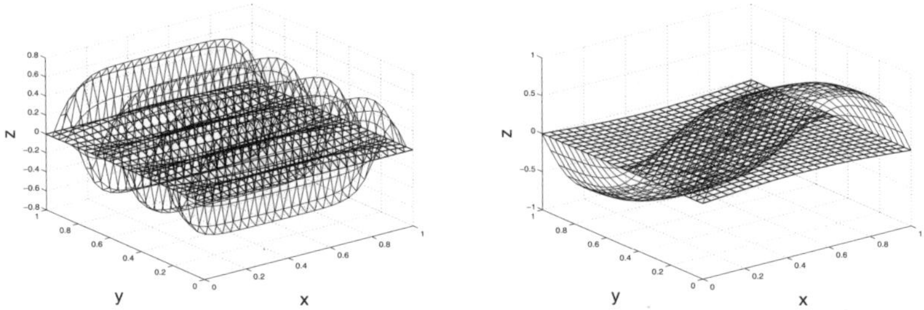

Figure 5.16. Case 1, (c), Case 2 (a)

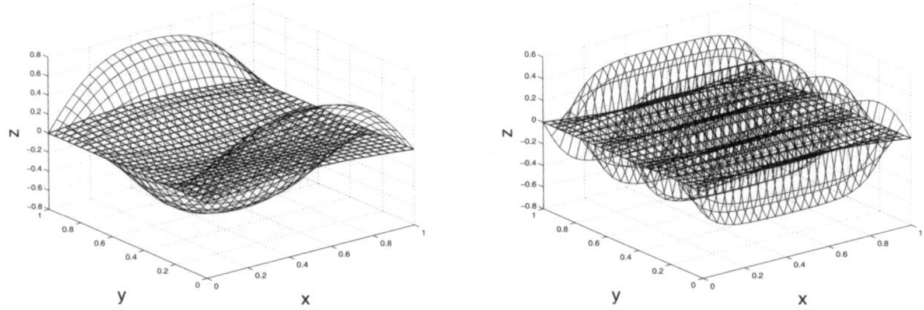

Figure 5.17. Case 2, (b), (c)

Example 5.18. See also (Keller72). Consider once more the problem $Lx + \mu Bx + C_r\langle x\rangle = 0$ where L is a *selfadjoint* operator such that $V^d = U^d$ and B symmetric and regular. Then the algebraic bifuraction equation is $\mu U^d B\langle U\zeta\rangle + U^d C_r\langle U\zeta\rangle = 0$. Suppose that $C_r\langle y\rangle = \operatorname{grad}\Phi(y)$ for some scalar function Φ as in the above examples. Let $y = U\zeta$ then $\zeta = U^T y$. Because Φ continuous, it attains its maximum and minimum at two distinct points y^* and y_* on the set $S = \{y\,,\ y^T By = 1\}$, and μ is the Lagrange multiplier of the problem $\Phi(y) = \text{extr!}$, $y^T By/2 = 1$. If r even then Φ odd and $\Phi(y) = -\Phi(-y)$ so that $y_* = -y^*$ and we have at least two distinct solutions (μ^*, y^*) and $(-\mu^*, -y^*)$. For r odd we have $\Phi(y) = \Phi(-y)$ and so $y^* \neq y_*$ unless $\Phi(y)$ konstant on S and every point on S is a solution. If not, we obtain at least four distinct solutions in this case.

5.7 Continuation

(a) Formulation of the Problem Let $F : \mathbb{R}^{n+1} \to \mathbb{R}^n$ be sufficiently smooth and let $x_0 \in \mathbb{R}^{n+1}$ be a point satisfying $F(x_0) = 0$ and $\dim \operatorname{Ker} F_x^0 = 1$. By the Implicit Function Theorem then there exists an open interval $\mathcal{I}$ with $0 \in \mathcal{I}$ and a uniquely determined curve $x : \mathcal{I} \ni s \mapsto x(s) \in \mathbb{R}^{n+1}$ such that $x(0) = x_0$. The path x is as smooth as F as well as

$$\forall\, s \in \mathcal{I} : F(x(s)) = 0 \in \mathbb{R}^n\,,\ \dim\ker \nabla F(x(s)) = 1\,,\ x'(s) \neq 0 \in \mathbb{R}^{n+1}\,;$$

without loss of generality let the path parameter s be the arc length. At the terminal point of the interval $\mathcal{I}$ either the condition of smoothness or the rank condition $\dim \operatorname{Ker} F_x^0 = 1$ is violated, the latter case suggests that there is a bifurcation point. A continuation method then has, say, the following form:

START: Specify x such that $F(x) = 0$; choose $N \in \mathbb{N}$. FOR i = 1:N *Predictor step:* Find a tangent t in x, choose suitable step length σ; set $$\widetilde{x} = x + \sigma\, t\,.$$ *Corrector step:* Find a new point y on the path such that $F(y) = 0$ *near* the predictor point $\widetilde{x}$; set $x = y$. END

The *length* of the continuation is specified by the number N if the pursued curve is *open*. If the curve is *closed*, i.e., if initial point and terminal point coincide, then of course an other stopping criterium may be selected. The step length near the terminal point has to be modified suitably if overlapping (overshooting) shall be avoided. The indispensable step length control in the predictor step is managed normally by a local extrapolation but some additional *security bounds* must guarantee that pathfollowing works also in less agreeable situations. At beginning of the procedure, step length is estimated coarsely by lack of appropriate data. A more thorough discussion must be renounced here, see however `SUPPLEMENT\chap05e`.

(b) Predictor Step The implicite representation $F(x(s)) = 0$ of the curve yields immediately $\nabla F(x(s))x'(s) = 0 \in \mathbb{R}^n$ which says that the tangent $x'(s)$ is perpendicular to the rows of $\nabla F(x(s))$ hence independent of these vectors, and they altogether span the entire $\mathbb{R}^{n+1}$. The *augmented* Jacobian

$$J(x(s)) := \begin{bmatrix} \nabla F(x(s)) \\ x'(s)^T \end{bmatrix} \in {\mathbb{R}^{n+1}}_{n+1}$$

is therefore regular in the basic interval $\mathcal{I}$; we call the curve x *positive oriented* if $\det(J(x(s)) > 0$. Let $A \in {\mathbb{R}^n}_{n+1}$ be rank-maximal then there exists a unique vector $t(A)$ with the following properties:

$$\boxed{A\,t(A) = 0\,,\ \ |t(A)| = 1\,,\ \ \det\begin{bmatrix} A \\ t(A)^T \end{bmatrix} > 0}\,. \tag{5.90}$$

This *direction vector* plays a crucial role in the sequel. Since $F(x(s))x'(s) = 0$, the curve x is solution of the initial value problem

$$x'(s) = t(\nabla F(x(s)))\,,\ \ x(0) = x_0\,,\ \ s \in \mathcal{I}\,.$$

Note, however, that $t(A)$ is not given explicitly. By consequence, we may *not* simply apply a numerical method of Sect. 2.2 for solving that initial value problem.

Let us now consider the QR decomposition of A^T,

$$Q\,R = A^T\,,\ \ Q = [q, \ldots, q_{n+1}]\,,\ q_i \in \mathbb{R}^{n+1}\,,$$

then the vector q_{n+1} has the properties enumerated in (5.90). This device for computing the tangent $t(\nabla F(x))$ is however less suited in large systems of equations since the orthogonal matrix Q is fully occupied but employed explicitly here. Therefore we choose simply an auxiliary vector $d \in \mathbb{R}^{n+1}$ such that $d^T t > 0$ where t denotes the tangent hitherto being unknown. Then that tangent follows uniquely by

$$\begin{bmatrix} \nabla F(x) \\ d \end{bmatrix} u = e_{n+1}\,,\ \ t = u/\|u\|_2\,, \tag{5.91}$$

where e_{n+1} denotes the $(n+1)$-th unit vector. The choice of d for t being unknown is only a small disfigurement being of no drawback numerically since $d^T t \neq 0$ does hold for almost all $d \neq 0$ (change sign possibly). The vector d is chosen arbitrarily in starting the procedure and is afterwards specified by using the known tangent.

It would be advantageous for the condition of the matrix in (5.91) if the vector d were as parallel as possible to the tangent vector t. On the other side, the i-th component x^i of x becomes directly the local curve parameter in the case where $d = e_i$. By this reason (Rheinboldt86) has proposed to choose $d = e_i$ and

$$i = \max_k\{|e_k^T t|\}\,,\ \ j = \max_{k \neq i}\{|e_k^T t|\} \tag{5.92}$$

then t^i is the largest and t^j the second-largest component of t in absolute value. In this choice of t however turning points are to be regarded especially by the following device:

Choose $\gamma > 0$ (e.g. $\gamma = 1.05$ and at beginning $\gamma = 2$), choose i as in (5.92). But if $$\lvert t^i\rvert < \lvert {t_{\rm old}}^i\rvert\,,\ \ \lvert t^j\rvert > \lvert {t_{\rm old}}^j\rvert\,,\ \ \lvert t^j\rvert \geq \mu\lvert t^i\rvert\,,$$ then choose $i = j$ by (5.92)

.

(c) Corrector Step For computation of a new path point x_{new}, after (Rheinboldt86) the system

$$\widetilde{F}(x_{\text{new}}) = \begin{bmatrix} F(x_{\text{new}}) \\ e_i^T(x_{\text{new}} - \widetilde{x}) \end{bmatrix} = 0 \tag{5.93}$$

has to be solved by a damped NEWTON method but it suffices also to evaluate the gradient of $\widetilde{F}$ *once* at the beginning (modified method). Of course $x_{\text{new}}^i = \widetilde{x}^i$ directly which may be used for a reduction of the system of equations in each iteration. By (5.93) the correcture $x_{\text{new}} - \widetilde{x}$ then is perpendicular to e_i in each step; cf. Fig. 5.18.

Also, in smaller problems, x_{new} is determined frequently by

$$|x_{\text{new}} - \widetilde{x}| = \arg\min_y\{|y - \widetilde{x}|\,;\; F(y) = 0\} \tag{5.94}$$

instead of (5.93); cf. (Allgower90). The matrix $\nabla F(y)$ is rank-maximal by assumption therefore, by Corollary 3.4, the multiplier rule

$$\nabla_y\left[|y - \widetilde{x}|^2\right] - z\nabla F(y) = 0 \in \mathbb{R}_{n+1}\,,\; F(y) = 0$$

($z \in \mathbb{R}_n$) is a necessary condition. This equation implies $2(y - \widetilde{x})^T = z\nabla F(y)$, hence by the Range Theorem 1.2,

$$y - \widetilde{x} \in \text{Range}(\nabla F(y)^T) = \text{Ker}(\nabla F(y))^{\perp} \iff t(\nabla F(y))^T(y - \widetilde{x}) = 0\,.$$

By this way the system (5.94) can replaced by the system

$$\begin{aligned} F(x_{\text{new}}) \qquad\qquad\quad &= 0 \\ t(\nabla F(x_{\text{new}}))^T(x_{\text{neu}} - \widetilde{x}) &= 0 \end{aligned}$$

which is solved by NEWTON's method:

done $= 0$, $x = \widetilde{x}$
WHILE NOT done
Solve the linear system w.r.t. y

$$\begin{aligned} F(x) + \nabla F(x)(y - x) &= 0 \\ t(\nabla F(x))^T(y - x) &= 0\,. \end{aligned} \tag{5.95}$$

Set $x = y$.
done $=$ convergence
END

Lemma 5.19. *Let $A \in \mathbb{R}^n{}_{n+1}$ be rank-maximal then the following statements are equivalent:*
(1°) $Ax = b\,,\; t(A)^T x = 0\,,$
(2°) $x = A^+b, A^+ = A^T(AA^T)^{-1}$ MOORE-PENROSE *inverse,*
(3°) $x = \arg\min\{|w|\,;\; Aw = b\}$ *minimal solution.*

Proof. The equivalence of (2°) and (3°) follows directly from Lemma 1.4. For the equivalence of (1°) and (2°) we observe that

$$\begin{bmatrix} A \\ t(A)^T \end{bmatrix} [A^+, t(A)] = I \,,$$

hence

$$(1^\circ) \iff \begin{bmatrix} A \\ t(A) \end{bmatrix} x = \begin{bmatrix} b \\ 0 \end{bmatrix} \iff x = [A^+, t(A)] \begin{bmatrix} b \\ 0 \end{bmatrix} \iff x = A^+ b \,.$$

□

By (5.95) using Lemma 5.19, finally

$$\boxed{y = x - [\nabla F(x)]^+ F(x)} \,. \tag{5.96}$$

So we are eventually faced with the problem to find the vector $y = A^+ b$ where $A \in \mathbb{R}^n{}_{n+1}$ is a matrix of rank n. To this end the algorithm proposed in Sect. 1.1**(h3)** can be employed.

In practice the modified NEWTON method is applied where $\nabla F(x) = \nabla F(\widetilde{x})$ remains fixed. Then the corrector $x_{\text{new}} - \widetilde{x}$ as well as all intermediary differences of the iteration stand perpendicular on the tangent of $F(x(s))$ in $\widetilde{x}$; see Fig. 5.19.

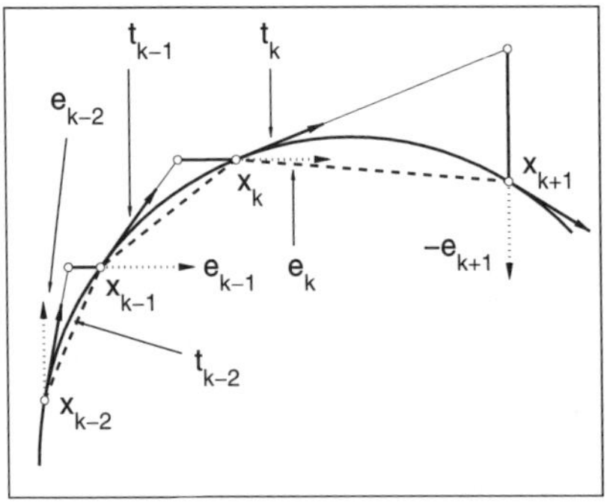

Fig. 5.18. Corrector (Rheinboldt)

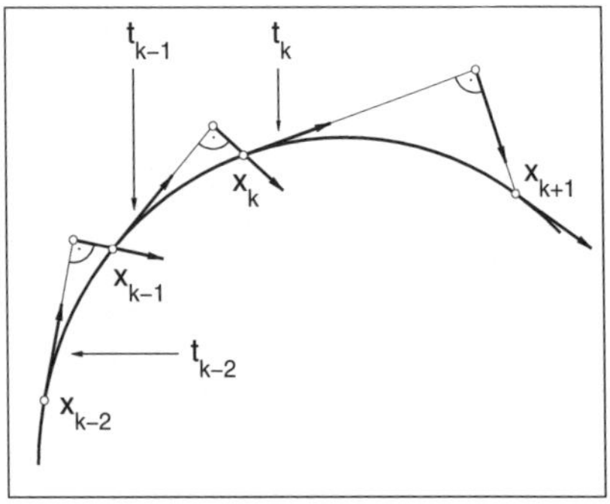

Fig. 5.19. Corrector (Allgower)

(d) Examples

Example 5.19. In most cases the MATLAB command `contour` suffices to illustrate the niveau lines of a scalar function $f : \mathbb{R}^2 \ni x \mapsto f(x) \in \mathbb{R}$. The exact computation however may be managed by a continuation method for $F(x) = f(x) - f(x_0) = 0$. We consider the simple example (Fig. 5.20)

$$f(x_1, x_2) = -\left[x_1^4 + x_2^4 + 2x_1^2 x_2^2 - 2x_1^2 + 2x_2^2\right] .$$

Example 5.20. Chemical reaction model (Kubicek), (Bank), (Deuflhard87) (Fig. 5.21).

$$\begin{aligned}
&\mu(1-x_3)E(x_1)-x_3 &&=0\\
&\mu\alpha(1-x_3)E(x_1)+\gamma\sigma-10(1+\gamma)x_1 &&=0\\
&x_3-x_4+\mu(1-x_4)E(x_1) &&=0\\
&10x_1-10(1+\gamma)x_2+\mu\alpha(1-x_4)E(x_1)+\delta\varrho &&=0\\
&\exp(10x/(1+\beta x)) &&=E(x)
\end{aligned}$$

where $\alpha = 22$, $\beta = 1.0E^{-2}$, $\gamma = \delta = 2$, $\sigma = \varrho = 0$, hence

$$\begin{aligned}
&\mu(1-x_3)E(x_1)-x_3 &&=0\\
&\mu 22(1-x_3)E(x_1)-30x_1 &&=0\\
&x_3-x_4+\mu(1-x_4)E(x_2) &&=0\\
&10x_1-30x_2+\mu 22(1-x_4)E(x_2) &&=0\\
&\exp(10x/(1+\beta x)) &&=E(x)
\end{aligned}$$

Initial point is $(\mu, x) = (0, 0)$.

Example 5.21. Classical test problem (Rheinboldt86), p. 146 (Figs. 5.22, 5.23).

$$\begin{aligned}
x_1 - x_2^3 + 5x_2^2 - 2x_2 + 34x_3 - 47 &= 0\\
x_1 + x_2^3 + x_2^2 - 14x_2 + 10x_3 - 39 &= 0
\end{aligned}$$

Initial point: $x_0 = [15, -2, 0]^T$. The solution passes through $x = [5, 4, 1]^T$ (terminal point). Exact solution to compare with:

$$x_1(t) = -\frac{11}{6}t^3 + \frac{2}{3}t^2 + 19t + \frac{107}{3}, \quad x_2(t) = t, \quad x_3(t) = \frac{1}{12}t^3 - \frac{1}{6}t^2 - \frac{1}{2}t + \frac{1}{3}.$$

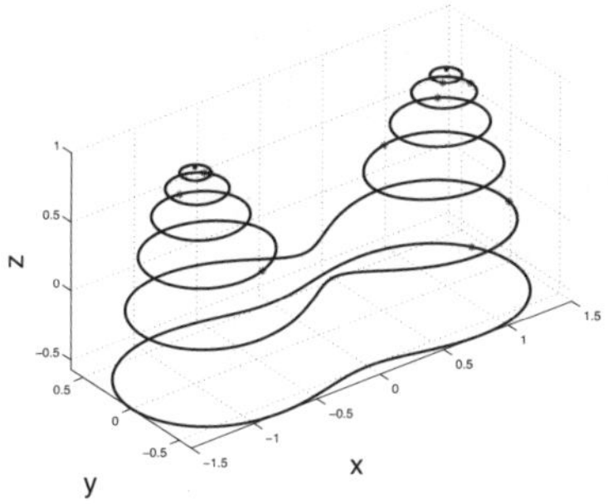

Figure 5.20. Example 5.19

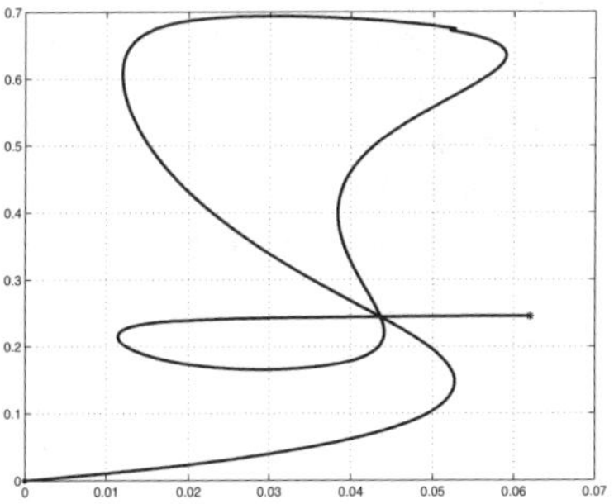

Figure 5.21. Example 5.20

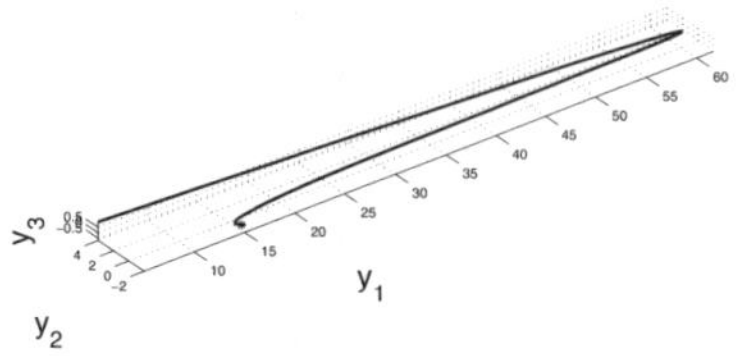

Figure 5.22. Example 5.21

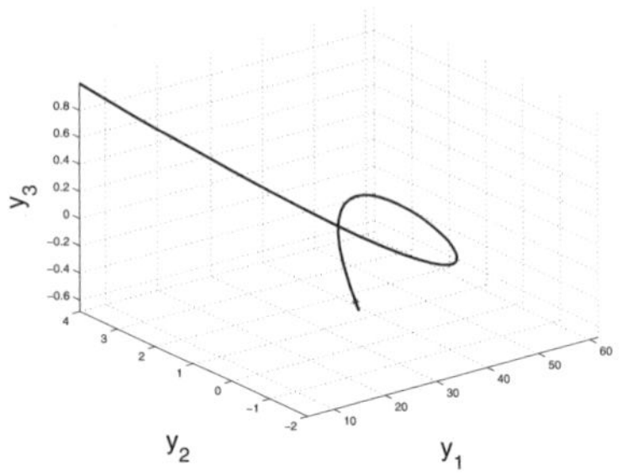

Figure 5.23. Example 5.21 scaled

5.8 Hints to the MATLAB Programs

```
KAPITEL05/SECTION_5, Hopf Bifurcation
conjgrad.m    Method of conjugate gradients after Stoer
cg_lq.m       Method of conjugate gradients after Allgower/Georg
demo1.m       Masterfile for HOPF bifurcation with backward
              differentiation or with trig. collocation
demo2.m       Masterfile for continuation in DEMO1.m only for
              backward differentiation
hopf_bdf.m    Hopf bifurcation with backward differentiation
hopf_trig.m   Hopf bifurcation with trig. collocation
hopf_contin   Simple continuation after HOPF.M
KAPITEL05/SECTION_6, Numerical Bifurcation
demo1.m    Pitchfork bifurcation
demo2.m    Example of Crandall (4 branching points)
demo3.m    Poisson's equation in unit square
           Newton's method (6 examples)
demo4.m    Poisson's equation in unit square
           direct iteration (6 examples)
demo5.m    Continuation by MU for Poisson's equation
bif.m      Direct iteration method
KAPITEL05/SECTION_7, Continuation Methods
demo1.m    Masterfile for continuation after Allgower/Georg
demo2.m    Masterfile for continuation after Rheinboldt
cont.m     Continuation after Allgower/Georg
pitcon1.m -- pitcon5.m  Continuation after Rheinboldt
newton.m   Newton's method for PITCON.M
```

6

Mass Points and Rigid Bodies

The author — as befits his area of expertise — has placed the mathematical part of this volume at the beginning in the conviction that a solid mathematical foundation can be only beneficial in understanding mechanics and its application to the computer. As mentioned in the preface, the following three chapters evolved from texts accompanying various lecture series on *higher mathematics* in calculus. They were meant to serve as motivation and pave the way to advanced studies. However, they was also supposed to connect to parallel lectures on mechanics, whose lecturers often took little note of what the mathematicians were teaching their students.

> To be able to distinguish the contributions of mechanics and those of mathematics as such, all results which can be verified (on the classical level) by experimentation only are referred to from now on as *axioms*, while mathematical conclusions from them are formulated in the form of *theorems* or the like. The scope of the latter is thus exactly defined and changes naturally if the foundation of the axioms is replaced.

Hint: All vectors — also point vectors — are underlined in this chapter.

6.1 The Force and its Moment

Let $\underline{a} \in \mathbb{R}^3$ be any *fixed point*, preferably the origin or the gravity center of a *rigid body* with geometrical configuration $\Omega \subset \mathbb{R}^3$.

Axiom 6.1. *Let $\underline{k}$ be a force attacking at a point $\underline{x}$ in Ω.*
(1°) *$\underline{k}$ can be moved arbitrarily in direction of its action line, i.e., the* pair *$(\underline{k}, \underline{x})$ has the property $\forall\, \tau \in \mathbb{R} : (\underline{k}, \underline{x}) = (\underline{k}, \underline{x} + \tau\, \underline{k})$.*
(2°) *$\underline{k}$ generates the moment $\underline{p}_a = \overrightarrow{\underline{a}\,\underline{x}} \times \underline{k}$ at point $\underline{a}$ (moment of force).*

Also, $\underline{p}_a = (\overrightarrow{\underline{a}\,\underline{x}} + \tau\, \underline{k}) \times \underline{k} = \overrightarrow{\underline{a}\,\underline{x}} \times \underline{k}$ holds for all $\tau \in \mathbb{R}$ because $\underline{k} \times \underline{k} = \underline{0}$; hence the force $\underline{k}$ is denoted as *aligned* vector in this context. The moment $\underline{p}_a$

depends on $\underline{a}$ ($\underline{x}$ fixed); hence it is to be understood as *fixed* vector at present. If now an additional force $-\underline{k}$ attacks at point $\underline{a}$ then both pairs, $(\underline{k},\underline{x})$ and $(-\underline{k},\underline{a})$, generate the moment

$$\begin{aligned}&(\overrightarrow{\underline{y}\,\underline{x}} \times \underline{k}) + (\overrightarrow{\underline{y}\,\underline{a}} \times (-\underline{k})) = -(\overrightarrow{\underline{x}\,\underline{y}} \times \underline{k}) - (\overrightarrow{\underline{y}\,\underline{a}} \times \underline{k})\\ &= -\overrightarrow{\underline{x}\,\underline{a}} \times \underline{k} = \overrightarrow{\underline{a}\,\underline{x}} \times \underline{k} = \underline{p}_a \perp \underline{k}\end{aligned}$$

at an *arbitrary point* $\underline{y}$ and this moment is now indepent of $\underline{y}$. After this consideration, $\underline{p}_a$ may be conceived as free moment which may be shifted everywhere. Consequently, in adding the pairs $(\underline{k},\underline{a})$ and $(-\underline{k},\underline{a})$ to the pair $(\underline{k},\underline{x})$, a new *pair* $(\underline{k},\underline{p}_a)$ is generated in $\underline{a}$ consisting of force $\underline{k}$ and free moment $\underline{p}_a$ with the property $\underline{k}\cdot\underline{p}_a = 0$ (*dyname* in $\underline{a}$). In other words, a dyname is a pair of force and moment acting perpendicular to each other.

If n forces $\underline{k}_i$ attack at n points $\underline{x}_i$ of the body then we proceed in the same way with all of them and obtain for each pair $(\underline{k}_i,\underline{x}_i)$ the dyname $(\underline{k}_i,\underline{p}_{a,i})$ at point $\underline{a}$.

Axiom 6.2. *(Superposition Principle) Let n forces $\underline{k}_i$ attack at n points $\underline{x}_i$ of a rigid body. Then*

$$\underline{k}_S = \sum_{i=1}^{n} \underline{k}_i \quad \text{and} \quad \underline{p}_S = \sum_{i=1}^{n} \underline{p}_{a,i}$$

is the resulting force and the resulting moment at the point $\underline{a}$.

But $\underline{k}_S \cdot \underline{p}_S = 0$ does not hold necessarily here; hence this pair is not a *dyname* at $\underline{a}$ in general. However it may decomposed at once in a *sum of two dynames,* e.g., into the unique decomposition

$$(\underline{k}_S\,,\,\underline{p}_S) = (\underline{k}_S\,,\,\underline{0}) + (\underline{0}\,,\,\underline{p}_S)\,,$$

where both terms are dynames in trivial way, or one chooses a vector $\underline{l} \neq \underline{0}$ satisfying $\underline{l}\cdot\underline{p}_S = 0$ and writes

$$(\underline{k}_S\,,\,\underline{p}_S) = (\underline{l}\,,\,\underline{p}_S) + (\underline{k}_S - \underline{l}\,,\,\underline{0})\,.$$

If a pair $(\underline{k}\,,\,\underline{p})$ acts at point $\underline{y}$ and the moment $\underline{p}$ points into the same direction as the force $\underline{k}$, i.e., $\underline{p} = \gamma\,\underline{k}$, one speaks of a *vector screw* with slope γ in $\underline{y}$. Obviously, this pair cannot be a dyname because $\underline{p} \neq \underline{0}$ cannot be parallel and perpendicular to $\underline{k}$ at the same time. The question is now whether there exists a point $\underline{y}$ such that the pair $(\underline{k}_S\,,\,\underline{p}_S)$ at point $\underline{a}$ acts as a vector screw at that point $\underline{y}$.

Lemma 6.1. *The pair $(\underline{k}_S\,,\,\underline{p}_S)$ at point $\underline{a}$ where $\underline{k}_S \neq \underline{0}$ acts as vector screw at every point of the straight line*

$$G = \left\{\underline{y} \in \mathbb{R}^3\,,\ \overrightarrow{\underline{a}\,\underline{y}} = \frac{\underline{k}_S \times \underline{p}_S}{\underline{k}_S \cdot \underline{k}_S} + \tau\,\underline{k}_S\,,\ \tau \in \mathbb{R}\right\}$$

with slope $\gamma = (\underline{k}_S \cdot \underline{p}_S)/(\underline{k}_S \cdot \underline{k}_S)$.

Proof. Using the projection of $\underline{p}_S$ onto $\underline{k}_S$, cf. Example 1.2,

$$[\underline{p}_S]_{\underline{k}} := \frac{\underline{k}_S \cdot \underline{p}_S}{\underline{k}_S \cdot \underline{k}_S}\, \underline{k}_S\,, \tag{6.1}$$

we obtain the decomposition

$$\underline{p}_S = [\underline{p}_S]_{\underline{k}} + \left(\underline{p}_S - [\underline{p}_S]_{\underline{k}}\right) = [\underline{p}_S]_{\underline{k}} + [\underline{p}_S]_{\underline{n}}\,, \quad [\underline{p}_S]_{\underline{k}} \cdot [\underline{p}_S]_{\underline{n}} = 0\,.$$

The pair $(\underline{k}_S\,, \underline{p}_S)$ generates the moment

$$\boxed{\underline{p}_y = \underline{p}_S + \overrightarrow{\underline{y}\,\underline{a}} \times \underline{k}_S = [\underline{p}_S]_{\underline{k}} + \left([\underline{p}_S]_{\underline{n}} - \overrightarrow{\underline{a}\,\underline{y}} \times \underline{k}_S\right)}\,,$$

at an arbitrary point $\underline{y}$ because $\underline{p}_S$ as sum of free moments is likewise a free moment and thus may be shifted everywhere. Consequently $\underline{p}_y = [\underline{p}_S]_{\underline{k}}$ and hence parallel to $\underline{k}_S$ if

$$\overrightarrow{\underline{a}\,\underline{y}} \times \underline{k}_S = [\underline{p}_S]_{\underline{n}}\,.$$

Because of $\left(\overrightarrow{\underline{a}\,\underline{y}} + \tau\,\underline{k}_S\right) \times \underline{k}_S = \overrightarrow{\underline{a}\,\underline{y}} \times \underline{k}_S\,,\ \tau \in \mathbb{R}$, this equation describes a straight line G with direction $\underline{k}_S$. Obviously $\overrightarrow{\underline{a}\,\underline{y}} - \overrightarrow{\underline{a}\,\underline{s}} = \tau\,\underline{k}_S$ does hold for the foot point $\underline{s}$ of the lot of $\underline{a}$ onto G. But, by Example 1.2,

$$\overrightarrow{\underline{a}\,\underline{s}} = \frac{1}{\underline{k}_S \cdot \underline{k}_S}\,\underline{k}_S \times \left(\overrightarrow{\underline{a}\,\underline{y}} \times \underline{k}_S\right) = \frac{1}{\underline{k}_S \cdot \underline{k}_S}\,\underline{k}_S \times [\underline{p}_S]_{\underline{n}} = \frac{1}{\underline{k}_S \cdot \underline{k}_S}\,\underline{k}_S \times \underline{p}_S$$

because $\underline{k}_S \times [\underline{p}_S]_{\underline{n}} = \underline{k}_S \times \underline{p}_S$. The slope γ follows from (6.1). □

6.2 Dynamics of a Mass Point

Let a cartesian coordinate system (COS) $\mathcal{E} = \{\mathcal{O}; \underline{e}_1, \underline{e}_2, \underline{e}_3\}$ be chosen in an EUKLIDian point space of dimension three; cf. Sect. 10.4**(a)**. Let $\mathcal{I} \subset \mathbb{R}$ be an open interval and $\Omega \subset \mathbb{R}^3$ an open set. Further, let $0 < m$ be a scalar called *mass* and $\underline{f} : \mathcal{I} \times \Omega \times \Omega \ni (t, \underline{x}, \underline{y}) \mapsto \underline{f}(t, \underline{x}, \underline{y}) \in \mathbb{R}^3$ a sufficiently smooth vector field called *force*. We consider a solution $\underline{x}$ of the initial value problem (IVP) (conservation law of momentum, NEWTON's axiom, equation of motion)

$$\boxed{\begin{aligned} &\frac{d}{dt} m\dot{\underline{x}}(t) = m\ddot{\underline{x}}(t) = \underline{f}(t, \underline{x}(t), \dot{\underline{x}}(t)) \\ &(t_0, \underline{x}(t_0), \dot{\underline{x}}(t_0)) \quad = (t_0, \underline{x}_0, \underline{v}_0) \in D_f := \mathcal{I} \times \Omega \times \Omega \in \mathbb{R} \times \mathbb{R}^6 \end{aligned}} \tag{6.2}$$

called *trajectory of a point with mass m* passing through $(t_0, \underline{x}_0, \underline{v}_0)$ (6 initial conditions). By assuming that $\underline{f}$ is sufficiently smooth (e.g., continuously differentiable), there exists a unique global solution of the IVP which cannot be continued further in D_f, i.e., runs up to the boundary of D_f for $t > t_0$ and for $t < t_0$ (which however does not mean that the solution exists on the entire interval $\mathcal{I}$).

> In the sequel let $\underline{x}$ be always a solution of (6.2).

(a) To characterize the motion (6.2) in $\mathbb{R}^n$ by means of a *single* scalar function, it suggests itself to multiply by a suitable vector. Velocity $\dot{\underline{x}}(t)$ is here the correct choice and yields after integration from t_0 to t a working integral or better the potential energy on the *right side*:

$$\frac{1}{2}m\dot{\underline{x}}(t)\cdot\dot{\underline{x}}(t) - \frac{1}{2}m\dot{\underline{x}}(t_0)\cdot\dot{\underline{x}}(t_0) = \int_{t_0}^{t} \underline{f}(\tau,\underline{x}(\tau),\dot{\underline{x}}(\tau))\cdot\dot{\underline{x}}(\tau)\,d\tau\,. \tag{6.3}$$

The corresponding result on the *left side* is consistently called *kinetic energy* and of course the difference must be the zero (but note the constants of integration in both the definite integrals).

Definition 6.1. (1°) *Relative to the basic COS,*

$E_{\text{kin}}(\underline{x}(t)) := \frac{1}{2}m\,\lvert\dot{\underline{x}}(t)\rvert^2$	is the kinetic energy
$E_{\text{pot}}(\underline{x},t,t_0) := -\int_{t_0}^{t} \underline{f}(\tau,\underline{x}(\tau),\dot{\underline{x}}(\tau))\cdot\dot{\underline{x}}(\tau)\,d\tau$	is the potential energy
$E(\underline{x}(t),t_0) := E_{\text{kin}}(\underline{x}(t)) + E_{\text{pot}}(\underline{x},t,t_0)$	is the total energy

of a mass point on its trajectory at time t.
(2°) *The vector field $\underline{f}$ is a* central field *if $\underline{f}(t,\underline{x},\underline{y}) = f(\underline{x})\underline{x}$ with a scalar function $f(\underline{x}) > 0$.*
(3°) *The vector field $\underline{f}$ is a* gradient field *or* conservative, *if $\underline{f}(t,\underline{x},\underline{y}) = \underline{f}(\underline{x})$ and a* potential $U : \underline{x} \mapsto U(\underline{x}) \in \mathbb{R}$ *exists such that*

$$\boxed{\operatorname{grad} U(\underline{x}) = -\underline{f}(\underline{x})}\,.$$

Remarks and addenda:

(1°) In many physical applications the proper choice of the basic COS is of crucial importance and must be frequently adapted a-posteriori.
(2°) The above introduction of the trajectory of a mass point shows that the *force* $\underline{f}$ does not depend on the *mass* m, and that NEWTON's axiom cannot serve for *definition* of $\underline{f}$.
(3°) A central field is *conservative* if and only if

$$\underline{f}(\underline{x}) = F'(\lvert\underline{x}\rvert)\frac{\underline{x}}{\lvert\underline{x}\rvert} \implies U(\underline{x}) = -F(\lvert\underline{x}\rvert)\,. \tag{6.4}$$

(4°) Let $E(t, \underline{x})$ and $B(t, \underline{x})$ be two electromagnetic fields and let $q > 0$ be the load of an electric particle with mass m, then the LORENTZ force $\underline{f}(t, \underline{x}, \dot{\underline{x}}) = q\,E(t, \underline{x}) + q(\dot{\underline{x}} \times B(t, \underline{x}))$ is an example for a velocity-dependent field of force. A further example is the damped oscillator of Sect. 1.4**(d)** with *damping energy* $\kappa \int_{t_0}^{t} |\dot{\underline{x}}(\tau)|^2\, d\tau$ and damping constant κ $[kg/s]$.

(5°) The potential energy $E_{\text{pot}}(\underline{x}, t, t_0)$ describes the work being performed by the force $\underline{f}$ along the path $\underline{x}(t)$ in time $T = t - t_0$. Its sign is chosen in a way that $\underline{f}$ points into direction of the greatest *descent* of E_{pot}. In general, this integral of potential energy depends on the way of the particle. If it shall be way-independent, i.e., if $E_{\text{pot}}(\underline{x}, t, t_0) = \widetilde{E}_{\text{pot}}(\underline{x}(t))$ shall hold for fixed (t_0, x_0) then $\underline{f}$ has to fulfill additional criteria. Letting $\underline{f}(t, \underline{x}, \underline{v}) = \underline{f}(\underline{x})$ then $\underline{f}$ has to satisfy $\operatorname{rot} \underline{f} = \underline{0}$ by the first and second main theorems on line integrals in order that $\widetilde{E}_{\text{pot}}(\underline{x}(t))$ is *locally* well-defined, hence is in fact a *function* of its argument $\underline{x}(t) \in \mathbb{R}^n$; this condition is fulfilled in particular if $\underline{f}$ is a gradient field.

(6°) The kinetic energy is invariant under rotations of the coordinate system. The potential energy is invariant under rotations if $D\underline{f}(t, \underline{x}, \underline{v})) = \underline{f}(t, D\underline{x}, D\underline{v})$ for every orthogonal matrix D with positive determinant (rotation matrix).

(7°) From Definition 6.1 we obtain by using (6.3)

$$\widetilde{E}(\underline{x}, t, t_0) := E_{\text{kin}}(\underline{x}(t)) + E_{\text{pot}}(\underline{x}, t, t_0); = E_{\text{kin}}(\underline{x}(t_0))\,, \tag{6.5}$$

therefore $\widetilde{E}(\underline{x}, t, t_0)$ is constant along the trajectory $\underline{x}$ through $\underline{x}(t_0)$. As consequence we can write $E(\underline{x}(t), t_0)) := \widetilde{E}(\underline{x}, t, t_0)$ and have $E(\underline{x}(t), t_0) = E(\underline{x}(t_0), t_0)$. Obviously the equation of motion (6.2) is found by differentiation of (6.5) w.r.t. t again. The total energy is therefore an *invariant* of the equation of motion (one of the *laws of conservation of energy*).

(8°) If the force $\underline{f}$ is a *gradient field*, then $E_{\text{pot}}(\underline{x}(t)) = U(\underline{x}(t))$, and every solution of $\boxed{m\,\ddot{\underline{x}}(t) = -\operatorname{grad} U(\underline{x}(t))}$ is given *implicitely* by

$$\boxed{m\,\frac{\dot{\underline{x}} \cdot \dot{\underline{x}}}{2} + U(\underline{x}) = m\,\frac{\dot{\underline{x}}_0 \cdot \dot{\underline{x}}_0}{2} + U(\underline{x}_0)}\,. \tag{6.6}$$

(b) By multiplication of (6.2) *vectorially* from left by $\underline{x}(t)$ one obtains

$$m\underline{x} \times \ddot{\underline{x}} = \underline{x} \times \underline{f}(\underline{x}, \dot{\underline{x}})\,. \tag{6.7}$$

Definition 6.2. *Let $\underline{c}$ be a fixed point in the basic COS. Then*

$\underline{d}(t) = m(\underline{x}(t) - \underline{c}) \times \dot{\underline{x}}(t)$	*is the angular momentum*
$\underline{p}(t) = (\underline{x}(t) - \underline{c}) \times \underline{f}(t, \underline{x}(t), \dot{x}(t), t)$	*is the moment of force*

w.r.t. $\underline{c}$ of the mass point on its path at time t.

Remarks and addenda:

(1°) $D\underline{d}(t) = D(\underline{x}(t) - \underline{c}) \times D\dot{\underline{x}}(t)$ by Lemma 1.5 for every rotation matrix D, and an analogous equation does hold for $\underline{p}(t)$.

(2°) It follows from (6.2) and Definition 6.2 directly that (without loss of generality $\underline{c} = \underline{0}$)

$$\frac{d}{dt}\underline{d}(t) = \frac{d}{dt}(m\underline{x} \times \dot{\underline{x}}) = m(\dot{\underline{x}} \times \dot{\underline{x}}) + m(\underline{x} \times \ddot{\underline{x}}) = \underline{x} \times \underline{f}(\underline{x}, \dot{\underline{x}}) = \underline{p}(t) . \quad (6.8)$$

(3°) Let further $\underline{c} = \underline{0}$. Then $\underline{p}(t) \equiv \underline{0}$ in a *central field* because $\underline{f}(\underline{x}) = f(\underline{x})\underline{x}$ and thus $\underline{d}(t) = \underline{d}(t_0)$ constant (*law of conservation of angular momentum*).

(4°) We have $\det(A) = \underline{a}_1 \cdot (\underline{a}_2 \times \underline{a}_3)$ for $A = [\underline{a}_1, \underline{a}_2, \underline{a}_3] \in \mathbb{R}^3{}_3$, cf. Sect. 1.1**(b)**, hence always $\underline{x}(t) \cdot \underline{d}(t) = 0$. Then $\underline{x}(t) \cdot \underline{d}(t_0) = 0$ in a central field and all trajectories are perpendicular to $\underline{d}(t_0)$.

(c) Let the force $\underline{f}$ be shifted to the side left of (6.2) and let the result be *multiplied by itself.* Then we obtain a functional $J(\underline{x}(t)) = \big[(m\ddot{\underline{x}}(t) - \underline{f}(t, \underline{x}(t), \dot{\underline{x}}(t))\big]^2$ which takes its absolute *minimum* for all solutions of the differential system (6.2) in a trivial way. But, under additional side conditions there results an extremal problem with constraints that may be treated further. If it is supposed, e.g., that $\underline{x} = \underline{x}^*$ and $\dot{\underline{x}} = \dot{\underline{x}}^*$ are already in optimum and it is only varied about acceleration then

$$\partial J(\underline{x}; \partial\ddot{\underline{x}}) = 2m\big[(m\ddot{\underline{x}} - \underline{f}(\underline{x}^*, \dot{\underline{x}}^*)\big] \cdot \partial\ddot{\underline{x}} = 0$$

constitutes a necessary condition for a stationary point. In this GAUSS' *principle of least constraint* one supposes that $\underline{x}^*$ und $\dot{\underline{x}}^*$ are determined already by constraints which are not necessarily holonomic here.

(d) When the equation of motion (6.2) shall be EULER equation of a *variational problem* then necessarily $\underline{f}(t, \underline{x}, \dot{\underline{x}}) = \underline{f}(\underline{x})$ (and $\underline{f}$ must be *conservative*) hence in particular $\nabla\underline{f}$ symmetric.

Lemma 6.2. *(*HAMILTON's principle, LAGRANGE principle of least action*)* *Let $\underline{f}$ be conservative and let $\mathcal{V} = \{\underline{x} \in \mathcal{C}^2[t_0, t_1], \underline{x}(t_0) = \underline{a}, \underline{x}(t_1) = \underline{b}\}$. Then $\underline{x} \in \mathcal{V}$ is solution of the equation of motion $m\,\ddot{\underline{x}}(t) = \underline{f}(\underline{x}(t))$ if and only if $\underline{x}$ is a stationary point of the* action integral

$$\mathcal{A}(\underline{x}) := \mathcal{A}_{[t_0,t_1]}(\underline{x}) = \int_{t_0}^{t_1} [\mathcal{E}_{\text{kin}}(\underline{x}(t)) - U(\underline{x}(t))]\, dt\,, \quad \underline{x} \in \mathcal{V}\,.$$

Proof. Setting the first variation of $\mathcal{A}$ equal to zero and integrating partially yields

$$0 = \delta\mathcal{A}(\underline{x}; \underline{v}) := \frac{d}{d\varepsilon}\mathcal{A}(\underline{x} + \varepsilon\underline{v})\Big|_{\varepsilon=0}$$

$$= \int_{t_0}^{t_1} [m\,\dot{\underline{x}} \cdot \dot{\underline{v}} - \operatorname{grad} U(\underline{x}) \cdot \underline{v}]\, dt = m\,\dot{\underline{x}} \cdot \underline{v}\Big|_{t_0}^{t_1} + \int_{t_0}^{t_1} \underline{v} \cdot [-m\,\ddot{\underline{x}} + \underline{f}]\, dt\,.$$

For $\underline{x} + \varepsilon \underline{v} \in \mathcal{V}$ necessarily $\underline{v} = 0$ at the terminal points and the assertion follows in both directions. □

The proof is managed here by a simple sign change due to partial integration. HAMILTON's principle in its *strong* form is however much more than that namely a fundamental extremal principle of physics; see Sect. 8.4.

The formula $U(\underline{x}(t)) - U(\underline{x}(t_0)) = \int_{t_0}^{t} \operatorname{grad} U(\underline{x}(\tau)) \cdot \dot{\underline{x}}(\tau)\, d\tau$ has for consequence that $\underline{f}(\underline{x}) = -\operatorname{grad} U(\underline{x})$ must hold at least locally hence $\underline{f}$ must be a local gradient field; then obviously $\delta U(\underline{x}; \underline{v}) = \operatorname{grad} U(\underline{x}) \cdot \underline{v}$. This formula must also hold for the representation (6.3). For the proof we apply partial integration and note that ∇f symmetric then

$$\begin{aligned}
&-\delta U(\underline{x}; \underline{v}) \\
&:= \frac{d}{d\varepsilon} \int_{t_0}^{t} \underline{f}(\underline{x} + \varepsilon \underline{v})(\dot{\underline{x}} + \varepsilon \dot{\underline{v}})\, d\tau \Big|_{\varepsilon=0} = \int_{t_0}^{t} [(\nabla \underline{f}(\underline{x})\, \underline{v}) \cdot \dot{\underline{x}} + \underline{f}(\underline{x}) \cdot \dot{\underline{v}}]\, d\tau \\
&= \int_{t_0}^{t} (\nabla \underline{f}(\underline{x})\, \underline{v}) \cdot \dot{\underline{x}}\, d\tau + \underline{f}(\underline{x}) \cdot \underline{v} \Big|_{t_0}^{t} - \int_{t_0}^{t} (\nabla \underline{f}(\underline{x})\, \dot{\underline{x}}) \cdot \underline{v}\, d\tau = \underline{f}(\underline{x}(t)) \cdot \underline{v}(t)
\end{aligned}$$

where the test function has to satisfy $\underline{v}(t_0) = \underline{0}$.

Example 6.1. In coordinate space $\mathbb{R}^n$, we have for $i = 1 : n$

$$\frac{\partial}{\partial x_i} \left(\frac{1}{|\underline{x}|} \right) = -\frac{x_i}{|\underline{x}|^3} .$$

Let a point of mass M lie in origin and let a further point with mass $m \ll M$ be given. Then NEWTON's law of gravitation yields for the gravitational force $\underline{f}$

$$U(\underline{x}) = -\frac{\gamma\, m\, M}{|\underline{x}|}, \quad \operatorname{grad} U(\underline{x}) = \frac{\gamma\, m\, M}{|\underline{x}|^3} \underline{x} = -\underline{f}(\underline{x})$$

with gravitation constant $\gamma = 6.67 \cdot 10^{-11}$ $[m^3/(kg \cdot s^2)]$ (earth).

(e) In **systems with one degree of freedom** we consider the *one-dimensional* motion $\mathcal{I} \ni t \mapsto x(t) \in \mathbb{R}$. Then, for $\underline{x} = (x, y) \in \mathbb{R}^2$, NEWTON's equation is equivalent to the two-dimensional system

$$\dot{\underline{x}} = \underline{v}(\underline{x}) = \begin{bmatrix} y \\ f(x)/m \end{bmatrix} \in \mathbb{R}^2, \quad y = \dot{x}, \tag{6.9}$$

and a solution $t \mapsto (x(t), \dot{x}(t)) \in \mathbb{R}^2$ of (6.4) is called *phase curve.* Recall that a phase curve is called *separatrix* if it tends to a singular point $\underline{x}$ with $\underline{v}(\underline{x}) = \underline{0}$ for $|t| \to \infty$ and that singular point is not a center. A separatrix is called *homoclinic orbit* if it tends to the same singular point for $|t| \to \infty$ and *heteroclinic orbit* if it tends to two different singular points for $t \to \infty$ and for $t \to -\infty$.

Example 6.2. (Small oscillation of a pendulum) Mass $m = 1$.

$$\ddot{x} = -x\,,\quad U(x) = x^2/2\,,\quad E(x,y) = (y^2 + x^2)/2 \implies y^2 + x^2 =: c \ \text{(constant)}\,.$$

Example 6.3. Mass $m = 1$ (Fig. 6.1).

$$\ddot{x} = x - 3x^2/2,,\quad U(x) = (x^3 - x^2)/2\,,\quad E(x,y) = y^2/2 + (x^3 - x^2)/2 =: c\,.$$

Then $y^2 - x^2 + x^3 = d$ and the homoclinic orbit is given by $y = \pm(x^2 - x^3)^{1/2}$ $= x(1-x)^{1/2}$.

Example 6.4. Mass $m = 1$ (Fig. 6.2).

$$\ddot{x} = x - 2x^3\,,\quad U(x) = -(x^2 - x^4)/2\,,\quad E(x,y) = y^2/2 + (x^4 - x^2)/2 =: c\,.$$

Then $y^2 - x^2 + x^4 = d$ and the both homoclinic orbits are given by $y^2 = x^2 - x^4$ for $x > 0$ and $x < 0$.

Example 6.5. (Mathematical pendulum) Mass $m = 1$ (Fig. 6.3).

$$\ddot{x} = -\omega^2 \sin x\,,\quad U(x) = -\omega^2 \cos x\,,\quad E(x,y) = y^2/2 - \omega^2 \cos x =: c\,.$$

Then $y = \pm(2\omega^2 \cos x + c)^{1/2}$ and singular points are $(x,y) = (k\pi, 0)$, $k \in \mathbb{Z}$. By substitution into E one obtains the corresponding constants $c = \pm 1$. An heteroclinic orbit is given, e.g., by $y = 2\omega^2 \cos x - 1$, $-\pi \le x \le \pi$.

In *Examples* 6.3 – 6.5, the separatrices separate regions with positive and negative total energy E.

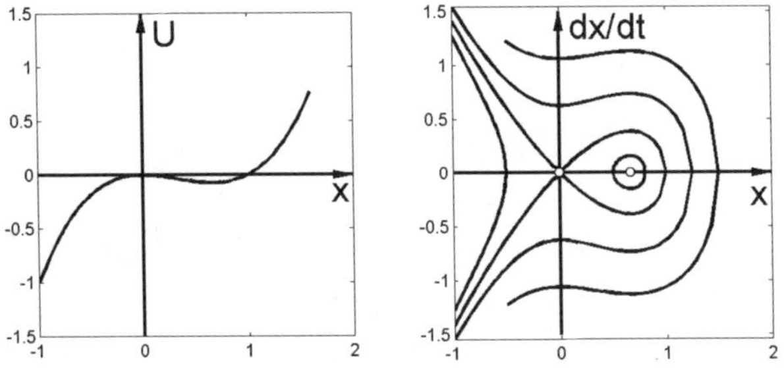

Figure 6.1. Ex. 6.3, potential and phase portrait

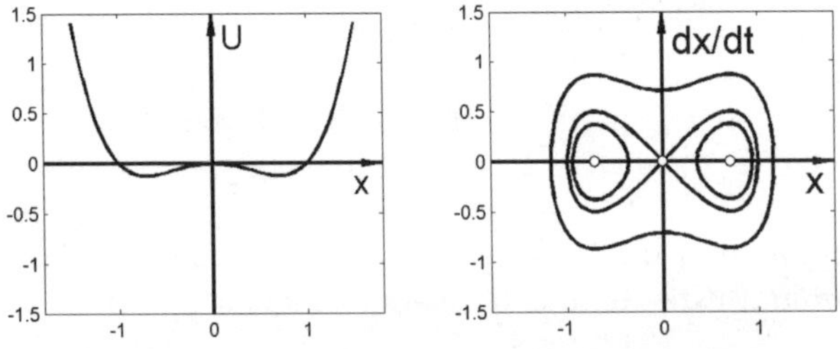

Figure 6.2. Ex. 6.4, potential and phase portrait

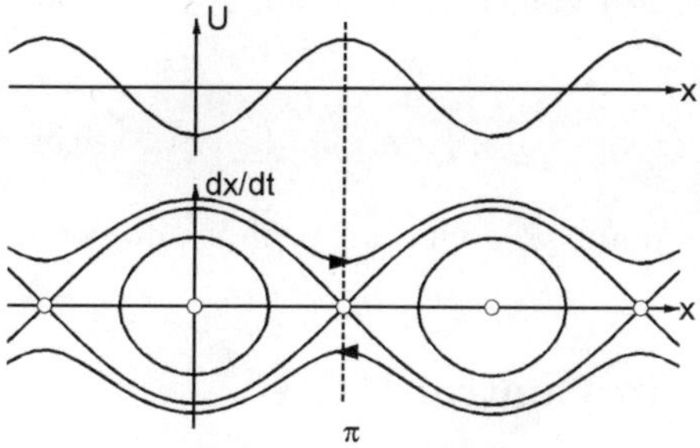

Figure 6.3. Ex. 6.5, potential and phase portrait

(f) Consider a **rigid rotation** in $\mathbb{R}^3$ with *fixed* rotation axis $\underline{a} \in \mathbb{R}^3$, $|\underline{a}| = 1$ (point vector) and *rotation matrix* $D(t)$ of (1.13) then $\underline{x}(t) = D(t)\underline{x}(t_0)$. Recall that

$\underline{\varphi}(t) = \varphi(t)\underline{a}$	vector of rotation angle
$\underline{\omega}(t) = \dot{\varphi}(t)\underline{a}$	angular velocity
$\dot{\underline{x}}(t) = \underline{\omega}(t) \times \underline{x}(t)$	velocity

.

by Sect. 1.1**(i)**. By the Expansion Theorem (1.2) we obtain for the angular moment

$$\begin{aligned} \underline{d} = m[\underline{x} \times (\underline{\omega} \times \underline{x})] &= m[(\underline{x}^T\underline{x})\underline{\omega} - (\underline{x}^T\underline{\omega})\underline{x}] \\ = m[(\underline{x}^T\underline{x})\underline{\omega} - (\underline{x}\,\underline{x}^T)\underline{\omega}] &= m[\underline{x}^T\underline{x}I - \underline{x}\,\underline{x}^T]\underline{\omega} =: T(\underline{x})\,\underline{\omega} \end{aligned} \tag{6.10}$$

with inertia tensor $T(\underline{x}) \in \mathbb{R}^3{}_3$. Rigid rotation has *one* constraint in $\mathbb{R}^2$ and *two* constraints in $\mathbb{R}^3$: The projection $\underline{x}_a$ of $\underline{x}$ onto the axis $\underline{a}$ — cf. Sect. 1.1**(a)** — is a *constant* vector and the *radius vector* $\underline{r} := \underline{x} - \underline{x}_a$ being *perpendicular* to $\underline{a}$, has a *constant* absolute value $|\underline{r}(t)| = r$. Therefore $\dot{\underline{x}} = \dot{\underline{r}}$, and by consequence

$$\dot{\underline{x}} = \dot{\underline{r}} = \underline{\omega} \times \underline{x} = \underline{\omega} \times (\underline{r} + \underline{x}_a) = \underline{\omega} \times \underline{r}\,, \quad |\dot{x}| = r|\dot{\varphi}(t)|\,. \tag{6.11}$$

Accordingly the point vector $\underline{x}$ can be replaced by the radius vector $\underline{r}$ (both are equal in $\mathbb{R}^2$) which corresponds to the introduction of *polar coordinates.* Then $T(\underline{x}) = m\,r^2\,I$ for the inertia tensor, $m\,r^2$ is the *moment of inertia* and

$$E_{\text{kin}} = \frac{mr^2|\underline{\omega}|^2}{2}\,, \quad E_{\text{pot}} = -\int_{t_0}^{t} \underline{\omega} \cdot \underline{p}\,d\tau\,,$$

because

$$\begin{aligned} |\dot{\underline{x}}|^2 &= |\underline{\omega} \times \underline{x}|^2 = |\underline{\omega} \times \underline{r}|^2 = |\underline{\omega}|^2|\underline{r}|^2 - (\underline{\omega} \cdot \underline{r})^2 = |\underline{\omega}|^2 r^2 \\ \underline{f} \cdot \dot{\underline{x}} &= \underline{f} \cdot (\underline{\omega} \times \underline{x}) = \underline{\omega} \cdot (\underline{x} \times \underline{f}) = \underline{\omega} \cdot \underline{p}\,. \end{aligned}$$

By (6.7) and using the Expansion Theorem 1.3, the *central acceleration* becomes

$$\ddot{\underline{r}} = \dot{\underline{\omega}} \times \underline{r} + \underline{\omega} \times \dot{\underline{r}} = \underline{\omega} \times (\underline{\omega} \times \underline{r}) = (\underline{\omega} \cdot \underline{r})\,\underline{\omega} - |\underline{\omega}|^2 \underline{r} = -|\underline{\omega}|^2 \underline{r}\,,$$

to which every mass point on a circular path is subjected.

6.3 Mass Point in Central Field

(a) Equation of motion (KEPLER 1571 – 1630, NEWTON 1642 – 1727).

(a1) Consider the trajectory $\underline{x} : t \mapsto \underline{x}(t) \in \mathbb{R}^3$ of a particle with mass m in a *central gradient field* with potential $U(|\underline{x}|)$:

$$m\ddot{\underline{x}} = -\operatorname{grad}_x U(|\underline{x}|) = -a(|\underline{x}|)\underline{x}\,, \quad a(|\underline{x}|) > 0 \tag{6.12}$$

(NEWTON's axiom). Then $\ddot{\underline{x}}$ is antiparallel to $\underline{x}$ and

$$\frac{d}{dt}\underline{d}(t) = \frac{d}{dt}(m\underline{x} \times \dot{\underline{x}}) = m\underline{x} \times \ddot{\underline{x}} + m\dot{\underline{x}} \times \dot{\underline{x}} = 0$$

yields immediately a *law of conservation of angular momentum*

$$\boxed{\underline{d}(t) := m\underline{x}(t) \times \dot{\underline{x}}(t) \in \mathbb{R}^3 \ \text{ constant}}\,. \tag{6.13}$$

Consequently the trajectory remains always in a plane perpendicular to the constant vector $\underline{d}$ and we may consider motion in a plane, $\underline{x}(t) \in \mathbb{R}^2$, without loss of generality. Then the orbit is conveniently written in polar coordinates with *radius vector* $\underline{r}(t) = \big(r(t)\cos\varphi(t)\,,\ r(t)\sin\varphi(t)\big)$, $r(t) := |\underline{r}(t)| > 0$ is the absolute value, and $|\underline{d}(t)| = mD_1$ constant where, by Sect. 6.2**(f)**,

$$\boxed{D_1 := r(t)^2\dot{\varphi}(t) \ \text{constant} \ \ \text{(angular moment for mass } m = 1)} \tag{6.14}$$

instead of (6.13). For $D_1 = 0$ straight motion is obtained with constant angle $\varphi(t)$. If however $D_1 \neq 0$, then φ is monotone in t hence invertible, $\underline{r}(t) = \underline{\widetilde{r}}(\varphi) = \big(\widetilde{r}(\varphi)\cos\varphi\,, \widetilde{r}(\varphi)\sin\varphi\big)$.

(a2) Sectorial area swept out by the radius vector:

$$S(t) = \frac{1}{2}\int_{\varphi(0)}^{\varphi(t)} \widetilde{r}^2(\varphi)\,d\varphi = \frac{1}{2}\int_0^t r^2(t)\dot{\varphi}(t)\,dt = \frac{D_1\,t}{2}\,. \tag{6.15}$$

Since the *sectorial velocity* $D_1/2 = r(t)^2\dot{\varphi}(t)/2$ is constant, we have:

Second KEPLER*'s law:* The radius vector $\underline{r}(t)$ sweeps out equal areas in equal times (Fig. 6.4).

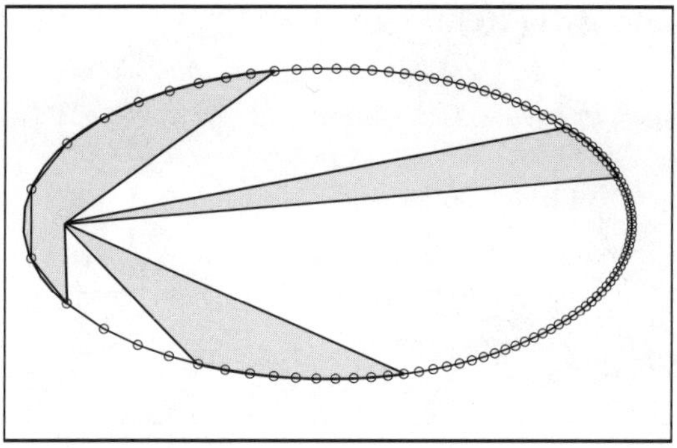

Figure 6.4. 2nd Kepler's law

Note that this law is valid for every motion of a particle in plane central gradient field. Let now T be the *time of revolution* then $S(T) = D_1 T/2$ is obviously the area of the ellipse in case of elliptic motion and we get :

Third KEPLER*'s law:* Revolution time = surface divided by sectorial velocity for an elliptic orbit.

(a3) Scalar equation of motion for $r(t) = |\underline{r}(t)|$ by a famous result of NEWTON:

Theorem 6.1. *Let $D_1 = r^2\dot{\varphi}$ and $r(t) = |\underline{r}(t)|$ then*

$$\boxed{m\,\ddot{r} = -\frac{d}{dr}V(r)\,,\quad V(r) = U(r) + \frac{m\,D_1^2}{2r^2}} \tag{6.16}$$

where $V(r)$ is the effective potential energy *(and D_1 is a constant).*

Proof. Let $\underline{e}_r(t) = \big(\cos\varphi(t),\, \sin\varphi(t)\big)\,,\ \underline{e}_\varphi(t) = \big(-\sin\varphi(t),\, \cos\varphi(t)\big)$ be a natural coordinate system (moving frame) on the trajectory then $\underline{x}(t) = r(t)\underline{e}_r(t)$, $\dot{\underline{e}}_r = \underline{e}_\varphi\dot{\varphi}\,,\ \dot{\underline{e}}_\varphi = -\underline{e}_r\dot{\varphi}$ hence

$$\dot{\underline{x}} = \dot{r}\,\underline{e}_r + r\dot{\varphi}\,\underline{e}_\varphi\,,\quad \ddot{\underline{x}} = (\ddot{r} - r\dot{\varphi}^2)\underline{e}_r + (2\dot{r}\dot{\varphi} + r\ddot{\varphi})\underline{e}_\varphi\,,\quad \boxed{\operatorname{grad}_x U = U_r(r)\underline{e}_r} \tag{6.17}$$

(HUYGEN's decomposition). Substitution into the original equation of motion (6.12) yields

$$\boxed{m(\ddot{r} - r\dot{\varphi}^2) = -U_r(r)\,,\quad 2\dot{r}\dot{\varphi} + r\ddot{\varphi} = 0}\,. \tag{6.18}$$

The second equation is equivalent to the condition $D_1 = r^2\dot{\varphi} =$ constant. Substitution of $\dot{\varphi} = D_1/r^2$ in the first equation yields

$$m\,\ddot{r} = m\,r\dot{\varphi}^2 - U_r(r) = r\frac{m\,D_1^2}{r^4} - U_r(r) = -\frac{d}{dr}\left(\frac{m\,D_1^2}{2\,r^2} + U(r)\right)\,. \ \square$$

(a4) Total energy of (6.12) resp. of (6.16):

$$E_1 = m\frac{\dot{\underline{x}}\cdot\dot{\underline{x}}}{2} + U(|\underline{x}|) \text{ resp. } E_2 = m\frac{\dot{r}^2}{2} + V(r)\,,$$

and of course $E_1 = E_2$ but also by

$$m\frac{|\dot{\underline{x}}|^2}{2} = m\frac{\dot{r}^2 + r^2\dot{\varphi}^2}{2} = m\frac{\dot{r}^2}{2} + \frac{m\,D_1^2}{2\,r^2}\,,$$

hence the total energy is

$$\boxed{E := E_1 = E_2 = m\frac{\dot{r}^2}{2} + V(r) = \text{ constant}}\,. \tag{6.19}$$

This equation leads immediately to a nonlinear differential equation of first order for r in dependence of time t and vice versa:

$$\frac{dr}{dt} = \left(\frac{2}{m}(E - V(r))\right)^{1/2}\,,\quad t(r) - t(r_0) = \int_{r_0}^{r} \frac{m^{1/2}}{\big(2(E - V(r))\big)^{1/2}}\,dr\,. \tag{6.20}$$

The angle φ is introduced as independent variable by $\dot{\varphi}(t) = D_1/r^2(t)$:

$$\frac{dr}{d\varphi} = \frac{dr}{dt}\cdot\frac{dt}{d\varphi} = \frac{r^2}{D_1}\left(\frac{2}{m}(E - V(r))\right)^{1/2}$$

and conversely

$$\frac{d\varphi}{dr} = \left(\frac{dr}{d\varphi}\right)^{-1} \Longrightarrow \varphi(r) - \varphi(r_0) = \int_{r_0}^{r} \frac{m^{1/2}D_1}{r^2\big(2(E - V(r))\big)^{1/2}}\,dr\,. \tag{6.21}$$

(As usual in physics, the different functions $r(t)$ and $\widetilde{r}(\varphi)$ are denoted by the same letter r.) By (6.20) or (6.21), the total energy E and the angular moment D_1 are the *invariants of motion* (6.12) for constant mass m.

(a5) Shape of the orbit The orbits in a central field depend only on the constants, D_1 and E (besides the mass m). They lie in the region $B_{D,E} := \{(r,\varphi) \in \mathbb{R}^2\,,\ V(r) \le E\}$ and on the boundary we have $\dot{r} = 0$; hence $V(r) = E$. By this result the region $B_{D,E}$ is composed of one or several annular domains with some $0 \le r_{\min} \le r \le r_{\max} \le \infty$.

Case 1: The domain $B_{D,E}$ is a circular orbit for $E - V(r_0) = 0$ and $V'(r_0) = 0$.

Case 2: The orbit is bounded by (6.20) when the integral

$$r(\infty) - r(t_0) = \int_{t_0}^{\infty} \frac{dr}{dt}\,dt = \int_{t_0}^{\infty} \left(\frac{2}{m}(E - V(r(\tau)))\right)^{1/2} d\tau$$

is convergent else unbounded, $r_{\max} = \infty$. If $\lim_{r\to\infty} U(r) =: U_\infty$ ($= \lim_{r\to\infty} V(r)$) is finite and $E - U_\infty > 0$, the orbit is obviously unbounded

— see the *hyperbola curve* in Example 6.6. If however $E - U_\infty = 0$, then $\lim_{|t|\to\infty}\dot r(t) = 0$, and the orbit *may be* unbounded — see the *parabola curve* in Example 6.6.
Case 3: If $r_{\min} = 0$, then obviously $\lim_{r\to 0} V(r) = E$ finite. By the energy equation (6.19) then

$$m\,\frac{\dot r^2}{2} = E - U(r) - \frac{m\,D^2}{2r^2} \ge 0 \iff r^2 U(r) + \frac{m\,D^2}{2} \le r^2 E\,.$$

Thus the condition

$$\lim_{r\to 0} r^2 U(r) < -\frac{m\,D^2}{2}$$

must be satisfied for $r \to 0$. The distance $r_{\min} = 0$ can be reached in *finite* time if $U(r)$ behaves like $-1/r^n$, $n > 2$ for $r \to 0$.

Let now $r_0 > 0$ be a stationary point of the effective potential $V(r)$, i.e., $V'(r_0) = 0$, let moreover $E \ne V(r_0)$ and let $V'(r)$ increase monotonically. Then $B_{D,E}$ is a single annulus where $0 < r_{\min} \le r(t) \le r_{\max}$, $r_{\min} < r_{\max}$. A point on the orbit with $r(t) = r_{\min}$ is called *pericenter*, and *apocenter* if $r(t) = r_{\max}$ (resp. *perihelion* and *apohelion* if sun is the center, and *perigee* and *apogee* if earth is the center). The orbit oscillates periodically between the extremal points, and each ray from the origin to the apocenter is a symmetry axis of the orbit. The angle

$$\Delta\Phi = 2\int_{r_{\min}}^{r_{\max}} \frac{m^{1/2} D_1}{r^2\big(2(E - V(r))\big)^{1/2}}\,dr \tag{6.22}$$

between two successive apocenters follows immediately from (6.21). The orbit is *closed* if $\Delta\Phi$ is a rational multiple of 2π otherwise the orbit is *dense* in annulus $B_{D,E}$.

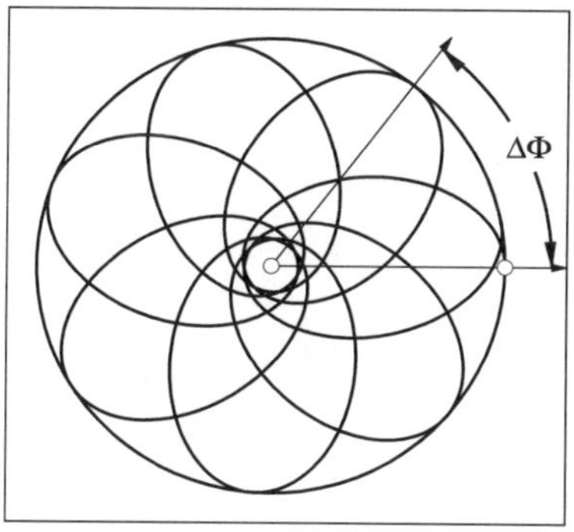

Figure 6.5. Motion in central field

Theorem 6.2. *Let the orbit of a mass point in central field be bounded then it is a closed curve exactly for two potentials, namely*

$$U(r) = -\frac{k}{r} \quad \textit{or} \quad U(r) = k\,r^2\,,\; k > 0\,.$$

In the first case $U(r)$ is NEWTON's gravitation potential and in the second case we have an harmonic oscillator without damping. The original proof of (Bertrand) (1873) is also found, e.g., in (Perelomov).

(b) Kepler's Problem

(b1) Focal equation We remember $D_1 = r^2\dot{\varphi} =$ constant and use the effective potential V further on but with NEWTON's gravitation potential $\boxed{U(r) = -k/r}$ where $k = \gamma\, m\, M$ in a physical two-body problem with masses $M \gg m > 0$ and gravitational constant γ. To get expressions for orbits in central fields being simpler as in **(a4)**, observe that

$$\frac{dr}{dt} = \frac{dr}{d\varphi}\frac{d\varphi}{dt} = \frac{D_1}{r^2}\frac{dr}{d\varphi} = -\,D_1\frac{d}{d\varphi}\left(\frac{1}{r}\right) \tag{6.23}$$

and therefore, by using (6.16),

$$\frac{d^2r}{dt^2} = \frac{d\dot{r}}{d\varphi}\frac{d\varphi}{dt} = -\,\frac{D_1^2}{r^2}\frac{d^2}{d\varphi^2}\left(\frac{1}{r}\right) = -\,\frac{d}{r}V(r)\,.$$

This relation suggests the transformation $r = 1/u$. Then

$$-\,\frac{d}{dr}V(r) = -\,\frac{1}{r^2}\frac{d}{du}V\left(\frac{1}{u}\right) \Longrightarrow m\frac{d^2u}{d\varphi^2} = -\,\frac{1}{D_1^2}\frac{d}{du}V\left(\frac{1}{u}\right)\,. \tag{6.24}$$

But

$$V(r) = -\,\frac{k}{r} + \frac{m\,D_1^2}{2r^2} \Longrightarrow V\left(\frac{1}{u}\right) = -\,k\,u + \frac{m\,D_1^2}{2}u^2\,,$$

and a substitution into (6.24) yields the simple linear differential equation

$$m\,u''(\varphi) + m\,u(\varphi) = \frac{k}{D_1^2}$$

with well-known solution

$$u(\varphi) = A\cos(\varphi - \varphi_0) + \frac{k}{m\,D_1^2} = \frac{1}{p}\Big(1 + \varepsilon\,\cos(\varphi - \varphi_0)\Big)\,,\quad p = \frac{m\,D_1^2}{k} > 0\,, \tag{6.25}$$

defining $\varepsilon := A\,p > 0$ as arbitrary constant of integration for the present. Returning to $r = 1/u$ again we obtain the so-called *focal equation* for a *conic section*

$$\boxed{r(\varphi) = \frac{p}{1 + \varepsilon\,\cos\varphi}} \tag{6.26}$$

where one focus lies in the origin. For $0 < \varepsilon < 1$ the conic section is an ellipse. The number p is called *parameter* of the ellipse and ε *(numerical) eccentricity.*

KEPLER*'s first law* says that planets describe ellipses with one focus in the sun. He discovered this law empirically by observations of the planet Mars and by applying measured data of the Danish astronomer TYCHO BRAHE (1546 – 1601). The above computations show that this law is a *mathematical* inference of NEWTON's axiom if NEWTON's gravitational potential is inserted. On the other side, if we adopt that planets move in a central field of gravity, KEPLER's first law implies NEWTON's law of gravity $U = -k/r$.

(b2) Recovery of eccentricity from the invariants D_1 and E. By substitution of (6.23) into (6.19), $E = m\dot{r}^2/2 + V(r)$, we obtain $E = m\, D_1^2(u'(\varphi)^2 + u^2(\varphi))/2 - k\, u(\varphi)$. Substitution of the solution (6.25) then yields

$$0 \le \varepsilon = \left(\frac{2pE}{k} + 1\right)^{1/2} \begin{cases} > 1 \iff E > 0 \\ = 1 \iff E = 0 \\ < 1 \iff E < 0\,. \end{cases} \tag{6.27}$$

In the present case $U(r) = -k/r$, $k > 0$, we have $\lim_{r\to\infty} V(r) = 0$; hence $\lim_{t\to\infty} \dot{r}(t) = (2E/m)^{1/2}$ by (6.20). It is shown below that $E = 0$ for a parabola then the velocity of a mass point tends to zero for $|t| \to \infty$ whereas it remains positive in case of a hyperbola; cf. Example 6.6. A mass point must have at least energy $E = 0$ to overcome the gravitational field of the center. Recall that $E = m|\dot{\underline{x}}|^2/2 - \gamma\, m\, M/|\underline{x}|$ therefore $|\dot{\underline{x}}|_E = (2\gamma\, M/|\underline{x}|)^{1/2}$ is the *escape velocity.*

(b3) Interplanetary orbits are frequently described by the distance $r(t)$ to the center and the velocity $\underline{v}(t) = (u(t)\,,\ v(t))$ with components $u(t) := \dot{r}(t)$ in direction to the center and the *track velocity* $v(t) = r(t)\,\dot{\varphi}(t)$ perpendicular to that direction as in (6.17); see, e.g., Example 4.10. Then $\dot{v} = -u\,v/r$ by (6.18) and the equations of motion become

$$\boxed{\dot{r} = u\,,\quad \dot{u} = \frac{v^2}{r} - \frac{\gamma M}{r^2}\,,\quad \dot{v} = -\frac{u\,v}{r}\,,\quad v_0 := \left(\frac{\gamma M}{r}\right)^{1/2}}\,. \tag{6.28}$$

For $\dot{r} = \ddot{r} = 0$ we get again the *track velocity* v_0 *on a circular orbit* which is also found by setting centripetal force equal to gravitational force, i.e., $\gamma m M/r^2 = m v^2/r$.

(c) Geometry

(c1) Recovery of Axes a and b from p and ε. For an *ellipse* (Fig. 6.6) we have

$$2a = \frac{p}{1+\varepsilon} + \frac{p}{1-\varepsilon} \implies a = \frac{p}{1-\varepsilon^2}\,. \tag{6.29}$$

For the further investigation we replace the *true anomaly* φ by the *the eccentric anomaly* ψ using the relation

$$\boxed{\begin{aligned} r(\varphi)\cos\varphi &= a\cos\psi - a\,\varepsilon \\ r(\psi) &= a - a\,\varepsilon\cos\psi \end{aligned}}\,; \tag{6.30}$$

see Fig. 6.7. The second equation is found by substituting $\cos\varphi = (a\cos\psi - a\varepsilon)/r$ into the focal equation (6.26). Let $\varphi = \varphi_0$ for $\psi = \pi/2$ then $r(\varphi_0)\cos\varphi_0 = -a\varepsilon =: e$ is the focal distance (*linear eccentricity*) and $r(\varphi_0) = a$ is found by inserting (6.29) into the polar equation (6.26) for $\varphi = \varphi_0$. Together

$$e = a\varepsilon = \sqrt{a^2 - b^2} \Longrightarrow b = p(1-\varepsilon^2)^{-1/2} \Longrightarrow p = b^2/a\,.$$

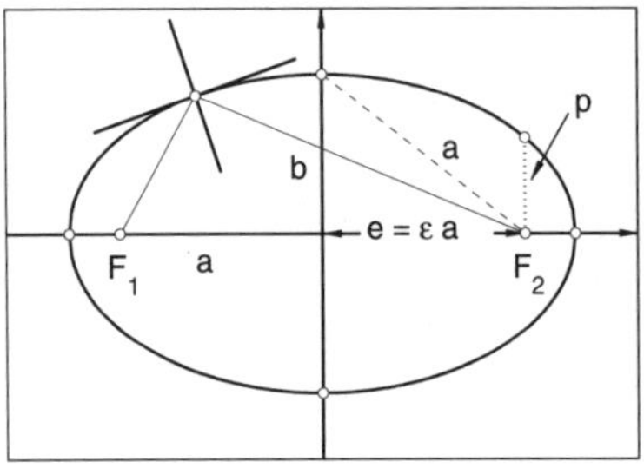

Figure 6.6. Ellipse

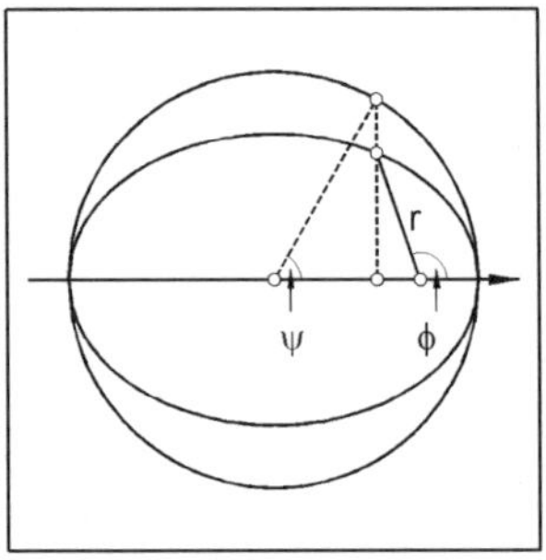

Figure 6.7. True and eccentric anomaly

Summary:

$$\boxed{\begin{array}{lll} \text{Ellipse} & 0 \le \varepsilon < 1\,, & a = p(1-\varepsilon^2)^{-1},\ b = p(1-\varepsilon^2)^{-1/2} \\ \text{Parabola} & \varepsilon = 1\,, & r(0) = p/2 \\ \text{Hyperbola} & 1 < \varepsilon\,, & a = p(\varepsilon^2-1)^{-1},\ b = p(\varepsilon^2-1)^{-1/2} \end{array}}\,. \qquad (6.31)$$

Call also `parabel.m` and `hyperbel.m` in `KAPITEL06\SECTION_2_3_4` for detailed illustration.

(c2) Recovery of space coordinates and time By using (6.27) and (6.31) one computes

$$a = \frac{k}{2|E|}\,,\quad b = D_1\left(\frac{m}{2|E|}\right)^{1/2}\,,\quad p = \frac{D_1^2 m}{k}$$

for the semi-axes and the parameter of an *elliptic* orbit. Then, recalling that $S(T) = D_1 T/2$ is the area of the ellipse (6.15) with revolution time T but also $S(T) = \pi ab$, we obtain

$$T = \frac{2\,\pi\,ab}{D_1} = 2\pi\,\frac{k}{2|E|}\,\frac{D_1\sqrt{m}}{\sqrt{2|E|}}\,\frac{1}{D_1} = \pi k\left(\frac{m}{2|E|^3}\right)^{1/2} = 2\pi a^{3/2}\sqrt{m/k}\,.$$

On the other side, substitution of the focal equation (6.26) into the first equation of (6.30) yields after simple computation

$$\cos\varphi = \frac{\cos\psi - \varepsilon}{1 - \varepsilon\cos\psi}\,, \quad \sin\varphi = \frac{(1-\varepsilon^2)^{1/2}\sin\psi}{1-\varepsilon\cos\psi} =: g(\psi)$$

using $\cos^2\varphi + \sin^2\varphi = 1$ and then

$$\frac{d\sin\varphi(\psi)}{d\psi} = \frac{dg(\psi)}{d\psi} = \frac{d\sin\varphi}{d\varphi}\frac{d\varphi}{d\psi} \implies \frac{d\varphi}{d\psi} = \frac{(1-\varepsilon^2)^{1/2}}{1-\varepsilon\cos\psi}\,.$$

Now recall (6.15) and use once more (6.30) then

$$\begin{aligned} D_1 t &= \int_{\varphi(0)}^{\varphi(t)} r^2(\varphi)\,d\varphi = \int_{\psi(0)}^{\psi(t)} a^2(1-\varepsilon\cos\psi)^2 \frac{(1-\varepsilon^2)^{1/2}}{1-\varepsilon\cos\psi}\,d\psi \\ &= C(\psi - \varepsilon\sin\psi)\Big|_{\psi_0}^{\psi} \\ C &= a^2\sqrt{1-\varepsilon^2} = a\sqrt{a^2(1-\varepsilon^2)} = a\sqrt{ap} = \left(\frac{a^3 D_1^2 m}{k}\right)^{1/2} = D_1\frac{T}{2\pi}\,; \end{aligned}$$

therefore

$$\boxed{\omega\, t = \psi - \varepsilon\sin\psi}$$

writing $\omega = 2\pi/T$ for the circular frequency and T for the time of revolution.

By applying the focal equation (6.26) in combination with (6.30) we obtain also for cartesian coordinates $x = r\,\cos\varphi\,,\ y = r\,\sin\varphi = (r^2 - x^2)^{1/2}$

$$\varepsilon\, x = p - r = a(1-\varepsilon^2) - a(1-\varepsilon\,\cos\psi) = a\,\varepsilon(\cos\psi - \varepsilon)\,;$$

hence together for *ellipse* and *hyperbola* (Landau) vol. I, Sect. 15,

$$\begin{aligned} &r = a(1-\varepsilon\,\cos\psi)\,, \quad t = \sqrt{m\,a^3/k}(\psi - \varepsilon\,\sin\psi) \\ &x = a(\cos\psi - \varepsilon)\,, \quad y = a\sqrt{1-\varepsilon^2}\,\sin\psi\,,\ 0 \le \psi < 2\pi\,, \qquad 0 < \varepsilon < 1 \\ &r = a(\varepsilon\,\cosh\psi - 1)\,, \ t = \sqrt{m\,a^3/k}(\varepsilon\,\sinh\psi - \psi) \\ &x = a(\cosh\psi - \varepsilon)\,, \quad y = a\sqrt{\varepsilon^2 - 1}\,\sinh\psi\,,\ -\infty < \psi < \infty\,, 1 < \varepsilon\,. \end{aligned} \tag{6.32}$$

(c3) Recovery of geometric data from physical data and vice versa

Case 1: In an inertial cartesian coordinate system, let the focus be the origin and let the initial position $\underline{x}_0$ relative to the focus and the initial velocity $\underline{v}_0$ be specified. Then the polar coordinates $(r_0,\,\varphi_0)$ of $\underline{x}_0$ can be determined and the initial values $\dot{r}_0$ and $\dot{\varphi}_0$ follow from

$$\underline{v}_0 = \dot{r}_0\underline{e}_{r_0} + r_0\dot{\varphi}_0\underline{e}_{\varphi_0} \implies \dot{r}_0 = \underline{v}_0 \cdot \underline{e}_{r_0}\,,\ \dot{\varphi}_0 = \underline{v}_0 \cdot \underline{e}_{\varphi_0}/r_0\,. \tag{6.33}$$

Then compute the invariants E and D_1 and thereafter the axes a and b by **(c2)**; for their position use φ_0.

Case 2: Conversely, let the *shape* of the ellipse be fixed by the axes a and b in an inertial system; then $e = \sqrt{a^2 - b^2}$ and, say, starting in the apocenter,

$r_0 = a + e$, $\varphi_0 = 0$, $\dot{r}_0 = 0$ and thus $\dot{\varphi}_0 = |\underline{v}_0|/r_0$ by (6.33). Therefore the initial velocity must be found. To this end we observe that $|E| = k/2a$ and $|E| = m|\underline{v}_0|^2/2 + U(r_0)$, $U(r) = -k/r$. For the time of revolution $T = 2\pi ab/D_1$ we need also $D_1 = b\sqrt{2|E|}$, cf. **(c1)**.

(d) Examples

Example 6.6. $D^2 = 1$, $k = 1$, $\mathrm{m} = 1$, hence $p = 1$, $\varepsilon = (2E+1)^{1/2}$;

$$V(r) = -\frac{1}{r} + \frac{1}{2r^2}\,,\ V(r_{\min}) = E\,,$$

ε	1/2	1	3/2
E	−3/8	0	5/8
$r_{\min}$	2/3	1/2	2/5

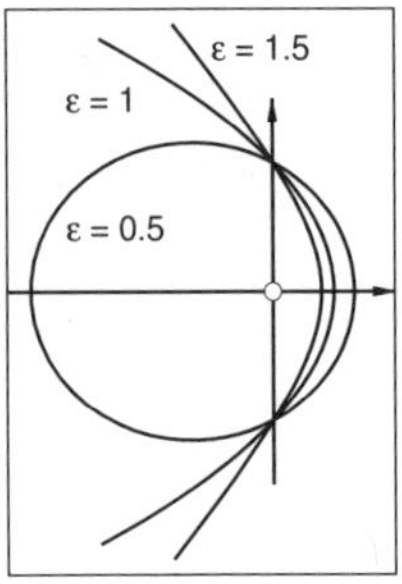

Figure 6.8. Conic sections

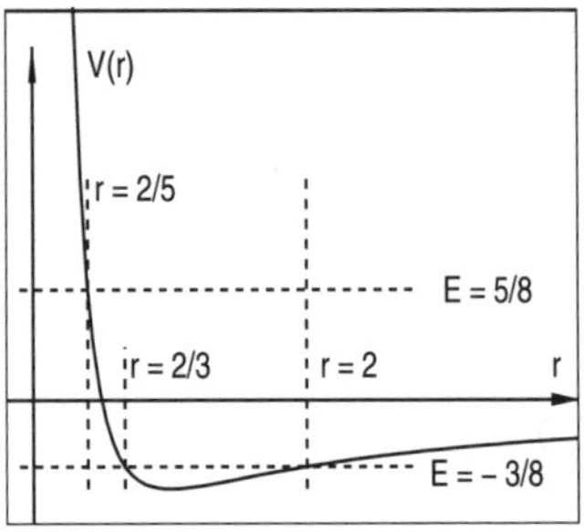

Figure 6.9. Effective potential

Example 6.7. An orbit passing through a point $\underline{x}_0$ is uniquely determined by the velocity vector $\underline{v}_0$ at that point if orientation of the orbit is fixed. If the conic sections are to be displayed in closed form then the symmetry axes must be known. In Figure 6.10 $\underline{v}_0 = 0.8(-3, 1)/\sqrt{10}$, $x_0 = \alpha(1, 1)/\sqrt{2}$ and $\alpha = 2, 2.5, 2.75, 3, 3.5, 4$; $\gamma = m = M = 1$; in Figure 6.11 $\underline{x}_0 = (1, 1)$, $\underline{v}_0 = \beta(-3, 1)/\sqrt{10}$ and $\beta = 0.4, 0.6, 0.8, 1, 1.1892$ (parabola), 1.5.

In *numerical approach*, the system (6.12) can be solved directly of course but the system (6.18) illustrates the time proportions more properly. Letting $\varphi = x_1$, $r = x_2$, $\dot{\varphi} = x_3$, $\dot{r} = x_4$ we have

$$\dot{x}_1 = x_3\,,\ \dot{x}_2 = x_4\,,\ \dot{x}_3 = -2x_3x_4/x_2\,,\ \dot{x}_4 = x_2x_3^2 - k/(mx_2^2)\,.$$

Recall that $k = \gamma mM$ therefore these three values have to be specified together with the four initial conditions.

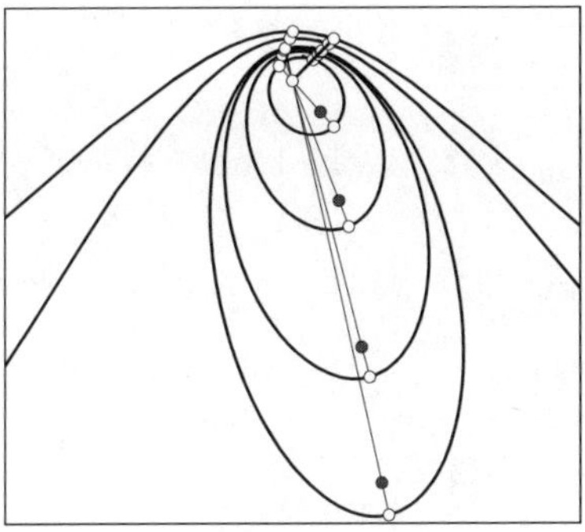

Figure 6.10. Several values of $|\underline{x}_0|$, $\underline{v}_0$ fixed

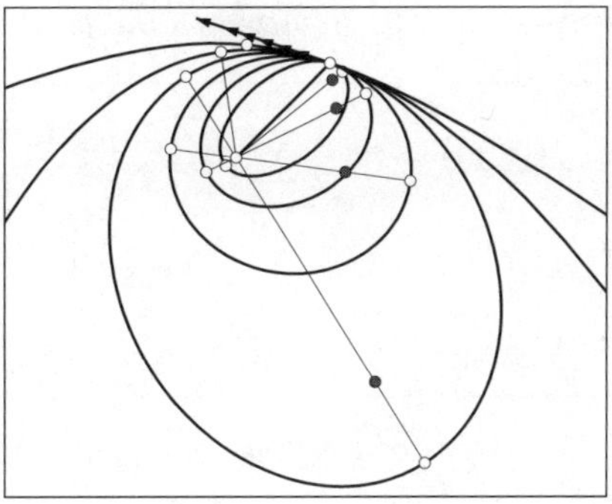

Figure 6.11. Several values of $|\underline{v}_0|$, x_0 fixed

References: (Arnold78), (Arnold90), (French), (Landau), (Perelomov), (Schneider)

6.4 Systems of Mass Points

NEWTON's mechanics deals with systems of mass points without constraints.

Under the same assumptions as in Sect. 6.2 let $\underline{x}_i : \mathcal{I} \ni t \mapsto \underline{x}_i(t) \in \mathbb{R}^3$, $i = 1 : n$, be the trajectories of n points of mass m_i, i.e., solutions of the initial value problem with a differential system

$$m_i\ddot{\underline{x}}_i = \underline{f}_i(t, \underline{x}_1, \ldots, \underline{x}_n, \dot{\underline{x}}_1, \ldots, \dot{\underline{x}}_n)\,, \quad i = 1 : n \tag{6.34}$$

and $6n$ suitable initial conditions. For brevity we write

$$\begin{aligned} \underline{X}(t) &= [\underline{x}_1(t), \ldots, \underline{x}_n(t)]^T \in \mathbb{R}^{3n} = \mathbb{R}^3 \times \ldots \times \mathbb{R}^3 \\ \underline{F}(t, \underline{X}(t), \dot{\underline{X}}(t)) &= [\underline{f}_1(t, \underline{X}(t), \dot{\underline{X}}(t)), \ldots, \underline{f}_n(t, \underline{X}(t), \dot{\underline{X}}(t))]^T \in \mathbb{R}^{3n} \end{aligned}$$

and then obtain

$$\boxed{m_i\ddot{\underline{x}}_i = \underline{f}_i(t, \underline{X}(t), \dot{\underline{X}}(t))\,, \quad i = 1 : n}\,.$$

(a) If each equation is multiplied by $\dot{\underline{x}}_i$, ensuing integrated from t_0 to t, and the results are all sumed up, we obtain

$$\sum_{i=1}^{n} \frac{1}{2} m_i |\dot{\underline{x}}_i(t)|^2 - \sum_{i=1}^{n} \frac{1}{2} m_i |\dot{\underline{x}}_i(t_0)|^2 - \int_{t_0}^{t} \underline{F}(\tau, \underline{X}(\tau), \dot{\underline{X}}(\tau)) \cdot \dot{\underline{X}}(\tau)\, d\tau = 0\,. \tag{6.35}$$

Definition 6.3. *Relative to the basic cartesian coordinate system (COS)*

$$\begin{array}{lll} E_{\text{kin}}(\underline{X}(t)) & := \dfrac{1}{2}\sum\limits_{i=1}^{n} m_i\,|\underline{\dot{x}}_i(t)|^2 = \sum\limits_{i=1}^{n} E_{\text{kin}}(\underline{x}_i(t)) & \text{is the kinetic energy} \\ E_{\text{pot}}(\underline{X},t,t_0) & := -\displaystyle\int_{t_0}^{t} \underline{F}(\tau,\underline{X}(\tau),\underline{\dot{X}}(\tau))\cdot\underline{\dot{X}}(\tau)\,d\tau & \text{is the potential energy} \\ \widetilde{E}(\underline{X},t,t_0) & := E_{\text{kin}}(\underline{X}(t)) + E_{\text{pot}}(\underline{X},t,t_0) & \text{is the total energy} \end{array}$$

of the system of mass points on their orbits at time t.

Remarks and addenda of Sect. 6.2 hold likewise. Note that $E_{\text{pot}}(\underline{X},t,t_0) = \sum_{i=1}^{n} E_{\text{pot}}(\underline{x}_i,t,t_0)$ and that the *conservation law of energy* does hold again,

$$\widetilde{E}(\underline{X},t,t_0) = E(\underline{X}(t),t_0) = E_{\text{kin}}(\underline{X}(t_0))\,. \tag{6.36}$$

In the sequel the fields $\underline{f}_i$ of force have to be partitioned in a slightly restricting way:

Axiom 6.3. (1°) *The forces $\underline{f}_i$ satisfy*

$$\underline{f}_i(t,\underline{X}(t),\underline{\dot{X}}(t)) = \underline{f}_{ii}(t,\underline{x}_i(t),\underline{\dot{x}}_i(t)) + \underline{g}_i(t,\underline{X}(t),\underline{\dot{X}}(t))\,,\quad i=1:n\,.$$

(2°) *Exactly the vector fields $\underline{f}_{ii}$ depend only of the mass point $\underline{x}_i$.*
(3°) *The generalized principle of* actio equals reactio,

$$\forall\, t\in\mathcal{I}: \sum_{i=1}^{n}\underline{g}_i(t,\underline{X}(t),\underline{\dot{X}}(t)) = 0 \tag{6.37}$$

does hold for all times t .
(4°) *The sum of interior angular momenta is zero,*

$$\forall\, i\in\mathcal{I}: \sum_{i=1}^{n}(\underline{x}_i(t)-\underline{c})\times\underline{g}_i(t,\underline{X}(t),\underline{\dot{X}}(t)) = 0 \tag{6.38}$$

where $\underline{c}=0$ without loss of generality by (6.37).

Remarks and Addenda:

(1°) The forces $\underline{f}_{ii}$ are called *external* forces and the forces $\underline{g}_i$ *internal* forces.
(2°) In almost all cases

$$\underline{g}_i(t,\underline{X}(t),\underline{\dot{X}}(t)) = \sum_{k=1,k\neq i}^{n}\underline{f}_{ik}(t,\underline{x}_i(t),\underline{x}_k(t),\underline{\dot{x}}_i(t),\underline{\dot{x}}_k(t))$$

(*two-particle forces*) and $\underline{f}_{ik} = -\underline{f}_{ki}$, $i\neq k$, (principle of *actio* = *reactio*).

(3°) Suppose that each mass point $\underline{x}_i$ generates a central field with $\underline{x}_i$ as origin then

$$\boxed{\underline{f}_{ik}(\underline{X}(t)) = f_{ik}(|\underline{x}_i - \underline{x}_k|)\frac{\underline{x}_i - \underline{x}_k}{|\underline{x}_i - \underline{x}_k|}\,, \quad i \neq k} \tag{6.39}$$

(*central two-particle forces of interaction*).

(4°) By definition of the *gravity center*,

$$\underline{x}_S(t) = M^{-1}\sum_{i=1}^{n} m_i\,\underline{x}_i(t)\,,\ M = \sum_{i=1}^{n} m_i\,,$$

we obtain the *crucial decomposition*

$$\underline{x}_i(t) = \underline{x}_S(t) + \underline{y}_i(t) \Longrightarrow \sum_{i=1}^{n} m_i\,\underline{y}_i(t) = \sum_{i=1}^{n} m_i\,\underline{x}_i(t) - M\underline{x}_S(t) = \underline{0} \tag{6.40}$$

and thus directly

$$M\dot{\underline{x}}_S(t) = \underline{I}(t)\,,\quad M\underline{x}_S(t)\times\dot{\underline{x}}_S(t) = \underline{x}_S(t)\times\underline{I}(t)\,, \tag{6.41}$$

and further, by (6.34) and Axiom 6.3(3°),

$$\boxed{M\,\ddot{\underline{x}}_S(t) = \frac{d}{dt}\underline{I}(t) = \sum_{i=1}^{n}\underline{f}_{ii}(\underline{X}(t))}\,. \tag{6.42}$$

In particular, the gravity center of a *closed* system moves on a straight line with constant velocity.

By the decomposition (6.40) one obtains also a decomposition in *external* und *internal* kinetic energy

$$\boxed{E_{\rm kin}(\underline{X}(t)) = \frac{1}{2}M\,|\dot{\underline{x}}_S(t)|^2 + \frac{1}{2}\sum_{i=1}^{n} m_i|\dot{\underline{y}}_i(t)|^2}\,. \tag{6.43}$$

(b) Vectorial multiplication of (6.34) from left by $\underline{x}_i$ and summing up over i yields

$$\sum_{i=1}^{n} m_i\underline{x}_i(t)\times\ddot{\underline{x}}_i(t) = \sum_{i=1}^{n}\underline{x}_i\times\underline{f}_i(t,\underline{X}(t),\dot{\underline{X}}(t))\,. \tag{6.44}$$

Definition 6.4. (1°) *Relative to a fixed point* $\underline{c}$ *of the basic COS*

$\underline{I}(t) = \sum_{i=1}^{n} m_i\,\dot{\underline{x}}_i(t)$	*is the total momentum*
$\underline{D}_{\underline{c}}(t) = \sum_{i=1}^{n} m_i(\underline{x}_i(t)-\underline{c})\times\dot{\underline{x}}_i(t)$	*is the total angular momentum*
$\underline{P}_{\underline{c}}(t) = \sum_{i=1}^{n}(\underline{x}_i(t)-\underline{c})\times\underline{f}_i(t,\underline{X}(t),\dot{\underline{X}}(t))$	*is the total moment*

of the system of mass points on their trajectories at time t *.*
(2°) *The system is* closed *if no external forces appear.*

Remarks and Addenda:

(1°) $\underline{D}_{\underline{c}}(t) = \underline{D}_0 - \underline{c} \times \underline{I}(t)$.
(2°) By (6.34) and the definition we obtain again

$$\boxed{\frac{d}{dt}\underline{D}_{\underline{c}}(t) = \underline{P}_{\underline{c}}(t)} \, . \tag{6.45}$$

(3°) The rate of change of the total momentum and of the total angular momentum depends only on the external forces $\underline{f}_{ii}$ under Axiom 6.3 (3°) and (4°):

$$\frac{d}{dt}\underline{I}(t) = \sum_{i=1}^{n} \underline{f}_{ii}(t, \underline{x}_i(t), \dot{\underline{x}}_i(t)) \,, \quad \frac{d}{dt}\underline{D}_{\underline{c}}(t) = \sum_{i=1}^{n} (\underline{x}_i(t) - \underline{c}) \times \underline{f}_{ii}(t, \underline{x}_i(t), \dot{\underline{x}}_i(t)) \, . \tag{6.46}$$

Consequently, the total momentum and the total angular momentum are constant in a *closed* system (*law of conservation of momentum resp. angular momentum*). This law does hold also for the angular momentum if all external forces are central forces relative to the origin of the basic COS.
(4°) Likewise, by definition of the gravity center and decomposition (6.40) we obtain a decomposition into *external* and *internal* angular momentum

$$\underline{D}(t) = M \, \underline{x}_S(t) \times \dot{\underline{x}}_S(t) + \sum_{i=1}^{n} m_i \underline{y}_i(t) \times \dot{\underline{y}}_i(t) \, . \tag{6.47}$$

By Axiom 6.3 (3°), (4°)

$$\begin{aligned} \frac{d}{dt} \sum_{i=1}^{n} m_i \underline{y}_i \times \dot{y}_i \overset{(!)}{=} \sum_{i=1}^{n} m_i \underline{y}_i \times \ddot{\underline{x}}_i &= \sum_{i=1}^{n} (\underline{x}_i - \underline{x}_S) \times \underline{f}_i(\underline{X}, \dot{\underline{X}}) \\ &= \sum_{i=1}^{n} (\underline{x}_i - \underline{x}_S) \times \underline{f}_{ii}(\underline{x}_i, \dot{\underline{x}}_i) \, ; \end{aligned}$$

hence the second term on the right side of (6.47 is constant in a *closed* system and $\underline{D}(t)$ constant thus also, by (6.40) ,

$$M \underline{x}_S(t) \times \dot{\underline{x}}_S(t) = \underline{x}_S(t) \times \underline{I}(t) = \text{ constant} \, . \tag{6.48}$$

Example 6.8. The motion of N points $\underline{x}_i = (x_i^1, x_i^2, x_i^3) \in \mathbb{R}^3$, $i = 1 : N$, with mass m_i under mutual gravitational attraction is called *"N-body problem"*. By NEWTON's law of gravitation the total gravitational potential is the *sum* of the individual potentials,

$$U(\underline{X}(t)) = - \sum_{i,k=1, i \neq k} \frac{\gamma \, m_i \, m_k}{|\underline{x}_i - \underline{x}_k|} \, , \quad \frac{\partial U}{\partial {x^j}_i} = \sum_{i \neq k} \frac{\gamma \, m_i \, m_k}{|\underline{x}_i - \underline{x}_k|^3} (x_i^j - x_k^j)$$

where γ is the gravitational constant. Then NEWTON's axiom yields immediately the equations of motion

$$m_i\ddot{\underline{x}}_i = -\operatorname{grad}_{\underline{x}_i} U(\underline{X}(t)) = \gamma\, m_i\, m_k \sum_{k\neq i} \frac{\underline{x}_k - \underline{x}_i}{|\underline{x}_k - \underline{x}_i|^3}\,,$$

and the total energy E of the (closed) system is obviously

$$E(t) = \sum_{i=1}^{n} \frac{|\dot{\underline{x}}_i(t)|^2}{2} + U(\underline{X}(t)) = \text{ const.}$$

Note also the double negative sign on the right side of the equation of motion.

(c) Mass Points with Constraints Consider again a system of n mass points $\underline{x}_i \in \mathbb{R}^3$ and adopt the above notations. Suppose that the system is momentarily subjected to some holonomic constraints.

Assumption 6.1. (1°) *Let hold the law of conservation of energy (6.36),*

$$\frac{d}{dt}E(\underline{X}(t), t_0) = 0 \in \mathbb{R}\,. \tag{6.49}$$

(2°) *There exist $m < 3n$ sufficiently smooth holonomic-scleronomic side conditions*

$$\underline{g}(\underline{X}) = \underline{0} \in \mathbb{R}^m \tag{6.50}$$

satisfying the regularity condition $\operatorname{rank} \operatorname{grad}_{\underline{X}}\, g(\underline{X}(t)) = m$.

Note that the energy does not depend explicitely on t. By Definition 6.3 there exists a vector field $\widehat{\operatorname{grad}}E(\underline{X}(t)t_0)$ in $\mathbb{R}_{3n}$ such that

$$\frac{d}{dt}E(\underline{X}(t), t_0) = \widehat{\operatorname{grad}}E(\underline{X}(t), t_0) \cdot \dot{\underline{X}}(t) = 0 \in \mathbb{R}\,, \tag{6.51}$$

and (6.50) yields likewise

$$\operatorname{grad} \underline{g}(\underline{X}(t)) \cdot \dot{\underline{X}}(t) = \underline{0} \in \mathbb{R}^m\,. \tag{6.52}$$

The linear system (6.52) for $\dot{\underline{X}}(t)$ has $3n - m$ linearly independent solutions $\underline{U}_i(t, \underline{X}(t)) \in \mathbb{R}^{3n}$, $i = 1 : 3n - m$, by assumption. Substitution into (6.51) leads to the D'ALEMBERT-LAGRANGE *system* (system of virtual displacements)

$$\widehat{\operatorname{grad}}E(\underline{X}(t), t_0) \cdot \underline{U}_i(t, \underline{X}(t)) = 0\,,\;\; i = 1 : 3n - m\,. \tag{6.53}$$

The system (6.53) is linear in the highest derivative $\ddot{\underline{X}}(t)$ if the system (6.51) has this property cf. also the EULER-LAGRANGE equations in Sect. 4.1.

In this formulation of the principle of D'ALEMBERT-LAGRANGE, the "virtual displacements" $\dot{\underline{X}}(t)$ have become *tangents* at the hyper-surfaces represented by the implicit side conditions; see also (Arnold78). However, m dependent variables in (6.53) have still to be eliminated for the final solution by applying once more the system of side conditions $\underline{g}(\underline{X}) = 0$. This is a serious drawback in *nonlinear* constraints and prohibits this device for large mechanical problems. Rather one considers both systems (6.34) and (6.50) together as differential algebraic system and applies the relating numerical approaches for this type of problems. One the other side, the equations of motion can be conceived by HAMILTON's principle as EULER-LAGRANGE equations i.e. variational equations of an extremal problem. Then equality constraints can be incorporated via local LAGRANGE theory see Sect. 3.6 but the result is also an differential-algebraic system enhanced by the LAGRANGE multipliers of which however no derivatives are involved. The latter system seems to have some advantages because it contains the side conditions explicitely whereas D'ALEMBERT's *principle* applies them first implicitely and then explicitely again. Also the author does not know an algorithmic implementation of D'ALEMBERT's principle but two involving LAGRANGE multipliers, BRASEY's `HEM5` and MURUA's `PHEM56`. For further investigation we refer to Sect. 6.8, for some examples to § 11.3, and for additional remarks to Sect. 6.9.

(d) In the subsequent **Examples** in $\mathbb{R}^2$, *discs* with inertia moment appear also besides mass points; therefore additional rotation energy must be regarded.

Example 6.9. (Szabo76). A cylinder $\mathcal{W}$ with radius R and a disc $\mathcal{S}$ with radius $r < R$ are concentric connected and have together the weight $m_2 g$ where g is the gravitaional acceleration. Two ropes (A) and (B) without weight are spooled up on cylinder and disc. Rope (B) carries the weight $m_1 g$ whereas the entire system is suspended at rope (A) see Fig. 6.12. Compute the accelerations!

By D'ALEMBERT's principle of virtual displacement and some intuition we obtain

$$(m_1 g - m_1 \ddot{q})\,\partial q + (m_2 g - m_2 \ddot{s})\,\partial s - \Theta \ddot{\varphi}\,\partial \varphi = 0\,.$$

$\Theta = \displaystyle\int_{\varrho \in \mathcal{W}-\mathcal{S}} \varrho^2\, dm$ is the geometrical moment of inertia and the last term called D'ALEMBERT's force of inertia appears somewhat abruptly at first but supplies the crucial contribution. Now

$$q + (r - R)\varphi = 0\,, \quad s + r\varphi = 0 \tag{6.54}$$

are the side conditions; hence $\partial s = -r\partial\varphi$ und $\partial q = (R - r)\partial\varphi$ and a substitution of $\ddot{s} = -r\ddot{\varphi}$ and $\ddot{q} = (R - r)\ddot{\varphi}$ leads to

$$\{[m_1 g - m_1(R - r)\ddot{\varphi}]\,(R - r) - (m_2 g + m_2 r\ddot{\varphi})\,r - \Theta\ddot{\varphi}\}\,\partial\varphi = 0\,. \tag{6.55}$$

By this way we obtain

$$\ddot{\varphi} = g\frac{m_1(R - r) - m_2 r}{m_1(R - r)^2 + m_2 r^2 + \Theta}\,, \quad \ddot{s} = -r\ddot{\varphi}\,, \quad \ddot{q} = (R - r)\ddot{\varphi}\,.$$

This example shall now be considered in the above more geometrical context. To this end we observe that the problem has three variables $\underline{x}(t) = [q(t), s(t), \varphi(t)]^T$ and two constraints (6.54). Following the concept of LAGRANGE mechanics two artificial constraint forces $\underline{z}_1, \underline{z}_2$ are introduced which shall keep the "mass points" on the surfaces described by (6.54). The total energy E is constant in a closed system,

$$\begin{aligned} E(\underline{x}(t)) &= \frac{m_1}{2}\dot{q}^2 - m_1 g\, q + \frac{m_2}{2}\dot{s}^2 - m_1 g\, s + \frac{\Theta}{2}\dot{\varphi}^2 = E(\underline{x}(t_0)) \\ \frac{dE(\underline{x}(t))}{dt} &= \begin{bmatrix} m_1\ddot{q} - m_1 g \\ m_2\ddot{s} - m_2 g \\ \Theta\ddot{\varphi} \end{bmatrix}\begin{bmatrix} \dot{q} \\ \dot{s} \\ \dot{\varphi} \end{bmatrix} =: \widehat{\mathrm{grad}} E(\underline{x}(t)) \cdot \dot{\underline{x}}(t) = 0\,. \end{aligned}$$

By this way we obtain the augmented NEWTON's equations of motion

$$\widehat{\mathrm{grad}} E(\underline{x}(t)) = \underline{z}_1 + \underline{z}_2\,. \tag{6.56}$$

The constraint forces are supposed to be perpendicular to the surfaces (6.54) and thus point up to sign into the direction of the normals of the surfaces

$$\underline{z}_1 = \lambda_1 \underline{n}_1\,,\ \underline{n}_1 = [1, 0, r - R]^T\,,\ \ \underline{z}_2 = \lambda_2 \underline{n}_2\,,\ \underline{n}_2 = [0, 1, r]^T\,.$$

The tangent vector of *both* implicit constraints is the cross product of the normals

$$\underline{T} = \underline{n}_1 \times \underline{n}_2 = \begin{vmatrix} \underline{e}_1 & 1 & 0 \\ \underline{e}_2 & 0 & 1 \\ \underline{e}_3 & r - R & r \end{vmatrix} = \begin{bmatrix} R - r \\ -r \\ 1 \end{bmatrix}.$$

Now the virtual constraint forces $\underline{z}_1$ and $\underline{z}_2$ in the system (6.56) disappear again ("do not perform real work") if we form the *scalar product* with the vector $\underline{T}$. *Ensuing* a substitution of $\ddot{s} = -r\ddot{\varphi}$, $\ddot{q} = (R - r)\ddot{\varphi}$ supplies also equation (6.55).

Example 6.10. A simple pulley has four independent variables $\underline{x} = [s_1, s_2, \varphi_1, \varphi_2]^T$ and three constraints

$$s_1 + r\varphi_1 = 0\,,\ \ s_2 + r\varphi_2\,,\ \ \varphi_1 + 2\varphi_2 = 0 \tag{6.57}$$

see Fig. 6.13. The total energy is constant in a closed system,

$$\begin{aligned} E(\underline{x}(t)) &= \frac{1}{2}[\Theta\dot{\varphi}_1^2 + \Theta\dot{\varphi}_2^2 + m_1\dot{s}_1^2 + (m_2 + m)\dot{s}_2^2] + m_1 g s_1 + (mg + m_2 g)s_2 \\ &= E(\underline{x}(t_0)) \\ \frac{dE(\underline{x}(t))}{dt} &= [m_1\ddot{s}_1 + m_1 g, (m_2 + m)\ddot{s}_2 + (m_2 g + mg), \Theta\ddot{\varphi}_1, \Theta\ddot{\varphi}_2] \cdot \dot{\underline{x}}(t) \\ &=: \widehat{\mathrm{grad}} E(\underline{x}(t)) \cdot \dot{\underline{x}}(t) = 0 \end{aligned}$$

where m, r, Θ are mass, radius and inertia moment of the discs. The normals of the surface (6.57) are

$$\underline{n}_1 = [1,0,r,0]^T\,,\quad \underline{n}_2 = [0,1,0,r]^T\,,\quad \underline{n}_3 = [0,0,1,2]^T\,,$$

and a tangent $\underline{T}$ with property $\underline{T}\cdot\underline{n}_i = 0\,,\ i = 1:3$ is $\underline{t} = [2r, -r, -2, 1]^T$. By the scalar product $\operatorname{grad} E\cdot\underline{T} = 0$ we obtain

$$0 = 2r\big[m_1\ddot{s}_1 + m_1 g\big] - r\big[(m_2+m)\ddot{s}_2 + (m_2 g + mg)\big] - 2\Theta\ddot{\varphi}_1 + \Theta\ddot{\varphi}_2$$

and, by substitution, see also (Szabo76),

$$\boxed{[r^2(4m_1 + m_2 + m) + 5\Theta]\ddot{\varphi}_1 = 2gr[2m_1 - m_2 - m]}\,.$$

In the following examples with pendulum the x-axis points to right and the y-axis to above (COS).

Example 6.11. The double physical pendulum has six independent variables $\underline{x} = [x_1,\ldots,x_4,\varphi_1,\varphi_2]^T$ and four constraints, see Fig. 6.14,

$$\begin{array}{ll} x_1 - \ell_1\sin\varphi_1 = 0\,, & x_2 + \ell_1\cos\varphi_1 = 0\,,\\ x_3 - \ell\sin\varphi_1 - \ell_2\sin\varphi_2 = 0\,, & x_4 + \ell\cos\varphi_1 + \ell_2\cos\varphi_2 = 0\,. \end{array} \tag{6.58}$$

(x_1, y_1) are the coordinates of the gravity center S_1 and the inertia moment of the first body relative to the rotational axis is Θ_1, (x_3, x_4) are the coordinates of the second gravity center S_2 and Θ_2 is the corresponding inertia moment w.r.t. S_2.(!) Kinetic energy T, potential energy U and total energy $E = T+U$ are

$$\begin{aligned} T(t) &= \frac{1}{2}\Theta_1\dot{\varphi}_1^2 + \frac{1}{2}\Theta_2\dot{\varphi}_2^2 + \frac{1}{2}m_2(\dot{x}_3^2 + \dot{x}_4^2)\\ U(t) &= m_1 g x_2 + m_2 g x_4\\ E(t) &= \frac{1}{2}\Theta_1\dot{\varphi}_1^2 + \frac{1}{2}\Theta_2\dot{\varphi}_2^2 + \frac{1}{2}m_2(\dot{x}_3^2 + \dot{x}_4^2) + m_1 g x_2 + m_2 g x_4 = E(t_0)\\ \frac{dE(t)}{dt} &= [0,\, m_1 g,\, m_2\ddot{x}_3,\, m_2\ddot{x}_4 + m_2 g,\, \Theta_1\ddot{\varphi}_1,\, \Theta_2\ddot{\varphi}_2]\cdot\underline{\dot{x}}\\ &= \widehat{\operatorname{grad}E}\cdot\underline{\dot{x}} = 0\,. \end{aligned} \tag{6.59}$$

Both the desired tangents $\underline{t}_1$, $\underline{t}_2$ are two linearly independent solutions of the homogeneous linear system

$$\begin{bmatrix} 1 & 0 & 0 & 0 & -\ell_1\cos\varphi_1 & 0\\ 0 & 1 & 0 & 0 & -\ell_1\sin\varphi_1 & 0\\ 0 & 0 & 1 & 0 & -\ell\cos\varphi_1 & -\ell_2\cos\varphi_2\\ 0 & 0 & 0 & 1 & -\ell\sin\varphi_1 & -\ell_2\sin\varphi_2 \end{bmatrix}\underline{t} = 0\,;$$

hence

$$\begin{bmatrix}\underline{t}_1^T\\ \underline{t}_2^T\end{bmatrix} = \begin{bmatrix} \ell_1\cos\varphi_1 & \ell_1\sin\varphi_1 & \ell\cos\varphi_1 & \ell\sin\varphi_1 & 1 & 0\\ 0 & 0 & \ell_2\cos\varphi_2 & \ell_2\sin\varphi_2 & 0 & 1 \end{bmatrix}.$$

The both equations $\widehat{\text{grad}}E \cdot \underline{t}_1 = 0$ and $\widehat{\text{grad}}E \cdot \underline{t}_2 = 0$ lead to

$$\begin{aligned} &m_1\ell_1 g \sin\varphi_1 + m_2\ell\ddot{x}_3\cos\varphi_1 + m_2\ell\sin\varphi_1(\ddot{x}_4 + g) + \Theta_1\ddot{\varphi}_1 = 0 \\ &m_2\ell_2\ddot{x}_3\cos\varphi_2 + m_2\ell_2\sin\varphi_2(\ddot{x}_4 + g) + \Theta_2\ddot{\varphi}_2 = 0\,. \end{aligned}$$

In this way we obtain

$$\boxed{\begin{aligned} &\begin{bmatrix} \Theta_1 + \ell^2 m_2 & \ell\ell_2 m_2\cos(\varphi_1-\varphi_2) \\ \ell\ell_2 m_2\cos(\varphi_1-\varphi_2) & \Theta_2 + \ell_2^2 m_2 \end{bmatrix}\begin{bmatrix}\ddot{\varphi}_1\\ \ddot{\varphi}_2\end{bmatrix} \\ + &\quad \ell\ell_2 m_2\begin{bmatrix} \sin(\varphi_1-\varphi_2)(\dot{\varphi}_2)^2 \\ -\sin(\varphi_1-\varphi_2)(\dot{\varphi}_1)^2\end{bmatrix} \\ = &-\begin{bmatrix} (\ell_1 m_1 + \ell m_2)g\sin\varphi_1 \\ \ell_2 m_2 g\sin\varphi_2 \end{bmatrix} \end{aligned}}$$

cf. `SUPPLEMENT\chap06`. The same result is obtained by applying HAMILTON's principle and the corresponding EULER equations (Szabo77), p. 89.

Example 6.12. The double mathematical pendulum follows from the double physical pendulum by setting $\ell_1 = \ell$, $\Theta_1 = m_1\ell^2$, $\Theta_2 = 0$; hence

$$\boxed{\begin{aligned} &\begin{bmatrix} (m_1 + m_2)\ell^2 & \ell\ell_2 m_2\cos(\varphi_1-\varphi_2) \\ \ell\ell_2 m_2\cos(\varphi_1-\varphi_2) & \ell_2^2 m_2 \end{bmatrix}\begin{bmatrix}\ddot{\varphi}_1\\ \ddot{\varphi}_2\end{bmatrix} \\ + &\quad \ell\ell_2 m_2\begin{bmatrix} \sin(\varphi_1-\varphi_2)(\dot{\varphi}_2)^2 \\ -\sin(\varphi_1-\varphi_2)(\dot{\varphi}_1)^2\end{bmatrix} \\ = &-\begin{bmatrix} (m_1 + m_2)\ell g\sin\varphi_1 \\ \ell_2 m_2 g\sin\varphi_2 \end{bmatrix} \end{aligned}}\,.$$

Alternative derivation of double mathematical pendulum, cf. Sect. 11.3, Example 11.1. The double mathematical pendulum has four independent variables $\underline{x} = [x_1, x_2, x_3, x_4] := [x_1, y_1, x_2, y_2]$, see Fig. 6.14), and two constraints

$$x_1^2 + x_2^2 = \ell^2\,,\quad (x_3 - x_1)^2 + (x_4 - x_2)^2 = \ell_2^2\,. \tag{6.60}$$

The total energy is

$$\begin{aligned} E(t) &= \frac{m_1}{2}(\dot{x}_1^2 + \dot{x}_2^2) + \frac{m_2}{2}(\dot{x}_3^2 + \dot{x}_4^2) + m_1 g x_2 + m_2 g x_4 = E(t_0) \\ \frac{dE(t)}{dt} &= [m_1\ddot{x}_1,\ m_1\ddot{x}_2 + m_1 g,\ m_2\ddot{x}_3,\ m_2\ddot{x}_4 + m_2 g]\cdot\underline{\dot{x}}(t) \\ &= \widehat{\text{grad}}E\cdot\underline{\dot{x}}(t) = 0 \end{aligned} \tag{6.61}$$

which coincides with (6.59) under Assumption (6.60) because $\Theta_1(\dot{\varphi}_1)^2 = m_1(\dot{x}_1^2 + \dot{x}_2^2)$. The normals of the surfaces (6.60) are

$$\underline{n}_1 = [x_1, x_2, 0, 0]^T\,,\quad \underline{n}_2 = [x_1 - x_3,\ x_2 - x_4,\ x_3 - x_1,\ x_4 - x_2]^T$$

and two tangents perpendicular to the both normals are

$$\begin{aligned}\underline{t}_1 &= [x_2(x_4 - x_2),\ -x_1(x_4 - x_2),\ 0,\ x_2x_3 - x_1x_4]^T\\ \underline{t}_2 &= [0,\ 0,\ x_4 - x_2,\ x_1 - x_3]^T\,.\end{aligned}$$

Multiplication of (6.61) by $\underline{t}_1$ resp. $\underline{t}_2$ leads to the system

$$\begin{aligned}&m\ddot{x}_1x_2(x_4 - x_2) - (m\ddot{x}_2 - mg)x_1(x_4 - x_2)\\ &+(m\ddot{x}_4 - mg)(x_2x_3 - x_1x_4) &= 0\\ &m\ddot{x}_3(x_4 - x_2) + (m\ddot{x}_4 - mg)(x_1 - x_3) &= 0\,.\end{aligned} \tag{6.62}$$

The rest follows in the same way as above to the same result.

An example with holonomic-rheonomic constraints is given in `SUPPLEMENT\chap06b`.

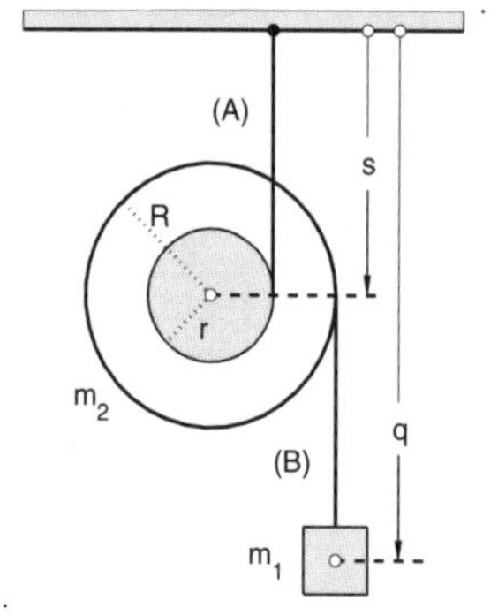

Figure 6.12. Ex. 6.9

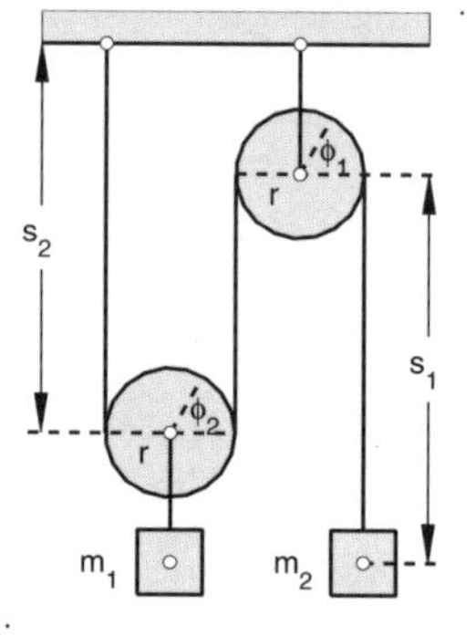

Figure 6.13. Pulley

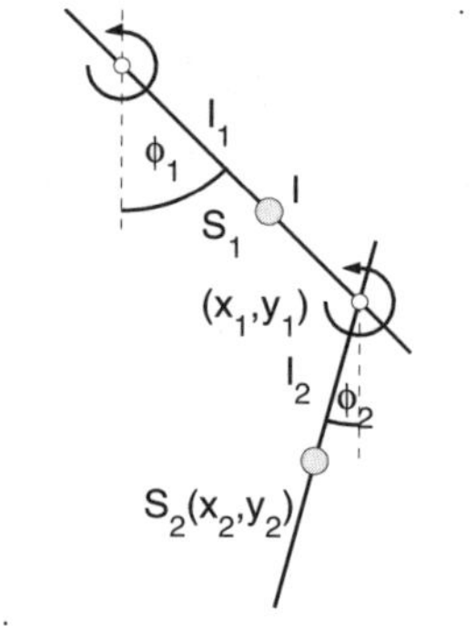

Figure 6.14. Pendulum

6.5 The Three-Body Problem

(a) Formulation of the Problem Let three points P_i with masses m_i , $i = 1, 2, 3$, move in a *plane* under mutual attraction, and let $(x_i,\ y_i)$ be the coordinates of P_i in a fixed global COS. Then NEWTON's axiom and the law of gravitation lead to a system of six differential equations

$$\left.\begin{aligned}m_i\ddot{x}_i &= \frac{\gamma\, m_i m_{i+1}}{r_{i,i+1}^3}(x_{i+1} - x_i) + \frac{\gamma\, m_i m_{i+2}}{r_{i,i+2}^3}(x_{i+2} - x_i)\\ m_i\ddot{y}_i &= \frac{\gamma\, m_i m_{i+1}}{r_{i,i+1}^3}(y_{i+1} - y_i) + \frac{\gamma\, m_i m_{i+2}}{r_{i,i+2}^3}(y_{i+2} - y_i)\end{aligned}\right\},\quad i = 1:3, \tag{6.63}$$

(indices modulo 3, no summation) where

$$r_{i,k} = r_{k,i} = \big((x_i - x_k)^2 + (y_i - y_k)^2\big)^{1/2}$$

and the masses m_i are canceled out, see also Example 6.8. The total energy

$$E = \sum_{i=1}^{3} \frac{1}{2} m_i \left(\dot{x_i}^2 + \dot{y_i}^2 \right) - \frac{\gamma m_1 m_2}{r_{1,2}} - \frac{\gamma m_2 m_3}{r_{2,3}} - \frac{\gamma m_3 m_1}{r_{3,1}}$$

is again constant and hence an invariant of the system. We introduce a *characteristic unit of mass* M and a *characteristic unit of length* L (both relative to the given problem) then

$$\widetilde{m}_i = \frac{m_i}{M}\,, \quad \widetilde{x}_i = \frac{x_i}{L}\,, \quad \widetilde{t} = t \left(\frac{\gamma M}{L^3} \right)^{1/2}, \quad \widetilde{v} = \left(\frac{L}{\gamma M} \right)^{1/2} v \tag{6.64}$$

are quantities without dimension (v velocity). The units of mass, length, and time in dimensionless system are then M, L, $(L^3/\gamma M)^{1/2}$, and $v_0 = (\gamma M)/L)^{1/2}$ is the unit of velocity. By (6.28), this is exactly the velocity which keeps a point of arbitrary mass m on a circular orbit with radius L. Applying the transformation (6.64) the differential system (6.63) becomes

$$\boxed{\begin{aligned} \ddot{x}_i &= \frac{m_{i+1}(x_{i+1} - x_i)}{r_{i,i+1}^3} + \frac{m_{i+2}(x_{i+2} - x_i)}{r_{i,i+2}^3} \\ \ddot{y}_i &= \frac{m_{i+1}(y_{i+1} - y_i)}{r_{i,i+1}^3} + \frac{m_{i+2}(y_{i+2} - y_i)}{r_{i,i+2}^3} \end{aligned}\,, \quad i = 1:3 \text{ modulo } 3} \tag{6.65}$$

where a point denotes differentiation w.r.t. $\widetilde{t}$ but afterwards we write again t instead $\widetilde{t}$, m_i instead $\widetilde{m}_i$ etc.. In *relative motion* the orbits of the mass points are displayed relative to the gravity center S of the system which has the coordinates

$$x_s = \frac{m_1 x_1 + m_2 x_2 + m_3 x_3}{m_1 + m_2 + m_3}\,, \quad y_s = \frac{m_1 y_1 + m_2 y_2 + m_3 y_3}{m_1 + m_2 + m_3}\,.$$

The gravity center moves on a straight line which is given by the initial conditions because no external energy is pumped into the system.

(b) Of course, the differential system for the **Two-Body Problem** is (6.63) for $i = 1:2$. In general, there arise ellipses which are neither concentric nor stationary. But suppose that the two mass points m_1 and m_2 rotate about their common mass center with constant distance d which it also supposed in **(c)**. Conveniently we choose $L = d$ in (6.64) and remember that centripetal force equals gravitational force (v scalar velocity),

$$\frac{m_1 v_1^2}{r_1} = \frac{\gamma m_1 m_2}{d^2} = \frac{m_2 v_2^2}{r_2}\,;$$

then

$$\begin{aligned}
r_1 &= \frac{m_2}{M}\,d \Longrightarrow v_1 = m_2\left(\frac{\gamma}{M\,d}\right)^{1/2} = \frac{m_2}{M}\left(\frac{\gamma M}{d}\right)^{1/2} \Longrightarrow \widetilde{v}_1 = \frac{m_2}{M}\\
r_2 &= \frac{m_1}{M}\,d \Longrightarrow v_2 = m_1\left(\frac{\gamma}{M\,d}\right)^{1/2} = \frac{m_1}{M}\left(\frac{\gamma M}{d}\right)^{1/2} \Longrightarrow \widetilde{v}_2 = \frac{m_1}{M}\\
\frac{v_1}{r_1} &= m_2\left(\frac{\gamma}{M\,d}\right)^{1/2}\frac{M}{m_2 d} = \left(\frac{\gamma M}{d^3}\right)^{1/2} = \frac{v_2}{r_2} \quad \text{common angular velocity}\,.
\end{aligned}$$

Summary: Two mass points m_1 and m_2 rotate in *dimensionless system* about their common mass center in origin if

$$r_1 = -\frac{m_2}{m_1+m_2}\,,\; r_2 = \frac{m_1}{m_1+m_2}\,,\; v_1 = r_1\,,\; v_2 = r_2\,.$$

(c) In the **Restricted Three-Body Problem**, the gravitational constant is likewise normed to one and the partition of mass is chosen such that

$$m_1 = 1-\mu =: \mu'\,,\;\; m_2 = \mu\,,\;\; m_3 = 0\,,\;\; 0 \le \mu \le 1\,. \tag{6.66}$$

Then the motion of P_1 and P_2 does not depend on P_3, and one supposes that both move on circular orbits with angular velocity one relative to their *common gravity center*. This simplification leads to a *reduced differential system*

$$\begin{aligned}
\ddot{x} &= x + 2\dot{y} - \mu\frac{x-\mu'}{\left[(x-\mu')^2+y^2\right]^{3/2}} - \mu'\frac{x+\mu}{\left[(x+\mu)^2+y^2\right]^{3/2}}\\
\ddot{y} &= y - 2\dot{x} - \mu\frac{y}{\left[(x-\mu')^2+y^2\right]^{3/2}} - \mu'\frac{y}{\left[(x+\mu)^2+y^2\right]^{3/2}}
\end{aligned} \tag{6.67}$$

for the orbit of $P_3(x,y) \in \mathbb{R}^2$.

For the sake of completeness we note LAGRANGE function L of the action integral (of which these equations are the EULER equations), total energy E and HAMILTONian H. For a unified representation we write $\underline{q}$ instead $\underline{x}$ and $\underline{p} = \operatorname{grad}_{\dot{\underline{q}}} L$ as usual; let also for brevity

$$\varrho(\underline{q}) = \left[(q_1-\mu')^2+q_2^2\right]^{1/2}\,,\;\; \sigma(\underline{q}) = \left[(q_1+\mu)^2+q_2^2\right]^{1/2}\,.$$

$$\begin{aligned}
L(\underline{q},\dot{\underline{q}}) &= \frac{|\dot{\underline{q}}|^2}{2} + \dot{q}_1 q_2 - \dot{q}_2 q_1 + \frac{\mu}{\varrho(\underline{q})} + \frac{\mu'}{\sigma(\underline{q})}\\
E(\underline{q},\dot{\underline{q}}) &= \frac{|\dot{\underline{q}}|^2}{2} - \frac{|\underline{q}|^2}{2} - \frac{\mu}{\varrho(\underline{q})} - \frac{\mu'}{\sigma(\underline{q})}\\
H(\underline{p},\underline{q}) &= \frac{|\underline{p}|^2}{2} + q_1 p_2 - q_2 p_1 - \frac{\mu}{\varrho(\underline{q})} - \frac{\mu'}{\sigma(\underline{q})}\,.
\end{aligned}$$

Recall that a differential system $\dot{\underline{u}} = \underline{f}(\underline{u})$ has singular points $\underline{u}^*$ where $\underline{f}(\underline{u}^*) = \underline{0}$. The first order system associated to (6.67) is

$$\begin{aligned}\dot{\underline{x}} &= \underline{y}\\ \dot{\underline{y}} &= 2\begin{bmatrix}0 & 1\\ -1 & 0\end{bmatrix}\underline{y} + \underline{x}\left(1 - \mu'\frac{1}{\varrho(\underline{x})^3} - \mu\frac{1}{\sigma(\underline{x})^3}\right) + \mu\mu'\begin{bmatrix}\varrho(\underline{x})^{-3} - \sigma(\underline{x})^{-3}\\ 0\end{bmatrix}.\end{aligned} \tag{6.68}$$

But, for $\underline{y} = 0$, the second equation can be written as gradient of a scalar function V where

$$V(x_1, x_2) = \frac{1}{2}\left[x_1^2 + x_2^2\right] + \frac{\mu}{\varrho(\underline{x})} + \frac{\mu'}{\sigma(\underline{x})}. \tag{6.69}$$

Accordingly, the singular points of the first order system (6.68) are the points satisfying

$$\dot{\underline{x}} = 0\,, \quad \text{grad}_x V(\underline{x}) = 0\,.$$

For $x_2 = 0$ we find *three* collinear singular points and for $x_2 \neq 0$ the lower second equation in (6.68) shows that further singular points occur when $\varrho = \sigma$ and this equation is fulfilled for LAGRANGE*'s equilateral triangle solutions* $x_1 = \frac{1}{2} - \mu$ and $x_2 = \pm 3^{1/2}/2$. For *Earth* and *Moon* ($\mu = 0.012277471$), $V(x, 0)$ is sketched in Fig. 6.21 and a graph of the potential $V(x_1, x_2)$ in Figure 6.22.

(d) Periodic Solutions The trajectories of the general as well as the restricted three-body problem develop *chaotically* in normal case; cf. e.g., (Acheson). Following (Arenstorf), the system (6.67) for the *restricted* problem is written as *complex* differential equation for the computation of special periodic solutions,

$$\ddot{u} + 2i\,\dot{u} - u = -\frac{\mu}{|u + \mu - 1|^3}(u + \mu - 1) - \frac{1 - \mu}{|u + \mu|^3}(u + \mu) \tag{6.70}$$

where $i = \sqrt{-1}$ and $u = x + i\,y$. The solutions of this equation are well-known for $\mu = 0$ and may be written as $u(t) = e^{-i\,t}z(t)$ where $z(t) \in \mathbb{C}$ denotes the KEPLER motion, i.e., is the solution of $\ddot{z} = -z\,|z|^{-3}$. Choosing, e.g., the initial conditions

$$z(0) = a(1 + \varepsilon)\,, \quad \dot{z} = \frac{i\,c}{z(0)}\,, \quad c^2 = a(1 - \varepsilon^2)\,, \quad 0 < a\,,\ 0 < \varepsilon < 1\,,$$

the orbit z is an *ellipse* with large semi-axis a, numerical eccentricity ε, one focus in origin, apocenter in $z(0)$ and time of circulation $T_0 = 2\pi|a^{3/2}|$; hence by (6.32)

$$u(t) = e^{-i\,t}z(t)\,, \ z(t) = a(\varepsilon + \cos\psi + i\sqrt{1 - \varepsilon^2}\sin\psi)\,, \ t = a^{3/2}(\psi + \varepsilon\sin\psi)\,.$$

The orbit $u(t)$ is periodic if T_0 is a rational multiple of 2π resp. if $a^{3/2} = m/k$ with relatively prime $k, m \in \mathbb{N}$ ($\text{sign}(k) = \text{sign}(c)$). The function u then decribes in $\mathbb{C}$ a *rotating ellipse* with total circulation time $T = 2\pi\,m$ which becomes a closed curve after $k - m$ rotations about the origin. (Arenstorf) has proved that this periodicity (with different circulation time) is preserved for small parameter $\mu > 0$ which is an astonishing result in view of the fragility of periodic solutions.

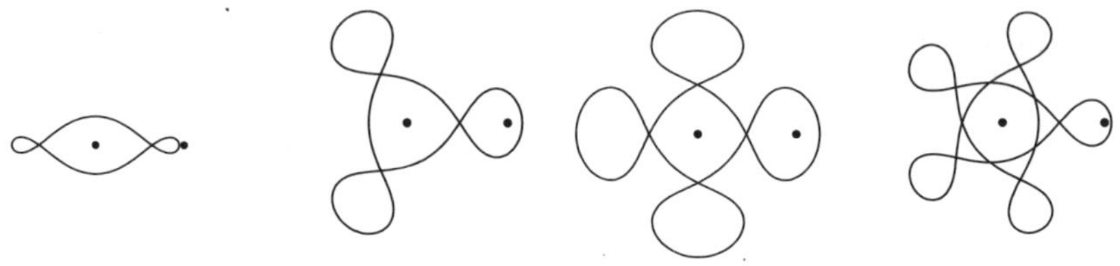

(m,k) = (1,2) (m,k) = (2,3) (m,k) = (3,4) (m,k)= (3,5)

Figure 6.15. Arenstorf orbits

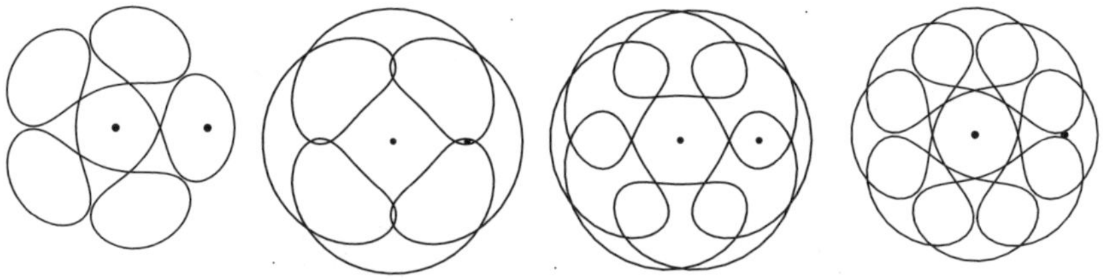

(m,k) = (4,5) (m,k) = (5,4) (m,k) = (7,6) (m,k)= (7,8)

Figure 6.16. Arenstorf orbits cont.

Figures 6.17 and 6.18 show an ARENSTORF orbit in absolute and relative frame of reference; cf. *Example* 2.19. In Figures 6.19 and 6.20 $m_1 = m_2 = 0.5$, both masses rotate with same velocity $v = 0.5$ about the origin; the mass point P_3 with $m_3 = 0$ leaves here the system also if its velocity becomes too large. In all cases the dimensionless differential system has been solved by `ode45.m` and (6.65). Call also `KAPITEL06\SECTION_5\demo2.m` for some further interesting examples.

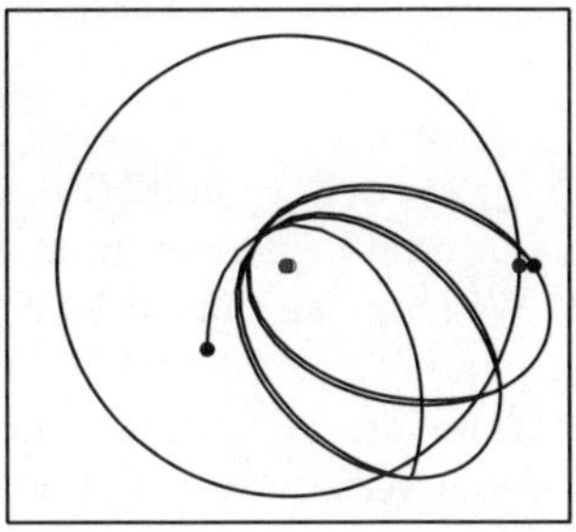

Figure 6.17. Absolute motion

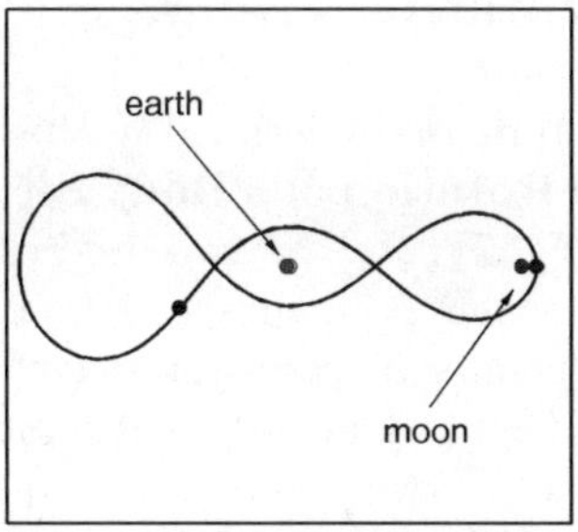

Figure 6.18. Relative motion

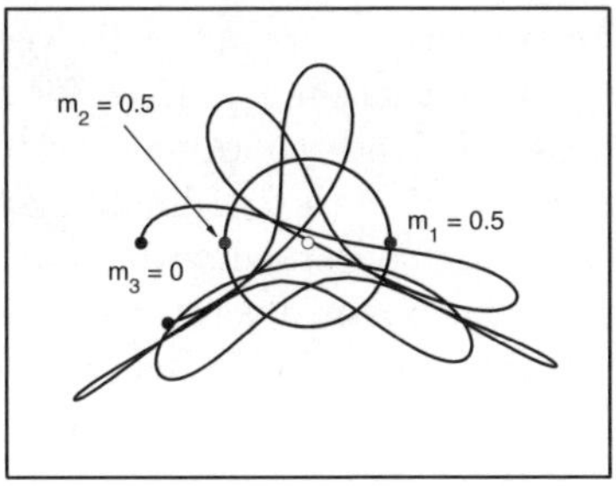

Figure 6.19. Absolute motion

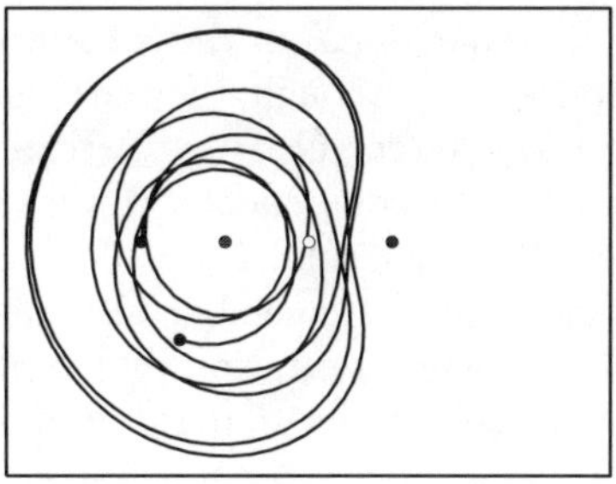

Figure 6.20. Relative motion

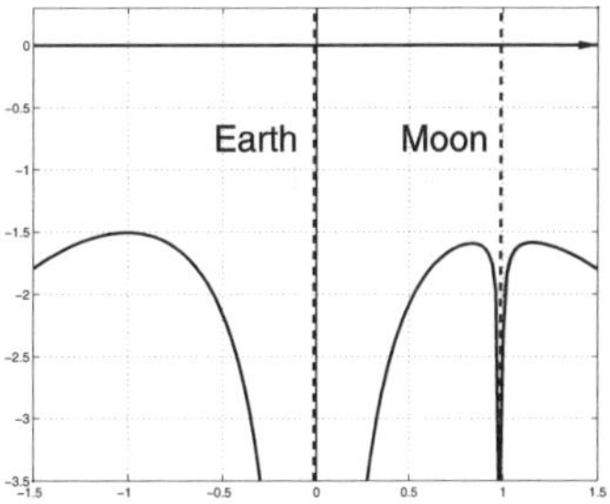

Figure 6.21. Potential $V(x, 0)$ for Earth and Moon

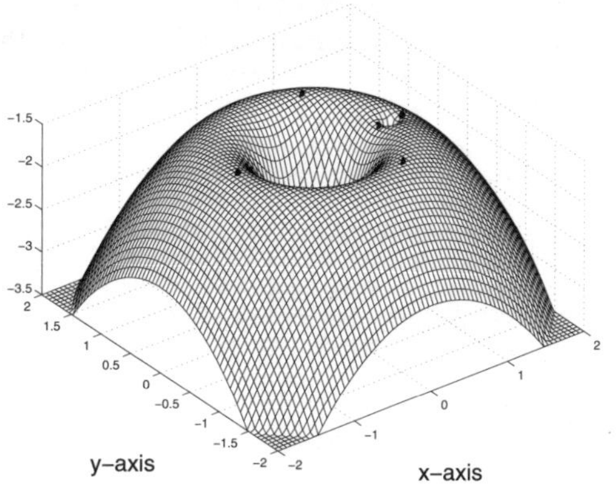

Figure 6.22. Potential V for Earth and Moon

6.6 Rotating Frames

All vectors are underlined in this section.

(a) Rotation of a Body Let $\{\mathcal{O}_\mathcal{E};\, \mathcal{E}\} = \{\mathcal{O}_\mathcal{E};\, \mathcal{E}_1,\, \mathcal{E}_2,\, \mathcal{E}_3\}$ and $\{\mathcal{O}_\mathcal{F};\, \mathcal{F}\} = \{\mathcal{O}_\mathcal{F};\, \mathcal{F}_1,\, \mathcal{F}_2,\, \mathcal{F}_3\}$ be two different cartesian coordinate systems in an EUKLIDian scalar product space E, cf. Sect. 10.4. We suppose that $\{\mathcal{O}_\mathcal{E};\, \mathcal{E}\}$ is a space-fixed or *inertial* system (reference system) and $\{\mathcal{O}_\mathcal{F}(t);\, \mathcal{F}(t)\}$ is a body-fixed COS of some body which moves relative to the inertial system. Since the difference of two points in E is defined and is also a vector, we can describe a *rigid body motion* per definitionem by

$$\mathcal{F} = (\mathcal{O}_\mathcal{F}(t) - \mathcal{O}_\mathcal{E}) + \mathcal{E}D(t) \tag{6.71}$$

in *formal* notation where $D(t) \in \mathbb{R}^3{}_3$ is an orthogonal matrix with positive determinant. "Formal" means here again that we can operate with the elements $\mathcal{E}_i$ of $\mathcal{E}$ and $\mathcal{F}_k$ of $\mathcal{F}$ in the same way as with *column vectors* in matrix-vector computation although they are really column vectors only if $E = \mathbb{R}^3$. Of course the vector $\mathcal{O}_\mathcal{F}(t) - \mathcal{O}_\mathcal{E}$ can be expressed in both coordinate systems with different components. However we suppose that both coordinate system have the same origin, $\mathcal{O}_\mathcal{F} = \mathcal{O}_\mathcal{E}$, since we are only concerned with rotations in this section. Remember also that the rotation axis is always eigenvector of the associated rotation matrix; hence it is the one and only vector which remains fixed under that rotation.

An arbitrary vector $\underline{v}$ in the EUCLIDian space E has the form

$$\boxed{\underline{v} = \mathcal{F}\underline{x}_\mathcal{F} = \mathcal{E}\underline{x}_\mathcal{E} = \mathcal{E}DD^T\underline{x}_\mathcal{E} \quad\Longrightarrow\quad \underline{x}_\mathcal{F} = D^T\underline{x}_\mathcal{E}\,,\ \underline{x}_\mathcal{E} = D\underline{x}_\mathcal{F}}\,; \tag{6.72}$$

see also Sect. 1.1**(a6)**. (In technical sciences the transformation matrix D is replaced frequently by D^T according to the opinion that coordinate transformation is more important than basis transformation.)

Example 6.13. Let

$$D(\underline{e}_1,\alpha) = \begin{bmatrix} 1 & 0 & 0\\ 0 & \cos\alpha & -\sin\alpha\\ 0 & \sin\alpha & \cos\alpha \end{bmatrix},\quad D(\underline{e}_2,\alpha) = \begin{bmatrix} \cos\alpha & 0 & -\sin\alpha\\ 0 & 1 & 0\\ \sin\alpha & 0 & \cos\alpha \end{bmatrix},\quad D(\underline{e}_3,\alpha) = \begin{bmatrix} \cos\alpha & -\sin\alpha & 0\\ \sin\alpha & \cos\alpha & 0\\ 0 & 0 & 1 \end{bmatrix}. \tag{6.73}$$

be the *fundamental rotation matrices* where $I = [\underline{e}_1, \underline{e}_2, \underline{e}_3] \in \mathbb{R}^3{}_3$ is the unit matrix. Then, e.g., a new basis $\mathcal{F} = \mathcal{E}D(\underline{e}_i, \alpha)$ results from $\mathcal{E}$ by rotation about the basis vector $\mathcal{E}_i$ with angle α in mathematically *positive* sense.

Example 6.14. Let $\underline{a} \in \mathbb{R}^3$ be a normed rotation axis, $|\underline{a}| = 1$, and let

$$D(\underline{a}, \varphi) = \cos(\varphi)I + (1 - \cos(\varphi))\underline{a}\,\underline{a}^T + \sin(\varphi)C\,,\quad C\underline{x} = \underline{a} \times \underline{x}\,,$$

be the rotation matrix of Sect. 1.1(**i**). Then $\mathcal{F} = \mathcal{E}D(\underline{a}, \varphi)$ is rotated about $\mathcal{E}\underline{a}$ with angle φ in *positive* direction; and as in (1.14)

$$\dot{D}(\underline{a}, \omega t) = \omega C D(\underline{a}, \omega t) = \omega D(\underline{a}, \omega t) C \,. \tag{6.74}$$

Example 6.15. In KARDAN *angles* (G. CARDANO, 1501-1576), at first $\mathcal{E}_1$ is rotation axis with angle α then the *image* of $\mathcal{E}_2$ is rotation axis with angle β and finally the *image* of $\mathcal{E}_3$ is rotation axis with angle γ, therefore

$$\mathcal{F} = \Big[\big[\mathcal{E}D(\underline{e}_1, \alpha)\big]D(\underline{e}_2, \beta)\Big]D(\underline{e}_3, \gamma) = \mathcal{E}D(\underline{e}_1, \alpha)D(\underline{e}_2, \beta)D(\underline{e}_3, \gamma)\,,$$

In order to prevent overlapping and singularities, the rotation angles have to be sufficiently small or *complementary rotation angles* have to introduced; cf., e.g., (Schiehlen86).

Example 6.16. In EULER-*angles* (L. EULER, 1707-1783), at first $\mathcal{E}_3$ is rotation axis with angle φ, then the *image* of $\mathcal{E}_1$ is rotation axis with angle ϑ and finally the *image* $\mathcal{F}_3$ of $\mathcal{E}_3$ is axis with angle ψ,

$$\mathcal{F} = \mathcal{E}D(\underline{e}_3, \varphi)D(\underline{e}_1, \vartheta)D(\underline{e}_3, \psi)\,,$$
$$0 \le \varphi < 2\pi\,, \quad 0 \le \vartheta < \pi\,, \quad 0 \le \psi < 2\pi\,.$$

In the theory of top there is a local COS $\mathcal{F}$ on the body and a global COS $\mathcal{E}$ in space with common origin. $\mathcal{F}_3$ is then the symmetry axis of, say, a symmetric heavy top, the angle φ describes the *precession*, i.e., the declination of the top under influence of gravity force, the angle ϑ describes the *nutation*, i.e., the periodic change in inclination, and the angle ψ the rotation around the top's axis $\mathcal{F}_3$, however the notations are not always the same. (In EUKLIDian space $\mathbb{R}^3$ the axes $\mathcal{E}_i = \underline{e}_i$ may be chosen conveniently.) We then obtain

$$D(\varphi, \vartheta, \psi) :=$$
$$\begin{bmatrix} \cos\psi\cos\varphi - \sin\psi\cos\vartheta\sin\varphi & -\sin\psi\cos\varphi - \cos\psi\cos\vartheta\sin\varphi & \sin\vartheta\sin\varphi \\ \cos\psi\sin\varphi + \sin\psi\cos\vartheta\cos\varphi & -\sin\psi\sin\varphi + \cos\psi\cos\vartheta\cos\varphi & -\sin\vartheta\cos\varphi \\ \sin\psi\sin\vartheta & \cos\psi\sin\vartheta & \cos\vartheta \end{bmatrix} \tag{6.75}$$

for the matrix of the resulting total rotation with unnormed rotation axis $\mathcal{E}\underline{a}$ and

$$\underline{a} = [\sin\vartheta(\sin\varphi - \sin\psi)\,,\ \sin\vartheta(\cos\varphi + \cos\psi)\,,\ \sin(\varphi + \psi)(\cos\vartheta + 1)]^T \in \mathbb{R}^3\,.$$

Example 6.17. Writing $\underline{u} = (r, \varphi, \vartheta)$, the *spherical coordinates* are defined by

$$\underline{f}(\underline{u}) = \big[r\cos\varphi\,\cos\vartheta,\ r\sin\varphi\,\cos\vartheta,\ r\sin\vartheta\big]^T\,,$$
$$0 \le r\,,\quad 0 \le \varphi < 2\pi\,,\quad -\pi/2 < \vartheta < \pi/2\,.$$

(Sometimes the range of ϑ is translated by $\pi/2$.) For instance on the surface of earth, the basis vector $\mathcal{E}_3$ is the axis of the earth pointing to the *North Pole* and $\mathcal{E}_1$ has the null-length of GREENWICH. Then the angle φ is the geographic length being positive in direction *East*, and ϑ the geographic latitude being positive in direction *North*. After normalization by the metric tensor $M(\underline{u})$ the columns of the matrix

$$\begin{aligned} F(\underline{u}) = \operatorname{grad} \underline{f}(\underline{u}) M(\underline{u})^{-1/2} &= \begin{bmatrix} \cos\varphi\,\cos\vartheta & -\sin\varphi & -\cos\varphi\,\sin\vartheta \\ \sin\varphi\,\cos\vartheta & \cos\varphi & -\sin\varphi\,\sin\vartheta \\ \sin\vartheta & 0 & \cos\vartheta \end{bmatrix} \\ &= D(\underline{e}_3, \varphi) D(\underline{e}_2, \vartheta) \end{aligned} \tag{6.76}$$

form an orthogonal *moving frame* on the surface of earth, where the origin as well as North and South pole are *singularities* to be excluded; cf. Sect. 10.5.

In the present notation $F(\underline{u}) = [\mathcal{F}_1, \mathcal{F}_2, \mathcal{F}_3]$, the vector $\mathcal{F}_1$ points to the zenith, $\mathcal{F}_2$ in direction of increasing φ and $\mathcal{F}_3$ in direction of increasing ϑ. The system moves on a circle of latitude in direction East if ϑ is kept fixed and φ increases monotonically. The system moves on a meridian to North for fixed φ and increasing ϑ. But it is of advantage to use instead of $\mathcal{F}$ the likewise positive oriented system

$$\mathcal{G} = \{\mathcal{G}_1,\, \mathcal{G}_2,\, \mathcal{G}_3\} = \{\mathcal{F}_2,\, \mathcal{F}_3,\, \mathcal{F}_1\} \tag{6.77}$$

then $\mathcal{G}_1$ points in direction of increasing degree of longitude φ, $\mathcal{G}_2$ in direction of increasing degree of altitude ϑ, and $\mathcal{G}_3$ points to the zenith.

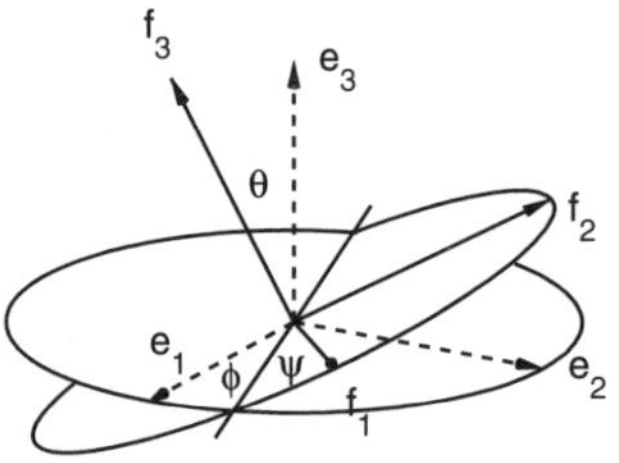

Figure 6.23. EULER angles

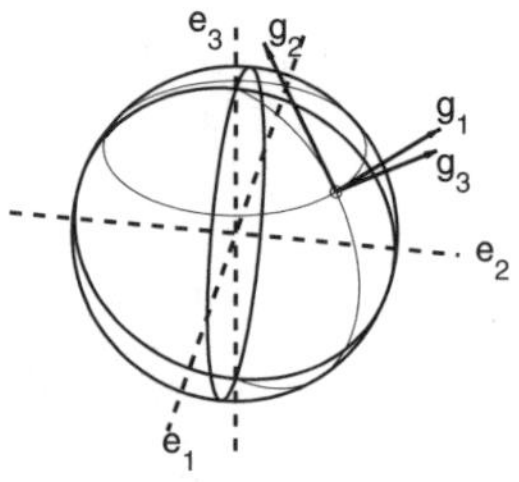

Figure 6.24. Spherical coordinates

(b) Two Rotations in succession with angle velocities $\omega_1\underline{a}_1$ and $\omega_2\underline{a}_3$ give again a rotation with an angle velocity $\omega_3\underline{a}_3(t)$ in $\mathcal{E}$ $(|\underline{a}|_i = 1)$,

$$\begin{aligned} &\mathcal{F} = \mathcal{E}D(\underline{a}_1, \omega_1 t)\,, \quad \mathcal{G} = \mathcal{F}D(\underline{a}_2, \omega_2 t)\,, \\ &\mathcal{H} = \mathcal{E}D(\underline{a}_3(t), \omega_3 t) = \mathcal{E}D(\underline{a}_1, \omega_1 t) D(\underline{a}_2, \omega_2 t)\,. \end{aligned}$$

The axis $\underline{a}_1$ shall be *fixed* in $\mathcal{E}$ and $\underline{a}_2$ *fixed* in $\mathcal{F}$. For brevity let $D_1(t) = D(\underline{a}_1, \omega_1 t)$, $D_2(t) = D(\underline{a}_2, \omega_2 t)$, $D_3(t) = D(\underline{a}_3(t), \omega_3 t)$ then $D_3(t) =$

$D_1(t)D_2(t)$ and $\omega_1 C_1 = \dot{D}_1 D_1^T$, $\omega_2 C_2 = \dot{D}_2 D_2^T$ by (6.74), (1.15). We thus obtain

$$\begin{aligned}\dot{D}_3 D_3^T &= \dot{D}_1 D_2 D_2^T D_1^T + D_1 \dot{D}_2 D_2^T D_1^T = \dot{D}_1 D_1^T D_1 D_2 D_2^T D_1^T + \omega_2 D_1 C_2 D_1^T \\ &= \omega_1 C_1 + \omega_2 D_1 C_2 D_1^T =: \omega_3 C_3 .\end{aligned}$$

The matrix C_3 is likewise skew-symmetric and

$$\begin{aligned}\omega_3 C_3 \underline{x} &= \omega_1 C_1 \underline{x} + \omega_2 D_1 C_2 D_1^T \underline{x} = \omega_1 (\underline{a}_1 \times \underline{x}) + \omega_2 D_1 (\underline{a}_2 \times D_1^T \underline{x}) \\ &= (\omega_1 \underline{a}_1 + \omega_2 D_1(t) \underline{a}_2) \times \underline{x} =: \omega_3 \underline{a}_3(t) \times \underline{x} ,\end{aligned}$$

where $D_1(t)\underline{a}_2$ is the axis pulled back from the system $\mathcal{F}$ into the system $\mathcal{E}$; the rotation axis $\underline{a}_3(t)$ is also called *instantaneous rotation axis* of the body or *instantaneous pole* in $\mathbb{R}^2$. Since $\underline{a}_1^T D_1 = \underline{a}_1^T$ we have altogether

$$\boxed{\omega_3 \underline{a}_3(t) = \omega_1 \underline{a}_1 + \omega_2 D_1(t) \underline{a}_2 , \quad |\omega_3|^2 = \omega_1^2 + 2\omega_1 \omega_2 \underline{a}_1^T \underline{a}_2 + \omega_2^2} \, .$$

(c) Consider now a rotation $\mathcal{F} = \mathcal{E} D(t) D_0$ where a fixed rotation is premultiplied for later application. The angular velocity shall be constant with normed rotation axis $\underline{a}$, i.e., $\underline{\omega}_{\mathcal{E}} = \omega \underline{a}$, $|\underline{a}| = 1$ then

$$\underline{\omega}_{\mathcal{F}} = D_0^T D(t)^T \underline{\omega}_{\mathcal{E}} = D_0^T \underline{\omega}_{\mathcal{E}} .$$

Furthermore, suppose that $\underline{y}_{\mathcal{F}}(t) = \underline{\varphi}_{\mathcal{F}}(t, \underline{x}_{\mathcal{F}})$ is a motion in *rotating* system $\mathcal{F}(t)$ then the image of $\underline{y}_{\mathcal{F}}(t)$ in system $\mathcal{E}$ satisfies

$$\underline{y}_{\mathcal{E}}(t) = D(t) D_0 \underline{y}_{\mathcal{F}}(t) = D(t) D_0 \underline{\varphi}_{\mathcal{F}}(t, \underline{x}_{\mathcal{F}}) . \tag{6.78}$$

The product rule yields by using (6.74)

$$\dot{\underline{y}}_{\mathcal{E}}(t) = D(t) D_0 \dot{\underline{y}}_{\mathcal{F}}(t) + \dot{D}(t) D_0 \underline{y}_{\mathcal{F}}(t) = D(t) D_0 \dot{\underline{y}}_{\mathcal{F}}(t) + \omega C D(t) D_0 \underline{y}_{\mathcal{F}}(t) \tag{6.79}$$

where

$$\begin{aligned}\omega C D(t) D_0 \underline{y}_{\mathcal{F}} &= \omega D(t) C D_0 \underline{y}_{\mathcal{F}} &&= D(t)(\underline{\omega}_{\mathcal{E}} \times D_0 \underline{y}_{\mathcal{F}}) \\ &= D(t) D_0 (D_0^T \underline{\omega}_{\mathcal{E}} \times \underline{y}_{\mathcal{F}}) &&= D(t) D_0 (\underline{\omega}_{\mathcal{F}} \times \underline{y}_{\mathcal{F}}) .\end{aligned} \tag{6.80}$$

By this way we obtain

$$\boxed{\dot{\underline{y}}_{\mathcal{E}}(t) = D(t) D_0 \big[\dot{\underline{y}}_{\mathcal{F}}(t) + \underline{\omega}_{\mathcal{F}} \times \underline{y}_{\mathcal{F}}(t)\big] := D(t) D_0 \big[\underline{v}_{\mathcal{F},r}(t) + \underline{v}_{\mathcal{F},\ell}(t)\big]} \, . \tag{6.81}$$

Fazit: The absolute velocity $\dot{\underline{y}}_{\mathcal{E}}(t)$ in system $\mathcal{E}$ is composed of the "pulled back" *relative velocity* $\underline{v}_{\mathcal{F},r}(t)$ in system $\mathcal{F}(t)$ and the "pulled back" *leading velocity* $\underline{v}_{\mathcal{F},\ell}(t)$ of rotation in system $\mathcal{F}(t)$. (Component vectors of different coordinate system shall never be mixed.) Further, we obtain for acceleration in system $\mathcal{E}$ by (6.74), (6.79) und (6.80)

$$\begin{aligned}\ddot{\underline{y}}_{\mathcal{E}}(t) &= \dot{D}(t)D_0\dot{\underline{y}}_{\mathcal{F}}(t) + D(t)D_0\ddot{\underline{y}}_{\mathcal{F}}(t) + \omega C\big[D(t)D_0\dot{\underline{y}}_{\mathcal{F}}(t) + \dot{D}(t)D_0\underline{y}_{\mathcal{F}}(t)\big]\\ &= \dot{D}(t)D_0\dot{\underline{y}}_{\mathcal{F}}(t) + D(t)D_0\ddot{\underline{y}}_{\mathcal{F}}(t) + \omega D(t)CD_0\dot{\underline{y}}_{\mathcal{F}}(t) + \omega^2 D(t)CCD_0\underline{y}_{\mathcal{F}}(t)\\ &= D(t)D_0\big[\ddot{\underline{y}}_{\mathcal{F}}(t) + 2\omega D_0^T CD_0\dot{\underline{y}}_{\mathcal{F}}(t) + \omega^2 D_0^T CCD_0\underline{y}_{\mathcal{F}}(t)\big]\\ &= D(t)D_0\big[\ddot{\underline{y}}_{\mathcal{F}}(t) + 2\underline{\omega}_{\mathcal{F}} \times \dot{\underline{y}}_{\mathcal{F}}(t) + \underline{\omega}_{\mathcal{F}} \times (\underline{\omega}_{\mathcal{F}} \times \underline{y}_{\mathcal{F}}(t))\big]\end{aligned}$$

and conversely

$$\begin{aligned}\ddot{\underline{y}}_{\mathcal{F}}(t) &= D_0^T D(t)^T \ddot{\underline{y}}_{\mathcal{E}}(t) - 2\,\underline{\omega}_{\mathcal{F}} \times \dot{\underline{y}}_{\mathcal{F}}(t) - \underline{\omega}_{\mathcal{F}} \times (\underline{\omega}_{\mathcal{F}} \times \underline{y}_{\mathcal{F}}(t))\\ &=: D_0^T D(t)^T \ddot{\underline{y}}_{\mathcal{E}}(t) + \underline{b}_{\mathcal{F},c}(t) + \underline{b}_{\mathcal{F},f}(t)\,.\end{aligned} \tag{6.82}$$

$$\boxed{\underline{b}_{\mathcal{F},c}(t) = -2\,\underline{\omega}_{\mathcal{F}} \times \dot{\underline{y}}_{\mathcal{F}}(t) \ \text{ and } \ \underline{b}_{\mathcal{F},f}(t) = -\underline{\omega}_{\mathcal{F}} \times (\underline{\omega}_{\mathcal{F}} \times \underline{y}_{\mathcal{F}}(t))}$$

are the CORIOLIS-*acceleration* and the *leading acceleration* (centrifugal force) (note the negative signs).

Fazit: The absolute acceleration $\ddot{\underline{y}}_{\mathcal{E}}$ in system $\mathcal{E}$ is composed of $-b_{\mathcal{F},f}$, $-\underline{b}_{\mathcal{F},c}$ and the *relative acceleration* $\ddot{\underline{y}}_{\mathcal{F}}$ (all in system $\mathcal{F}(t)$ and pulled back to $\mathcal{E}$). Conversely, let there be given a particle in system $\mathcal{F}$ with position vector $\underline{y}_{\mathcal{F}}(t)$, unit mass, and velocity $\underline{v}_{\mathcal{F}}$, and let $\underline{k}_{\mathcal{E}}$ be a global force acting in $\mathcal{E}$ then the relative force $D_0^T D(t)^T \underline{k}_{\mathcal{E}}$ acts on that particle in $\mathcal{F}$.

For instance, the angle velocity of earth is a vector pointing from center to the *North Pole* with length $\omega_{\text{earth}} = 2\pi/(3600\cdot 24) \sim 7.3\cdot 10^{-5}\,\text{sec}^{-1}$ (rotation in mathematical positive direction) where all other influences are neglected. The leading acceleration $\underline{b}_{\mathcal{F},f}$ behaves proportionally to ω^2 and is thus often neglected in context with rotation of earth.

(d) Choose the basis $\{\mathcal{F}_1, \mathcal{F}_2, \mathcal{F}_3\}$ of Example 6.17 for fixed system on surface of earth and let (6.76) be the rotation matrix,
$D(t)D_0 = D(\underline{e}_3, \varphi(t))D(\underline{e}_2, \vartheta)$, $\varphi(t) = \omega_{\text{earth}} t$, then we obtain

$$\underline{\omega}_{\mathcal{F}} = D^T(\underline{e}_2, \vartheta) D^T(\underline{e}_3, \omega_{\text{earth}} t)\underline{\omega}_{\mathcal{E}}$$

But $\underline{\omega}_{\mathcal{E}} = \omega_{\text{earth}}\underline{e}_3$ therefore $D^T(\underline{e}_3, \omega_{\text{earth}} t)\underline{\omega}_{\mathcal{E}} = \omega_{\text{earth}}\underline{e}_3$ since $D\underline{x} = \underline{x}$ implies $\underline{x} = D^T\underline{x}$ for orthogonal D. Accordingly

$$\underline{\omega}_{\mathcal{F}} = D^T(\underline{e}_2, \vartheta)\underline{\omega}_{\mathcal{E}} = \omega_{\text{earth}}[\sin\vartheta,\ 0,\ \cos\vartheta]^T$$

for the rotational axis in $\mathcal{F}$ and in system
$\mathcal{G} = \{\mathcal{G}_1, \mathcal{G}_2, \mathcal{G}_3\} = \{\mathcal{F}_2, \mathcal{F}_3, \mathcal{F}_1\}$ we have

$$\underline{\omega}_{\mathcal{G}} = \omega_{\text{earth}}[0,\ \cos\vartheta,\ \sin\vartheta]^T\,. \tag{6.83}$$

(d1) Let ψ be the angle of *East* to North (North $\psi = \pi/2$) and $\underline{v}$ a velocity vector in system $\mathcal{G}$ where

$$\underline{v} = [\widetilde{\underline{v}},\ 0]^T = v[\cos\psi,\ \sin\psi,\ 0]^T\,,\ \ 0 < v \in \mathbb{R}\,.$$

Then $\underline{v}$ is *tangential* to the surface of earth and we obtain the CORIOLIS acceleration

$$\underline{b}_C = -2\,\underline{\omega}_\mathcal{G} \times \underline{v} = -2\,v\,\omega_{\text{earth}} \begin{bmatrix} 0 \\ \cos\vartheta \\ \sin\vartheta \end{bmatrix} \times \begin{bmatrix} \cos\psi \\ \sin\psi \\ 0 \end{bmatrix} = 2\,v\,\omega_{\text{earth}} \begin{bmatrix} \sin\vartheta\sin\psi \\ -\sin\vartheta\cos\psi \\ \cos\vartheta\cos\psi \end{bmatrix}$$

in system $\mathcal{G}$. The projection of $\underline{b}_C$ onto the surface of earth ($\{\mathcal{G}_1, \mathcal{G}_2\}$-system) is

$$\underline{b} := 2\,v\,\omega_{\text{earth}} \sin\vartheta\,[\sin\psi,\, -\cos\psi]^T ,\ \underline{b} \perp \underline{\widetilde{v}} ,$$

and $\det[\underline{b}, \underline{\widetilde{v}}] = 2\,v\omega_{\text{earth}} \sin\vartheta > 0$ for $0 < \vartheta \leq \pi/2$. Relative to $\underline{v}$, the vector $\underline{b}$ points always to *right* on the northern hemisphere and to *left* on the southern hemisphere taking, in absolute value, the maximum at the poles and the minimum at the equator.

(d2) Let $\underline{v}$ be a velocity vector in system $\mathcal{G}$ where

$$\underline{v} = v[0,\, 0,\, -1]^T ,\ \ 0 < v \in \mathbb{R} .$$

Then $\underline{v}$ is *perpendicular* to the surface of earth pointing to the center and we obtain the Coriolis acceleration

$$\underline{b}_C = -2\,\underline{\omega}_\mathcal{G} \times \underline{v} = -2\,v\,\omega_{\text{earth}} \begin{bmatrix} 0 \\ \cos\vartheta \\ \sin\vartheta \end{bmatrix} \times \begin{bmatrix} 0 \\ 0 \\ -1 \end{bmatrix} = 2\,v\,\omega_{\text{earth}} \begin{bmatrix} \cos\vartheta \\ 0 \\ 0 \end{bmatrix} .$$

in system $\mathcal{G}$. The projection of $\underline{b}_C$ onto the surface of earth ($\{\mathcal{G}_1, \mathcal{G}_2\}$-system) is

$$\underline{b} := 2\,v\,\omega_{\text{earth}}[\cos\vartheta,\, 0]^T \geq \underline{0} ,\ \ -\pi/2 < \vartheta < \pi/2 .$$

Now the vector $\underline{b}$ points to *East* and attains its maximum at the equator.

6.7 Inertia Tensor and Top

In this section, coordinate vectors in the body-fixed coordinate system (COS) are denoted by *large* and coordinate vectors in space-fixed COS by *small* letters.

(a) Inertia Tensor We consider a body with local (body-own) cartesian coordinate system $\mathcal{F}$. Let $K \subset \mathbb{R}^3$ be the geometic form of the body in $\mathcal{F}$ and ϱ the mass density then

$$\mathbf{T} = \int_K \varrho(\underline{X})[\underline{X}^T \underline{X}\,\boldsymbol{\delta} - \underline{X}\,\underline{X}^T]\,d\underline{X} \in \mathbb{R}^3{}_3 \tag{6.84}$$

($\boldsymbol{\delta}$ unit tensor) is the inertia tensor where points (point vectors) in $\mathcal{F}$ are written again as capitals. The diagonal elements of the matrix $\mathbf{T}$ are the *moments of inertia* relative to the body-fixed X_1-, X_2-, and X_3-axis, and the off-diagonal elements are the *deviation moments.* The inertia tensor is symmetric hence orthogonally diagonalizable, and by Schwarz's inequality one verifies easily that it is at least positive semi-definite.

The *principle axes* of a tensor are its eigenvectors $\underline{U}_i$, $i = 1, 2, 3$. It is well-known from the theory of quadratic forms that the moments of deviation disappear if the principle axes are chosen for rotation axes. By normalization, suitable ordering and orientation, they form a cartesian COS and the eigenvalues satisfy $\lambda_1 \geq \lambda_2 \geq \lambda_3 \geq 0$. We denote the vector of angular velocity in system $\mathcal{F}$ by $\underline{\Omega}(t) = \dot{\varphi}(t)\underline{A} \in \mathbb{R}^3$, $|\underline{A}| = 1$ then

$$E_{\rm rot}(t) = \frac{1}{2}\underline{\Omega}(t)^T \mathbf{T} \underline{\Omega}(t) \in \mathbb{R}$$

is the instantaneous rotation energy. Again by the theory of quadratic forms

$$E_{\rm rot} \text{ is maximum for } \underline{A} = \underline{U}_1, \quad E_{\rm rot} \text{ ist minimum for } \underline{A} = \underline{U}_3.$$

Let K be a *surface* in (X_1, X_2)-plane of the system $\mathcal{F}$ then the components $T_{1,3} = T_{3,1}$ und $T_{2,3} = T_{3,2}$ disappear and $T_{3,3}$ is called *polar moment of inertia.* Frequently $T_{3,3}$ is dropped such that T becomes a $(2, 2)$-tensor where

$$\begin{aligned}
T_{1,1} &= \int_K X_2^2 \, dX_1 dX_2 \,, && \text{moment of inertia w.r.t. to } X_1\text{-axis} \\
T_{2,2} &= \int_K X_1^2 \, dX_1 dX_2 \,, && \text{moment of inertia w.r.t. } X_2\text{-axis} \\
T_{1,2} &= \int_K X_1 X_2 \, dX_1 dX_2 && \text{moment of deviation}\,.
\end{aligned}$$

The change of the inertia tensor under translation of the body-own COS is ruled by the parallel axes theorem of STEINER:

Theorem 6.3. *Let the origin of the body-own COS $\mathcal{F}$ be the gravity center S of a body with mass M. If $\mathcal{F}$ is translated by the vector $\underline{D}$ then the inertia tensor $\widetilde{\mathbf{T}}$ relative to the shifted system is*

$$\widetilde{\mathbf{T}} = \mathbf{T} + M(\underline{D}^T \underline{D}\, \boldsymbol{\delta} - \underline{D}\,\underline{D}^T)\,.$$

The sign of $\underline{D}$ is cancelled out and the axial moments (diagonal elements of T) are minimal if the coordinate axes are gravity axes, i.e., pass through the gravity center.

Let now the system $\mathcal{F}$ be rotated by the matrix D. Then $Y = D^T X$ follows for the components in the new system $\mathcal{G}$ by Sect. 6.6**(a)**. Substitution into (6.84) yields

$$X^T X\, \boldsymbol{\delta} - XX^T = Y^T D D^T Y\, \boldsymbol{\delta} - D^T Y Y^T D = D^T [Y^T Y\, \boldsymbol{\delta} - YY^T] D\,,$$

hence

$$\boxed{\mathbf{T}(Y) = D\mathbf{T}(X)D^T}\,. \tag{6.85}$$

The nonlinear inertia tensor behaves consequently like an ordinary matrix under rotations of the COS!

(b) Rigid Body with Stationary Point We consider a homogenous body with constant mass density and mass m. The space-fixed COS $\mathcal{E}$ and the body-fixed COS $\mathcal{F}$ shall have the common origin $\mathcal{O}$ for *stationary point.* The body rotates with angular velocity $\omega > 0$ about the axis $\mathcal{E}\underline{\omega}_{\mathcal{E}}(t) := \omega\mathcal{E}\underline{a}(t)$, $|\underline{a}(t)| = 1$, and $D(t) := D(\underline{a}(t), \omega(t))$ is the corresponding rotation matrix. Then, by NEWTON's axiom,

$$\frac{d}{dt}\underline{d}(t) = \underline{p}(t) \in \mathbb{R}^3 \tag{6.86}$$

does hold for the angular momentum $\underline{d}(t)$ and the moment $\underline{p}(t)$ in $\mathcal{E}$.

A point $\underline{x}(t)$ of the body satisfies the equation of motion $\dot{\underline{x}}(t) = \omega(t)\underline{a}(t) \times \underline{x}(t)$ and one computes the angular momentum relative to the stationary point by the representation formula (1.3),

$$\begin{aligned}
\underline{d}(t) &= m\omega \int_{K(t)=D(t)K} \underline{x} \times \underline{v}(x)\, d\underline{x} = m\omega \int_K D(t)\underline{x} \times (\underline{a} \times D(t)\underline{x})\, d\underline{x} \\
&= D(t)m\omega \int_K \underline{X} \times (D(t)^T\underline{a} \times \underline{X})\, d\underline{X} = D(t)m\omega \int_K \underline{X} \times (\underline{A} \times \underline{X})\, d\underline{X} \\
&= D(t)m\omega \int_K \left(\underline{X}^T\underline{X}\underline{A} - (\underline{X}^T\underline{A})\underline{X}\right) d\underline{X} \\
&= D(t)m\omega \int_K \left(\underline{X}^T\underline{X}\,\boldsymbol{\delta} - \underline{X}\underline{X}^T\right) d\underline{X}\underline{A} = D(t)\mathbf{T}\omega\underline{A} =: D(t)\mathbf{T}\underline{\Omega}_{\mathcal{F}}(t)\,.
\end{aligned}$$

$$\begin{aligned}
\underline{D}(t) &= \mathbf{T}\,\underline{\Omega}_{\mathcal{F}}(t) \in \mathbb{R}^3 && \text{is called instantaneous angular momentum} \\
\underline{P}(t) &= \frac{d}{dt}\underline{D}(t) && \text{is called instantaneous moment}
\end{aligned} \tag{6.87}$$

(both in COS $\mathcal{F}$). Hence $\underline{d}(t) = D(t)\underline{D}(t)$. Relation (6.81) thus follows for the angular momentum by the formal similarity with (6.78),

$$\dot{\underline{d}}(t) = D(t)\big[\dot{\underline{D}}(t) + \underline{\Omega}_{\mathcal{F}}(t) \times \underline{D}(t)\big] = D(t)D(t)^T\underline{p}(t) = D(t)\underline{P}(t)\,,$$

or the *dynamic* EULER *equations* in body-fixed COS,

$$\boxed{\underline{P}(t) = \mathbf{T}\,\dot{\underline{\Omega}}_{\mathcal{F}}(t) + \underline{\Omega}_{\mathcal{F}}(t) \times \mathbf{T}\,\underline{\Omega}_{\mathcal{F}}(t)}\,. \tag{6.88}$$

Note that $\underline{P}(t) = \underline{0}$ and $\underline{p}(t) = \underline{0}$ if no *external* moment exists. If $\dot{\underline{D}} = 0$, i.e., if the rate of change of $\underline{D}$ in the body-fixed COS disappears,

$$\dot{\underline{d}} = D(t)\big(\underline{\Omega}_{\mathcal{F}} \times \underline{D}\big) = \underline{\omega}_{\mathcal{E}} \times D(t)\underline{D} = \underline{\omega}_{\mathcal{E}} \times \underline{d} = \underline{p}\,. \tag{6.89}$$

Let especially the inertia tensor have diagonal form in COS $\mathcal{F}$,

$$\mathbf{T} = \begin{bmatrix} T_1 & 0 & 0 \\ 0 & T_2 & 0 \\ 0 & 0 & T_3 \end{bmatrix},$$

where T_i are the eigenvalues of $\mathbf{T}$. Then these equation have the more simple form

$$\boxed{\begin{aligned} T_1\dot{\Omega}_1 - (T_2 - T_3)\Omega_2\Omega_3 &= P_1 \\ T_2\dot{\Omega}_2 - (T_3 - T_1)\Omega_3\Omega_1 &= P_2 \\ T_3\dot{\Omega}_3 - (T_1 - T_2)\Omega_1\Omega_2 &= P_3 \end{aligned}} \; . \tag{6.90}$$

(c) Rotors Suppose that $\dot{\underline{\Omega}} = \underline{0}$, i.e., rotational axis and rotational velocity are *constant* in body-fixed COS $\mathcal{F}$. If then, e.g.,

$$\boxed{T_1 = T_2\,, \quad \underline{\Omega} = [0\,,\ \omega \sin\alpha\,,\ \omega\cos\alpha]^T}\,,$$

the rotational axis lies in the (X_2, X_3)-plane and, by (6.90),

$$\begin{aligned} P_1 &= (T_3 - T_2)\omega^2 \sin\alpha\cos\alpha = \frac{1}{2}(T_3 - T_2)\omega^2 \sin(2\alpha) \\ P_2 &= T_1\dot{\omega}\,\sin\alpha = 0\,,\quad P_3 = T_3\dot{\omega}\,\cos\alpha = 0\,. \end{aligned}$$

Only the external moment P_1 about the X_1-axis remains non-zero and we obtain for the *internal* moment $P_{K,X_1} = -P_1$

$$P_{K,X_1} = (T_2 - T_3)\omega^2 \sin(2\alpha)/2\,;$$

hence the force

$$\underline{K} = |\underline{K}|[0, -\cos\alpha, \sin\alpha]^T\,, \quad |\underline{K}| = \frac{(T_2 - T_3)\omega^2 \sin(2\alpha)}{2d}$$

applies to the rotational axis in distance d of the origin since $d[0, \sin\alpha, \cos\alpha] \times \underline{K} = \underline{P}_K$. The triple $\{\underline{\Omega},\ \underline{K},\ \underline{P}_K\}$ forms a right-oriented system for $T_2 > T_3$ and $\underline{K}$ points away from $\underline{\Omega}$ in distance d. This is the case in a long narrow rotor with symmetry axis X_3. In the other case $\underline{K}$ points to $\underline{\Omega}$ and tries to draw the rotor into the direction of the rotational axis (flat rotor or disc). It is to be regarded in the computation of support forces that the body-fixed COS $\mathcal{F}$ rotates; hence $\underline{K}$ has to be projected on the fixed support. Then, for the support forces,

$$\underline{K}_L = |\underline{K}|(\cos(\omega t), \sin(\omega t))$$

in a plane perpendicular to the rotational axis at distance d from the origin.

(d) Top without External Forces A body in rotation with one stationary point is called *top*. Let the *body-fixed* COS $\mathcal{F}$ be chosen such that the inertia tensor has *normal form* and the origin is the *stationary point*. Moreover, let no moment act on the top, i.e., $\underline{P} = \underline{0}$. This is the case, e.g., if the gravity center lies in the origin or if no external forces act on the top. Then the rotational energy E_{rot} is constant, and the surface in implicit representation

$$\underline{\Omega}(t) \cdot \underline{D}(t) = \underline{\Omega}(t)^T \mathbf{T} \underline{\Omega}(t) = \Omega_1^2 T_1 + \Omega_2^2 T_2 + \Omega_3^2 T_3 = 2E_{\text{rot}} \tag{6.91}$$

resp. $$\frac{X_1^2}{2E_{\rm rot}/T_1}+\frac{X_2^2}{2E_{\rm rot}/T_2}+\frac{X_3^2}{2E_{\rm rot}/T_3}=1$$

is called *ellipsoid of energy* $\mathcal{E}_E$ in body-fixed COS $\mathcal{F}$. The smallest axis of $\mathcal{E}_E$ corresponds to the largest moment of inertia whereby $\mathcal{E}_E$ reflects roughly the shape of the body. The point (point vector) $\underline{\Omega}(t)$ moves on $\mathcal{E}_E$. The tangent plane $\mathcal{T}(t)$ of $\mathcal{E}_E$ at point $\underline{\Omega}(t)$ has in *space-fixed* COS $\mathcal{E}$ the implicit representation

$$\underline{x}_{\mathcal{E}}\cdot\underline{d}_{\mathcal{E}}(t)=[D(t)X_{\mathcal{F}}]^TD(t)\underline{D}(t)=X_{\mathcal{F}}\cdot\underline{D}(t)=\underline{\Omega}(t)\cdot\underline{D}(t)=2E_{\rm rot}\,.$$

This plane is also constant in system $\mathcal{E}$ since $\underline{d}_{\mathcal{E}}(t)=\underline{d}_{\mathcal{E}}$ is constant in $\mathcal{E}$. Consequently the ellipsoid $\mathcal{E}_E$ rolls on this plane and the vertex of $\underline{\Omega}_{\mathcal{F}}(t)$ is the point of contact (Poinsot motion). The trajectory desribed by this point in the plane $\mathcal{E}_E$ is called *polhode*, and the trajectory desribed in the plane $\mathcal{T}$ is called *herpolhode.*

The absolute value $|\underline{D}(t)|^2=\underline{D}(t)^T\underline{D}(t)$ is constant in both systems and $\underline{D}(t)=\mathbf{T}\,\underline{\Omega}(t)$. The surface with implicit form

$$(\mathbf{T}\,\underline{\Omega}(t))^T\mathbf{T}\,\underline{\Omega}(t)=\Omega_1^2T_1^2+\Omega_2^2T_2^2+\Omega_3^2T_3^2=|\underline{D}|^2 \tag{6.92}$$

resp. $$\frac{X_1^2}{(|\underline{D}|/T_1)^2}+\frac{X_2^2}{(|\underline{D}|/T_2)^2}+\frac{X_3^2}{(|\underline{D}|/T_3)^2}=1$$

is called *ellipsoid of moment of momentum* $\mathcal{E}_I$ in COS $\mathcal{F}$. The point $\underline{\Omega}(t)$ moves also on $\mathcal{E}_I$.

Fazit: If the tops is free of moments, the trajectory of intersection of $\mathcal{E}_E$ and $\mathcal{E}_I$ describes the path of $\underline{\Omega}(t)$ in *body-fixed* COS $\mathcal{F}$, and the rolling of $\mathcal{E}_E$ onto $\mathcal{T}$ the path of $\underline{\Omega}(t)$ in space-fixed COS $\mathcal{E}$. Obviously both paths may degenerate to a point.

(e) Symmetric Top without External Forces Let further no moment act on the top. Then the top is called *symmetric* if two principal moments of inertia coincide. If, e.g., $T_1=T_2$ then the body-own X_3-axis is called *axis of the top.* Then we are faced with three axes through the origin of both COS, the *body-fixed* axis of the top, the *space-fixed* vector of angular moment $\underline{d}$ and the instantaneous rotary axis $\underline{\Omega}(t)$. The motion of these three axes relative to each other is described by three cones

the X_3-axis about	$\underline{d}$	*precession cone,*
the instantaneous rotary axis $\underline{\Omega}(t)$ about	$\underline{d}$	space-fixed *herpolhode cone*
the instantaneous rotary axis $\underline{\Omega}(t)$ about	X_3-axis	body-fixed *polhode cone.*

Polhode cone and herpolhode cone touch each other at the instantaneous rotary axis. In general there are three different cases of motion:

$T_3<T_1=T_2$ polhode cone rolls with exterior side on herpolhode cone,
$T_3=T_1=T_2$ $\underline{\Omega}(t)$ and $\underline{d}$ parallel,
$T_3>T_1=T_2$ polhode cone rolls with interior side on herpolhode cone.

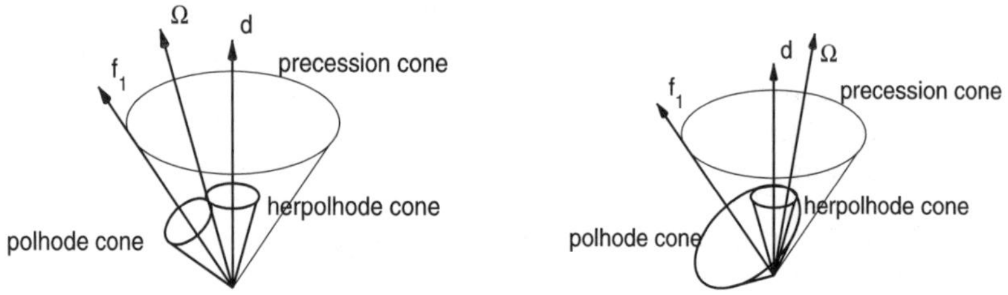

Figure 6.25. Polhode and herpolhode cone

(Of course, the the angular moment may point into every other direction and not necessarily into the direction of the x_3-axis.)

(f) Leaded Symmetric Top Let the top be symmetric, $T_1 = T_2$, and let the top's axis be rotary axis and X_3-axis in both COS. If no moment applies to the stationary point then $\underline{d} = \underline{D} = [0,\, 0,\, D_3]^T = [0,\, 0,\, \Omega_3 T_3]^T$ in both COS. Let now an *external* angular moment attack on the top, e.g., without restriction in direction of the x_2-axis of the *space-fixed* COS with vector of angular velocity $\underline{\omega}_{\mathcal{E}} = [0,\, \omega_2,\, 0]^T$. Then

$$\underline{d} = \underline{D} = [0,\, D_2,\, D_3]^T = [0,\, \omega_2 T_2,\, \Omega_3 T_3]^T .$$

Since the relative rate of change $\dot{\underline{D}}$ of $\underline{D}$ in *body-fixed* COS is now zero, we obtain by (6.89)

$$\underline{p} = \underline{\omega}_{\mathcal{E}} \times \underline{d} = \begin{vmatrix} \underline{e}_1 & 0 & 0 \\ \underline{e}_2 & \omega_2 & \omega_2 T_2 \\ \underline{e}_3 & 0 & \Omega_3 T_3 \end{vmatrix} = \begin{bmatrix} \omega_2 \Omega_3 T_3 \\ 0 \\ 0 \end{bmatrix}$$

in *space-fixed COS*. Let $\kappa > 0$ be the distance to the stationary point then the following pair of forces acts on the support axes $a(0,\, 0,\, \kappa)$ and $b(0,\, 0,\, -\kappa)$ in *space-fixed* COS

$$\underline{k}_A = -\frac{1}{2}\, [0,\, \omega_2 \Omega_3 T_3/\kappa,\, 0]^T , \quad \underline{k}_B = \frac{1}{2}\, [0,\, \omega_2 \Omega_3 T_3/\kappa,\, 0]^T ,$$

which generates a moment in direction of the x_1-axis. The opposite *moment of top* $-\underline{p}$ has to be overcome in order that the rotation about the x_2-axis in *space-fixed* COS is maintained. If the top's axis is suspended at one point only then the top responses to the moment $\underline{p}$ by a lateral evasive movement.

(g) Kinematic Euler Equations Let $\underline{u}(t) = (\varphi(t),\, \vartheta(t),\, \psi(t))$ be the vector of EULER angles and $D(\underline{u})$ the rotation matrix (6.75). The *kinematic* EULER *equations* supply a relation between $\dot{\underline{u}}$ and the vector of angular velocity $\underline{\omega}_{\mathcal{E}}(t)$ in *space-fixed* COS. The components may be computed by

$\dot{D}(\underline{u}) = \omega C(\underline{u})D(\underline{u})$; hence $\omega C(\underline{u}) = \dot{D}(\underline{u})D^T(\underline{u})$ using the representation of Sect. 1.1**(i)**, i.e.,

$$\omega C(\underline{u}) = \begin{bmatrix} 0 & -\omega_3(\underline{u}) & \omega_2(\underline{u}) \\ \omega_3(\underline{u}) & 0 & -\omega_1(\underline{u}) \\ -\omega_2(\underline{u}) & \omega_1(\underline{u}) & 0 \end{bmatrix},$$

and the result is

$$\begin{bmatrix} \omega_1 \\ \omega_2 \\ \omega_3 \end{bmatrix} = \begin{bmatrix} 0 & \cos\varphi & \sin\vartheta\sin\varphi \\ 0 & \sin\varphi & -\sin\vartheta\cos\varphi \\ 1 & 0 & \cos\vartheta \end{bmatrix} \begin{bmatrix} \dot{\varphi} \\ \dot{\vartheta} \\ \dot{\psi} \end{bmatrix}. \tag{6.93}$$

Then the vector of angular velocity in system $\mathcal{F}$ follows by $\underline{\Omega}_{\mathcal{F}} = D(\underline{u})^T \underline{\omega}_{\mathcal{E}}$ and has the components

$$\begin{bmatrix} \Omega_1 \\ \Omega_2 \\ \Omega_3 \end{bmatrix} = \begin{bmatrix} \sin\vartheta\sin\psi & \cos\psi & 0 \\ \sin\vartheta\cos\psi & -\sin\psi & 0 \\ \cos\vartheta & 0 & 1 \end{bmatrix} \begin{bmatrix} \dot{\varphi} \\ \dot{\vartheta} \\ \dot{\psi} \end{bmatrix} \tag{6.94}$$

where, moreover,

$$\Omega_1^2 + \Omega_2^2 = \dot{\varphi}^2 \sin^2\vartheta + \dot{\vartheta}^2 , \quad \Omega_3^2 = (\dot{\varphi}\cos\vartheta + \dot{\psi})^2 . \tag{6.95}$$

(h) Heavy Symmetric Top We consider a symmetric top with inertia tensor in normal form and $T_1 = T_2$, and suppose that the gravity centre lies at distance $l > 0$ of the origin on the top's axis.

Then the kinetic energy E_{rot}, potential energy E_{pot} and LAGRANGE function L

$$\begin{aligned} &E_{\text{rot}}(\underline{u}, \dot{\underline{u}}) = \frac{1}{2}\left[T_1(\Omega_1^2 + \Omega_2^2) + T_3\Omega_3^2\right] = \\ &\frac{1}{2}T_1\left(\dot{\vartheta}^2 + \dot{\varphi}^2\sin^2\vartheta\right) + \frac{1}{2}T_3\left(\dot{\psi} + \dot{\varphi}\cos\vartheta\right)^2 \\ &E_{\text{pot}}(\underline{u}, \dot{\underline{u}}) = mgl\,\cos\vartheta \\ &L(\underline{u}, \dot{\underline{u}}) = E_{\text{rot}}(\underline{u}, \dot{\underline{u}}) - E_{\text{pot}}(\underline{u}, \dot{\underline{u}}) \end{aligned} \tag{6.96}$$

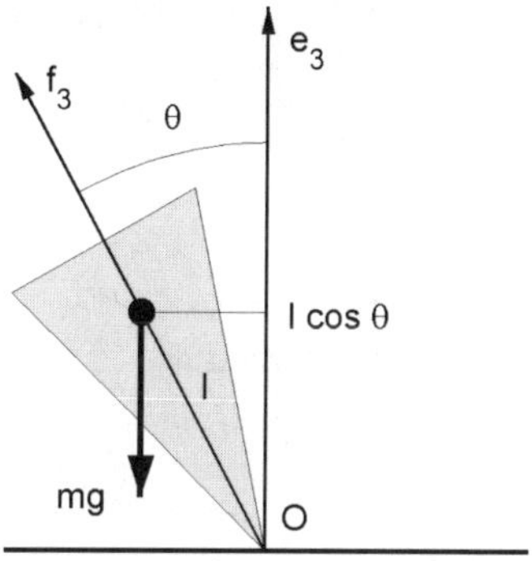

Figure 6.26. Potential energy

By HAMILTON's principle the top behaves in a way that the integral of action, $\int_0^T L(\underline{u}(t), \dot{\underline{u}}(t))\,dt$, has a stationary value in $\underline{u}$; hence the EULER-LAGRANGE equations hold; cf. Sect. 4.1,

$$0 = \frac{d}{dt}\operatorname{grad}_{\dot{u}} L(\underline{u}, \dot{\underline{u}}) - \operatorname{grad}_u L(\underline{u}, \dot{\underline{u}}) =: A(\underline{u})\ddot{\underline{u}} + b(\underline{u}, \dot{\underline{u}}) \tag{6.97}$$

where

$$A(\underline{u}) = \begin{bmatrix} T_1 \sin^2\vartheta + T_3\cos^2\vartheta & 0 & T_3\cos\vartheta \\ 0 & T_1 & 0 \\ T_3\cos\vartheta & 0 & T_3 \end{bmatrix}, \quad \det(A) = T_1^2 T_3 \sin^2\vartheta\,,$$

and

$$b(\underline{u}, \dot{\underline{u}}) = \sin\vartheta \begin{bmatrix} 2\dot{\varphi}\dot{\vartheta}(T_1 - T_3)\cos\vartheta - \dot{\vartheta}\dot{\psi}T_3 \\ \dot{\varphi}^2(T_3 - T_1)\cos\vartheta + \dot{\varphi}\dot{\psi}T_3 - mgl \\ -\dot{\varphi}\dot{\vartheta}T_3 \end{bmatrix}.$$

These equations do not depend directly on the angles φ and ψ but only on their derivatives. The matrix A is invertible as long as ϑ is non-zero; or the system may be solved by the MATLAB program `ode23t.m`; the trajectory of the top's axis in space-fixed system is then given by the third column of the matrix $D(\varphi, \vartheta, \psi)$ in (6.75) which is likewise independent of the EULER angle ψ.

(i) Energy of Heavy Top We observe that the first and third component of $\operatorname{grad}_u L$ are zero therefore the first and third component of $\operatorname{grad}_{\dot{u}} L$ must be *invariants* of the system by (6.97) (*cyclic* variable; cf. Sect. 4.1**(f)**). Let $d_3 := [\underline{d}]_3$ be the third component of angular moment in space-fixed COS $\mathcal{E}$ and $D_3 := [\underline{D}]_3$ the third component of angular moment in body-fixed COS $\mathcal{F}$. Then by theory or by direct verification,

$$\begin{aligned} \partial L/\partial\dot{\varphi} &= \dot{\varphi}(T_1\sin^2\vartheta + T_3\cos^2\vartheta) + \dot{\psi}T_3\cos\vartheta = d_3 \\ \partial L/\partial\dot{\psi} &= T_3(\dot{\psi} + \dot{\varphi}\cos\vartheta) = D_3\,. \end{aligned}$$

Resolution w.r.t. $\dot{\varphi}$ and $\dot{\psi}$ yields

$$\dot{\varphi} = \frac{d_3 - D_3\cos\vartheta}{T_1\sin^2\vartheta}\,, \quad \dot{\psi} = \frac{D_3}{T_3} - \frac{d_3 - D_3\cos\vartheta}{T_1\sin^2\vartheta}\cos\vartheta\,, \tag{6.98}$$

and substitution into the total energy $E = E_{\text{rot}} + E_{\text{pot}}$ yields the *third invariant* E of the system as a function of only the angle ϑ and its derivative,

$$E = \frac{1}{2}T_1\dot{\vartheta}^2 + V(\vartheta)\,, \quad T_1\dot{\vartheta}\ddot{\vartheta} = -\frac{dV}{d\vartheta}(\vartheta)\dot{\vartheta} \tag{6.99}$$

where the *effective potential energy* $V(\vartheta) \le E$ is

$$\begin{aligned} V(\vartheta) &= \frac{(d_3 - D_3\cos\vartheta)^2}{2T_1\sin^2\vartheta} + mgl\,\cos\vartheta + \frac{D_3^2}{2T_3} \quad \text{for } d_3^2 \neq D_3^2 \\ &= \frac{D_3^2(1 - \cos\vartheta)}{2T_1(1 + \cos\vartheta)} + mgl\cos\vartheta + \frac{D_3^2}{2T_3} \quad \text{for } d_3 = D_3 \\ &= \frac{D_3^2(1 + \cos\vartheta)}{2T_1(1 - \cos\vartheta)} + mgl\cos\vartheta + \frac{D_3^2}{2T_3} \quad \text{for } d_3 = -D_3\,. \end{aligned}$$

Observe the formal similarity with Theorem 6.1 with the difference that $0 \leq \vartheta \leq \pi$ must be satisfied. By cancelling out $\dot{\vartheta}$ one obtains from the second equation in (6.99) together with the first equation in (6.98) a differential system which describes the motion of the top's axis completely since the angle ψ in the third column of (6.75), being the top's symmetry axis in space-fixed COS, does not appear. The azimuthal motion of the top w.r.t. the EULER angle φ is called *precession* and the periodic motion w.r.t. the angle ϑ is called *nutation*; if the angle ϑ is constant, nutation disappears and the precession is *regular.*

For the further investigation commonly the substitution $\cos\vartheta = u$ is used in the theory of top. Then by (6.99)

$$\dot{u}^2 = f(u) = (\alpha - \beta u)(1 - u^2) - (a - bu)^2 , \tag{6.100}$$

$a = d3/T1\,,\ b = D3/T_1\,,\ \alpha = 2E'/T_1\,,\ \beta = 2\,mgl/T_1 > 0\,,\ E' = E - D_3^2/2\,T_3\,.$ After having specified the physical data of the top, the polynomial $f(u)$ of third degree has one double or two different real roots $u_1 = \cos\vartheta_1$ and $u_2 = \cos\vartheta_2$ in interval $-1 \leq u \leq 1$. $V(\vartheta)$ has exactly one minimum between the poles $\vartheta = 0$ and $\vartheta = \pi$ for $d_3^2 \neq D_3^2$ and, in this case, the angle ϑ between the top's symmetry axis and the space-fixed $\underline{e}_3$-axis varies between the two extremal ϑ_1 and ϑ_2 with property $V(\vartheta_i) = E$. Otherwise the minimum of $V(\vartheta)$ lies at the boundary of the intervall $[0,\,\pi]$, and ϑ varies between this boundary point and the value ϑ_0 with $V(\vartheta_0) = E$.
Case 1: $d_3^2 \neq D_3^2$ and $\cos\vartheta_1 < d_3/D_3$; then $\dot{\varphi} > 0$ in (6.98) and φ increases strictly monotonically.
Case 2: $d_3^2 \neq D_3^2$ and $d_3/D_3 < \cos\vartheta_2$; then $\dot{\varphi} < 0$ in (6.98) and φ decreases strictly monotonically.
Case 3: $d_3^2 \neq D_3^2$ and $\cos\vartheta_1 = d_3/D_3$; then $\dot{\varphi} = 0$ for $\vartheta = \vartheta_1$.
Case 4: $d_3^2 \neq D_3^2$ and $\cos\vartheta_2 = d_3/D_3$; then $\dot{\varphi} = 0$ for $\vartheta = \vartheta_2$.
Case 5: $d_3^2 \neq D_3^2$ and $\cos\vartheta_2 < d_3/D_3 < \cos\vartheta_1$; then $\dot{\varphi}$ changes sign in $(\vartheta_1,\,\vartheta_2)$, and $\dot{\varphi}$ has opposite signs in ϑ_1 und ϑ_2.
Case 6: $d_3^2 \neq D_3^2$ and $E = \min_\vartheta V(\vartheta)$; then $\vartheta_1 = \vartheta_2$ and $\dot{\varphi}$, $\dot{\psi}$ are constant; the result is again pure nutation (regular precession).
Case 7: $d_3 = -D_3 \neq 0$. Then $\vartheta_2 = \pi$ because $\underline{d}(t) = D(t)\underline{D}(t)$ and (6.88). $V(\vartheta)$ decreases monotonically in interval $(0,\,\pi)$ therefore the top's axis remains stable in the negative $\underline{e}_3$-Achse (sleeping top) or oscillates between ϑ_1 and π if $\vartheta(0) \neq \pi$.
Case 8: $d_3 = D_3 \neq 0$ and $D_3 \geq 2(T_1 mgl)^{1/2}$. Again $\vartheta_1 = 0$ because $\underline{d}(t) = D(t)\underline{D}(t)$ and (6.88). $V(\vartheta)$ increases monotonically in interval $(0,\,\pi)$, therefore the top's axis remains stable in the $\underline{e}_3$-Achse (sleeping top) or oscillates between 0 and ϑ_2 if $\vartheta(0) \neq 0$.
Case 9: $d_3 = D_3 \neq 0$ and $D_3 < 2(T_1 mgl)^{1/2}$. $V(\vartheta)$ has a minimum in interval $(0,\,\pi)$. The angle ϑ oscillates between $\vartheta_1 = 0$ and ϑ_2 where $V(\vartheta_2) = E$. Because of $D_3 = T_3\Omega_3$, $\Omega_3^* = 2(T_1 mgl)^{1/2}/T_3$ is the *critical* velocity of rotation, below of which the sleeping top wakes up and begins to tumble under small perturbations.

(j) Examples $T_1 = 1\,,\ T_3 = 2\,,\ m = 1\,.$
By having specified d_3 und $D_3 > 0$ one obtains α and β by setting (6.100) equal to zero for $u_1 = \cos\vartheta_1$ and $u_2 = \cos\vartheta_2$. Unphysical data are sorted out by the condition $\beta \geq 0$. The both invariants d_3 and D_3 are connected with the initial conditions for $\dot{\varphi}$ and $\dot{\psi}$ as displayed in (6.98).

Table 6.1. $T_1 = 1\,,\ T_3 = 2\,,\ m = 1$

Case	ϑ_1	ϑ_2	d_3	D_3	α	$\beta \simeq$
1	$\pi/6$	$\pi/2$	3	2	9	2.9667
2	$\pi/2$	$3\pi/4$	-3	1	9	2.1421
3	$\pi/6$	$\pi/2$	$\sqrt{3}$	2	3	3.4641
5	$\pi/4$	$\pi/2$	1	2	1	0.9289
6	$\pi/4$	$\pi/4$	1.2	1	~ 0.9	0.5973
7	$\pi/2$	π	-1	1	1	2
8	0	$\pi/2$	3	3	9	1
9	0	$\pi/2$	1	1	1	1

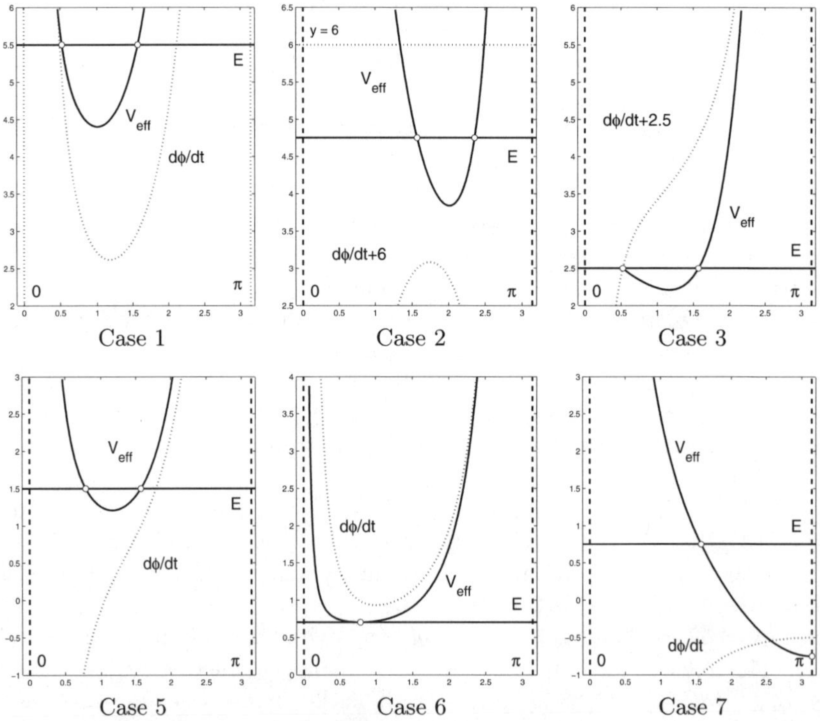

Figure 6.27. Effective potential energy V and $d\varphi/dt$ in a heavy top

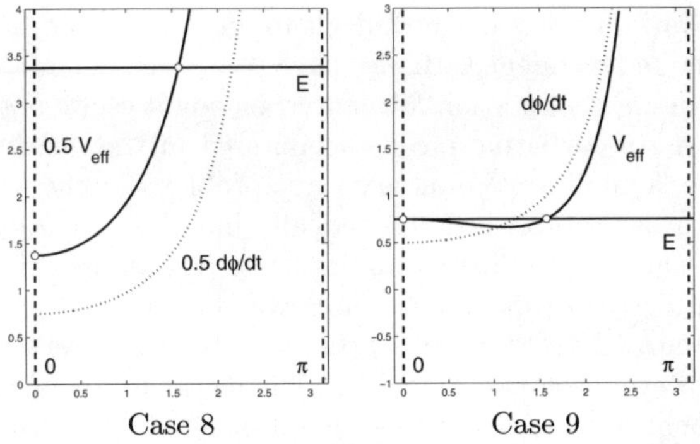

Figure 6.28. Effective potential energy V and $d\varphi/dt$ (contd)

Trace of top's axis : initial values $(\vartheta(0),\,\dot\vartheta(0),\,\varphi(0)) = (\vartheta_i,\,0,\,0)$, $i = 1$ or 2; in Case 9 $\vartheta(0) = \pi/2$ and $\vartheta(0) = \pi/4$:

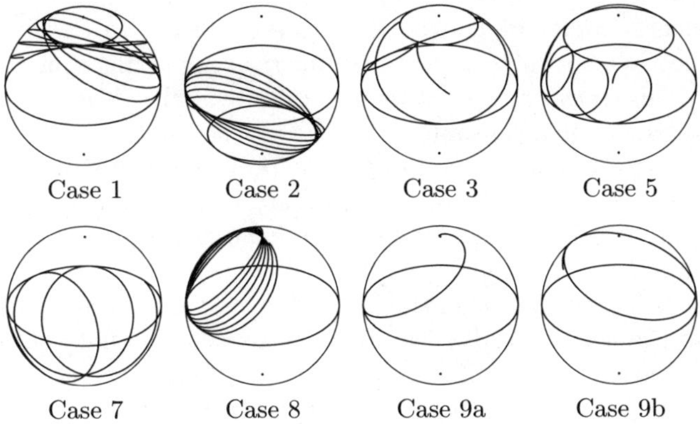

Figure 6.29. The relevant cases

Plot also from north pole by `demo3.m` in `KAPITEL06\SECTION_6_7`.

6.8 On Multibody Problems

The *dynamic* behavior of a system of *rigid bodies* under constraints is called *multibody problem* or *mechanical system* in technical applications. Analytic mechanics after LAGRANGE and HAMILTON describes, e.g., the general motion of mass points under equality constraints resp. on differentiable manifolds; cf., e.g., (Heil) and (Arnold78). It provides a closed framework but is less recommended for technical applications; cf., e.g., (Schiehlen86). One reason may be that the way from the generalized coordinates $\underline{p}$ and $\underline{q}$ back to the state and velocity variables $\underline{x}$ and $\dot{x}$ is difficult to manage numerically and the other that lastly the independent variables have to be selected out which is likewise problematic in implementation. Moreover, inequalities are not assigned in this calculation. A similar situation is encountered in the nearly related control theory. There, a nonlinear boundary value problem for the pair $(\underline{x}, \underline{y})$ of state and costate is but established theoretically but the numerical approach fails frequently because of the high complexity such that the original problems is discretized directly in this volume; cf. Sects. 4.3, 4.4.

In mechanics, D'ALEMBERT*'s principle of virtual displacement* is clearly preferred. However each variational problem has an extremal problem, the (energy) functional of which is (at least) stationary at the position in question. It is not explicitly set up in many cases; rather, the relevant variational equations are found directly. As previously mentioned in Sect. 4.1, one must strictly differentiate between *essential* and *natural* boundary conditions to avoid incorrect approximation results. The essential boundary conditions have to be respected by the comparing functions (that is, the test functions or virtual displacements), whereas homogenous natural boundary conditions do not appear explicitely in the variational problem (as their name already shows) and by consequence cannot be gauged from that problem. In the finite element method, they are not met exactly by numerical approximation; rather, they are met only in transition to the limit of infinitisimal refinement. To counter this, *all* boundary conditions must be taken into account and discretized when directly solving the related boundary value problem. The exact derivation of the variational problem with all its boundary conditions can only occur via the extremal problem because its variation is necessary to determine the two types of boundary conditions.

HAMILTON's principle takes the place of the extremal principle in the case of dynamic problems, although it must be considered that the forces have to be conservative and the LAGRANGE function must not *explicitely* depend on time. The variational equations of the action integral are the *dynamic* EULER-LAGRANGE equations from Theorem 4.1 with the boundary conditions described therein. On the other side, the conservation equations can be directly set up when evading the extremal principle and can depend on time in any form. Because the desired state function of the dynamic system no longer consists of space variables only but can also contain rotational angles, for example, it is applied as a *generalized* state function $\underline{q}(t)$ with values in the *configuration* space.

Notations; cf. Sects. 4.1 and 8.4:

$t \in [0, T]$	time interval
$\underline{q}(t) \in \mathbb{R}^m$	generalized coordinates
$\dot{\underline{q}}(t) \in \mathbb{R}^m$	generalized velocity
$E_{\text{kin}}(\dot{\underline{q}}(t)) = \frac{1}{2}\dot{\underline{q}}(t)^T A\dot{\underline{q}}(t)$	kinetic energy
$E_{\text{pot}}(\underline{q}(t))$	potential energy

$L(\underline{q}(t), \dot{\underline{q}}(t)) = E_{\text{kin}}(\dot{\underline{q}}(t)) - E_{\text{pot}}(\underline{q}(t))$	LAGRANGE function
$G(t, \underline{q}(t), \dot{\underline{q}}(t)) = 0 \in \mathbb{R}^r$	constraints

Constraints G are distinguished according to the variables of which they depend explicitely:

$G(\underline{q})$	$= 0$	holonomic-skleronomic
$G(t, \underline{q})$	$= 0$	holonomic-rheonomic
$G(\underline{q}, \dot{\underline{q}})$	$= 0$	non-holonomic skleronomic
$G(t, \underline{q}, \dot{\underline{q}})$	$= 0$	non-holonomic rheonomic

(*skleronom*: rigid, *rheonom*: flowing). Sometimes constraints G are also called non-holonomic if they cannot be written as total t-derivative of a holonomic-rheonomic constraint, i.e., if there does not exist a constraint $H(t, \underline{q}) = 0$ such that $G(t, \underline{q}, \dot{\underline{q}}) = d\,H(t, \underline{q})/dt$.

Consider now the constrained extremal problem

$$\int_0^T L(t, \underline{q}(t), \dot{\underline{q}}(t))\, dt = \text{extr!}\,,\ \underline{q}(0) = \underline{a}\,,\ \underline{q}(T) = \underline{b}\,,\ G(t, \underline{q}(t), \dot{\underline{q}}(t)) = 0 \tag{6.101}$$

for the state function $[0, T] \ni t \mapsto \underline{q}(t) \in \mathbb{R}^m$ where $L \in \mathcal{C}^2[0, T]$ and $G \in \mathcal{C}^2([0, T]; \mathbb{R}^m)$. This problem becomes a *control problem* of the form dealed with in Sect. 4.3 if $\underline{u} = \dot{\underline{q}}$ is introduced as further dependent variable. In this section however a direct method shall be discussed which leads to a differential-algebraic problem.

Solution curves of EULER's equations (4.2) are usually called *extremals* although they are not necessarily extremas of the associated integral. In the following theorem of (Gelfand), L is allowed to be an *abitrary* scalar function (with the above smoothness). In particular, L may depend explicitely on the independent time variable t.

Theorem 6.4. *(Existence of* LAGRANGE *multiplier)* (1°) *Let* $\underline{q}^*$ *be an extremum of (6.101).*
(2°) *For* $t \in [0, T]$ *let* $\operatorname{rank} \operatorname{grad}_{\dot{q}} G(t, \underline{q}^*, \dot{\underline{q}}^*) = m$ *or* $\operatorname{rank} \operatorname{grad}_{q} G(t, \underline{q}^*) = m$ *when* G *does not depend on* $\dot{q}$.
Then there exists *a function* $y \in \mathcal{C}^2([0, T]; \mathbb{R}^m)$ *such that* $\underline{q}^*$ *is* extremal *of the augmented* LAGRANGE *functional* $\int (L + y^T G)\, dt$, *i.e.,*

$$\operatorname{grad}_q (L + y^T G) - \frac{d}{dt} \operatorname{grad}_{\dot{q}} (L + y^T G) = 0\,.$$

Notations for numerical solution:

$\underline{x}$, $\underline{y} := \dot{\underline{x}}$, $\underline{z}$:	generalized space coordinates, their derivatives, LAGRANGE multipliers
$(\underline{u}, \underline{v}, \underline{w})$:	numerical approximation of $(\underline{x}, \underline{y}, \underline{z})$

.

We consider a mechanical system with constraints in the form

$$M(t,\underline{x}(t))\ddot{\underline{x}}(t) = \underline{f}(t,\underline{x}(t),\dot{\underline{x}}(t)) \in \mathbb{R}^n\,, \quad \underline{g}(t,\underline{x}(t)) = \underline{0} \in \mathbb{R}^m\,. \tag{6.102}$$

The differential system is to be understood as variational equation of an extremal problem; see Sect. 4.1. Therefore the transformation into a system of first order and an application of the *multiplier rule* of Sect. 3.2**(b)** leads to the system

$$\boxed{\begin{array}{ll} \dot{\underline{x}} & = \underline{y} \\ M(t,\underline{x})\dot{\underline{y}} & = \underline{f}(t,\underline{x},\underline{y}) - \nabla\underline{g}(t,\underline{x})^T\underline{z} \\ \underline{0} & = \underline{g}(t,\underline{x}) \end{array}}\,. \tag{6.103}$$

For transformation into a system of index 1 — cf. Sect. 2.7**(d)** — the constraint is differentiated w.r.t. t:

$$\boxed{\begin{array}{lll} \dot{\underline{x}} & & = \underline{y} \\ M(t,\underline{x})\dot{\underline{y}} & + \nabla\underline{g}(t,\underline{x})^T\underline{z} & = \underline{f}(t,x,y) \\ \nabla\underline{g}(t,\underline{x})\dot{\underline{x}} & & = -\dfrac{\partial g}{\partial t}(t,\underline{x}) \end{array}}\,. \tag{6.104}$$

The LAGRANGE matrix

$$L = \begin{bmatrix} M & [\nabla\underline{g}]^T \\ \nabla\underline{g} & O \end{bmatrix}$$

of this system must be regular in a neighborhood of the solution in order that a semi-explicit RUNGE-KUTTA method can be applied; cf. Sect. 2.7**(e)**. For this condition it suffices however by Lemma 1.2, that the *generalized mass matrix* M is regular on the kernel of $\nabla\underline{g}$.

Using the data $(A,\, b,\, c)$ of a semi-explicit method let

$$\begin{array}{llll} \underline{u}_i(t) = \underline{u}(t) + \tau\displaystyle\sum_{j=1}^{i-1}\alpha_{ij}\underline{v}_j\,, & & \underline{v}_i(t) = \underline{v}(t) + \tau\displaystyle\sum_{j=1}^{i-1}\alpha_{ij}\dot{\underline{v}}_j \\ M_i \;\; = M(t+\gamma_i\tau,\underline{u}_i)\,, & \underline{f}_i & = \underline{f}(t+\gamma_i\tau,\underline{u}_i,\underline{v}_i) \\ G_i \;\; = \nabla\underline{g}(t+\gamma_i\tau,\underline{u}_i)\,, & \underline{h}_i & = \dfrac{\partial g}{\partial t}(t+\gamma_i\tau,\underline{u}_i) \end{array}$$

for $i = 1:r$ where $\dot{\underline{v}}$ is to be understood as *approximation* of the derivative of $\underline{v}$. Then we obtain a system

$$\begin{array}{ll} \dot{\underline{u}}_i & = \underline{v}_i \\ M_i\dot{\underline{v}}_i + G_i^T\underline{w}_i & = \underline{f}_i \\ G_i\underline{v}_i & = -\underline{h}_i\,. \end{array} \tag{6.105}$$

for the intermediate steps $i = 1:r$ by (6.104). Now, because of the triangular form of the matrix A, the index i in the third row of this system may be enhanced by one and

$$\begin{aligned}\underline{v}_{i+1} &= \underline{v}(t) + \tau \sum_{j=1}^{i-1} \alpha_{i+1,j}\dot{\underline{v}}_j + \tau\alpha_{i+1,i}\dot{\underline{v}}_i\,, \; i = 1 : r-1 \\ \underline{v}(t+\tau) &= \underline{v}(t) + \tau \sum_{j=1}^{r-1} \beta_j\dot{\underline{v}}_j + \tau\beta_r\dot{\underline{v}}_r\,, \qquad i = r\end{aligned}$$

can be inserted in (6.105). Then, by resolution w.r.t. $\dot{v}_i$, a *linear system*

$$\boxed{\begin{bmatrix} M_i & G_i^T \\ G_{i+1} & 0 \end{bmatrix}\begin{bmatrix} \dot{\underline{v}}_i \\ \underline{w}_i \end{bmatrix} = \begin{bmatrix} \underline{f}_i \\ \underline{r}_i \end{bmatrix}}\,. \tag{6.106}$$

is obtained for each step $i = 1 : r$. By this way, semi-explicit RUNGE-KUTTA methods reveal to be especially well-suited in solving mechanical systems being not stiff in general. The initial values however have to satisfy the condition of consistence

$$\underline{g}(t_0, \underline{u}_0) = 0\,, \;\; G_0\underline{v}_0 + \underline{h}_0 = \underline{0}\,.$$

Examples in Sect. 11.3.

6.9 On Some Principles of Mechanics

A mechanical principle can be an *axiom* in mathematical sense or a *computational device* as LAGRANGE's multiplier method or D'ALEMBERT's principle; the notation is not applied in unique way and sometimes handled generously. The classical principles of mechanics concern mainly mass points or rigid bodies in which thermal transport, i.e., the energy law remains disregarded and solely the balance laws of momentum and of angular momentum in stationary or instationary form are involved.

First rate principles are:

(a) The *Balance Laws* of Sect. 8.3.

(b) The *Energy Principle* says that the sum of kinetic and potential energy is kept constant in a closed mechanical system (serves also for the definition of a *Closed System* in mechanics). This law holds in general as long as the kinetic energy is a quadratic form in $\dot{\underline{x}}$ and the potential energy does not contain velocities (there are potentials which do contain velocities); moreover, both must be scleronomic.

(c) The *Extremal Principle of Energy* says that a stationary mechanical system is in equilibrium if the total potential energy takes a stationary value (ordinarily a minimum). Its variational form is called *Principle of Virtual Work.*

(d) D'ALEMBERT and LAGRANGE. It is not the intention here to develop a full theory of variational principles. Rather we consider a single particle $\underline{x} \in \mathbb{R}^n$ obeying NEWTON's axiom (law of momentum)

$$A\ddot{\underline{x}} = \underline{f}(\underline{x}) \tag{6.107}$$

where $A \in \mathbb{R}^n{}_n$ is a symmetric, positive definite matrix, $\underline{f} = -\operatorname{grad} U$, and U is a potential function. Then kinetic energy T and total energy E of the *orbit* points $\underline{x}(t)$ are

$$T(\dot{\underline{x}}(t)) = \frac{1}{2}\dot{\underline{x}}(t)^T A\dot{\underline{x}}(t)\,, \quad E(\underline{x}(t)) = T(\dot{\underline{x}}(t)) + U(\underline{x}(t)) \tag{6.108}$$

and, by Sect. 6.2,

$$\frac{d}{dt}E = [A\ddot{\underline{x}} - \underline{f}(\underline{x})] \cdot \dot{\underline{x}} =: \widehat{\operatorname{grad}} E(\underline{x})\,\dot{\underline{x}} = 0\,.$$

Therefore $E(\underline{x}(t))$ is constant on the orbit $\underline{x}(t)$ if and only if (6.107) does hold. D'ALEMBERT's *Principle of Virtual Displacement* says in this unconstrained case simply that

$$(\forall\, \underline{u} \in \mathbb{R}^n : [A\ddot{\underline{x}} - \underline{f}(\underline{x})] \cdot \underline{u} = 0) \implies A\ddot{\underline{x}} - \underline{f}(\underline{x}) = 0\,.$$

But suppose now that there are m holonomic-scleronomic constraint conditions

$$\underline{g}(\underline{x}) = 0 \in \mathbb{R}^m\,, \quad \operatorname{rank} \operatorname{grad} \underline{g} = m\,.$$

Following LAGRANGE's *Multiplier Method*, the rows of $\operatorname{grad} \underline{g}(\underline{x})$ are considered as *constraint forces* which are multiplied by some factor y_i, the LAGRANGE multipliers, and added to the *impressed* force $\underline{f}$. The result is an augmented *differential-algebraic system*

$$[\operatorname{grad} \underline{g}(\underline{x})]^T \underline{y} = [\widehat{\operatorname{grad}} E(\underline{x})]^T\,, \quad \underline{g}(\underline{x}) = 0$$

which is solved for $\underline{x}$ and $\underline{y}$. Following D'ALEMBERT's *Principle*, we calculate $\operatorname{Ker} \operatorname{grad} \underline{g}(\underline{x}) = \operatorname{span}\{\underline{u}_1(\underline{x}), \ldots, \underline{u}_{n-m}(\underline{x})\}$ and multiply $\widehat{\operatorname{grad}} E(\underline{x}) = \underline{y}^T \operatorname{grad} \underline{g}(\underline{x})$ from right by $\underline{u}_i$. The result is again a *differential-algebraic system*

$$\widehat{\operatorname{grad}} E(\underline{x}) \cdot \underline{u}_i(\underline{x}) = 0\,,\; i = 1 : n - m\,, \;\; \underline{g}(\underline{x}) = 0\,, \tag{6.109}$$

see Sect. 6.4**(c)**, which is however not solved in this form. Rather the m surplus variables are eliminated by means of $\underline{g}(\underline{x}) = 0$ to obtain a *pure* differential system for $n - m$ independent variables.

By the Range Theorem 1.5 and Corollary 1.8, both devices are equivalent,

$$[\operatorname{grad} \underline{g}]^T \underline{y} = [\widehat{\operatorname{grad}} E]^T \iff \big(\operatorname{grad} \underline{g}\, \underline{u} = 0 \implies \widehat{\operatorname{grad}} E\, \underline{u} = 0\big)\,.$$

Accordingly, LAGRANGE's multiplier method and D'Alembert's principle reveal to be dual to each other from geometrical point of view.

In case the system is stationary, we have $T = 0$ and both devices lead to an algebraic system for the unknown solution $\underline{x} \in \mathbb{R}^n$. D'ALEMBERT's principle becomes the (generalized) principle of virtual work again from which it is was obtained by introducing the *inertia force* $m\,\ddot{\underline{x}}$ as dynamic component (scalar mass m instead A).

In solving large constrained mechanical systems, LAGRANGE's method is always preferred today because appropriate codes are available and the algebraic solution of the implicit side conditions in D'ALEMBERT's method causes some difficulties in case these constraints are nonlinear. The case of general non-holonomic rheonomic side conditions $\underline{g}$ in LAGRANGE's multiplier method is ruled by GELFAND's Theorem 6.4. D'ALEMBERT's principle can be generalized to holonomic-rheonomic constraints $\underline{g}(t,\underline{x}) = 0$, at least theoretically. Then

$$\frac{d}{dt}\underline{g}(t,\underline{x}) = \underline{g}_t(t,\underline{x}) + \operatorname{grad}\underline{g}(t,\underline{x})\,\underline{u} = 0\,,\ \underline{u} = \dot{\underline{x}}\,.$$

Under the above rank condition for $\operatorname{grad}\underline{g}$, there are again $n-m$ linear independent solutions $\underline{u}_i$ of that system to be substituted into (6.109).

(e) HAMILTON's *Principle.* Recall that the EULER equations

$$\operatorname{grad}_x L(t,\underline{x},\dot{\underline{x}}) - \frac{d}{dt}\operatorname{grad}_{\dot{x}} L(t,\underline{x},\dot{\underline{x}}) = 0$$

are a necessary and sufficient condition for a stationary orbit $\underline{x}$ of the extremal problem

$$\mathcal{A}(\underline{x}) := \int_{t_0}^{t_1} L(t,\underline{x},\dot{\underline{x}})dt = \text{extr!}\,,\ \underline{x}(t_0) = \underline{a}\,,\ \underline{x}(t_1) = \underline{b}\,.$$

Let now $L = T - U$ be the LAGRANGE function. Then $\mathcal{A}(\underline{x})$ is called *action integral* (action = work · time) and HAMILTON's *Principle* says that any orbit $\underline{x}$ with kinetic energy T and potential energy U as in (6.108) is a stationary orbit of $\mathcal{A}(\underline{x})$; but note that L must not depend explicitly on time. HAMILTON's principle is well established in physics even in his stronger form as *minimum principle.* Then it suffices to consider the action integral for a single interval because the minimum property holds for every subinterval by the *principle of optimality*; see Sect. 4.2**(e)**. The EULER equations of this integral are (6.107) and conversely, the equations of motion $A\ddot{\underline{x}} = \underline{f}(\underline{x})$ lead to HAMILTON's principle in weak form by Sect. 6.2**(e)**. Therefore the weak principle is equivalent to NEWTON's axiom (and thus also equivalent to D'ALEMBERT's principle).

Recall that $L - \operatorname{grad}_{\dot{x}} L\dot{\underline{x}}$ is a first integral (invariant) of EULER's equations in case L does not depend explicitly on time t (DUBOIS-REYMOND condition). But $\operatorname{grad}_{\dot{x}} L\dot{\underline{x}} = \operatorname{grad}_{\dot{x}} T\dot{\underline{x}} = 2T$ hence

$$L - 2T = \text{const.} \iff E = T + U = \text{const.}$$

Now, as E constant, it can be added to the action integral without changing its stationary values. Therefore we obtain an alternative formulation of HAMILTON's principle:

> If $\underline{x}$ is a stationary point of the action integral $\mathcal{A}(\underline{x})$ then $\underline{x}$ is a stationary point of the action integral $\widetilde{\mathcal{A}}(\underline{x}) := 2\int_a^b T\,dt$.
> Conversely, if $\underline{x}$ is a stationary point of $\widetilde{\mathcal{A}}(\underline{x})$ and E constant then $\underline{x}$ is stationary point of $\mathcal{A}(\underline{x})$.

(f) JACOBI's *Principle.* Recall that $ds = |\dot{\underline{x}}(t)|\,dt$ holds for the arc length s. If $|\dot{\underline{x}}| \neq 0$ on the orbit $\underline{x}$ then we can substitute s for t and obtain

$$T(\dot{\underline{x}}(t)) = T(\widetilde{\underline{x}}'(s))\frac{1}{(t'(s))^2} \tag{6.110}$$

$$\widetilde{\mathcal{A}}(\underline{x}) = 2\int_{s_1}^{s_2} T(\dot{\underline{x}}(t(s))t'(s)\,ds = 2\int_{s_1}^{s_2} \frac{T(\widetilde{\underline{x}}'(s))}{t'(s)}\,ds\,. \tag{6.111}$$

Inserting (6.110) in $E = T + U$ yields $t'(s) = \big(2(E-U)\big)^{-1/2}$. Substituting once more $T = E - U$ in (6.111), we obtain

$$\widehat{\mathcal{A}}(\underline{x}) = \int_{s_1}^{s_2} \sqrt{(2(E-U)}\,ds = \int_{s_1}^{s_2} \sqrt{2T(\widetilde{\underline{x}}'(s))}\,ds$$

(of course, the factor $\sqrt{2}$ can be cancelled.) JACOBI's *principle of least action* says that the motion $\widetilde{\underline{x}}(s)$ of a particle between $\widetilde{\underline{x}}(s_1)$ and $\widetilde{\underline{x}}(s_2)$ makes $\widehat{\mathcal{A}}(\underline{x})$ stationary (or more strongly a minimum). For $A = I$ identity we have

$$\sqrt{2T(\widetilde{\underline{x}}'(s))} = |\widetilde{\underline{x}}'(s)| = 1$$

since velocity with respect to arc length is always one. Then JACOBI's principle says simply that an orbit $\underline{x}$ with kinetic and potential energy as above is always the *shortest path* between two of its points under the auxiliary condition that E constant. A general fixed matrix A leads only to an other metric.

Conversely, consider this problem as classical constrained extremal problem

$$\widetilde{\mathcal{A}}(\underline{x}) = \text{extr!}\,,\quad E = \frac{T(\widetilde{\underline{x}}'(s))}{t'(s)^2} + U(\widetilde{\underline{x}}(s)) = \kappa \ \text{ constant.}$$

Then LAGRANGE's multiplier method, Theorem 3.26, leads to the modified action integral

$$\widetilde{\widetilde{\mathcal{A}}}(\underline{x}) = \int_{s_1}^{s_2} \left[2\frac{T(\widetilde{\underline{x}}'(s))}{t'(s)} + \lambda(s)\left(\frac{T(\widetilde{\underline{x}}'(s))}{t'(s)^2} + U(\widetilde{\underline{x}}(s))\right)\right] ds$$

Since t' is a dependent variable now, we obtain by minimizing with respect to t'

$$-2\frac{T(\widetilde{\underline{x}}'(s))}{t'(s)^2} - \frac{2\lambda T(\widetilde{\underline{x}}'(s))}{t'(s)^3} = 0$$

which gives $\lambda(s) = -t'(s)$. Inserting yields

$$\widetilde{\widetilde{\mathcal{A}}}(\underline{x}) = \int_{s_1}^{s_2} \left[\frac{T(\widetilde{\underline{x}}'(s))}{t'(s)^2} - U(\widetilde{\underline{x}}(s))\right] t'(s)\,ds$$

The new variational problem is a *free* problem without auxiliary condition therefore we can use t as independent variable again and recover HAMILTON's action integral by this way,

$$\mathcal{A}(\underline{x}) = \int_{t_0}^{t_1} (T - U)\, dt\,,$$

which leads back to HAMILTON's principle.

The above results hold also when $\underline{x}(t) \in \mathbb{R}^n$ is replaced by generalized coordinates $\underline{q}$ in some *configuration space*, therefore they hold also under side conditions of various types or, more generally, on manifolds. (The side conditions are "solved" to find the independent (generalized) variables $\underline{q}$.) So *canonical transformations* take a large place in a more general discussion. Also the generalized momentum $\underline{p} = \text{grad}_{\dot{\underline{q}}}\, L$ is introduced to obtain the simple form $\dot{\underline{p}} = \text{grad}_{\underline{q}}\, L(t, \underline{q}, \dot{\underline{q}})$ for the EULER equations; see Sect. 4.1(**e**). But state $\underline{x}$ and velocity $\dot{\underline{x}}$ remain the primary quantities in technical applications and it can be difficult to gain them back from the transformed system. Therefore we do not pursue this matter here and refer to the respective literature on analytical mechanics for further studies, e.g., (Lanczos).

6.10 Hints to the MATLAB Programs

```
KAPITEL06/SECTION_2_3_4, Central Fields
demo1.m    Graphics for Kepler's second law
demo2.m    Motion in central field, different potentials
demo3.m    Arbitrary conic sections under different
           initial positions and velocities
kepler.m   Computes conic section by intial data
ellipse.m  Draws ellipse with data
parabel.m  Draws parabola with data
hyperbel.m Draws hyperbola with data
KAPITEL06/SECTION_5, Three-Body Problem
arenstorf.m Different Arenstorf orbits
demo1.m Two-body problem, trajectories by differential system
demo2.m Three-body problem, trajectories by differential system
KAPITEL06/SECTION_6_7, Top
demo1.m  Computes EULER angles for top and trajectory
         of top's axis directly by EULER-LAGRANGE equations
demo2.m  Top demo, the 7 examples
demo3.m  Computes EULER angles phi and theta by initial data of
         DEMO2.M with differential system and trajectory of
         top's axis
demo4.m  Draws curve of Euler angle theta and curve of
         derivation of phi
demo5.m  Movie for top
```

7

Rods and Beams

7.1 Bending Beam

A beam or bar is a long narrow plate. In studying its deformation one confines to the *"neutral fibre"* which shall be the x-axis in unloaded and the *bending line* in loaded state. The approximation of a three-dimensional elastic body by a one-dimensional curve has many benefits but entails also severe restrictions such that many additional assumptions are tacitly introduced to overcome the apparent inconsistencies. An imbedding into three-dimensional continuum theory without contradictions is impossible. On the other side, many phenomena are explained in satisfying way by this model, and the accuracy suffices in most cases within the somewhat nebulous "engineering exactness". In order to circumvent the entire difficulties, bending and torsion are introduced in this chapter more or less *axiomatically*. Moreover, the bending beam is mainly studied in the plane. One-dimensional beam models in space suffer from the lack of a sufficient number of independent variables and thus can be handled only under restrictions of freedom in motion and loading.

Notations: ([N] (Newton) unit of force, [L] unit of length)

ℓ	length	$A(x)$ $(area)$	cross-section
V	volume		
F	force	M	moment $[N\,L]$
$E(x) > 0$	elasticity modul $[N/L^2]$	$\nu \in (0\,,\,1/2)$	Poisson number
$I(x)$	moment of inertia $[L^4]$	$p(x) = E(x)I(x)$	flexural rigidity $[N\,L^2]$
$\varepsilon = \Delta\ell/\ell$	strain	$\sigma = F/A$	stress
$\kappa = E\,A/\ell$	spring constant in tension rod		

Additional notations for a beam with constant rectangular cross-section:

$$b \text{ width in } y\text{-direction}\,,\quad h \text{ height in } z\text{-direction}\,;\quad b, h \ll \ell\,.$$

$$I_z = \int_{-h/2}^{h/2} \left(\int_{-b/2}^{b/2} y^2\, dy \right) dz = \frac{1}{12} bh^3 \quad \text{moment of inertia w.r.t. } z\text{-axis}$$

$$I_p = \int_A (y^2 + z^2)\, dydz = \frac{1}{12} bh(b^2 + h^2) \text{ polar moment of inertia}$$

$$\text{w.r.t. cross-section } A\,.$$

Recall that stress σ is positive under tension.

Some stipulations:

(1°) The neutral fibre passes through the barycenter of the area of cross-section in unloaded and loaded state, and the gravity center coincides with the origin of the local system of (y, z)-coordinates.

(2°) Shear forces in direction of the beam axis appear only as moments, shear forces in other directions are assembled into the torsion of the beam.

(3°) Bending about x-axis is neglected because of small moment.

(4°) In the BERNOULLI beam or *shear-rigid beam*, the angle φ between bending line and cross-section is *constant*, $\varphi = \pi/2$, before and after bending, whereas it is an additional independent variable in the TIMOSHENKO beam or *shear-soft* beam. BERNOULLI's beam returns the real situation with sufficient accuracy hence we consider only this type in the sequel.

(a) In a **Tension Rod**, the scalar function $u(x)$ describes the displacement in x-direction and $\kappa = AE/\ell$ is the *spring constant.*

Axiom 7.1. (1°) Adopt the *linear* HOOKE's law,

$$\boxed{\sigma(x) = E(x)\varepsilon(x)\,, \quad \varepsilon(x) = u'(x)}\,.$$

(2°) The *total energy* = *interior energy* + *exterior energy* of the tension rod is

$$\boxed{\Pi_S = \frac{1}{2}\int_0^\ell E(x)A(x)u'(x)^2 dx - F_1 u_1 - F_2 u_2\,, \quad u_1 = u(0),\ u_2 = u(\ell)} \tag{7.1}$$

where F_1, $F_2 \in \mathbb{R}$ are the exterior forces in direction of the rod.

In compression of the rod and equilibrium, we have $F_1 > 0$ and $F_2 = -F_1$.

(b) In the plane **Bending Beam**, the x-axis is again the axis of the beam and the y-axis shall point to "below" by clearness such that a positive load (to below) produces a positive displacement. Bending about the z-axis in mathematical positive sense then leads to a positive displacement in the (x, y)-plane. Let $y = u(x)$ be the *bending line* again, then a shortening of the beam in x-direction is given by $u_1(x, y) = yu'(x)$ in first order approximation (linearization such that $\tan(\beta) = u'(x)/1$). The linearization thus yields the displacements

$$\begin{aligned} u_1(x,y) &= yu'(x) && \text{in } x\text{-direction} \\ u_2(x) &= u(x) && \text{in } y\text{-direction} \\ u_3(x) &= 0 && \text{in } z\text{-direction} \end{aligned}$$

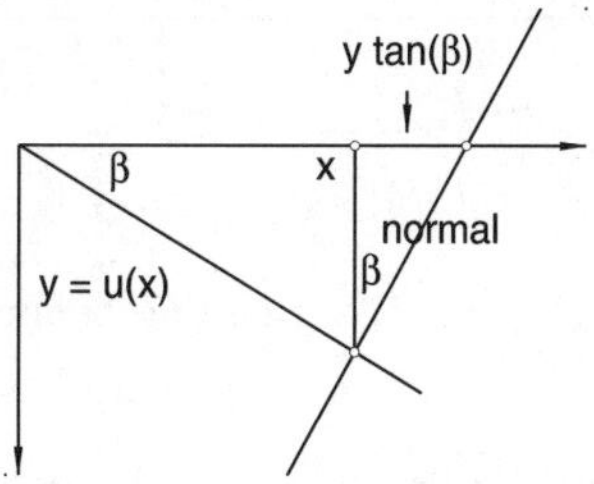

Figure 7.1. Displacement by KIRCHHOFF

Now a stress can be related again to the displacement by bending:

Axiom 7.2. (1°) HOOKE's law

$$\boxed{\sigma(x,y) = E(x)\varepsilon(x,y)\,, \quad \varepsilon(x,y) = yu''(x)}\,.$$

(2°) In bending about the z-axis, the *bending energy* of the beam is

$$\boxed{\begin{aligned} \Pi_B &= \frac{1}{2}\int_V \sigma(x,y)\,\varepsilon(x,y)\,dv = \frac{1}{2}\int_V E(x)y^2u''(x)^2\,dxdydz \\ &= \frac{1}{2}\int_0^\ell E(x)I_z(x)u''(x)^2\,dx \end{aligned}}\,.$$

(3°) Exterior energy, interior energy by bending, and interior energy by tension/compression are *summed up* to total energy.

(c) Energy In bending about the z-axis, volume forces and surface forces are combined to a stress $r(x)$ [force/(cross-section-)area] to establish the *total energy* in a way that the length force density $r(x)u(x)$ operates in *negative* y-direction for $r(x) > 0$ (e.g., elastic support), and in positive y-direction for $r(x) < 0$ (e.g., self-weight).

As is seen for instance in the first case of Figure 7.3, an axially acting force may lead to a stress in direction of the beam axis *and* to a crosswise displacement; cf. Example 7.2. This situation shall be regarded by an additional term $q(x)u'(x)$ with *force* $q(x) > 0$. Then

$-\,p(x)u''(x)$	bending moment (flexural rigidity · approximated curvature) $[N\,L]$
$q(x)u'(x)$	axial force $[N]$
$-\,r(x)u(x)$	length-force density $[N/L]$
$f(x) > 0$	continuously distributed load density $[N/L]$.

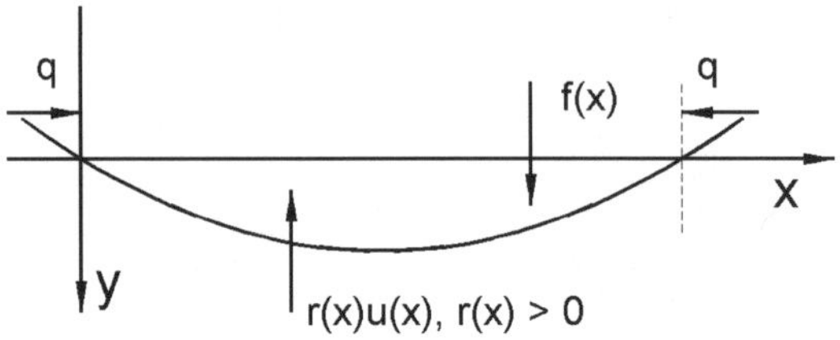

Figure 7.2. Bending beam

In equilibrium, the total interior energy of the beam is

$$\Pi_B = \frac{1}{2}\int_0^l \left(p(x)u''(x)^2 - q(x)u'(x)^2 + r(x)u(x)^2\right)dx\,.$$

by Axiom 7.2. Besides, pointwise loads F_i and pointwise *bending moments* M_k are now to be added, and the total energy is sumed up to

$$\mathcal{E}(u) = \frac{1}{2}\int_0^\ell \left(p(x)u''(x)^2 - q(x)u'(x)^2 + r(x)u(x)^2\right)dx - \int_0^\ell f(x)u(x)\,dx$$
$$- \sum_{i=1}^{I} F_i u(x_i) - \sum_{k=1}^{K} M_k \varphi(x_k)\,;$$

u is the displacement, and the positive y-direction points to "below". Because $\tan\varphi(x) = u'(x)$ and the small displacements supposed here, we may write also approximatively $\varphi(x_k) = u'(x_k)$ in the sum of moments.

(d) Variational Problem and Boundary Value Problem In order to derive conditions for the existence of a stationary solution u in equilibrium, we form again the directional derivative (first variation) relative to a *test function* (virtual displacement) v and set the result equal zero:

$$G(\varepsilon) := \mathcal{E}(u+\varepsilon v) \Longrightarrow G'(0) = \frac{d}{d\varepsilon}\mathcal{E}(u+\varepsilon v)\Big|_{\varepsilon=0} =: \delta\mathcal{E}(u;v) \overset{!}{=} 0\,,$$

then we obtain the EULER or variational equations of the extremal problem

$$\boxed{\mathcal{E}(u) = \min!\,,\quad u \in \mathcal{U}}\,. \tag{7.2}$$

The vector space $\mathcal{U}$ has now to be specified more exactly for the proof of existence of a solution, on the one side because of the attained smoothness of the desired solution and on the other side because the test functions $v \in \mathcal{U}$ have to regard the *essential boundary conditions* which depend on the individual problem formulation. In the present case we obtain

$$
\begin{aligned}
\delta\mathcal{E}(u;v) &:= \int_0^\ell p(x)u''(x)v''(x)\,dx - \int_0^\ell q(x)u'(x)v'(x)\,dx \\
&+ \int_0^\ell \Big(r(x)u(x) - f(x)\Big)v(x)\,dx - \sum_{i=1}^{I} F(x_i)v(x_i) - \sum_{k=1}^{K} M(x_k)v'(x_k) = 0\,.
\end{aligned}
\tag{7.3}
$$

Twofold partial integration yields *under sufficient smoothness* of the solution u

$$
\begin{aligned}
&\int_0^\ell p(x)u''(x)v''(x)\,dx = \Big[p(x)u''(x)v'(x)\Big]_0^\ell - \int_0^\ell \big(p(x)u''(x)\big)'v'(x)\,dx \\
&= \Big[p(x)u''(x)v'(x)\Big]_0^\ell - \Big[\big(p(x)u''(x)\big)'v(x)\Big]_0^\ell + \int_0^\ell \big(p(x)u''(x)\big)''v(x)\,dx\,.
\end{aligned}
$$

In disregarding pointwise loads and moments, (7.2) reveals to be equivalent to the relation

$$
\begin{aligned}
&\Big[p(x)u''(x)v'(x) - \Big((p(x)u''(x))' + q(x)u'(x)\Big)v(x)\Big]_0^\ell \\
&+ \int_0^\ell \Big[\big(p(x)u''(x)\big)'' + \big(q(x)u'(x)\big)' + r(x)u(x) - f(x)\Big]v(x)\,dx = 0\,,
\end{aligned}
\tag{7.4}
$$

where the solution u is supposed to be *fourtimes* continuously differentiable. Choosing at first an *arbitrary* test function v with $v(0) = v(\ell) = 0$, we obtain by this way a differential equation of order four,

$$
\boxed{\big(p(x)u''(x)\big)'' + \big(q(x)u'(x)\big)' + r(x)u(x) = f(x)\,,\quad 0 < x < \ell}\,. \tag{7.5}
$$

Choosing now either $v(0) = 0$ and $v(\ell)$ free, or $v(0)$ free and $v(\ell) = 0$, we obtain the boundary conditions for $\xi = 0$ and $\xi = \ell$:

$$
\begin{array}{lll}
v(\xi)\ \text{free} \Longrightarrow \big[(p(x)u''(x))' + q(x)u'(x)\big]_{x=\xi} & = & 0 \\
v(\xi)\ = 0 \Longrightarrow \big[(p(x)u''(x))' + q(x)u'(x)\big]_{x=\xi} & \text{free} & \\
v'(\xi)\ \text{free} \Longrightarrow p(x)u''(x)\big|_{x=\xi} & = & 0 \\
v'(\xi) = 0 \Longrightarrow p(x)u''(x)\big|_{x=\xi} & \text{free.} &
\end{array}
\tag{7.6}
$$

If boundary conditions $u(\xi) = a$ and/or $u'(\xi) = b$ are given, then all test functions have to regard the boundary conditions as follows:

$$
\begin{aligned}
u(\xi) = a &\Longrightarrow u(\xi) + \varepsilon v(\xi) = a \ \Longrightarrow v(\xi) = 0 \\
u'(\xi) = b &\Longrightarrow u'(\xi) + \varepsilon v'(\xi) = b \Longrightarrow v'(\xi) = 0
\end{aligned}
\qquad \xi \in \{0,\, \ell\}\,,
$$

therefore boundary conditions of this type are called *essential* or *geometric.* The remaining (homogenous) boundary conditions

$$
\begin{aligned}
u(\xi)\ \text{free} &\Longrightarrow \big[(p(x)u''(x))' + q(x)u'(x)\big]_{x=\xi} = 0 \\
u'(\xi)\ \text{free} &\Longrightarrow \qquad\qquad\qquad p(x)u''(x)\big|_{x=\xi} = 0
\end{aligned}
\qquad \xi \in \{0, \ell\}
$$

are satisfied by the exact solution of the variational problem (7.4) because the corresponding values $v(\xi)$ resp. $v'(\xi)$ of the test functions are free; hence they are called *natural* or *dynamic* boundary conditions. However, in solving the *boundary value problem* with equation (7.5), all boundary conditions — together four — are to be regarded explicitly; for instance the following possibilities do exist ((e) essential,(n) natural):

$$\boxed{\begin{array}{lrr} \text{simple supported end:} & u(\xi)=0\ (\text{e})\,, & pu''(\xi)=0\ (\text{n}) \\ \text{clamped fixed end:} & u(\xi)=0\ (\text{e})\,, & u'(\xi)=0\ (\text{e}) \\ \text{clamped movable end:} & u'(\xi)=0\ (\text{e})\,, & [(pu'')'+qu'](\xi)=0\ (\text{n}) \\ \text{free end, } q=0 & pu''(\xi)=0\ (\text{n})\,, & (pu'')'(\xi)=0\ (\text{n}) \end{array}}\ .$$

It is seen in most bending problems directly which *combination* of boundary conditions allows an equilibrium position of the beam and which combination has no physical meaning.

(e) The differential equation (7.5) may be derived also by a **Balance of Moments**. To this end let Q be a constant axial force having the same direction before and after bending (*dead force or load*) then the following relation does hold at point x in equilibrium for the bending moment and the moment belonging to Q,

$$\boxed{-p(x)u''(x) = Q\,u(x) = Q\int_0^\ell u'(\xi)\,d\xi}\ ;$$

cf. Fig. 7.3, *Case 1.* If the pointwise axial forces are replaced by a uniformly distributed force density as in **(c)**, this balance equation has the form

$$-p(x)u''(x) = \int_0^x q(\xi)u'(\xi)\,d\xi\,.$$

The length-force density $-r(x)u(x)$ as well as the load density $f(x)$ entail further moments and the addition of all moments at point x leads to

$$p(x)u''(x) + \int_0^x q(\xi)u'(\xi)\,d\xi + \int_0^x r(\xi)u(\xi)(x-\xi)\,d\xi - \int_0^x f(\xi)(x-\xi)\,d\xi = 0\,.$$

Twofold differentiation and an application of Leibniz' rule then yields (7.5) again. But the *definiteness of the quadratic form* in the extremal problem (7.2) remains always the crucial criterium for suitable boundary conditions.

(f) Further Boundary Conditions We neglect pointwise loads and moments as before and write for *brevity*, cf. also Sect. 7.3,

$$\boxed{\begin{array}{ll} b(u,v) := \displaystyle\int_0^\ell p(x)u''(x)v''(x)\,dx\,, & c(u,v) := \displaystyle\int_0^\ell q(x)u'(x)v'(x)\,dx \\ d(u,v) := \displaystyle\int_0^\ell r(x)u(x)v(x)\,dx\,, & f(u) := \displaystyle\int_0^\ell f(x)u(x)\,dx \\ a(u,v) := b(u,v) - c(u,v) + d(u,v) & \end{array}}\ .$$

The *bilinear form a* has to be equipped with a suitable additional boundary term for the consideration of further *dynamic* boundary conditions but, for brevity, we confine ourselves to the homogenous case here. Let

$$R(\xi,u)=\alpha_\xi p(\xi)u'(\xi)^2+\beta_\xi p(\xi)u(\xi)^2+\gamma_\xi q(\xi)u(\xi)^2,\ \ \xi\in\{0,\,\ell\}\,,$$

hence either $\xi=0$ or $\xi=\ell$. Then

$$\delta R(\xi,u;v)=2\alpha_\xi p(\xi)u'(\xi)v'(\xi)+2\beta_\xi p(\xi)u(\xi)v(\xi)+2\gamma_\xi q(\xi)u(\xi)v(\xi)\,,$$

and the modified minimum problem (7.2) has the form

$$2\,\widetilde{\mathcal{E}}(u)=\left[b(u,u)+d(u,u)-c(u,u)+R(\xi,u)\Big|_{\xi=0}^{\ell}\right]-2f(u)=\min!\ \ u\in\mathcal{U}\,. \tag{7.7}$$

By setting the first variation equal to zero, one obtains the differential equation (7.4) in the same way as above, but the boundary terms now have a different form, namely,

$$\begin{aligned}&\big[p(x)u''(x)+\alpha_x p(x)u'(x)\big]v'(x)\Big|_0^\ell\\ -\ &\big[(p(x)u''(x))'+q(x)u'(x)-\beta_x p(x)u(x)+\gamma_x q(x)u(x)\big]v(x)\Big|_0^\ell=0\,.\end{aligned} \tag{7.8}$$

Besides others, the following combinations are possible:
(1°) Combination of fixed support and elastic clamping left/right with spring constant $\alpha>0$:

$$\boxed{\begin{array}{l}u(0)=0\ (\mathrm{e})\,,\ \ pu''(0)-\alpha pu'(0)=0\ (\mathrm{n})\\ u(\ell)=0\ (\mathrm{e})\,,\ \ pu''(\ell)+\alpha pu'(\ell)=0\ (\mathrm{n})\end{array}}\ ;$$

boundary term $R(\xi,u)=\pm\,\alpha\,p(u')^2(0)$, ("$-$" for $\xi=0$).
(2°) Combination of movable support and elastic clamping left/right with spring constant $\beta>0$:

$$\boxed{\begin{array}{l}u'(0)=0\ (\mathrm{e})\,,\ \ (pu'')'(0)+qu'(0)-\beta pu(0)+\gamma qu(0)=0\ (\mathrm{n})\\ u'(\ell)=0\ (\mathrm{w})\,,\ \ (pu'')'(\ell)+qu'(\ell)+\beta pu(\ell)-\gamma qu(\ell)=0\ (\mathrm{n})\end{array}}\ ;$$

boundary term $R(\xi,u)=\pm\,[\beta\,pu^2(\xi)+\gamma\,qu^2(\xi)]$, ("$-$" for $\xi=0$).
(3°) Elastic supported end left/right with spring constant $\beta>0$:

$$\boxed{\begin{array}{l}pu''(0)=0\ (\mathrm{n})\,,\ \ (pu'')'(0)-\beta pu(0)+\gamma qu^2(0)=0\ (\mathrm{n})\\ pu''(\ell)=0\ (\mathrm{n})\,,\ \ (pu'')'(\ell)+\beta pu(\ell)-\gamma qu^2(\ell)=0\ (\mathrm{n})\end{array}}\ ;$$

boundary term $R(\xi,u)=\pm\,\beta\,pu^2(\xi)$, ("$-$" for $\xi=0$).
(4°) Combination of elastic support and elastic clamping left/right with spring constants $\alpha>0$ and $\beta>0$:

$$\boxed{\begin{array}{ll} pu''(0) - \alpha pu'(0) = 0\ (\mathrm{n})\,, & (pu'')'(0) + qu'(0) - \beta pu(0) + \gamma qu(0) = 0\ (\mathrm{n}) \\ pu''(\ell) + \alpha pu'(\ell) = 0\ (\mathrm{n})\,, & (pu'')'(\ell) + qu'(\ell) + \beta pu(\ell) - \gamma qu(\ell) = 0\ (\mathrm{n}) \end{array}}\;;$$

boundary term $R(\xi, u) = \pm[\alpha\, p(u')^2(0) + \beta\, pu^2(\xi) + \gamma\, qu^2(\xi)]$, (" $-$ " for $\xi = 0$). It remains to be verified in this case whether a nonzero parameter γ has any physical meaning.

(g) To prove the **Existence** of solutions, we introduce a vector space $\mathcal{H}$ with scalar product and norm,

$$(u, v) = \int_0^\ell u(x)v(x)\, dx\,, \quad \|u\|^2 = \int_0^\ell u(x)^2\, dx\,.$$

Further, let $\mathcal{H}$ be closed w.r.t. $\|\cdot\|$ hence a HILBERT space, and let $\mathcal{U} \subset \mathcal{H}$ be a closed subspace. A direct application of Theorem 1.25 then yields

Theorem 7.1. *(Existence and Uniqueness) Let the bilinear form a be symmetric, $a(u, v) = a(v, u)$, and let*

$$\forall\, u \in U:\ \kappa\|u\| \le a(u, u) \le \|a\|\,\|u\|^2 \tag{7.9}$$

where $\|a\|$ is finite and $\kappa > 0$. Then the minimum problem (7.7),

$$2\,\widetilde{\mathcal{E}}(u) = \widetilde{a}(u, u) - 2f(u) = \min!\,, \quad u \in \mathcal{U}\,,$$

has a unique solution.

This theorem warrants only the existence of a *weak* solution in HILBERT space $\mathcal{H}$, its smoothness has to be verified by other means; if the functions p, q, r together with their derivatives are continuous then the bilinear form a is uniformly bounded and the right side of the inequality (7.9) is fulfilled. The left inequality has to be verified for each boundary term $R(\xi, u)$. Among others, it constitutes a condition for an appropriate choice of the parameters α, β and γ.

Lemma 7.1. *(Elementary RAYLEIGH-RITZ Inequality)* $\|u\|^2 \le \dfrac{\ell^2}{2}\|u'\|^2$ *if* $u(0) = 0$.

Proof. We have $u(x) = \int_0^\ell u'(x)\, dx = \int_0^\ell 1 \cdot u'(x)\, dx$ by assumption. By CAUCHY-SCHWARZ' inequality, $(u, v)^2 \le (u, u)\cdot(v, v)$, we obtain $u(x)^2 \le x\int_0^\ell u'(x)^2\, dx$. Integration of both sides yields the assertion. □

For instance, let $0 \le q(x) \le q_0$ and let $u(0) = 0$ be boundary condition then, by Lemma 7.1,

$$-c(u, u) \ge -\frac{2q_0}{\ell^2}\|u\|^2\,,$$

which may be used for the proof of definiteness.

If e.g. $p(x) \geq p_0 > 0$ and $u(0) = u'(0) = 0$ (beam clamped at left end), a twofold application of Lemma 7.1 yields

$$\int_0^\ell p(x)(u''(x))^2\,dx \geq p_0 \int_0^\ell (u''(x))^2\,dx \geq \kappa \int_0^\ell u^2(x)\,dx = \kappa \|u\|^2\,,\quad \kappa > 0\,.$$

If now e.g. $\alpha = \beta = \gamma = 0$, $q = 0$ and $r(x) > 0$ (elastic support), then Theorem 7.1 guarantees the existence of a weak solution.

7.2 Eigenvalue Problems

(a) The **Minimum Problem**

$$2\,\mathcal{E}(u) = a(u,u) - 2f(u) = \min!\,,\quad u \in \mathcal{U}$$

has a unique solution under the assumptions discussed in the preceding section. Let now two *continuous* and *symmetric* bilinear forms b and c be given and let both be uniformly positive definite on a closed subspace $\mathcal{U} \subset \mathcal{H}$:

$$\boxed{\begin{array}{l} \exists\, 0 < \beta \leq \|b\| \;\; \forall\, u \in \mathcal{U}:\; \beta\|u\|^2 \leq b(u,u) \leq \|b\|\|u\|^2 \\ \exists\, 0 < \gamma \leq \|c\| \;\; \forall\, u \in \mathcal{U}:\; \gamma\|u\|^2 \leq c(u,u) \leq \|c\|\|u\|^2 \end{array}}\,. \tag{7.10}$$

Then a bilinear form a being defined on $\mathcal{U}$ by

$$a(u,u;\lambda) := b(u,u) - \lambda\, c(u,u)\,,\quad \lambda \in \mathbb{R}\,,$$

is certainly positive definite if

$$\beta - \lambda\,\|c\| > 0\,.$$

Accordingly, the minimum problem $a(u,u;\lambda) = \min!\,,\ u \in \mathcal{U}$, has only the *trivial* (but not less important) solution $u = 0$ for these values of λ which corresponds to zero displacement in a problem of bending beam. If however the bilinear form a is only *positive semidefinite*, there exists a $0 \neq u \in \mathcal{U}$ such that $a(u,u;\lambda) = 0$. In this case also $\kappa\, u$ is a solution for $\kappa \in \mathbb{R}$, which corresponds to an indifferent solution in bending beam. The smallest value of the parameter λ satisfying

$$b(u,u) - \lambda\, c(u,u) = 0\,,\quad 0 \neq u \in \mathcal{U}\,, \tag{7.11}$$

is the smallest *eigenvalue* with appertaining *eigensolution* u of the *eigenvalue problem* (7.11). It is the global *minimum* point of the Rayleigh *quotient*

$$Q(u) = b(u,u)/c(u,u)\,,\quad u \in \mathcal{U}\,;$$

the further eigenvalues are *local* minimum points of $Q(u)$.

Theorem 7.2. *(Characterization Theorem) Let $\mathcal{U} \subset \mathcal{H}$ be a closed subspace. The function $u \in \mathcal{U}$ is a stationary point of the* RAYLEIGH *quotient if and only if*

$$\boxed{\forall\, v \in \mathcal{U} : b(u,v) = \lambda\, c(u,v)}\,. \tag{7.12}$$

In particular, we have $\lambda = b(u,u)/c(u,u)$ for $v = u$.
Proof. An expansion shows that

$$Q(u + \varepsilon v) = \frac{b(u,u) + 2\varepsilon b(u,v) + \varepsilon^2 b(v,v)}{c(u,u) + 2\varepsilon c(u,v) + \varepsilon^2 c(v,v)}\,.$$

As condition for a stationary value we set the first variation of the RAYLEIGH quotient equal to zero and obtain

$$0 = \frac{d}{d\varepsilon} Q(u + \varepsilon v)\Big|_{\varepsilon=0} = 2\,\frac{c(u,u)b(u,v) - b(u,u)c(u,v)}{c(u,u)^2}\,. \tag{7.13}$$

Substitution of $b(u,u) = \lambda\, c(u,u)$ and cancelling of $c(u,u) \neq 0$ yields the necessary condition (7.12). On the other side, (7.13) follows from (7.12) in simple way by substituting (7.12) and then substituting $b(u,u) = \lambda\, c(u,u)$ once more. □

The system (7.12) is called *generalized eigenvalue problem* and constitutes the basis for many numerical methods of approximation (RITZ-GALERKIN method). In particular, an eigenvalue problem *must be a homogenous problem*, i.e., all the exterior loads must be zero. For λ, e.g., the volume force density is substituted in applications, $r(x) = -\lambda$, or the axially acting force, $q(x) = \lambda$. When the above introduced boundary term R appears, it has to be partitioned into a part belonging to b and a part belonging to c. The modified bilinear forms have to satisfy again Assumption (7.10) on a suitable subspace.

Example 7.1. Bending oscillations without axial loads. Cf. (Collatz63), 2. ed. p. 24. We choose $q = 0$ in (7.13) and $r(x) = -\lambda s(x)$, $s(x) = \varrho(x)A(x)$ (ϱ mass density, A area of cross-section). Then (7.11) reads:

$$\left[\int_0^\ell p(x)u''(x)^2\,dx + R_b(\xi,u)\Big|_{\xi=0}^{\ell}\right] - \lambda \int_0^\ell s(x)u(x)^2\,dx = 0$$
$$R_b(\xi,u) = \alpha_\xi p(\xi)u'(\xi)^2 + \beta_\xi p(\xi)u(\xi)^2\,.$$

The associated differential equation reads:

$$(p(x)u'')'' - \lambda s(x)u = 0\,.$$

Suitable boundary conditions are for instance:
(1.) Left end clamped, right end free:

$$u(0) = u'(0) = 0\,,\ \ u''(\ell) = (pu'')'(\ell) = 0\,.$$

(2.) Both ends hinged:

$$u(0) = u''(0) = 0\,,\ \ u(\ell) = u''(\ell) = 0\,.$$

(3.) Both ends elastically supported :

$$u''(0) = (pu'')'(0) + \gamma u(0) = 0\,,\ \ u''(\ell) = (pu'')'(\ell) - \gamma u(\ell) = 0\,.$$

γ spring constant. The eigenvalues are negative for $\gamma < 0$ and the problem becomes unstable in this case.

(4.) Left end hinged, right end elastically supported:

$$u(0) = u''(0) = 0\,,\ \ u''(\ell) = (pu'')'(\ell) - \gamma u(\ell) = 0\,.$$

Example 7.2. Oscillation of a bar with self-weight. (Gravitational force acts in direction of the negative x-axis.) We choose

$$q(x) = g\int_0^x s(\xi)\,d\xi\,,\ \ r(x) = -\lambda s(x)\,,\ \ s(x) = \varrho(x)A(x)$$

where g is the gravitational acceleration. Equation (7.11) then reads:

$$\left[\int_0^\ell [p(x)u''(x)^2 + q(x)u'(x)^2]\,dx + R_b(\xi,u)\Big|_{\xi=0}^{\ell}\right] - \lambda\int_0^\ell s(x)u(x)^2\,dx = 0\,,$$
$$R_b(\xi,u) = \alpha_\xi p(\xi)u'(\xi)^2 + \beta_\xi p(\xi)u(\xi)^2 + \gamma_\xi q(\xi)u(\xi)^2\,.$$

The associated differential equation reads:

$$[p(x)u'']'' + [q(x)u']' - \lambda s(x)u = 0\,.$$

Suitable boundary conditions are, e.g., left end clamped, right end free:

$$u(0) = u'(0) = 0\,,\ \ u''(\ell) = (pu'')'(\ell) = 0\,.$$

(b) Buckling of a Beam Cf. (Collatz63), pp. 46–66, and 2. ed. p. 9, 435 ff. Here $q(x) = \widetilde{\lambda} = F$ with force F. (7.11) has the form

$$\left[\int_0^\ell [p(x)u''(x)^2 + r(x)u(x)^2]\,dx + R_b(\xi,u)\Big|_{\xi=0}^{\ell}\right]$$
$$-\widetilde{\lambda}\left[\int_0^\ell u'(x)^2\,dx + R_c(\xi,u)\Big|_{\xi=0}^{\ell}\right] = 0\,,$$

$$R_b(\xi,u) = \alpha_\xi p(\xi)u'(\xi)^2 + \beta_\xi p(\xi)u(\xi)^2,\ \ R_c(\xi,u) = \gamma_\xi u(\xi)^2,\ \ \xi\in\{0,\,\ell\}\,,$$

(means $\xi = 0$ or $\xi = \ell$). The associated differential equation reads:

$$\boxed{(p(x)u''(x))'' + r(x)u + \widetilde{\lambda}\,u''(x) = 0}\,. \qquad (7.14)$$

The smallest eigenvalue is called EULER*'s buckling load.* We consider more exactly the simplest case where $p(x) = EI$ constant and $r(x) \equiv 0$. By writing $u'' = v$, one then obtains the differential equation

$$v'' + \lambda v = 0\,, \quad \lambda = \widetilde{\lambda}/EI$$

with general solution $(a, b, c, d \in \mathbb{R})$

$$\boxed{\begin{aligned} v(x) = u''(x) &= a\,\sin(\sqrt{\lambda}x) + b\,\cos(\sqrt{\lambda}x) \\ u'(x) &= -\frac{a}{\sqrt{\lambda}}\,\cos(\sqrt{\lambda}x) + \frac{b}{\sqrt{\lambda}}\,\sin(\sqrt{\lambda}x) + d \\ u(x) &= -\frac{a}{\lambda}\,\sin(\sqrt{\lambda}x) - \frac{b}{\lambda}\,\cos(\sqrt{\lambda}x) + c + d\,x \end{aligned}}\,.$$

Case 1: left end clamped, right end free. Boundary conditions are

$$u(0) = 0\ (\text{e})\,, \quad u'(0) = 0\ (\text{e})\,, \quad u(\ell) = 0\ (\text{n})\,, \quad u'''(\ell) + \lambda u'(\ell) = 0\ (\text{n})\,.$$

By using the left boundary conditions we obtain

$$\begin{aligned} u(x) &= \frac{a}{\lambda}[\sqrt{\lambda}x - \sin(\sqrt{\lambda}x)] + \frac{b}{\lambda}[1 - \cos(\sqrt{\lambda}x)], \\ u'(x) &= \frac{a}{\sqrt{\lambda}}[1 - \cos(\sqrt{\lambda}x)] + \frac{b}{\sqrt{\lambda}}\sin(\sqrt{\lambda}x)\,. \end{aligned}$$

The right boundary condition $u'''(\ell) + \lambda u'(\ell) = 0$ yields

$$a\sqrt{\lambda}\cos(\sqrt{\lambda}\ell) - b\sqrt{\lambda}\sin(\sqrt{\lambda}\ell) + \lambda\left[\frac{a}{\sqrt{\lambda}}[1 - \cos(\sqrt{\lambda}\ell)] + \frac{b}{\sqrt{\lambda}}\sin(\sqrt{\lambda}\ell)\right] = 0\,,$$

hence $a = 0$. Because $b \neq 0$, the right condition $u''(\ell) = v(\ell) = 0$ yields $\cos(\sqrt{\lambda}\ell) = 0$, and then

$$\sqrt{\lambda}\ell = \frac{\pi}{2} + 2k\pi, \quad k \in \mathbb{Z}\,,$$

hence

$$\boxed{\begin{aligned} &\text{Solution:} && u(x) = c(1 - \cos(\sqrt{\lambda}x))\,,\ c \in \mathbb{R} \\ &\text{EULER's buckling load: } \lambda_1 && = \frac{\pi^2}{4\ell^2} \Longrightarrow F = \frac{\pi^2}{4}\frac{EI}{\ell^2} \end{aligned}}\,.$$

Case 2: Both ends hinged. The boundary conditions are

$$u(0) = 0\ (\text{e})\,, \quad u''(0) = 0\ (\text{n})\,, \quad u(\ell) = 0\ (\text{e})\,, \quad u''(\ell) = 0\ (\text{n})\,.$$

One obtains easily $\sqrt{\lambda}\ell = k\pi$, $k \in \mathbb{Z}$.

Solution:	$u(x) = c\sin(\sqrt{\lambda}x)\,,\ c \in \mathbb{R}$
EULER's buckling load: λ_1	$= \dfrac{\pi^2}{\ell^2} \Longrightarrow F = \pi^2\,\dfrac{EI}{\ell^2}$.

Case 3: Left end clamped, right end hinged. The boundary conditions are

$$u(0) = 0\ (\text{w})\,,\quad u'(0) = 0\ (\text{w})\,,\quad u(\ell) = 0\ (\text{w})\,,\quad u''(\ell) = 0\ (\text{n})\,.$$

At first, the left condition yield as in *Case 1*

$$u(x) = \frac{a}{\lambda}[\sqrt{\lambda}x - \sin(\sqrt{\lambda}x)] + \frac{b}{\lambda}[1 - \cos(\sqrt{\lambda}x)]\,.$$

Substitution of $x = \ell$ and the condition $u''(\ell) = v(\ell) = 0$ yield the system

$$\begin{aligned} &a\sin(\sqrt{\lambda}\ell) && + b\cos(\sqrt{\lambda}\ell) && = 0 \\ &a[\sqrt{\lambda}\ell - \sin(\sqrt{\lambda}\ell)] && + b[1 - \cos(\sqrt{\lambda}\ell)] && = 0\,. \end{aligned}$$

The determinant of the associated matrix must be zero in order that a non-zero solution $[a,\ b]^T$ exists: $\sin(\sqrt{\lambda}\ell) - \sqrt{\lambda}\ell\cos(\sqrt{\lambda}\ell) = 0$.

For *approximation* of the first eigenvalue one chooses here frequently $\sqrt{\lambda}\ell = \sqrt{2}\cdot\pi$. Solution and EULER's buckling load:

$u(x) = c\big[\sqrt{\lambda}x - \sin(\sqrt{\lambda}x) + \tan(\sqrt{\lambda}\ell)(\cos(\sqrt{\lambda}x) - 1)\big]$	
$\lambda_1 \simeq \dfrac{2\pi^2}{\ell^2} \Longrightarrow F \simeq 2\pi^2\,\dfrac{EI}{\ell^2}$	.

Case 4: Both ends clamped (one movable in x-direction). The boundary conditions are

$$u(0) = 0\ (\text{e})\,,\quad u'(0) = 0\ (\text{e})\,,\quad u(\ell) = 0\ (\text{e})\,,\quad u'(\ell) = 0\ (\text{e})\,.$$

At first, the condition at the left end yields as in *case 1*

$$u(x) = \frac{a}{\lambda}[\sqrt{\lambda}x - \sin(\sqrt{\lambda}x)] + \frac{b}{\lambda}[1 - \cos(\sqrt{\lambda}x)]\,.$$

Substitution of the right terminal conditions yields the system

$$\begin{aligned} &a[\sqrt{\lambda}\ell - \sin(\sqrt{\lambda}\ell)] && + b[1 - \cos(\sqrt{\lambda}\ell)] && = 0 \\ &a[1 - \cos(\sqrt{\lambda}\ell)] && + b\sin(\sqrt{\lambda}\ell) && = 0 \end{aligned}$$

of which the determinant must be zero again in order that a non-zero solution $[a,\ b]^T$ exists: $\sqrt{\lambda}\ell\sin(\sqrt{\lambda}\ell) + 2\cos(\sqrt{\lambda}\ell) - 2 = 0$ and consequently $\sqrt{\lambda}\ell = 2k\pi\,,\ k \in \mathbb{Z}$.

Solution:	$u(x) = c(1-\cos(\sqrt{\lambda}x))\,,\ c \in \mathbb{R}$
EULER's buckling load: λ_1	$= \dfrac{4\pi^2}{\ell^2} \Longrightarrow F = 4\pi^2 \dfrac{EI}{\ell^2}$

.

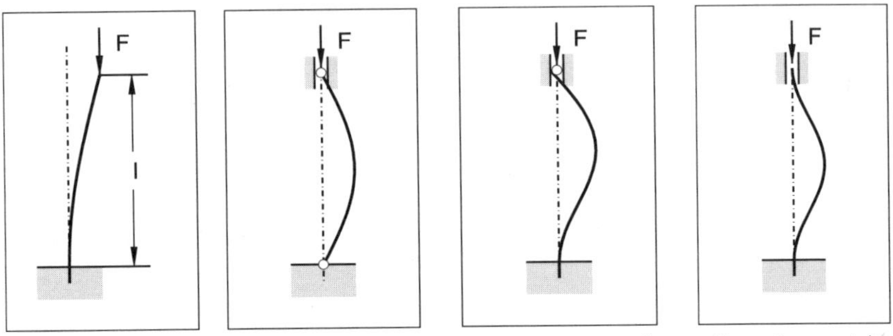

Case 1, $\kappa = 1$ Case 2, $\kappa = 1$ Case 3, $\kappa = -1/4$ Case 4, $\kappa = 1/2$

Figure 7.3. EULER's buckling loads

(c) The equations of motion of an **Oscillating Beam** can be found by the local balance theorem of momentum or by HAMILTON's principle (8.31),

$$\mathcal{W}(u; t_1, t_2) := \int_{t_1}^{t_2} [\mathcal{E}_k(u) - \mathcal{E}_p(u)]\, dt = \text{ stationary in } u$$

where $\mathcal{E}_k(u)$ is the kinetic energy and $\mathcal{E}_p(u)$ is the potential energy of the beam. $\mathcal{E}_p(u)$ may be one of the above considered bilinear forms but we choose here simply the bending energy then

$$\mathcal{E}_k(u) = \frac{1}{2}\int_0^\ell r(x)\dot{u}^2(t,x)\, dx\,,\quad \mathcal{E}_p(u) = \frac{1}{2}\int_0^\ell p(x)u''^2(t,x)\, dx\,, \qquad (7.15)$$

$r(x) = \varrho(x)A(x)\,,\ p(x) = E(x)I(x)$. Other forms of the potential energy are treated in a similar way. We set the variation equal to zero, $\partial\mathcal{W}(u;v) = 0$, integrate $\partial\mathcal{E}_k$ partially over t and $\partial\mathcal{E}_p$ two-times partially over x, the result reads:

$$\int_{t_1}^{t_2} \left[\int_0^\ell [r\ddot{u} + (pu'')'']v\, dx - (pu'')'v\Big|_0^\ell + pu''v'\Big|_0^\ell \right] dt = 0\,. \qquad (7.16)$$

The square-bracketed term must disappear since t_1 and t_2 are arbitrary. By the usual argumentation of variational calculus we obtain

$$\begin{aligned} r(x)\ddot{u}(t,x) + [p(x)u''(t,x)]'' = 0 \\ [p(x)u''(t,x)]'v(x)\Big|_0^\ell - p(x)u''(t,x)v'(x)\Big|_0^\ell = 0\,. \end{aligned} \qquad (7.17)$$

The diversification of boundary conditions is the same as in (7.6) for $q = 0$. The solution of the resulting hyperbolic boundary value problem follows by *separation of variables* leading to an *eigenvalue problem* of the above form: Inserting $u(t, x) = v(t)w(x)$ into (7.17) we obtain for a constant $\lambda \in \mathbb{R}$

$$\frac{\ddot{v}}{v} = \lambda = \frac{1}{r}\frac{(pw'')''}{w} \Longrightarrow \ddot{v} = \lambda\, v\,,\ (p(x)w'')'' - \lambda\, r(x)w = 0\,.$$

So the associated eigenvalue problem reveals to be the dominant part of an oscillation problem.

7.3 Numerical Approximation

(a) The **Tension Rod** is first rescaled here to unit intervall $[0\,,\,1]$ because of later application in the method of finite elements, see Chap. 9,

$$x = \xi\,\ell\,,\ u(x) = v(\xi)\,,\ u'(x) = v'(\xi)(d\xi/dx)\,,\ 0 \leq \xi \leq 1\,.$$

Then we obtain

$$\int_0^\ell u'(x)^2 dx = \frac{1}{l}\int_0^1 v'(\xi)^2 d\xi\,.$$

By supposing constant stress $\sigma = E\varepsilon(x) = Eu'(x)$, the *linear ansatz*

$$\boxed{v(\xi) = \alpha_1 + \alpha_2\xi}$$

is exact. The relation between coefficients with and without physical meaning, namely $u_1 = v(0) = \alpha_1$, $u_2 = v(1) = \alpha_1 + \alpha_2$, follows from the relation $u(x(\xi)) = v(\xi)$; or in matrix-vector notation

$$\begin{bmatrix}\alpha_1\\ \alpha_2\end{bmatrix} = \begin{bmatrix}1 & 0\\ -1 & 1\end{bmatrix}\begin{bmatrix}u_1\\ u_2\end{bmatrix} =: B\underline{u}\,.$$

Inserting yields

$$\int_0^\ell u'(x)^2 dx \simeq \underline{u}^T\widetilde{S}\underline{u},\ \ \widetilde{S} = \frac{1}{\ell}\begin{bmatrix}1 & -1\\ -1 & 1\end{bmatrix},$$

and, by this way,

$$\boxed{\widehat{\Pi} = \frac{\kappa}{2}\underline{u}^T\widetilde{S}\underline{u} - \underline{f}^T\underline{u}\,,\ \ \underline{f} = \begin{bmatrix}f_1\\ f_2\end{bmatrix}} \tag{7.18}$$

is a suitable approximation of interior energy in (7.1) where $\kappa = E\,A/\ell$ denotes the spring constant.

(b) Bending Beam Using the notations of Sect. 7.1, we consider a beam element with constant rectangular cross-section and constant flexural rigidity

EI dropping the axial tension $q(x)u'(x)$. Further, we suppose that pointwise loads appear only at the ends of the beam element. Then the total energy of the beam element is

$$\mathcal{E}(u) = \frac{EI}{2}\int_0^\ell u''(x)^2\,dx - \int_0^\ell f(x)u(x)\,dx - \sum_{i=1}^2 F_i u_i - \sum_{k=1}^2 M_k u_k' \qquad (7.19)$$

where $u(0) = u_1$, $u(\ell) = u_2$, etc.. The numerical approximation is here properly managed by a Hermitian interpolating polynomial of degree three, cf. Sect. 2.1, **(e)** whose coefficients are uniquely determined by the values of the *node vector* $U = [u_1, u_1', u_2, u_2']^T$,

$$\begin{aligned} p(x,U) &= a + b(x - x_1) + c(x - x_1)^2 + d(x - x_1)^3 \\ p(x_1;U) &= u_1\,,\ \ p'(x_1;U) = u_1'\,,\ \ p(x_2;U) = u_2\,,\ \ p'(x_2;U) = u_2'\,. \end{aligned}$$

By $\ell = x_2 - x_1$, $x_1 = 0$ we obtain $a = u_1$, $b = u_1'$, and

$$c = \frac{3(u_2 - u_1) - \ell(2u_1' + u_2')}{\ell^2}\,,\quad d = \frac{2(u_1 - u_2) + \ell(u_1' + u_2')}{\ell^3}\,.$$

Now the *unknown* function u is replaced approximatively by the polynomial $p(\circ\,;U)$, then an integration w.r.t. x leads to

$$\frac{EI}{2}\int_0^\ell u''(x)^2\,dx \sim \frac{EI}{2}\int_0^\ell p''(x;U)^2\,dx = \frac{1}{2}U^TKU$$

where the *stiffness matrix* of the beam element is

$$K = \frac{2EI}{\ell^3}\begin{bmatrix} 6 & 3\ell & -6 & 3\ell \\ 3\ell & 2\ell^2 & -3\ell & \ell^2 \\ -6 & -3\ell & 6 & -3\ell \\ 3\ell & \ell^2 & -3\ell & 2\ell^2 \end{bmatrix}. \qquad (7.20)$$

For the second integral in (7.19) it is advantageous to approximate $f(x)$ likewise by a polynomial $p(x;F)$ of degree three with node vector $F = [f_1, f_1', f_2, f_2']^T$. Then we obtain in the same way

$$\int_0^\ell f(x)u(x)\,dx \sim F^TMU$$

where the *mass matrix* of the beam element has the form

$$M = \frac{\ell}{420}\begin{bmatrix} 156 & 22\ell & 54 & -13\ell \\ 22\ell & 4\ell^2 & 13\ell & -3\ell^2 \\ 54 & 13\ell & 156 & -22\ell \\ -13\ell & -3\ell^2 & -22\ell & 4\ell^2 \end{bmatrix}. \qquad (7.21)$$

Now the approximation of potential energy of the beam element reads:

$$\boxed{P(U) = \frac{1}{2}U^T KU - F^T MU - R^T U} \tag{7.22}$$

where $R = [F_1, M_1, F_2, M_2]^T$ is the vector of pointwise loads at the ends.

Let us now suppose that the beam consists of several beam elements. Then the individual equations (7.22) are summed up to a global quadratic form,

$$\mathbf{P}([U]) := \frac{1}{2}[U]^T[K][U] - [F]^T[M][U] - [R]^T[U], \tag{7.23}$$

where $[U]$ is the global node vector of unknowns.

The total energy of a beam takes a minimum in equilibrium, but the computation of $\min\{\mathbf{P}([U])\}$ does not lead to a proper result without additional *support conditions* which are written here in the form of linear side conditions $[B][U] = [C]$. The solution of the linear system

$$\begin{bmatrix} [K] & [B]^T \\ [B] & [O] \end{bmatrix} \begin{bmatrix} [U] \\ [V] \end{bmatrix} = \begin{bmatrix} [M][F] + [R] \\ [C] \end{bmatrix} \tag{7.24}$$

then supplies an approximation of the bending line u and the accuracy increases with the number of beam elements proportionally to $1/\ell^3$ (ℓ maximum length of all elements).

Example 7.3. For constant axial load $q(x) = q$, TIMOSHENKO and GOODIER (1951) have found the exact bending line of a simply supported bending beam with rectangular cross-section,

$$u(x) = d - \frac{q\ell^4}{64EI}\left[1 + \left(\frac{8}{5} + \nu\right)\lambda^2 - \frac{1}{6}\xi^2\right]\xi^2,\ d = \frac{5q\ell^4}{384EI}\left[1 + \frac{6}{5}\lambda^2\left(\frac{8}{5} + \nu\right)\right]$$

Here $\lambda = h/\ell$, $\xi = 2x/\ell$, and ν denotes again POISSON's ratio.

Example 7.4. (Schwarz80), p. 38. Input data: (length in $[cm]$),
$\mathrm{E} = 2 \cdot 10^7\ [N/cm^2]$, $\mathrm{I} = 16\ [cm^4]$, $\mathrm{F} = 100\ [N]$, $\mathrm{f} = 2\ [N/cm]$.
Nodes: $[P_1,\ P_2,\ P_3,\ P_4] = [0,\ 150,\ 200,\ 300]$,
constant load densities in the three subintervals: $f = [0,\ 0,\ 2]$, supports in nodes P_i:

$$\begin{bmatrix} u_i \\ u_i' \end{bmatrix} = \begin{bmatrix} 0 & - & 0 & 0 \\ 0 & - & - & - \end{bmatrix}$$

Supports more detailed: $u(0) = u(200) = u(300) = 0,\ u'(0) = 0$.
Loads and moments in nodes P_i:

$$\begin{bmatrix} F_i \\ M_i \end{bmatrix} = \begin{bmatrix} 0 & 100 & 0 & 0 \\ 0 & 0 & 0 & 0 \end{bmatrix}$$

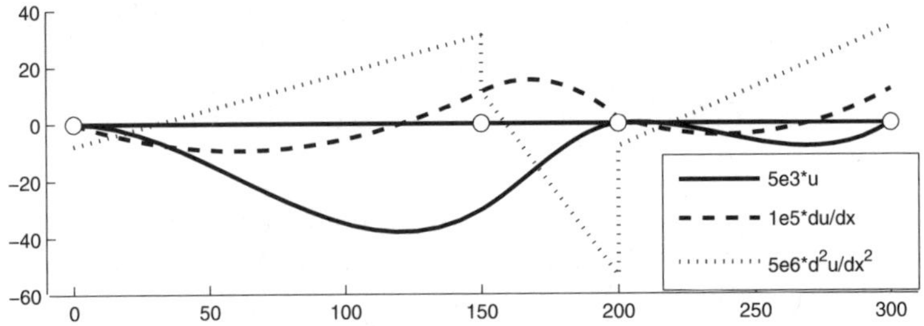

Figure 7.4. Example 7.4 scaled

7.4 Frameworks of Rods

(a) Tension Rod in General Position Let $P_i = P_i(x, y, z)$, $i = 1, 2$, be initial and terminal point of a rod of length ℓ in general position. Further, let

$\widetilde{u}_i \in \mathbb{R}$ displacement of P_i in direction of the rod,
$\underline{u}_i = (u_i, v_i, w_i)$ displacement of P_i in general position.

Introducing the *cosinus values of direction*

$$c_1 = \frac{x_2 - x_1}{\ell}, \quad c_2 = \frac{y_2 - y_1}{\ell}, \quad c_3 = \frac{z_2 - z_1}{\ell},$$

the projection of $\widetilde{u}_i$ onto the axes of the coordinate system yields

$$u_i = c_1 \widetilde{u}_i\,, \quad v_i = c_2 \widetilde{u}_i\,, \quad w_i = c_3 \widetilde{u}_i\,.$$

Conversely, because $c_1^2 + c_2^2 + c_3^2 = 1$,

$$\widetilde{u}_i = c_1 u_i + c_2 v_i + c_3 w_i\,, \quad i = 1, 2, \implies \boxed{\underline{\widetilde{u}} = C\underline{u}}$$

where

$$\underline{\widetilde{u}} = [\widetilde{u}_1, \widetilde{u}_2]^T, \; C = \begin{bmatrix} c_1 & c_2 & c_3 & 0 & 0 & 0 \\ 0 & 0 & 0 & c_1 & c_2 & c_3 \end{bmatrix}, \; \underline{u} = [u_1, v_1, w_1, u_2, v_2, w_2]^T.$$

Substitution into the *stiffness matrix* $\widetilde{S}$ of (7.18) supplies a $(6, 6)$-Matrix

$$S = \frac{1}{\ell} EAC^T \begin{bmatrix} 1 & -1 \\ -1 & 1 \end{bmatrix} C\,. \tag{7.25}$$

The external forces do no longer point into direction of the rod therefore the potential of external forces has to be replaced by a scalar product $\underline{f}^T \underline{u}$ where

$$\underline{f} = [f_{11},\, f_{12},\, f_{13},\, f_{21},\, f_{22},\, f_{23}]^T, \quad \underline{u} = [u_1,\, v_1,\, w_1,\, u_2,\, v_2,\, w_2]^T \in \mathbb{R}^6.$$

(b) Framework of Rods Summing up the modified equations (7.18) for approximation of the total energy supplies a minimum problem for the energy of the framework in equilibrium by the *extremal principle*,

$$Q([U]) := \frac{1}{2}\,[U]^T[S][U] - [F]^T[U] = \min! \tag{7.26}$$

where $Q([U])$ is the *quadratic form* for the global node vector $[U] = [(u_i, v_i, w_i)]_{i=1}^N$. Derivation w.r.t. $[U]$ leads to a linear system again,

$$\boxed{[S][U] = [F]} \tag{7.27}$$

which constitutes a necessary and here also sufficient condition for the existence of a solution of (7.26). Note that the displacements (not the positions) of the node points are computed in equilibrium if that exists. The individual displacements shall be "small" however by the linearizing assumption in Axiom 7.1(1°). The internal forces are cancelled in normal case by the principle of *actio* = *reactio* such that only external forces come into question for possible *loads.*

(c) Support Conditions The global *stiffness matrix* $[S]$ is symmetric and positive semi-definite but never definite whence the extremal problem does not have a unique solution. Boundary conditions (supports) are to be chosen in a way that the solution exists uniquely (and the problem has a solution at all). In a spatial framework, a support point can have null, one, or two degrees of freedom. Thus, recalling that we always suppose *small* displacements, it can be either fixed, or move on a straight line, or move in a plane; the latter possibility drops naturally in *plane* frameworks. Altogether, the support conditions constitute a linear (underdetermined) system of equations for the global node vector $[U]$,

$$[P][U] = [H]\,.$$

By applying Lemma 1.2 and Theorem 3.8 we obtain the following result.

Lemma 7.2. *Assumption:* (1°) *$[P]$ is a (m, n)-matrix with $m < n$.*
(2°) *$[P]$ has maximum rank, i.e.* $\text{rank}[P] = m$.
(3°) *$[S]$ is positive definite on the kernel of $[P]$, i.e.,*

$$[X] \neq [0] \text{ und } [P][X] = [0] \implies [X]^T[S][X] > 0\,.$$

Then the linear system

$$\begin{bmatrix} [S] & [P]^T \\ [P] & [O] \end{bmatrix}\begin{bmatrix} [V] \\ [Z] \end{bmatrix} = \begin{bmatrix} [S][U]_0 - [F] \\ [P][U]_0 - [H] \end{bmatrix}$$

has a unique solution $([V], [Z])$ for arbitrary $[U]_0$, and $[U]^* = [V] - [U]_0$ is a solution of the problem

$$Q([U]) = \min!\,, \quad [P][U] = [H]\,. \tag{7.28}$$

The first and second condition says that no "superfluous" boundary conditions are allowed. The third condition selects the admissible conditions but cannot be easily verified in advance. The *condition* of the matrix is a measure for the *stability* of the entire system and is supplied by MATLAB on demand. It indicates the accuracy with which condition (3°) is satisfied.

Obviously the solution $[U]^*$ in (7.28) depends among others from the right side $[H]$ of the side conditions. In a general linear-quadratic optimization problem, the i-th component H_i of H represent the amount of the i-th available resource. The LAGRANGE multiplier $[Z]$ supplies the *sensitivity* of the solution w.r.t the resources applied (*shadow price*); more exactly, we have

$$\frac{\partial Q([U]^*([H]))}{\partial H_i} = -Z_i\,,$$

where Z_i has the negative sign because a minimum problem is under consideration.

(d) Support Loads If a framework is in equilibrium, the forces acting on an individual node balance each other. This condition yields two equations for each node in plane and three equations in spatial frameworks. The direction of force $\underline{k}$ in a tension rod is obviously given up to sign by the initial and terminal point of the rod itself. Thus we have to work either with *normed* rods or the calculated absolute value $|\underline{k}|$ has to be multiplied ensuing by the *length* of the rod. The support forces must be *perpendicular* to the (linear) side conditions in equilibrium. More exactly, let $P(\underline{u}) = P(u_1, u_2, u_3)$ be a support point and let $\underline{k} = [k_1, k_2, k_3]^T$ be an appertaining unknown supporting force, then Table 7.1 gives the different possibilities:

Table 7.1.

Degr. of freedom of $\underline{u}$	Cond. for $\underline{u}$	Cond. for $\underline{k}$	Degr. of freedom of $\underline{k}$
	plane framework		
0	$\underline{u} = \underline{0}$	$\underline{k}$ free	2
1	$\underline{a} \cdot \underline{u} = 0$	$\underline{k} = \alpha\,\underline{a}$, α free	1
	spatial framework		
0	$\underline{u} = \underline{0}$	$\underline{k}$ free	3
1	$\underline{a} \cdot \underline{u} = 0$	$\underline{k} = \alpha\,\underline{a} + \beta\,\underline{b}$	2
	$\underline{b} \cdot \underline{u} = 0$	α, β free	
2	$\underline{a} \cdot \underline{u} = 0$	$\underline{k} = \alpha\,\underline{a}$, α free	1

The resulting linear system of equations is called

statically determined,	if the solution exists uniquely,
statically undetermined,	if several solutions exist,
kinematically undetermined,	if no solution exists.

Example 7.5. Plane framework. As pattern model for later implementation by finite element methods, input data and their arrangement are described in detail here (z-coordinates being dropped naturally in plane frameworks).

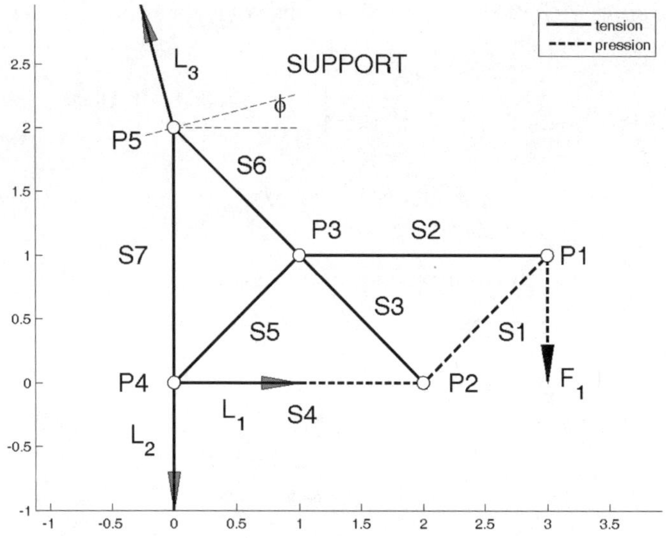

Figure 7.5. Ex. 7.5, plane framework

(1°) The i-th column of the *node matrix* P contains the coordinates of the i-th node point in arbitrary succession of nodes. The i-th column of the *load matrix* F contains the coordinates of the *load vector* attacking at the i-th node.
(2°) The k-th column of the *element matrix* S contains the *numbers* of the nodes of the k-th rod in arbitrary succession of rods whereby the *direction* of the k-th rod is fixed and has to be regarded later. Accordingly, in the present plane example:

$$P = \begin{bmatrix} 3 & 2 & 1 & 0 & 0 \\ 1 & 0 & 1 & 0 & 2 \end{bmatrix}, \quad F = \begin{bmatrix} 0 & 0 & 0 & 0 & 0 \\ f & 0 & 0 & 0 & 0 \end{bmatrix}, \quad S = \begin{bmatrix} 2 & 3 & 3 & 4 & 4 & 5 & 4 \\ 1 & 1 & 2 & 2 & 3 & 3 & 5 \end{bmatrix}.$$

(3°) A support condition for the displacement $\underline{u}$ has the form $\underline{a} \cdot \underline{u} = 0$, $|\underline{a}| = 1$, $\underline{a} = [\alpha_1, \alpha_2, \alpha_3]^T$. The *support matrix* L contains in the first row the *number* of node points in which support conditions are specified and in the remaining rows the components α_i of the corresponding support vector. Thus we have 0, 1, 2 or 3 conditions for each node in spatial frameworks. In the present (plane) example:

$$L = [\ell^i{}_k] = \begin{bmatrix} 4 & 4 & 5 \\ 1 & 0 & -\sin\varphi \\ 0 & -1 & \cos\varphi \end{bmatrix}$$

(in a fixed point, only two resp. three *linearly independent* vectors have to be supporting forces hence the sign of $\ell^3{}_2$ does not play any role). Unnormed rod directions, support force and load directions with corresponding lengths read in the present example:

$$[\underline{S}_1, \underline{S}_2, \underline{S}_3, \underline{S}_4, \underline{S}_5, \underline{S}_6, \underline{S}_7 | \underline{L}_4, \underline{L}_5 | \underline{F}_1] = \left[\begin{array}{ccccccc|cc|c} 1 & 2 & 1 & 2 & 1 & 1 & 0 & L_1 & -\sin\varphi & 0 \\ 1 & 0 & -1 & 0 & 1 & -1 & 2 & L_2 & \cos\varphi & -1 \end{array}\right]$$

$$\begin{aligned} &[s_1, s_2, s_3, s_4, s_5, s_6, s_7 \mid l_4, l_5 \mid f] \\ &= \left[\sqrt{2}, 2, \sqrt{2}, 2, \sqrt{2}, \sqrt{2}, 2 \mid (L_1^2 + L_2^2)^{1/2}, 1 \mid 1\right] \end{aligned}$$

Relations for the forces attacking in nodes 1 : 5:

$$\left.\begin{array}{|ccccccc|cc|} \widetilde{k}_1\underline{S}_1 & +\widetilde{k}_2\underline{S}_2 & & & & & & & = \underline{F}_1 \\ \widetilde{k}_1\underline{S}_1 & & -\widetilde{k}_3\underline{S}_3 & -\widetilde{k}_4\underline{S}_4 & & & & & = \underline{0} \\ & -\widetilde{k}_2\underline{S}_2 & +\widetilde{k}_3\underline{S}_3 & & -\widetilde{k}_5\underline{S}_5 & -\widetilde{k}_6\underline{S}_6 & & & = \underline{0} \\ & & & \widetilde{k}_4\underline{S}_4 & +\widetilde{k}_5\underline{S}_5 & & +\widetilde{k}_7\underline{S}_7 & -\underline{L}_4 & = \underline{0} \\ & & & & & \widetilde{k}_6\underline{S}_6 & -\widetilde{k}_7\underline{S}_7 & -l_5\underline{L}_5(\varphi) & = \underline{0} \end{array}\right. .$$

This system consists of ten equations for ten unknowns $\widetilde{k}_1, \ldots, \widetilde{k}_7, l_5$, and the both components of $\underline{L}_4$. The resulting linear system is uniquely solvable hence statically determined for $0 < \varphi < 2\pi$ and unsolvable hence dynamically undetermined fore $\varphi = 0$. Absolute values of forces in direction of the rods are $k_i = \widetilde{k}_i \cdot s_i$, $i = 1 : 7$.

Example for $\varphi = \pi/12$, $E = 0.2e^9$, $A = 0.5e^{-3}$, $f = 10$:

rod	1	2	3	4	5	6	7
force	-1.1412	1.0000	1.4142	-2.0000	0.7071	2.1213	4.0981
stress $\times 1.0e^4$	2.8284	-2.0000	-2.8284	4.0000	-1.4142	-4.2426	-8.1962

Example 7.6. Plane truss, left support fixed, right support hinged.

$$\text{Nodes:} \quad P = \begin{bmatrix} 0\,2\,4\,6\,4\,2 \\ 0\,0\,0\,0\,1\,2 \end{bmatrix}, \quad \text{Rods:} \; S = \begin{bmatrix} 1\,2\,3\,4\,5\,6\,6\,5\,6 \\ 2\,3\,4\,5\,6\,1\,2\,3\,3 \end{bmatrix}$$

$$\text{Supports:} \; L = \begin{bmatrix} 1 & 1 & 4 \\ -1 & 0 & 0 \\ 0 & -1 & -1 \end{bmatrix}, \quad \text{Loads:} \; F = \begin{bmatrix} 0 & 0\,0\,0\,f\,0 \\ 0 & -2f\,0\,0\,0\,0 \end{bmatrix}.$$

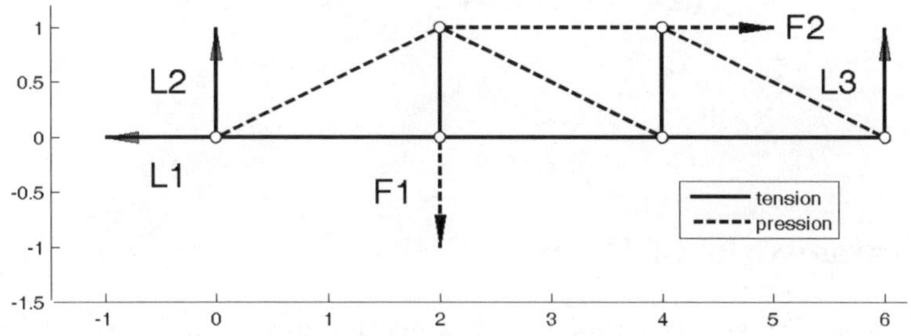

Figure 7.6. Ex. 7.6, plane truss

Further examples with frameworks shown in Figures 7.7 and 7.8 (Schwarz91), (Schwarz80) are studied in directory `KAPITEL07\SECTION_4`. The displacements are scaled with factor κ for better visibility.

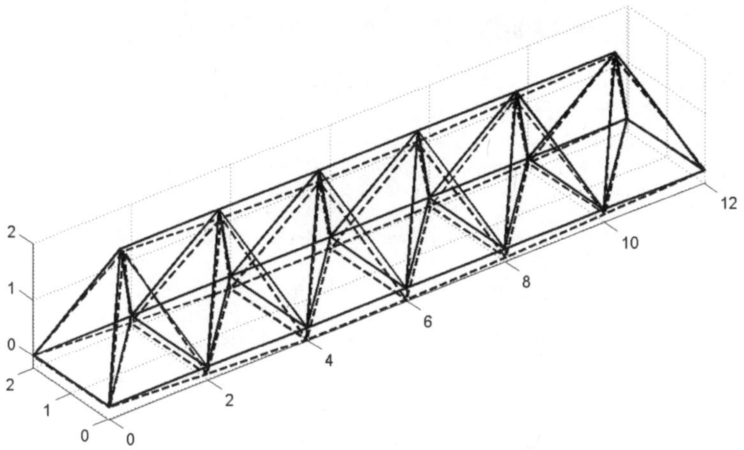

Figure 7.7. Cantilever, $\kappa = 50$

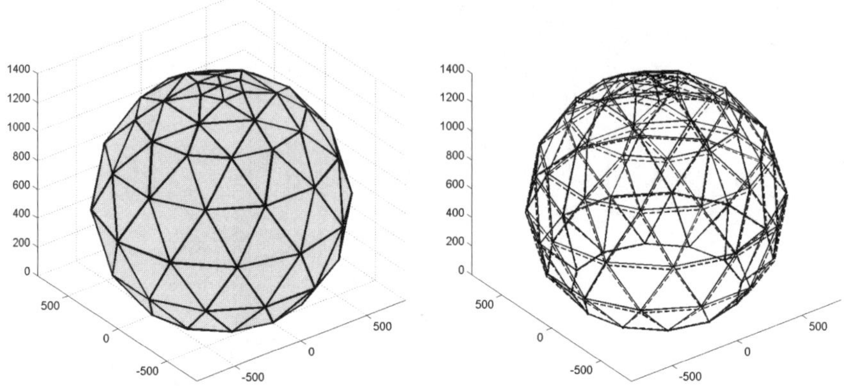

$\kappa = 0$ $\kappa = 1000$

Figure 7.8. Radar dome

7.5 Frameworks of Beams

(a) Torsion We suppose that the cross-section of a beam or bar rotates in (y, z)-plane with angle $\varphi(x)$, then displacements arise only in y, z-direction and we have

$$\begin{bmatrix} \widetilde{y} \\ \widetilde{z} \end{bmatrix} = \left[\begin{bmatrix} \cos\varphi & -\sin\varphi \\ \sin\varphi & \cos\varphi \end{bmatrix} - I\right]\begin{bmatrix} y \\ z \end{bmatrix}$$

(I unit matrix) hence

$$\begin{aligned} u_1(x,y,z,\varphi) &= 0 \\ u_2(x,y,z,\varphi) &= y(\cos\varphi - 1) - z\sin\varphi \\ u_3(x,y,z,\varphi) &= y\sin\varphi + z(\cos\varphi - 1)\,. \end{aligned}$$

Supposing small amounts of $|\varphi|$, linearization yields $\cos\varphi \sim 1$, $\sin\varphi \sim \varphi$, hence

$$\boxed{u_1(x,y,z,\varphi) = 0\,,\;\; u_2(x,y,z,\varphi) = -z\varphi(x)\,,\;\; u_3(x,y,z,\varphi) = y\varphi(x)}\,.$$

Thus the *strain vector* $\underline{\varepsilon} = [u_x, v_y, w_z, u_y + v_x, v_z + w_y, w_x + u_z]^T$ has the form

$$\underline{\varepsilon} = [0,\, 0\,, 0, -z\varphi'(x),\, 0,\, y\varphi'(x)]^T.$$

We suppose also that the *shear modulus* $G = E/(2(1+\nu))$ is constant then $\underline{\sigma} = G\underline{\varepsilon}$ is the *stress vector*. The stress energy of a *circular* torsion bar is then given by the volume integral

$$\Pi_S = \frac{1}{2}\int_V \underline{\sigma}\cdot\underline{\varepsilon}\,dv = \frac{1}{2}G\int_V (y^2 + z^2)\varphi'(x)^2\,dxdydz\,,$$

and the total energy of a circular torsion bar is

$$\Pi = \frac{1}{2}GI_p\int_0^l \varphi'(x)^2\,dx - M_1\varphi_1 - M_2\varphi_2\,.$$

In torsion bars with non-circular cross-section *warpings* of the cross-section area do appear which cannot be neglected in computation, and the polar moment of inertia I_p must be replaced by the torsion moment of inertia I_t whose values are mostly given in tabular form. For instance, let the bar have rectangular area of cross-section $A = h \cdot b$ and $h > b$ then

$$I_t \simeq \eta_2 h\, b^3\,, \quad \eta_2 = \frac{2.370592\, q^2 - 2.486211\, q + 0.826518}{7.111777\, q^2 - 3.057824\, q + 1}\,, \quad q = \frac{h}{b} > 1\,;$$

cf. (Holzmann), vol. 3, Sect. 7.1.4; (Szabo77).

(b) Total Energy We suppose in the sequel that *no shear forces* act on the beam and that no bending occurs about the x-axis in the (y, z)-plane because h and b are small relative to the length l of the beam. Then the total energy is composed of bending about the z-axis, bending about the y-axis, elongation/shortening in direction of the x-axis and the torsion energy. In the *most simple model of beam* we then have

$$\begin{aligned} \Pi_B = \frac{1}{2}E\Big\{ & \frac{bh^3}{12}\int_0^l w''(x)^2 dx + \frac{b^3h}{12}\int_0^l v''(x)^2 dx \\ & + bh\int_0^l u'(x)^2 dx + \frac{I_t}{2(1+\nu)}\int_0^l \varphi'(x)^2 dx \Big\} \end{aligned} \tag{7.29}$$

where the beam shall be rectangular with x-axis for symmetry axis.

(c) Beam with Bending and Torsion in (nearly) general position (Schwarz80). The local values of the beam are denoted by capitals and the values in the global coordinate system by small letters, i.e., big letters before bending/rotation and small letters thereafter. Moreover, we consider again "small" bendings and rotations such that $\tan\varphi \sim \varphi$ can be written for the rotation angle.

Remember: The slope angle φ of the tangent of a function $y = f(x)$ at point x satisfies $\tan\varphi = f'(x)$ hence approximatively $\varphi \sim \tan\varphi = f'(x)$ if φ is small.

Let P_1 and P_2 be initial and terminal point of the beam then

V_i' slope of bending line in P_i, (X, Y)-plane $\sim$ rotary angle about Z-axis
W_i' slope of bending line in P_i, (X, Z)-plane $\sim$ rotary angle about Y-axis

cf. (7.31). Succession by (7.29)

$$U_e = [W_1, W_1', W_2, W_2'; V_1, V_1', V_2, V_2'; U_1, U_2; \Phi_1, \Phi_2]^T . \tag{7.30}$$

However, it is more advantageous for the subsequent operations to assemble displacements pointwise:

$$\widetilde{U}_e = \Big[[U_1, V_1, W_1, \Phi_1, W_1', V_1'] ,\ [U_2, V_2, W_2, \Phi_2, W_2', V_2']\Big]^T .$$

The spatial position of the beam is determined by *three cosinus values* of direction in the same way as in the rod:

$$c_{Xx} = \frac{x_2 - x_1}{l}\,, \quad c_{Xy} = \frac{y_2 - y_1}{l}\,, \quad c_{Xz} = \frac{z_2 - z_1}{l}\,;$$

these are the cosinus values for rotation about the X-axis. Likewise one has 3 cosinus values c_{Yx}, c_{Yy}, c_{Yz} for rotation about the Y-axis,
3 cosinus values c_{Zx}, c_{Zy}, c_{Zz} for rotation about the Z-axis.

Because of the presentation of the beam as one-dimensional system, *one* of the 6 cosinus values for the rotation of the local Y-axis and the local Z-axis must be specified in advance: We suppose here that the Y-axis corresponding to the width b of the beam shall remain always perpendicular to the global z-axis. More precisely, if the X-axis is not parallel to the z-axis then the Y-axis shall remain parallel to the global (x, y)-plane otherwise the Y-axis shall remain parallel to the x-axis. The direction of the "width-side" of the beam does not change by this way and $c_{Yz} = 0$ does hold in advance.

By this way, the Y-axis is orthogonal to the projection of the X-axis onto the (x, y)-plane and

$$L = (c_{Xx}^2 + c_{Xy}^2)^{1/2}$$

is the length of the projection of the unit vector in X-direction onto the (x, y)-plane. Because $(-v, u) \perp (u, v)$ we obtain

$$\boxed{\begin{array}{llll} c_{Yx} = -\dfrac{c_{Xy}}{L}\,, & c_{Yy} = \dfrac{c_{Xx}}{L}\,, & c_{Yz} = 0\,, & L \neq 0 \\ c_{Yx} = 0\,, & c_{Yy} = 0\,, & c_{Yz} = 0\,, & L = 0 \end{array}}\,.$$

The cosinus values of the Z-axis are given by the cross product

$$\begin{bmatrix} c_{Zx} \\ c_{Zy} \\ c_{Zz} \end{bmatrix} = \begin{bmatrix} c_{Xx} \\ c_{Xy} \\ c_{Xz} \end{bmatrix} \times \begin{bmatrix} c_{Yx} \\ c_{Yy} \\ c_{Yz} \end{bmatrix}\,, \quad c_{Yz} = 0\,.$$

Remember that $U_i = c_{Xx}u_i + c_{Xy}v_i + c_{Xz}w_i$, for $i = 1, 2$ in a rod element. Now in the same way

$$\begin{aligned} U_i &= c_{Xx}u_i + c_{Xy}v_i + c_{Xz}w_i \\ V_i &= c_{Yx}u_i + c_{Yy}v_i + c_{Yz}w_i \\ W_i &= c_{Zx}u_i + c_{Zy}v_i + c_{Zz}w_i\,. \end{aligned} \tag{7.31}$$

Likewise, for the rotation angle by linearization

$$\begin{aligned} \Phi_i &= c_{Xx}\varphi_i + c_{Xy}w_i' + c_{Xz}v_i'\,, && \text{rotation about } X\text{-axis} \\ W_i' &= c_{Yx}\varphi_i + c_{Yy}w_i' + c_{Yz}v_i'\,, && \text{rotation about } Y\text{-axis} \\ V_i' &= c_{Zx}\varphi_i + c_{Zy}w_i' + c_{Zz}v_i'\,, && \text{rotation about } Z\text{-axis}\,. \end{aligned}$$

The result reads now in matrix-vector notation:

$$C = \begin{bmatrix} c_{Xx} & c_{Xy} & c_{Xz} \\ c_{Yx} & c_{Yy} & c_{Yz} \\ c_{Zx} & c_{Zy} & c_{Zz} \end{bmatrix}, \quad D = \begin{bmatrix} C & 0 & 0 & 0 \\ 0 & C & 0 & 0 \\ 0 & 0 & C & 0 \\ 0 & 0 & 0 & C \end{bmatrix}, \tag{7.32}$$

$$\begin{aligned} \widetilde{U}_e &= \Big[[U_1, V_1, W_1, \Phi_1, W_1', V_1'],\ [U_2, V_2, W_2, \Phi_2, W_2', V_2']\Big]^T \\ \widetilde{u}_e &= \Big[[u_1, v_1, w_1, \varphi_1, w_1', v_1'],\ [u_2, v_2, w_2, \varphi_2, w_2', v_2']\Big]^T \\ \widetilde{U}_e &= D\widetilde{u}_e\,. \end{aligned}$$

(d) In **Numerical Approximation**, the linear ansatz of Sect. 7.3**(a)** is chosen for u and φ, and the cubic basic approach of Sect. 7.3**(b)** for v und w; cf. (Schwarz80), (Schwarz91). Writing again the local node vector as in (7.30), the stiffness matrix of a beam element is a block diagonal matrix of dimension $4+4+2+2=12$. The individual stiffness matrices in the diagonal are the matrices S of (7.25) and K of (7.20) with different pre-factors where, in addition, K is subjected to the orthogonal transformation with (permutation) matrix D of (7.32).

Example 7.7. (Schwarz91) in `KAPITEL07\SECTION_5`. The displacements in Figure 7.9 are multiplied by the factor two for better illustration.
Input data:

```
NODES = [...
  1  0.0  0.0  0.0;  2  0.0  0.0  4.0;  3  1.0  1.0  5.0;
  4  1.0  4.0  5.0;  5  0.0  5.0  4.0;  6  0.0  5.0  0.0;
  7  5.0  0.0  0.0;  8  5.0  0.0  4.0;  9  5.0  1.0  5.0;
 10  5.0  4.0  5.0; 11  5.0  5.0  4.0; 12  5.0  5.0  0.0;
 13 10.0  0.0  0.0; 14 10.0  0.0  4.0; 15  9.0  1.0  5.0;
 16  9.0  4.0  5.0; 17 10.0  5.0  4.0; 18 10.0  5.0  0.0];
% LOADS: column(i) = [FX; FY; FZ; MX; MY; MZ]; i node nr.
LOADSF = zeros(3,size(p,2));
LOADSF(:,3)  = [0.0;0.0;-20.0]; LOADSF(:,4)  = [0.0;0.0;-20.0];
LOADSF(:,9)  = [0.0;0.0;-25.0]; LOADSF(:,10) = [0.0;0.0;-25.0];
LOADSF(:,15) = [0.0;0.0;-30.0]; LOADSF(:,16) = [0.0;0.0;-30.0];
LOADSG = zeros(3,size(p,2)); LOADS = [LOADSF;LOADSG];
% SUPPORT: column(i) = [NODE Nr.; U; V; W; Th; WS; VS];
% U = 1/0 : U fixed/free, etc.
SUPPORTF = [...
1, 1, 1, 6, 6, 6, 7, 7, 7, 12, 12, 12, 13, 13, 13, 18, 18, 18;
1, 0, 0, 1, 0, 0, 1, 0, 0,  1,  0,  0,  1,  0,  0,  1,  0,  0;
0, 1, 0, 0, 1, 0, 0, 1, 0,  0,  1,  0,  0,  1,  0,  0,  1,  0;
0, 0, 1, 0, 0, 1, 0, 0, 1,  0,  0,  1,  0,  0,  1,  0,  0,  1];
SUPPORTG = zeros(3,size(SUPPORTF,2)); SUPPORT = [SUPPORTF;SUPPORTG];
```

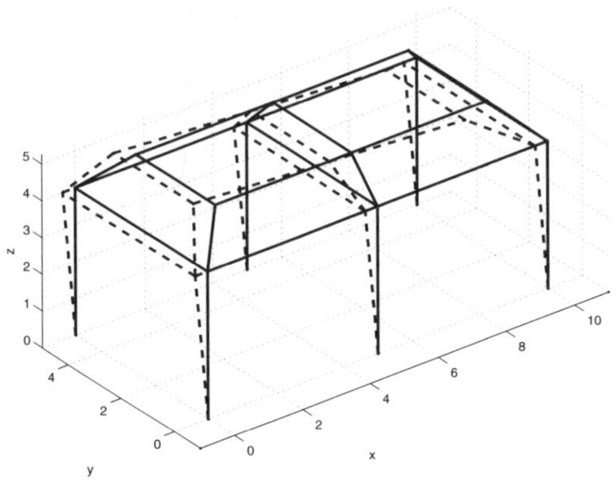

Figure 7.9. Example 7.7, spatial framework (scaled)

7.6 Hints to the MATLAB Programs

```
KAPITEL07/SECTION_3 Beam in Special Position
demo4.m         Masterfile, bending beam
balkelement1.m  Beam element
balken1.m       Beam in special position
balken2.m       Beam in general plane position
KAPITEL07/SECTION_4, Frameworks of Rods
demo1.m        Masterfile, forces in plane framework
demo2.m        Masterfile, displacements in plane framework
               with image sequence
demo3.m        Masterfile, displacements in spatial framework
stabelement1.m Tension rod in plane position
stabelement2.m Tension rod in spatial position
stabwerk1.m    Forces in plane framework
stabwerk2.m    Displacements in plane framework
stabwerk3.m    Displacements in spatial framework
KAPITEL07/SECTION_5, Spatial Frameworks
demo1.m     Masterfile for spatial frameworks
balken2.m   Beam element, nearly general position
rahmen2.m   Displacements in spatial frameworks
```

8

Continuum Theory

8.1 Deformations

(a) Deformation In a fixed cartesian coordinate system, let $\emptyset \neq \Omega \subset \mathbb{R}^3$ be an open bounded domain with sufficiently smooth boundary $\partial\Omega =: \Gamma$, and let $\mathcal{I} \subset \mathbb{R}$ be an open interval with $0 \in \mathcal{I}$; Ω shall describe the geometric shape of a liquid or solid "body". We consider a mapping

$$\Phi : \mathcal{I} \times \Omega \ni (t, X) \mapsto \Phi(t, X) =: x(t) \in \mathbb{R}^3$$

with the following properties:

(1°) Φ is two times continuously differentiable;
(2°) $\Phi(0, X) = X$;
(3°) $\det \mathrm{Grad}_X\, \Phi(t, X) > 0$.

By the last assumption no volume element can degenerate. Together with the first assumption it implies local reversibility of Φ for fixed t. In dependence of the individual problem the mapping Φ is called *deformation*, *motion*, or *flux* and $\Upsilon(t, X) := \Phi(t, X) - X =: \underline{u}$ is the corresponding *displacement*; their computation resp. approximation is the goal of many problems in physics and engineering sciences. If a system is in static equilibrium, time t is omitted. The points (point vectors) $X \in \Omega$ are called *material points* and their components material coordinates or LAGRANGE coordinates. The points $x \in \Phi(t, \Omega)$ are called *space points* and their components space coordinates or EULER coordinates. The material point X is transferred to the space point x by the motion Φ or in other words: the material point X is located at position x after time t. In the view of LAGRANGE, X is the present position of the material point X and x its future position. In other words, the observer is placed at the point X and considers the "world" from this point. In the today preferred view of EULER, x is the present position of the fictive particle and X the former position; hence the observer sees the world passing by from this point. In this approach all x-coordinates share a physical meaning whereas the corresponding "pulled back" equations in X-coordinates are necessary for the mathematical inferences. The material points X are commonly understood as fixed points in a fixed EUCLIDian space (reference space) being identified with its coordinate space $\mathbb{R}^3$. So we do not discuss the problematic nature of choice of a "observer standpoint" but take this for given in the one or other way. The rule

from large to small

has always to be regarded in the sequel. All functions and operators with material points as arguments are written as capital letters, and all functions and operators with space points for arguments are written as small letters. This notation after (Marsden) leads however to some difficulties if several deformations are considered in composed form, therefore all terms are written sometimes as small letters but elements and operators in the image domain are indicated by Φ after (Ciarlet93). Differential operators as "divergence" and "gradient" relate always to space variables only and not to time t; tensor fields are written boldface.

Point vectors X, x and the both basic vector fields Φ and Υ are not underlined.

Let $\underline{v}$ be the velocity and $\underline{b}$ the acceleration of a material point X being at position x at time t, i.e.,

$$\dot{x}(t) := \frac{D}{Dt}x(t) := \frac{\partial \Phi}{\partial t}(t,X) =: \underline{V}(t,X) =: \underline{v}(t,x)\,, \quad x = \Phi(t,X)\,, \tag{8.1}$$

then by the chain rule, cf. (Marsden), p. 3,

$$\begin{aligned}\ddot{x}(t) &:= \frac{D^2}{Dt^2}x(t) = \frac{\partial^2 \Phi}{\partial t^2}(t,X) =: \underline{B}(t,X) \equiv \frac{\partial}{\partial t}\underline{V}(t,X) = \frac{d}{dt}\underline{v}(t,\Phi(t,X)) \\ &= \frac{\partial \underline{v}}{\partial t}(t,x) + [\operatorname{grad}\underline{v}(t,x)]\underline{v}(t,x) =: \frac{D}{Dt}\underline{v}(t,x) =: \dot{\underline{v}}(t,x)\,,\end{aligned} \tag{8.2}$$

and accordingly

$$\boxed{\dot{\underline{v}}(t,x) = \frac{D}{Dt}\underline{v}(t,x) = \frac{\partial}{\partial t}\underline{V}(t,X)}\,.$$

The *absolute* acceleration $\underline{B}$ is additively composed of the *relative* acceleration $\partial\underline{v}/\partial t$ and the *convective* acceleration $(\operatorname{grad}\underline{v})\underline{v}$!

Example 8.1. Let $Q(t)$ be an orthogonal matrix, $Q(0) = I$, and $\Phi(t,X) = Q(t)X = x(t)$ then

$$\begin{aligned}\underline{V}(t,X) &= \dot{Q}X = \dot{Q}Q^T x = \underline{v}(t,x) \\ \underline{B}(t,X) &= \ddot{Q}X = \ddot{Q}Q^T x = \underline{b}(t,x)\,.\end{aligned}$$

On the other side, $Q\dot{Q}^T = -\dot{Q}Q^T$ by $QQ^T = I$ and the product rule, hence $\dot{Q}Q^T\dot{Q}Q^T = -\dot{Q}\dot{Q}^T$ and then

$$\begin{aligned}\frac{D}{Dt}\underline{v}(t,x) &= \underline{v}_t(t,x) + \operatorname{grad}\underline{v}(t,x)\underline{v}(t,x)\\ &= \ddot{Q}Q^Tx + \dot{Q}\dot{Q}^Tx + \dot{Q}Q^T\dot{Q}Q^Tx = \ddot{Q}Q^Tx = \ddot{Q}X\,.\end{aligned}$$

The *trajectory* or *path* $\Phi(\,\cdot\,,X) : t \mapsto \Phi(t,X)$ of a particle X is the unique solution of the initial value problem

$$\dot{x}(t) = \frac{\partial\Phi}{\partial t}(t,X) = \underline{v}(t,x(t)) = \underline{v}(t,\Phi(t,X))\,,\quad \Phi(0,X) = X\,. \tag{8.3}$$

If $\underline{v}$ is a stationary velocity field being independent of time t then

$$\Phi(s+t,\,\cdot\,) = \Phi(s,\,\cdot\,)\circ\Phi(t,\,\cdot\,) = \Phi(s,\Phi(t,\,\cdot\,)) = \Phi(t,\Phi(s,\,\cdot\,))\,.$$

The fundamental mapping Φ is called *flux* in this context, cf. § 1.7, and it is an one-parametric group of transformations in the present case of stationary velocity field.

(b) Derivation of the Gradient w.r.t. the variable t yields

$$\begin{aligned}\frac{\partial}{\partial t}[\operatorname{Grad}\Phi(t,X)] &= \operatorname{Grad}\left[\frac{\partial}{\partial t}\Phi(t,X)\right] = \operatorname{Grad}\underline{V}(t,X)\\ &= \operatorname{Grad}\underline{v}(t,\Phi(t,X)) = \operatorname{grad}\underline{v}(t,x)\ \operatorname{Grad}\Phi(t,X)\,;\end{aligned}$$

hence

$$\boxed{\frac{\partial}{\partial t}[\operatorname{Grad}\Phi(t,X)] = \operatorname{grad}\underline{v}(t,x)\ \operatorname{Grad}\Phi(t,X)}\,. \tag{8.4}$$

Derivation of

$$[\operatorname{Grad}\Phi(t,X)][\operatorname{Grad}\Phi(t,X)]^{-1} = I$$

w.r.t. time t then yields by using the product rule

$$\operatorname{grad}\underline{v}(t,x) = -[\operatorname{Grad}\Phi(t,X)]\frac{\partial}{\partial t}[(\operatorname{Grad}\Phi(t,X))^{-1}]$$

or

$$\frac{\partial}{\partial t}[(\operatorname{Grad}\Phi(t,X))^{-1}] = -[\operatorname{Grad}\Phi(t,X)]^{-1}\operatorname{grad}\underline{v}(t,x)\,. \tag{8.5}$$

Since $\operatorname{grad}\Phi^{-1}(t,X) = [\operatorname{Grad}\Phi(t,X)]^{-1}$ we obtain by this result also

$$\boxed{\frac{\partial}{\partial t}[\operatorname{grad}\Phi^{-1}(t,x)]^T = -[\operatorname{grad}\,\underline{v}(t,x)]^T[\operatorname{Grad}\Phi(t,X)]^{-T}}\,. \tag{8.6}$$

(c) Material Derivatives (Substantial Derivatives) The material derivation is the total derivation w.r.t. t where thereafter frequently $\Phi(t,X) = x$ is written again as argument. Let more generally $\underline{w}$ be a vector field in space coordinates x and let

$$\underline{W}(t,X) = \underline{w}(t,x) = \underline{w}(t,\Phi(t,X))\,, \quad x = \Phi(t,X)\,,$$

be the same vector field as function of the material points X. Then $(\partial\underline{w}/\partial t)(t,x)$ is the rate of change of $\underline{w}$ for fixed x, i.e., the change of $\underline{w}$ seen from x, and

$$\dot{\underline{w}} := \frac{D}{Dt}\underline{w}(t,x) := \frac{d}{dt}\underline{w}(t,\Phi(t,X))$$

is the change of $\underline{w}$ for fixed X, i.e. the change of $\underline{w}$ seen form the material point X. Using $x = \Phi(t,X)$ and the chain rule we obtain

$$\frac{d}{dt}\underline{w}(t,\Phi(t,X)) = \frac{\partial \underline{w}}{\partial t}(t,x) + \operatorname{grad}\underline{w}(t,x)\underline{v}(t,x) =: \left[\frac{\partial}{\partial t} + \underline{v}\cdot\nabla_x\right]\underline{w}(t,x)\,.$$

Definition 8.1. *The operator (or the operation)*

$$\boxed{\frac{D}{Dt} := \frac{\partial}{\partial t} + \underline{v}\cdot\nabla_x}$$

is the material derivative (w.r.t. t).

Normally the material derivative is different from zero also if $\underline{w}$ does not depend on t explicitely:

$$\frac{D}{Dt}\underline{w}(x) = 0 + \operatorname{grad}\underline{w}(x)\underline{v}(t,x)\,, \quad x = \Phi(t,X)\,.$$

In particular, we obtain for the velocity vector

$$\frac{D}{Dt}\underline{v}(t,x) = \frac{\partial}{\partial t}\underline{v}(t,x) + \nabla\underline{v}(t,x)\underline{v}(t,x)\,.$$

For the identity $\underline{w}(t,x) = x$ we obtain again $\dot{x}(t) = (D/Dt)x(t) = \underline{v}(t,x)$, and for a scalar function φ

$$\frac{D}{Dt}\varphi(t,x) = \frac{\partial}{\partial t}\varphi(t,x) + \operatorname{grad}\varphi(t,x)\underline{v}(t,x)\,.$$

The convective derivative is zero here if $\operatorname{grad}\varphi(t,x)$ stands perpendicularly on the velocity $\underline{v}(t,x)$.

In material derivatives it is always to be regarded that the point vector x depends normally on t and X although this fact is not always expressed explicitly. One goes back to the material coordinates, forms the derivative w.r.t. t and passes again to the space coordinates.

(d) Piola Transformation In continuum mechanics one works frequently with normals of *infinitesimal surfaces* consisting of a point and the normal vector only. Therefore the behavior of a normal vector of a surface under the mapping Φ is of interest.

Let $D \subset \mathbb{R}^2$ be open, $F := \{X(u)\,,\ u = (u^1, u^2) \in D\} \subset \mathbb{R}^3$ a surface in reference space (material coordinates) and $\Phi(F) = \{x = (\Phi \circ X)(u),\ u \in D\} \subset \mathbb{R}^3$ the transformed (transported) surface in space coordinates which shall not depend here on t for simplicity. *Remember* the scalar and vectorial surface elements in detailed representation:

$$\begin{aligned}
dO &= \left|\left(\frac{\partial X}{\partial u^1} \times \frac{\partial X}{\partial u^2}\right)(u^1, u^2)\right| du^1\, du^2 && \text{material coordinates,}\\
d\underline{O} &= \underline{N} dO = \left(\frac{\partial X}{\partial u^1} \times \frac{\partial X}{\partial u^2}\right)(u^1, u^2)\, du^1\, du^2 && \text{material coordinates,}\\
do &= \left|\left(\frac{\partial x}{\partial u^1} \times \frac{\partial x}{\partial u^2}\right)(u^1, u^2)\right| du^1\, du^2\,,\ x = \Phi(u) && \text{space coordinates,}\\
d\underline{o} &= \underline{n} do = \left(\frac{\partial x}{\partial u^1} \times \frac{\partial x}{\partial u^2}\right)(u^1, u^2)\, du^1\, du^2 && \text{space coordinates}
\end{aligned}$$

cf. §1.2(**c1**); remember also that $\operatorname{cof} A = \det(A)A^{-T}$. In the sequel let $\underline{w}$ be a vector field in space coordinates x and $\underline{W} = \underline{w} \circ \Phi$ the same (pulled back) vector field as function of material coordinates X. Then, by using the normalized normal vectors $\underline{n}$ and $\underline{N}$,

$$\int_{\Phi(F)} \underline{w} \cdot \underline{n}\, do = \int_F (\mathbf{A}\underline{W}) \cdot \underline{N}\, dO = \int_F \underline{W}^T \mathbf{A}^T \underline{N}\, dO\,, \tag{8.7}$$

where $\mathbf{A} \in \mathbb{R}^3{}_3$ is to be specified, because $\underline{W}$ has to be rotated in the same way as $\underline{N}$ arises from $\underline{n}$.

Lemma 8.1.

$$\boxed{\mathbf{A}(X) = [\operatorname{Cof} \nabla\Phi(X)]^T \equiv \det(\nabla\Phi(t, X))[\nabla\Phi(t, X)]^{-1}}\,.$$

Proof. Let $\underline{a}\,, \underline{b}\,, \underline{c} \in \mathbb{R}^3$, let $\det(\underline{a}, \underline{b}, \underline{c}) = <\underline{a}, \underline{b}, \underline{c}> = \underline{a}^T(\underline{b} \times \underline{c})$ be the spat product and let $C \in \mathbb{R}^3{}_3$. Then

$$\begin{aligned}
&\det(\underline{a}, C\underline{b}, C\underline{c}) = \det(CC^{-1}\underline{a}, C\underline{b}, C\underline{c}) = \det(C)\det(C^{-1}\underline{a}, \underline{b}, \underline{c})\\
&= <\det(C)C^{-1}\underline{a}, \underline{b}, \underline{c}> = <(\operatorname{cof} C^T)\underline{a}, \underline{b}, \underline{c}> = \underline{a}^T(\operatorname{cof} C)(\underline{b} \times \underline{c})\,.
\end{aligned}$$

Note that $x_{u^1} = \nabla\Phi(X)X_{u^1}$, $x_{u^2} = \nabla\Phi(X)X_{u^2}$ and take $C = \nabla\Phi$, then the assertion follows from

$$\begin{aligned}
&\int_{\Phi(F)} \underline{w} \cdot \underline{n}\, do = \int_D (\underline{w} \circ \Phi \circ X(u)) \cdot (x_{u^1} \times x_{u^2})(u)\, du^1\, du^2\\
&= \int_D (\underline{W} \circ X)(u)) \cdot (\operatorname{Cof} \nabla\Phi(X(u)))(X_{u^1} \times X_{u^2})(u)\, du^1\, du^2\,.
\end{aligned}$$

□

Since relation (8.7) does hold for an arbitrary vector field $\underline{w}$, Lemma 8.1 leads to the *substitution rules* for normal vector and surface area $|F|$:

$$\boxed{\begin{aligned} \underline{n}(x) &= \frac{\operatorname{Cof}\nabla\Phi(X)\underline{N}(X)}{|\operatorname{Cof}\nabla\Phi(X)\underline{N}(X)|} \\ |\Phi(F)| &= \int_{\Phi(F)} do = \int_F |\operatorname{Cof}\nabla\Phi(X)\underline{N}(X)|\, dO \end{aligned}}\,,$$

or, in detailed parameter representation of F,

$$|\Phi(F)| = \int_D \left|\operatorname{Cof}\nabla\Phi(X(u))\left(\frac{\partial X}{\partial u^1}\times\frac{\partial X}{\partial u^2}\right)(u)\right| du^1\, du^2 .$$

So we have altogether the general transformation rule

$$do = |\operatorname{Cof}\nabla\Phi\underline{N}|dO,\quad d\underline{o} = \operatorname{Cof}\nabla\Phi\, d\underline{O}\,,$$

or in differential form

$$\begin{bmatrix} dx^2\wedge dx^3 \\ dx^3\wedge dx^1 \\ dx^1\wedge dx^2 \end{bmatrix} = \operatorname{Cof}\nabla\Phi(X)\begin{bmatrix} dX^2\wedge dX^3 \\ dX^3\wedge dX^1 \\ dX^1\wedge dX^2 \end{bmatrix}\,,\quad x = \Phi(X)\,.$$

By these results it becomes evident that the matrix $\operatorname{Cof}\nabla\Phi(X)$ plays a crucial role whenever normal vectors come into play.

(e) Pull Back of Divergence Theorem Let $U \subset \Omega$ be an arbitray subvolume with sufficiently smooth surface ∂U. Further, let $\mathbf{t} : x \mapsto \mathbf{t}(x)$ be a tensor field in space points x. Then there follows from Lemma 8.1 and the divergence theorem of § 1.1.2**(c2)** by row-wise application

$$\begin{aligned} \int_{\Phi(U)} \operatorname{div}\mathbf{t}\, dv &= \int_{\partial\Phi(U)} \mathbf{t}\,\underline{n}\, do = \int_{\partial U} (\mathbf{t}\circ\Phi)(\operatorname{Cof}\nabla\Phi)\underline{N}\, dO \\ &= \int_U \operatorname{Div}[(\mathbf{t}\circ\Phi)\operatorname{Cof}\nabla\Phi)]\, dV\,. \end{aligned}$$

Definition 8.2. *The tensor*

$$\mathbf{T}(X) := \mathbf{t}(\Phi(X))\operatorname{Cof}\nabla\Phi(X)$$

is the PIOLA *transformation of* $\mathbf{t}(x)$.

If now $\mathbf{t}(x)$ is the stress tensor then $\mathbf{T}(\mathbf{X})$ is called PIOLA-KIRCHHOFF *stress tensor*. Since $U \subset \Omega$ is an arbitrary subvolume in the above equation we obtain as goal of the entire action

$$\boxed{\begin{aligned} &\mathbf{t}\,\underline{n}\, do = \mathbf{T}\,\underline{N}\, dO\,,\quad x = \Phi(X)\,,\quad X\in\Omega \\ &\int_{\Phi(U)} \operatorname{div}\,\mathbf{t}\, dv = \int_U \operatorname{Div}\,\mathbf{T}\, dV \quad \text{(column vector)} \end{aligned}}\,. \tag{8.8}$$

Lemma 8.2. *(*PIOLA *Identity)* $\operatorname{Div}\operatorname{Cof}\nabla\Phi(t,X) = \underline{0} \in \mathbb{R}^3$.

Since $U \subset \Omega$ arbitrary, this follows immediately from (8.8) for $\mathbf{t}(x) = I$ using Definition 8.2. □

Lemma 8.3. $\operatorname{Div}\mathbf{T}(X) = \det[\nabla\Phi(X)]\operatorname{div}\mathbf{t}(x)\,,\ \ x = \Phi(X)$.

This is likewise an inference to (8.8) because $dv = \det[\nabla\Phi(X)]\,dV$. □

Let now $\dot{x} = \underline{v}(t,x)$ be the velocity vector again.

Lemma 8.4. *Let* $J(t,X) := \det[\nabla\Phi(t,X)]$ *then*

$$\frac{\partial}{\partial t}J(t,X) = J(t,X)\ \operatorname{div}\underline{v}(t,x)\,,\ \ x = \Phi(t,X)\,.$$

Proof. Remember $\nabla\underline{V}(t,X) = \nabla\Phi_t(t,X) = [\nabla\Phi(t,X)]_t$ then

$$\begin{aligned}
&\det[\nabla\Phi(t,X)]\operatorname{div}\underline{v}(t,x) = \det[\nabla\Phi(t,X)]\operatorname{div}\underline{v}(t,x)^T\\
&= \operatorname{Div}[\underline{v}(t,\Phi(t,X))^T\operatorname{Cof}\nabla\Phi(t,X)]\ \ \text{(Lemma 8.3)}\\
&= \Phi_t(t,X)^T\operatorname{Div}\operatorname{Cof}\nabla\Phi(t,X) + \operatorname{Grad}\Phi_t(t,X) : \operatorname{Cof}\nabla\Phi(t,X)\ \ \text{(product rule)}\\
&= \operatorname{Grad}\Phi_t(t,X) : \operatorname{Cof}\nabla\Phi(t,X)\ \ \text{(Lemma 8.2)}\\
&= \left[\frac{\partial}{\partial t}\nabla\Phi(t,X)\right] : \operatorname{Cof}\nabla\Phi(t,X) = \frac{\partial}{\partial t}J(t,X)\ \ \text{(determinant rule).}
\end{aligned}$$

□

Lemma 8.4 describes the infinitesimal change of volume w.r.t. time t.

Lemma 8.5. *(*HELMHOLTZ *Identity)*

$$\frac{\partial}{\partial t}\operatorname{Cof}\nabla\Phi(t,X) = [\operatorname{div}\underline{v}(t,x)I - \operatorname{grad}\underline{v}(t,x)]^T[\operatorname{Cof}\nabla\Phi(t,X)]\,,\quad x = \Phi(t,X)\,.$$

Proof. Note that

$$[\operatorname{Cof}\nabla\Phi]_t = [\det(\nabla\Phi)]_t(\nabla\Phi)^{-T} + \det(\nabla\Phi)[(\nabla\Phi)^{-T}]_t\,.$$

Then, by Lemma 8.4 and (8.6),

$$\begin{aligned}
[\operatorname{Cof}\nabla\Phi]_t &= \det(\nabla\Phi)\,(\operatorname{div}\underline{v})(\nabla\Phi)^{-T} - \det(\nabla\Phi)(\operatorname{grad}\underline{v})^T(\nabla\Phi)^{-T}\\
&= [\operatorname{div}\underline{v}(t,X)I - \operatorname{grad}\underline{v}(t,X)]^T[\operatorname{Cof}\nabla\Phi(t,X)]\,.
\end{aligned}$$

□

8.2 The Three Transport Theorems

Let again $\Phi : \mathbb{R}\times\Omega \ni (t,X) \mapsto \Phi(t,X) = x \in \mathbb{R}^3$ be a deformation, $\underline{v}$ the velocity field and $\underline{w}$ an arbitrary vector field in space coordinates, $\underline{w}(t,x) = \underline{w}(t,\Phi(t,X)) = \underline{W}(t,X)$. Further, let $U \subset \Omega$ be an open subset with sufficiently smooth boundary.

Theorem 8.1. *(Transport Theorem for Volume Integrals)*

$$\boxed{\frac{D}{Dt}\int_{\Phi(t,U)} \underline{w}(t,x)\,dv = \int_{\Phi(t,U)} \left[\frac{D}{Dt}\underline{w}(t,x) + \underline{w}(t,x)\,\mathrm{div}\,\underline{v}(t,x)\right] dv} \,.$$

Proof. Recalling $dv = J(t,X)dV$, $J(t,X) = \det[\nabla\Phi(t,X)]$, we obtain by application of Lemma 8.4

$$\begin{aligned}
&\frac{D}{Dt}\int_{\Phi(t,U)} \underline{w}(t,x)\,dv = \int_U \frac{d}{dt}\left[\underline{w}(t,\Phi(t,X))J(t,X)\right] dV \\
&= \int_U \left[J(t,X)\frac{d}{dt}\underline{w}(t,\Phi(t,X)) + \underline{w}(t,\Phi(t,X))J(t,X)\,\mathrm{div}\,\underline{v}(t,\Phi(t,X))\right] dV \\
&= \int_{\Phi(t,U)} \left[\frac{D}{Dt}\underline{w}(t,x) + \underline{w}(t,x)\,\mathrm{div}\,\underline{v}(t,x)\right] dv\,.
\end{aligned}$$

□

In this result, named after REYNOLDS, the domain of integration $\Phi(t,U)$ depends on t as before but the material derivative is passed under the integral sign.

Let now $a = a(t,x)$ be any scalar function (e.g., mass density) then

$$\mathrm{div}(a\,\underline{v}) = (\mathrm{grad}\,a)\underline{v} + a\;\mathrm{div}\,\underline{v}$$

and

$$\frac{D}{Dt}(a\,\underline{w}) + a\,\underline{w}\,\mathrm{div}\,\underline{v} = \frac{Da}{Dt}\underline{w} + \frac{D\underline{w}}{Dt}a + a\,\underline{w}\,\mathrm{div}\,\underline{v} = \underline{w}\left(\frac{Da}{Dt} + a\;\mathrm{div}\,\underline{v}\right) + a\,\frac{D\underline{w}}{Dt}\,.$$

Therefore Theorem 8.1 has two important special cases:

$$\boxed{\begin{aligned}
\frac{D}{Dt}\int_{\Phi(t,U)} a\,dv &= \int_{\Phi(t,U)} \left[\frac{\partial a}{\partial t} + \mathrm{div}(a\,\underline{v})\right] dv \\
\frac{D}{Dt}\int_{\Phi(t,U)} a\,\underline{w}\,dv &= \int_{\Phi(t,U)} \left[\underline{w}(\frac{\partial a}{\partial t} + \mathrm{div}(a\,\underline{v})) + a\frac{D\underline{w}}{Dt}\right] dv
\end{aligned}}\,. \tag{8.9}$$

In the sequel the arguments (t,x) resp. (t,X) of the integrands are omitted partly for simplicity. Let $D \subset \mathbb{R}^2$ be open, $F := \{X(u),\ u = (u_1,u_2) \in D\} \subset \mathbb{R}^3$ a surface in reference space (material coordinates) and $\Phi(t,F) = \{x = (\Phi(t,X(u)),\ u \in D\} \subset \mathbb{R}^3$ the transformed (transported) surface in space coordinates.

Theorem 8.2. *(Transport Theorem for Surface Integrals)*

$$\boxed{\frac{D}{Dt}\int_{\Phi(t,F)} \underline{w}\cdot\underline{n}\,do = \int_{\Phi(t,F)} \left[\frac{D\underline{w}}{Dt} + (\mathrm{div}\,\underline{v}\,I - \mathrm{grad}\,\underline{v})\underline{w}\right]\cdot\underline{n}\,do}\,.$$

Proof. At first, substitution and product rule yield

$$\frac{D}{Dt}\int_{\Phi(t,F)} \underline{w}(t,x)\cdot\underline{n}(t,x)\,do = \int_F \frac{d}{dt}\underline{w}(t,\Phi(t,X))\cdot \mathrm{Cof}\,\nabla\Phi(t,X)\underline{N}(X)\,dO$$
$$= \int_{\Phi(t,F)} \frac{D\underline{w}(t,x)}{Dt}\cdot\underline{n}(t,x)\,do + \int_F \underline{W}(t,X)\cdot\frac{d}{dt}[\mathrm{Cof}\,\nabla\Phi(t,X)]\underline{N}(X)\,dO\,.$$

By means of the HELMHOLTZ identity,

$$\frac{d}{dt}\mathrm{Cof}\,\nabla\Phi = [(\mathrm{div}\,\underline{v})I - \mathrm{grad}\,\underline{v}]^T\,\mathrm{Cof}\,\nabla\Phi\,,$$

we obtain

$$\int_F \underline{W}(t,X)\cdot\frac{d}{dt}[\mathrm{Cof}\,\nabla\Phi(t,X)]\underline{N}(X)\,dO$$
$$= \int_{\Phi(t,F)} \underline{w}(t,x)\cdot[\mathrm{Cof}\,\nabla\Phi(t,X)]_t[\mathrm{Cof}\,\nabla\Phi(t,X)]^{-1}\underline{n}\,do$$
$$= \int_{\Phi(t,F)} [\mathrm{div}\,\underline{v}\,I - \mathrm{grad}\,\underline{v}]\underline{w}\cdot\underline{n}\,do\,.$$

□

Finally let $\mathcal{I}\subset\mathbb{R}$ be an open interval, $C := \{X(u)\,,\ u\in\mathcal{I}\}\subset\mathbb{R}^3$ a line segment in reference space (material coordinates) and $\Phi(t,C) = \{x = (\Phi(t,X(u)), u\in\mathcal{I}\}\subset\mathbb{R}^3$ the transformed (transported) line in space coordinates.

Theorem 8.3. *(Transport Theorem for Line integrals)*

$$\boxed{\frac{D}{Dt}\int_{\Phi(t,C)} \underline{w}(t,x)\cdot d\underline{x} = \int_{\Phi(t,C)}\left[\frac{D\underline{w}(t,x)}{Dt}\cdot d\underline{x} + \underline{w}(t,x)\cdot d\underline{v}\right]}\,.$$

The *proof* of this theorem follows directly from the rule for derivations of products observing that $d\underline{x} = (\Phi\circ X)_u\,du$ and

$$\frac{\partial}{\partial t}\frac{\partial}{\partial u}\Phi(t,X(u)) = \frac{\partial}{\partial u}\frac{\partial}{\partial t}\Phi(t,X(u))$$
$$= \frac{\partial}{\partial u}\underline{V}(t,X(u)) = \frac{\partial}{\partial u}\underline{v}(t,\Phi(t,X(u))) =: d\underline{v}\,.$$

□

If $\underline{w} = \underline{v}$ is the velocity field, then

$$\int_{\Phi(t,C)} \underline{w}\cdot d\underline{v} = [\underline{v}(t,B)\cdot\underline{v}(t,B) - \underline{v}(t,A)\cdot\underline{v}(t,A)]/2$$

where A is the initial point and B the terminal point of the line C; therefore a *closed* curve C satisfies

$$\frac{D}{Dt}\oint_{\Phi(t,C)} \underline{v}\cdot d\underline{x} = \oint_{\Phi(t,C)} \frac{D\underline{v}}{Dt}\cdot d\underline{x}\,.$$

8.3 Conservation Laws

Conservation laws are also called balance theorems. The conservation *theorems* of physics have to be conceived as *axioms* in mathematical sense. On the level of the present volume they cannot be derived from other results by pure *mathematical* conclusions but verified only in an experimental way. They are defined here at first *relative to space coordinates* where altogether the following *space-related* physical quantities are involved (*specific* quantities relate to mass unit):

$\varepsilon(x,t)$	spec. energy density	[energy/mass]
$\vartheta(t,x)$	abs. temperature (>0)	[Kelvin]
$\varrho(t,x)$	mass denisty	[masse/volume]
$h(t,x;\underline{n}(t,x))$	thermal flux density	[energy/(area· time)]
$p(t,x)$	pressure	[force/area]
$r(x,t)$	spez. thermal source density	[heat/(time· mass)]
$s(x,t)$	spez. entropy	[heat/(temperature·mass)]
$\underline{f}(x,t)$	spez. volume-force density	[force/mass]
$\underline{g}(t,x;\underline{n}(t,x))$	surface-force density	[force/area]
$\underline{k}(t,x)$	volume-force density	[force/volume]
$\underline{q}(t,x)$	thermal flux vector	[energy/(area·time)]
$\underline{v}(t,x)$	velocity field	[space/time]
$\psi = \varepsilon - \vartheta\, s$	free energy density	
$e := \varepsilon + \underline{v}\cdot\underline{v}/2$	abbreviation.	

(Frequently $\varepsilon = c\vartheta$ with specific heat c.) Further, let

$\boldsymbol{\delta}$	unit tensor,
$\mathbf{t}(t,x) \in \mathbb{R}^3{}_3$	stress tensor after CAUCHY depending on material [force/area],
$\varepsilon(\underline{u}) = \dfrac{1}{2}\left[\operatorname{grad}\underline{u} + (\operatorname{grad}\underline{u})^T\right]$	linearized strain tensor without dimension for displacement field.

Commonly the pressure p is neglected in mechanics of *solid media.* In mechanics of *fluids*, the pressure p is separated from the stress tensor $\mathbf{t}$ by $\mathbf{t} = \boldsymbol{\sigma} - p\,\boldsymbol{\delta}$, and the tensor $\boldsymbol{\sigma}$ depends on *velocity* $\underline{v}$ instead of displacement.

Let $\{\Omega, \varrho, \mathbf{t}, \ldots\}$ be a body and Φ a deformation; the transformed body is briefly denoted by $\Phi(t,\Omega)$. The mechanical and thermodynamical properties are defined by a specification of the above quantities and thus play here again the role of *axioms* in mathematical sense. All quantities shall be two times continuously differentiable henceforth, and the boundaries of all considered domains shall be continuous and piecewise continuously differentiable. One then says also that the body and the motion Φ are *simple.* Further, $U \subset \Omega$ shall be an *arbitrary* subset with likewise continuous and piecewise continuously differentiable boundary.

(a) Conservation Law of Mass Let

$$\mathcal{M}(t,U) := \int_{\Phi(t,U)} \varrho(t,x)dv$$

be the mass of the "moving" volume $\Phi(t,U)$.

Axiom 8.1. (Conservation Law of Mass) $\mathcal{M}(t,U)$ is constant or, in other words,

$$\boxed{\forall\, U \subset \Omega : \frac{D}{Dt}\mathcal{M}(t,U) = 0}\,.$$

(b) Conservation Law of Momentum Let

$$\mathcal{I}(t,U) = \int_{\Phi(t,U)} \varrho(t,x)\,\underline{v}(t,x)dv \in \mathbb{R}^3$$

be the total momentum of volume $\Phi(t,U)$. The remaining moments in interior of $\Phi(t,U)$ cancel out each other (also axiom).

Axiom 8.2. (Conservation Law of Momentum) $\forall\, U \subset \Omega$:

$$\boxed{\frac{D}{Dt}\mathcal{I}(t,U) = \int_{\Phi(t,U)} \varrho(t,x)\,\underline{f}(t,x)\,dv + \int_{\partial\Phi(t,U)} \underline{g}(t,x;\underline{n}(t,x))\,do}\,.$$

The stress vector $\underline{g}$ is not a vector field, but the following fundamental theorem of CAUCHY does hold in case of sufficient smoothness:

Theorem 8.4. *Under Axiom 8.2 there exists a tensor field (stress tensor)*

$$\mathbf{t} : (t,x) \mapsto \mathbf{t}(t,x) \in {\mathbb{R}^3}_3\,,$$

such that

$$\underline{g}(t,x;\underline{n}(t,x)) = \mathbf{t}(t,x)\underline{n}(t,x)\,.$$

Proof see e.g. (Ciarlet93).

(c) Conservation Law of Angular Momentum For arbitrary x_0 let

$$\mathcal{L}(t,U) = \int_{\Phi(t,U)} \varrho(t,x)[(x-x_0)\times \underline{v}(t,x)]\,dv$$

be the total angular momentum of the subvolume $\Phi(t,U)$ relative to x_0. The other angular momentums cancel out each other (axiom); without loss of generality let also $x_0 = 0$ below.

Axiom 8.3. (Conservation Law of Angular Momentum) $\forall\, U \subset \Omega$:

$$\boxed{\begin{aligned}&\frac{D}{Dt}\mathcal{L}(t,U)\\ &= \int_{\Phi(t,U)} \varrho(t,x)[x\times \underline{f}(t,x)]\,dv + \int_{\partial\Phi(t,U)} [x \times \underline{g}(t,x;\underline{n}(t,x))]\,do\end{aligned}}\,.$$

Theorem 8.5. *Adopt Axiom 8.1 and 8.2. Then Axiom 8.3 does hold if and only if the stress tensor* $\mathbf{t}(t,x)$ *is symmetric.*

Proof SUPPLEMENT\chap08a.

Suppose Axiom 8.1 and 8.2. Then Axiom 8.3 yields $\forall\, U \subset \Omega$:

$$\frac{D}{Dt}\mathcal{L}(t,U) = \int_{\Phi(t,U)} \varrho(t,x)[x \times \underline{f}(t,x)]\, dv + \int_{\partial\Phi(t,U)} [x \times \mathbf{t}(t,x)\,\underline{n}(t,x)]\, do\,. \tag{8.10}$$

Since all axioms are supposed to hold in the sequel, the balance theorem of angular momentum is no longer mentioned but equated with the symmetry of CAUCHY's stress tensor $\mathbf{t}$ by Theorem 8.5.

(d) Conservation Law of Energy Let $h(t,x;\underline{n}(t,x))$ be the flux of energy *from interior to exterior* through the surface of each considered subvolume where $\underline{n}$ denotes the normal vector (mostly thermal flux). The sum of interior and kinetic energy in subvolume $\Phi(t,U)$ without potential energy is

$$\mathcal{E}(t,U) = \int_{\Phi(t,U)} \varrho\left[\varepsilon + \frac{\underline{v}\cdot\underline{v}}{2}\right] dv\,.$$

Axiom 8.4. (Conservation Law of Energy) $\forall\, U \subset \Omega$:

$$\boxed{\begin{aligned}&\frac{D}{Dt}\mathcal{E}(t,U) = \\ &\int_{\Phi(t,U)} \varrho(\underline{f}\cdot\underline{v} + r)\, dv + \int_{\partial\Phi(t,U)} \left[\underline{g}(t,x;\underline{n}(t,x))\cdot\underline{v} - h(t,x;\underline{n}(t,x))\right] do\end{aligned}}\,.$$

Theorem 8.6. *Suppose Axiom 8.1 to 8.4, then there exists an energy-flux vector (thermal-flux vector)* $\underline{q}(t,x)$ *such that*

$$\forall\, \underline{n} : h(t,x;\underline{n}(t,x)) = \underline{q}(t,x)\cdot\underline{n}(t,x)\,.$$

Motions of rigid bodies have the form $\Psi(t,X) = \underline{c}(t) + Q(t)X$ where $\underline{c}(t)$ is an arbitrary translation vector and $Q(t)$ is an arbitrary rotation matrix. The set of these motions forms a *group* $\mathcal{S}$, i.e., for each motion there exists the inverse, and two successive motions of $\mathcal{S}$ are again a rigid motion (belonging to $\mathcal{S}$). The conservation law of energy is called *invariant* under (transformations of) the group $\mathcal{S}$ if it does hold for arbitrary $\Psi \circ \Phi$ yielding always the same result. This invariance is a simple and evident condition for this balance theorem excluding some pathological cases.

Theorem 8.7. (1°) *If Axiom 8.4 holds and is invariant under* $\mathcal{S}$ *then Axioms 8.1, 8.2 and 8.3 do hold.*
(2°) *If all four axioms do hold then Axiom 8.4 is invariant under* $\mathcal{S}$.

Cf. (Marsden), Theorem 3.8. By this result, the balance theorem of energy plays an exceptional role under all four balance theorems.

(e) Conservation Laws in Differential Form We suppose that all four axioms apply. If Axiom 8.2 is considered componentwise then all axioms have the form

$$\boxed{\frac{D}{Dt}\int_{\Phi(t,U)} a(t,x)\,dv = \int_{\Phi(t,U)} b(t,x)\,dv + \int_{\partial\Phi(t,U)} \underline{w}(t,x)\cdot\underline{n}(t,x)\,do} \tag{8.11}$$

where a, b are scalar functions and $\underline{w}$ is a vector field.

Theorem 8.8. (Localization) *The scalar fields a, b and the vector field $\underline{w}$ obey the law of conservation (8.11) for arbitrary subsets $U \subset \Omega$ if and only if*

$$\frac{\partial a}{\partial t} + \operatorname{div}(a\,\underline{v}) = b + \operatorname{div}\underline{w}$$

where $\underline{v}$ is the velocity field of Φ.

Proof. It follows by (8.9) and the Divergence Theorem directly that (8.11) is equivalent to

$$\int_{\Phi(t,U)} \left(\frac{\partial a}{\partial t} + \operatorname{div}(a\,\underline{v})\right) dv = \int_{\Phi(t,U)} b\,dv + \int_{\Phi(t,U)} \operatorname{div}\underline{w}\,dv\,,$$

by REYNOLDS' Transport Theorem 8.1 and the Divergence Theorem. Thus divergence theorem and the differential form imply the integral form. Conversely, the integral form implies the differential form because $U \subset \Omega$ is an arbitrary subset. □

Note however that the differential form demands higher smoothness of the employed functions than the integral form.

At first, this result or (8.9) leads to the law of conservation of mass in differential form (*equation of continuity*)

$$\boxed{\frac{\partial \varrho}{\partial t} + \operatorname{div}(\varrho\,\underline{v}) = 0}\,. \tag{8.12}$$

Then, by integration and application of the divergence theorem,

$$\int_{\Phi(t,U)} \frac{\partial \varrho}{\partial t}\,dv = -\int_{\Phi(t,U)} \operatorname{div}(\varrho\underline{v})\,dv = -\int_{\partial\Phi(t,U)} \varrho\,\underline{v}\cdot\underline{n}\,do\,.$$

The *number* $-\displaystyle\int_{\partial\Phi(t,U)} \varrho\,\underline{v}\cdot\underline{n}\,do$ describes the flux of mass in direction from interior to exterior (decreasing).

Next, equation (8.12) is applied to a general vector field $\underline{w}$ and yields

$$\frac{D}{Dt}\int_{\Phi(U,t)} \varrho\,\underline{w}\,dv = \int_{\Phi(U,t)} \left[\underline{w}\Big(\frac{\partial \varrho}{\partial t} + \operatorname{div}(\varrho\,\underline{v})\Big) + \varrho\frac{D\underline{w}}{Dt}\right] dv = \int_{\Phi(U,t)} \varrho\frac{D\underline{w}}{Dt}\,dv\,. \tag{8.13}$$

Then the balance law of momentum together with Theorem 8.4 and the Divergence Theorem supply the law in differential form

$$\boxed{\varrho\frac{D\underline{v}}{Dt} - \operatorname{div}\mathbf{t} = \varrho\,\underline{f}}\,. \tag{8.14}$$

Finally (8.13), with scalar $w = e$, and the Divergence Theorem are applied to the law of energy. Together we have the conservation laws in differential form writing the material derivative explicitely:

$$\boxed{\begin{array}{llll} \dfrac{\partial\varrho}{\partial t} & + & \operatorname{div}(\varrho\,\underline{v}) & = 0 \\ \varrho\,\dfrac{\partial\underline{v}}{\partial t} + \varrho(\operatorname{grad}\underline{v})\underline{v} & - & \operatorname{div}\mathbf{t} & = \varrho\,\underline{f} \\ & & \mathbf{t} & = \mathbf{t}^T \\ \varrho\,\dfrac{\partial e}{\partial t} + \varrho\operatorname{grad} e\cdot\underline{v} & - & \operatorname{div}\mathbf{t}\underline{v} + \operatorname{div}\underline{q} & = \varrho\,\underline{f}\cdot\underline{v} + \varrho\,r\,, \quad e = \varepsilon + \dfrac{\underline{v}\cdot\underline{v}}{2} \end{array}}\,. \tag{8.15}$$

This representation of the balance theorems is called *non-conservative* form. For the *conservative* form we add the continuity equation (8.12), multiplied by $\underline{v}$ and e respectively, to the second and fourth equation. Then, after some simple transformations,

$$\boxed{\begin{array}{llll} \dfrac{\partial(\varrho\,\underline{v})}{\partial t} + \operatorname{div}(\varrho\,\underline{v}\,\underline{v}^T) & - & \operatorname{div}\mathbf{t} & = \varrho\,\underline{f} \\ \dfrac{\partial(\varrho\,e)}{\partial t} + \operatorname{div}(\varrho\,e\,\underline{v}) & - & \operatorname{div}\mathbf{t}\underline{v} + \operatorname{div}\underline{q} & = \varrho\,\underline{f}\cdot\underline{v} + \varrho\,r \end{array}}\,. \tag{8.16}$$

The conservative form is strongly recommended for solving *compressible* flow problems (Zienkiewicz); note that $\underline{v}\,\underline{v}^T$ is a dyadic product (matrix).

The balance theorem of energy in (8.15) can be transformed further by using the above *strain tensor* $\boldsymbol{\varepsilon}(\underline{v})$ where the volume-force density is cancelled out:

Lemma 8.6.

$$\boxed{\varrho\frac{\partial\varepsilon}{\partial t} + \varrho\,(\operatorname{grad}\varepsilon)\cdot\underline{v} - \boldsymbol{\varepsilon}(\underline{v}) : \mathbf{t} + \operatorname{div}\underline{q} = \varrho\,r}\,.$$

Proof. Note that e is replaced here by the intrinsic energy ε. By (1.2) and §1.2 **(b)**

$$\operatorname{div}(\mathbf{t}\underline{v}) = \underline{v}\cdot\operatorname{div}\mathbf{t} + \operatorname{grad}\underline{v} : \mathbf{t} = \operatorname{div}\mathbf{t}\cdot\underline{v} + \mathbf{t} : \boldsymbol{\varepsilon}(\underline{v})$$

because of the symmetry of $\mathbf{t}$. On the other side, multiplying the equation of momentum by $\underline{v}$,

$$\underline{v}\cdot\operatorname{div}\mathbf{t} = \varrho\,\underline{v}\cdot\frac{D\underline{v}}{Dt} - \varrho\,\underline{f}\cdot\underline{v}\,.$$

Substitution into the non-conservative equation of energy yields the assertion because

$$\frac{D}{Dt}\frac{\underline{v}\cdot\underline{v}}{2} = \underline{v}\cdot\frac{D\underline{v}}{Dt}\,.$$

□

Axiom 8.4 is also called *first law of thermodynamics* in the above general form. Remember the general notations in integral form

$\mathcal{K} = \dfrac{1}{2}\displaystyle\int_{\Phi(t,U)} \varrho\,\underline{v}\cdot\underline{v}\,dv$	kinetic energy
$\mathcal{E}_{\text{int}} = \displaystyle\int_{\Phi(t,U)} \varrho\,\varepsilon\,dv$	interior energy
$\dfrac{D\mathcal{W}}{Dt} = \displaystyle\int_{\Phi(t,U)} \varrho\,\underline{f}\cdot\underline{v}\,dv + \int_{\partial\Phi(t,U)} \underline{v}\cdot\mathbf{t}(t,x)\,\underline{n}\,do$	mechanical power
$\dfrac{D\mathcal{Q}}{Dt} = \displaystyle\int_{\Phi(t,U)} \varrho\,r\,dv - \int_{\partial\Phi(t,U)} \underline{q}\cdot\underline{n}\,do$	non-mechanical power .

By these notations this first main theorem obtains the more stringent form

$$\boxed{\frac{D}{Dt}(\mathcal{K}+\mathcal{E}_{\text{int}}-\mathcal{W}-\mathcal{Q}) = 0}$$

where $\mathcal{W}$ denotes the mechanical energy and $\mathcal{Q}$ the non-mechanical energy. Thus Axiom 8.4 says also that the individual energies may be added together with the properly chosen sign and that the sum is *constant.* Cf. (Marsden), p. 144. (Of course total energy must be constant in a closed system.)

(f) Second Law of Thermodynamics Let

$$\mathcal{S}(t,U) = \int_{\Phi(t,U)} \varrho(t,x)\,s(t,x)\,dv$$

be the *entropy* of the subvolume $\Phi(t,U)$.

Axiom 8.5. (Second Law of Thermodynamics, Entropy Inequality) $\forall\, U \subset \Omega$:

$$\boxed{\begin{aligned}&\frac{D}{Dt}\int_{\Phi(t,U)} \varrho(t,x)\,s(t,x)\,dv\\ &\geq \int_{\Phi(t,U)} \frac{\varrho(t,x)\,r(t,x)}{\vartheta(t,x)}\,dv - \int_{\partial\Phi(t,U)} \frac{\underline{q}(t,x)\cdot\underline{n}(t,x)}{\vartheta(t,x)}\,do\end{aligned}}\,. \tag{8.17}$$

If thermal source density r and energy-flux vector $\underline{q}$ are both zero then Axiom 8.5 has the more simple and well-known form

$$\frac{D}{Dt}\int_{\Phi(t,U)} \varrho(t,x)s(t,x)\,dv \geq 0\,.$$

This inequality says that the entropy of a subvolume $\Phi(t,U)$ increases weakly monotone with time t under the above assumptions.

Let us transform the left term of the inequality (8.17) by (8.13),

$$\frac{D}{Dt}\int_{\Phi(t,U)} \varrho(t,x)s(t,x)\,dv = \int_{\Phi(t,U)} \varrho(t,x)\frac{D}{Dt}s(t,x)\,,$$

apply the divergence theorem to the second term on the right side of (8.17) and remember

$$\operatorname{div}(\varphi\,\underline{w}) = \varphi\,\operatorname{div}\underline{w} + \operatorname{grad}\varphi\cdot\underline{w}\,.$$

Then, because U arbitrary again, we obtain the second law in differential form

$$\varrho\frac{Ds}{Dt} \geq \frac{\varrho\,r}{\vartheta} - \operatorname{div}\left(\frac{1}{\vartheta}\underline{q}\right) = \frac{\varrho\,r}{\vartheta} - \frac{1}{\vartheta}\operatorname{div}\underline{q} + \frac{1}{\vartheta^2}\operatorname{grad}\vartheta\cdot\underline{q}\,,$$

resp., after multiplication by $\vartheta > 0$,

$$\boxed{\varrho\,\vartheta\,\frac{Ds}{Dt} \geq \varrho\,r - \operatorname{div}\underline{q} + \frac{1}{\vartheta}\operatorname{grad}\vartheta\cdot\underline{q}}\,. \tag{8.18}$$

Theorem 8.9. *Let $\psi = \varepsilon - \vartheta\,s$ be the free specific energy then*

$$\boxed{\varrho\left(s\,\dot{\vartheta} + \dot{\psi}\right) - \operatorname{div}(\mathbf{t}\underline{v}) + \operatorname{div}\mathbf{t}\cdot\underline{v} + \frac{1}{\vartheta}\operatorname{grad}\vartheta\cdot\underline{q} \leq 0}\,.$$

Proof. Observe

$$\dot{\psi} = \dot{\varepsilon} - \vartheta\,\dot{s} - \dot{\vartheta}\,s \implies \vartheta\,\dot{s} = \dot{\varepsilon} - \dot{\vartheta}\,s - \dot{\psi}\,.$$

Substitution into (8.18) yields

$$\varrho\left(\dot{\varepsilon} - \dot{\vartheta}\,s - \dot{\psi}\right) \geq \varrho\,r - \operatorname{div}\underline{q} + \frac{1}{\vartheta}\operatorname{grad}\vartheta\cdot\underline{q}\,.$$

Substitution of the differential form of the energy law,

$$\varrho\,\dot{\varepsilon} = \varrho\,r - \operatorname{div}\underline{q} + \boldsymbol{\varepsilon}(\underline{v}) : \mathbf{t} = \varrho\,r - \operatorname{div}\underline{q} + \operatorname{div}(\mathbf{t}\underline{v}) - \operatorname{div}\mathbf{t}\cdot\underline{v}\,,$$

by Lemma 8.6, then yields the assertion. □

Remember: A thermodynamic process is called

(1°) adiabatic if $\underline{q} = \underline{0}$, (2°) isentropic if $\dot{s} = 0$,
(3°) homentropic if $\operatorname{Grad} S = 0$, (4°) isothermic if $\vartheta = \text{const}$.

Finally it should be remarked that the initial quantities $M(0,\Omega)$, $\mathcal{L}(0,\Omega)$, $\mathcal{I}(0,\Omega)$ and $\mathcal{E}(0,\Omega)$ — being constants of integration — may be chosen arbitrarily in all Axioms 8.1 to 8.4.

References: Mainly (Marsden).

8.4 Material Forms

The conservation laws have to be *pulled back*, i.e., be transformed into material form for further considerations and for *computing* solutions. In particular the domain $\Phi(t,\Omega)$ is unknown and one has to integrate over the known domain Ω at last. Also in *Material Theory* the *response functions* characterizing the elastic material are developed at first relative to material coordinates. They describe the stress of a material as *response* to the strain and are indispensable for the construction of the stress tensor. Thereafter it is the question whether and which terms of higher order may be neglected in further computation since, finally, a *linear* relation between strain and stress tensor is aimed at (Altenbach), (Ciarlet93). One supposes in *linear elasticity* that the considered body changes its form in a negligible way under deformation (!). As a consequence, the external forces retain their direction before and after deformation ("dead loads") and are thus independent of displacement. By this way the difference between material and space coordinates is vanished to some degree and is then no longer accentuated in notation.

The distinction between conservative and non-conservative representation disappears obviously in material forms. Let

$$\begin{aligned} &A(t,X) := a(t,\Phi(t,X))\,,\ B(t,X) := b(t,\Phi(t,X))\\ &E(t,X) := e(t,\Phi(t,X))\,,\ \underline{F}(t,X) := \underline{f}(t,\Phi(t,X))\\ &R(t,X) := r(t,\Phi(t,X)) \end{aligned}$$

where the quantities on left side are to be understood as abbreviations. They have *no* explicit physical meaning as functions of their arguments; only e.g. $A(0,X)=a(0,X)$ for $t=0$ because $\Phi(0,X)=X$. Further, let

$$J(t,X) := \det\nabla\Phi(t,X)\,,\quad \varrho_{\rm ref}(X) = \text{mass density in } \Omega\,.$$

Moreover, it is frequently operated in elastic bodies with *dead masses* and *dead loads* as mentioned above, i.e. the same values

$$\varrho(t,x) = \varrho_{\rm ref}(X)\,,\quad \underline{f}(t,x) := \underline{f}(t,X)\ \ (=\underline{F}(t,X))\,,\quad x=\Phi(t,X)\,,$$

are taken before and after displacement.

(a) We need the PIOLA transformations of the vectors $\underline{q}$ and $\underline{w}$, and of the tensor $\mathbf{t}$,

$$\begin{aligned} \underline{Q}(t,X) &= [\mathrm{Cof}\,\nabla\Phi(t,X)]^T\,\underline{q}(t,x)\,,\quad x=\Phi(t,X)\\ \underline{W}(t,X) &= [\mathrm{Cof}\,\nabla\Phi(t,X)]^T\,\underline{w}(t,x)\\ \mathbf{T}(t,X) &= \mathbf{t}(t,x)\,\mathrm{Cof}\,\nabla\Phi(t,X)\,. \end{aligned}$$

The *first* PIOLA-KIRCHHOFF *stress tensor* $\mathbf{T}$ is not symmetric hence the *second* PIOLA-KIRCHHOFF *stress tensor* $\mathbf{S}(t,X) = \nabla\Phi(t,X)^{-1}\mathbf{T}(t,X)$ is introduced. If the CAUCHY stress tensor $\mathbf{t}$ is symmetric then also $\mathbf{S}$ is *symmetric* because

$$\mathbf{S}(t,X) = J(t,X)\nabla\Phi(t,X)^{-1}\mathbf{t}(t,\Phi(t,X)[\nabla\Phi(t,X)]^{-T}\,. \tag{8.19}$$

(b) The continuity equations now reads simply

$$\begin{aligned}\mathcal{M}(t,U) &= \int_{\Phi(t,U)} \varrho(t,x)\,dv = \int_U \varrho(t,\Phi(t,X))J(t,X)\,dV\\ &= \int_U \varrho_{\rm ref}(X)\,dV = \text{constant},\end{aligned}$$

hence in material form

$$\boxed{\varrho_{\rm ref}(X) = \varrho(t,\Phi(t,X))J(t,X)}\,.$$

(c) Applying the results of the preceding section, in place of the general rule (8.11) now the representation in material form

$$\frac{d}{dt}\int_U A(t,X)J(t,X)\,dV = \int_U B(t,x)J(t,X)\,dV + \int_{\partial U} \underline{W}(t,X)\cdot\underline{N}(t,X)\,dO\,,$$

is obtained by substitution, resp. the differential form

$$\boxed{\frac{\partial}{\partial t}(A\,J) = B\,J + \operatorname{Div}\underline{W}}\,. \tag{8.20}$$

Then the conservation law of momentum reads in material form

$$\begin{aligned}\frac{d}{dt}\int_U \varrho(t,\Phi(t,X))\underline{V}(t,X)J(t,X)\,dV &= \int_U \varrho_{\rm ref}\frac{\partial}{\partial t}\underline{V}(t,X)\,dV\\ = \int_U \varrho_{\rm ref}\underline{F}(t,X)\,dV + \int_{\partial U} \mathbf{T}(t,X)\cdot\underline{N}(t,X)\,dO\,,&\end{aligned} \tag{8.21}$$

and, after application of the divergence theorem, in local form

$$\boxed{\varrho_{\rm ref}(X)\frac{\partial}{\partial t}\underline{V}(t,X) - \operatorname{Div}\mathbf{T}(t,X) = \varrho_{\rm ref}(X)\underline{F}(t,X)}\,. \tag{8.22}$$

(d) The conservation law of angular momentum is the same in material form and in space-related form, namely $\mathbf{t} = \mathbf{t}^T$, resp.

$$\mathbf{S}(t,X) = \mathbf{S}^T(t,X)\,.$$

(e) Regarding Lemma 8.6, the conservation law of energy reads in material form

$$\varrho_{\rm ref}\left(\frac{\partial E}{\partial t} + (\operatorname{grad} E)\cdot\underline{V}\right) - \operatorname{Grad}\underline{V} : \mathbf{T} - \operatorname{Div}\underline{Q} = \varrho_{\rm ref}\,R\,. \tag{8.23}$$

But observe

$$\operatorname{Grad}\underline{V} : \mathbf{T} = \operatorname{Grad}\underline{V} : [\nabla\Phi\,\mathbf{S}] = \operatorname{trace}\left([\operatorname{Grad}\underline{V}]^T\nabla\Phi\,\mathbf{S}\right) = (\nabla\Phi)^T\operatorname{Grad}\underline{V} : \mathbf{S}\,,$$

hence because of the symmetry of $\mathbf{S}$

$$\operatorname{Grad}\underline{V} : \mathbf{T} = \frac{1}{2}\left[(\nabla\Phi)^T\operatorname{Grad}\underline{V} + (\operatorname{Grad}\underline{V})^T\nabla\Phi\right] : \mathbf{S} =: \widetilde{\mathbf{E}}(\underline{V}) : \mathbf{S}$$

therefore (8.22) yields

$$\boxed{\varrho_{\text{ref}}\left(\frac{\partial E}{\partial t} + (\operatorname{grad} E)\cdot\underline{V}\right) - \widetilde{\mathbf{E}}(\underline{V}) : \mathbf{S} - \operatorname{Div}\underline{Q} = \varrho_{\text{ref}}\,R} \tag{8.24}$$

where $\mathbf{S}$ and $\widetilde{\mathbf{E}}$ are *symmetric* tensors.

(f) Also the second law of thermodynamics is transformed by applying Theorem 8.9 and reads in material form

$$\boxed{\varrho_{\text{ref}}\left(S\frac{\partial\Theta}{\partial t} + \frac{\partial\Psi}{\partial t}\right) - \widetilde{\mathbf{E}} : \mathbf{S} + \frac{1}{\Theta}\operatorname{Grad}\Theta\cdot\underline{Q} \leq 0}\,. \tag{8.25}$$

(g) Let us now write $x = X + \underline{u}$ where $\underline{u}$ is a (small) *displacement* then

$$\nabla\Phi(t,X) = I + \nabla\underline{u}(t,X)\,,\quad \det\nabla\Phi(t,X) = 1 + \operatorname{trace}\nabla\underline{u}(t,X) + \text{h.o.t.}\,,$$

and

$$\begin{aligned}[\operatorname{Cof}\nabla\Phi(t,X)]^T &= \det(\nabla\Phi(t,X))[\nabla\Phi(t,X)]^{-1}\\ &= 1 + \operatorname{trace}\nabla\underline{u}(t,X) + \nabla\underline{u}(t,X) + \text{h.o.t.}\,.\end{aligned}$$

In *linear* theory of elasticity terms of higher order in $\underline{u}$ are neglected and

$$\begin{aligned}\mathbf{T}(t,X) &= \quad \nabla\Phi(t,X)\mathbf{S}(t,X) &&\simeq \mathbf{S}(t,X)\\ \mathbf{E}(\underline{V}) &= \quad \frac{1}{2}\left[\operatorname{Grad}\underline{V} + (\operatorname{Grad}\underline{V})^T\right] &&\simeq \widetilde{\mathbf{E}}(\underline{V})\\ \mathbf{S}(t,X;\underline{u}) &= \mathbf{C}(t,X)\mathbf{E}(\underline{u}(t,X))\,,\ \mathbf{C} = \mathbf{C}^T\,,\end{aligned} \tag{8.26}$$

are taken approximatively where the last equation is called *linear law of material* with the *elasticity matrix* $\mathbf{C}$. Then we obtain the *linearized* equations

$$\boxed{\begin{aligned}&\varrho_{\text{ref}}(X) &&= \varrho(t,\Phi(t,X))J(t,X)\\ &\varrho_{\text{ref}}(X)\frac{\partial}{\partial t}\underline{V}(t,X) - \operatorname{Div}\mathbf{S}(t,X;\underline{u}(t,X)) &&= \varrho_{\text{ref}}(X)\underline{F}(t,X)\\ &\varrho_{\text{ref}}(X)\left(\frac{\partial E}{\partial t}(t,X) + (\operatorname{Grad}E(t,X))\cdot\underline{V}(t,X)\right)\\ &-\mathbf{E}(\underline{V}(t,X)) : \mathbf{S}(t,X;\underline{u}(t,X)) - \operatorname{Div}\underline{Q}(t,X) &&= \varrho_{\text{ref}}(X)\,R(t,X)\end{aligned}}\,. \tag{8.27}$$

The corresponding equations in *fluids* are linearized in a similar way, but the stress tensor **S** then depends on the velocity $\underline{V}$.

It is customary to use the stress tensor (8.19) resp. its linearization in literature but otherwise to start out from the conservation equations (8.15) in space-related form; if necessary, it is integrated over the domain Ω in material space (reference space).

(h) Variational Problem We multiply the second equation of (8.27) by a test function $\underline{W}(t, X) =: \partial\Phi(t, X)$ and integrate the result over the domain Ω. Then, in a trivial way,

$$\int_\Omega \underline{W}(t,X) \cdot \Big[\varrho_{\rm ref}(X)\frac{\partial}{\partial t}\underline{V}(t,X) - \operatorname{Div} \mathbf{S}(t,X;\underline{u}(t,X)) - \varrho_{\rm ref}(X)\underline{F}(t,X)\Big]\, dV = 0\,. \tag{8.28}$$

An application of formula (1.21),

$$\int_V \left[\underline{v} \cdot \operatorname{div} T + \operatorname{grad} \underline{v} : T\right] dV = \oint_{\partial V} \underline{v} \cdot T\, \underline{n}\, dO\,,$$

writing $\underline{V}_t = \underline{\ddot{u}}$ yields

$$\int_\Omega \left[\varrho_{\rm ref}\, \underline{W} \cdot \underline{\ddot{u}} + \mathbf{E}(\underline{W}) : \mathbf{S}(\underline{u})\right] dV - \int_{\partial\Omega} \underline{W} \cdot \mathbf{S}(\underline{u})\, \underline{N}\, dO = \int_\Omega \varrho_{\rm ref}\, \underline{W} \cdot \underline{F}\, dV \tag{8.29}$$

in no longer entirely trivial way. This GALERKIN *form* constitutes the basis for numerical treatment; see also § 1.11**(e)**. But, as mentioned repeatedly, the test functions may not be chosen arbitrarily but have to regard the *essential boundary conditions* of $\underline{u}$ in the concrete problem. Note also that the requirements on smoothness are weakened in passing from (8.28) to (8.29), resp. have to be required additionally in the other direction.

(i) Extremal Principle If (8.29) is stationary, i.e., independent of time t, then this equation is *variational equation* of the extremal problem $\forall\ \underline{u} \in \underline{u}_0 + \mathcal{V}$

$$\boxed{\mathcal{E}(\underline{u}) := \int_\Omega \frac{1}{2}\mathbf{E}(\underline{u}) : \mathbf{S}(\underline{u})\, dV - \int_\Omega \underline{u} \cdot \underline{K}\, dV - \int_{\partial\Omega} \underline{u} \cdot \underline{G}\, dO = \text{extr!}}\ . \tag{8.30}$$

$\underline{K}$ and $\underline{G}$ denote here the pulled back non-specific volume-force and surface-force densities. The function $\underline{u}_0$ and the vector space $\mathcal{V}$ are problem-related and are to be specified more precisely in the individual case.

(j) Hamilton's Principle comes into play here by integration of (8.29) w.r.t. time and ensuing simple partial integration; but stress tensor **S** and volume-force density $\underline{F}$ must not depend explicitely on time (*conservative system*). Permutation of integration in (8.29) yields at first

$$\int_\Omega \int_0^\tau [-\varrho_{\rm ref}\, \underline{W} \cdot \underline{\ddot{u}} - \mathbf{E}(\underline{W}) : \mathbf{S}(\underline{u}) + \varrho_{\rm ref}\, \underline{W} \cdot \underline{F}]\, dt\, dV + \int_{\partial\Omega} \int_0^\tau \underline{W} \cdot \mathbf{S}(\underline{u})\, \underline{N}\, dt\, dO = 0\,. \tag{8.31}$$

If the test functions $\underline{W}$ satisfy the boundary condition $\underline{W}(0, X) = \underline{W}(\tau, X) = 0$, partial integration leads to

$$\int_0^\tau \varrho \underline{\dot{W}} \cdot \underline{\dot{u}}\, dt = \varrho \underline{W} \cdot \underline{\dot{u}}\Big|_0^\tau - \int_0^\tau \varrho \underline{W} \cdot \underline{\ddot{u}}\, dt = -\int_0^\tau \varrho \underline{W} \cdot \underline{\ddot{u}}\, dt\,,$$

hence, by (8.31),

$$0 = \partial\mathcal{A}(\underline{u}\,; \underline{W}) := \int_0^\tau \Big[\int_\Omega \Big[\varrho_{\rm ref}\, \underline{\dot{W}} \cdot \underline{\dot{u}} - \mathbf{E}(\underline{W}) : \mathbf{S}(\underline{u}) + \varrho_{\rm ref} \underline{W} \cdot \underline{F}\Big]\, dV + \int_{\partial\Omega} \underline{W} \cdot \mathbf{S}(\underline{u})\, \underline{N}\, dO\Big]\, dt\,. \tag{8.32}$$

Let now

$$\mathcal{L}(\underline{u}) = \frac{1}{2}\int_\Omega [\varrho_{\rm ref}\, \underline{\dot{u}} \cdot \underline{\dot{u}} - \mathbf{E}(\underline{u}) : \mathbf{S}(\underline{u})]\, dV + \int_\Omega \varrho_{\rm ref}\, \underline{W} \cdot \underline{F}\, dV + \int_{\partial\Omega} \underline{u} \cdot \mathbf{S}(\underline{u})\, \underline{N}\, dO$$

be the LAGRANGE *function* and observe that $\mathbf{E}(\underline{W}) : \mathbf{S}(\underline{u}) = \mathbf{E}(\underline{u}) : \mathbf{S}(\underline{W})$ by (8.26) then (8.32) is variational equation of the *integral of action*

$$\mathcal{A}(\underline{u}, [0, \tau]) = \int_0^\tau \mathcal{L}(\underline{u})\, dt\,. \tag{8.33}$$

Theorem 8.10. *(HAMILTON's principle) Let the system be conservative then the solution $\underline{u}$ of*

$$\varrho_{\rm ref}\, \underline{\ddot{u}} - {\rm Div}\, \mathbf{S}(\underline{u}) = \varrho_{\rm ref}\, \underline{F}$$

runs between two fixed points $(t_0, \underline{u}_0)$ and $(t_1, \underline{u}_1)$ such that the integral of action $\mathcal{A}(\underline{v}, [t_0, t_1])$ has a stationary point $\underline{v} = \underline{u}$.

HAMILTON's principle is well established in physics even in its stronger form as minimum principle but a stringent quantum-mechanical proof of this strong form is still lacking (Denninger).

By using the interpretation

$$\mathcal{E}_k(\underline{u}) := \frac{1}{2}\int_\Omega \varrho_{\rm ref}\, \underline{\dot{u}} \cdot \underline{\dot{u}}\, dV \quad \text{kinetic energy}$$

$$\mathcal{E}_d(\underline{u}) := \frac{1}{2}\int_\Omega \mathbf{E}(\underline{u}) : \mathbf{S}(\underline{u})\, dV \quad \text{deformation energy}$$

$$\mathcal{E}_v(\underline{u}) := -\int_\Omega \varrho_{\rm ref}\, \underline{F} \cdot \underline{u}\, dV \quad \text{potential energy of volume forces}$$

$$\mathcal{E}_s(\underline{u}) := \int_{\partial\Omega} \underline{u} \cdot \mathbf{S}(\underline{u})\, \underline{N}\, dO \quad \text{potential energy of surface forces}$$

the LAGRANGE function obtains the form

$$\boxed{\mathcal{L} = \mathcal{E}_k - \mathcal{E}_d - \mathcal{E}_v - \mathcal{E}_s}\ .$$

cf. (Marsden). Essentially, HAMILTON's principle is equivalent here to the balance law of momentum by which the goal of the operation, namely finding a proper boundary or initial value problem is faster attained in most cases. On the other side, note that exactly the solutions of the EULER equations (4.2) with $q = \widetilde{\mathcal{L}}$ are stationary point of the integral of action.

8.5 Linear Elasticity Theory

> Hint: 1. All scalar products $\underline{v} \cdot \underline{w}$ below may be understood also as matrix product $\underline{v}\,\underline{w}$ where $\underline{v}$ on the left is a *row vector* $\underline{v}$ and $\underline{w}$ on the right is a *column vector*.
> 2. By optical reasons and to take care of the reader, some concessions to the classical notation are made in this chapter.

Applied constants:

$\nu = \dfrac{\lambda}{2(\lambda+\mu)}\,,\ 0 < \nu < 1/2$	Poisson's ratio
$E = \dfrac{\mu(3\lambda+2\mu)}{\lambda+\mu} > 0$	elasticity modul
h	thickness of plate
$\lambda = \dfrac{E\nu}{(1+\nu)(1-2\nu)}$	Lamé constant
$\mu = \dfrac{E}{2(1+\nu)}$	Lamé constant
$\widetilde{\lambda} = \dfrac{\lambda\mu}{\lambda+2\mu} = \dfrac{E\nu}{2(1-\nu^2)}$	$\widetilde{\lambda} + \mu = \dfrac{E}{2(1-\nu^2)}$
$\kappa = \dfrac{Eh^3}{12(1-\nu^2)} = \dfrac{h^3}{4}\dfrac{2}{3}(\widetilde{\lambda}+2\mu)$	plate rigidity
$\chi = \dfrac{3}{3\lambda+2\mu}$	compressibility

(a) Strain- and Stress Tensor Let $\Omega \subset \mathbb{R}^3$ be a "body" with sufficiently smooth boundary $\partial\Omega = \Gamma$, let $X = (x, y, z)$ be a point of Ω and $\underline{u} = (u, v, w) : \Omega \to \mathbb{R}^3$ a sufficiently smooth function (displacement).

The *strain tensor* is defined in *linear* elasticity theory by

$$\varepsilon := \varepsilon(\underline{u}) := \mathbf{E}(\underline{u}) = \frac{1}{2}\left[\operatorname{grad}\underline{u} + (\operatorname{grad}\underline{u})^T\right]$$

$$= \begin{bmatrix} u_x & (u_y+v_x)/2 & (u_z+w_x)/2 \\ (v_x+u_y)/2 & v_y & (v_z+w_y)/2 \\ (w_x+u_z)/2 & (w_y+v_z)/2 & w_z \end{bmatrix} =: \begin{bmatrix} \varepsilon_x & \gamma_{xy}/2 & \gamma_{xz}/2 \\ \gamma_{xy}/2 & \varepsilon_y & \gamma_{yz}/2 \\ \gamma_{xz}/2 & \gamma_{yz}/2 & \varepsilon_z \end{bmatrix}$$

where ε_x , ε_y , ε_z are the *stretches* and γ_{xy} , γ_{xz} , γ_{yz} are the *shifts*. The *stress tensor* has to be defined by a *law of material* being an *axiom* in mathematical sense. We write

$$\boldsymbol{\sigma} := \boldsymbol{\sigma}(\underline{u}) := \mathbf{S}(t, X; \underline{u}) =: \begin{bmatrix} \sigma_x & \tau_{xy} & \tau_{xz} \\ \tau_{xy} & \sigma_y & \tau_{yz} \\ \tau_{xz} & \tau_{yz} & \sigma_z \end{bmatrix}$$

where σ_x , σ_y , σ_z are the *normal stresses* and τ_{xy} , τ_{xz} , τ_{yz} the *shear stresses*. Shifts and shear stresses differ *formally* by the factor two because of VOIGT's representation in **(e)**.

(b) Extremal Problem and Variational Problem The displacement $\underline{u}$ of a solid body $(\Omega, \varrho, \boldsymbol{\sigma})$ is solution of a *extremal problem* regarding the total potential energy

$$\boxed{\forall\, \underline{v} \in \underline{u}_0 + \mathcal{U} : \int_\Omega \frac{1}{2}\boldsymbol{\varepsilon}(\underline{v}) : \boldsymbol{\sigma}(\underline{v})\, d\Omega - \int_\Omega \underline{v} \cdot \underline{k}\, d\Omega - \int_\Gamma \underline{v} \cdot \underline{g}\, d\Gamma = \min!} \tag{8.34}$$

where $\underline{k}$ is the volume-force density and $\underline{g}$ the surface-force density. The function space $\mathcal{U}$ is to be specified more precisely. By §§ 1.11, 4.1, a variational problem (EULER equation) is associated to the extremal problem (8.34) which is called *weak problem* in the sequel,

$$\boxed{\begin{gathered} \exists\, \underline{u} \in \underline{u}_0 + \mathcal{U}\ \ \forall\, \underline{v} \in \mathcal{U} : \\ \int_\Omega \boldsymbol{\varepsilon}(\underline{v}) : \boldsymbol{\sigma}(\underline{u})\, d\Omega = \int_\Omega \underline{v} \cdot \underline{k}\, d\Omega + \int_\Gamma \underline{v} \cdot \underline{g}\, d\Gamma \end{gathered}}\ . \tag{8.35}$$

This problem constitutes the weak form of the law of conservation of momentum and is the basis of numerical approach by the finite element method.

(c) Again an application of the divergence theorem leads to corresponding boundary value problem. By (1.2) and (1.21), and because $\boldsymbol{\sigma}(\underline{u})$ is symmetric

$$\begin{aligned} &\int_\Omega \boldsymbol{\varepsilon}(\underline{v}) : \boldsymbol{\sigma}(\underline{u})\, d\Omega = \int_\Omega \frac{1}{2}\left(\operatorname{grad}\underline{v} + (\operatorname{grad}\underline{v})^T\right) : \boldsymbol{\sigma}(\underline{u})\, d\Omega \\ &= \int_\Omega \operatorname{grad}\underline{v} : \boldsymbol{\sigma}(\underline{u})\, d\Omega = -\int_\Omega \underline{v} \cdot \operatorname{div}\boldsymbol{\sigma}(\underline{u})\, d\Omega + \oint_\Gamma \underline{v} \cdot \boldsymbol{\sigma}(\underline{u})\underline{n}\, d\Gamma \end{aligned}$$

if $\underline{u}$ satisfies the required smoothness. By this way one obtains the weak form

$$\int_\Omega \underline{v} \cdot [-\operatorname{div}\boldsymbol{\sigma}(\underline{u}) - \underline{k}]\, d\Omega = \oint_\Gamma \underline{v} \cdot [\underline{g} - \boldsymbol{\sigma}(\underline{u})\underline{n}]\, d\Gamma\,,\ \ \underline{u} \in \underline{u}_0 + \mathcal{U}\,, \tag{8.36}$$

where $\underline{v} \in \mathcal{U}$ is at present an arbitrary test function. We suppose that the boundary Γ of Ω is partitioned into DIRICHLET boundary and CAUCHY boundary following (9.1) and that $\underline{u}$ is specified at the boundary Γ_D by $\underline{u} = \underline{h}$.

The function u_0 in V then has to satisfy $\underline{u}_0 = \underline{h}$ on Γ_D and consequently $\underline{v} = 0$ on Γ_D in order that every variation $\underline{u} + \varepsilon\underline{v}$ is an element of the affine space $u_0 + V$ for sufficiently small $|\varepsilon|$; cf. §§ 1.11, 9.1. On choosing first $\underline{v}$ arbitrary with $\underline{v} = 0$ on the entire boundary it follows that the square-bracketed part on the left side must be zero. Thereafter, on choosing $\underline{v}$ on Γ_C arbitrary the square-bracketed term on the right side must be zero. Together we obtain by this way the *boundary value problem*

$$\boxed{\begin{aligned} -\operatorname{div}\boldsymbol{\sigma}(\underline{u}) &= \underline{k} \text{ in } \Omega \\ \underline{u} &= \underline{h} \text{ on } \Gamma_D \\ \boldsymbol{\sigma}(\underline{u})\underline{n} &= \underline{g} \text{ on } \Gamma_C \end{aligned}} \tag{8.37}$$

where the first equation represents the local law of conservation of momentum. *Fazit:* The solution of the weak problem (8.35) fulfils the dynamic boundary condition $\boldsymbol{\sigma}(\underline{u})\underline{n} = \underline{g}$ on boundary Γ_C. Contrary to the DIRICHLET conditions, this boundary condition is regarded by finite element solutions only in passing to the limit of infinitesimal approximation.

Further combinations of boundary conditions are deduced if the boundary term of the waek form is partitioned in the same way as in 8.10**(b)**.

(d) St Venant-Kirchhoff Material of linear elasticity has by definition the stress tensor

$$\boxed{\boldsymbol{\sigma}(\underline{u}) := \mathbf{S}(\underline{u}) = 2\mu\,\boldsymbol{\varepsilon}(\underline{u}) + \lambda \operatorname{trace}(\boldsymbol{\varepsilon}(\underline{u}))\boldsymbol{\delta}} \tag{8.38}$$

(HOOKE's law) where μ and λ are the LAMÉ constants; cf. (Ciarlet93), p. 130. We also may write instead

$$\boldsymbol{\sigma} = \frac{E}{1+\nu}\left(\boldsymbol{\varepsilon} + \frac{\nu}{1-2\nu}\operatorname{trace}(\boldsymbol{\varepsilon})\,\boldsymbol{\delta}\right) \tag{8.39}$$

which allows the inversion

$$\boxed{\boldsymbol{\varepsilon} = \frac{1+\nu}{E}\boldsymbol{\sigma} - \frac{\nu}{E}\operatorname{trace}(\boldsymbol{\sigma})\,\boldsymbol{\delta}} \tag{8.40}$$

by $\operatorname{trace}\boldsymbol{\delta} = 3$. Recalling

$$\boldsymbol{\varepsilon} : \boldsymbol{\sigma} = 2\mu\,\boldsymbol{\varepsilon} : \boldsymbol{\varepsilon} + \lambda(\operatorname{trace}\boldsymbol{\varepsilon})\,\boldsymbol{\varepsilon} : \boldsymbol{\delta}\,, \quad \boldsymbol{\varepsilon} : \boldsymbol{\delta} = \operatorname{trace}(\boldsymbol{\varepsilon}) = \operatorname{div}\underline{u}\,,$$

the associated extremal problem (8.34) obtains the form

$$\boxed{\begin{gathered} \int_\Omega \frac{1}{2}\left[2\mu\,\boldsymbol{\varepsilon}(\underline{v}) : \boldsymbol{\varepsilon}(\underline{v}) + \lambda\,(\operatorname{div}\underline{v})^2\right] d\Omega - \int_\Omega \underline{v}\cdot\underline{k}\,d\Omega - \int_{\Gamma_C} \underline{v}\cdot\underline{g}\,d\Gamma = \min! \\ \forall\,\underline{v} \in \mathcal{U} \text{ such that } \underline{v} = \underline{h} \text{ on } \Gamma_D \end{gathered}} \tag{8.41}$$

Further, recall

$$\begin{aligned} \operatorname{div}(\operatorname{div}(\underline{u})\boldsymbol{\delta}) &= [\operatorname{grad}\operatorname{div}\underline{u}]^T \quad \text{(column vector)} \\ \operatorname{div}([\operatorname{grad}\underline{u}]^T) &= [\operatorname{grad}\operatorname{div}\underline{u}]^T \quad \text{(column vector)}\,, \end{aligned}$$

hence, writing $\operatorname{grad}\varphi$ as column vector,

$$\begin{aligned} \operatorname{div}\boldsymbol{\sigma}(\underline{u}) &= \mu\,\left[\operatorname{div}\operatorname{grad}\underline{u} + \operatorname{div}[\operatorname{grad}\underline{u}]^T\right] + \lambda\,[\operatorname{grad}\operatorname{div}\underline{u}]^T \\ &= \mu\,\Delta\underline{u} + (\lambda+\mu)\operatorname{grad}\operatorname{div}\underline{u}\,. \end{aligned} \tag{8.42}$$

Substitution into (8.37) yields the boundary value problem

$$\boxed{\begin{aligned} -\mu\Delta\underline{u} - (\lambda+\mu)\operatorname{grad}\operatorname{div}\underline{u} &= \underline{k} \ \text{ in } \Omega \\ \underline{u} &= \underline{h} \ \text{ on } \Gamma_D \\ [2\mu\,\boldsymbol{\varepsilon}(\underline{u}) + \lambda(\operatorname{div}\underline{u})\boldsymbol{\delta}]\underline{n} &= \underline{g} \ \text{ on } \Gamma_C \end{aligned}}\,. \tag{8.43}$$

(e) Notation after (Voigt). Stress tensor and strain tensor are both determined by six components respectively because of symmetry, therefore a *strain vector* $\underline{\varepsilon} \in \mathbb{R}^6$ and a *stress vector* $\underline{\sigma} \in \mathbb{R}^6$ may be associated to the tensor $\boldsymbol{\varepsilon}$ resp. to the tensor $\boldsymbol{\sigma}$,

$$\boxed{\begin{aligned} \underline{\varepsilon} &= [\varepsilon_x,\, \varepsilon_y,\, \varepsilon_z,\, \gamma_{xy},\, \gamma_{xz},\, \gamma_{yz}]^T \\ \underline{\sigma} &= [\sigma_x,\, \sigma_y,\, \sigma_z,\, \tau_{xy},\, \tau_{xz},\, \tau_{yz}]^T \end{aligned}}$$

such that $\boldsymbol{\varepsilon} : \boldsymbol{\sigma} = \underline{\varepsilon}\cdot\underline{\sigma}$. By this way the uncomfortable scalar product of tensors may be replaced by a simple scalar product of vectors. Furthermore, a *operator matrix* D is introduced conveniently and a symmetric *elasticity matrix* C by

$$\underline{\sigma} = C\underline{\varepsilon}\,, \quad \underline{\varepsilon} = D\underline{u}\,, \quad D = \begin{bmatrix} \partial_x & 0 & 0 \\ 0 & \partial_y & 0 \\ 0 & 0 & \partial_z \\ \partial_y & \partial_x & 0 \\ \partial_z & 0 & \partial_x \\ 0 & \partial_z & \partial_y \end{bmatrix}\,. \tag{8.44}$$

These notions allow a more lucid representation

$$\boxed{\boldsymbol{\varepsilon}(\underline{v}) : \boldsymbol{\sigma}(\underline{u}) = \underline{\varepsilon}(\underline{v})\cdot\underline{\sigma}(\underline{u}) = \underline{v}^T D^T C D\underline{u}}\,, \tag{8.45}$$

which is most suitable for numerical approach.

By (8.35) and (8.41) we now obtain the representation

$$\int_\Omega \underline{v}^T D^T C D\underline{u}\,d\Omega = \int_\Omega \underline{v}\cdot\underline{k}\,d\Omega + \int_{\Gamma_C} \underline{v}\cdot\underline{g}\,d\Gamma \ \ \forall\,\underline{v}\in V \text{ such that } \underline{v} = 0 \text{ on } \Gamma_D$$

where C is the *elasticity matrix* and its inverse C^{-1} the *compliance matrix*,

$$C = \frac{E}{(1+\nu)(1-2\nu)}\begin{bmatrix} 1-\nu & \nu & \nu & 0 & 0 & 0 \\ \nu & 1-\nu & \nu & 0 & 0 & 0 \\ \nu & \nu & 1-\nu & 0 & 0 & 0 \\ 0 & 0 & 0 & \frac{1}{2}(1-2\nu) & 0 & 0 \\ 0 & 0 & 0 & 0 & \frac{1}{2}(1-2\nu) & 0 \\ 0 & 0 & 0 & 0 & 0 & \frac{1}{2}(1-2\nu) \end{bmatrix},$$

$$C^{-1} = \frac{1}{E}\begin{bmatrix} 1 & -\nu & -\nu & 0 & 0 & 0 \\ -\nu & 1 & -\nu & 0 & 0 & 0 \\ -\nu & -\nu & 1 & 0 & 0 & 0 \\ 0 & 0 & 0 & 2(1+\nu) & 0 & 0 \\ 0 & 0 & 0 & 0 & 2(1+\nu) & 0 \\ 0 & 0 & 0 & 0 & 0 & 2(1+\nu) \end{bmatrix}.$$

The boundary problem (8.37) may also be derived via (8.45) for checking. Using the matrices

$$D_1 = \begin{bmatrix} 1 & 0 & 0 \\ 0 & 0 & 0 \\ 0 & 0 & 0 \\ 0 & 1 & 0 \\ 0 & 0 & 1 \\ 0 & 0 & 0 \end{bmatrix}, \quad D_2 = \begin{bmatrix} 0 & 0 & 0 \\ 0 & 1 & 0 \\ 0 & 0 & 0 \\ 0 & 0 & 1 \\ 0 & 0 & 0 \\ 0 & 0 & 1 \end{bmatrix}, \quad D_3 = \begin{bmatrix} 0 & 0 & 0 \\ 0 & 0 & 0 \\ 0 & 0 & 1 \\ 0 & 0 & 0 \\ 1 & 0 & 0 \\ 0 & 1 & 0 \end{bmatrix}.$$

we have

$$\underline{\varepsilon} = D_1[\operatorname{grad} u]^T + C\, D_2[\operatorname{grad} v]^T + C\, D_3[\operatorname{grad} w]^T,$$

hence

$$\begin{aligned} \underline{\varepsilon}\cdot\underline{\sigma} &= \left[(\operatorname{grad} u)D_1^T + (\operatorname{grad} v)D_2^T + (\operatorname{grad} w)^T D_3^T\right]\underline{\sigma} \\ &= \left[(\operatorname{grad} u)\underline{\sigma}_1 + (\operatorname{grad} v)\underline{\sigma}_2 + (\operatorname{grad} w)\underline{\sigma}_3\right] \end{aligned}$$

$$\underline{\sigma}_1 = [\sigma_x\,,\ \tau_{xy}\,,\ \tau_{xz}]^T,\quad \underline{\sigma}_2 = [\tau_{xy}\,,\ \sigma_y\,,\ \tau_{yz}]^T,\quad \underline{\sigma}_3 = [\tau_{xz}\,,\ \tau_{yz}\,,\ \sigma_z]^T.$$

Consequently

$$\underline{\varepsilon}\cdot\underline{\sigma} = [\operatorname{grad} u\cdot\underline{\sigma}_1 + \operatorname{grad} v\cdot\underline{\sigma}_2 + \operatorname{grad} w\cdot\underline{\sigma}_3]\,.$$

Finally, the relation

$$\int_\Omega \operatorname{grad}\varphi\cdot\underline{w}\,d\Omega = -\int_\Omega \varphi\,\operatorname{div}\underline{w}\,d\Omega + \oint_\Gamma \varphi\,\underline{w}\cdot\underline{d\Gamma}\,,\ \ \underline{d\Gamma} := \underline{n}\,d\Gamma\,,$$

yields after suming up

$$\boxed{\int_\Omega \underline{\varepsilon}(\underline{v})\cdot\underline{\sigma}(\underline{u})\,d\Omega = -\int_\Omega \underline{v}\cdot\operatorname{div}\boldsymbol{\sigma}(\underline{u})\,d\Omega + \oint_\Gamma \underline{v}\cdot\boldsymbol{\sigma}(\underline{u})\underline{n}\,d\Gamma}\ .$$

8.6 Discs

Let $\widetilde{\Omega}(x,y,z) = \Omega(x,y) \times [-h/2\,,\, h/2]$ be the geometric shape of a disc (flat body with constant thickness h) where the boundary is partitioned again by $\Gamma = \Gamma_D \oplus \Gamma_C$ as proposed in (9.1); let the thickness h of the disc be small against the diameter.

(a) Plane Stress A deformation in z-direction is allowed, but external forces $\underline{k}$ and $\underline{g}$ act only in (x,y)-plane hence have no z-component. Here we suppose that

$$\boxed{\begin{array}{c} u = u(x,y)\,, \quad v = v(x,y)\,, \quad w(x,y,z) = z\varepsilon_{33}(x,y) \\ \sigma_{3i} = \sigma_{i3} = 0\,, \quad i = 1,2,3 \end{array}}$$

does hold for the vector of displacement $\underline{u} = (u,v,w)$ and the stress tensor $\boldsymbol{\sigma}$. Then the general linear law of material (8.40) becomes the law of material for plane stress. Note first that $\varepsilon_{i3} = \varepsilon_{3i} = 0\,,\ i = 1,2$, is a direct consequence and that $\sigma_{33} = 0$ yields by resolution

$$\varepsilon_{33} = -\frac{\nu}{1-\nu}(\varepsilon_{11} + \varepsilon_{22})\,.$$

Now ε_{33} may be cancelled out and we obtain a law of material in $\mathbb{R}^2$ for the vector $\underline{u} = (u,v)$

$$\begin{aligned} \boldsymbol{\sigma} &= \frac{E}{1+\nu}\left[\boldsymbol{\varepsilon} + \frac{\nu}{1-\nu}(\varepsilon_{11} + \varepsilon_{22})\boldsymbol{\delta}\right] = 2\mu\,\boldsymbol{\varepsilon} + 2\widetilde{\lambda}\ \mathrm{trace}(\boldsymbol{\varepsilon})\boldsymbol{\delta} \\ \boldsymbol{\varepsilon} &= \frac{1+\nu}{E}\left[\boldsymbol{\sigma} - \frac{\nu}{1+\nu}(\sigma_{11} + \sigma_{22})\boldsymbol{\delta}\right]\,. \end{aligned} \tag{8.46}$$

The space variable z does no longer occur therefore this equation and (8.38) leads to a boundary problem for $\underline{u} = (u,v)$ in the same way as in (8.43)

$$\boxed{\begin{aligned} -\mu\Delta\underline{u} - (2\widetilde{\lambda} + \mu)\,\mathrm{grad}\,\mathrm{div}\,\underline{u} &= \underline{k} \ \text{ in } \Omega \\ \underline{u} &= \underline{h} \ \text{ on } \Gamma_D \\ [2\mu\,\boldsymbol{\varepsilon}(\underline{u}) + 2\widetilde{\lambda}(\mathrm{div}\,\underline{u})\boldsymbol{\delta}]\underline{n} &= \underline{g} \ \text{ on } \Gamma_C \end{aligned}}\,. \tag{8.47}$$

Further combinations of boundary conditions are possible here, too. By introducing the vectors

$$\underline{\varepsilon} = [\varepsilon_x, \varepsilon_y, \gamma_{xy}]^T, \quad \underline{\sigma} = [\sigma_x, \sigma_y, \tau_{xy}]^T,$$

we have again $\boldsymbol{\varepsilon} : \boldsymbol{\sigma} = \underline{\varepsilon} \cdot \underline{\sigma}$, and obtain by (8.46)

$$\underline{\sigma} = \frac{E}{1-\nu^2}\begin{bmatrix} 1 & \nu & 0 \\ \nu & 1 & 0 \\ 0 & 0 & (1-\nu)/2 \end{bmatrix}\underline{\varepsilon} =: C_S\underline{\varepsilon}\,. \tag{8.48}$$

The matrix C_S is the elasticity matrix of plane stress. Thereby the conservation law of momentum leads to the total energy of a disc in plane stress after integration over the space variable z,

$$\mathcal{E}(\underline{u}) = h\int_\Omega\left[\frac{1}{2}\,\varepsilon(\underline{u}):\sigma(\underline{u}) - \underline{k}\cdot\underline{u}\right]d\Omega - h\int_{\Gamma_C}\underline{g}\cdot\underline{u}\,d\Gamma - \sum_i \underline{F}_i\cdot\underline{u} \quad (8.49)$$

where still some pointwise external forces $\underline{F}_i$ are added. More detailed, for $\underline{u} = (u, v)$,

$$\begin{aligned}\varepsilon(\underline{u}):\sigma(\underline{u}) &= \underline{\varepsilon}\cdot\underline{\sigma} = [\varepsilon_x, \varepsilon_y, \gamma_{xy}]\,\frac{E}{1-\nu^2}\begin{bmatrix}1 & \nu & 0\\ \nu & 1 & 0\\ 0 & 0 & (1-\nu)/2\end{bmatrix}\begin{bmatrix}\varepsilon_x\\ \varepsilon_y\\ \gamma_{xy}\end{bmatrix}\\ &= \frac{E}{1-\nu^2}\left[\varepsilon_x^2 + 2\nu\varepsilon_x\varepsilon_y + \varepsilon_y^2 + \frac{1}{2}(1-\nu)\gamma_{xy}^2\right]\\ &= \frac{E}{1-\nu^2}\left[u_x^2 + 2\nu u_x v_y + v_y^2 + \frac{1}{2}(1-\nu)(u_y+v_x)^2\right]\\ &= \underline{u}^T D^T C_S D\underline{u}\end{aligned} \quad (8.50)$$

where C_S is the elasticity matrix of (8.48) and D is the operator matrix,

$$D = \begin{bmatrix}\partial_x & 0\\ 0 & \partial_y\\ \partial_y & \partial_x\end{bmatrix}.$$

(b) Plane Strain A deformation in z-direction is *not* possible here. It is supposed that external forces on the disc do prevent this. Also we suppose here

$$\boxed{\begin{array}{c} u = u(x,y)\,,\quad v = v(x,y)\,,\quad w(x,y) = 0\\ \varepsilon_{3i} = \varepsilon_{i3} = 0\,,\quad i = 1,2,3\end{array}}\,.$$

Then the general law of material (8.39) leads to the law of material for plane strain as well. Note that $\sigma_{33} = \nu(\sigma_{11} + \sigma_{22})$ because $\varepsilon_{33} = 0$. Therefore σ_{33} may be eliminated and we obtain for $\underline{u} = (u, v)$:

$$\begin{aligned}\sigma &= \frac{E}{1+\nu}\left[\varepsilon + \frac{\nu}{1-2\nu}(\varepsilon_{11}+\varepsilon_{22})\delta\right] = 2\mu\varepsilon + \lambda\,\mathrm{trace}(\varepsilon)\delta\\ \varepsilon &= \frac{1+\nu}{E}\left[\sigma - \nu(\sigma_{11}+\sigma_{22})\delta\right].\end{aligned} \quad (8.51)$$

The associated boundary problem reads:

$$\boxed{\begin{aligned}-\mu\Delta\underline{u} - (\lambda+\mu)\,\mathrm{grad}\,\mathrm{div}\,\underline{u} &= \underline{k} \text{ in } \Omega\\ \underline{u} &= \underline{h} \text{ on } \Gamma_D\\ [2\mu\varepsilon(\underline{u}) + \lambda(\mathrm{div}\,\underline{u})\delta]\underline{n} &= \underline{g} \text{ on } \Gamma_C\end{aligned}}\,, \quad (8.52)$$

and the law of material for stress and strain vector is by (8.51)

$$\underline{\sigma} = \frac{E}{(1+\nu)(1-2\nu)}\begin{bmatrix} 1-\nu & \nu & 0 \\ \nu & 1-\nu & 0 \\ 0 & 0 & (1-2\nu)/2 \end{bmatrix} \underline{\varepsilon} =: C_V \underline{\varepsilon}\,.$$

Again the total energy of a disc under plane strain is found after integration over z and has the same form as (8.49) with elasticity matrix C_S replaced by the matrix C_V. Let $\underline{u} = (u, v)$ and let D be the operator matrix of 8.50) again. Then

$$\begin{aligned}
\boldsymbol{\varepsilon} : \boldsymbol{\sigma} = \underline{\varepsilon} \cdot \underline{\sigma} &= [\varepsilon_x, \varepsilon_y, \gamma_{xy}] \frac{E}{(1+\nu)(1-2\nu)}\begin{bmatrix} 1-\nu & \nu & 0 \\ \nu & 1-\nu & 0 \\ 0 & 0 & (1-2\nu)/2 \end{bmatrix}\begin{bmatrix} \varepsilon_x \\ \varepsilon_y \\ \gamma_{xy} \end{bmatrix} \\
&= \frac{E}{(1+\nu)(1-2\nu} \left[(1-\nu)\varepsilon_x^2 + 2\nu\varepsilon_x\varepsilon_y + (1-\nu)\varepsilon_y^2 + \frac{1}{2}(1-2\nu))\gamma_{xy}^2\right] \\
&= \frac{E}{(1+\nu)(1-2\nu)} \left[(1-\nu)u_x^2 + 2\nu u_x v_y + (1-\nu)v_y^2 + \frac{1-2\nu}{2}(u_y + v_x)^2\right] \\
&= \underline{u}^T D^T C_V D \underline{u}\,.
\end{aligned} \tag{8.53}$$

8.7 Kirchhoff's Plate

(a) Extremal Problem and Variational Problem Let again $\widetilde{\Omega}(x, y, z) = \Omega(x, y) \times [-h/2, h/2]$ and let h be small against the diameter of Ω. In a point (x, y) of the middle plane $z = 0$, let $\mathcal{N}_x$ be the *image* of the cross-section perpendicular to the x-axis and $\mathcal{N}_y$ the image of the cross-section perpendicular to the y-axis. Then the intersection of $\mathcal{N}_x$ and the (x, z)-plane has an angle $\varphi_1(x, y)$ relative to the z-axis and the intersection of $\mathcal{N}_y$ and the (y, z)-plane has an angle $\varphi_2(x, y)$ relative to the z-axis, where we suppose that $\mathcal{N}_x$ and $\mathcal{N}_y$ are planes again. Let $\underline{u} = (u_1\,, u_2\,, u_3)$ denote the displacement. Then we have by this assumption

$$\boxed{u_1 = -z\,\varphi_1(x, y)\,,\ \ u_2 = -z\,\varphi_2(x, y)\,,\ \ u_3 = w(x, y)}\ ; \tag{8.54}$$

cf. the considerations in § 7.1. It is now supposed in KIRCHHOFF's plate or *shear-rigid* plate that $\mathcal{N}_x$ and $\mathcal{N}_y$ remain perpendicular to the middle plane before and after deformation. Moreover, it is supposed that

$$\sigma_{33} \equiv \sigma_z = 0 \ \text{ und } \ k^i = 0\,,\ \ i = 1, 2\,,\ \ k^3 =: f\,, \tag{8.55}$$

for the volume-force densities $\underline{k}$ since h is small, where the last condition says that no bending forces appear in (x, y)-direction. Recalling that

$$\tan\varphi_1 = w_x \implies \varphi_1 = \arctan w_x = w_x + \text{h.o.t.}\,,$$

we may write $w_x = \varphi_1$ and $w_y = \varphi_2$ in first order approximation and obtain

$$u_1 = -z\,w_x(x,y)\,,\ \ u_2 = -z\,x_y(x,y)\,,\ \ u_3 = w(x,y) \tag{8.56}$$

by (8.54). Thereby $\sigma_{13} = \tau_{xz} = 0\,,\ \sigma_{23} = \tau_{yz} = 0$ by (8.44), and altogether

$$\begin{aligned}
&\varepsilon_x \ = u_{1,x} = -z\,w_{xx}, && \varepsilon_y \ = u_{2,y} = -z\,w_{yy},\ \varepsilon_z \ = 0,\\
&\gamma_{xy} = u_{1,y} + u_{2,x} = -2zw_{xy}, && \gamma_{yz} = u_{2,z} + u_{3,y} = 0,\ \gamma_{zx} = u_{3,x} + u_{1,z} = 0\,.
\end{aligned}$$

So we may write with slight modification

$$z\boldsymbol{\varepsilon}(w) = -z\begin{bmatrix} w_{xx} & 2w_{xy}\\ 2w_{xy} & w_{yy}\end{bmatrix}\,,\ \ z\underline{\varepsilon}(w) = -z[w_{xx}\,,\, w_{yy}\,,\, w_{xy}]^T\,.$$

The general law of material (8.38) applied to this strain tensor yields *formally* the same law of material as in plane stress

$$z\boldsymbol{\sigma}(w) = \frac{zE}{1+\nu}\left[\boldsymbol{\varepsilon}(w) + \frac{\nu}{1-\nu}\,\text{trace}(\boldsymbol{\varepsilon}(w))\boldsymbol{\delta}\right] = z\left[2\mu\boldsymbol{\varepsilon}(w) + 2\widetilde{\lambda}\,\text{trace}(\boldsymbol{\varepsilon}(w))\boldsymbol{\delta}\right]\,, \tag{8.57}$$

hence

$$\underline{\sigma}(w) = zC_S\underline{\varepsilon}(w) = -zC_S Dw\,,\ \ D = [\partial_{xx}\,,\,\partial_{yy}\,,\,\partial_{xy}]^T$$

by (8.48), resp.

$$z^2\boldsymbol{\varepsilon}(v) : \boldsymbol{\sigma}(w) = z^2\underline{\varepsilon}(v)\cdot\underline{\sigma}(w) = z^2 vD^TC_S Dw\,.$$

By integration w.r.t. z from $-h/2$ to $h/2$

$$\int_{\widetilde{\Omega}} z^2\boldsymbol{\varepsilon}(v) : \boldsymbol{\sigma}(w)\,d\widetilde{\Omega} = \frac{h^3}{12}\int_\Omega \boldsymbol{\varepsilon}(v) : \boldsymbol{\sigma}(w)\,d\Omega \equiv \frac{h^3}{12}\int_\Omega \underline{\varepsilon}(v)\cdot\underline{\sigma}(w)\,d\Omega\,.$$

The general extremal problem for KIRCHHOFF's plate now reads

$$\boxed{\begin{aligned}
\mathcal{E}(w) = &\int_\Omega \left[\frac{1}{2}\frac{h^3}{12}\,\boldsymbol{\varepsilon}(w) : \boldsymbol{\sigma}(w) - h\,w\,f\right]\,d\Omega\\
&+ h\oint_\Gamma \left(\frac{1}{2}\alpha\,w^2 + \beta w\right)\,d\Gamma + h\oint_\Gamma \left(\frac{1}{2}\gamma\,w_n^2 + \delta w_n\right)\,d\Gamma\\
&= \min!
\end{aligned}} \tag{8.58}$$

where $\alpha\,,\,\beta\,,\,\gamma\,,\,\delta : \Gamma \ni s \to \mathbb{R}\,, \alpha(s) \geq 0\,,\ \gamma(s) \geq 0$. The associated variational problem reads for present

$$\begin{aligned}
&\frac{h^3}{12}\int_\Omega \boldsymbol{\varepsilon}(v) : \boldsymbol{\sigma}(w)\,d\Omega \equiv \frac{h^3}{12}\int_\Omega \underline{\varepsilon}(v) : \underline{\sigma}(w)\,d\Omega\\
&= h\int_\Omega v\,f\,d\Omega - h\oint_\Gamma v(\alpha w + \beta)\,d\Gamma - h\oint_\Gamma v_n(\gamma w_n + \delta)\,d\Gamma\,,
\end{aligned}$$

where $v = v(x, y) \in \mathbb{R}$ is a test function (virtual displacement), and suitable boundary conditions are to be specified.

(b) Transformation In detail

$$\sigma_x = \frac{E}{1+\nu}\left[-zw_{xx} + \frac{\nu}{1-\nu}(-z(w_{xx} + w_{yy}))\right] = -\frac{Ez}{1-\nu^2}\left[w_{xx} + \nu w_{yy}\right],$$

$$\sigma_y = -\frac{Ez}{1-\nu^2}[\nu w_{xx} + w_{yy}], \quad \tau_{xy} = -\frac{Ez}{1-\nu^2} w_{xy},$$

hence

$$\begin{aligned}&\int_{\tilde{\Omega}} \underline{\sigma}(w)\cdot\underline{\varepsilon}(w)\, d\tilde{\Omega} \\ &= \frac{E}{1-\nu^2}\int_{\tilde{\Omega}} z^2\left[(w_{xx} + \nu w_{yy})w_{xx} + (\nu w_{xx} + w_{yy})w_{yy} + 2(1-\nu)w_{xy}^2\right] d\Omega \\ &= \frac{h^3}{12}\int_\Omega \frac{E}{1-\nu^2}\left[w_{xx}^2 + 2\nu w_{xx}w_{yy} + w_{yy}^2 + 2(1-\nu)w_{xy}^2\right] d\Omega\end{aligned}$$

or

$$\boxed{\frac{h^3}{12}\int_\Omega \underline{\sigma}(w)\cdot\underline{\varepsilon}(w)\, d\Omega = \kappa\int_\Omega \left[\Delta^2 w + 2(1-\nu)(w_{xy}^2 - w_{xx}w_{yy})\right] d\Omega} \tag{8.59}$$

where

$$\boxed{\kappa = \frac{Eh^3}{12(1-\nu^2)}}$$

is the *plate rigidity*. Then, after simple computation,

$$\underline{\varepsilon}(v)\cdot\underline{\sigma}(w) = \frac{E}{1-\nu^2}\left[\Delta v\Delta w + (1-\nu)\left[(v_y w_{xy} - v_x w_{yy})_x + (v_x w_{xy} - v_y w_{xx})_y\right]\right]. \tag{8.60}$$

On using the common abbreviations $v_n = \nabla v \cdot \underline{n}$, $v_{nn} = \nabla\nabla v[\underline{n}, \underline{n}] = \underline{n}^T[\nabla\nabla v]\underline{n}$, etc., remember that

$$\Delta v \Delta w = \operatorname{div}(\Delta w \operatorname{grad} v) - \operatorname{grad} v \operatorname{grad} \Delta w,$$

$$\begin{aligned}\int_\Omega \operatorname{div}(\Delta w \operatorname{grad} v)\, d\Omega &= \oint_\Gamma \Delta w(\operatorname{grad} v\cdot\underline{n})\, d\Gamma = \oint_\Gamma \Delta w v_n\, d\Gamma, \\ \int_\Omega \operatorname{grad} v \operatorname{grad}\Delta w\, d\Omega &= \int_\Omega \operatorname{div}(v \operatorname{grad}\Delta w)\, d\Omega - \int v\Delta^2 w\, d\Omega \\ &= \oint_\Gamma v(\Delta w)_n\, d\Gamma - \int_\Omega v\Delta^2 w\, d\Omega.\end{aligned}$$

So, altogether, we obtain

$$\int_\Omega \Delta v\Delta w\, d\Omega = \int_\Omega v\Delta^2 w\, d\Omega - \oint_\Gamma v(\Delta w)_n\, d\Gamma + \oint_\Gamma \Delta w v_n\, d\Gamma.$$

for the first term in (8.59). As concerns the second term, observe the representations

$$\begin{aligned} d\underline{n} &= n_1 dx + n_2 dy, \quad dx = n_1 d\underline{n} - n_2 d\underline{t}\,, \\ d\underline{t} &= -n_2 dx + n_1 dy, \; dy = n_2 d\underline{n} + n_1 d\underline{t}\,, \end{aligned}$$

where $\underline{n}$ denotes the normal and $\underline{t}$ the tangent on the boundary. Application of the divergence theorem yields

$$\begin{aligned} &\int_\Omega \left[(v_y w_{xy} - v_x w_{yy})_x + (v_x w_{xy} - v_y w_{xx})_y\right] d\Omega \\ &= \oint_\Gamma \left[n_1(v_y w_{xy} - v_x w_{yy}) + n_2(v_x w_{xy} - v_y w_{xx})\right] d\Gamma \\ &= \oint_\Gamma v_n \left[2n_1 n_2 w_{xy} - n_2^2 w_{xx} - n_1^2 w_{yy}\right] d\Gamma \\ &+ \oint_\Gamma v_t \left[n_1 n_2(w_{yy} - w_{xx}) + (n_1^2 - n_2^2) w_{yy}\right] d\Gamma \\ &= - \oint_\Gamma v_n w_{tt}\, d\Gamma + \oint_\Gamma v_t w_{nt}\, d\Gamma\,. \end{aligned}$$

After substitution into (8.59), the weak problems now reads as follows,

$$\begin{aligned} &\frac{h^3}{12} \int_\Omega \underline{\varepsilon}(v) \cdot \underline{\sigma}(w)\, d\Omega \\ &= \kappa \left[\int_\Omega v \Delta^2 w\, d\Omega - \oint_\Gamma v[\Delta w - (1-\nu) w_{tt}]_n\, d\Gamma\right] \\ &+ \kappa \oint_\Gamma v_n [\Delta w - (1-\nu) w_{tt}]\, d\Gamma \\ &= h \int_\Omega v\, f\, d\Omega - h \oint_\Gamma v(\alpha w + \beta)\, d\Gamma\,. - h \oint_\Gamma v_n(\gamma w_n + \delta)\, d\Gamma\,. \end{aligned} \tag{8.61}$$

(c) By the usual argumentation, this problem leads to a **Boundary Value Problem** of order *four*,

$$\boxed{\begin{array}{ll} \kappa\, \Delta^2 w & = h\, f \text{ in } \Omega \\ v\Big[\kappa[\Delta w - (1-\nu) w_{tt}]_n - h(\alpha w + \beta)\Big] & = 0 \quad \text{on } \Gamma \\ v_n\Big[\kappa[\Delta w - (1-\nu) w_{tt}] - h(\gamma w_n + \delta)\Big] & = 0 \quad \text{on } \Gamma \end{array}} \tag{8.62}$$

When w resp. w_n are given on some part of the boundary, the test function v resp. v_n must vanish on that part. This leads altogether to a diversification of the boundary conditions into four different types:

$$\begin{array}{ll} w \;\text{ fixed}, & \kappa[\Delta w - (1-\nu) w_{tt}]_n - h(\alpha w + \beta) \text{ free} \\ w \;\text{ free}, & \kappa[\Delta w - (1-\nu) w_{tt}]_n - h(\alpha w + \beta) \text{ zero} \\ w_n \text{ fixed}, & \kappa[\Delta w - (1-\nu) w_{tt}] - h(\gamma w_n + \delta) \;\text{ free} \\ w_n \text{ free}, & \kappa[\Delta w - (1-\nu) w_{tt}] - h(\gamma w_n + \delta) \;\text{ zero}\,. \end{array}$$

More exactly, let $\varGamma_1$, $\varGamma_2$, $\varGamma_A$, $\varGamma_B$, $\varGamma_C$, $\varGamma_D$ be open subsets of $\varGamma$, and let

$$\varGamma = \overline{\varGamma_1 \cup \varGamma_2}, \quad \varGamma_1 \cap \varGamma_2 \text{ arbitrary},$$

$$\varGamma_1 = \varGamma_A \cup \varGamma_B\,,\ \varGamma_A \cap \varGamma_B = \emptyset\,,\ \ \varGamma_2 = \varGamma_C \cup \varGamma_D\,,\ \varGamma_C \cap \varGamma_D = \emptyset.$$

Then the inhomogenous boundary problem (8.62) has the form

$$\begin{aligned}
&\kappa\Delta^2 w &&= h\,f &&\text{in } \varOmega,\\
&w &&= p &&\text{on } \varGamma_A \text{ essential BC}\\
&\kappa[\Delta w - (1-\nu)w_{tt}]_n &&= h(\alpha w + \beta) &&\text{on } \varGamma_B \text{ natural BC}\\
&w_n &&= q &&\text{on } \varGamma_C \text{ essential BC}\\
&\kappa[\Delta w - (1-\nu)w_{tt}] &&= h(\gamma w_n + \delta) &&\text{on } \varGamma_D \text{ natural BC}\,.
\end{aligned} \tag{8.63}$$

Thus, in case $\alpha = \beta = \gamma = \delta = 0$, the boundary conditions $w = p$ and $w_n = q$ are essential and $[\Delta w - (1-\nu)w_{tt}]_n = 0$, $\Delta w - (1-\nu)w_{tt} = 0$ are natural boundary conditions.

Consider, e.g., the *clamped* plate then $\varGamma = \varGamma_A = \varGamma_C$, $\varGamma_B = \varGamma_D = \emptyset$ and

$$\boxed{\begin{aligned}
&\kappa\Delta^2 w = h\,f \text{ in } \varOmega,\\
&w \quad\ = p \quad \text{on } \varGamma, \text{ essential BC}\\
&w_n \quad = q \quad \text{on } \varGamma, \text{ essential BC}
\end{aligned}}\,.$$

On the other side, $\varGamma = \varGamma_A = \varGamma_D$, $\varGamma_B = \varGamma_C = \emptyset$ in the *simply supported* plate, and

$$\boxed{\begin{aligned}
&\kappa\Delta^2 w &&= h\,f &&\text{in } \varOmega,\\
&w &&= p &&\text{on } \varGamma, \text{ essential BC}\\
&\kappa[\Delta w - (1-\nu)w_{tt}] &&= h(\gamma w_n + \delta) &&\text{on } \varGamma, \text{ natural BC}
\end{aligned}}\,.$$

(d) Let the **Boundary** $\varGamma$ is parametrized w.r.t. arc length $s \mapsto x(s)$. The normal $\underline{n}$ pointing to the exterior of the plate and the tangent $\underline{t}$ in $x(s)$ form a cartesian coordinate system. Then, noting that Δw remains invariant under rotations,

$$\Delta w = w_{xx} + w_{yy} = w_{tt} + w_{nn}\,,\quad \Delta w - (1-\nu)w_{tt} = w_{nn} + \nu w_{tt}\,.$$

Twofold differentiation w.r.t. the arc length yields

$$w(x(s)) = p(s)\,,\quad \nabla w(x)x' = p'\,,\quad \nabla w(x)x'' + \nabla\nabla w(x)[x', x'] = p''.$$

Now $x''(s) = -\chi(s)\underline{n}(s)$ where $\chi(s)$ is the curvature by FRENET's formulas of § 1.3 hence

$$\boxed{\nabla\nabla w(x(s))[x'(s), x'(s)] = w_{tt}(x(s)) = \chi(s)w_n(x(s)) + p''(s)}\,.$$

By consequence on $\Gamma_A \cap \Gamma_D$

$$w_{nn} + \nu w_{tt} = 0\,, \quad \Delta w = (1-\nu)(\chi w_n + p'')$$

whereby the curvature is introduced explicitely into the boundary conditions.

The clamped plate has only essential boundary conditions and enjoys therefore a preferential treatment; in contrast, $w = p$ is essential and $w_{nn} + \nu(\chi w_n + p'') = 0$ in the simply supported plate. If now a curvilinear boundary is replaced by a polygon then obviously all terms with nonzero curvature disappear. Therefore a convergent approximation is impossible in this case (BABUSKA *paradoxon*; cf., e.g., (Babuska63), (Babuska90)).

(e) Example To show the influence of curvature in the simply supported plate we introduce $-\Delta w = v$ and replace the boundary value problem (8.63) of order four by two boundary problems of order two,

$$\begin{aligned} -\kappa\Delta v &= f && \text{in } \Omega \\ v &= (\nu-1)w_{tt} = (\nu-1)(\chi w_n + p'') && \text{on } \Gamma \end{aligned}$$

$$\begin{aligned} -\Delta w &= v && \text{in } \Omega \\ w &= p && \text{on } \Gamma\,. \end{aligned}$$

Both systems are solved alternating in several iteration steps, the exact solution is introduced via the right side f, and w_n is approximated numerically. For starting value we choose $v = 0$ on Γ. In the following three examples, the unit circle is the basic domain Ω where the curvature on the boundary is $\chi = 1$. We set $\kappa = 1$ and POISSON's ratio ν is fixed on the boundary by the relation $v = (\nu - 1)[\chi w_n + p'']$. The triangulation is generated by

```
[p,e,t]  = initmesh('circleg','Hmax',0.2);
p        = jigglemesh(p,e,t);.
```

The solution is computed by the program `assempde.m` of the MATLAB PDE-TOOLBOX i.e. with simple linear triangular elements.

Example 8.2. $w(r,\varphi) = (1-r^2)(5-r^2)\,,\ w(1,\varphi) = 0\,,\ \nu = 0\,,\ \Delta^2 w = 64\,.$

Example 8.3. $w(r,\varphi) = (1-2r^2)(5-r^2)\,,\ w(1,\varphi) = -4\,,\ \nu = 1/7\,,$
$\Delta^2 w = 128\,.$

Example 8.4. $w(r,\varphi) = (r^6 - 2r^4 + 16r^2)\cos(2\varphi)/10\,,\ w(1,\varphi) = -\cos(2\varphi)/2\,,$
$\nu = 1/5\,,\ \Delta^2 w = 384r^2\cos(2\varphi)/10$ (Figs. 8.1 and 8.2).

Example 8.5. $w(r,\varphi) = (2r^6 - 5r^4)\cos(2\varphi)/12\,,\ w(1,\varphi) = -\cos(2\varphi)/4\,,$
$\nu = 0\,,\ \Delta^2 w = 64r^2\cos(2\varphi)\,.$

Table 8.1. Maximum Error:

	$\chi = 0$	$\chi = 1$, 5 steps	$\chi = 1$, 10 steps
Ex. 8.2	2.0534	0.1692	0.0812
Ex. 8.3	3.1077	0.1865	0.1133
Ex. 8.4	0.0190	0.0061	0.0061
Ex. 8.5	0.0188	0.0076	0.0076

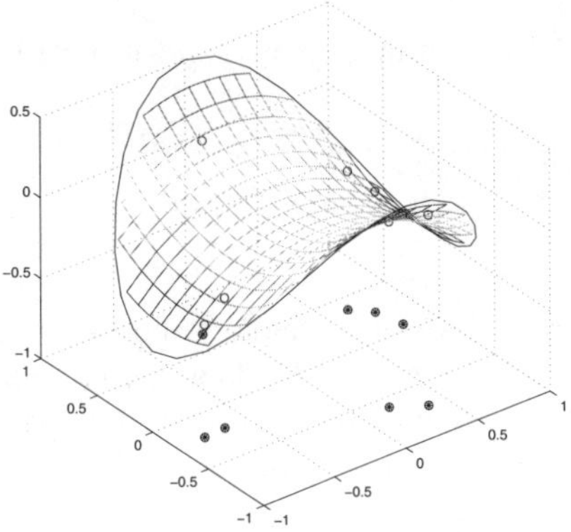

Figure 8.1. Example 8.4, solution

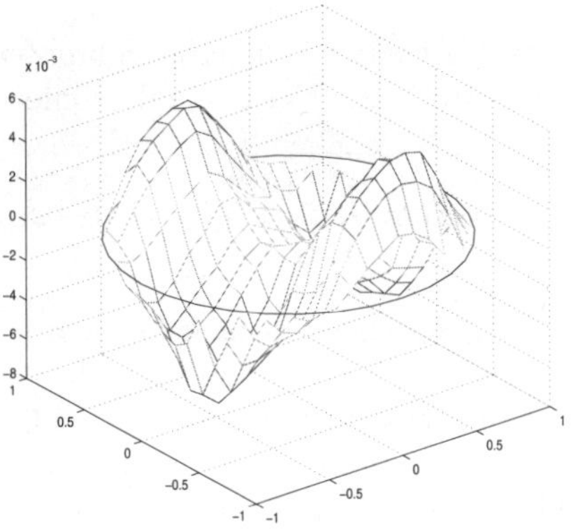

Figure 8.2. Example 8.4, error scaled ($\times 10^{-3}$)

8.8 Von Karman's Plate and the Membrane

We proceed at first as in KIRCHHOFF's plate. Let the thickness h be small again and let volume force density $\underline{k}$ act only in z-direction. Then $\underline{k} = [0,\ 0,\ f]^T$ is the external force and

$$u_1 = -z\, w_x(x,y),\ u_2 = -z\, w_y(x,y),\ u_3 = w(x,y),\ \underline{\widetilde{u}} = (u_1, u_2),\ \underline{u} = (\underline{\widetilde{u}}, w)$$

where $\underline{u} \in \widetilde{\Omega} = \Omega \times [-h/2\,,\, h/2] \subset \mathbb{R}^3$. But in addition to the bending energy now a *strain energy* is introduced with the nonlinear strain tensor

$$\begin{aligned}\widetilde{\varepsilon} := \varepsilon(\underline{u}) &= \frac{1}{2}\left[\operatorname{grad}\underline{\widetilde{u}} + (\operatorname{grad}\underline{\widetilde{u}})^T + (\operatorname{grad} w)^T \operatorname{grad} w\right] \\ &= \begin{bmatrix} u_{1,x} + w_x^2/2 & (u_{1,x} + u_{2,y} + w_x w_y)/2 \\ (u_{1,x} + u_{2,y} + w_x w_y)/2 & u_{2,y} + w_y^2/2 \end{bmatrix}\end{aligned}$$

where grad w shall be a row vector hence $(\text{grad}\, w)^T \text{grad}\, w$ is a dyadic product. Therefore

$$\underline{\varepsilon} = [\varepsilon_x,\, \varepsilon_y,\, \gamma_{xy}]^T = [u_{1,x} + w_x^2/2\,,\, u_{2,y} + w_y^2/2\,,\, u_{1,x} + u_{2,y} + w_x w_y]^T\,.$$

Also, in the same way as in KIRCHHOFF's plate

$$\underline{\sigma} = C\,\underline{\varepsilon}\,,\;\; \underline{\widetilde{\sigma}} = C\,\underline{\widetilde{\varepsilon}}\,,$$

where C is the elasticity matrix of (8.48) and C^{-1} the compliance matrix,

$$C = \frac{E}{1-\nu^2}\begin{bmatrix} 1 & \nu & 0 \\ \nu & 1 & 0 \\ 0 & 0 & (1-\nu)/2 \end{bmatrix}\,,\quad C^{-1} = \frac{1}{E}\begin{bmatrix} 1 & -\nu & 0 \\ -\nu & 1 & 0 \\ 0 & 0 & 2(1+\nu) \end{bmatrix}.$$

The total energy is linearly composed of the bending energy and the strain energy and leads to extremal problem

$$\boxed{\mathcal{E}(\underline{u}) = \int_{\widetilde{\Omega}} \left[\frac{1}{2}[\boldsymbol{\varepsilon}(\underline{u}) : \boldsymbol{\sigma}(\underline{u}) + \widetilde{\boldsymbol{\varepsilon}}(\underline{u}) : \widetilde{\boldsymbol{\sigma}}(\underline{u})] - f\, u_3\right] d\widetilde{\Omega} = \min!} \tag{8.64}$$

in the case where surface forces are disregarded; cf. (Landau). For a justification of this choice we refer to (Ciarlet97). We consider further only the boundary condition $w = 0\,,\ w_n = 0$ on Γ such that boundary integrals disappear. Because of the nonlinearity of $\widetilde{\boldsymbol{\varepsilon}}$, the variational problem reads now under observation of symmetry

$$\int_{\widetilde{\Omega}} [\boldsymbol{\varepsilon}(\underline{v}) : \boldsymbol{\sigma}(\underline{u}) + \delta\widetilde{\boldsymbol{\varepsilon}}(\underline{u};\underline{v}) : \widetilde{\boldsymbol{\sigma}}(\underline{u}) - f v_3]\, d\widetilde{\Omega} = 0$$

where $\underline{v}$ is a test function and

$$\delta\widetilde{\boldsymbol{\varepsilon}}(\underline{u};\underline{v}) = \frac{1}{2}[\text{grad}\,\underline{\widetilde{v}} + (\text{grad}\,\underline{\widetilde{v}})^T] + (\text{grad}\, w)^T \text{grad}\, v_3\,,\;\; \underline{\widetilde{v}} = [v_1, v_2]^T$$

is the first variation of $\widetilde{\boldsymbol{\varepsilon}}$. The first part stems form pure bending and yields in the same way as in the clamped KIRCHHOFF plate

$$\int_{\widetilde{\Omega}} \boldsymbol{\varepsilon}(\underline{v}) : \boldsymbol{\sigma}(\underline{u})\, d\widetilde{\Omega} \simeq \int_{\Omega} \boldsymbol{\varepsilon}(v_3) : \boldsymbol{\sigma}(w)\, d\Omega = \kappa \int_{\Omega} v_3 \Delta^2 w\, d\Omega\,.$$

Regarding the symmetry of $\widetilde{\boldsymbol{\sigma}}$ we obtain for the second part after integrating over z

$$\int_{\widetilde{\Omega}} \delta\widetilde{\boldsymbol{\varepsilon}}(\underline{u};\underline{v}) : \widetilde{\boldsymbol{\sigma}}(\underline{u})\, d\widetilde{\Omega} = \frac{h}{2}\int_{\Omega} \Big[\text{grad}(\underline{\widetilde{v}}) + (\text{grad}(\underline{\widetilde{v}}))^T$$
$$+ (\text{grad}\, w)^T \text{grad}\, v_3 + (\text{grad}\, v_3)^T \text{grad}\, w\Big] : \widetilde{\boldsymbol{\sigma}}(\underline{u})\, d\Omega$$
$$= h\int_{\Omega} \left[(\text{grad}(\underline{\widetilde{v}}))^T + (\text{grad}\, v_3)^T \text{grad}\, w\right] : \widetilde{\boldsymbol{\sigma}}(\underline{u})\, d\Omega\,.$$

Partial integration yields because of the zero boundary conditions

$$\int_\Omega \delta\widetilde{\varepsilon}(\underline{u};\underline{v}) : \widetilde{\boldsymbol{\sigma}}(\underline{u})\, d\Omega = -h \int_\Omega \left[\widetilde{\underline{v}} \cdot \operatorname{div} \widetilde{\boldsymbol{\sigma}}(\underline{u}) + v_3 \operatorname{div}\{\widetilde{\boldsymbol{\sigma}}(\underline{u})(\operatorname{grad} w)^T\}\right] d\Omega\,.$$

Together the variational equation of the extremal problem (8.64) has the form

$$\int_\Omega \Big[v_3 \left(\kappa \Delta^2 w - h \operatorname{div}\{\widetilde{\boldsymbol{\sigma}}(\underline{u})(\operatorname{grad} w)^T\} - h\, f\right) - h\, \widetilde{\underline{v}} \cdot \operatorname{div} \widetilde{\boldsymbol{\sigma}}(\underline{u})\Big]\, d\Omega = 0\,.$$

Accordingly, the boundary value problem

$$\kappa \Delta^2 w - h \operatorname{div}(\widetilde{\boldsymbol{\sigma}}(\underline{u})(\operatorname{grad} w)^T) = h\, f\,, \quad \operatorname{div} \widetilde{\boldsymbol{\sigma}} = 0\,, \quad \kappa = \frac{Eh^3}{12(1-\nu)^2} \tag{8.65}$$

is associated to the extremal problem (8.64) and the above boundary conditions remain the same. Recall from § 1.2**(c2)** that

$$\operatorname{div}(S\underline{v}) = \underline{v} \cdot \operatorname{div} S + \operatorname{grad} \underline{v} : S \ \text{ for } \ S = S^T \in {\mathbb{R}^3}_3\,.$$

Now one introduces the scalar AIRY stress function q by

$$\widetilde{\sigma}_x = \frac{\partial^2 q}{\partial y^2}\,, \quad \widetilde{\sigma}_y = \frac{\partial^2 q}{\partial x^2}\,, \quad \widetilde{\tau}_{xy} = -\frac{\partial^2 q}{\partial x \partial y}$$

(note that $\nabla^2 q \neq \boldsymbol{\sigma}$) and the symmetric product

$$[u, v] = u_{xx} v_{yy} + u_{yy} v_{xx} - 2u_{xy} v_{xy}$$

then $\operatorname{grad}(\operatorname{grad} w)^T : \widetilde{\boldsymbol{\sigma}}(\underline{u}) = [q, w]$ and (8.65) reads simply

$$\boxed{\kappa\, \Delta^2 w - h\,[q, w] = h\, f}\,. \tag{8.66}$$

Conversely the relations

$$\widetilde{\varepsilon}_x = \frac{1}{E}(\widetilde{\sigma}_x - \nu \widetilde{\sigma}_y)\,, \quad \widetilde{\varepsilon}_y = \frac{1}{E}(\widetilde{\sigma}_y - \nu \widetilde{\sigma}_x)\,, \quad \widetilde{\gamma}_{xy} = 2\frac{1+\nu}{E}\widetilde{\tau}_{xy}$$

follow from $\underline{\widetilde{\varepsilon}} = C^{-1}\underline{\widetilde{\sigma}}$ hence with AIRY's stress function q

$$u_{1,x} + \frac{w_x^2}{2} = \frac{1}{E}(q_{yy} - \nu q_{xx})\,, \quad u_{2,y} + \frac{w_y^2}{2} = \frac{1}{E}(q_{xx} - \nu q_{yy})\,,$$
$$u_{1,y} + u_{2,x} + w_x\, w_y = -2\frac{1+\nu}{E} q_{xy}\,.$$

The first equation is differentiated two-times w.r.t. y, the second two-times w.r.t. x and the third w.r.t. x and y. Thereafter the first and second equation are added and then the third is subtracted. Together we obtain VON KARMAN*'s equations* of nonlinear plate theory,

$$\boxed{\begin{aligned} \kappa\,\Delta^2 w - h\,[q,w] &= h\,f \\ \Delta^2 q + \frac{E}{2}\,[w,w] &= 0 \end{aligned}}\;. \tag{8.67}$$

A *membrane* is to be understood as a thin plate which is strongly stretched by plane forces attacking at boundary. Moreover one supposes that the components of the stress tensor are *constant* throughout. Then the first term in (8.65) may be neglected and one obtains the condition for equilibrium

$$h\,\mathrm{div}(\widetilde{\boldsymbol{\sigma}}\,\mathrm{grad}\,w) + hf = 0\,. \tag{8.68}$$

Note that hf has the dimension of pressure. If now the $h\widetilde{\boldsymbol{\sigma}} = \mu\boldsymbol{\delta}$ holds for the stress tensor with unity tensor $\boldsymbol{\delta}$ and the *boundary-force density* μ [force/length] is constant then we get eventually the LAPLACE equation

$$\boxed{\mu\Delta w + hf = 0}\;. \tag{8.69}$$

8.9 On Fluids and Gases

Here the velocity field $\underline{u}$ is the unknown vector field to be computed and no longer the displacements. Let therefore $\underline{v}$ or $\delta\underline{u}$ be the associated test function (virtual displacement of $\underline{u}$).

The tension in a fluid depends only on the velocity

$$\underline{u}(t,x) := \frac{\partial\Phi}{\partial t}(t,X)\,,\quad x = \Phi(t,X)\,,$$

(besides t) and the stress tensor has per definitionem the form

$$\mathbf{t}(t,x) = \boldsymbol{\sigma}(t,\underline{u}(x,t)) - p(t,x)\boldsymbol{\delta}\,,\ \text{or briefly}\ \mathbf{t} = \boldsymbol{\sigma}(\underline{u}) - p\boldsymbol{\delta} \tag{8.70}$$

where the scalar function p denotes the initial hydrostatic pressure independent of the strain rate and $\boldsymbol{\sigma}(t,\underline{0}) = \mathbf{0}$.

(a) Conservation Laws Recalling $\mathrm{div}(p(t,x)\boldsymbol{\delta}) = \mathrm{grad}\,p(t,x)$, the differential (local) laws of conservation read now in non-conservative form

$$\boxed{\begin{aligned} &\frac{\partial\varrho}{\partial t} + \mathrm{div}(\varrho\,\underline{u}) = 0 \\ &\varrho\frac{\partial u}{\partial t} + \varrho(\mathrm{grad}\,\underline{u})\underline{u} - \mathrm{div}\,\boldsymbol{\sigma} + \mathrm{grad}\,p = \varrho\underline{f} \\ &\varrho\frac{\partial e}{\partial t} + \varrho\,\mathrm{grad}\,e\cdot\underline{u} - \mathrm{div}(\boldsymbol{\sigma}\,\underline{u}) + \mathrm{div}(p\,\underline{u}) + \mathrm{div}\,q = \varrho\underline{f}\cdot\underline{u} + \varrho\,r \end{aligned}} \tag{8.71}$$

where $e = \varepsilon + \underline{u} \cdot \underline{u}/2$ is to be inserted and sometimes also

$$(\operatorname{grad} \underline{u})\underline{u} = \frac{1}{2} \operatorname{grad}(\underline{u} \cdot \underline{u}) - \underline{u} \times \operatorname{rot} \underline{u} \,.$$

(b) Notations (also to remember):

(1°) A *perfect* fluid (inviscid fluid) is defined by $\boldsymbol{\sigma} = 0$, then there exist no shear forces at the surface of the body hence also no vorticities.

(2°) A fluid is called *incompressible* if the deformation Φ satisfies

$$\Phi(t, U) \equiv \int_{\Phi(t,U)} dv = \int_U \det[\operatorname{grad} \Phi(t, X)] \, dU = \text{constant}$$

for all subvolumes U and all times t. By Lemma 8.4

$$\frac{d}{dt}\Phi(t, U) = \int_U \det[\operatorname{grad} \Phi(t, X)] \operatorname{div} \underline{u}(t, \Phi(t, X)) \, dU = 0 \,,$$

hence a fluid is incompressible if and only if $\operatorname{div} \underline{u}(t, x) = 0$, i.e., if the *velocity field* has no sources. By the so-called equation of continuity,

$$\frac{D\varrho}{Dt} + \varrho \operatorname{div} \underline{u} = 0 \,, \tag{8.72}$$

this is the case if and only if $D\varrho/Dt = 0$.

(3°) A fluid is *homogenous* if ϱ does not depend on the space variable x, then $D\varrho/Dt = d\varrho/dt$. By (8.72) a fluid is homogenous and incompressible if and only if the mass density ϱ is *constant.*

(4°) A compressible fluid is a *gas.*

(5°) A fluid is *isentropic* if its entropy is constant. This means that thermal source density and thermal flux vector in the balance law of energy vanish. Then interior energy depends only on ϱ and $\underline{u}$ besides time and space.

(6°) Recall that a body or fluid is said to *isotropic* if its material properties are the same in all directions. A (viscous) fluid or gas is called (linear isotropic) NEWTONIAN fluid if GREEN-LAGRANGE strain tensor and stress tensor have per definitionem the form

$$\boxed{\begin{aligned}
\boldsymbol{\varepsilon}(\underline{u}) &= \frac{1}{2}\left[\operatorname{grad} \underline{u} + (\operatorname{grad} \underline{u})^T\right] \\
\boldsymbol{\sigma}(\underline{u}) &= 2\mu_1 \boldsymbol{\varepsilon}(\underline{u}) + \mu_2 \operatorname{trace} \boldsymbol{\varepsilon}(\underline{u}) \boldsymbol{\delta} \\
&= 2\mu_1 \boldsymbol{\varepsilon}(\underline{u}) - \frac{2}{3}\mu_1 (\operatorname{div} \underline{u})\boldsymbol{\delta} + \left(\frac{2}{3}\mu_1 + \mu_2\right) \operatorname{div} \underline{u}\boldsymbol{\delta} \\
&= 2\mu_1 \left[\boldsymbol{\varepsilon}(\underline{u}) - \frac{1}{3}(\operatorname{div} \underline{u})\boldsymbol{\delta}\right] + \mu_3 (\operatorname{div} \underline{u})\boldsymbol{\delta} \,, \quad \mu_3 := (2\mu_1 + 3\mu_2)/3
\end{aligned}}$$

(8.73)

μ_1	first coeff. of viscosity, shear viscosity or simply viscosity
μ_2	second coeff. of viscosity, volume viscosity
μ_3	volumetric viscosity, bulk viscosity, bulk modulus
$\nu = \mu_1/\varrho$	kinematic viscosity

(8.74)

The trace of [....] in $\boldsymbol{\sigma}$ in the last equation of (8.73) is zero hence the first term represents the infinitisimal rate of change of shape and the second the infinitisimal rate of change of volume. Therefore $\widetilde{p} = -\mu_3 \operatorname{div} \underline{u} + p$ is the strain-dependent pressure but frequently the term $\mu_3 \operatorname{div} \underline{u}$ is neglected.

(7°) In a *non-Newtonian* fluid, the stress tensor $\boldsymbol{\sigma}$ depends in non-linear way of the strain tensor $\boldsymbol{\varepsilon}$, e.g. μ_1 is a function of $\boldsymbol{\varepsilon}$. Then the scalar p may lose its physical interpretation as a normal stress and plays the role of a LAGRANGE multiplier.

(c) Conservation Laws of Viscous Fluids We suppose that the constants of material μ_1 and μ_2 are *constant* in time and space, and $\operatorname{grad} p$ shall be a column vector below as is also seen from the context. By (8.42)

$$\begin{aligned}\operatorname{div} \boldsymbol{\sigma}(\underline{u}) &= \mu_1 \Delta \underline{u} + (\mu_1 + \mu_2) \operatorname{grad} \operatorname{div} \underline{u} \\ &= \mu_1 \left[\Delta \underline{u} - \frac{1}{3} \operatorname{grad} \operatorname{div} \underline{u} \right] + \mu_3 \operatorname{grad} \operatorname{div} \underline{u} \,.\end{aligned}$$

Substitution into the local conservation law of momentum (8.71), an application of Lemma 8.6 onto the local conservation law of energy and the continuity equation then yield together the conservation laws for viscous fluids in non-conservative form

$$\boxed{\begin{aligned}&\frac{D\varrho}{Dt} + \varrho \operatorname{div} \underline{u} &&= 0\\ &\varrho \frac{D\underline{u}}{Dt} - \mu_1 \Delta \underline{u} - (\mu_1 + \mu_2) \operatorname{grad} \operatorname{div} \underline{u} + \operatorname{grad} p &&= \varrho \underline{f}\\ &\varrho \frac{De}{Dt} - \boldsymbol{\varepsilon}(\underline{u}) : \boldsymbol{\sigma}(\underline{u}) + \operatorname{div}(p\,\underline{u}) + \operatorname{div} \underline{q} &&= \varrho \underline{f} \cdot \underline{u} + \varrho r\end{aligned}} \,. \tag{8.75}$$

By FOURIER's law, the conductive heat flux for an isotropic material is $\underline{q} = -\kappa \operatorname{grad} \vartheta$ where κ is the *thermal conductivity* and ϑ the absolute temperature.

(d) As an example we consider a *homogenous* fluid, $\operatorname{grad} \varrho = 0$.
(1°) The fluid is *perfect, compressible* and *isentropic*. Then the law of energy in (8.71) yields, because $\underline{q} = \underline{0}$ and $r = 0$ (recall $e = \varepsilon + \underline{u} \cdot \underline{u}/2$),

$$\varrho \frac{De}{Dt} + \operatorname{div}(p\,\underline{u}) = 0 \,. \tag{8.76}$$

But

$$\frac{De}{Dt} = \frac{d}{dt} e\left(\varrho(t, \Phi(t, X))\right) = \frac{\partial e}{\partial t} + \frac{\partial e}{\partial \varrho}\left(\frac{D\varrho}{Dt}\right) = \frac{\partial e}{\partial t} - e_\varrho\, \varrho \operatorname{div} \underline{u}$$

by application of the so-called continuity equation, i.e. $D\varrho/Dt + \varrho \operatorname{div} \underline{u} = 0$. Substitution into (8.76) yields

$$\boxed{\varrho \frac{\partial e}{\partial t} - \frac{\partial e}{\partial \varrho} \varrho^2 \operatorname{div} \underline{u} + (\operatorname{grad} p)\underline{u} + p \operatorname{div} \underline{u} = 0} \; .$$

If the interior energy e does not depend on time t and $\operatorname{grad} p = 0$, this equation yields $p - \varrho^2(\partial e/\partial \varrho) = 0$ after cancelling of $\operatorname{div} \underline{u}$. That equation however can be integrated separately if the function $p = p(\varrho)$ is known. Gases satisfy frequently the isentropic equation $p(\varrho) = A\varrho^\gamma$, $A \geq 0$, where $\gamma \geq 1$ is the exponent of adiabatics.

(2°) The fluid is *homogenous* and *incompressible*, $\operatorname{div} \underline{u} = 0$, and the interior energy is *constant* (temperature constant, thermal flux vector $\underline{q} = 0$, r dropped). Then the second and third equation of (8.75) are equivalent for $\operatorname{div} \underline{u} = 0$ and the balance equations read in non-conservative form

$$\boxed{\begin{array}{rl} \partial\varrho/\partial t + (\operatorname{grad} \varrho)\underline{u} + \varrho \operatorname{div} \underline{u} = 0 & \text{(mass)} \\ \varrho\partial\underline{u}/\partial t + \varrho(\operatorname{grad} \underline{u})\underline{u} - \mu_1 \Delta \underline{u} + \operatorname{grad} p = \varrho \underline{f} & \text{(momentum)} \\ \operatorname{div} \underline{u} = 0 & \text{(energy)} \end{array}} \; . \qquad (8.77)$$

(The term $\mu_3 \operatorname{grad} \operatorname{div} \underline{u}$ is omitted here and below such that only the hydrostatic pressure is involved.) Because $\partial\varrho/\partial t = 0$ by assumption on homogenous incompressible fluid, the first equation can be cancelled. Then this *"differential-algebraic"* system of two equations is called NAVIER-STOKES *equations.* (Gresho), p. 451: *"The combination of the non-linear* convective *term and the pressure-velocity coupling makes the NS equations difficult to solve. If either is absent, the equations are much simpler and are known to have solutions — the limiting cases being* STOKES *flow and the so-called* BURGER*'s equations* $\varrho(\partial\underline{u}/\partial t + (\operatorname{grad} \underline{u})\underline{u} - \mu_1 \Delta \underline{u} = 0$ *respectively".*

8.10 Navier-Stokes Equations

Notation and Assumptions: See also (8.74).

ϱ	mass density	$\varrho =$ *constant*
$\underline{u}$	solution (column vector)	velocity field
$\underline{v}$	test function	velocity field
$\nu = \mu/\varrho$	kinematic viscosity	[area/time]
	for simplicity	$\nu =$ *constant*
$p_{\text{spec}} = p/\varrho$	specific pressure	
$\underline{f}$	specific volume-force density	

.

(a) Velocity Pressure Form (Direct form) We reconsider the system (8.77) under these assumptions and notations then after division by ϱ

$$\boxed{\frac{\partial \underline{u}}{\partial t} + (\operatorname{grad} \underline{u})\underline{u} - \nu \Delta \underline{u} + \operatorname{grad} p_{\text{spez}} = \underline{f}\,, \quad \operatorname{div} \underline{u} = 0\,.} \tag{8.78}$$

Scalar products are written again such that $\underline{v} \cdot \underline{u} = \underline{v}^T \underline{u}$ for possible MATLAB implementations. Remember and write

$$\begin{aligned} \int_\Omega \underline{v} \cdot \operatorname{grad} p \, d\Omega &= -\int_\Omega p \operatorname{div} \underline{v} \, d\Omega + \int_\Gamma p \, \underline{v} \cdot \underline{n} \, d\Omega\,, \\ \int_\Omega \underline{v} \cdot \Delta \underline{u} \, d\Omega &= -\int_\Omega \operatorname{grad} \underline{v} : \operatorname{grad} \underline{u} \, d\Omega + \int_\Gamma \underline{v} (\operatorname{grad} \underline{u}) \underline{n} \, d\Omega \end{aligned}$$

$$\begin{aligned} a(\underline{v}, \underline{u}) &= \nu \int_\Omega \operatorname{grad} \underline{v} : \operatorname{grad} \underline{u} \, d\Omega\,, \quad b(\underline{v}, q) = \int_\Omega \operatorname{div} \underline{v} \, q \, d\Omega\,, \\ c(\underline{v}, \underline{u}, \underline{w}) &= \int_\Omega \underline{v} \cdot (\operatorname{grad} \underline{u}) \underline{w} \, d\Omega\,, \\ (\underline{v}, \underline{u}) &= \int_\Omega \underline{v} \cdot \underline{u} \, d\Omega\,, \quad (\underline{v}, \underline{u})_\Gamma = \int_\Gamma \underline{v} \cdot \underline{u} \, d\Gamma\,. \end{aligned} \tag{8.79}$$

Then, multiplying the two equations (8.78) by test functions $\underline{v}$ and q respectively, we obtain the *weak* non-linear velocity-pressure form

$$\boxed{\begin{array}{lll} (\underline{v}, \partial \underline{u}/\partial t) & & \\ +a(\underline{v}, \underline{u}) + c(\underline{v}, \underline{u}, \underline{w}) - (\operatorname{div} \underline{v}, p_{\text{spec}}) & = (\underline{v}, \underline{f}) + (\underline{v}, \underline{\sigma}_n(\underline{u}, p_{\text{spec}}))_\Gamma\,, & \underline{v} \in \mathcal{V} \\ \qquad - (\operatorname{div} \underline{v}, q) & = 0\,, & q \in \mathcal{Q} \\ \qquad \underline{\sigma}_n(\underline{u}, p_{\text{spec}}) & = (\nu \operatorname{grad} \underline{u} - p_{\text{spec}}) \underline{n} & \end{array}} \tag{8.80}$$

where $\mathcal{V}$, $\mathcal{Q}$ are appropriate function spaces to be specified later. This weak form of the NAVIER-STOKES equations requires the lowest smoothness of the solution compared with subsequent forms.

(b) Boundary Value Problem for Velocity Pressure Form We subdivide the boundary Γ of the basic domain Ω into four different parts,

$$\Gamma = \partial \Omega = \Gamma_D \oplus \Gamma_G \oplus \Gamma_H \oplus \Gamma_N$$

where however some may be empty; cf. (9.1). Let $\{\underline{t}_1(x), \underline{t}_2(x)\}$ be an orthogonal system in each point x on smooth parts of boundary; in particular but not only we may choose $\{\underline{t}_1, \underline{t}_2\} = \{\underline{n}, \underline{t}\}$ or $\{\underline{t}_1, \underline{t}_2\} = \{\underline{t}, -\underline{n}\}$. Now $I = \underline{t}_1 \underline{t}_1^T + \underline{t}_2 \underline{t}_2^T$ in $\mathbb{R}^2$ by assumption (I identity) and for the *inhomogeneous* boundary condition $(\underline{v}, \underline{\sigma}_n(\underline{u}, p))_\Gamma = (\underline{v}, \underline{g})_\Gamma$ follows the decomposition

$$\int_\Gamma \underline{v} \cdot \underline{\sigma}_n(\underline{u}, p) \, d\Gamma = \int_\Gamma \underline{v} \cdot (\underline{t}_1 \underline{t}_1^T + \underline{t}_2 \underline{t}_2^T) \underline{\sigma}_n(\underline{u}, p) \, d\Gamma\,.$$

The general DIRICHLET boundary condition

$$\underline{u} = \underline{g} \text{ on } \Gamma_D\,, \quad \underline{u} \cdot \underline{t}_1 = g_1 \text{ on } \Gamma_G$$

requires that

$$\underline{v} = \underline{0} \text{ on } \Gamma_D\,, \quad \underline{v} \cdot \underline{t}_1 = 0 \text{ on } \Gamma_G$$

for the test function and thus

$$\int_\Gamma \underline{v} \cdot \underline{\sigma}_n(\underline{u}, p)\, d\Gamma = \int_{\Gamma_H} g_2\, \underline{v} \cdot \underline{t}_2\, d\Gamma + \int_{\Gamma_N} \underline{v} \cdot \underline{g}\, d\Gamma\,.$$

Therefore the general boundary value problem for the stationary NAVIER-STOKES equations reads:

$$\boxed{\begin{aligned} -\nu \Delta \underline{u} + (\operatorname{grad} \underline{u})\underline{u} + \operatorname{grad} p_{\text{spec}} &= \underline{f} \text{ in } \Omega \\ -\operatorname{div} \underline{u} &= 0 \quad \text{in } \Omega \\ \underline{u} &= \underline{g} \quad \text{on } \Gamma_D \quad \text{(essential BC)} \\ \underline{t}_1 \cdot \underline{u} &= g \quad \text{on } \Gamma_G \quad \text{(essential BC)} \\ \underline{t}_2 \cdot \underline{\sigma}_n(\underline{u}, p_{\text{spec}}) &= g \quad \text{on } \Gamma_H \quad \text{(natural BC)} \\ \underline{\sigma}_n(\underline{u}, p_{\text{spec}}) &= \underline{g} \quad \text{on } \Gamma_N \quad \text{(natural BC)} \end{aligned}}\,. \tag{8.81}$$

The essential boundary condition $\underline{t} \cdot \underline{u} = g$ leads to the boundary condition $\underline{t} \cdot \underline{v} = 0$ for the test functins which is difficult to realize by the finite element method.

In order that a unique solution of the problem exists, some conditions on the smoothness of the involved data are to be required but also some *regularity conditions* for the right sides $\underline{f}$ in Ω and $\underline{g}$ on Γ. Cf. e.g. (Gresho).

Example 8.6. (Orlt) Let $\underline{t}(\underline{x})$ be the tangential vector on the boundary Γ, let

$$\begin{aligned} &\mathcal{C}^{0,1} = \text{set of LIPSCHITZ-continuous functions in } \Omega, \\ &\mathcal{W} = \left\{ \underline{v} \in \mathcal{C}^{0,1}(\Omega)^2, \ \forall\, \underline{x} \in \Gamma : \underline{v}(\underline{x}) \cdot \underline{t}(\underline{x}) = 0\,, \ \underline{v}(\underline{x}) = \underline{a} + b \begin{bmatrix} -x_2 \\ x_1 \end{bmatrix} \right\}, \\ &\Gamma_0 = \Gamma_N \cup (\Gamma_G \backslash \overline{\{\underline{x} \in \Gamma_G\,,\ \underline{t}_1(\underline{x}) = \pm \underline{n}(\underline{x})\}})\,. \end{aligned}$$

The space $\mathcal{W}$ consists of all rigid motions in $\mathbb{R}^2$ (translation and rotation) which are not prevented by homogeneous essential boundary conditions, and Γ_0 is that subset of boundary on which the normal component of $\underline{u}$ is not specified. Then a solution of the boundary value problem exists in weak sense if the right sides satisfy the following equilibrium conditions:

$$\boxed{\begin{array}{ll} (\underline{f}, \underline{v})_\Gamma + (\underline{g}, \underline{v})_{\Gamma_N} = 0 & \text{for } \underline{v} \in \mathcal{W} \\ (\underline{n}, \underline{g})_{\Gamma_D} = 0 & \text{if } \Gamma_0 = \emptyset \\ \text{one value of } p \text{ fixed} & \text{or } \displaystyle\int_\Omega p\, d\Omega = 0 \end{array}}\;;$$

for $\Gamma_0 \neq \emptyset$ the second condition is dropped. The third condition on the pressure p is easily implemented but leads to serious instabilities in numerical solution because the approximation depends on one single value of p whereas the total numbers of unknown (discrete) values of p may be very large.

Further results are found, e.g., in (Gresho) or in the comprehensive work of (Orlt).

(c) Non-dimensional System Reconsider the original momentum equation of (8.77) with constant mass density ϱ and $p_{\rm spec} = p/\varrho$,

$$\partial \underline{u}/\partial t + (\mathrm{grad}\,\underline{u})\underline{u} - \nu \Delta \underline{u} + \mathrm{grad}\, p_{\rm spec} = \underline{f}\,. \tag{8.82}$$

It is customary and advisable to transpose this equation into non-dimensional form before numerical approach. To this end the user has to specify a *characteristic length* L and a *characteristic velocity* U which depend both on the individual physical model. Then the REYNOLDS number R_e is introduced as ratio of advective to diffusive momentum transport such that

$$\underline{x} = L\,\underline{\widetilde{x}}\,,\ \ \underline{u} = U\,\underline{\widetilde{u}}\,,\ \ R_e = \frac{UL}{\nu}\,. \tag{8.83}$$

Regard also that the current flow must be *laminar* which means that the REYNOLDS number remains below some critical value.

Case 1: We introduce a *characteristic time* L^2/ν (sometimes called FOURIER time) and a *characteristic pressure* $\mu U/L$,

$$t = \frac{L^2}{\nu}\,\widetilde{t}\,,\ \ p = \frac{\varrho \nu U}{L}\,\widetilde{p}\,. \tag{8.84}$$

Multiplying (8.82) by $L^2/\nu U$ then leads to the non-dimensional form

$$\frac{\partial \underline{\widetilde{u}}}{\partial \widetilde{t}} + R_e(\mathrm{grad}\,\underline{\widetilde{u}})\underline{\widetilde{u}} - \Delta \underline{\widetilde{u}} + \mathrm{grad}\,\widetilde{p} = \underline{\widetilde{f}} \ \text{ where } \ \underline{\widetilde{f}} = \frac{L^2}{\nu U}\,\underline{f}\,. \tag{8.85}$$

As the tangential vector is non-dimensional, both *natural* boundary conditions in (8.81) must be *multiplied* by $L/\nu U$ such that

$$\underline{\sigma}_n(\underline{\widetilde{u}}, \widetilde{p})\underline{n} = (\mathrm{grad}\,\underline{\widetilde{u}} - \widetilde{p})\underline{n} = \frac{L}{\nu U}\underline{g} =: \underline{\widetilde{g}}\,.$$

Case 2: We introduce a *characteristic time* L/U and a *characteristic pressure* ϱU^2,

$$t = \frac{L}{U}\,\widetilde{t}\,,\ \ p = \varrho U^2\,\widetilde{p}\,. \tag{8.86}$$

Multiplication of (8.82) by L/U^2 then leads to the alternative non-dimensional form

$$\frac{\partial \underline{\widetilde{u}}}{\partial \widetilde{t}} + (\mathrm{grad}\,\underline{\widetilde{u}})\underline{\widetilde{u}} - \frac{1}{R_e}\Delta \underline{\widetilde{u}} + \mathrm{grad}\,\widetilde{p} = \underline{\widetilde{f}} \ \text{ where } \ \underline{\widetilde{f}} = \frac{L}{U^2}\,\underline{f}\,. \tag{8.87}$$

The *natural* boundary conditions in (8.81) must now be *divided* by U^2 such that

$$\underline{\sigma}_n(\underline{\widetilde{u}},\widetilde{p})\underline{n} = \left(\frac{1}{R_e}\operatorname{grad}\underline{\widetilde{u}} - \widetilde{p}\right)\underline{n} = \frac{1}{U^2}\underline{g} =: \underline{\widetilde{g}}\,.$$

In both cases, the *essential* boundary conditions in (8.81) must be divided by U such that e.g. $\underline{\widetilde{u}} = \underline{\widetilde{g}}$, $\underline{\widetilde{g}} = \underline{g}/U$.

In consequence, The non-dimensional system (8.85) is more appropriate for low REYNOLDS numbers, and linear STOKES flow is recovered for $R_e = 0$, whereas (8.87) is to be preferred in the advection-dominated domain of high R_e.

(d) Stream-Function Vorticity Form Let in $\mathbb{R}^2$

$$z(x,y) = c\,,\quad c\in\mathbb{R}\,,$$

be an invariant of the differential system $\dot{\underline{x}} = \underline{v}(\underline{x})$ where $\underline{x} = [x\,,\,y]^T$ and $\underline{v} = (u,v)$, i.e., a solution in implicit form, then z is called *stream function* of $\underline{v}$. If now $\underline{x}(t) = (x(t), y(t))$ is a stream-line, i.e., an individual solution of $\dot{\underline{x}} = \underline{v}(\underline{x})$ then

$$\frac{dz}{dt} = z_x\dot{x} + z_y\dot{y} = 0\,.$$

After having fixed the sign we obtain

$$\boxed{z_y = u\,,\quad z_x = -v\,,\quad \underline{v} =: \operatorname{curl} z}\,. \tag{8.88}$$

The *vorticity* w [1/time] of the velocity field $\underline{v}$ is defined in $\mathbb{R}^2$ by

$$\boxed{w = v_x - u_y}\,, \tag{8.89}$$

again and, by (8.89), we have the direct connection

$$\Delta z = -w\,. \tag{8.90}$$

The vorticity w of an ideal fluid vanishes hence $\Delta z = 0$ in this case. Also, because $w = 0$, the velocity field $\underline{v}$ is a gradient field in this case with potential Ψ and $\operatorname{grad}\Psi = \underline{v}$.

Now consider the NAVIER-STOKES equation in $\mathbb{R}^2$ again,

$$\underline{u}_t + (\operatorname{grad}\underline{u})\underline{u} - \nu\Delta\underline{u} + \operatorname{grad} p = \underline{f} \in \mathbb{R}^2\,.$$

The second equation is differentiated w.r.t. x, the first w.r.t. y and thereafter the first equation is subtracted from the second. The result of this manipulation represents together with the above equation $\Delta z = -w$ the NAVIER-STOKES equations in *stream-function vorticity form*

$$\boxed{\begin{array}{c} w_t - \nu\Delta w + \operatorname{curl} z\cdot\operatorname{grad} w = f_{2,x} - f_{1,y} \equiv \operatorname{rot}\underline{f} \\ -\Delta z - w = 0 \end{array}} \tag{8.91}$$

where $\operatorname{curl} z \cdot \operatorname{grad} w = z_y w_x - z_x w_y$ is the convective nonlinear term and the proper goal of the operation namely the elimination of the pressure p is attained. The *stationary* equations (8.91) constitute a *nonlinear elliptic system*

$$\begin{aligned} -\Delta z - w &= 0 \\ -\Delta w + \nu^{-1} z_y w_x - \nu^{-1} z_x w_y &= g\,, \quad g = \nu^{-1} \operatorname{rot} \underline{f} \end{aligned}$$

hence

$$\begin{bmatrix} \operatorname{div}(C_{11} \operatorname{grad} z + C_{12} \operatorname{grad} w) \\ \operatorname{div}(C_{21} \operatorname{grad} z + C_{22} \operatorname{grad} w) \end{bmatrix} + \begin{bmatrix} 0 & 1 \\ 0 & 0 \end{bmatrix} \begin{bmatrix} z \\ w \end{bmatrix} = \begin{bmatrix} 0 \\ g \end{bmatrix},$$

$$C_{11} = \begin{bmatrix} 1 & 0 \\ 0 & 1 \end{bmatrix},\ C_{12} = \begin{bmatrix} 0 & 0 \\ 0 & 0 \end{bmatrix},\ C_{21} = \begin{bmatrix} 0 & \nu^{-1} w \\ -\nu^{-1} w & 0 \end{bmatrix},\ C_{22} = \begin{bmatrix} 1 & 0 \\ 0 & 1 \end{bmatrix}.$$

The stream-function vorticity form requires higher smoothness of the data which in particular becomes apparent at possible corners of the domain Ω. On the other side, it allows the application of *simple* finite elements with linear triangular elements.

(e) Connection with the Plate Equation We consider the linear STOKES equation in $\mathbb{R}^2$ with DIRICHLET *boundary condition*

$$\boxed{\begin{aligned} -\Delta \underline{u} + \operatorname{grad} p &= \underline{f} \ \text{ in } \Omega \\ \operatorname{div} \underline{u} &= 0 \ \text{ in } \Omega \\ \underline{u} &= \underline{g} \ \text{ on } \Gamma = \partial\Omega \end{aligned}}\ . \tag{8.92}$$

(1°) HELMHOLTZ' decomposition theorem 1.7 supplies $\underline{u} = \operatorname{grad} \varphi + \operatorname{rot} \underline{v}$ for every (smooth) vector field $\underline{u}$. Since grad φ is cancelled out later, we write $\underline{u}$ as solenoidal field $\underline{u} = \operatorname{rot} \underline{v}$ such that $\operatorname{div} \underline{u} = 0$,

$$-\Delta \operatorname{rot} \underline{v} + \operatorname{grad} p = \underline{f}\,, \quad \operatorname{div} \operatorname{rot} \underline{v} = 0\,, \tag{8.93}$$

where the latter equation is *always* fulfilled for smooth fields.

(2°) Recall that $\operatorname{rot} \operatorname{rot} \underline{u} = \operatorname{grad} \operatorname{div} \underline{u} - \Delta \underline{u}$ in $\mathbb{R}^3$, apply it to (8.93) and insert the result into (8.92) then

$$\begin{aligned} -\Delta \operatorname{rot} \underline{v} = \operatorname{rot}^3 \underline{v} - \operatorname{grad} \operatorname{div} \operatorname{rot} \underline{v} &= \underline{f} \\ \operatorname{rot}^3 \underline{v} + \operatorname{grad} p &= \underline{f}\,. \end{aligned} \tag{8.94}$$

(3°) Apply "rot" to (8.94), $\operatorname{rot}^4 \underline{v} + \operatorname{rot} \operatorname{grad} p = \operatorname{rot} \underline{f}$, then the pressure is cancelled out because a gradient field has no vorticities,

$$\boxed{\begin{aligned} \operatorname{rot}^4 \underline{v} &= \operatorname{rot} \underline{f} \ \text{ in } \Omega \\ \operatorname{rot} \underline{v} &= \underline{u}_0 \quad \text{ on } \Gamma \end{aligned}}\ . \tag{8.95}$$

(4°) Let now $u_3 = 0$, i.e. $\underline{u}$ is a plane vector field in $\mathbb{R}^2$. Then, because $\underline{u} = \operatorname{rot} \underline{v}$, $u_3 = v_{2,x} - v_{1,y} = 0$. Since $\underline{u}$ is a plane field, we also may set $v_1 = v_2 = 0$ then

$$\begin{aligned}
\operatorname{rot} \underline{v} &= [v_{3,y} - v_{2,z}, v_{1,z} - v_{3,x}, v_{2,x} - v_{1,y}]^T = [v_{3,y}, -v_{3,x}, 0]^T \\
\operatorname{rot}^2 \underline{v} &= [0, 0, -v_{3,xx} - v_{3,yy}]^T \\
\operatorname{rot}^3 \underline{v} &= [-v_{3,xxy} - v_{3,yyy}, v_{3,xxx} + v_{3,yyx}, 0]^T \\
\operatorname{rot}^4 \underline{v} &= [0, 0, v_{3,xxxx} + v_{3,yyxx} + v_{3,xxyy} + v_{3,yyyy}]^T .
\end{aligned}$$

Now $v_3 =: z$ is the stream-function because $\underline{u} = \operatorname{rot} \underline{v}$ and $v_1 = v_2 = 0$, and moreover $u_1 = v_{3,y} = z_y$, $u_2 = -v_{3,x} = -z_x$, as well as $\Delta^2 z = (\operatorname{rot}^4 \underline{v})_3 = z_{xxxx} + 2z_{xxyy} + z_{yyyy}$. In plane no forces act in z-direction, $f_3 = 0$, thus finally the following form of the *Stokes* equation is obtained from (8.95) for z:

$$\Delta^2 z = (\operatorname{rot} \underline{f})_3 = f_{2,x} - f_{1,y} .$$

(5°) We write $\underline{t} = (dx, dy)$ on the boundary Γ for the tangent vector and $\underline{n} = (dy, -dx)$ for the normal vector. Note that $z_y = u_{0,1}$ and $z_x = -u_{0,2}$ on Γ because $\operatorname{rot} \underline{v} = \underline{u}_0$ and $u_{0,3} = 0$. Further, $z_n = z_x n_1 + z_y n_2 = z_x dy - z_y dx = -u_{0,2} dy - u_{0,1} dx = -\underline{u}_0 \cdot \underline{t}$ on Γ and

$$\begin{aligned}
z(\underline{x}(s)) &= \int_{s_0}^{s} (z_x dx + z_y dy) + c = \int_{s_0}^{s} (-u_{0,2} dx + u_{0,1} dy) + c \\
&= \int_{s_0}^{s} \underline{u}_0 \cdot \underline{n} \, ds + c ,
\end{aligned}$$

hence on Γ together $z_n = -\underline{u}_0 \cdot \underline{t}$ and $z = 0$ for $\underline{u}_0 \cdot \underline{n} = 0$. Altogether a linear boundary value problem of *order four* is derived by this way for z:

$$\boxed{\begin{aligned}
\Delta^2 z &= f_{2,x} - f_{1,y} && \text{in } \Omega \\
z(\underline{x}) &= \int_{C(\underline{x}) \subset \Gamma_D} \underline{g} \cdot \underline{n} \, d\Gamma + \kappa && \text{on } \Gamma \\
z_n &= \underline{g} \cdot \underline{t} && \text{on } \Gamma
\end{aligned}} ,$$

where $C(\underline{x}) \subset \Gamma$ is a line segment from $\underline{x}_0$ to $\underline{x}$ and κ is an arbitrary constant (e.g. $\kappa = 0$).

(f) Pressure Poisson equation In solving NAVIER-STOKES equations, the results for pressure p are frequently not very convincing. Rather it is calculated by a separate POISSON problem which however has only NEUMANN boundary conditions such that a reference value of p must be specified somewhere in the domain Ω or $\int_\Omega p \, dV = 0$ must be required for normalization but nevertheless the problem is and remains *unstable*.

(f1) For the computation of p by the flow field $\underline{u}$, remember that

$$\Delta \underline{v} = \operatorname{div}(\operatorname{grad} \underline{v})^T = [\operatorname{grad} \operatorname{div} \underline{v}]^T .$$

and apply divergence to the *linear* STOKES equation, $-\Delta \underline{u} + \operatorname{grad} p = \underline{f}$, yielding

$$-\operatorname{div} \Delta \underline{u} + \operatorname{div} \operatorname{grad} p = \operatorname{div} \underline{f} .$$

But

$$\operatorname{div} \Delta \underline{u} = \operatorname{div} \operatorname{div} \operatorname{grad} \underline{u} = \operatorname{div} \operatorname{grad} \operatorname{div} \underline{u} = \Delta \operatorname{div} \underline{u} = 0\,,$$

because $\operatorname{div} \underline{u} = 0$ by assumption hence

$$\boxed{\Delta p = \operatorname{div} \underline{f}}\,.$$

On the other side, multiplying the STOKES equation on the boundary Γ by the normal vector $\underline{n}$, yields the condition $\Delta \underline{u} \cdot \underline{n} + \partial p / \partial \underline{n} = \underline{f} \cdot \underline{n}$. On summarizing we obtain a NEUMANN problem for the pressure p,

$$\boxed{\begin{array}{ll} \Delta p = \operatorname{div} \underline{f} & \text{in } \Omega \\ \dfrac{\partial p}{\partial n} = \Delta \underline{u} \cdot \underline{n} + \underline{f} \cdot \underline{n} & \text{on } \partial\Omega \end{array}}\,.$$

(**f2**) (Sohn). Reconsider the homogeneous non-dimensional NAVIER-STOKES equation (8.87),

$$\frac{\partial \underline{u}}{\partial t} + (\nabla \underline{u})\underline{u} - \frac{1}{R_e} \Delta \underline{u} + \nabla p = \underline{0}\,, \quad \operatorname{div} \underline{u} = 0\,. \tag{8.96}$$

An alternative pressure POISSON equation is obtained by differentiating the first momentum equation in (8.96) with respect to x and the second with respect to y, and adding them. The result can be written as

$$\Delta p = -\nabla[(\nabla \underline{u})\underline{u}] + R_e^{-1}[(\Delta u)_x + (\Delta v)_y]\,. \tag{8.97}$$

Let q be an arbitrary test function for p and apply GREEN's theorem then we obtain the weak form where $\underline{n}$ is the unit vector normal to boundary $\Gamma = \partial\Omega$ pointing outward from Ω,

$$\begin{aligned} \int_\Omega \nabla q \cdot \nabla p \, d\Omega = &- \int_\Omega \nabla q \cdot (\nabla \underline{u})\underline{u} \, d\Omega + R_e^{-1} \int_\Omega \nabla q \cdot \Delta \underline{u} \, d\Omega \\ &+ \int_\Gamma q\underline{n} \cdot [(\nabla \underline{u})\underline{u} + \nabla p - R_e^{-1} \Delta \underline{u}] \, d\Gamma\,. \end{aligned} \tag{8.98}$$

Substituting once more the momentum equation (8.96) into the line integral yields

$$\int_\Omega \nabla q \cdot \nabla p \, d\Omega = - \int_\Omega \nabla q \cdot (\nabla \underline{u})\underline{u} \, d\Omega + R_e^{-1} \int_\Omega \nabla q \cdot \Delta \underline{u} \, d\Omega - \int_\Gamma q \frac{\partial \underline{u}}{\partial t} \cdot \underline{n} \, d\Gamma\,. \tag{8.99}$$

The line integral involving the time rate of change of the velocities will vanish at all solid boundaries or when steady-state solutions are sought. It is non-zero only along open boundaries in time-dependent flows, or if the flow is excited by the time-varying motion of a wall (Sohn). On the other side, the evaluation of the right side of (8.99) requires the second derivatives of the velocity components (which, by the way, are known globally only as continuous functions). To overcome this difficulty at least locally, (Sohn) proposes a least squares approximation of the first derivatives on the ansatz functions in case they are linear or bilinear and gets appealing results for the latter case.

9

Finite Elements

9.1 Elliptic Boundary Value Problems

Usually, an elliptic problem has three different "faces" namely extremal problem, variational or weak problem, and boundary value problem for a differential system, and either face illuminates the problem from a different point of view. Since ancient times the classical potential equation i.e., POISSON's problem with LAPLACE equation, has served as that basic model to consider the different aspects and to develop appropriate numerical approaches. One encounters it in many physical applications as for instance in bending of an elastic membrane, stationary heat distribution of a plate, or in the computation of minimal surfaces, nothing to say of the representation of electric potential fields.

Let $\Omega \subset \mathbb{R}^3$ be an open connected set (domain) with piecewise smooth boundary $\Gamma := \partial\Omega$. The boundary has to obey certain *regularity conditions* which shall not be discussed here in detail; cf. e.g., (Braess), (Velte). All involved functions shall be defined on the closure $\overline{\Omega}$ and shall be sufficiently smooth. Because of the two different types of boundary conditions (BC) let the boundary be decomposed into DIRICHLET boundary Γ_D and CAUCHY boundary Γ_C:

$$\begin{aligned} \emptyset \neq \Gamma_D \subset \Gamma, \;\; \Gamma_C \subset \Gamma, \;\; \Gamma_D\,, \Gamma_C \text{ open in } \Gamma\,, \\ \Gamma_D \cap \Gamma_C = \emptyset\,, \;\; \Gamma = \overline{\Gamma_D \cup \Gamma_C} \end{aligned} \tag{9.1}$$

which we write for brevity as $\Gamma = \Gamma_D \oplus \Gamma_C$. Moreover, a normal vector $\underline{n}$ of unit length shall exist in every point of Γ_C pointing to the exterior of the domain Ω.

(a) Extremal Problem Point of departure here and later on is the extremal problem for energy of the closed system as it has already been considered to some extent in Sect. 1.11. The "energy functional" of a scalar linear elliptic problem has the general form

$$J(u) = \int_\Omega \left[\frac{1}{2}\nabla u \cdot A(x)\nabla u + \varrho(x)u^2 - f(x)u\right] d\Omega + \int_{\Gamma_C} \left[\beta(x)u^2 - \gamma(x)u\right] d\Gamma \tag{9.2}$$

where $A : \Omega \to \mathbb{R}^3{}_3$, $\varrho, f : \Omega \to \mathbb{R}$ and $\beta, \gamma, \delta : \Gamma \to \mathbb{R}$ shall be continuous functions. The matrix A is symmetric and positive definite and $\varrho \geq 0$; corresponding modifications for two-dimensional problems are evident. Some physical or technical applications are presented later. The domain of definition of $J(u)$, called also objective function, is an affine subspace $u_0 + \mathcal{U} \subset \mathcal{V}$ with a function space $\mathcal{V}$. In detail

$\mathcal{V}$	vector space of all functions u for which $J(u)$ exists, $J(u)$ "lives" on $\mathcal{V}$,	
$\mathcal{U} = \{v \in \mathcal{V}, v = 0 \text{ on } \Gamma_D\}$	subspace of all functions satisfying the specified homogeneous DIRICHLET boundary conditions	(9.3)
$u_0 + \mathcal{U} \subset \mathcal{V}$, $u_0 = \delta$ on Γ_D	affine subspace of all functions satisfying the inhomogeneous DIRICHLET boundary conditions.	

Note that $u_0 = 0$ in case where no *inhomogen ous* DIRICHLET conditions are given, elsewhere the function $u_0 \in \mathcal{V}$ is arbitrary up to the mentioned boundary values on Γ_D.

(b) Weak Form By the Characterization Theorem 1.21 a variational problem (with EULER equations) is associated to the extremal problem (9.2) called weak form of the problem in the sequel,

$$\boxed{\begin{aligned}&\exists\, u \in u_0 + \mathcal{U}\ \ \forall\, v \in \mathcal{U}:\\ &\int_\Omega [\nabla v \cdot A(x)\nabla u + \varrho\, v\, u]\, d\Omega + \int_{\Gamma_C} \beta\, v\, u\, d\Gamma = \int_{\Gamma_C} v\,\gamma\, d\Gamma + \int_\Omega v\, f\, d\Omega\end{aligned}}\,. \tag{9.4}$$

(c) Boundary Value Problem An application of GREEN's formula to (9.4) yields

$$\int_\Omega v\,[-\operatorname{div} A(x)\operatorname{grad} u + \varrho u - f]\, d\Omega + \int_{\Gamma_C} v\,[\operatorname{grad} u \cdot \underline{n} + \beta\, u - \gamma]\, d\Gamma = 0\,. \tag{9.5}$$

The test functions v have to vanish on Γ_D in order that the variations $u + \varepsilon v$ are contained in $u_0 + \mathcal{V}$ as well for $|\varepsilon| \ll 1$. On choosing first v such that $v = 0$ on the entire boundary, the first square bracket reveals to be zero. Choosing thereafter v on Γ_C arbitrary the second square bracket is also zero. By this common argumentation we obtain the boundary problem

$$\boxed{\begin{aligned}-\operatorname{div}(A(x)\operatorname{grad} u) + \varrho u &= f\,, & x &\in \Omega\\ u &= \delta\,, & x &\in \Gamma_D \text{ DIRICHLET BC}\\ \operatorname{grad} u \cdot \underline{n} + \beta u &= \gamma\,, & x &\in \Gamma_C \text{ CAUCHY BC}\end{aligned}}\,. \tag{9.6}$$

The CAUCHY condition (also called third boundary condition) is called NEUMANN condition in case $\beta = 0$.

Obviously every solution of the boundary value problem (9.6) is also solution of the weak problem but the converse does hold only if the solution of (9.5) is sufficiently smooth. All variational functions $u + \varepsilon v$ in the weak problem (9.4) have to satisfy the DIRICHLET boundary conditions, whereas the CAUCHY boundary condition does not occur explicitely in the weak problem resp. only via the functions β and γ. Therefore DIRICHLET conditions are *essential* or *geometrical* boundary conditions and CAUCHY conditions are *natural* or *dynamic* conditions; cf. Sect. 4.1**(a)**.

(d) To prove the **Existence of Solutions**, the problem (9.2) has to be stated in the context of quadratic functionals for application of the Existence Theorem 1.25. To this end the components, namely the vector space $\mathcal{U}$, the bilinear form a and the function $u_0 : \Omega \to \mathbb{R}$ satisfying $u_0 = \delta$ on Γ_D, are to be specified properly. We suppose that Ω is a *bounded domain* with the above mentioned but not more detailed regularity conditions, and that the above function u_0 does exist at all. If continuity of the bilinear form is guaranteed and only DIRICHLET boundary conditions appear then we may restrict ourselves to the homogeneous case following (1.62). Finally, let the matrix $A(x)$ be *uniformly* positive definite on Ω,

$$\exists\, \alpha > 0 \;\; \forall\, x \in \Omega \;\; \forall\, 0 \neq y \in \mathbb{R}^3 : y^T A(x) y \geq \alpha\, y^T y\,.$$

(d1) The problem with DIRICHLET boundary conditions,

$$-\operatorname{div}(A \operatorname{grad} u) + \varrho u = f\,, \quad x \in \Omega; \quad u = 0\,, \quad x \in \Gamma\,, \tag{9.7}$$

$\varrho, f \in \mathcal{C}(\overline{\Omega})$, $u \in \mathcal{C}^2(\overline{\Omega})$, has the weak form $\forall\, v \in \mathcal{H}_0^1(\Omega)$:

$$a(v,u) := \int_\Omega (\nabla v \cdot A \nabla u + \varrho v \cdot u)\, d\Omega = \int_\Omega v \cdot f\, d\Omega =: (v,f)\,, \tag{9.8}$$

$\varrho, f \in \mathcal{L}_2(\Omega)$, $u \in \mathcal{H}_0^1(\Omega)$. The ellipticity of the bilinear form a on HILBERT space $\mathcal{U} = \mathcal{H}_0^1(\Omega)$ follows for $\varrho \geq 0$ immediately by the POINCARÉ-FRIEDRICHS inequality, Lemma 1.13. Therefore the weak problem has a unique solution $u \in \mathcal{H}_0^1$ by the Existence Theorem 1.25 for quadratic functionals.

(d2) The problem with mixed boundary conditions,

$$\begin{aligned} -\operatorname{div}(A \operatorname{grad} u) + \varrho u = f\,, \;\; x \in \Omega; \;\; u = 0\,, \;\; x \in \Gamma_D\,, \\ \operatorname{grad} u \cdot \underline{n} + \beta u = \gamma\,, \;\; x \in \Gamma_C\,, \;\; \beta, \gamma \in \mathcal{C}(\overline{\Gamma}_C)\,, \end{aligned} \tag{9.9}$$

has the weak form $\forall\, v \in \mathcal{U}$:

$$\begin{aligned} & a^*(v,u) := a(v,u) + \int_{\Gamma_C} \beta v u\, d\Gamma = (v,f) + \int_{\Gamma_C} \gamma v\, d\Gamma =: f^*(v) \\ & \varrho, f \in \mathcal{L}_2(\Omega)\,, \;\; \beta, \gamma \in \mathcal{L}_2(\Gamma_C)\,, \;\; u \in \mathcal{U}\,, \end{aligned} \tag{9.10}$$

on HILBERT space $\mathcal{U}$, $\mathcal{H}_0^1(\Omega) \subset \mathcal{U} := \{v \in \mathcal{H}^1(\Omega),\ v = 0 \text{ on } \Gamma_D\} \subset \mathcal{H}^1(\Omega)$. If $\int_{\Gamma_D} d\Gamma > 0$ then $\mathcal{U} \in \mathcal{H}^1(\Omega)$ is a closed subspace by Corollary 1.5. Again the

$\mathcal{U}$-ellipticity of the bilinear form a^* has to be verified for the proof of unique existence of a solution of the weak problem (9.10). For $\varrho \geq 0$ and $\beta \geq 0$

$$\forall\, v \in \mathcal{U} : \int_\Omega \varrho\, v^2\, d\Omega \geq 0\,, \quad \int_{\Gamma_C} \beta\, v^2\, d\Gamma \geq 0\,,$$

hence by Sect. 1.7, Example 1.15(2°)

$$a^*(v,v) \geq |v|^2 \geq m(\Omega)\, \|v\|_1^2\,, \quad m(\Omega) > 0\,.$$

The estimation $a^*(v,v) \leq M(\Omega)\, \|v\|_1^2$ needs

$$\int_{\Gamma_C} \beta\, v^2\, d\Gamma \leq \beta_{\max} \int_\Gamma v^2\, d\Gamma \leq \text{const}\, \|v\|_{1,\Omega}\,,$$

by the Trace Theorem 1.15 and likewise for f^*

$$\int_{\Gamma_C} \gamma\, v\, d\Gamma \leq \left[\int_{\Gamma_C} \gamma^2\, d\Gamma\right]^{1/2} \left[\int_\Gamma v^2\, d\Gamma\right]^{1/2} \leq \text{const}\, \|v\|_0 \leq \text{const}\, \|v\|_1\,.$$

Consequently a unique solution $u \in \mathcal{U}$ of (9.10) exists again by the Existence Theorem 1.25.

(d3) The problem with CAUCHY condition on the entire boundary,

$$\begin{aligned} -\operatorname{div}(A \operatorname{grad} u) + \varrho\, u &= f\,, \\ \operatorname{grad} u \cdot \underline{n} + \beta\, u = \gamma\,, \quad x \in \Gamma;& \quad \beta, \gamma \in C(\overline{\Gamma})\,, \end{aligned} \tag{9.11}$$

has the weak form

$$\forall\, v \in \mathcal{U} \,:\, a^*(v,u) := a(v,u) + \int_\Gamma \beta\, v u\, d\Gamma = (v,f) + \int_\Gamma \gamma\, v\, d\Gamma =: f^*(v)\,. \tag{9.12}$$

where $\mathcal{U} = \mathcal{H}^1(\Omega)$ is the associated HILBERT space.

Case 1: If ϱ is uniformly positive in Ω, $\varrho(x) \geq \varrho_0 > 0$, and $\beta \geq 0$ then, by Example 1.15(1°),

$$a^*(v,v) \geq \min(1,\varrho_0)\left[|v|_0^2 + |v|_1^2\right] \geq m(\Omega)\|v\|_1^2\,, \quad m(\Omega) > 0\,,$$

which proves the ellipticity of a^* on $\mathcal{H}^1(\Omega)$.

Case 2: Let $\varrho \geq 0$ and $\beta \geq 0$ then the bilinear form a^* is elliptic on HILBERT space $\mathcal{U}$ in Corollary 1.5 (3°) by Example 1.15(3°). But the right side f^* has to be an element of the dual space $\mathcal{U}_d$ for an application of the Existence Theorem. By RIESZ' Representation Theorem, a HILBERT space may be identified canonically with its dual space. Therefore the elements $f \in \mathcal{U}_d$ have to satisfy likewise a condition of the form $\int_\Omega v\, d\Omega = 0$. This leads to the condition

$$\int_\Omega f\, d\Omega + \int_\Gamma \gamma\, d\Gamma = 0$$

for the above $f^* \in \mathcal{U}_d$. The same condition is also obtained by the variational equation $a^*(v,u) = f^*(v)$, if $\varrho \equiv 0$ and $\beta \equiv 0$ and an arbitrary constant is chosen for test function v.

The solution of the weak problem does not necessarily satisfy the smoothness required in the differential equation of the associated boundary value problem. This *regularity* may be proved by application of SOBOLEV's Imbedding Theorem together with further estimations. Furthermore, it is also to be shown that the solution u depends continuously on the right side f, the problem is then said to be "well-defined". However, for this deeper results of functional analysis we have to refer to the respective monographs.

See also Sect. 4.1(**e4**) for the dual DIRICHLET problem.

References: (Braess), (Ciarlet79), (Michlin), (Velte).

9.2 From Formula to Figure, Example

Two ways lead to numerical approach of boundary value problems: Either the domain of definition of the *solution* is replaced by a finite-dimensional mesh or the space of functions on which the *differential equation* lives, is replaced by a finite-dimensional subspace where the latter device is rather generously handled sometimes (non-conforming finite elements). *Difference methods* are obtained in the first case as dealed with in Section 2.2 to some part. In the second case numerical devices are the result being named today after RITZ and GALERKIN; cf. also Section 1.11.

Difference methods are used always in initial value problems and for discretization of initial boundary value problems in "time" direction. The example in Sect. 2.2(**h**) shows that they may be used also for discretization of boundary problems and provide acceptable results but their application remains limited to *simple* boundaries and boundary conditions. The formula of differences reaches beyond the boundary in more complicated cases or approximations of higher order and, by consequence, artificial boundary layers are to be introduced. However if the differential equation contains *higher* derivatives of the space variables then difference methods may become attractive again (X.Chen) since finite elements are forced to use very expending approximations here.

The method of RITZ being briefly described in Sect. 1.11 is not applied normally to solve initial value problems but is the first choice today in solving boundary value problems or to discretize initial boundary value problems in *space* direction. In its original form, the unknown solution has been replaced by a polynomial which leads to astonishing exact results if these polynomials are of comparatively low degree. It however leads to systems being difficult to handle if polynomials of higher degree are used hoping to reach better approximation. Therefore *piecewise* polynomial approximations have been proposed here since more accurate numerical approaches became at all possible. Also the resulting systems of equations have the property of *sparse occupancy* which can be exploited fully by using *sparse computation* algorithms.

By their way of construction, finite element approximations are no longer as smooth as the solution of the basic partial differential equation, say (9.6), but, in the *conforming* case, still as smooth as the solution of the *weak problem* $a(v,u) = (v,f)\,,\ v \in \mathcal{V}$ of (9.8). In other words, *conforming* approximations are contained in a *finite subspace* of that vector space $\mathcal{V}$ on which the weak problem lives. Therefore we say briefly that a (piecewise polynomial) finite element approximation is *conforming* with respect to *second* order problems if it forms an overall *continuous* function, i.e., it passes continuously over the element boundaries in all directions. Likewise a finite element approximation is conforming with respect to *fourth* order problems as, e.g., plate problems if it represents a overall *continuously differentiable* function. Conforming elements facilitate considerably the proof of convergence (to the analytic solution for mesh width tending to zero) nevertheless more recently proposed elements for elastic bodies and fluids are mostly non-conforming by various reasons. So, e.g., the smallest conforming triangular element for plate problems has 18 DOFs (degrees of freedom), see BELL's triangle in **(c)**.

(a) Problem In this subsection the individual steps leading to the numerical solution of a boundary value problem shall be presented completely for a simple example that may serve for pattern later on in more complicated applications. To this end we consider the boundary value problem (9.6) in weak form (9.4) and assume that the boundary Γ of the domain $\Omega \in \mathbb{R}^2$ is a polygon. Further, the domain Ω is partitionend into triangles T_i without "pending" nodes and the edges at the boundary are denoted by R_j briefly for the present. We suppose also for simplicity that A is the unit matrix in (9.4), that ϱ is constant on the triangles T_i and that β, γ are constant on the boundary segments R_j . Then the partitioned representation

$$\boxed{\sum_i \int_{T_i} [\operatorname{grad} v \cdot \operatorname{grad} u + \varrho_i\, v\, u - v\, f]\, dxdy + \sum_j \int_{R_j} v\, [\beta_j\, u - \gamma_j]\, ds = 0} \,. \tag{9.13}$$

follows from (9.4). Note however that discontinuities between elements are completely ignored in (9.13). Therefore non-conforming elements must pass IRONS' patch test for correct application (Ciarlet79), Sect. 4.2; see also Sect. 9.4**(f)**.

The triangulation is commonly specified in MATLAB by three matrices, the node or **p**oint matrix "p", the **e**dge matrix "e" and the **t**riangle matrix "t". The point matrix contains the (x, y)-coordinates of the nodes in arbitrary but *strongly fixed* succession. The edge matrix contains the numbers of the terminal points of each boundary edge in counterclockwise order for outer boundary segments and clockwise order for interior boundary segments; additional rows may contain further attributes of each edge as, e.g., parameter intervals and the segment number. The columns of the triangle matrix contain the numbers of the vertex points of every triangle in counterclockwise order; the succession of the columns of "t" may be changed. For further details see the MATLAB suite of this volume.

Example 9.1. Simple triangulation of a square, cf. Fig. 9.2. The three matrices p, e, t associated to the triangulation read

$$p = \begin{bmatrix} 0\,1\,1\,0\,0.5 \\ 0\,0\,1\,1\,0.5 \end{bmatrix}, \quad e = \begin{bmatrix} 1\,2\,3\,4 \\ 2\,3\,4\,1 \end{bmatrix}, \quad t = \begin{bmatrix} 1\,2\,3\,4 \\ 2\,3\,4\,1 \\ 5\,5\,5\,5 \end{bmatrix}.$$

Overlapping of entries in the global stiffness matrix $[K]$ and likewise in the global mass matrix $[M]$ of (9.17):

	1	2	3	4	5
1	A	A			A
2	A	A			A
3					
4					
5	A	A			A

	1	2	3	4	5
1					
2		B	B		B
3		B	B		B
4					
5		B	B		B

	1	2	3	4	5
1					
2					
3			C	C	C
4			C	C	C
5			C	C	C

	1	2	3	4	5
1	D			D	D
2					
3					
4	D			D	D
5	D			D	D

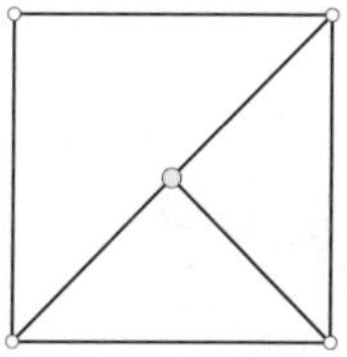

Figure 9.1. A pending node

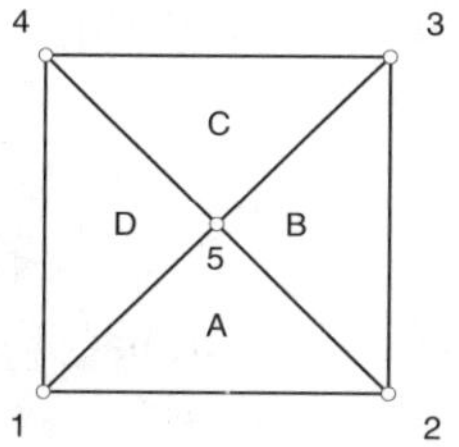

Figure 9.2. Example 9.1

(b) Approximation *The exact solution and its numerical approximation are denoted below by the* same *letter u again!*

Let $T \subset \mathbb{R}^2$ be an arbitrary element of the triangulation then the following integrals have to be evaluated by (9.13):

$$K(v,u) = \int_T (v_x\, u_x + v_y\, u_y)\; dxdy\,,\quad M(v,u) = \int_T v\, u\, dxdy \;\text{ area integrals,}$$
$$P(v,u) = \int_R v\, u\, ds\,,\quad Q(v) = \int_R v\, ds \;\text{ line integrals on boundary.} \tag{9.14}$$

Choose now (at this first stage) a set of mutually distinct node points $P(x_i, y_i)$, $i = 1 : n$, on T and a *two-dimensional interpolating polynomial in* LAGRANGE *form*

$$u(x,y) = u^1\varphi_1(x,y) + \ldots + u^n\varphi_n(x,y)\,,\;\; \underline{u} = [u^1,\ldots,u^n]^T,$$

vanishing outside of T, where the basis polynomials φ_i are shape functions with the *interpolation property* $\varphi_i(x_k, y_k) = \delta^i{}_k$. Then substitution into the integrals (9.14) provides *quadratic rsp. linear forms* for numerical approximation, e.g., with same representation for v,

$$\begin{aligned} M(v,u) &\approx \underline{v}^T M \underline{u}\,,\ M = [m^i{}_j]_{i,j=1}^n\,,\ m^i{}_j = \int_T \varphi_i \varphi_j \, dxdy\,,\\ K(v,u) &\approx \underline{v}^T K \underline{u}\,,\ \ K = [k^i{}_j]_{i,j=1}^n\,,\ \ k^i{}_j = \int_T [\varphi_{i,x}\varphi_{j,x} + \varphi_{i,y}\varphi_{j,y}]\, dxdy\,. \end{aligned} \tag{9.15}$$

The right side f of the differential equation is appropriately replaced in the same way as u by

$$f(x,y) \approx f^1\varphi_1(x,y) + \ldots + f^n\varphi_n(x,y)\,,\ \ \underline{f} = [f^1, \ldots, f^n]^T\,,$$

as far as f is not constant. For a fixed triangle in the sum (9.13) we then obtain in case $\beta = \gamma = 0$

$$\int_T [\operatorname{grad} v \cdot \operatorname{grad} u + \varrho_i\, v\, u - v\, f]\, dxdy \approx \underline{v}^T \left[(K + \varrho_i\, M)\underline{u} - M\underline{f}\right].$$

Let, e.g., ϱ be constant in Ω then the summation of all these equations yields an approximation of (9.13),

$$[V]^T \Big[[[K] + \varrho\,[M]]\,[U] - [M][F] \Big] = 0\,. \tag{9.16}$$

Since $[V]$ is an arbitrary vector up to now, (9.16) leads immediately to the linear system of equations

$$\boxed{[A][U] = [R]\,,\ \ [A] = [K] + \varrho\,[M]\,,\ \ [R] = [M][F]\ \ \text{(or the like)}} \tag{9.17}$$

for the (unknown) *global node vector* $[U]$.

(c) Linear Triangular Elements (COURANT's element). In the *most simple case* a *linear ansatz*

$$u(x,y) = u^1\varphi_1(x,y) + u^2\varphi_2(x,y) + u^3\varphi_3(x,y)$$

is chosen in triangle T where $\varphi_i = \alpha_i + \beta_i x + \gamma_i y$ are again three linearly independent functions vanishing outside of T. The components are chosen in a way that these functions are *shape functions* relative to the three vertices $P_k(x_k, y_k)$, $k = 1:3$, of T numerated counterclockwise. Write briefly $x_{ik} = x_i - x_k$, $y_{ik} = y_i - y_k$ then $|T| = [x_{21}y_{31} - x_{31}y_{21}]/2$ is the area of triangle of T. The desired result for the above area integrals (9.15) resp. the associated stiffness matrix K and mass matrix M then reads:

$$\frac{1}{4|T|}\begin{bmatrix} x_{32}^2 + y_{32}^2 & y_{23}y_{31} + x_{23}x_{31} & x_{32}x_{21} + y_{32}y_{21} \\ & x_{31}^2 + y_{31}^2 & y_{13}y_{21} + x_{13}x_{21} \\ \text{symm.} & & x_{21}^2 + y_{21}^2 \end{bmatrix},\ \ \frac{|T|}{12}\begin{bmatrix} 2 & 1 & 1 \\ 1 & 2 & 1 \\ 1 & 1 & 2 \end{bmatrix}. \tag{9.18}$$

The resulting global *mass matrix* $[M]$ of (9.17) is symmetric, positive definite and well-conditioned, the resulting global *stiffness matrix* $[K]$ however is symmetric and positive semi-definite but ill-conditioned as already has been shown in the similar situated example of Sect. 2.2**(h)** where the matrix $[A]$ in (9.16) is ill-conditioned as well and is even singular for $\varrho = 0$.

(d) Implementation of Dirichlet Boundary Conditions

(d1) *Direct Method.* The system (9.17) is *modified* by means of the boundary conditions such that the resulting system enjoys a positive definite matrix $[A]$ also in case $\varrho = 0$. Then a CHOLESKY decomposition may be applied for solution being advantageous from numerical point of view. Let there be, e.g., the boundary condition $U^i = d^i \in \mathbb{R}$ for the value U^i of the node vector $[U]$ at boundary and let $[A] = [a^i{}_k] = [\underline{a}_1, \ldots, \underline{a}_N]$ with columns $\underline{a}_k$, then one has to proceed as follows:
(1°) Form $[G] = [R] - d^i\,\underline{a}_i$,
(2°) Replace the component g^i of $[G]$ by d^i,
(3°) Replace row i and column i von $[A]$ by the zero vector,
(4°) Set $a^i{}_i = 1$.

This operation has to be carried out for *each* corner point on the DIRICHLET boundary (polygon). The result is a modified system $[\widetilde{A}][\widetilde{U}] = [\widetilde{G}]$ with the mentioned properties of which the solution takes the desired boundary values.

(d2) *Direct Method with Reduction* The succession of nodes in the *node matrix* p is not allowed to be permuted since each position is used in the *element matrix* t for identification of the vertices of the triangularization. But obviously the points in the system (9.17) may be separated by permutation into interior points and boundary points without DIRICHLET condition $\mathcal{I}$ on the one side, and boundary points $\mathcal{R}$ with DIRICHLET condition on the other side by using numbers of boundary nodes specified in the *edge matrix* e; cf. Sect. 9.7. After permutation of rows and corresponding columns the following system is obtained in place of (9.17)

$$\begin{bmatrix} [A]_{\mathcal{I},\mathcal{I}} & [A]_{\mathcal{I},\mathcal{R}} \\ [A]_{\mathcal{R},\mathcal{I}} & [A]_{\mathcal{R},\mathcal{R}} \end{bmatrix} \begin{bmatrix} [U]_{\mathcal{I}} \\ [U]_{\mathcal{R}} \end{bmatrix} = \begin{bmatrix} [R]_{\mathcal{I}} \\ [R]_{\mathcal{R}} \end{bmatrix}.$$

After having inserted the DIRICHLET boundary conditions $[U]_{\mathcal{R}} = [D]_{\mathcal{R}}$, the lower block row may be cancelled and the first row leads to the reduced system

$$[A]_{\mathcal{I},\mathcal{I}}[U]_{\mathcal{I}} = [R]_{\mathcal{I}} - [A]_{\mathcal{I},\mathcal{R}}[D]_{\mathcal{R}}$$

for the values of displacement u at the node points with index in the index set $\mathcal{I}$. The reduced matrix $[A]_{\mathcal{I},\mathcal{I}}$ is positive definite in a properly posed elliptic problem but continues to be ill-conditioned as well. After solution, the node vector $[U]_{\mathcal{I}}$ has to be completed again by $[U]_{\mathcal{R}}$ regarding the former permutation. Especially, this method of implementing DIRICHLET boundary conditions is applied in *eigenvalue problems* (oscillation problems) where the boundary conditions are homogeneous, $[D]_{\mathcal{R}} = \underline{0}$.

(d3) *Indirect Method after* LAGRANGE. The system (9.17) is *augmented* to a system with LAGRANGE matrix, cf. Sect. 1.1**(e)**, which is still symmetric and regular but no longer definite (in normal case). To this end, let the DIRICHLET conditions be written as $[C][U] = [D]$ where $[C]$ is a (P, N)-matrix of *maximum* rank $P < N$, and let $[U]_0$ be an arbitrary vector, e.g., the null-vector. By Theorem 3.8 a vector $[U] = [U]_0 - [X]$ with solution $[X]$ of

$$\begin{bmatrix} [A] & [C]^T \\ [C] & [O] \end{bmatrix} \begin{bmatrix} [X] \\ [Y] \end{bmatrix} = \begin{bmatrix} [A][U]_0 - [R] \\ [C][U_0] - [D] \end{bmatrix}$$

is a solution of the original system (9.17) and satisfies the boundary conditions $[C][U] = [D]$. The matrix Matrix $[A]$ must however be positive definite on the kernel of the matrix $[C]$ in order that the LAGRANGE matrix is regular. This condition represents the discrete analogon to the possible choice of suitable boundary conditions in Sect. 9.1. This method is more flexible than the direct method and allows incorporating of more complicated boundary conditions as, e.g., $\underline{u} \cdot \underline{n} = 0$ in a flow problem where $\underline{u}$ is the velocity and $\underline{n}$ the normal of the boundary, or $\int_\Omega p\, d\Omega = 0$ where p is the pressure.

(e) Implementation of Cauchy Boundary Conditions The boundary integrals in (9.13) have to be approximated numerically *only* in the case where the coefficients β_i and/or γ_i are non-zero. In the *most simple case* a *linear* ansatz supplies the same result as in a linear rod element, cf. Sect. 7.1,

$$P(v, u) \approx \underline{v}^T P_R \underline{u}\,, \quad Q(v) \approx Q_R^T \underline{v}\,, \quad P_R = \frac{L}{6} \begin{bmatrix} 2 & 1 \\ 1 & 2 \end{bmatrix}\,, \quad Q_R^T = \frac{L}{2}\,[1\,,\ 1]\,.$$

The shape of the specific boundary segment R is involved here only by its length L. Summarizing of the matrices $\beta_i\, P_{R_i}$ and the vectors $\gamma_i\, Q_{R_i}$ in (9.13) yields, in place of (9.17), the linear system

$$\boxed{\Big[[K] + \varrho\,[M] + [M_\beta]_R\Big][U] = [R] + [B_\gamma]_R}\ .$$

ATTENTION: This system has to be formed *before* the implementation of DIRICHLET boundary conditions.

Example 9.2. By Section 9.7**(a)**, the *stationary heat distribution* u in a disc Ω satisfies the elliptic differential equation $-\operatorname{div}(\lambda \operatorname{grad} u) = f$. In Figure 9.4 we have chosen $\lambda = 1$ and $f = 10$ and four different types of boundary conditions.

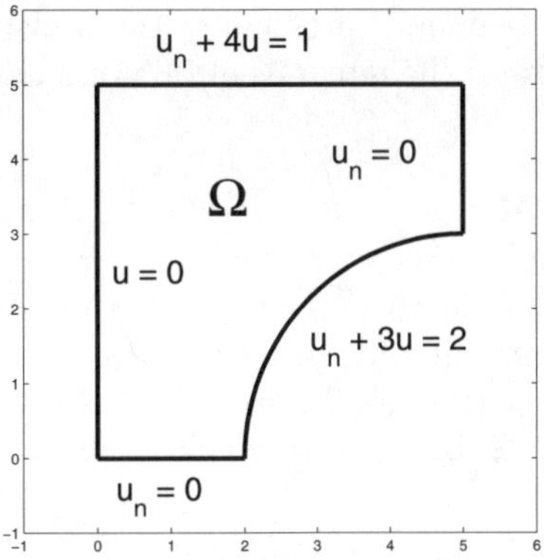

Figure 9.3. Ex. 9.2, geometry

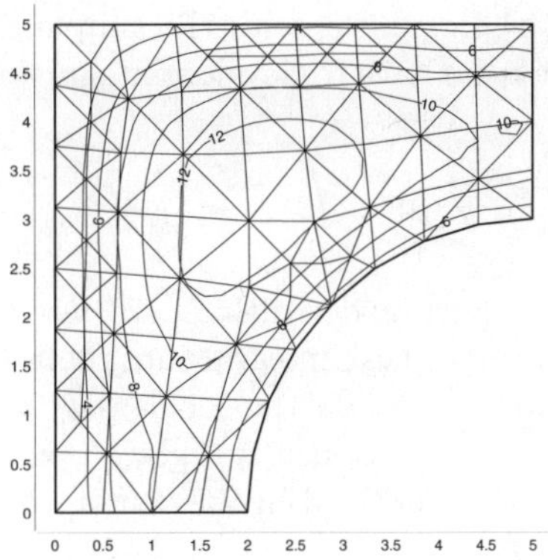

Figure 9.4. Ex. 9.2, isothermes

9.3 Constructing Finite Elements

(a) Problem Consider an elliptic boundary problem of type (9.13) and let the triangle T be an element of a triangular decomposition of the basic domain Ω (with polygonal boundary). The unknown solution u is replaced on T by an *interpolating polynomial* being denoted by the same letter u for simplicity.

Notations for triangle $T \subset \mathbb{R}^2$ in general position and *global* (x, y)-coordinates:

(1°) $P_i(x_i, y_i)$, $i = 1, 2, 3$, *counterclockwise* vertices of T .
(2°) $\mathcal{U}$ vector space of polynomials on T with n degrees of freedom.
(3°) $\Theta(x, y) = [1, x, y, \ldots]^T$ *column* vector of algebraic basis of $\mathcal{U}$,
$\underline{c} = [\gamma^1, \ldots, \gamma^n]^T$ *column* vector of coefficients,
$u(x, y) = \Theta(x, y)^T \underline{c} \in \mathcal{U}$ polynomial ansatz on T .
(4°) $K_j(x_j, y_j)$, $j = 1 : n$, support *nodes* in T (mutually distinct for the present).
(5°) $u^j = u(K_j) = u(x_j, y_j)$, $i = 1 : n$, support *values* of u at points K_j ,
$\underline{u} = [u^1, \ldots, u^n]^T$ column vector of support ordinates.
(6°) $x_{21} = x_2 - x_1$ etc. short-cuts for coordinate differences in vertices, $J := x_{21}y_{31} - x_{31}y_{21} > 0$ double area of T.

Hint. We suppose first that the elements of $\Theta(x, y)$ form a *complete* basis of the vector space $\Pi_n(x, y)$ of polynomials for some degree n but there are many interesting exceptions. For instance, the complete cubic polynomial has ten degrees of freedom (DOF) but in solving plate problems diverse nonconforming modifications with nine DOFs are generally preferred. Also the vertices of the triangle (quadrangle) are not always support points of the interpolating polynomial, for instance in some elements for the NAVIER-STOKES equation; cf. (Turek) and `KAPITEL09\FEM_1`.

(b) Beginning with the simple elliptic boundary problem of the previous section we have to develop integration rules for the integrals of (9.14), namely

$$K(v,u) = \int_T (v_x\, u_x + v_y\, u_y)\, dxdy\,, \quad M(v,u) = \int_T v\, u\, dxdy\,; \tag{9.19}$$

higher partial derivatives are later involved in the approximation of plate problems. These integrals are replaced numerically by *quadratic forms* $\underline{u}^T A_i \underline{u}$ where $\underline{u} \in \mathbb{R}^n$ is the local *node* vector on triangle T. In a *first step*, the unknown function u is replaced on triangle T by a linear combination of the *monoms* contained in the vector $\Theta(x,y)$,

$$u(x,y) = \Theta(x,y)^T \underline{a} = \alpha^1 \vartheta_1(x,y) + \ldots + \alpha^n \vartheta_n(x,y)\,,$$

e.g., $u(x,y) = \alpha^1 + \alpha^2 x + \alpha^3 y$ (same letter u in approximation). Then we have to change that basis into $u = \Phi(x,y)^T \underline{u}$ to obtain the today commonly used representation by *shape functions* shape functions $\Phi = (\varphi_1, \ldots \varphi_n)$ which are the straightforward generalization of the well-known LAGRANGE basis functions (2.7) to the present two-dimensional case. To this end we use the otherwise less popular *direct interpolation* $u^i = \Theta(x_i, y_i)\underline{a}\,,\ i = 1:n$, of § 2.1**(a)** because the dimension of that system remains moderate in the present context.

> To express the algebraic vector $\underline{a}$ by the node vector $\underline{u}$ we need the *inverse* B_T^{-1} of the matrix $B_T = [\Theta(x_i, y_i)]_{i=1}^n$ which therefore must be regular.

This inverse matrix, depending here on the individual triangle T, is frequently called *design matrix* of the element. There are however elements violating the regularity condition for some triangle constellations (e.g., TOUCHER's element) so that they require some specific *regularity* from the triangular domain decomposition. Also the inversion of $[\Theta(x_i, y_i)]_{i=1}^n$ limits the present direct way of construction in some applications. (But for checking it is always advantageous to have an alternative way of calculation.)

Now we obtain the basis Φ of general shape functions from the algebraic basis Θ by a simple transformation,

$$\Phi(x,y)^T \underline{u} = \Theta(x,y)^T \underline{a} = \Theta(x,y)^T B_T^{-1} \underline{u}\,,\ \varphi_k(x,y) = \Theta(x,y)^T \underline{\widetilde{b}}_k\,,\ i = 1:n\,,$$

where $\underline{\widetilde{b}}_k$ are the *columns* of the design matrix B_T^{-1}.

Example 9.3. (COURANT's triangle) Consider the linear approximation, $u = \Theta^T \underline{a} = \alpha^1 + x\alpha^2 + y\alpha^3$, and suppose that the vertices $P_i(x_i, y_i)\,,\ i = 1:3$, are node points with node values u^i. Then $u^i = \alpha^1 + \alpha^2 x_i + \alpha^3 y_i\,,\ i = 1:3$,

$$\underline{u} = \begin{bmatrix} 1 & x_1 & y_1 \\ 1 & x_2 & y_2 \\ 1 & x_3 & y_3 \end{bmatrix} \underline{a}\,,\ \underline{a} = \frac{1}{J} \begin{bmatrix} x_2 y_3 - x_3 y_2 & x_3 y_1 - x_1 y_3 & x_1 y_2 - x_2 y_1 \\ y_{23} & y_{31} & y_{12} \\ x_{32} & x_{13} & x_{21} \end{bmatrix} \underline{u} =: B_T^{-1} \underline{u}\,.$$

Then $\Phi(x,y)^T = \Theta(x,y)^T B_T^{-1}$ and, e.g., $\varphi_1(x,y) = J^{-1}(x_2y_3 - x_3y_2 + y_{23}x + x_{32}y)$ whereas φ_2, φ_3 follow by cyclic permutation of indices $i = 1:3$.

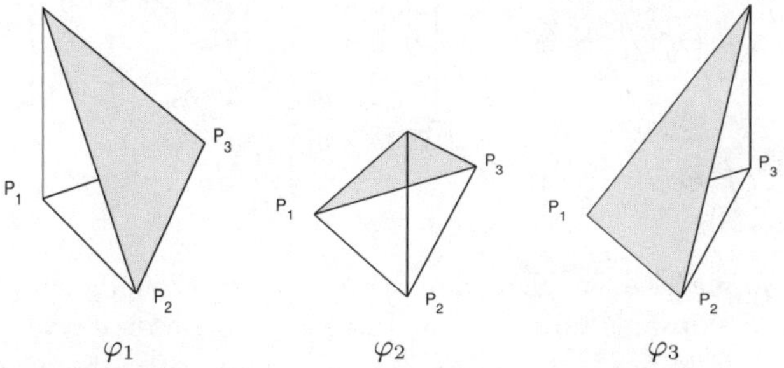

Figure 9.5. Shape functions of COURANT's triangle

If once the shape functions Φ are found, we insert $u(x,y) = \Phi^T \underline{u}$ into (9.19) as *second step* of construction,

$$K(v,u) = \int_T \left(\underline{v}^T \Phi_x \Phi_x^T \underline{u} + \underline{v}^T \Phi_y \Phi_y^T \underline{u}\right) dxdy\,, \quad M(v,u) = \int_T \underline{v}^T \Phi\, \Phi^T \underline{u}\, dxdy\,; \tag{9.20}$$

where $\Phi\Phi^T \in \mathbb{R}^n{}_n$, etc., denotes a *dyadic product.* Regarding once more the important relation $\Phi^T = \Theta^T B_T^{-1}$ between the algebraic basis Θ and the LAGRANGE basis Φ of shape functions, we obtain, e.g.,

$$\begin{aligned} K_1(v,u) &= \underline{v}^T B_T^{-T} \left[\int_T \Theta_x \Theta_x^T\, dxdy\right] B_T^{-1} \underline{u}\,, \\ K_2(v,u) &= \underline{v}^T B_T^{-T} \left[\int_T \Theta_y \Theta_y^T\, dxdy\right] B_T^{-1} \underline{u}\,, \end{aligned} \tag{9.21}$$

and a similar result for $M(v,u)$. In a *third step* we then have to calculate the matrix-valued domain integrals

$$\int_T \Theta_x \Theta_x^T dxdy\,, \int_T \Theta_y \Theta_y^T dxdy\,, \int_T \Theta\, \Theta^T dxdy \in \mathbb{R}^n{}_n$$

by going back to a *reference configuration* (e.g., a *unit triangle* in triangular decompositions) or by evaluating the integrals $\int_T x^r y^s\, dxdy$ directly, see § 2.3(**f3**).

All three steps together constitute a "road map" for making finite elements from a pure computational point of view. Requirements on physical and mechanical properties or even on convergence and stability are completely neglected at this first glance but must be regarded later.

Example 9.4. In the most simple case of COURANT's triangle $\Theta(x,y) = (1,\, x,\, y)$,

$$
\begin{aligned}
\int_T \Theta_x\,\Theta_x^T\,dxdy &= \int_T \begin{bmatrix} 0 & 0 & 0 \\ 0 & 1 & 0 \\ 0 & 0 & 0 \end{bmatrix} dxdy = \begin{bmatrix} 0 & 0 & 0 \\ 0 & |T| & 0 \\ 0 & 0 & 0 \end{bmatrix} \\
\int_T \Theta_y\,\Theta_y^T\,dxdy &= \int_T \begin{bmatrix} 0 & 0 & 0 \\ 0 & 0 & 0 \\ 0 & 0 & 1 \end{bmatrix} dxdy = \begin{bmatrix} 0 & 0 & 0 \\ 0 & 0 & 0 \\ 0 & 0 & |T| \end{bmatrix} \\
\int_T \Theta\,\Theta^T\,dxdy &= \int_T \begin{bmatrix} 1 & x & y \\ x & x^2 & xy \\ y & xy & y^2 \end{bmatrix} dxdy\,.
\end{aligned}
\tag{9.22}
$$

The first two matrices in (9.20) are of course exceptionally simple. The third matrix may be obtained directly by Table 2.3 but observe that first the local coordinate system has to be moved into the center of T. So we obtain, e.g.,

$$
\int_T xy\,dxdy = \frac{|T|}{24}[2x_1y_1+x_1y_2+x_2y_1+2x_2y_2+2x_3y_3+x_3y_1+x_3y_2+x_1y_3+x_2y_3]
\tag{9.23}
$$

but the *complete* result of this example is more simple again and already given in (9.18).

(c) Reduction to Unit Triangle Theoretically we may always manage step three of the above computational device by using SYMBOLIC MATHEMATICS and "copy and paste" but Example 9.4 shows that the resulting formulas blow up extraordinarily although the final result is of moderate complexity. Therefore the *reference configuration* is commonly introduced already at the beginning and not only in the last step. Of course all above considerations remain true in this case and we have only to study the transformation. In the sequel we restrict ourselves again to triangles in the plane.

Notations for the unit triangle S in *local* (ξ,η)-coordinates:

(7°) $Q_1(0,0)$, $Q_2(1,0)$, $Q_3(0,1)$ vertices of triangle S.
(8°) $\mathcal{V}$ vector space of polynomials with n degrees of freedom on S.
(9°) $\Theta(\xi,\eta) = [1,\xi,\eta,\ldots]^T$ column vector of algebraic basis of $\mathcal{V}$,
$\underline{a} = [\alpha^1,\ldots,\alpha^n]^T$ column vector of coefficients,
$v(\xi,\eta) = \Theta(\xi,\eta)^T\underline{a} \in \mathcal{V}$ polynomial ansatz in S.
(10°) Affine linear mapping transfering S into T:

$$
\begin{aligned}
\begin{bmatrix} x \\ y \end{bmatrix} &= \underline{g}(\xi,\eta) \quad := \begin{bmatrix} x_1 \\ y_1 \end{bmatrix} + \begin{bmatrix} x_{21} & x_{31} \\ y_{21} & y_{31} \end{bmatrix}\begin{bmatrix} \xi \\ \eta \end{bmatrix} \\
\begin{bmatrix} \xi \\ \eta \end{bmatrix} &= \underline{g}^{-1}(x,y) = \frac{1}{J}\begin{bmatrix} y_{31} & -x_{31} \\ -y_{21} & x_{21} \end{bmatrix}\begin{bmatrix} x-x_1 \\ y-y_1 \end{bmatrix}
\end{aligned}
\tag{9.24}
$$

with partial derivatives — $J = \det \operatorname{grad} \underline{g}(\xi,\eta)$, cf. (6°),

$$
\xi_x = y_{31}/J,\quad \xi_y = x_{13}/J,\quad \eta_x = y_{12}/J,\quad \eta_y = x_{21}/J\,.
\tag{9.25}
$$

(11°) $L_i = L_i(\xi_i,\eta_i) = \underline{g}^{-1}(K_i)$, $i = 1:n$, support points in S.
(12°) Simple but crucial relation: $u(x,y) = v(\xi,\eta)$, $(x,y) = \underline{g}(\xi,\eta)$.

The support points L_i in S are given in applications by the type of element; moreover, the formal vectors $\Theta(x,y)$ and $\Theta(\xi,\eta)$ shall have the same components as the notation indicates, however x, y are *global* coordinates and ξ, η *local* coordinates.

Of course the design matrix B_S^{-1} relative to the unit triangle S does not contain data of the varying triangle T and becomes much simpler by this way. This remains also true if (later on) partial derivatives of the unknown function u are chosen for interpolation but no longer in the case of normal derivatives; see MORLEY's triangle below. Now remember the crucial relation $\Psi(\xi,\eta)^T = \Theta(\xi,\eta)^T B_S^{-1}$ between the LAGRANGE basis Ψ and the algebraic basis Θ in the same way as in **(b)**:

$$v(\xi,\eta) = \Theta(\xi,\eta)^T \underline{a} = \Theta(\xi,\eta)^T B_S^{-1}\underline{u} =: \Psi(\xi,\eta)^T \underline{u} \equiv \sum_{k=1}^{n} u^k \psi_k(\xi,\eta)\,.$$

Let $\underline{\widetilde{b}}_k$ be the columns of the design matrix B_S^{-1} again, i.e. $B_S^{-1} = [\underline{\widetilde{b}}_1, \ldots, \underline{\widetilde{b}}_n]$, then

$$\boxed{u^k\psi_k(\xi,\eta) = \Theta(\xi,\eta)^T\underline{\widetilde{b}}_k u^k\,,\quad \psi_k(\xi,\eta) = \Theta(\xi,\eta)^T\underline{\widetilde{b}}_k\,,\quad k = 1:n}\,.$$

Then, by the rules of substitution (12°) for arbitrary $\underline{u} \in \mathbb{R}^n$, e.g.

$$\begin{aligned}
\int_T \underline{u}^T \Phi(x,y)\,\Phi^T(x,y)\underline{u}\,dxdy &= \underline{u}^T \left[\int_S \Psi(\xi,\eta)\,\Psi(\xi,\eta)^T J\,d\xi d\eta\right]\underline{u} \\
\int_T \underline{u}^T \Phi_x\,\Phi_y^T \underline{u}\,dxdy &= \underline{u}^T \left[\int_S \left(\Psi_\xi\xi_x + \Psi_\eta\eta_x\right)\left(\Psi_\xi\xi_y + \Psi_\eta\eta_y\right)^T J\,d\xi d\eta\right]\underline{u}
\end{aligned} \qquad (9.26)$$

where all *geometric data* of the varying triangle T are condensed in the *numbers* ξ_x, η_x, ξ_y, η_y and J; cf. Notations (10°). So we have to evaluate *once for all* the much simpler matrices

$$\int_S \Psi\Psi^T\,d\xi d\eta\,,\quad \int_S \Psi_\xi\Psi_\xi^T\,d\xi d\eta\,,\quad \int_S \Psi_\xi\Psi_\eta^T\,d\xi d\eta\,,\quad \int_S \Psi_\eta\Psi_\eta^T\,d\xi d\eta \qquad (9.27)$$

The *fundamental* or basic matrices (9.27) are first calculated for a specific finite element and stored up. On demand they are called up and equipped with the geometric data of the individual triangle T. The result is particularly simple representation in which analytic and geometric data are strictly separated.

As long as we deal with *affine-equivalent* LAGRANGE elements where a single transformation $\underline{g}$ suffices for reduction to reference configuration and no derivatives are interpolated, we may also recover the general shape functions Φ on triangle T from the shape functions Ψ on unit triangle S by simple insertion $\Phi = \Psi \circ \underline{g}^{-1}$, i.e.

$$u(x,y) = \Phi(x,y)^T\underline{u} = \Psi(\underline{g}^{-1}(x,y))\underline{u} \equiv \sum_{k=1}^{n} \varphi_k(x,y)u^k\,.$$

Example 9.5. (COURANT's triangle) as in Example 9.3 but w.r.t. to unit triangle S. Define the vector $\Theta(\xi,\eta)^T = [1\,,\,\xi\,,\,\eta]$ then the interpolation condition on the vertices of S reads

$$\begin{array}{l} u^1 = \alpha^1 + 0\cdot\alpha^2 + 0\cdot\alpha^3 \\ u^2 = \alpha^1 + 1\cdot\alpha^2 + 0\cdot\alpha^3 \\ u^3 = \alpha^1 + 0\cdot\alpha^2 + 1\cdot\alpha^3\,, \end{array} \Longrightarrow B_S = \begin{bmatrix} 1 & 0 & 0 \\ 1 & 1 & 0 \\ 1 & 0 & 1 \end{bmatrix}\,,\quad B_S^{-1} = \begin{bmatrix} 1 & 0 & 0 \\ -1 & 1 & 0 \\ -1 & 0 & 1 \end{bmatrix}\,.$$

$$\Psi(\xi,\eta)^T = \Theta(\xi,\eta)^T B_S^{-1} = [1\,,\,\xi\,,\,\eta]\begin{bmatrix} 1 & 0 & 0 \\ -1 & 1 & 0 \\ -1 & 0 & 1 \end{bmatrix} = [1-\xi-\eta\,,\,\xi\,,\,\eta]$$

Then, for instance,

$$A := \int_S \Theta\,\Theta^T\,d\xi d\eta = \int_{\eta=0}^{1}\int_{\xi=0}^{1-\eta}\begin{bmatrix} 1 & \xi & \eta \\ \xi & \xi^2 & \xi\eta \\ \eta & \xi\eta & \eta^2 \end{bmatrix}\,d\xi d\eta = \frac{1}{24}\begin{bmatrix} 12 & 4 & 4 \\ 4 & 2 & 1 \\ 4 & 1 & 2 \end{bmatrix}$$

and the mass matrix $M = JB_S^{-T}AB^{-1}$ is again that of (9.18).

Example 9.6. In the *mini element* for NAVIER-STOKES equations, *pressure* is approximated by a constant in triange T whereas both the components of the *flow velocity* vector are linear interpolated as in Example 9.3 augmented by the cubic "bubble" function which vanishes at the edges and has value one at the center of T (Fig. 9.8). So we have together a basis $\widetilde{\Psi}(\xi,\eta) = [1-\xi-\eta, \xi, \eta, 27\xi\eta(1-\xi-\eta)]^T$ for the interpolation space of a flow component in unit triangle. But this basis is not a basis of shape functions since the first three functions do not vanish in center $(1/3, 1/3)$ being now an additional support point. Therefore a renewed transformation $\Psi(\xi,\eta)^T = \widetilde{\Psi}(\xi,\eta)^T B_S^{-1}$ becomes necessary,

$$\Psi(\xi,\eta)^T = [1-\xi-\eta,\,\xi,\,\eta,\,27\xi\eta(1-\xi-\eta)]\begin{bmatrix} 1 & 0 & 0 & 0 \\ 0 & 1 & 0 & 0 \\ 0 & 0 & 1 & 0 \\ -1/3 & -1/3 & -1/3 & 1 \end{bmatrix}\,.$$

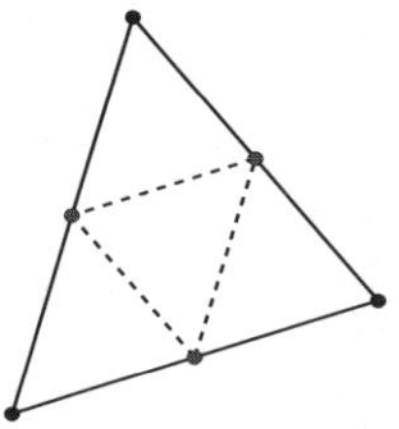

Figure 9.6. Quadratic LAGRANGE element

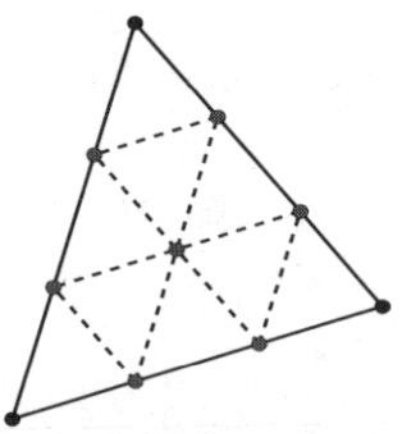

Figure 9.7. Cubic LAGRANGE element

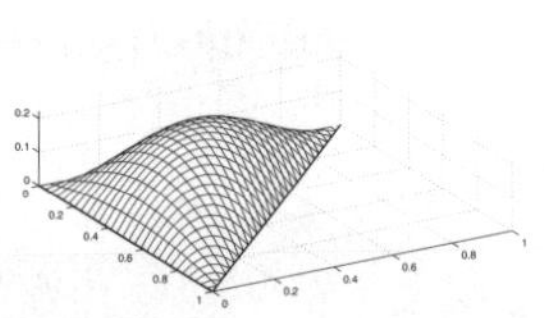

Figure 9.8. "Bubble" function

Barycentric coordinates are nearly always the natural choice in triangles (and higher-dimensional simplices). They allow a very concise and appealing representation which can reduce the computational amount of integration to some extent by its *symmetry.*

Example 9.7. Consider the quadratic ansatz $\Theta = (1, \xi, \eta, \xi^2, \xi\eta, \eta^2)$ in unit triangle then $\Psi^T = \Theta^T B_S^{-1}$ is found again in the same way as above,

$$\begin{aligned}
&\psi_1 = (1-\xi-\eta)(1-2\xi-2\eta) = \zeta_1(2\zeta_1 - 1)\,, && \psi_2 = \xi(2\xi-1) = \zeta_2(2\zeta_2-1)\\
&\psi_3 = \eta(2\eta-1) = \zeta_3(2\zeta_3-1)\,, && \psi_4 = 4\xi(1-\xi-\eta) = 4\zeta_1\zeta_2\\
&\psi_5 = 4\xi\eta = 4\zeta_2\zeta_3\,, && \psi_6 = 4\eta(1-\xi-\eta) = 4\zeta_1\zeta_3
\end{aligned}$$

($\psi_k\,,\ k = 4, 5, 6\,,$ for mid-points of edges). Writing $u^4 = u^{12}\,,\ u^5 = u^{23}\,,$ $u^6 = u^{13}$, we obtain for interpolating polynomial, using again the transformation $(x, y) = \underline{g}(\xi, \eta)\,,$

$$\begin{aligned}
u(x,y) &= \Phi(x,y)^T\underline{u} = \Psi(\xi,\eta)^T\underline{u} = Z(\zeta_1,\zeta_2,\zeta_3)^T\underline{u}\\
&= \sum_{i=1}^{3} u^i\zeta_i(2\zeta_i - 1) + \sum_{1\le i<j\le 3} u^{ij}4\zeta_i\zeta_j\,;
\end{aligned}$$

see also Example 2.6. Now mass matrix, stiffness matrix and other components can be calculated in *schematic form* by applying the integration rule (2.44),

$$\int_T \zeta_1^m\zeta_2^n\zeta_3^p\, d\zeta_1 d\zeta_2 = 2|T|\,\frac{m!n!p!}{(m+n+p+2)!}\,,$$

holding for arbitrary triangles T. This way of proceeding is of special advantage in elements having many degrees of freedom. For instance, the general rule

$$\int_T u\, dxdy \approx \left[\int_T \Phi(x,y)^T\, dxdy\right]\underline{u} = 2|T|\left[\int_S \Phi(\zeta_1,\zeta_2,\zeta_3)^T\, d\zeta_1 d\zeta_2\right]\underline{u} = \underline{r}^T\underline{u}$$

does hold where $r_k = 2|T|\displaystyle\int_S \varphi_k(\zeta_1,\zeta_2,\zeta_3)\, d\zeta_1 d\zeta_2$. For instance, the above quadratic ansatz yields

$$\begin{aligned}
r_1 = r_2 = r_3 &= 2|T|\int_S \zeta_1(2\zeta_1-1)\, d\zeta_1 d\zeta_2 = 2|T|\left[2\frac{2!}{4!} - \frac{1}{3!}\right] = 0\,,\\
r_4 = r_5 = r_6 &= 2|T|\int_S 4\zeta_1\zeta_2\, d\zeta_1 d\zeta_2 \qquad = 2|T|\,4\frac{1!1!}{4!} = \frac{2|T|}{6}\,,
\end{aligned}$$

hence $\underline{r}^T = 2|T|[0\,,\,0\,,\,0\,,\,1\,,\,1\,,\,1]/6\,.$

Further Examples will hardly be necessary at the present time.

9.4 Further Topics

The discretization error of finite element approximations depends strongly on the ratio of longest and shortest triangular edge in a triangulation. Therefore "flat" triangles with small angles should be avoided as far as possible. To improve an initial triangulation, long common edges may be replaced by smaller common edges which infers a renumeration of the *element matrix*. Also interior points may be moved into the center of the surrounding polygon. Both modifications are applied repeatedly in ascending and descending order (relative to the sequence of elements in the element matrix). But a domain decomposition regards in normal case only this geometric characteristic and *not* possible principal stress directions, wave fronts, characteristic directions or stream lines. Such an adaption to the analytic structure of a continuum problem would, however, be of great advantage under various aspects. Therefore parallelogram elements or even general quadrilateral elements are employed also, or, beside other procedures, the decomposition regards "characteristic directions" iteratively during computation. General quadrilateral elements allow no longer a transformation to unit square by an *affine linear* mapping similar to the mapping g in (9.20). Therefore they belong to the class of *isoparametric* elements of which we describe one example below; see also for instance the RANNACHER-TUREK element of the MATLAB suite. In contrast, the mapping (9.20) may be used directly in parallelogram elements because one corner of the parallelogram depends linearly on the other three corners. The corners of the parallelogram are numbered by $1-2-3-4$ counterclockwise and the affine linear mapping $\underline{g}$ applies to the corners $1-2-4$. The remaining modifications versus a triangle are of purely technical nature; hence it is referred to the respective programs of the MATLAB suite for details.

The elements considered hitherto are elements of LAGRANGE type where only *values of the function* u are interpolated and no derivatives are employed for approximation. The number of degrees of freedom — strongly connected with the number of support points — may be enhanced arbitrarily with limited numerical improvement, however. One the one side, stiffness and mass matrix become fuller and fuller occupied and, on the other side, global smoothness on Ω cannot be improved by this way. Therefore small elements and fine decompositions are preferred today.

(a) Hermitian Elements As an example for a finite element with derivatives, we consider the complete cubic triangular element with 10 degrees of freedom. It is especially well-suited for *disc problems* dealed with in Sect. 8.6 since stresses in vertices of the triangle are composed in simple way of the *partial derivatives* at those points. But derivatives need some additional transformations in passing from local (ξ, η)-coordinates to global (x, y)-coordinates. (The complete cubic element does however not pass (Irons)' *patch test* for plate problems; see below.)

(a1) Support values of the cubic polynomial in unit triangle S are the three *values* and the six *partial derivatives* of the unknown function u at the vertices of the triangle. In addition the function value is prescribed at the center $Q_{10} = (\xi_{10}, \eta_{10}) = (1/3, 1/3)$ of the unit triangle. We follow strictly the patterns of the previous section, choose the suitable algebraic basis

$$\Theta(\xi, \eta) = [1, \xi, \eta, \xi^2, \xi\eta, \eta^2, \xi^3, \xi^2\eta, \xi\eta^2, \eta^3]^T$$

but choose an *intermediate* node vector

$$\underline{\widetilde{u}} = [\underline{\widetilde{u}}^1, \underline{\widetilde{u}}^2, \underline{\widetilde{u}}^3, u^4]^T\,, \quad \underline{\widetilde{u}}^i = [u^i, v^i_\xi, v^i_\eta]\,,$$

in (ξ, η)-coordinates, where the local partial ξ, η-derivatives are not yet expressed by global x, y-derivatives. Then the *design matrix* B_S^{-1} of the transformation $\widetilde{u}(\xi, \eta) = \Psi(\xi, \eta)^T \underline{\widetilde{u}} = \Theta(\xi, \eta)^T B_S^{-1} \underline{\widetilde{u}}$ is calculated as in Sect. 9.3 by direct interpolation,

$$\begin{aligned} u^i = \Theta(\xi_i, \eta_i)\underline{a}\,, \quad v^i_\xi = \Theta_\xi(\xi_i, \eta_i)\underline{a}\,, \quad v^i_\eta = \Theta_\eta(\xi_i, \eta_i)\underline{a}\,, \quad i = 1:3 \\ \widetilde{u}(\xi_{10}, \eta_{10}) = \Theta(1/3, 1/3)\underline{a} \end{aligned}$$

(10 equations) where simply $(\xi_1, \eta_1) = (0,0)$, $(\xi_2, \eta_2) = (1,0)$, $(\xi_3, \eta_3) = (0,1)$. Recall that $\psi_k = \Theta^T \underline{\widetilde{b}}_k$, $k = 1:10$, where $\underline{\widetilde{b}}_k$ is the k-th column of B_S^{-1} so for instance — see Sect. 12.1(**c**)

$$\psi_1 = (1 - \xi - \eta)[(1 - \xi + 2\eta)(1 + 2\xi - \eta) - 16\xi\eta] = \zeta_1^2(3 - 2\zeta_1) - 7\zeta_1\zeta_2\zeta_3\,.$$

The shape function of the center is again the bubble function of Example 9.6, $\psi_{10} = 27\xi\eta(1 - \xi - \eta) = 27\zeta_1\zeta_2\zeta_3$ which disappears at the edges of the triangle S.

(a2) The partial derivatives of v w.r.t. ξ, η have now to be expressed by partial derivatives of u w.r.t. x, y in a second step. Using the transformation (9.24) we have $\text{grad}_\xi v = \text{grad}_{\underline{x}} u \, \text{grad}_{\underline{\xi}} \underline{g}$ where $\widetilde{C}_T := \text{grad}_\xi \underline{g}$ contains the geometric data of the varying triangle T, i.e.

$$\boxed{\begin{matrix} v_\xi = u_x x_\xi + u_y y_\xi = u_x x_{21} + u_y y_{21} \\ v_\eta = u_x x_\eta + u_y y_\eta = u_x x_{31} + u_y y_{31} \end{matrix} =: \widetilde{C}_T \begin{bmatrix} u_x \\ u_y \end{bmatrix}}\,. \qquad (9.28)$$

This transformation provides the transition from the preliminary node vector $\underline{\widetilde{u}}$ to the final local node vector

$$\underline{u} = [u^1, \ldots, u^{10}]^T := [u_1, u_x^{(1)}, u_y^{(1)}, u_2, u_x^{(2)}, u_y^{(2)}, u_3, u_x^{(3)}, u_y^{(3)}, u_4]^T$$

by means of a further matrix: Let $\widehat{C}_T := \begin{bmatrix} 1 & 0 \\ 0 & \widetilde{C}_T \end{bmatrix}$ and let

$C_T = \text{diag}[\widehat{C}_T\,, \widehat{C}_T\,, \widehat{C}_T\,, 1]$ be a block diagonal matrix then $\underline{\widetilde{u}} = C_T \underline{u}$. Let for instance $\widetilde{A}_i$ be the *preliminary* fundamental matrices for a simple elliptic problem,

$$\widetilde{A}_1 = \int_S \Theta_\xi \, \Theta_\xi^T \, d\xi d\eta \,, \quad \widetilde{A}_2 = \int_S \Theta_\xi \, \Theta_\eta^T \, d\xi d\eta$$
$$\widetilde{A}_3 = \int_S \Theta_\eta \, \Theta_\eta^T \, d\xi d\eta \,, \quad \widetilde{A}_4 = \int_S \Theta \, \Theta^T \, d\xi d\eta \,,$$

then the final *fundamental matrices* $A_i = B_S^{-T} \widetilde{A}_i B_S^{-1}$, $i = 1 : 4$ do not depend on the data of the varying triangle T and can be calculated once for all in advance. Now, e.g.,

$$\underline{u}^T \left[\int_S \Psi_\xi \Psi_\xi^T \, d\xi d\eta \right] \underline{u} = \underline{\widetilde{u}}^T B_S^{-T} \widetilde{A}_1 B_S^{-1} \underline{\widetilde{u}} = \underline{u}^T C_T^T B_S^{-T} \widetilde{A}_1 B_S^{-1} C_T \underline{u} \,, \quad (9.29)$$

and the matrices $\int_T \Phi_x \Phi_x^T dxdy$, $\int_T \Phi_y \Phi_y^T dxdy$ are found in exactly the same way as in (9.26). However, because of the matrix C_T, these calculations do no longer separate strictly analytic data of the element and geometric data of triangle T. The geometry-independent matrices $B_S^{-T} \widetilde{A}_i B_S^{-1}$ are listed up in program `fem_drksch.m` of the MATLAB suite. For an alternative representation of the cubic interpolating polynomial in barycentric coordinates see Sect. 2.3(**e**).

By contrast with the vertices, the center node is only once occupied in this element. Also an element corresponds with his neighbors only by the boundary nodes. Therefore various attempts have been made to cancel this tenth degree of freedom. One way is to reduce the associated final matrix by *condensation* as in `fem_drksch.m` of the MATLAB suite, see also Sect. 1.1(**e3**). But an element derived on this basis does not converge in *plate problems* (Zienkiewicz), vol. I, p. 28.

(b) Normal Derivatives Finite elements are called *conforming* if they enjoy the same global smoothness as the solution of the basic *weak* problem. For instance, solutions of the weak elliptic problem in Sect. 9.2 and of disc problems in Sect. 8.6 are generically continuous therefore elements for those problems are *conforming* if the are continuous everywhere on the domain Ω. Solutions of plate problems in Sect. 8.7 are generically continuously differentiable hence finite elements are *conforming* in this case if they are continuously differentiable ("$\mathcal{C}^1$ elements"). See also the remarks at the beginning of Sect. 9.2. However the requirement of conformity is too strong for tractable elements in plate problems therefore many attempts have been made to develop alternative convergent approaches without this property.

A very simple non-conforming element — originally for plates — was first proposed by (Morley). It is a quadratic triangular element with node values at the vertices $P(x_i, y_i)$, $i = 1 : 3$, and normal derivatives at the mid-points $P(x_i, y_i)$, $i = 4 : 6$, of the edges; cf. Fig. 9.13. Accordingly,

$$\Theta(\xi, \eta)^T \underline{a} = a_1 + a_2\xi + a_3\eta + a_4\xi^2 + a_5\xi\eta + a_6\eta^2 \,, \quad \underline{a} \in \mathbb{R}^6 \,,$$

and $\underline{u} = [u_1, u_2, u_3, u_{n,4}, u_{n,5}, u_{n,6}]^T \in \mathbb{R}^6$ is the local node vector on triangle T. Let $\underline{n}_i = (c_i, s_i)$, $i = 4 : 6$, be the (normed) normals at the midpoints

$\underline{x}_i$, $i = 4:6$ of the edges pointing outwards and let ℓ_i, $i = 1:3$, be the lengths of the edges then

$$c_4 = y_{21}/\ell_1\,,\ s_4 = -x_{21}/\ell_1\,,\ c_5 = y_{32}/\ell_2\,,\ s_5 = -x_{32}/\ell_2\,,$$
$$c_6 = y_{13}/\ell_3\,,\ s_6 = -x_{13}/\ell_3\,,$$

$$u_{n,i} = \left(\frac{\partial u}{\partial \underline{n}_i}\right)_i = c_i \left(\frac{\partial u}{\partial x}\right)_i + s_i \left(\frac{\partial u}{\partial y}\right)_i, \quad i = 4:6\,. \tag{9.30}$$

Substitution of $u_x = v_\xi \xi_x + v_\eta \eta_x$ and $u_y = v_\xi \xi_x + v_\eta \eta_x$ yields together with the relations (9.25)for ξ_x etc.

$$u_{n,i} = \alpha_i v_\xi(\underline{\xi}_i) + \beta_i v_\eta(\underline{\xi}_i) \tag{9.31}$$

$$\alpha_i = (c_i y_{31} + s_i x_{13})/2|T|\,,\ \ \beta_i = (c_i y_{12} + s_i x_{21})/2|T|\,,\ \ i = 4:6\,.$$

Because of these coefficients, the design matrix *relative to unit triangle S* does now contain already geometric data of the triangle T and must therefore be found for each triangle T by inversion of

$$B_T = \begin{bmatrix} 1 & 0 & 0 & 0 & 0 & 0 \\ 1 & 1 & 0 & 1 & 0 & 0 \\ 1 & 0 & 1 & 0 & 0 & 1 \\ 0 & \alpha_4 & \beta_4 & \alpha_4 & 0.5\beta_4 & 0 \\ 0 & \alpha_5 & \beta_5 & \alpha_5 & 0.5(\alpha_5+\beta_5) & \beta_5 \\ 0 & \alpha_6 & \beta_6 & 0 & 0.5\alpha_6 & \beta_6 \end{bmatrix}. \tag{9.32}$$

For instance, the fourth row is $\alpha_4 \Theta_\xi(1/2,0)^T + \beta_4 \Theta_\eta(1/2,0)^T$. An inversion by means of SYMBOLIC MATHEMATICS is not recommended. By this way we obtain the familiar relation $\Psi(\xi,\eta)^T \underline{u} = \Theta(\xi,\eta)^T \underline{a} = \Theta(\xi,\eta)^T B_T^{-1} \underline{u}$ for a representation of the shape functions in local ξ,η-coordinates. The result is inserted again into the integrals $\int_S \Psi\Psi^T d\xi d\eta$ etc. of (9.27) and it is proceeded in the same way for the higher partial derivatives involved in plate problems.

We calculate the design matrix B of (9.32) but using the normals of the unit triangle S, $\underline{n}_4 = (0, -1)$ in $\underline{\xi}_4 = (1/2, 0)$, $\underline{n}_5 = (1, 1)/\sqrt{2}$ in $\underline{\xi}_5 = (1/2, 1/2)$ and $\underline{n}_6 = (-1, 0)$ in $\underline{\xi}_6 = (0, 1/2)$. Then the shape functions Ψ are obtained via $\Psi(\xi,\eta)^T = \Theta(\xi,\eta)^T B$ (row vector) and expressed in barycentric coordinates by substituting $1 - \xi - \eta = \zeta_1$, $\xi = \zeta_2$ and $\eta = \zeta_3$. The shape functions relative to the normal derivatives at the midpoints of the edges,

$$\begin{aligned} \psi_4 &= \eta - \eta^2 &&= \zeta_3(\zeta_3 - 1) \\ \psi_5 &= (-\xi - \eta + \xi^2 + 2\xi\eta + \eta^2)/\sqrt{2} &&= \zeta_1(\zeta_1 - 1)/\sqrt{2} \\ \psi_6 &= \xi^2 - \xi &&= \zeta_2(\zeta_2 - 1) \end{aligned} \tag{9.33}$$

are correct for *all six node points* and must only be multiplied by a constant $2|T|/\ell_1$, $2\sqrt{2}|T|/\ell_2$, $2|T|\ell_3$ for ψ_4, ψ_5, ψ_6 respectively. But the first three shape functions ψ_i, $i = 1:3$, must be modified by adding a linear combination of (9.33) because they are not adjusted to normal derivatives in x, y-coordinates. It is convenient to choose

$$\widetilde{\psi}_i = \zeta_i + \zeta_i(\zeta_i - 1) + a_i\zeta_{i+1}(\zeta_{i+1} - 1) + b_i\zeta_{i+2}(\zeta_{i+2} - 1)\,,\ i = 1:3 \text{ modulo } 3\,.$$

Then there are nine conditions for six coefficients a_i, b_i , $i = 1:3$,

$$\frac{\partial\widetilde{\psi}_i}{\partial\underline{n}_k}(\underline{\xi}_k) = 0\,,\quad i = 1:3\,,\quad k = 4:6\,,$$

but three of them reveal to be trivial $(0 + 0 = 0)$. The final result is a representation of quadratic polynomials in Morley's form:

$$\boxed{\begin{gathered}\forall\, p \in \Pi_2 : p(\underline{x}) = \sum_{i=1}^{3} p(\underline{x}_i)\varphi_i(\underline{x}) + \sum_{i=4}^{6} \frac{\partial p}{\partial\underline{n}_i}(\underline{x}_i)\varphi_i(\underline{x})\\ \varphi_1 = \zeta_1 + \zeta_1(\zeta_1 - 1) + a_1\zeta_2(\zeta_2 - 1) + b_1\zeta_3(\zeta_3 - 1)\\ a_1 = (y_{31}y_{23} + x_{31}x_{23})/\ell_3^2\,,\ \ b_1 = (y_{12}y_{23} + x_{12}x_{23})/\ell_1^2\\ \varphi_4 = 2|T|\,\zeta_3(\zeta_3 - 1)/\ell_1\end{gathered}}\ .$$

The remaining shape functions φ_2 and φ_3 are obtained by cyclic permutation of indices. Note that the barycentric coordinates ζ_i are to be comprehended as functions of $\underline{x} = (x, y) \in \mathbb{R}^2$ by (2.40) and partial derivatives w.r.t. x and y are also calculated by means of (2.40). See also `SUPPLEMENT\chap09e`.

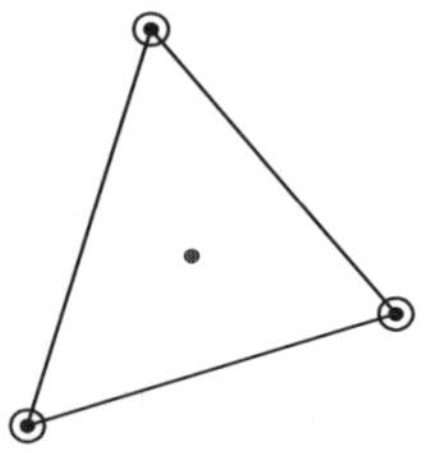

Figure 9.9. Hermitean cubic element

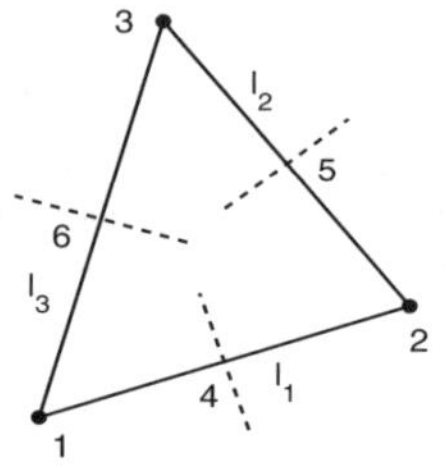

Figure 9.10. Morley's quadratic element

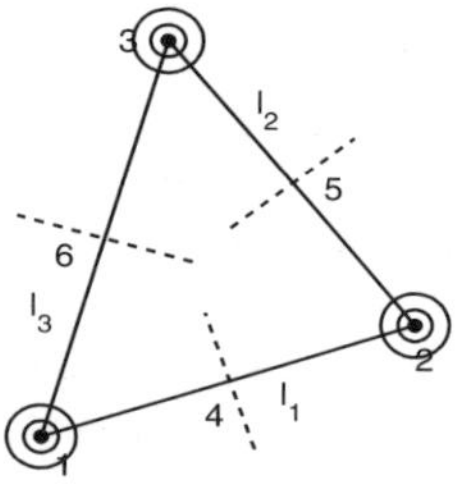

Figure 9.11. Argyris' quintic element

(c) Argyris' Triangle As already mentioned, the number of degrees of freedom has to be considerably enhanced for a C^1 element, i.e., an element yielding continuously differentiable approximation on the *entire* basic domain Ω. The complete quintic element with 21 degrees of freedom due to (Argyris) et al. enjoying this property is uniquely determined by the values as well as all partial derivatives of first and second order at the vertices (18 degrees of freedom) and in addition the normal derviatives at the midpoints of the edges. So we may rely on the above results on cubic elements and Morley's triangle and have only to incorporate the now occuring second partial derivatives.

The algebraic basis of the polynomial vector space over the unit triangle S is

$$\Theta(\xi,\eta) = \Big[1,\xi,\eta|,\ \xi^2,\xi\eta,\eta^2|,\ \xi^3,\xi^2\eta,\xi\eta^2,\eta^3|,\ \xi^4,\xi^3\eta,\xi^2\eta^2,\xi\eta^3,\eta^4|,\\ \xi^5,\xi^4\eta,\xi^3\eta^2,\xi^2\eta^3,\xi\eta^4,\eta^5\Big]^T .$$

The first derivatives are handled as in **(a2)**, and the second derivatives handled in a straight-forward generalization. They are transformed global-local by the chain rule and the *affine linear* mapping (9.20), $\nabla_\xi{}^2 v = [\operatorname{grad}_\xi \underline{g}]^T \nabla_x{}^2 u \operatorname{grad}_\xi \underline{g}$ or explicitely

$$\begin{bmatrix} v_{\xi\xi} \\ v_{\xi\eta} \\ v_{\eta\eta} \end{bmatrix} = \begin{bmatrix} x_{21}^2 & 2x_{21}y_{21} & y_{21}^2 \\ x_{21}x_{31} & x_{21}y_{31}+x_{31}y_{21} & y_{21}y_{31} \\ x_{31}^2 & 2x_{31}y_{31} & y_{31}^2 \end{bmatrix} \begin{bmatrix} u_{xx} \\ u_{xy} \\ u_{yy} \end{bmatrix} =: \widetilde{D}_T \begin{bmatrix} u_{xx} \\ u_{xy} \\ u_{yy} \end{bmatrix} .$$

The integrals (9.27) as, e.g., $\int_S \Psi_\xi \Psi_\xi^T \, d\xi d\eta$ are evaluated at first as in the cubic element by using the auxiliary vector

$$\underline{\widetilde{u}} = [\underline{\widetilde{u}}^1, \underline{\widetilde{u}}^2, \underline{\widetilde{u}}^3]^T \in \mathbb{R}^{21}\ ,\ \ \underline{\widetilde{u}}^i = [u^i, v_\xi^i, v_\eta^i, v_{\xi\xi}^i, v_{\xi\eta}^i, v_{\eta\eta}^i, u_n^i] \tag{9.34}$$

regarding the additional matrix $\widetilde{D}$. The normal derivatives u_n^i , $i = 1:3$, at the edges of triangle T can be handled as in **(b)** of course. This means that the global-local transformation

$$u_n^i = (\operatorname{grad} u)^i \cdot \underline{n}^i = (\operatorname{grad} v)^i \left[(\operatorname{grad} \underline{g})^i\right]^{-1} \underline{n}^i =: (\operatorname{grad} v)^i \cdot \underline{\widetilde{n}}^i$$

is already regarded in the construction of the design matrix $B^{-1} \in \mathbb{R}^{21}{}_{21}$ relative to *unit triangle* S which then depends on the data of varying triangle T. A computation of this parameter-dependent matrix as an *inverse* can but be performed theoretically by SYMBOLIC MATHEMATICS but the result is a tremendous data mismatch. So it remains to compute this matrix numerically for each triangle T as inverse of the matrix B of the (modified) interpolation conditions relative S. The auxiliary vector $\underline{\widetilde{u}}$ is thereafter expressed by the node vector

$$\underline{u} = [\underline{u}^1, \underline{u}^2, \underline{u}^3]^T\ ,\ \ \underline{u}^i = [u^i, u_x^i, u_y^i, u_{xx}^i, u_{xy}^i, u_{yy}^i, u_n^i] .$$

again by means of a matrix-vector multiplication: Let

$$\widehat{D}_T := \begin{bmatrix} 1 & 0 & 0 & 0 \\ 0 & \widetilde{C}_T & 0 & 0 \\ 0 & 0 & \widetilde{D}_T & 0 \\ 0 & 0 & 0 & 1 \end{bmatrix}\ ,\ \ \widetilde{C}_T = \begin{bmatrix} x_{21} & x_{31} \\ y_{21} & y_{31} \end{bmatrix} ,$$

$\widetilde{C}_T$ being the matrix of (9.28), and let $D_T = \operatorname{diag}[\widehat{D}_T , \widehat{D}_T , \widehat{D}_T]$ be a block diagonal matrix, then $\underline{\widetilde{u}} = D_T \underline{u}$. Now, e.g., instead of (9.29),

$$\underline{u}^T \left[\int_S \Psi_\xi \Psi_\xi^T \, d\xi d\eta \right] \underline{u} = \underline{u}^T D_T^T B_T^{-T} \widetilde{A}_1 B_T^{-1} D_T \underline{u} \,, \ \ \widetilde{A}_1 = \int_S \Theta_\xi \Theta_\xi^T \, d\xi d\eta \tag{9.35}$$

where $\underline{u} \in \mathbb{R}^{21}$ is the local node vector relative to global x, y-coordinates. The matrices of the other integrals have the same form with different basic matrix $\widetilde{A}_1$ however.

In spite of its magnitude, ARYRIS' element fascinates in many aspects. It may be applied in problems where higher than second derivatives can no longer be neglected as in the theory of surface waves (X.Chen), and it is also proposed for application in shell theory. The expensive calculation of the matrix inversions can be circumvented theoretically by using a representation of the basic polynomial in barycentric coordinates in a similar way as in MORLEY's triangle. See Sect. 12.1**(d)** and `SUPPLEMENT\chap09f`.

Following a proposition of (Bell), the number of degrees of freedom of the quintic element may be reduced to 18 *without loss of smoothness* (but with loss of accuracy) by the requirement that

| the normal derivative of u is a cubic polynomial on each edge of T | .

This condition allows to eliminate the values of the normal derivative at the mid-points of the edges in a similar way as in **(b)**. BELL's requirement implies that $u_n(s) = a + bs + cs^2 + ds^3$ on each edge after reparametrization. The coefficients of $u_n(s)$ are expressed by four boundary conditions at the terminal points of the individual edge involving mixed derivatives $u_{n,s}$ however but those can be expressed by available first and second partial derivatives of u. For details see `KAPITEL02\SECTION_1_2_3\bell.m` and also Sect. 7.3**(b)**.

(d) A Triangular Element with Curvilinear Edges Obviously the mapping $\underline{g} : S \to T$ of (9.20) must not necessarily be linear as long as it remains *invertible* on every triangle T. In choosing polynomials of higher degree for that mapping we obtain a further class of elements, in particular elements with curved boundary. However, for computational reasons, the edges shall be here polynomials of the *same* degree as the two-dimensional polynomial $\underline{g}$, e.g., quadratic if $\underline{g}$ is quadratic; therefore such elements are called *isoparametric.* The integration of the quadratic forms in (9.24) then has to be managed by a numerical integration rule, normally by a GAUSSian rule, since the weights ξ_x , ξ_y , η_x , η_y in (9.26) are no longer constants. The same holds also for the direct representation via shape functions and, besides, elements with derivatives are more difficult to be implemented here. Because of higher effort in construction and evaluation, this type is receded somewhat into the background due to computational speeds of today.

(d1) For example, we reconsider the quadratic triangular element of Example 9.7, and recall that every triangle T is specified here by *six* support nodes $P_i(x_i, y_i)$, $i = 1 : 6$, including midpoints of edges such that every edge can be quadratically interpolated. Also recall that $\Theta = (1, \xi, \eta, \xi^2, \xi\eta, \eta^2)$ is the algebraic basis in unit triangle and $\Psi^T = \Theta^T B_S^{-1}$ etc. where B_S^{-1} is the design matrix of that element relative to unit triangle. As the element shall be *isoparametric*, the mapping $\underline{g} : S \to T$ is supposed to be also *quadratic*, therefore the requirement $P(x_i, y_i) = \underline{g}(\xi_i, \eta_i)$, $i = 1 : 6$, for the specified support points of the curved triangle T on the one side and the support points of unit triangle S on the other side implies that $\underline{g}$ consists of two quadratic polynomials

$$\begin{bmatrix} x \\ y \end{bmatrix} = \begin{bmatrix} \gamma_1 + \gamma_2\xi + \gamma_3\eta + \gamma_4\xi^2 + \gamma_5\xi\eta + \gamma_6\eta^2 \\ \delta_1 + \delta_2\xi + \delta_3\eta + \delta_4\xi^2 + \delta_5\xi\eta + \delta_6\eta^2 \end{bmatrix} =: \underline{g}(\xi, \eta).$$

of which the coefficients γ_i and δ_i are to be determined by

(x, y)	(ξ, η)
$x_1 = \gamma_1$	$(0, 0)$
$x_2 = \gamma_1 + \gamma_2 + \gamma_4$	$(1, 0)$
$x_3 = \gamma_1 + \gamma_3 + \gamma_6$	$(0, 1)$
$x_4 = \gamma_1 + \frac{1}{2}\gamma_2 + \frac{1}{4}\gamma_4$	$(\frac{1}{2}, 0)$
$x_5 = \gamma_1 + \frac{1}{2}\gamma_2 + \frac{1}{2}\gamma_3 + \frac{1}{4}\gamma_4 + \frac{1}{4}\gamma_5 + \frac{1}{4}\gamma_6$	$(\frac{1}{2}, \frac{1}{2})$
$x_6 = \gamma_1 + \frac{1}{2}\gamma_3 + \frac{1}{4}\gamma_6$	$(0, \frac{1}{2})$
$y_1 = \delta_1$	$(0, 0)$
$y_2 = \delta_1 + \delta_2 + \delta_4$	$(1,0)$
... likewise	

.

Letting $\underline{x} = [x_1, \dots, x_6]^T$ and $\underline{y} = [y_1, \dots, y_6]^T$ we see that the interpolation conditions have again the familiar form $\underline{x} = B_S\underline{\gamma}$ and $\underline{y} = B_S\underline{\delta}$ and the global-local relation between x, y- and ξ, η-coordinates reads, e.g., $x = \Theta^T\underline{\gamma} = \Theta^T B_S^{-1}\underline{x} = \Psi^T\underline{x}$ for x. Therefore we may use the *same* shape functions as in the straight quadratic triangular element for x, y and the unknown function u:

$$u(x, y) = v(\xi, \eta) = \Psi(\xi, \eta)^T\underline{u}, \quad x = \Psi(\xi, \eta)^T\underline{x}, \quad y = \Psi(\xi, \eta)^T\underline{y}.$$

However the gradient of $\underline{g}$ is no longer a constant matrix (relative to triangle T) but varies now with ξ, η. The JACOBI determinant $J(\xi, \eta)$ must be nonzero on S in order that $\underline{g}$ is bijective on S which is apparently true because the elements of the algebraic basis Θ are linearly independent. More explicitely,

$$J(\xi, \eta) = \left|\frac{\partial(x, y)}{\partial(\xi, \eta)}\right| = \begin{vmatrix} x_\xi & x_\eta \\ y_\xi & y_\eta \end{vmatrix} = \begin{vmatrix} \Psi_\xi(\xi, \eta)^T\underline{x} & \Psi_\eta(\xi, \eta)^T\underline{x} \\ \Psi_\xi(\xi, \eta)^T\underline{y} & \Psi_\eta(\xi, \eta)^T\underline{y} \end{vmatrix}.$$

The partial derivatives of the inverse function $\underline{g}^{-1}$ are easily found by CRAMER's rule,

$$\frac{\partial(\xi, \eta)}{\partial(x, y)} = \begin{bmatrix} \xi_x & \xi_y \\ \eta_x & \eta_y \end{bmatrix} = \begin{bmatrix} x_\xi & x_\eta \\ y_\xi & y_\eta \end{bmatrix}^{-1} = \frac{1}{J}\begin{bmatrix} y_\eta & -x_\eta \\ -y_\xi & x_\xi \end{bmatrix}$$

but are now likewise *rational functions* in ξ, η. Consider, e.g., the stiff part of the model problem in Sect. 9.2 then

$$\int_T (u_x^2 + u_y^2)\,dxdy = \int_S (v_\xi \xi_x + v_\eta \eta_x)^2 + (v_\xi \xi_y + v_\eta \eta_y)^2\, J\, d\xi d\eta$$

$$= \underline{u}^T \Big\{ \int_S \big[(\Psi_\xi \xi_x + \Psi_\eta \eta_x)(\Psi_\xi \xi_x + \Psi_\eta \eta_x)^T$$

$$+ (\Psi_\xi \xi_y + \Psi_\eta \eta_y)(\Psi_\xi \xi_y + \Psi_\eta \eta_y)^T\big]\, J\, d\xi d\eta \Big\} \underline{u} = \underline{u}^T K \underline{u}\,.$$

The integrals $\int_S \Psi_\xi\, \Psi_\xi^T\, d\xi d\eta$ etc. can no longer be calculated once for all in advance but the composed integrals $\int_S \Psi_\xi\, \Psi_\xi^T \xi_x^2\, d\xi d\eta$ etc. must be calculated for each triangle T by a numerical integration rule, usually a two-dimensional GAUSSian rule. See also `fem_isodrqell.m` in the MATLAB suite.

(d2) Evaluation of Boundary Integrals

$$\int_{\Gamma \cap T} u^2(s)\,ds\,, \quad \int_{\Gamma \cap T} u(s)\,ds\,.$$

The quadratic shape functions in unit interval $[0,1]$ are

$$\psi_1 = (1-\sigma)(1-2\sigma),\ \psi_2 = 4\sigma(1-\sigma),\ \psi_3 = -\sigma(1-2\sigma)\,.$$

On a boundary edge $R = \Gamma \cap T$ we have as above but one-dimensinal

$$u(s) = v(\sigma) = \Psi(\sigma)^T \underline{u}\,,\ \ x(\sigma) = \Psi(\sigma)^T \underline{x},\ y(\sigma) = \Psi(\sigma)^T \underline{y}\,,\ \ \underline{x} = [x_1, x_2, x_3]^T\,.$$

Substitution yields

$$\int_{\Gamma \cap T} u^2(s)\,ds = \int_0^1 v^2(\sigma)(x'^2(\sigma) + y'^2(\sigma))^{1/2} d\sigma$$

$$= \int_0^1 [\Psi(\sigma)^T \underline{u}]^2 \Big((\Psi'(\sigma)^T \underline{x})^2 + (\Psi'(\sigma)^T \underline{y})^2 \Big)^{1/2} d\sigma$$

$$= \underline{u}^T \Big\{ \int_0^1 \Psi(\sigma)\Psi(\sigma)^T (\ldots)^{1/2} d\sigma \Big\} \underline{u} = \underline{u}^T M_R \underline{u}$$

$$\textstyle\int_{\Gamma \cap T} u(s)\,ds = \Big\{ \int_0^1 \Psi(\sigma)^T (\ldots)^{1/2}\, d\sigma \Big\} \underline{u} = \underline{r}^T \underline{u}\,.$$

The line integrals are commonly evaluated by application of one-dimensional GAUSSian rules again.

(e) Finite Elements for Discs Consider *plane stress* for discs. Then $\underline{u} = (u, v)$ denotes the displacement and, by (8.49) and (8.50),

$$\begin{aligned} \mathcal{E}_T(\underline{u}) &= \frac{h}{2} \int_T \underline{\varepsilon}(\underline{u}) \cdot \underline{\sigma}(\underline{u})\,dxdy\,, \\ \underline{\varepsilon}(\underline{u}) \cdot \underline{\sigma}(\underline{u}) &= \underline{u}^T D^T C_S D \underline{u} \\ &= \frac{E}{1-\nu^2} \left[u_x^2 + 2\nu u_x v_y + v_y^2 + \frac{1}{2}(1-\nu)(u_y + v_x)^2 \right] \end{aligned} \tag{9.36}$$

is the stress energy in an arbitrary triangle T where

$$C_S = \begin{bmatrix} 1 & \nu & 0 \\ \nu & 1 & 0 \\ 0 & 0 & (1-\nu)/2 \end{bmatrix}, \quad D = \begin{bmatrix} \partial_x & 0 \\ 0 & \partial_y \\ \partial_y & \partial_x \end{bmatrix} \tag{9.37}$$

is the elasticity resp. the operator matrix. In direct computation with shape functions we write

$$u = \Phi(x,y)^T \underline{U}\,, \;\; \underline{U} = [u^1, \ldots, u^n]^T\,; \;\; v = \Phi(x,y)^T \underline{V}\,, \;\; \underline{V} = [v^1, \ldots, v^n]^T$$

again. Substitution into (9.36) then yields

$$\frac{hE}{1-\nu^2}\,[\underline{U}^T, \underline{V}^T] \int_T \widetilde{K}\, dF \begin{bmatrix} \underline{U} \\ \underline{V} \end{bmatrix}$$

where $\underline{U}$ and $\underline{V}$ denote the local vector in global node variables. The stiffness matrix is now

$$\widetilde{K} =$$

$$\left[\begin{bmatrix} \partial_x & 0 \\ 0 & \partial_y \\ \partial_y & \partial_x \end{bmatrix}\begin{bmatrix} \Phi(x,y)^T & \mathrm{Null}_n \\ \mathrm{Null}_n & \Phi(x,y)^T \end{bmatrix}\right]^T \begin{bmatrix} 1 & \nu & 0 \\ \nu & 1 & 0 \\ 0 & 0 & (1-\nu)/2 \end{bmatrix}\begin{bmatrix} \partial_x & 0 \\ 0 & \partial_y \\ \partial_y & \partial_x \end{bmatrix}\begin{bmatrix} \Phi(x,y)^T & \mathrm{Null}_n \\ \mathrm{Null}_n & \Phi(x,y)^T \end{bmatrix}$$

The shape functions are expressed by barycentric coordinates, cf. Sect. 2.3**(f)**, **(g)**, and the derivatives Φ_x and Φ_y are calculated. The derivative of a single barycentric coordinate is a real number by (2.40) which depends only on the geometry of the triangle. The integration of the individual components may be managed ensuing by application of formula (2.44) without difficulties and, by the way, may be standardized to a large extent. For instance we obtain a matrix $\widetilde{K}$ of dimension twenty for *cubic elements* by this way. Integration over the general triangle T is reduced to integration over the unit triangle again as repeatedly done in this section,

$$\begin{aligned} \underline{\varepsilon}(\underline{u}) \cdot \underline{\sigma}(\underline{u}) &= \frac{E}{1-\nu^2}\Big[(u_\xi \xi_x + u_\eta \eta_x)^2 + 2\nu(u_\xi \xi_x + u_\eta \eta_x)(v_\xi \xi_y + v_\eta \eta_y) \\ &+ (v_\xi \xi_y + v_\eta \eta_y)^2 + \frac{1}{2}(1-\nu)(u_\xi \xi_y + u_\eta \eta_y + v_\xi \xi_x + v_\eta \eta_x)^2\Big] \\ &= \frac{E}{1-\nu^2}\Big[a_1 u_\xi^2 + 2b_1 u_\xi u_\eta + c_1 u_\eta^2 + a_2 v_\xi^2 + 2b_2 v_\xi v_\eta + c_2 v_\eta^2 \\ &+ 2a_3 u_\xi v_\xi + 2b_3 u_\xi v_\eta + 2c_3 u_\eta v_\xi + 2d_3 u_\eta v_\eta\Big]\,. \end{aligned} \tag{9.38}$$

The coefficients a_i , b_j , c_k depend again on the geometry of the triangle T and are listed in the appertaining programs of the MATLAB suite. Let

$$\widetilde{A}_1 = \int_S \Theta_\xi\, \Theta_\xi^T\, d\xi d\eta\,, \;\; \widetilde{A}_2 = \int_S \Theta_\xi\, \Theta_\eta^T\, d\xi d\eta\,, \;\; \widetilde{A}_3 = \int_S \Theta_\eta\, \Theta_\eta^T\, d\xi d\eta \tag{9.39}$$

be the fundamental matrices , let B_S^{-1} be the design matrix and $\widehat{A}_i = B_S^{-T}\widetilde{A}_i B_S^{-1}$ then

$$h\int_T \underline{\varepsilon}(\underline{u})\cdot\underline{\sigma}(\underline{u}) = [\underline{\widetilde{U}},\underline{\widetilde{V}}]\begin{bmatrix}\widehat{K}_{11} & \widehat{K}_{12}\\ \widehat{K}_{21} & \widehat{K}_{22}\end{bmatrix}\begin{bmatrix}\underline{\widetilde{U}}\\ \underline{\widetilde{V}}\end{bmatrix},$$

$$\begin{aligned}
\underline{\widetilde{U}}^T\widehat{K}_{11}\underline{\widetilde{U}} &= \frac{hE}{1-\nu^2}\int_S\left[a_1u_\xi^2 + 2b_1u_\xi u_\eta + c_1u_\eta^2\right]d\xi d\eta\\
\underline{\widetilde{V}}^T\widehat{K}_{22}\underline{\widetilde{V}} &= \frac{hE}{1-\nu^2}\int_S\left[a_2v_\xi^2 + 2b_2v_\xi v_\eta + c_2v_\eta^2\right]d\xi d\eta\\
\underline{\widetilde{U}}^T\widehat{K}_{12}\underline{\widetilde{V}} &= \frac{hE}{1-\nu^2}\int_S\left[a_3u_\xi v_\xi + b_3u_\xi v_\eta + c_3u_\eta v_\xi + d_3u_\eta v_\eta\right]d\xi d\eta
\end{aligned}$$

$$\begin{aligned}
\widehat{K}_{11} &= \frac{hE}{1-\nu^2}\left[a_1\widehat{A}_1 + b_1(\widehat{A}_2+\widehat{A}_2^T) + c_1\widehat{A}_3\right]\\
\widehat{K}_{22} &= \frac{hE}{1-\nu^2}\left[a_2\widehat{A}_1 + b_2(\widehat{A}_2+\widehat{A}_2^T) + c_2\widehat{A}_3\right]\\
\widehat{K}_{12} &= \frac{hE}{1-\nu^2}\left[a_3\widehat{A}_1 + b_3\widehat{A}_2 + c_3\widehat{A}_2^T + d_3\widehat{A}_3\right]
\end{aligned}\tag{9.40}$$

and $\widehat{K}_{21} = \widehat{K}_{12}^T$. Thereby we get along with three fundamental matrices (9.39) where A_2 is unsymmetric; the computational amount in (9.40) my be still reduced further.

We have $\underline{U} = \underline{\widetilde{U}}$ and $\underline{V} = \underline{\widetilde{V}}$ for the final global node vectors in elements of LAGRANGE type (without derivatives) whereas the additional transformation $\underline{\widetilde{U}} = C_T\underline{U}$, $\underline{\widetilde{V}} = C_T\underline{V}$ has to be regarded in elements of HERMITE type as, e.g., in the above cubic element. The corresponding modifications in parallelogram elements are of pure technical nature.

(f) On the Patch Test Reconsider the homogeneous DIRICHLET problem (9.4): Find $u \in \mathcal{U}$ such that

$$\forall\, v\in\mathcal{U}: \int_\Omega\left[\operatorname{grad} v\cdot\operatorname{grad} u - v\,f\right]d\Omega = 0\,,\quad v = 0 \text{ on } \Gamma = \partial\Omega \tag{9.41}$$

and $u = 0$ on $\Gamma = \partial\Omega$. Recall also the decomposed form (9.13)

$$\sum_i\int_{T_i}[\operatorname{grad} v\cdot\operatorname{grad} u - v\,f]\,dxdy = 0\,,\quad \Omega = \bigcup_i T_i\,, \tag{9.42}$$

being the basic formula for finite element approach. To recover the LAPLACE equation $-\Delta u - f = 0$ from (9.42) we apply GREEN's formula to every triangle T_i and obtain

$$\begin{aligned}
0 &= \sum_i\int_{T_i}[-v\Delta u - v\,f]\,dxdy + \sum_i\oint_{\partial T_i} v\operatorname{grad} u\cdot\underline{n}\,ds \quad\text{or}\\
0 &= -\int_\Omega v(\Delta u + f)\,dxdy + \sum_i\oint_{\partial T_i} v\operatorname{grad} u\cdot\underline{n}\,ds\,.
\end{aligned}\tag{9.43}$$

Obviously the line integrals run over all edges of all triangles T_i not being part of the boundary because $v = 0$ on Γ. The boundary ∂T_i of every triangle T_i is positively oriented such that the interior of T_i lies on the left of the boundary. Thus every *interior edge* E_j of the decomposition is run *twice* in opposite direction. But $\operatorname{grad} u \cdot \underline{n}$ may be *discontinuous* on interelement boundaries and thus may have a *jump* $\widetilde{\Delta} \operatorname{grad} u \cdot \underline{n}$ ($\widetilde{\Delta}$ shall denote the jump here and not a LAPLACE operator). Summing over the edges we can therefore write instead of (9.43)

$$0 = -\int_{\Omega} v(\Delta u + f)\, dxdy + \sum_j \int_{E_j} v\widetilde{\Delta} \operatorname{grad} u \cdot \underline{n}\, ds\,. \tag{9.44}$$

Let us quote (Strang), p. 178: *An essential feature of finite elements is their success on a coarse mesh; even some elements which fail the patch test and are nonconvergent give very satisfactory results for realistic mesh width.* But these jumps cannot be neglected in proving any form of *consistence, convergence* or even "robustness" of the approximation (9.42) for mesh width tending to zero.

Patch Test (Sufficient condition following (Zienkiewicz)). Let the decomposition of the basic domain have only straight edges, and let E denote an arbitrary interior edge. Let $\underline{n}$ and $\underline{t}$ denote the unit normal and tangent of each E, and let φ be an arbitrary shape function of the element under consideration. Then the element passes the patch test if

$$\forall\, E\ \forall\, \varphi : \int_E \widetilde{\Delta} \frac{\partial \varphi}{\partial \underline{t}}\, ds = 0\,, \quad \int_E \widetilde{\Delta} \frac{\partial \varphi}{\partial \underline{n}}\, ds = 0\,. \tag{9.45}$$

Following (Specht), the first condition can be replaced by $\displaystyle\int_E \widetilde{\Delta}\varphi\, ds = 0$ directly and is not the troublemaker. The patch test is explained very well in (Strang) SEct. 4.2 and p. 300 ff. But see also the *polynomial invariance criteria* of (Ciarlet79), Sect. 4.2, and (Taylor) where a counterexample is "countered" revealing that there are some ambiguities in the test being not removed up today. The test applies to triangular as well as quatrilateral decompositions and to problems of any order but concerns mainly non-conforming elements for *plates.* Besides some regularity conditions on the mesh refinement it (probably) suffices for *consistence.*

The complete cubic triangle of **(a)** with 10 degrees of freedom fails the patch test for plates and does not converge ((Irons), (Specht), (Zienkiewicz), p. 23) mainly because of the "bubble" function $\zeta_1\zeta_2\zeta_3$ being the shape function associated to the center of the triangle. The crucial second jump integral in (9.45) disappears of course for conforming elements in plates. But it disappears also if $\operatorname{grad} \varphi \cdot \underline{n}$ is constant on the edge with one value fixed as in Example 9.8 or if it is linear with two values specified at the vertices as in BATOZ' element below. Moreover it vanishes if $\operatorname{grad} \varphi \cdot \underline{n}$ is skew-symmetric with respect to the midpoint of the edge. Therefore (Specht) requires that the higher order shape functions of a *modified* cubic triangle have the property that $\varphi_{\underline{n}}(s) = \alpha_1 \cdot 1 + \alpha_2 s + \alpha_3 s^3$ on the edge which is fulfilled in case

$$\int_{-0.5}^{0.5} (12s^2 - 1)\varphi_{\underline{n}}\, ds = 0 \text{ because } \int_{-0.5}^{0.5} (12s^2 - 1)s^k\, ds = 0\,, \quad k = 0, 1, 3\,.$$

By this way he was able to develop an element with 9 DOFs passing the test. Let ℓ_i, $i = 1:3$, be the lengths of the edges and

$$\mu_1 = (\ell_3^2 - \ell_2^2)/\ell_1^2\,,\ \mu_2 = (\ell_1^2 - \ell_3^2)/\ell_2^2\,,\ \mu_3 = (\ell_2^2 - \ell_1^2)/\ell_3^2\,.$$

Then the shape functions of SPECHT are the components of $\widetilde{\Psi}_1$, $\widetilde{\Psi}_2$, $\widetilde{\Psi}_3$ where

$$\widetilde{\Psi}_1 = \left[\zeta_1\,,\ \zeta_1\zeta_2\,,\ \zeta_1^2\zeta_2 + \frac{1}{2}\zeta_1\zeta_2\zeta_3\big[3(1-\mu_3)\zeta_1 - (1+3\mu_3)\zeta_2 + (1+3\mu_3)\zeta_3\big]\right]$$

(with fourth order term). $\widetilde{\Psi}_2$, $\widetilde{\Psi}_3$ follow from $\widetilde{\Psi}_1$ by cyclic permutation of indices mod 3.

Example 9.8. Among many others, a non-conforming linear triangular element for second order problems has been proposed by CROUZEIX and RAVIART in (Crouzeix73). It has the *midpoints* of the edges for node points and the shape functions $\psi_i = 1 - 2\zeta_i$, $i = 1:3$. Interelement continuity is lost by this way (except at the midpoints) but the element passes the patch test even for irregular meshes as, e.g., shown very directly in (Strang), p. 178. RANNACHER and TUREK propose a corresponding quatrilateral element in (Rannacher) with local basis 1, ξ, η, $\xi^2 - \eta^2$, cf. Figure 9.13; see also `SUPPLEMENT\chap09e`. This isoparametric element passes the test for irregular (quadrilateral) mesh decompositions by the same reasons; it is rigorously examined and compared with other elements in (Turek). Both elements apply in particular to STOKES equations by properties being not discussed here whereby the first one is rather of theoretical interest. Today there exist valuable "meshers" for quadrilateral meshs with partial refinement following an idea of (Schneiders).

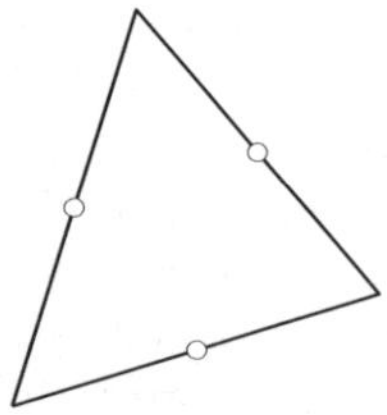

Figure 9.12. Simple CR element

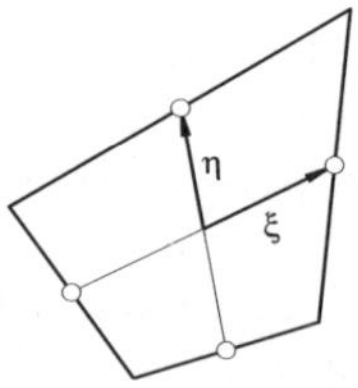

Figure 9.13. RT element

(g) A Cubic Triangular Element for Plates with nine degrees of freedom passing the patch test. In KIRCHHOFF's plate by Sect. 8.7**(b)**,

$$\mathcal{E}_T(w) = \frac{h^3}{12}\int_T \underline{\sigma}(w)\cdot\underline{\varepsilon}(w)\,dxdy\,,$$

$$\underline{\sigma}(w)\cdot\underline{\varepsilon}(w) = \frac{E}{1-\nu^2}\left[w_{xx}^2 + 2\nu\, w_{xx}w_{yy} + w_{yy}^2 + \frac{1}{2}(1-\nu)(w_{xy}+w_{yx})^2\right] \tag{9.46}$$

is the stress energy in an arbitrary triangle T. Let $\underline{u} = (u,v) = \operatorname{grad} w$ then by slight modification of the notation

$$\underline{\sigma}(\underline{u})\cdot\underline{\varepsilon}(\underline{u}) = \underline{u}^T D^T C_S D\underline{u}$$

where the matrices are the same as in (9.37). Following (Batoz) we choose the quadratic element of Example 9.7 for u and v and the reduced cubic element without "bubble" function for the bending w but, again following (Batoz) with the additional requirement that

$$\boxed{\text{the derivative of } w \text{ in normal direction of each edge is linear}}\;. \tag{9.47}$$

By this additional condition the values of u and v at the mid-points of the edges can be expressed by values of w, w_x, w_y at the vertices of the triangle T:

We consider an edge κ_{12} of length ℓ_{12} between the points $P_1(x_1,y_1)$ and $P_2(x_2,y_2)$ which shall be parametrized by the arc length s. The "bubble" function in the complete cubic ansatz for w vanishes on each edge therefore w must be a cubic polynomial on each edge also in the reduced cubic element. This implies that $w(s) = a + bs + cs^2 + ds^3$ after reparametrization. The coefficients of this one-dimensional polynomial are determined by four boundary conditions

$$w_1 := w(0)\,,\quad w_{s,1} := \frac{dw}{ds}(0)\,,\quad w_2 := w(\ell_{ij})\,,\quad w_{s,2} := \frac{dw}{ds}(\ell_{ij}) \tag{9.48}$$

at the ends of the considered edge: $a = w_1$, $b = w_{s,1}$,

$$c = \frac{3}{\ell_{12}^2}[w_2 - w_1] - \frac{1}{\ell_{12}}[2w_{s,1} + w_{s,2}]\,,\quad d = \frac{2}{\ell_{12}^3}[w_1 - w_2] + \frac{1}{\ell_{12}^2}[w_{s,1} + w_{s,2}]\,.$$

At the mid-point of the edge

$$\begin{aligned} w(1/2) &= a + b\frac{1}{2} + c\frac{1}{4} + d\frac{1}{8} = \frac{1}{2}[w_1 + w_2] + \frac{\ell_{12}}{8}[w_{s,1} - w_{s,2}] \\ w'(1/2) &= \frac{3}{2\ell_{12}}[w_2 - w_1] - \frac{1}{4}[w_{s,1} + w_{s,2}] \end{aligned}$$

and by (9.47)

$$w_n\big|_{s=1/2} := \operatorname{grad} w\cdot\underline{n}\big|_{s=1/2} = \frac{1}{2}\left[\operatorname{grad} w\cdot\underline{n}\big|_{s=0} + \operatorname{grad} w\cdot\underline{n}\big|_{s=l_{12}}\right].$$

Let now φ be the angle between the edge κ_{12} and the x-axis and let $x_{21} = x_2 - x_1$, $y_{21} = y_2 - y_1$, $\ell_{12}^2 = x_{21}^2 + y_{21}^2$,

$$c_{21} := \cos\varphi = \frac{x_{21}}{\ell_{21}}\,, \quad s_{21} := \sin\varphi = \frac{y_{21}}{\ell_{21}}\,, \quad \underline{t} = [c_{21}\,,\, s_{21}]^T\,, \quad \underline{n} = [s_{21}\,,\, -c_{21}]^T\,,$$

then $\operatorname{grad} w \cdot \underline{t} = c_{21}\, w_x + s_{21}\, w_y$, $\operatorname{grad} w \cdot \underline{n} = s_{21}\, w_x - c_{21}\, w_y$, hence at the mid-point of the edge

$$\begin{bmatrix} u^4 \\ v^4 \end{bmatrix} = \begin{bmatrix} w_x \\ w_y \end{bmatrix} = A_{21} \begin{bmatrix} w_t \\ w_n \end{bmatrix}, \quad A_{21} = A_{21}^{-1} = \begin{bmatrix} c_{21} & s_{21} \\ s_{21} & -c_{21} \end{bmatrix};$$

likewise we obtain

$$\begin{aligned} w_{s,1} &= c_{21} w_{x,1} + s_{21} w_{y,1} = c_{21} u_1 + s_{21} v_1 \\ w_{s,2} &= c_{21} w_{x,2} + s_{21} w_{y,2} = c_{21} u_2 + s_{21} v_2 \\ w_{n,1} &= s_{21} w_{x,1} - c_{21} w_{y,1} = s_{21} u_1 - c_{21} v_1 \\ w_{n,2} &= s_{21} w_{x,2} - c_{21} w_{y,2} = s_{21} u_2 - c_{21} v_2\,. \end{aligned}$$

By this way the unknown values at the mid-point can be expressed by the local node vector $\underline{U}$ in global (x, y)-coordinates at the vertices,

$$u^4 = \underline{P_4 U}\,, \quad v^4 = \underline{Q_4 U}\,, \quad \underline{U} = [w^1, v^1, u^1, w^2, u^2, v^2, w^3, u^3, v^3]^T \tag{9.49}$$

where $P_4 = [\underline{p}\,,\ 0\,,\ 0\,,\ 0]$, $Q_4 = [\underline{q}\,,\ 0\,,\ 0\,,\ 0]$ and

$$\underline{p} = \left[-\frac{3c_{21}}{2\ell_{12}}\,, \left[s_{21}^2 - \frac{1}{4} c_{21}^2 \right], -\frac{5}{4} c_{21} s_{21}\,, \frac{3c_{21}}{2} \ell_{12}\,, \left[s_{21}^2 - \frac{1}{4} c_{21}^2 \right], -\frac{5}{4} c_{21} s_{21} \right]$$

$$\underline{q} = \left[-\frac{3s_{21}}{2\ell_{12}}\,, -\frac{5}{4} c_{21} s_{21}\,, \left[c_{21}^2 - \frac{1}{4} s_{21}^2 \right], \frac{3s_{21}}{2} \ell_{12}\,, -\frac{5}{4} c_{21} s_{21}\,, \left[c_{21}^2 - \frac{1}{4} s_{21}^2 \right] \right].$$

The other two edges of the triangle are handled in the same way and finally one obtains a transformation

$$\underline{\widetilde{U}} = [u^1, \ldots, u^6]^T = P(x,y)\underline{U}\,, \quad \underline{\widetilde{V}} = [v^1, \ldots, v^6]^T = Q(x,y)\underline{U}$$

where P, $Q \in \mathbb{R}^6{}_9$ and, e.g.,

$$P = \begin{bmatrix} 0 & 1 & 0 & 0 & 0 & 0 & 0 & 0 & 0 \\ 0 & 0 & 0 & 0 & 1 & 0 & 0 & 0 & 0 \\ 0 & 0 & 0 & 0 & 0 & 0 & 0 & 1 & 0 \\ P_{41} & P_{42} & P_{43} & P_{44} & P_{45} & P_{46} & P_{47} & P_{48} & P_{49} \\ 0 & 0 & 0 & P_{54} & P_{55} & P_{56} & P_{57} & P_{58} & P_{59} \\ P_{61} & P_{62} & P_{63} & 0 & 0 & 0 & P_{67} & P_{68} & P_{69} \end{bmatrix}.$$

Now the unknown values at the mid-points of the edges are eliminated completely and we obtain in modification of (9.40)

$$\mathcal{E}_T(w) = \frac{h^3}{12} \int_T \underline{\varepsilon}(\underline{u}) \cdot \underline{\sigma}(\underline{u})\, dxdy = \frac{h^3}{12} [P\underline{U}, Q\underline{V}] \begin{bmatrix} \widehat{K}_{11} & \widehat{K}_{12} \\ \widehat{K}_{21} & \widehat{K}_{22} \end{bmatrix} \begin{bmatrix} P\underline{U} \\ Q\underline{V} \end{bmatrix} \tag{9.50}$$

where $\widehat{K}_{ij}$ are the same matrices as in (9.40); cf. the program `fem_batoz.m`.

9.5 On Singular Elements

All numerical devices require some smoothness of the unknown solution in dependence of the order (accuracy) of the individual method. But this smoothness is reduced or even lost in corners of the basic domain of definition, in particular when we are faced with *re-entrant* corners. The phenomenon may be observed directly in membranes, but also discs and plates crack most likely at these points. In order to attain an overall uniform quality of approximation, the exact solution has to be modelled with special care at "singular" points of the basic domain, or one accepts simply that loss of exactness. In this section we study the behavior of solutions at corners more thoroughly in the comparatively simple example of LAPLACE equation.

(a) Transition to polar coordinates. Remember:

$$\boxed{\begin{array}{l} x = r\cos\varphi\,,\;\; y = r\sin\varphi\,,\;\; r = (x^2+y^2)^{1/2}\,,\;\; -\pi \le \varphi < \pi \\ \varphi = \arccos(x/r) \text{ for } y \ge 0\,,\;\; \varphi = -\arccos(x/r) \text{ for } y < 0 \\ \text{resp. } \varphi = 2\pi - \arccos(x/r) \text{ for } y < 0\,,\; \text{if } 0 \le \varphi < 2\pi \end{array}}\,.$$

The angle φ is undetermined for $r = 0$. Further, remember the partial derivatives

$$r_x = \frac{x}{r},\;\; r_y = \frac{y}{r},\;\; r_{xx} = \frac{y^2}{r^3},\;\; r_{xy} = -\frac{xy}{r^3},\;\; r_{yy} = \frac{x^2}{r^3}\,;$$

$$\frac{d\varphi}{dx} = -\frac{y}{r^2},\;\; \frac{d\varphi}{dy} = \frac{x}{r^2} \text{ for } y \ge 0\,,\quad \frac{d\varphi}{dx} = \frac{y}{r^2},\;\; \frac{d\varphi}{dy} = -\frac{x}{r^2} \text{ for } y < 0\,.$$

By $F(r,\varphi) := f(r\cos\varphi, r\sin\varphi)$ we have $F_r = f_x\cos\varphi + f_y\sin\varphi$, $F_\varphi = -f_x r\sin\varphi + f_y r\cos\varphi$ and resolution yields

$$f_x = F_r\cos\varphi - \frac{1}{r}F_\varphi\sin\varphi\,,\quad f_y = F_r\sin\varphi + \frac{1}{r}F_\varphi\cos\varphi\,.$$

The normal vector $\underline{n}$ reads on the ray $\varphi = const$ as $\underline{n} = (-\sin\varphi, \cos\varphi)$ and therefore

$$\begin{aligned} f_n &= -f_x\sin\varphi + f_y\cos\varphi \\ &= \left(r^{-1}F_\varphi\sin\varphi - F_r\cos\varphi\right)\sin\varphi + \left(F_r\sin\varphi + r^{-1}F_\varphi\cos\varphi\right)\cos\varphi\,, \\ f_n &= r^{-1}F_\varphi. \end{aligned}$$

(b) We solve the boundary value problem with LAPLACE equation in polar coordinates

$$\boxed{\Delta u = u_{rr} + \frac{1}{r}u_r + \frac{1}{r^2}u_{\varphi\varphi} = 0} \tag{9.51}$$

on the pie-shaped domain $\Omega = \{(r,\varphi)\,,\ 0 < r \le R\,,\ \alpha < \varphi \le \beta\}\,, 0 < \beta - \alpha < 2\pi$. By substitution of $u(r,\varphi) = r^s F(\varphi)$ into (9.51) we obtain

$$s(s-1)r^{s-2}F(\varphi) + sr^{s-2}F(\varphi) + r^{s-2}F''(\varphi) = 0\,,$$

which leads to the ordinary differential equation

$$\boxed{F''(\varphi) + s^2F(\varphi) = 0}\,. \tag{9.52}$$

The general solution of (9.51) is a linear combination of fundamental solutions

$$u_s(r,\varphi) = r^s\Big(a\,\cos(s\varphi) + b\,\sin(s\varphi)\Big)$$

of (9.52) where the possible exponents $s > 0$ are still to be specified. Suppose now that there are $u(r,\alpha)$, $u(r,\beta)$, $0 < r \le R$ for possible homogeneous DIRICHLET boundary conditions (**D**) and that there are further homogeneous NEUMANN conditions (**N**).
Case D-D $u(r,\alpha) = u(r,\beta) = 0$.

$$\begin{aligned} u(r,\alpha) &= ar^s\cos(s\alpha) + br^s\sin(s\alpha) = 0 \\ u(r,\beta) &= ar^s\cos(s\beta) + br^s\sin(s\beta) = 0\,. \end{aligned}$$

A non-trivial solution exists if the determinant of the homogeneous system vanishes,

$$\cos(s\alpha)\sin(s\beta) - \cos(s\beta)\sin(s\alpha) = \sin(s(\beta-\alpha)) = 0\,,$$

which leads to the condition

$$\boxed{s_n = \frac{n\pi}{\beta-\alpha}\,,\quad n \in \mathbb{N}_0}\,.$$

In that case $a/b = -\sin(s\alpha)/\cos(s\alpha)$ and, recalling $\sin(\varphi)\cos(\alpha) - \cos(\varphi)\sin(\alpha) = \sin(\varphi-\alpha)$, we obtain

$$u_n(r,\varphi) = r^{s_n}\sin\big(s_n(\varphi-\alpha)\big)\,. \tag{9.53}$$

The general solution reads therefore

$$\boxed{u(r,\varphi) = \sum_{n=1}^{\infty} c_n r^{s_n}\sin\big(s_n(\varphi-\alpha)\big)}\,. \tag{9.54}$$

But consider a *re-entrant* corner where $\beta - \alpha > \pi$ hence $0 < s_1 < 1$. Then

$$\boxed{\lim{}_{r\to 0}\frac{\partial u}{\partial r}(r,\varphi) = \lim{}_{r\to 0}s_1\frac{1}{r^{1-s_1}}\sin\big(s_1(\varphi-\alpha)\big) = \infty}\,,$$

whereby the notion *singularity* is justified.

Case N-N $u_n(r,\alpha) = u_n(r,\beta) = 0$. Substitution of $u_n = u_\varphi/r$ leads to the system

$$-a\sin(s\alpha) + b\cos(s\alpha) = 0\,, \quad a\sin(s\beta) + b\cos(s\beta) = 0\,.$$

We thus obtain the same condition $\sin\big(s(\beta-\alpha)\big) = 0$ as in Case **D-D** and by consequence the same general solution.

Case D-N $u(r,\alpha) = u_n(r,\beta) = 0$. The boundary condition

$$a\cos(s\alpha) + b\sin(s\alpha) = 0\,, \quad a\sin(s\beta) - b\cos(s\beta) = 0$$

yields here $\cos\big(s(\beta-\alpha)\big) = 0$ hence

$$\boxed{s_n = \frac{(2n+1)\pi}{2(\beta-\alpha)}\,, \quad n \in \mathrm{N}_0}\,.$$

Case N-D Same as Case D-N.

(c) Example D-D Consider an approximation by finite elements and let T_S be a triangle containing the D-D singularity at a vertex. The singular corner shall have the local coordinates $(r,\varphi) = (0,0)$ in this approach. Then, in most simplest case, the first basis solution

$$w(r,\varphi) := \begin{cases} r^{s_1}\sin\big(s_1(\varphi-\alpha)\big)\,, & 0 < r \le r_0\,, \\ g(r)\sin(s_1(\varphi-\alpha))\,, & r_0 \le r \le r_1 \end{cases}$$

is taken for *additional* basis function to the basis of polynomials in T_S. (Further basis functions (9.53) may be added in case of higher order approximations.) The scalar function $g(r)$ shall be a cubic polynomial with the properties

$$g(r_0) = r_0^{s_1}\,, \quad g'(r_0) = s_1 r^{s_1-1}\,, \quad g(r_1) = g'(r_1) = 0\,,$$

where r_1 is sufficiently small such that v vanishes outside of T_S; cf. Sect. 2.1(**e**). Then

$$g(r) = g(r_0) + g'(r_0)(r-r_0) - \Big(3g(r_0) + 2g'(r_0)(r_1-r_0)\Big)\frac{(r-r_0)^2}{(r_1-r_0)^2}$$
$$+\Big(2g(r_0) + g'(r_0)(r_1-r_0)\Big)\frac{(r-r_0)^3}{(r_1-r_0)^3}$$

and the additional basis function w reads more explicitely

$$\boxed{\begin{aligned} w(r,\varphi) &= r^s\sin(s(\varphi-\alpha))\,, \quad 0 < r \le r_0 > 0\,, \quad w(0,\varphi) = 0 \\ w_r(r,\varphi) &= sr^{s-1}\sin(s(\varphi-\alpha)) \\ w_\varphi(r,\varphi) &= sr^s\cos(s(\varphi-\alpha)) \end{aligned}}$$

and shall be continued smoothly for $r_0 < r \le r_1$.

For instance, consider *linear elements in unit triangle with center* $S(1/3, 1/3)$. The singular point shall be the origin for simplicity; for r_0 we choose the distance of S to the origin hence $r_0 = \sqrt{2}/3$. Now the augmented representation in S has the form

$$v(\xi,\eta) = \alpha_1 + \alpha_2\xi + \alpha_3\eta + \alpha_4\widetilde{w}(\xi,\eta), \quad \widetilde{w}(\xi,\eta) = w(r(\xi,\eta),\varphi(\xi,\eta))$$

where, for the inverse of the design matrix,

$$\begin{aligned}
Q_1(0,0): \quad & u_1 = \alpha_1 + \alpha_4\widetilde{w}(0,0)\\
Q_2(1,0): \quad & u_2 = \alpha_1 + \alpha_2 + \alpha_4\widetilde{w}(1,0)\\
Q_3(0,1): \quad & u_3 = \alpha_1 + \alpha_3 + \alpha_4\widetilde{w}(0,1)\\
S(1/3,1/3): \; & u_4 = \alpha_1 + \frac{1}{3}\alpha_2 + \frac{1}{3}\alpha_3 + \alpha_4\widetilde{w}(1/3,1/3)\,.
\end{aligned}$$

$$\widetilde{w}(0,0) = 0\,,\ \ \widetilde{w}(1,0) = 0\,,\ \ \widetilde{w}(0,1) = 0\,,$$

$$\widetilde{w}(1/3,1/3) = \left(\frac{\sqrt{2}}{3}\right)^s \sin((s\pi - 4\beta)/4) =: a\,.$$

The design matrix is then found as usual

$$B^{-1} = \begin{bmatrix} 1 & 0 & 0 & 0\\ 1 & 1 & 0 & 0\\ 1 & 0 & 1 & 0\\ 1 & 1/3 & 1/3 & a\end{bmatrix}, \quad B = \frac{1}{3a}\begin{bmatrix} 3a & 0 & 0 & 0\\ -3a & 3a & 0 & 0\\ -3a & 0 & -3a & 0\\ -1 & -1 & -1 & 3\end{bmatrix}$$

Also, partial derivatives w.r.t. local (ξ,η)-coordinates are needed in applications. For $0 < r \le r_0$

$$\begin{aligned}
\widetilde{w} &= w(r(\xi,\eta),\varphi(\xi,\eta))\\
\widetilde{w}_\xi &= w_r r_\xi + w_\varphi \varphi_\xi\\
&= s\,r^{s-1}\sin(s(\varphi-\beta))\frac{\xi}{r} + s\,r^s\cos(s(\varphi-\beta))\left(-\frac{|\eta|}{r^2}\right)\\
&= s\,r^{s-2}[\sin(s(\varphi-\beta))\xi - \cos(s(\varphi-\beta))|\eta|]\\
\widetilde{w}_\eta &= w_r r_\eta + w_\varphi \varphi_\eta\\
&= s\,r^{s-1}\sin(s(\varphi-\beta))\frac{\eta}{r} + s\,r^s\cos(s(\varphi-\beta))\left(e\frac{\xi}{r^2}\right)\\
&= s\,r^{s-2}[\sin(s(\varphi-\beta))\eta + e\cos(s(\varphi-\beta))\xi]
\end{aligned}$$

where $e = 1$ for $y \ge 0$ and $e = -1$ for $y < 0$. Equivalently, for $r_0 < r \le r_1 = 1/\sqrt{2}$:

$$\begin{aligned}
\widetilde{w}_\xi &= s\,g'(r)\sin(s(\varphi-\beta))\frac{\xi}{r} + s\,g(r)\cos(s(\varphi-\beta))\left(-\frac{|\eta|}{r^2}\right)\\
\widetilde{w}_\eta &= s\,g'(r)\sin(s(\varphi-\beta))\frac{\eta}{r} + e\,s\,g(r)\cos(s(\varphi-\beta))\frac{\xi}{r^2}\,.
\end{aligned}$$

Recall that our intention was to solve POISSON's equation by linear triangular elements. Then

$$\Psi = [\psi_1, \psi_2, \psi_3]^T = [1 - \xi - \eta, \xi, \eta]^T$$

is the vector of shape functions in unit triangle and the following additional integrals appear at singularities by the above device

$$\int_S \widetilde{w}_\xi^2, \ \int_S \widetilde{w}_\eta^2, \ \int_S \widetilde{w}^2, \ \int_S \widetilde{w}, \ \int_S \psi_i \widetilde{w}, \ \int_S \psi_{i,\xi} \widetilde{w}_\xi, \ \int_S \psi_{i,\eta} \widetilde{w}_\eta \,.$$

These integrals are evaluated numerically by using a GAUSSian integration rule in unit triangle.

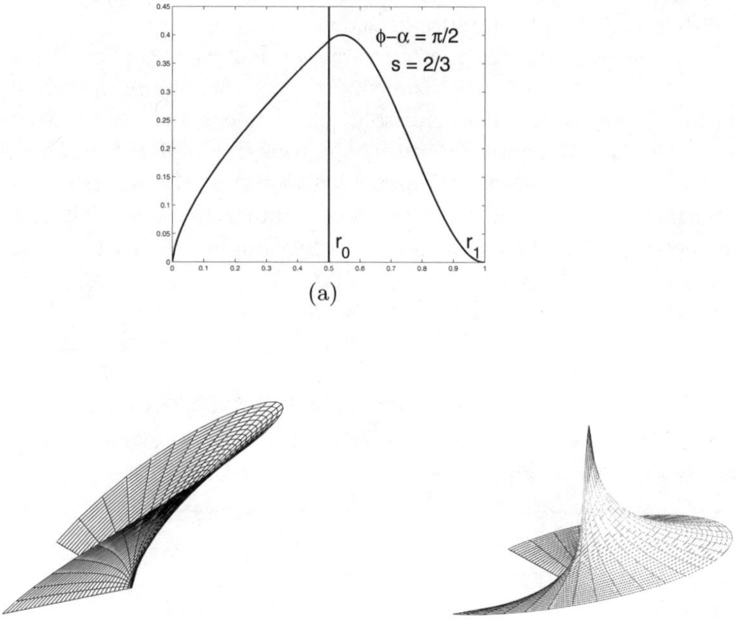

(a)

(b) (c)

Figure 9.14. Example

(a) $w(r, \pi/2) = r^s \sin(s\,\pi/2)$, (b) $w(r, \varphi) = r^s \sin(s\,\varphi)$,
(c) $w_r(r, \varphi) = s\,r^{s-1} \sin(s\,\varphi)$, $s = 2/3$.

9.6 Navier-Stokes Equations

At the beginning of the computer age, modeling fluid motions was a great challenge which later shifted to problems involving a lack of smoothness such as turbulences, shocks, cracks, compound materials, approximating the solution in the corners of the domain, etc. Nowadays anyone can simulate fluids, convection and related problems on his or her home computer and achieve perfectly respectable results for *simple* problems when all data exhibit adequate smoothness. Many methods are in noble rivalry over the visualization of problems with high REYNOLDS numbers whereby also difference methods play a substantial role, cf., e.g., (X.Chen), (Spotz), (Tanahashi).

There exists a large offer of finite elements for solving NAVIER-STOKES equations in their various different representations. In particular different nonconforming triangular and quadrilaterel elements are frequently proposed and very successful because they allow more regular domain decomposition; see, e.g, (Gresho), (Turek) and many others. By and large we restrict ourselves, however, in this section to the versatile stream-function vorticity form (where pressure is eliminated) and to COURANT's triangle as the most simple triangular element; some further methods are implemented in the MATLAB suite.

Whether velocity-pressure or stream-function vorticity form, the *stationary* versions sustain only limited values of REYNOLDS resp. RAYLEIGH numbers. Beyond, say, $Re = 10^3$ the condition of the leading matrix becomes too bad for the otherwise powerful solution of linear systems in MATLAB by *direct* methods. Alternative *iterative* methods like GMRES or the like are also offered by MATLAB but have not been tested in the MATLAB suite, also no stabilization effects. Higher REYNOLDS number are commonly handled by the time-dependent form and comparatively simple devices for time discretization as CRANK-NICHOLSON method ($\omega = 1/2$ in (2.50)) and *fractional-step-ϑ schemes*. But then some *stabilization effects* become necessary for convergence in time direction which can be realized in the most simple way by introducing an *artificial viscosity*; see (9.67).

(a) The *incompressible, stationary* NAVIER-STOKES equations in weak velocity-pressure form constitute a saddlepoint problem by (8.80): Find a pair $(\underline{u}, p) \in \mathcal{V} \times \mathcal{Q}$ such that

$$\boxed{\begin{aligned} &\forall\, \underline{v} \in \mathcal{V} : a(\underline{v}, \underline{u}) + c(\underline{v}, \underline{u}, \underline{u}) - b(\underline{v}, p) = (\underline{v}, \underline{f}) + (\underline{v}, \underline{\sigma}_n(\underline{u}, p))_\Gamma - a(v, u_0) \\ &\forall\, q \in \mathcal{Q} : -b(\underline{v}, q) \qquad\qquad\qquad = 0 \end{aligned}}\,, \tag{9.55}$$

$$\begin{aligned} a(\underline{v}, \underline{u}) &= \nu \int_\Omega \operatorname{grad} \underline{v} : \operatorname{grad} \underline{u}\, d\Omega\,, \quad b(\underline{v}, q) = \int_\Omega \operatorname{div} \underline{v}\, q\, d\Omega\,, \\ \underline{\sigma}_n &= (2\nu\varepsilon(\underline{u}) - p)\underline{n}\,, \quad c(\underline{v}, \underline{u}, \underline{w}) = \int_\Omega (\underline{v} \cdot (\operatorname{grad} \underline{u})\underline{w})\, d\Omega\,. \end{aligned} \tag{9.56}$$

For normalization either $\displaystyle\int_\Omega p\, d\Omega = 0$ has to be required or a value of pressure p has to be specified at a single point with the already mentioned numerical difficulties arising thereby. One speaks also of a *mixed* problem here since $\mathcal{V}$ and $\mathcal{Q}$ are vector spaces of different smoothness. Essential boundary conditions have to be regarded again in the choice of the vector space $\mathcal{V}$ but the abundance of possibilities shall not be discussed here once more; cf. however **(f)**. For instance, let the solution $\underline{u}$ be specified on the entire boundary, $\underline{u} = \underline{g}$ on Γ, and $\underline{u}_0 \in \mathcal{H}^1(\Omega; \mathbb{R}^n)$, $n = 2$ or 3, where $\underline{u}_0 = \underline{g}$ on Γ. In this case we have $\mathcal{V} = \mathcal{H}_0^1(\Omega; \mathbb{R}^n)$ and, say, $\mathcal{Q} = \{q \in \mathcal{L}_2(\Omega),\ \int_\Omega q\, d\Omega = 0\}$. Accordingly, the test functions $\underline{v}$ vanish on the boundary and the boundary integral (9.55) vanishes, too.

The following result provides an infinite-dimensional analogon to Lemma 1.2 in the present context of general linear-quadratic saddlepoint problems.

Theorem 9.1. *(Unique Existence) Let $\mathcal{V}$, $\mathcal{Q}$ be* HILBERT *spaces, let $\underline{f} \in \mathcal{V}$ and $g \in \mathcal{Q}$. The saddlepoint problem*

$$\begin{aligned} \forall\, \underline{v} \in \mathcal{V}: \ & a(\underline{v}, \underline{u}) - b(\underline{v}, p) = (\underline{v}, \underline{f}) \\ \forall\, q \in \mathcal{Q}: \ & -b(\underline{v}, q) \qquad\qquad = (g, q) \end{aligned} \tag{9.57}$$

has a unique solution $(\underline{u}, p) \in \mathcal{V} \times \mathcal{Q}$ if and only if the following four conditions are fulfilled:
(1°) The symmetric bilinear form $a : \mathcal{V} \times \mathcal{V} \to \mathbb{R}$ is bounded,

$$\exists\, \beta > 0\ \forall\, \underline{v} \in \mathcal{V}:\ a(\underline{v}, \underline{v}) \le \beta \|\underline{v}\|^2 .$$

(2°) The bilinear form $b : \mathcal{V} \times \mathcal{Q} \to \mathbb{R}$ is bounded,

$$\exists\, \kappa > 0\ \forall\, \underline{v} \in \mathcal{V}\ \forall\, q \in \mathcal{Q}:\ b(\underline{v}, q) \le \kappa \|\underline{v}\| \|q\| .$$

(3°) Let $\mathcal{W} := \{\underline{v} \in \mathcal{V};\ \forall\, q \in \mathcal{Q} : b(\underline{v}, q) = 0\}$ be the "kernel" of b. The bilinear form a is elliptic *on $\mathcal{W}$,*

$$\exists\, \alpha > 0\ \forall\, \underline{v} \in \mathcal{W}:\ a(\underline{v}, \underline{v}) \ge \alpha \|\underline{v}\|^2 .$$

(4°) The bilinear form b obeys the BABUSKA-BREZZI *condition (inf-sup condition)*

$$\exists\, \gamma > 0 :\ \inf_{q \in \mathcal{Q}} \sup_{\underline{v} \in \mathcal{V}} \frac{b(\underline{v}, q)}{\|\underline{v}\| \|q\|} \ge \gamma .$$

Proof see, e.g., (Braess), (Brenner).

The inf-sup condition replaces together with (3°) the assumption of Lemma 1.2 that the matrix A there is positive definite on the kernel of B *and* that B is rank-maximal. The celebrated result (4°) is also of crucial significance in finite-element methods where $\mathcal{V}$ and $\mathcal{Q}$ are finite-dimensional vector spaces. Among others it has for consequence that the flow vector $\underline{u}$ needs a *higher-order* approximation than pressure p for convergence; cf., e.g., (Braess).

(b) The general NAVIER-STOKES equation is augmented by the *convective* quadratic term $(\text{grad}\, \underline{u})\underline{u}$ resp. in the weak form by the *trilinear form* $c(\underline{v}, \underline{u}, \underline{w}) = (\underline{v}, (\text{grad}\, \underline{u})\underline{w})$ which infers that the proof of unique existence is rendered more difficult; cf., e.g., (Orlt). Adopt the notations of § 9.2 and let for the moment capitals denote local node vectors on triangle T, for instance $u_i = \Phi(x, y)^T \underline{U}_i$ then in numerical approximation

$$(\text{grad}\, \underline{u})\underline{w} \simeq \begin{bmatrix} \Phi_x^T \underline{U}_1 \Phi^T \underline{W}_1 + \Phi_y^T \underline{U}_1 \Phi^T \underline{W}_2 \\ \Phi_x^T \underline{U}_2 \Phi^T \underline{W}_1 + \Phi_y^T \underline{U}_2 \Phi^T \underline{W}_2 \end{bmatrix}$$

where Φ is a *column* vector, therefore on triangle T

$$\begin{aligned}c(\underline{v}, \underline{u}, \underline{w}) &= \int_T (\underline{v} \cdot (\operatorname{grad} \underline{u})\underline{w})\, dxdy \\ &\simeq \int_T \Big[\Phi^T \underline{V}_1 (\Phi_x^T \underline{U}_1 \Phi^T \underline{W}_1 + \Phi_y^T \underline{U}_1 \Phi^T \underline{W}_2) \\ &\qquad + \Phi^T \underline{V}_2 (\Phi_x^T \underline{U}_2 \Phi^T \underline{W}_1 + \Phi_y^T \underline{U}_2 \Phi^T \underline{W}_2)\Big]\, dxdy\,.\end{aligned}$$

But for instance $\Phi^T \underline{V}_1 \Phi_x^T \underline{U}_1 \Phi^T \underline{W}_1 = \underline{V}_1^T (\Phi_x^T \underline{U}_1)\Phi \Phi^T \underline{W}_1$ therefore

$$c(\underline{v}, \underline{u}, \underline{w}) \simeq \underline{V}_1^T C(\underline{U}_1)\underline{W}_1 + \underline{V}_1^T D(\underline{U}_1)\underline{W}_1 + \underline{V}_2^T C(\underline{U}_2)\underline{W}_1 + \underline{V}_2^T D(\underline{U}_2)\underline{W}_1\,,$$

$$C(\underline{U}) = \int_T (\Phi_x^T \underline{U})\Phi \Phi^T\, dxdy = \sum_{i=1}^n U^i C_i\,,\ C_i := \int_T \varphi_{i,x} \Phi \Phi^T\, dxdy \in \mathbb{R}^n{}_n$$

$$D(\underline{U}) = \int_T (\Phi_y^T \underline{U})\Phi \Phi^T\, dxdy = \sum_{i=1}^n U^i D_i\,,\ D_i = \int_T \varphi_{i,y} \Phi \Phi^T\, dxdy \in \mathbb{R}^n{}_n\,.$$

Example 9.9. For COURANT's triangle of § 9.3 we have by Example 9.3

$$J\varphi_1(x, y) = J + y_{23}(x - x_1) + x_{32}(y - y_1)\,,\ \varphi_{1,x} = y_{23}/J\,,\ \varphi_{1,y} = x_{32}/J$$

and the corresponding results for φ_2 and φ_3 are obtained by cyclic permutation of indices $i = 1 : 3$. Therefore, in this most simple case,

$$JC(\underline{U}) = M(y_{23}U^1 + y_{31}U^2 + y_{12}U^3)\,,\ JD(\underline{U}) = M(x_{32}U^1 + x_{13}U^2 + x_{21}U^3)$$

where $\underline{U} = [U^1, U^2, U^3]^T$, $J = 2|T|$ and M denotes the mass matrix of (9.18).

(c) For discretization of the velocity-pressure form, TAYLOR-HOOD elements are a good choice, cf., e.g., (Gresho), in particular the illustrative table in (Gresho), p. 552 ff.. Both components of the velocity vector are approximated here quadratically as displayed in § 9.3, Example 9.7, and pressure p is approximated linearly as in § 9.2, Example 9.3. In linear STOKES equation the viscosity is frequently scaled to one. Also, for improvement of the approximating qualities, the "bubble" function is added for basis function in velocity components and/or the above finite element basis is augmented by an additional constant in pressure (Gervais). Thereafter the dimension of the both velocity elements may be reduced from seven to six again by *static condensation* and in pressure from four to three (works however only in linear STOKES equation). Static condensation is for instance realized in the complete cubic element `fem_drksch.m` where the "bubble" functions plays a similar isolated role.

(d) The **Stream-Function Vorticity Form** in $\mathbb{R}^2$ is treated by a small part of the *community* as secret ideal way but justifiably viewed with skepticism by the majority; see however (Cheng), (Liao). By elimination of pressure the related numerical difficulties are entirely avoided and the computational amount is reduced by one third. Also the inf-sup condition must not be regarded and simple linear triangular elements can be used throughout. One the other side, no boundary conditions exist for vorticity w whereas the stream function z is double equipped with DIRICHLET and NEUMANN conditions. By consequence, *artificial* boundary conditions must be calculated for w in a susceptible process employing the NEUMANN conditions of z. The following Example 9.16 from the folder `KAPITEL09\FEM_3` exhibits however acceptable results like some other benchmark problems in § 9.8, although only the simple semi-implicit method (9.64) is applied for time discretization. One the other side, the test Example 9.18 shows (slow) convergence only under application of `ode23.m` (with step length control) in time direction so that (Spotz) proposes higher order approximations of w at the boundary.

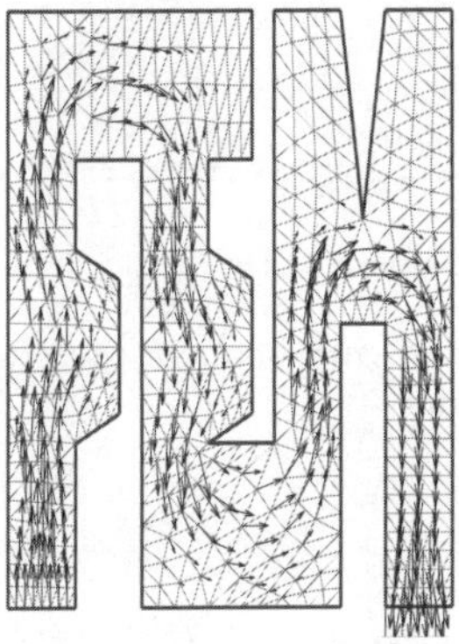

Figure 9.15. Flow velocity

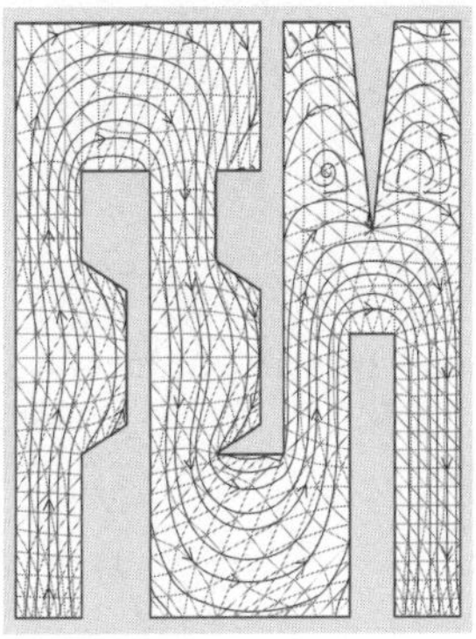

Figure 9.16. Streamlines

Example with linear STOKES equation, TAYLOR-HOOD element and mesh generation without MATLAB-PDE-TOOLBOX

Following (Stevens), the stream-function vorticity form leads to a *coupled system* in *stationary case* which tolerates the unbalanced distribution of boundary conditions between w and z without trouble; cf. **(e)**. Incidentally, both forms of the method, stationary and non-stationary, can be easily generalized to problems of *convection currents* and *mass transport* (of moderate complexity) and yield optically appealing results which provide a good overview at least with regard to quality.

Notations: L characteristic length, U characteristic velocity,

$\underline{u} = (u, v)$	velocity field
z	stream function, $\operatorname{grad} z = (-v,\, u)$
w	vorticity, $w = v_x - u_y\,,\ \Delta z = -w$
$\delta z\,,\ \delta w$	test functions for z and w
ν	kinematic viscosity $[m^2/s]$
$R_e = LU/\nu$	REYNOLDS number

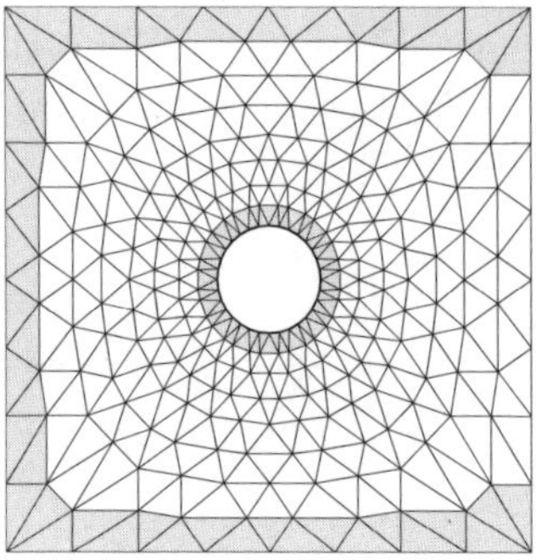

Figure 9.17. Boundary layer

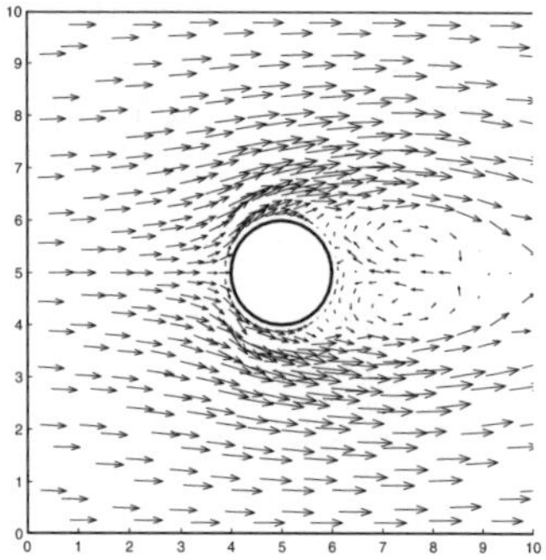

Figure 9.18. Velocity field

Example 9.16, stream function-vorticity method with boundary layer, linear triangular elements

Using the notations of § 1.7, let

$$\Gamma = \partial\Omega = \Gamma_D \cup \Gamma_N\,,\ \operatorname{int}(\Gamma_D) \cap \operatorname{int}(\Gamma_N) = \emptyset\,,\ \mathcal{H}_0^1 = \{v \in \mathcal{H}^1\,,\ v = 0 \text{ on } \Gamma_D\}$$

as in § 9.1. Recall the differential system (8.91) in *non-dimensional form*,

$$\begin{aligned} -\Delta z - w &= 0 \\ w_t - R_e^{-1}\Delta w + z_y w_x - z_x w_y &= f_{2,x} - f_{1,y} \equiv \operatorname{rot} \underline{f}\,, \end{aligned}$$

and write $\nu = 1/R_e$ instead $\nu = LU/R_e$. We multiply the first equation by the test function ∂z for z, the second by ∂w for w and apply GREEN's formula to both, then we obtain the weak form

$$\begin{aligned} &\forall\, \partial z \in \mathcal{H}^1 : \int_\Omega \nabla\delta z \cdot \nabla z \, d\Omega - \int_\Omega \delta z\, w \, d\Omega = \int_{\Gamma_D} \delta z\, z_{\underline{n}} \, d\Gamma + \int_{\Gamma_N} \delta z\, z_{\underline{n}} \, d\Gamma \\ &\forall\, \partial w \in \mathcal{H}_0^1 : \int_\Omega \delta w\, w_t \, d\Omega + \int_\Omega \delta w (z_y\, w_x - z_x\, w_y) \, d\Omega \\ &\qquad + \nu \int_\Omega \nabla\delta w \cdot \nabla w \, d\Omega = \int_\Omega \delta w \operatorname{rot} \underline{f} \, d\Omega + \nu \int_{\Gamma_N} \delta w\, w_{\underline{n}} \, d\Gamma \end{aligned} \tag{9.58}$$

The boundary conditions of the (direct or) velocity pressure form (8.78) are to be regarded also and lead to the specifications

$$\begin{aligned} & z = z_0 \text{ and } z_{\underline{n}} := \frac{\partial z}{\partial \underline{n}} = \underline{t} \cdot \underline{u}_0 \text{ on } \Gamma_D , \quad \delta w = 0 \text{ on } \Gamma_D ; \\ & \frac{\partial z}{\partial \underline{n}} \text{ and } \frac{\partial w}{\partial \underline{n}} \text{ on } \Gamma_N . \end{aligned} \tag{9.59}$$

The double condition for z and the empty condition for w on Γ_D constitutes the main drawback of the method and must be removed by an artificial adjustment. The further discussion of appropriate boundary conditions is however postponed to **(f)**.

The quadratic forms of (9.15) with stiffness matrix K and mass matrix M remain the same as before. For the additional terms we use representation by shape functions $z = \Phi^T \underline{z}$, $w = \Phi^T \underline{w}$ as in § 9.3**(b)**. Then

$$C = [c^i{}_j] = \int_T \Phi \Phi_x^T \, dxdy \, , \quad D = [d^i{}_j] = \int_T \Phi \Phi_y^T \, dxdy \in \mathbb{R}^n{}_n \, , \tag{9.60}$$

and the boundary integral gets the numerical equivalent

$$S(z) = [s^i(z)] = \int_{\Gamma \cap T} \Phi \, z_{\underline{n}} \, ds \in \mathbb{R}^n \, .$$

Also the trilinear form in (9.58) is written conveniently by using shape functions,

$$\begin{aligned} p(v, z, w) &= \int_T v(z_y w_x - z_x w_y) \, dxdy \simeq \int_T \Phi^T \underline{v} (\Phi_y^T \underline{z} \Phi_x^T \underline{w} - \Phi_x^T \underline{z} \Phi_y^T \underline{w}) \, dxdy \\ &= \underline{v}^T \left[\int_T \left[(\Phi_y^T \underline{z}) \Phi \Phi_x^T - (\Phi_x^T \underline{z}) \Phi \Phi_y^T \right] dxdy \right] \underline{w} =: \underline{v}^T P(\underline{z}) \underline{w} \end{aligned} \tag{9.61}$$

$$\begin{aligned} P(\underline{z}) = [p^i{}_k(\underline{z})]_{i,k=1}^n &= \int_T \left[(\Phi_y^T \underline{z}) \Phi \Phi_x^T - (\Phi_x^T \underline{z}) \Phi \Phi_y^T \right] dxdy = \sum_{i=1}^n z^i P_i \\ P_i &= \int_T \left[\varphi_{i,y} \Phi \Phi_x^T - \varphi_{i,x} \Phi \Phi_y^T \right] dxdy \in \mathbb{R}^n{}_n \, . \end{aligned} \tag{9.62}$$

Using EINSTEIN's convention, the equations for are single triangle T in local numbering read by (9.58)

$$\begin{aligned} & k^i{}_j z^j - m^i{}_j w^j && = s^i(\underline{z}) \\ & m^i{}_j \frac{dw^j}{dt} + p^i{}_j(z) w^j + \nu k^i{}_j w^j && = \nu s^i(\underline{w}) + c^i{}_j (f_2)^j - d^i{}_j (f_1)^j \end{aligned}$$

and summation over all triangles then leads to a differential-algebraic system of equations for the global node vectors $[Z]$ and $[W]$,

$$\boxed{\begin{aligned} & [K][[Z] - [M][W] && = [S][Z]_\Gamma \\ & [M][W]_t + [P([Z])][W] + \nu[K][W] && = \nu[S][W]_\Gamma + [C][F_2] - [D][F_1] \end{aligned}} \, . \tag{9.63}$$

This time-dependent system may be solved, e.g., by the semi-implicit iterational device

$$\boxed{\begin{aligned} &[K][[Z]^{n+1} &&= [M][W]^n + [S][Z]_\Gamma^n \\ &\left(\frac{1}{\Delta t}[M] + \nu[K]\right)[W]^{n+1} &&= \frac{1}{\Delta t}[M][W]^n - [P([Z]^{n+1})][W]^n \\ & &&\quad +\nu[S][W]_\Gamma^n + [C][F_2] - [D][F_1] \end{aligned}} \tag{9.64}$$

where $[Z]^n = [Z](n\Delta t)$ and Δt is the step length in time direction.

Example 9.10. Consider again the linear ansatz of § 9.3 Example 9.3 then

$$z(x,y) = \Phi(x,y)^T \underline{z}\,, \quad w(x,y) = \Phi(x,y)^T \underline{w}\,, \quad \Phi = [\varphi_1\,, \varphi_2\,, \varphi_3]^T\,, \quad \underline{w}, \underline{z} \in \mathbb{R}^3\,,$$

are the numerical approaches of z and w in a single triangle T carrying again the same notion as the unknown exact analogues. The matrix $P(z)$ of (9.62) then reads explicitely

$$P(z) = \frac{1}{6}\begin{bmatrix} z_{23} & z_{31} & z_{12} \\ z_{23} & z_{31} & z_{12} \\ z_{23} & z_{31} & z_{12} \end{bmatrix}, \quad \operatorname{grad}_z P(z)w = \frac{1}{6}\begin{bmatrix} w_{32} & w_{13} & w_{21} \\ w_{32} & w_{13} & w_{21} \\ w_{32} & w_{13} & w_{21} \end{bmatrix}, \tag{9.65}$$

and, for $i = 1:3$,

$$[c^i{}_1,\, c^i{}_2,\, c^i{}_3] = \frac{1}{6}[y_{23},\, y_{31},\, y_{12}]\,, \quad [d^i{}_1,\, d^i{}_2,\, d^i{}_3] = \frac{1}{6}[x_{32},\, x_{13},\, x_{21}]\,.$$

The numerical approximations $z_{\underline{n}}$ and $w_{\underline{n}}$ are constant on each triangular edge of length ℓ therefore, letting $v = z$ or $v = w$,

$$S(v) \in \left\{\frac{v_{\underline{n}}}{2\ell}[1\,,1\,,0]^T\,,\ \frac{v_{\underline{n}}}{2\ell}[1\,,0\,,1]^T\,,\ \frac{v_{\underline{n}}}{2\ell}[0\,,1\,,1]^T\right\}$$

in dependence of the edge of the triangle.

Components of the flow velocity $\underline{u} = (u, v)$ can be gained back by the components of z as constant in each triangle

$$\begin{aligned} u &= \varphi_{i,y} z^i \quad \left(= \frac{1}{2|T|}\left[x_{32}z_1 - x_{31}z_2 + x_{21}z_3\right]\right), \\ v &= -\varphi_{i,x} z^i \quad \left(= \frac{1}{2|T|}\left[y_{32}z_1 - y_{31}z_2 + y_{21}z_3\right]\right). \end{aligned} \tag{9.66}$$

It is commonly recommended to introduce an artificial viscosity in instationary Navier-Stokes equations of the form studied here. The positive effect on stability and numerical reliability may be verified theoretically but we omit such a discussion here and write simply

$$\nu_x = \nu + \frac{1}{2}u^2\Delta t\,, \quad \nu_y = \nu + \frac{1}{2}v^2\Delta t\,. \tag{9.67}$$

Then the viscosity-dependent term $\nu\Delta\underline{u}$ has to be modified by

$$\widetilde{\Delta}(\nu, \Delta t)\underline{u} = \begin{bmatrix} \nu_x u_{xx} + \nu_y u_{yy} \\ \nu_x v_{xx} + \nu_y v_{yy} \end{bmatrix} . \tag{9.68}$$

Example 9.11. The stiffness matrix (9.18) of linear triangular elements has to be modified in the corresponding way,

$$K = \frac{1}{4|T|}\begin{bmatrix} x_{32}^2\nu_y + y_{32}^2\nu_x & -x_{32}x_{31}\nu_y - y_{32}y_{31}\nu_x & x_{32}x_{21}\nu_y + y_{32}y_{21}\nu_x \\ & x_{31}^2\nu_y + y_{31}^2\nu_x & -x_{31}x_{21}\nu_y - y_{31}y_{21}\nu_x \\ \text{symm.} & & x_{21}^2\nu_y + y_{21}^2\nu_x \end{bmatrix}$$

and is used in this form by the programs of the MATLAB suite associated to this volume.

(e) Coupled System Reconsider the system (9.63) in the *non-dimensional stationary* form

$$\begin{aligned} &[M][W] - [K][Z] &&= -[S][Z]_\Gamma \\ &([K][W] + R_e[P([Z])])[W] &&= [S][W]_\Gamma + R_e[C][F_2] - R_e[D][F_1] \end{aligned} \tag{9.69}$$

where R_e is again the REYNOLDS number. Suppose that the components z_i of $Z = [z_1, \ldots, z_n]^T$ and likewise of W are *ordered* such that
$\mathcal{M} = \{1, \ldots, m\}$ is the index set of *all* node points *without* DIRICHLET boundary condition for z.
$\mathcal{R} = \{m+1, \ldots, n\}$ is the index set of all boundary points *with* DIRICHLET boundary condition for z.
$\mathcal{N} = \{\mathcal{M}, \mathcal{R}\} = \{1, \ldots, n\}$ is the index set of *all* node points.
Then the last $n - m$ equations of the second row in (9.69) can be cancelled because the corresponding components of $[Z]$ are known boundary values. The result is a system of $n + m$ equations for $n + m$ unknowns

$$\begin{bmatrix} M_{\mathcal{N},\mathcal{N}} & -K_{\mathcal{N},\mathcal{M}} \\ [K + R_eP(Z_\mathcal{N})]_{\mathcal{M},\mathcal{N}} & +\varepsilon M_{\mathcal{M},\mathcal{M}} \end{bmatrix}\begin{bmatrix} W_\mathcal{N} \\ Z_\mathcal{M} \end{bmatrix} - \begin{bmatrix} K_{\mathcal{N},\mathcal{R}}Z_\mathcal{R} - S_\mathcal{N}Z_\mathcal{R} \\ R_eM_{\mathcal{M},\mathcal{N}}G_\mathcal{N} + S_\mathcal{M}W_\mathcal{N} \end{bmatrix} = 0 \tag{9.70}$$

in a somewhat *shrinked* form not allowing the transient case. The additional term $\varepsilon M_{\mathcal{M},\mathcal{M}}$ with small $\varepsilon > 0$ is sometimes proposed for improving stability in high REYNOLDS numbers (low viscosity ν). M and K are the above mass and stiffness matrix whereas $S_\mathcal{N}Z_\mathcal{N}$ and $S_\mathcal{M}W_\mathcal{N}$ contain the boundary integrals,

$$S_\mathcal{N}Z_\mathcal{N} = \left[\int_\Gamma \varphi_i z_{\underline{n}}\, d\Gamma\right]_{i=1}^n , \quad S_\mathcal{M}W_\mathcal{N} = \left[\int_\Gamma \varphi_i w_{\underline{n}}\, d\Gamma\right]_{i=1}^m .$$

(The shape functions φ_i originally defined on their respective triangle are here tacitly continued by zero to the domain Ω.) All interior shape functions vanish on boundary therefore all *interior* node points satisfy

$$\int_\Gamma \varphi_i z_{\underline{n}}\, d\Gamma = 0\,, \quad \int_\Gamma \varphi_i w_{\underline{n}}\, d\Gamma = 0\,, \quad i = 1:m\,,$$

hence by consequence $S_\mathcal{N} Z_\mathcal{N} = [\underline{0}\,,\, S_\mathcal{R} Z_\mathcal{R}]$ und $S_\mathcal{M} W_\mathcal{N} = \underline{0}$. The mass matrix is regular and well-conditioned, and the matrix $K_{\mathcal{N},\mathcal{M}}$ has maximum rank because the matrix $K_{\mathcal{M},\mathcal{M}}$ enjoys this property. By consequence the total matrix of the *linear* system is regular, and the linear system (9.70) has a unique solution $(W_\mathcal{N}, Z_\mathcal{M})$ for every right side. Note also that vorticity is *computed everywhere* in solving (9.70). Therefore the coupled system does *not* suffer from the discrepancies in the boundary conditions being described in **(f)**. The *nonlinear* system (9.70) becomes unstable when the matrix $K_{\mathcal{M},\mathcal{N}} + R_e P(Z_\mathcal{N})_{\mathcal{M},\mathcal{N}}$ looses its rank-maximality by high values of REYNOLDS numbers; The nonlinear system of equations $\Phi(W_\mathcal{N}, Z_\mathcal{M}) = 0$ is solved by the globalized NEWTON method in normal case. The gradient of the system then reads

$$\nabla\Phi(W_\mathcal{N}, Z_\mathcal{M}) = \begin{bmatrix} M_{\mathcal{N},\mathcal{N}} & -K_{\mathcal{N},\mathcal{M}} \\ [K + R_e P(Z_\mathcal{N})]_{\mathcal{M},\mathcal{N}} & R_e \operatorname{grad}_Z [P(Z_\mathcal{N})_{\mathcal{M},\mathcal{N}} W_\mathcal{N}]_{\mathcal{M},\mathcal{M}} + \varepsilon M_{\mathcal{M},\mathcal{M}} \end{bmatrix}. \tag{9.71}$$

This matrix may be computed in simple way by use of (9.65) in present case where discretization is chosen following Example 9.10. For high REYNOLDS numbers however a continuation becomes necessary into direction of increasing values of that parameter until some value where the iteration breaks down in any case. This way to high REYNOLDS numbers is necessary because else the starting values for the otherwise "global" NEWTON method become to inexact. The nonlinear term may also be shifted to right side and thereafter a simple iteration method may be applied with step-wise updating. This method however breaks down earlier in higher REYNOLDS numbers.

(f) Boundary Conditions for Stream-Function Vorticity Form Let $\underline{n} = (n_1, n_2)$ be the unit normal vector and $\underline{t} = (-n_2, n_1)$ the unit tangential vector on the boundary $\Gamma = \partial\Omega$ of the basic domain Ω. Both vectors together form a cartesian coordinate system $\{\underline{n}, \underline{t}\}$ and by consequence

$$\underline{u} = (\underline{u}\cdot\underline{t})\underline{t} + (\underline{u}\cdot\underline{n})\underline{n}\,, \quad \Delta z = u_{xx} + u_{yy} = u_{\underline{t}\underline{t}} + u_{\underline{n}\underline{n}}\,. \tag{9.72}$$

Because $\underline{u} = (u, v)$, $u = z_y$, $v = -z_x$ also

$$z_{\underline{n}} := \frac{\partial z}{\partial \underline{n}} = -\underline{u}\cdot\underline{t}\,, \quad z_{\underline{t}} := \frac{\partial z}{\partial \underline{t}} = \underline{u}\cdot\underline{n}\,.$$

One speaks of a *slip condition* if $\underline{u}\cdot\underline{t} \neq 0$, e.g., on a free surface and of *no-slip condition* if $\underline{u}\cdot\underline{t} = 0$, e.g., on a solid boundary. We difer three types of boundary conditions:

(1°) *Inflow boundary.* Here $\underline{u}$ is specified on boundary hence also $\underline{u}\cdot\underline{t}$ and $\underline{u}\cdot\underline{n}$. But then also

$$z_{\underline{n}} = -\underline{u}\cdot\underline{t}\,,\quad w = \frac{\partial\underline{u}\cdot\underline{t}}{\partial\underline{n}} - \frac{\partial\underline{u}\cdot\underline{n}}{\partial\underline{t}}$$

is specified and z may be computed as primitive function of $\operatorname{grad} z = (-v, u)$ at least theoretically.
(2°) *Outflow boundary.* Here $z_{\underline{n}} = 0$ and $w_{\underline{n}} = 0$ are specified on boundary in normal case (or supposed). This type corresponds to Γ_N in the partition of (9.59).
(3°) *Slip and No-slip boundary.* Here z and $z_{\underline{n}} = -\underline{u}\cdot\underline{t}$ have to be specified because the boundary shall be a streamline. In no-slip case $z_{\underline{n}} = 0$ is supposed and sometimes also $w = 0$.

One states that $z_{\underline{n}}$ is specified on the entire boundary in all cases. In normal case, the different boundary conditions entail no difficulties in coupled systems (for *stationary* problems) also *not* the double occupancy in (3°). In instationary separated systems (9.63) however there are too many boundary conditions for stream function z and too few for vorticity w. The solution of $\Delta z = -w$ has overdetermined boundary conditions if z and $z_{\underline{n}}$ are given at same time. On the other side, the solution of the second equation in (9.63) has underdetermined boundary conditions on that part of boundary where neither w nor $w_{\underline{n}}$ are specified. If w is given here only then $w_{\underline{n}}$ acts as natural boundary condition being approximated in the limit by $w_{\underline{n}} = 0$. Therefore, in case (3°), boundary values z_n are *only* allowed to be used for providing *artificial* boundary values for w. In that construction however the relation $\Delta z = z_{\underline{t}\,\underline{t}} + z_{\underline{n}\,\underline{n}} = -w$ is involved on boundary whence this method falls under the rubric "variational crimes".

In order to make notations simple, we consider an impermeable piece of boundary parallel to y-axis where by consequence the component u_1 vanishes hence also the flow velocity perpendicular to boundary vanishes (in x-direction). Because $w = -z_{xx} - z_{yy}$, $z_x = -u_2$, $z_y = u_1$ let $f : x \mapsto f(x)$, $g : x \mapsto g(x)$ be two scalar functions and let $f = g_{xx}$; therefore, up to sign, f corresponds to vorticity w and g to stream function z in this situation. Noting that

$$g'''(x) = f'(x) \approx \frac{f(x+h) - f(x)}{h}\,,$$

a TAYLOR expansion yields

$$\begin{aligned} g(x+h) &= g(x) + hg'(x) + \frac{h^2}{2}g''(x) + \frac{h^3}{6}g'''(x) + \text{h.o.t.} \\ &\approx g(x) + hg'(x) + \frac{h^2}{2}f(x) + \frac{h^2}{6}\big(f(x+h) - f(x)\big) \\ &= g(x) + hg'(x) + \frac{h^2}{6}f(x+h) + \frac{2h^2}{6}f(x)\,. \end{aligned}$$

By this way we obtain the approximations

$$\begin{aligned}\frac{1}{3}h^2 f(x) &\approx g(x+h) - g(x) - hg'(x) - \frac{h^2}{6}f(x+h)\\ f(x) &\approx \frac{3}{h^2}\Big(g(x+h) - g(x) - hg'(x)\Big) - \frac{1}{2}f(x+h)\,.\end{aligned}$$

This result is transposed to an arbitrary point $\underline{x} \in \Gamma$ of the original boundary and then reads, applied to w and z,

$$\boxed{w(\underline{x}) = \frac{3}{h^2}\Big(z(\underline{x}+h\underline{n}) - z(\underline{x}) - h\frac{\partial z}{\partial \underline{n}}(\underline{x})\Big) - \frac{1}{2}w(\underline{x}+h\underline{n})}\,. \tag{9.73}$$

This formula supplies a first-order approximation in space-step h of the unknown value of w on boundary. However it needs the neighboring points $z(\underline{x}+h\underline{n})$ and $w(\underline{x}+h\underline{n})$ for practical computation which have to be supplied by interpolation of the basic triangulation. The corresponding value of $\partial z/\partial \underline{n}$ is either zero or specified; a TAYLOR espansion of $\partial z/\partial \underline{n}$ provides no improvement. Artificial boundary approximation of higher order for w has been studied by (Spotz).

For further details we refer to the respective programs in the MATLAB suite and to the subsequent examples.

9.7 Mixed Applications

Recall that $N = kg \cdot m/s^2$ (Newton), $J = N \cdot m$ (Joule), $Pa = N/m$ (Pascal), $K = 273.15 + {}^0C$ (Kelvin), $g = 9.81(m/s^2)$ gravity acceleration; $a/b \cdot c := a/(b \cdot c)$.

Physical constants for water and air at $20\,{}^0C$, viscosity and density of air at $0\,{}^0C$ and $101,3\,k\,Pa$:

Material	constants	Water	Air
$\varrho\,(kg/m^3)$	mass density	998.2	1.292
$\mu\,(Pa \cdot s)$	dynamic viscosity	$1.002 \cdot 10^{-3}$	$1.72 \cdot 10^{-5}$
$\nu\,(m^2/s)$	kinematic viscosity	$1.004 \cdot 10^{-6}$	$1.33 \cdot 10^{-5}$
$c\,(J/K \cdot kg)$	specific heat capacity	$4.182 \cdot 10^3$	$1.005 \cdot 10^3$
$\varrho c\,(J/m^3 \cdot K)$	heat capacity	$4.174 \cdot 10^5$	$1.298 \cdot 10^3$
$\kappa\,(J/m \cdot s \cdot K)$	thermal conductivity	0.598	0.0026
$\lambda\,(m^2/s)$	heat conduction coefficient	$1.40 \cdot 10^{-7}$	$2.00 \cdot 10^{-6}$
$\beta\,(1/K)$	thermal expansion coefficient	$2.07 \cdot 10^{-4}$	$3.66 \cdot 10^{-3}$

($\nu = \mu/\varrho$, $\lambda = \kappa/\varrho \cdot c$ and always $a/b \cdot c := a/(b \cdot c)$).

(a) Heat Conduction Physical quantities:

$\vartheta\,,\,\vartheta_{\rm ext}\,(K)$	temperature, neighboring temperature
$r\,(J/s\cdot kg)$	spezific thermal source density
$h\,(J/m^2\cdot s\cdot K)$	thermal transmission coefficient
$\underline{q}\,(J/m^2\cdot s)$	thermal flux vector

The thermal flux vector $\underline{q}$ of the law of conservation of energy in Sect. 8.3**(d)** satisfies FOURIER's law of heat conduction $\underline{q} = -\kappa\,\text{grad}\,\vartheta$, and the law of energy provides for $\underline{v} = \underline{0}$ the relation $\varrho\, c\, \vartheta_t + \text{div}\,\underline{q} = \varrho\, r$ or

$$\boxed{\varrho\, c\,\frac{\partial\vartheta}{\partial t} - \text{div}(\kappa\,\text{grad}\,\vartheta) = \varrho\, r} \quad [J/(m^3\cdot s)]\,. \tag{9.74}$$

Again a *characteristic length* L and a *characteristic difference of temperature* $\Delta\vartheta$ are introduced:

$$x = L\widetilde{x}\,,\quad y = L\widetilde{y}\,,\quad \widetilde{\vartheta} = (\vartheta - \vartheta_0)/\Delta\vartheta\,.$$

Then we obtain instead of (9.74) for *constant* κ the non-dimensional system

$$\frac{\partial\widetilde{\vartheta}}{\partial\widetilde{t}} = \left(\frac{\partial^2}{\partial\widetilde{x}^2} + \frac{\partial^2}{\partial\widetilde{y}^2}\right)\widetilde{\vartheta}$$

where $\widetilde{t}$ is the non-dimensional FOURIER *time*,

$$t = \frac{L^2}{\nu}\widetilde{t} \Longrightarrow u = \frac{\nu}{L}\widetilde{u}\,,\ v = \frac{\nu}{L}\widetilde{v}\,,$$

so that ν/L is the *characteristic velocity.*

Initial and boundary conditions are $\vartheta(t_0,x,y) = \vartheta_0(x,y)$ and ($\underline{n}$ being the non-dimensional normal vector), e.g.,:

$$\begin{array}{llll} (1^\circ)\ \text{specified temperature:} & \vartheta = \vartheta_D & \text{on } \Gamma_D & \\ (2^\circ)\ \text{specified heat flux:} & \underline{q}\cdot\underline{n} = -\kappa\vartheta_n & \text{on } \Gamma_C & \\ (3^\circ)\ \text{specified heat radiation:} & \underline{q}\cdot\underline{n} = h(\vartheta - \vartheta_{\rm ext}) & \text{on } \Gamma_S\,. & \end{array} \tag{9.75}$$

In two-dimensional problems the domain Ω is supposed to be a cross-section area of a medium parallel to (x,y)-plane and temperature ϑ shall be constant in z-direction. Cf. Example 9.2.

(b) Convection We consider a plane fluid of temperature ϑ (or the cross-plane of a fluid in $\mathbb{R}^3$). The buoyancy (volume-force density) depends only on ϑ by the *linearizing assumption* after BOUSSINESQ. The laws of conservation of momentum and energy supply the basic equations

$$\begin{aligned} &\underline{u}_t + (\text{grad}\,\underline{u})\underline{u} - \nu\Delta\underline{u} + \frac{1}{\varrho}\,\text{grad}\,p = \underline{f}(\vartheta)\,,\quad \text{div}\,\underline{u} = 0\\ &\vartheta_t + \text{grad}\,\vartheta\cdot\underline{u} - \lambda\Delta\vartheta \qquad\qquad\quad = 0 \end{aligned} \tag{9.76}$$

with same boundary conditions for ϑ as in (9.75). The transition of kinetic into thermal energy due to viscosity is neglected.

Example 9.12. $\varrho = \varrho_0[1 - \beta(\vartheta - \vartheta_0)]$, $f_1(\vartheta) = 0$, $f_2(T) = (\varrho_0 - \varrho)g/\varrho_0 = \beta g(\vartheta - \vartheta_0)$.

In this section we choose again the *stream-function vorticity form* for discretization of (9.76) but, of course, there are many other devices possible for numerical approach.

$$\begin{aligned} -\Delta z &= w \\ w_t + z_y w_x - z_x w_y - \nu \Delta w &= \operatorname{rot} \underline{f} \equiv f_{2,x} - f_{1,y} \\ \vartheta_t + z_y \vartheta_x - z_x \vartheta_y - \lambda \Delta \vartheta &= 0 \\ \operatorname{rot} \underline{f} &= \beta g \vartheta_x \ \text{ after Example 9.12}\,. \end{aligned} \tag{9.77}$$

RALEIGH number R_a, PRANDTL number P_r and GRASHOFF number G_r (all non-dimensional) are

Table 9.1.

Constant	$R_a = g\beta\Delta\vartheta L^3/\nu\lambda$	$P_r = \nu/\lambda$	$G_r = R_a/P_r$
Water	$1.444 \cdot 10^8 \cdot \Delta\vartheta \cdot L^3$	$0.717 \cdot 10^1$	$2.014 \cdot 10^7 \cdot \Delta\vartheta \cdot L^3$
Air	$1.35 \cdot 10^9 \cdot \Delta\vartheta \cdot L^3$	$0.665 \cdot 10^1$	$2.03 \cdot 10^8 \cdot \Delta\vartheta \cdot L^3$

.

Writing again u instead $\widetilde{u}$ etc., (9.77) leads to the non-dimensional system

$$\begin{aligned} -\Delta z &= w \\ w_t + z_y w_x - z_x w_y - \Delta w &= G_r\, \vartheta_x \\ \vartheta_t + z_y \vartheta_x - z_x \vartheta_y - P_r^{-1} \Delta \vartheta &= 0\,. \end{aligned} \tag{9.78}$$

A discretization via Sect. 9.6**(d)** yields in place of (9.63) the system

$$\boxed{\begin{aligned} &[K][[Z] - [M][W] &&= [S][Z]_\Gamma \\ &[M][W]_t + [P([Z])][W] + [K][W] &&= [S][W]_\Gamma + G_r[C][\Theta] \\ &[M][\Theta]_t + [P([Z])][\Theta] + P_r^{-1}[K][\Theta] &&= P_r^{-1}[S][\Theta]_\Gamma \end{aligned}}\,, \tag{9.79}$$

with the same matrices. The system is solved iteratively by a semi-implicite method of type (9.64) in the same way as (9.63). In stationary case the coupled nonlinear system (9.70) has to be modified likewise and reads now

$$\boxed{\begin{aligned} &[M]_{\mathcal{N},\mathcal{N}}[W]_{\mathcal{N}} - [K]_{\mathcal{N},\mathcal{M}}[Z]_{\mathcal{M}} = [K]_{\mathcal{N},\mathcal{R}}[Z]_{\mathcal{R}} - [S]_{\mathcal{N}}[Z]_\Gamma \\ &[K + P(Z_{\mathcal{N}})]_{\mathcal{M},\mathcal{N}}[W]_{\mathcal{N}} = G_r[C]_{\mathcal{M},\mathcal{N}}[\Theta]_{\mathcal{N}} + [S]_{\mathcal{M}}[W]_\Gamma \\ &[K + P_r P(Z_{\mathcal{N}})]_{\mathcal{N},\mathcal{N}}[\Theta]_{\mathcal{N}} = [S]_{\mathcal{N}}[\Theta]_\Gamma \end{aligned}}\,. \tag{9.80}$$

Possible boundary conditions of DIRICHLET type for the stream function have to be inserted into the first and third row; in normal case there are no boundary conditions for vorticity. Possible inhomogeneous CAUCHY conditions have to be regarded in the boundary integrals

$$[S]_{\mathcal{N}}[Z]_{\Gamma} = \left[\int_{\Gamma} \varphi_i z_{\underline{n}}\right]_{i=1}^{n}, \quad [S]_{\mathcal{M}}[W]_{\Gamma} = \left[\int_{\Gamma} \varphi_i w_{\underline{n}}\right]_{i=1}^{m} = \underline{0},$$
$$[S]_{\mathcal{N}}[\Theta]_{\Gamma} = \left[\int_{\Gamma} \varphi_i \vartheta_{\underline{n}}\right]_{i=1}^{n};$$

both types are already displayed in Sect. 9.6. Frequently, conditions of z and $z_{\underline{n}}$ are specified simultaneously on the entire boundary whereas specifications of ϑ and $\vartheta_{\underline{n}}$ are *not* allowed to overlap.

(c) Mass Transport Additional physical quantities:

η	diffusivity coefficient	m^2/s
σ	proportionality constant	
S	intensity of mass source	$kg/(m^3 \cdot s)$
C	concentration of mass of medium P	kg/m^3
$\underline{f}(C)$	specific volume-force density	m/s^2
$\underline{J}$	mass flux density of medium P	$kg/(m^2 \cdot s)$

We consider a substance P, e.g., a *pollution* in a plane fluid. Let the substance have concentration C and let intensity of discharge at world point $(t, \underline{x})$ be S units per second. The law of conservation of momentum has to be augmented here by an equation for concentration C. Conservation of mass for P is known here as FICK*'s law*

$$C_t + \operatorname{div} J = S \tag{9.81}$$

and the mass flow density of P consists of a diffusive and a convective term $J = -\eta \operatorname{grad} C + \underline{u}C$. Recalling that $\operatorname{div} \underline{u} = 0$, we have

$$\operatorname{div} J = -\eta \Delta C + \operatorname{grad} C \cdot \underline{u}$$

and subtitution into (9.81) yields

$$C_t + \operatorname{grad} C \cdot \underline{u} - \eta \Delta C = S .$$

We suppose again that *only* volume-force density depends on C by the *linearizing assumption after* BOUSSINESQ and then obtain the basis equations by joining with the NAVIER-STOKES equations

$$\begin{aligned} &\underline{u}_t + (\operatorname{grad} \underline{u})\underline{u} - \nu \Delta \underline{u} + \frac{1}{\varrho} \operatorname{grad} p = \underline{f}(C), \quad \operatorname{div} \underline{u} = 0 \\ &C_t + \operatorname{grad} C \cdot \underline{u} - \eta \Delta C \qquad\qquad = S . \end{aligned} \tag{9.82}$$

Example 9.13. $\varrho = \varrho_0[1 + \sigma(C - C_0)]$, $f_1(C) = 0$, $f_2(C) = (\varrho_0 - \varrho)g/\varrho_0 = -\sigma g(C - C_0)$.

In the same way as in **(b)**, we choose for approximation

$$\begin{aligned} -\Delta z &= w \\ w_t + z_y w_x - z_x w_y - \nu \Delta w &= \operatorname{rot} \underline{f} \equiv f_{2,x} - f_{1,y} \\ C_t + z_y C_x - z_x C_y - \eta \Delta C &= S \\ \operatorname{rot} \underline{f} &= -\sigma g\, C_x \text{ by Example 9.13}. \end{aligned}$$

Boundary conditions for C are, e.g.,

$$C = C_1 \text{ on } \Gamma_{DC}, \quad -\eta \frac{\partial C}{\partial n} = \underline{q}_C \cdot \underline{n} \text{ on } \Gamma_{NC},$$

where $\underline{q}_C \cdot \underline{n} = 0$ on a non-absorbing or reflecting boundary. From a formal point of view, the system (9.80) for convection flow (9.76) remains the same for mass transport (9.82) with corresponding modifications of coefficients:

$$S_c = \frac{\nu}{\eta}, \quad R_a = \frac{g\sigma \Delta C L^3}{\nu\eta}, \quad G_r = \frac{R_a}{S_c}$$

(PRANDTL number is called SCHMIDT number in mass transport).

(d) Shallow Water Problems Physical and other quantities in (x, y, z)-coordinate system:
h mean water depth, $w = h + z$ total water depth, $\widetilde{u}$ velocity to east, $\widetilde{v}$ velocity to north, $u = w^{-1} \int_{-h}^{z} \widetilde{u}\, d\zeta$ mean velocity to east, $v = w^{-1} \int_{-h}^{z} \widetilde{v}\, d\zeta$ mean velocity to north, $\vartheta\, [m/s]$ wind speed $10\,m$ over water surface, κ dimensionless coefficient of surface force due to wind, $\varphi\, [rad]$ latitude, ψ angle of wind direction from east, $\omega = 7.292 \times 10^{-5}\, [rad/s]$ angular velocity of terrestrial rotation, $f = 2\omega \sin\varphi\, [rad/s]$ CORIOLIS factor, $g = 9.81\, [m/s^2]$ gravity acceleration, $\mu_e\, [kg/s]$ eddy viscosity, $\gamma\, [m^{1/2}/s]$ CHEZY coefficient of friction on the sea bed, n MANNING coefficient of roughness of sea bed.
In computational examples we use for instance

$$\gamma = n^{-1} h^{1/6}, \; n = 0.025 \;\text{ or }\; \gamma = 1.486\, n^{-1} h^{1/6}, \; n = 0.0402 \;\text{(Peraire)}.$$

The coefficient κ depends on wind speed, e.g.,

$$\kappa = \begin{cases} 1.0 \times 10^{-3} \; (\vartheta \le 5) \\ 1.5 \times 10^{-3} \; (5 < \vartheta \le 15) \\ 2.0 \times 10^{-3} \; (15 < \vartheta \le 20). \end{cases}$$

The law of conservation of mass (continuity equation) yields

$$z_t + (wu)_x + (wv)_y = 0, \tag{9.83}$$

and the conservation of momentum yields approximatively for constant μ_e

$$\begin{aligned} & u_t + u\, u_x + v\, u_y - fv + gz_x \\ &= \frac{\mu_e}{\varrho w} \left[(wu_x)_x + (wu_y)_y \right] + \frac{\kappa \vartheta^2}{w} \cos\psi - \frac{gu(u^2 + v^2)^{1/2}}{w\, \gamma^2} \end{aligned}$$

$$v_t + u\,v_x + v\,v_y + fu + gz_y = \frac{\mu_e}{\varrho w}\big[(wv_x)_x + (wv_y)_y\big] + \frac{\kappa\,\vartheta^2}{w}\sin\psi - \frac{gv(u^2+v^2)^{1/2}}{w\,\gamma^2}\,.$$

When shear forces, surface wind and frictional forces are neglected, the equations of motion become

$$\boxed{\begin{aligned} z_t + (wu)_x + (wv)_y &= 0\\ u_t + u\,u_x + v\,u_y - f\,v + g\,z_x &= 0\\ v_t + u\,v_x + v\,v_y + f\,u + g\,z_y &= 0\end{aligned}} \tag{9.84}$$

being called *shallow water equations.* As usual in transition to the weak problem, the first equation is multiplied by a test function δz for z, the second by a test function δu for u, and the third by a test function δv for v. Thereafter all equations are integrated over the basis domain Ω; also the first equation is modified by writing

$$\int_\Omega \delta z\big[(wu)_x + (wv)_y\big] = \int_\Omega \delta z\big[w_x u + w_y v\big] + \int_\Omega \delta z\big[u_x + v_y\big]w\,.$$

Altogether we obtain the weak problem

$$\begin{aligned}\int_\Omega \delta z\, z_t &= -\int_\Omega \delta z\big[uw_x + vw_y\big] - \int_\Omega \delta z w\big[u_x + v_y\big]\\ \int_\Omega \delta u\, u_t &= -\int_\Omega \delta u\big(uu_x + vu_y\big) + \int_\Omega \delta u\big(fv - g\,z_x\big)\\ \int_\Omega \delta v\, v_t &= -\int_\Omega \delta v\big(uv_x + vv_y\big) - \int_\Omega \delta v\big(fu + g\,z_y\big)\end{aligned} \tag{9.85}$$

By consequence, the following tensors of second and third order are needed now in triangle T

$$\begin{aligned} M &= [m^i{}_j]\,,\; m^i{}_j = \textstyle\int_T \varphi_i\varphi_j\,dxdy\,, & C &= [c^i{}_j]\,,\; c^i{}_j = \textstyle\int_T \varphi_i\varphi_{j,x}\,dxdy\\ D &= [d^i{}_j]\,,\; d^i{}_j = \textstyle\int_T \varphi_i\varphi_{j,y}\,dxdy\,, & P &= [p^i{}_{jk}]\,,\; p^i{}_{jk} = \textstyle\int_T \varphi_i\varphi_j\varphi_k\,dxdy\\ Q &= [q^i{}_{jk}]\,,\; q^i{}_{jk} = \textstyle\int_T \varphi_i\varphi_j\varphi_{k,x}\,dxdy\,, & R &= [r^i{}_{jk}]\,,\; r^i{}_{jk} = \textstyle\int_T \varphi_i\varphi_j\varphi_{k,y}\,dxdy\,.\end{aligned}$$

Having assembled all ingredients for a discretization, the system (9.84) leads to the following result relative to a single triangle T (recalling EINSTEIN's conventions)

$$\begin{aligned} m^i{}_j z^j{}_t &= -(q^j{}_{ki}w^j u^k + r^j{}_{ki}w^j v^k)\\ m^i{}_j u^j{}_t &= -(q^i{}_{jk}u^j u^k + r^i{}_{jk}v^j u^k) + p^i{}_{jk}f^j v^k - g\,c^i{}_j z^j\\ m^i{}_j v^j{}_t &= -(q^i{}_{jk}u^j v^k + r^i{}_{jk}v^j v^k) - p^i{}_{jk}f^j u^k - g\,d^i{}_j z^j\,.\end{aligned} \tag{9.86}$$

Using local numbering in triangle T, the matrices

$$\begin{aligned} A(U,V) &= [a^i{}_k]\,,\; a^i{}_k = q^i{}_{jk}u^j + r^i{}_{jk}v^j\\ B(U,V) &= [b^i{}_k]\,,\; b^i{}_k = q^i{}_{kj}u^j + r^i{}_{kj}v^j\\ G(F) &= [g^i{}_k]\,,\; g^i{}_k = p^i{}_{jk}f^j\end{aligned}$$

are introduced, then equations (9.86) are assembled to a system for the global node vectors $[Z]$, $[U]$, $[V]$, $[W]$,

$$\begin{aligned}[M][Z]_t &= -[A(U,V)][W] - [B(U,V)][W]\\ [M][U]_t &= -[A(U,V)][U] + [G(F)][V] - g\,[C][Z]\\ [M][V]_t &= -[A(U,V)][V] - [G(F)][U] - g\,[D][Z]\end{aligned} \tag{9.87}$$

(global numbering).

The corresponding pure initial value problem may be solved simply by using a semi-implicit modification of HEUN's method; cf. Sect. 2.4 Example 2.9 (try also more sophisticated methods!) Thereby each of the three equations is solved separately with time step Δt and afterwards updated in the following way

$$\begin{aligned}[M][X]^{n+(1/2)} &= [M][X]^n + \frac{1}{2}\Delta t\,[M][X]_t^n\\ [M][X]^{n+1} &= [M][X]^n + \Delta t\,[M][X]_t^{n+(1/2)}\,;\end{aligned} \tag{9.88}$$

altogether, iteration runs for $n = 0, 1, \ldots$. The variables $[U]$, $[V]$, $[Z]$ are to be inserted for X one after other in that succession, and for $[M][X]_t$ the corresponding right side of (9.87) is to be chosen. The present method however suffers from spurious numerical damping therefore (Kawahara) proposes a decomposition of the mass matrix $M = [m^i{}_k] \in \mathbb{R}^n{}_n$ in (9.88) (selective lumping). To this end, let a *lumped* mass matrix $\widetilde{M}$ and a weighted matrix $\widehat{M}$ be defined by

$$\widetilde{M} = \text{diag}\left(\sum_{k=1}^{n} m^1{}_k\,, \ldots, \sum_{k=1} m^n{}_k\right) \in \mathbb{R}^n{}_n\,, \quad \widehat{M} = \varepsilon\widetilde{M} + (1-\varepsilon)\,M\,,$$

then, instead of (9.88), iterations runs by the rule

$$\begin{aligned}[\widetilde{M}][X]^{n+(1/2)} &= [\widehat{M}][X]^n + \frac{1}{2}\Delta t\,[M][X]_t^n\\ [\widetilde{M}][X]^{n+1)} &= [\widehat{M}][X]^n + \Delta t\,[M][X]_t^{n+(1/2)}\,.\end{aligned} \tag{9.89}$$

(Kawahara) also proposes the value $\varepsilon = 0.9$ for weighting and takes the COURANT-FRIEDRICHS-LEVY condition

$$\frac{\Delta t}{\Delta x} \le \frac{2-\varepsilon}{3\sqrt{2}} \cdot \frac{1}{(gH)^{1/2}}$$

for computation of time-step Δt relative to space-step Δx; Sect. 2.4**(g2)**.

We refer e.g. to (Ninomiya) for more comprehensive studies concerning the four examples for possible iteration presented here. The application of coupled stream-function vorticity equations to *stationary* problems of convection has been proposed by (Stevens). Finally, it should be remarked that solving shallow water problems by the system (9.84) and numerical device (9.88) resp. (9.89) remains unstable to some degree also if heuristical modifications as *mass lumping* are taken for improvement. The choice of time-step Δt as large as possible under observation of the COURANT-FRIEDRICHS-LEVY condition of Sect. 2.4**(g)** seems to provide the "best" results. Also, it has to be noted that the devices (9.88) and (9.89) are rather simple discretizations into direction of time.

9.8 Examples

In this section more or less well-known examples to the preceding section are considered, either in time-dependent *dynamical form* as in Sect. 9.6**(d)** or in time-independent *stationary form* as coupled system, cf. Sect. 9.6**(e)**. Discretization is performed by linear straight triangular elements as in Sect. 9.2**(c)** with the matrices described there; artificial boundary conditions for vorticity after (9.73) are denoted by w^*.

(a) Navier-Stokes Equations

Example 9.14. Lid Driven Cavity (Benchmark problem, cf., e.g., (Gresho)).
(1°) Transient problem, linear triangle elements, 2213 nodes; similar result also with bilinear quatrilateral elements, 289 nodes. **(A):** *slip boundary* with $u = 1\,[m/s]$, **(A)** – **(D):** $z = 0$, $w = w^*$. $\nu = 10^{-3}$, 500 time steps with $\Delta t = 0.05$. Cold start with $w_0 = 0$. Pressure p by SOHN's method; see Sect. 8.10, **(f2)**; $\int_\Omega p\,d\Omega = 0$ (Fig. 9.19).
(2°) Same result for stationary problem with coupled system, linear triangle or bilinear parallelogram elements, 1089 nodes. No boundary conditions for w. Cold start for $\nu = 10^{-1}$, five steps of continuation to $\nu = 10^{-3}$.

Example 9.15. Flow Past Half Cylinder (Benchmark problem)
(A): $z = 4.0$, $w = 0$, **(C):** $z = 0$, $w = w^*$, **(D):** *outflow boundary* free,
(B): *inflow boundary* with

$$u = -0.048\,y^2 + 0.48\,y\,, \quad v = 0\,, \quad z = \int_0^y u(y)\,dy\,, \quad w = -\frac{du}{dy}\,.$$

(a) (Ninomiya). 969 node points. $\nu = 0.01$, 200 time steps with $\Delta t = 0.1$.
(b) (Gresho). Modified geometry with 1647 node points. $\nu = 0.001$, 400 time steps with $\Delta t = 0.01$. (Figs. 9.20 and 9.21).

Example 9.16. Flow Past Cylinder (Benchmark problem (Ninomiya)). $\nu = 0.01$, 100 time steps with $\Delta t = 0.1$, start with $w_0 = 0$. **(A):** $z = 20$, $w = 0$, **(C):** $z = -20$, $w = 0$, **(E):** $z = 0$, $w = w^*$, **(D):** *outflow boundary* free, **(B):** *inflow boundary* with $z = y$, $w = 0$ (Fig. 9.22).

Example 9.17. Back Facing Step (Benchmark problem, cf., e.g., (Ninomiya), (Barton)). $\nu = 1.338 \cdot 10^{-5}$, $\Delta t = 7.5 \cdot 10^{-5}$, Cold start with $w_0 = 0$ (Figs. 9.23 and 9.24). **(A):** $z = 0.36$, $w = 0$, **(C):** $z = 0$, $w = w^*$, $v = 0$,
(D): *outflow boundary* free,

(B): *inflow boundary* with

$$u = 10\,,\quad v = 0\,,\quad z(y) = \int_{0.02}^{y} u\,dy = 10(y - 0.02)\,,\quad z(0.056) = 0.36\,,\quad w = 0\,.$$

In Figure 9.24 the problem is *scaled* choosing the step height $L = 0.02$ for characteristic length and $U = 10$ for characteristic velocity such that $R_e = 14950$ and $\widetilde{u} = 500$ corresponds to the velocity $u = 10$ at the inlet. For illustration however $\widetilde{u} = 1500$ has been choosed at the entrance corresponding to $u = 30$. Several meshes have been tested with the same data. The results agree near the step but differ to some extent in the far field. This effect is possibly due to the simple time iteration and the lack of better stabilizing procedures. The illustrations show also that the flow does not tend to a steady state but develops a periodic behavior in the present geometrical configuration.

Example 9.18. Exact Example for stream-function vorticity form (Fig. 9.25).

$$\begin{aligned} z &= -8(x-x^2)^2(y-y^2)^2\,, & w &= 16((6x^2-6x+1)(y-y^2)^2 \\ & & &\quad +(x-x^2)^2(6y^2-6y+1)) \\ u &= -16(x-x^2)^2(y-y^2)(1-2y)\,, & v &= 16(y-y^2)^2(x-x^2)(1-2x)\,. \end{aligned}$$

on unit square; $z_{\underline{n}} = 0$ on the entire boundary Γ, $\nu = 0.01$, $\Delta x = \Delta y = 1/16$ (545 nodes). Numerical solution by time-dependent problem with `ode23.m` and 10 time steps of time length 1.
(a) Vorticity for artificial *and* exact boundary values of w, (b) error of stream function $\times 10^2$ with $w = w^*$ on Γ, (c) error of stream function $\times 10^2$ with exact values of w on Γ.
This example shows even worse approximation for exact boundary values of w. Very similar result for $\nu = 0.001$ and 100 time steps of length 1.

(b) Convection

Example 9.19. Convection in a Cup (Benchmark problem (Ninomiya)). $\nu = 0.005$, 150 time steps with $\Delta t = 0.2$, start with $w_0 = 0$, $\vartheta_0 = 15\,^0C$. **(A):** $z = 0$, $w = 0$, $\underline{q}_n = h(\vartheta - \vartheta_u)$, $\vartheta_u = 15\,^0C$, **(B):** $z = 0$, $w = w^*$, $\partial\vartheta/\partial n = 0$ or $\vartheta = \vartheta_u = 15$, **(C):** $z = 0$, $w = w^*$, $\vartheta = 60\,^0C$ (Figs. 9.26 and 9.27).

Example 9.20. Convection in Unit Square (Benchmark problem). $R_a = 10^5$, $P_r = 1$, 800 time steps with $\Delta t = 10^{-4}$. Cold start with $w_0 = 0$, $\vartheta_0 = 0.5\,^0C$; **(A), (C):** $z = 0$, $\partial\vartheta/\partial n = 0$, **(B):** $z = 0$, $\vartheta = 0$, **(D):** $z = 0$, $\vartheta = 1\,^0C$. Pressure calculation by SOHN's method (Figs. 9.28, 9.29, 9.30, 9.31)

Example 9.21. BÉNARD *Cell.* Natural convection of water in a closed vessel with length $L = 0.12\,[m]$ and height $h = 0.01\,[m]$, temperature $\vartheta = 334\,[K]$ at bottom and $\vartheta = 333\,[K]$ at top. Scaled problem with $\Delta\vartheta = 1$ and $\Delta L = h$, $R_a = 1.444 \cdot 10^4$ (Figs. 9.32 and 9.33).

(c) Shallow Water Problems

Example 9.22. Tidal Current in a Bay with Island (Ninomiya)). Time periode $T = 12$ [*hours*], boundary conditions

$$\begin{aligned} \underline{u}_n &= un_1 + vn_2 = 0 \text{ in } \Gamma_K \text{ (coast)}, \\ \zeta(t) &= A \sin\left(\frac{2\pi}{T}t\right) \quad \text{in } \Gamma_S \text{ (open sea)}. \end{aligned} \tag{9.90}$$

$A = 1$ [m] amplitude, $\underline{n} = (n_1, n_2)$ normed normal vector. Initial conditions (cold start) for $t = 0$ in Ω are $\zeta = 0$, $u = 0$, $v = 0$ (Figs. 9.34 and 9.35).

Example 9.23. Travelling Waves in a Shallow Channel; cf. (Ninomiya). Length 1000 [m], width 1600 [m], depth $h = 20$ [m]. At left end a wave is generated as in (9.90),

$$\zeta = A \sin(2\pi t/T)\,, \quad A\,[m] \text{ amplitude}, \quad T\,[sec] \text{ period}.$$

(wave velocity $c = (g\,h)^{1/2}$ [m/s] and wave length $\lambda = c\,T$ [m] are not used.) Initial values $(U, V, \zeta) = (0, 0, 0)$ and $A = 1$. Water level is monitored at left and right end of the basin for the period of one hour. Regular triangular decomposition with 5-fold partitioning of x-axis and two-fold partitioning of y-axis; solving by `ode23.m` (image sequence) (Figs. 9.36, 9.37, 9.38, 9.39).

Example 9.24. Wave in a Channel; cf. (Petera). In a shallow basin of length 3000 [m], width 200 [m], water depth $h = 10$ [m], a wave is generated at left end as in (9.90),

$$\zeta = A \sin(2\pi t/T)\,, \quad A = 0.3\,[m] \text{ amplitude}, \quad T = 300\,[sec] \text{ period}.$$

As special feature the wave velocity $u = \zeta(g/(h+\zeta))^{1/2}$ is specified at right end to simulate an infinitely long channel (Petera). Using initial values $(U, V, \zeta) = (0, 0, 0)$, water level is monitored at *right* end of the channel after T seconds for a period of T seconds. The result is compared to the exact solution

$$\zeta(t, x) = A \sin\left(t - \frac{x}{(g(h + \zeta))^{1/2}}\right), \quad x = 3000\ [m]$$

(Petera). The channel is decomposed into uniform squares of edge length 100 [m] and the latter are decomposed into two triangles; solving by modified EULER method without *lumping* (image sequence) (Fig. 9.40).

Example 9.25. Solitary Wave on a Beach; cf. (Petera). A solitary wave is generated at the *right* end of a channel of length 40 [m], width 2 [m] and water depth $h = -x/30$ [m] (Figs. 9.41 and 9.42):

$$\begin{aligned} \zeta &= a_0 \left[\cosh\left(\frac{1}{2}(3a_0)^{1/2}(x - \alpha^{-1})\right)\right]^{-2}, \\ u &= -\left(1 + \frac{1}{2}a_0\right)\frac{\zeta}{\alpha x + \zeta}, \quad a_0 = 0.1\,, \ \alpha = 1/30\,. \end{aligned}$$

The shape of the wave is calculated each second over a period of 25 seconds.

(d) Discs and Plates

Example 9.26. Spanner. Length in $[cm]$, $E = 0.2 \cdot 10^8$ $[N/cm^2]$, $\nu = 0.3$, height $= 0.7$ $[cm]$, (A) support, (B) load $\sim 27\,[N/cm]$, total external force $\sim 218\,[N]$. $\sigma_1 > \sigma_2$ are the eigenvalues of the stress tensor $\boldsymbol{\sigma}$ at the nodes points; σ_1 is mainly positive and σ_2 mainly negative (Figs. 9.43, 9.44, 9.45, and 9.46).

Example 9.27. KIRCHHOFF*'s plate* after (Batoz). Length 30.48 $[cm]$, width 21.5526 $[cm]$, height 0.3175 $[cm]$, $E = 0.35153606 \cdot 10^7$ $[kp/cm^2]$, $\nu = 0.3$, uniform load density $q = 1.83262778799 \cdot 10^{-2}$ $[kp/cm^2]$; the plate is clamped at left end (Fig. 9.47).

Test values in cm:

Nr.	1	2	3	4	5	6
numerical	0.7540	0.5180	0.3070	0.3280	0.1420	0.0560
experimental	0.75438	0.51816	0.30734	0.32766	0.14224	0.05588

.

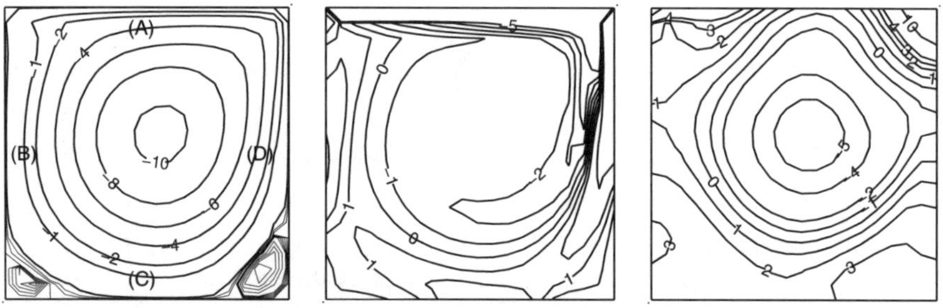

Figure 9.19. Ex. 9.14, stream lines, vorticity, pressure $(\times 10^2)$

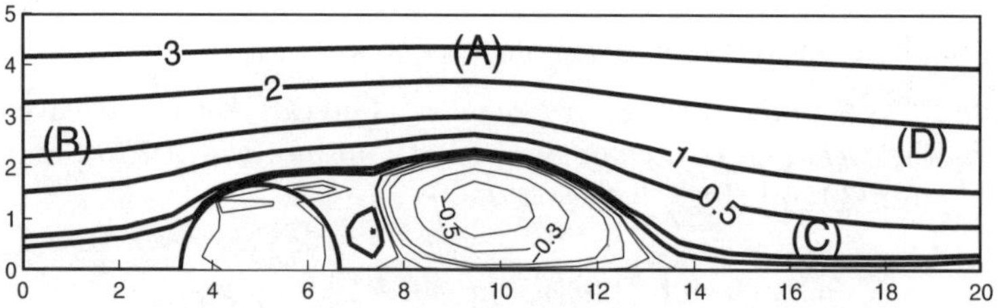

Figure 9.20. Ex. 9.15(a), stream lines, $t = 20$ $[s]$

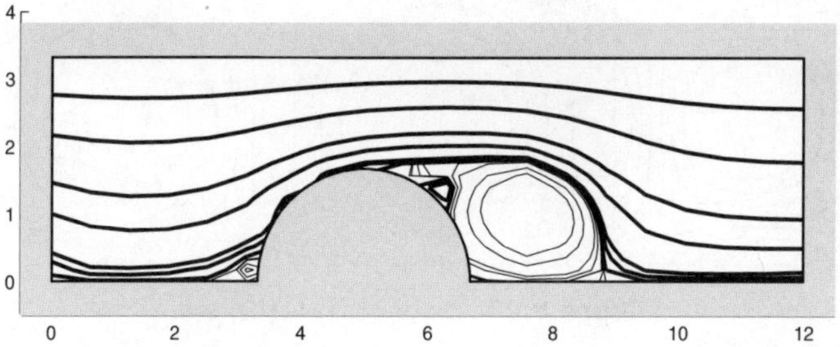

Figure 9.21. Ex. 9.15(b), stream lines, $t = 4$ $[s]$

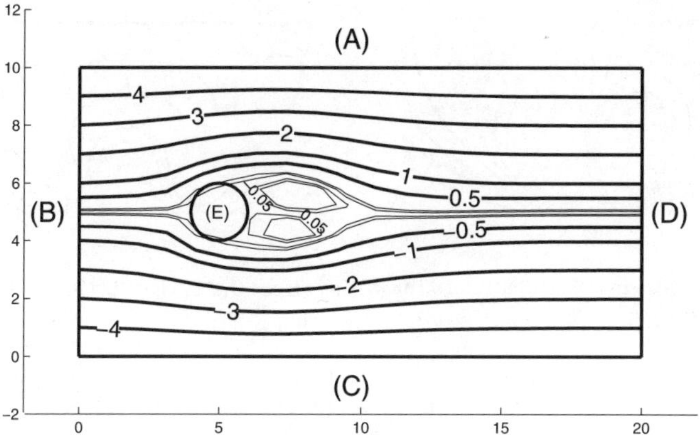

Figure 9.22. Ex. 9.16, stream lines

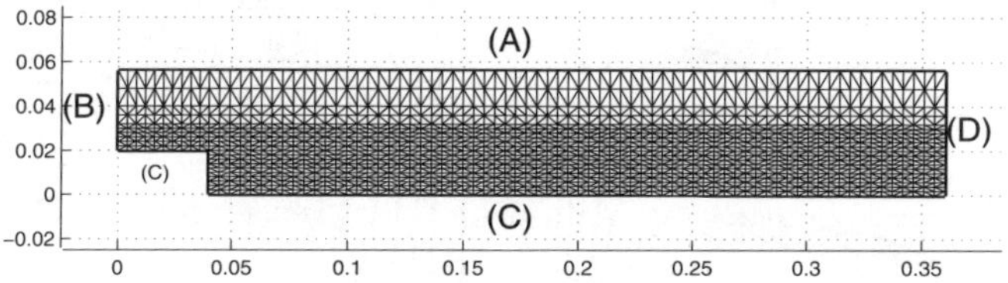

Figure 9.23. Ex. 9.17, unscaled geometry, mesh

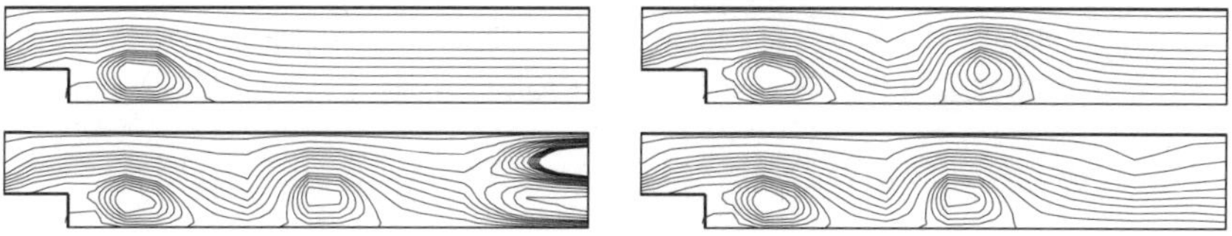

Figure 9.24. Ex. 9.17, scaled $t = k \cdot 9.375$ $[ms]$, $k = 1:4$

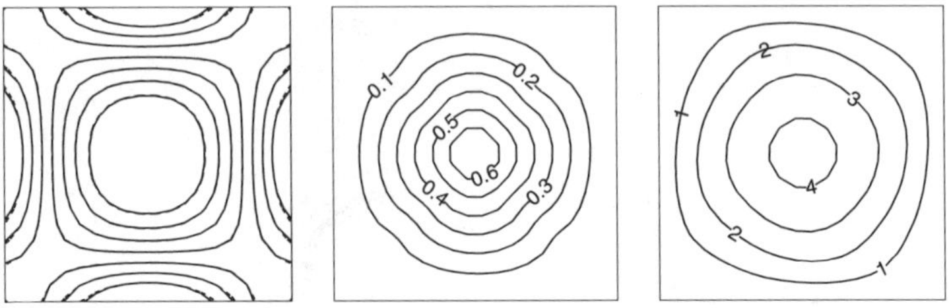

Figure 9.25. Ex. 9.18, (a), (b), (c)

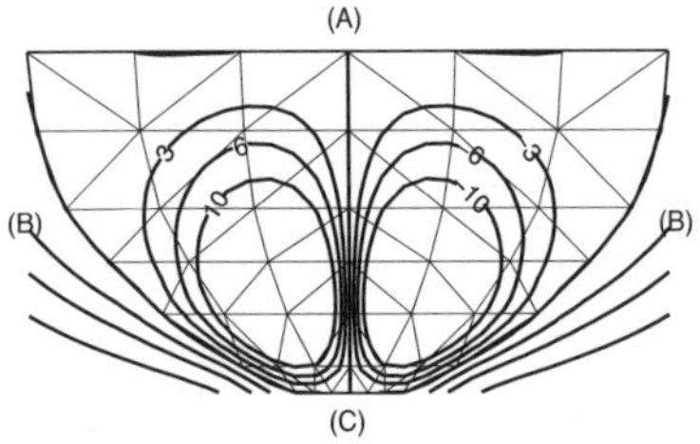

Figure 9.26. Ex. 9.19, stream lines unscaled ($\times 10^3$)

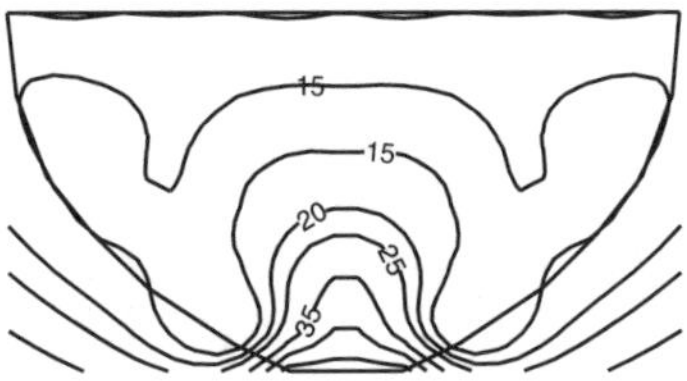

Figure 9.27. Ex. 9.19, isoterms unscaled

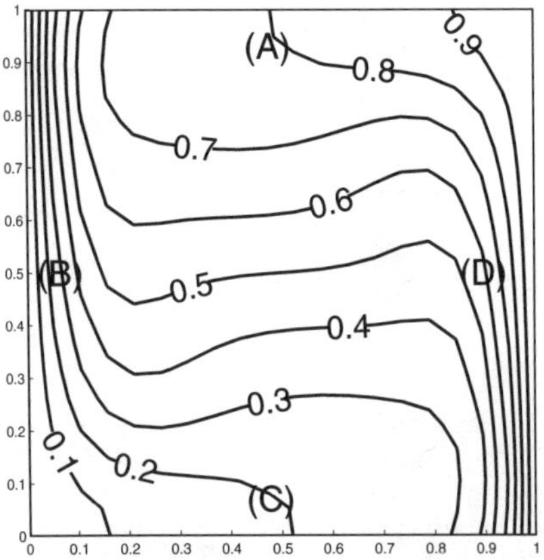

Figure 9.28. Ex. 9.20, isotherms

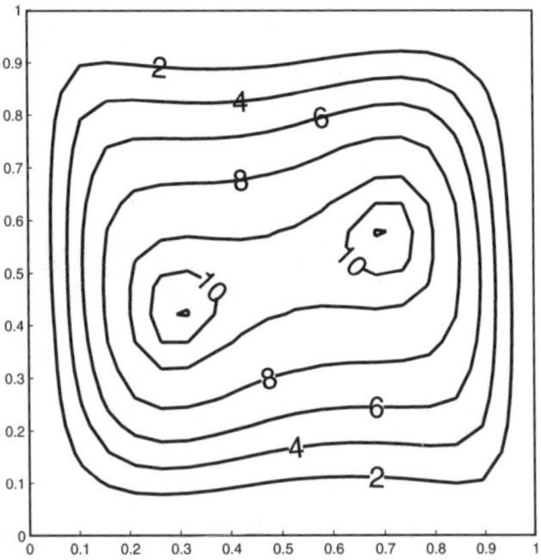

Figure 9.29. Ex. 9.20, streamlines

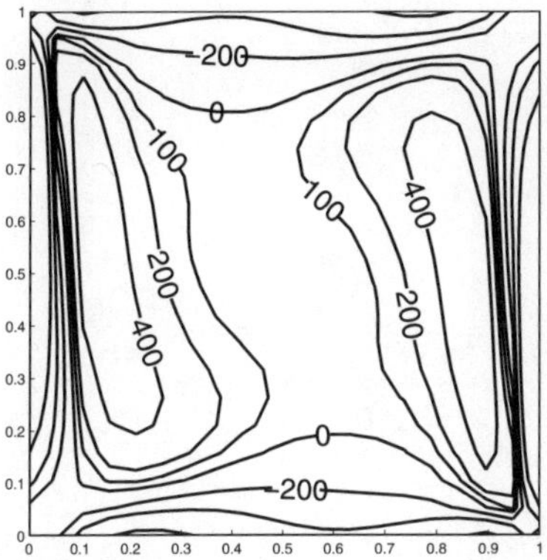

Figure 9.30. Ex. 9.20, vorticity

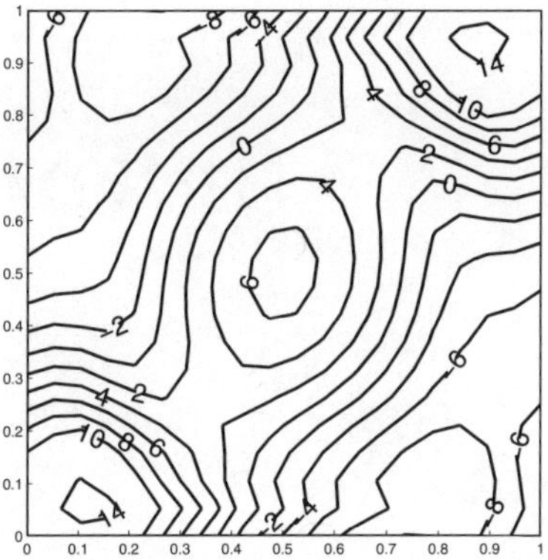

Figure 9.31. Ex. 9.20, pressure /100

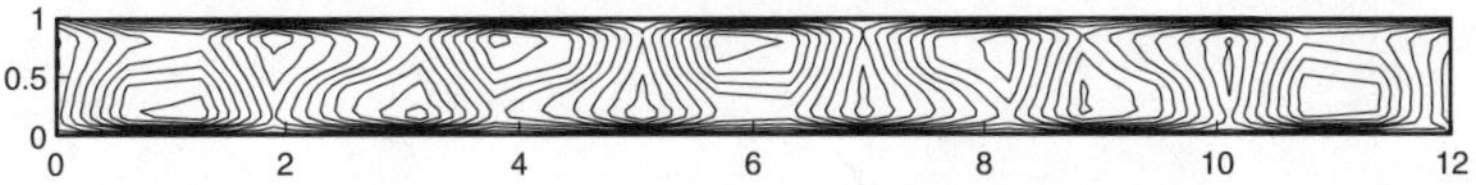

Figure 9.32. Ex. 9.21, Temperature

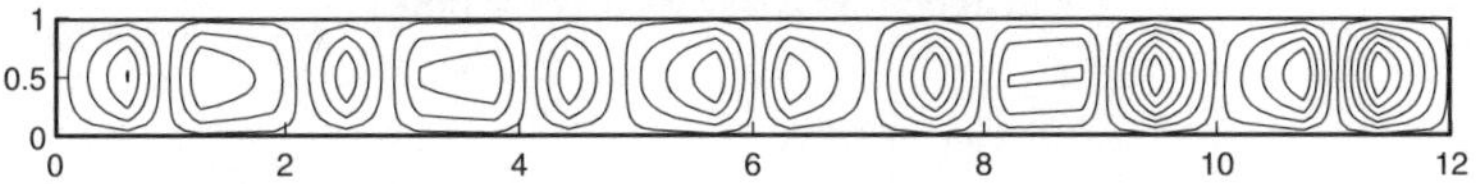

Figure 9.33. Ex. 9.21, Stream function

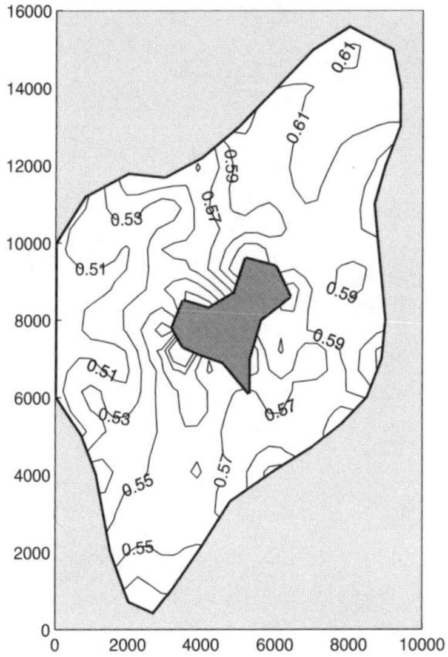

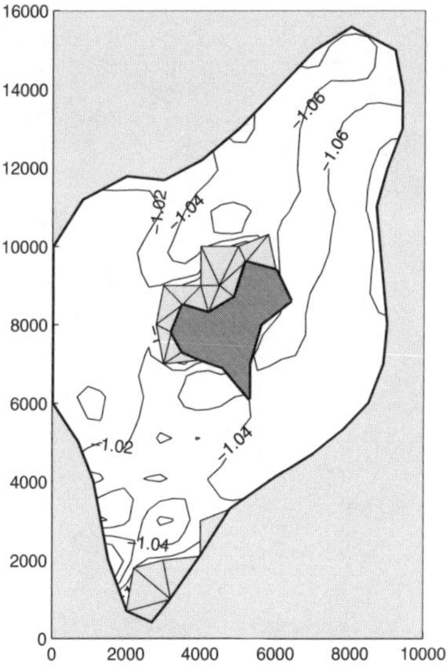

Figure 9.34. Ex. 9.22, 1^{fst} hour **Figure 9.35.** Ex. 9.22, 9nd hour

Water levels relative to sea level

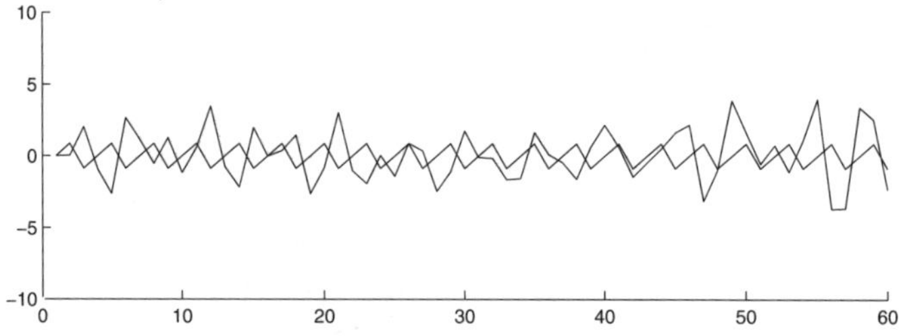

Figure 9.36. Ex. 9.23, $T = 180$

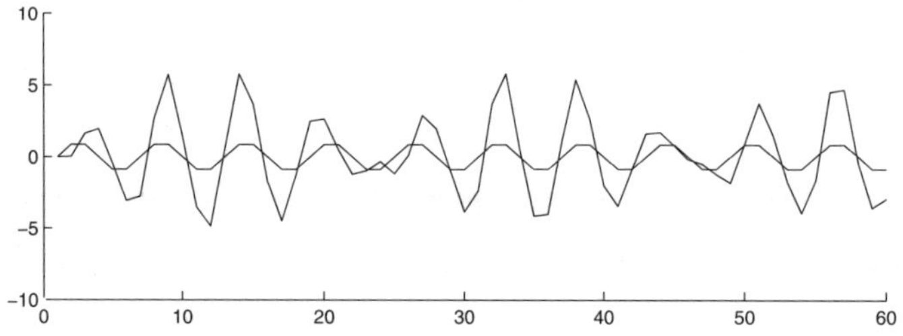

Figure 9.37. Ex. 9.23, $T = 360$

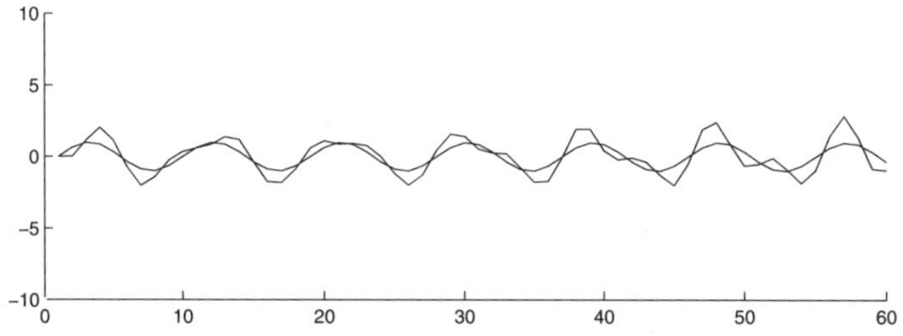

Figure 9.38. Ex. 9.23, $T = 540$

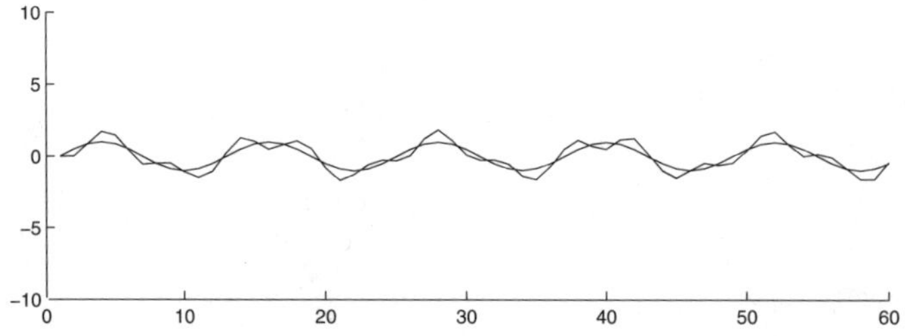

Figure 9.39. Ex. 9.23, $T = 720$

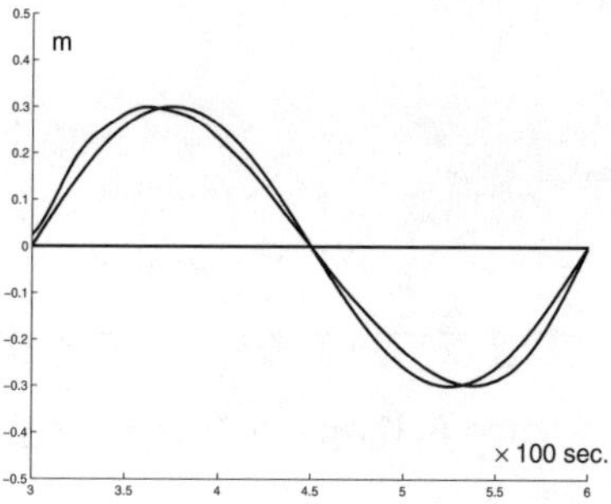

Figure 9.40. Ex. 9.24, water level exact and calculated

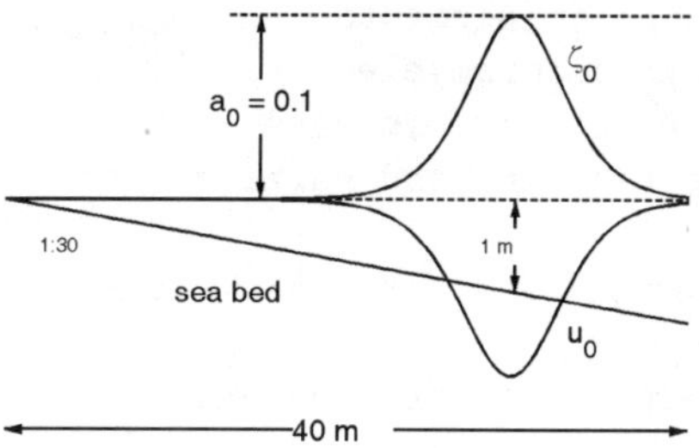

Figure 9.41. Ex. 9.25, boundary conditions

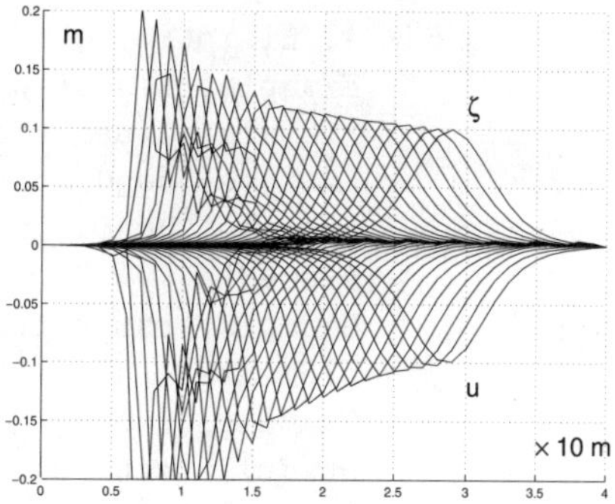

Figure 9.42. Ex. 9.25, water levels

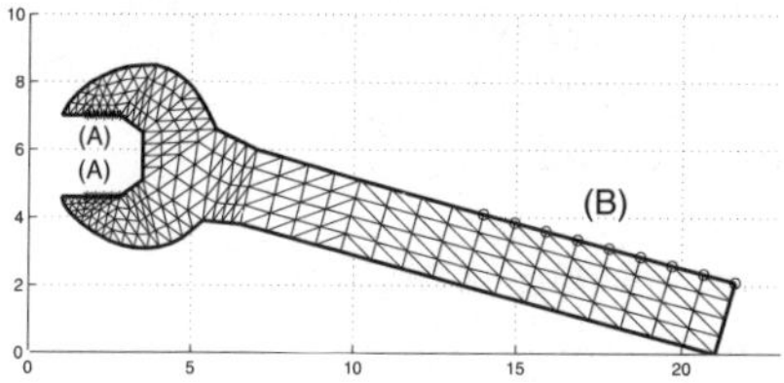

Figure 9.43. Ex. 9.26, geometry

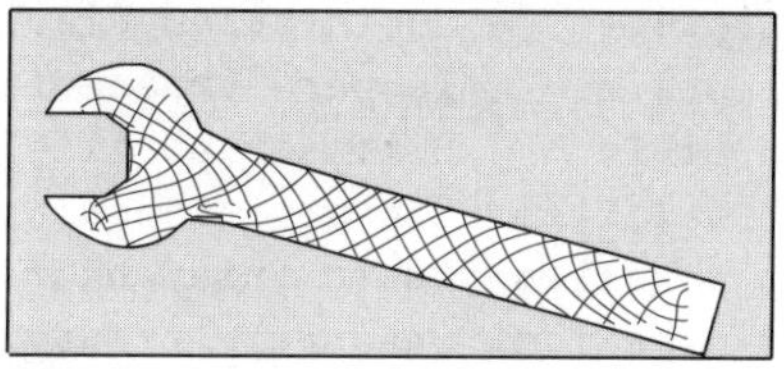

Figure 9.44. Ex. 9.26, principal stress lines

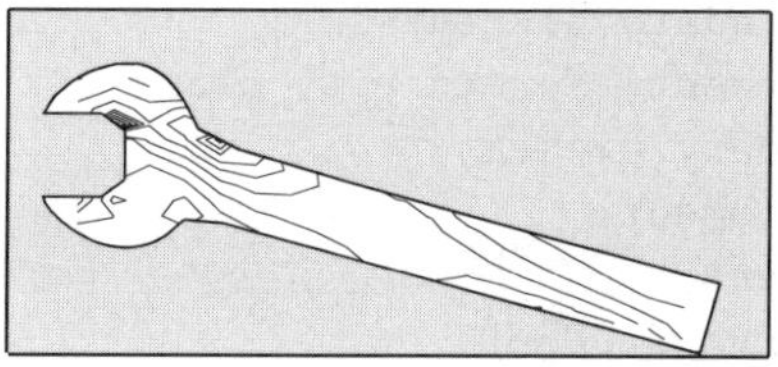

Figure 9.45. Ex. 9.26, contours of σ_1

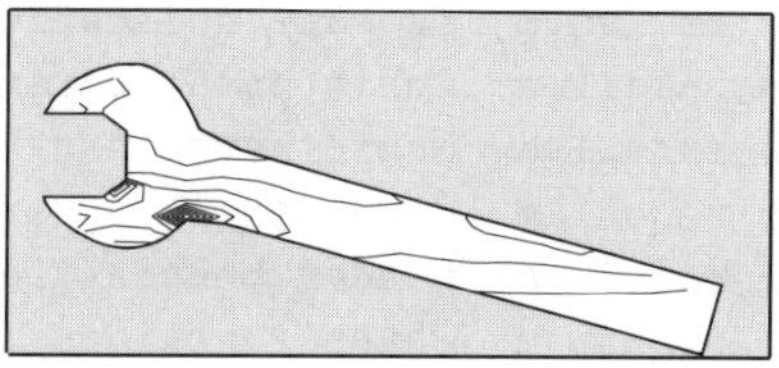

Figure 9.46. Ex. 9.26, contours of σ_2

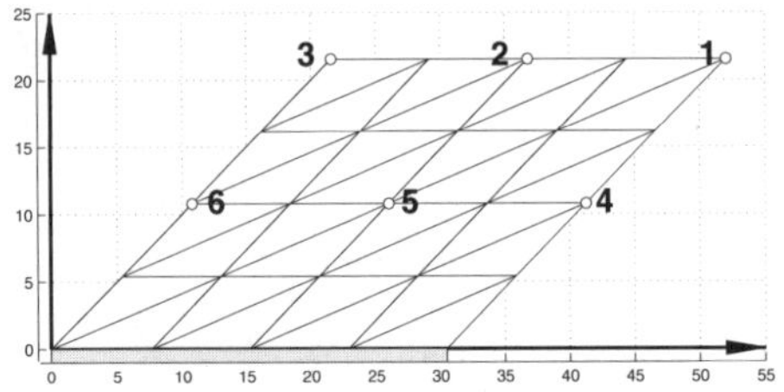

Figure 9.47. Ex. 9.27, geometry

9.9 Hints to MATLAB Programs

KAPITEL09/FEM_1, Elliptic boundary value problems

demo1.m	Example, linear triangular elements
demo2.m	Example, lineare parallelogram elements
demo3.m	Example, quadratic triangular elements
demo4.m	Example, quadratic triangular and parallelogram elements
demo5.m	Example, cubic triangular and parallelogram elements
demo6.m	Example, isopar. quadratic triangular and quadrilateral elements
ellipt1.m	Linear triangular element
ellipt2a.m	Linear parallelogram element
ellipt2b.m	Isopar. quadrilateral element
ellipt3.m	Quadratic triangular and parallelogram element
ellipt4.m	Cubic triangular and parallelogram element
ellipt5.m	Isopar. triangular and quadrilateral element
fem_bilin.m	Bilinear parallelogram element
fem_drlell.m	Linear triangular element
fem_drkell.m	Cubic triangular element after Zienkiewicz
fem_drqell.m	Quadratic triangular element
fem_isobil.m	Isopar. bilinear quadrilateral element
fem_isodrq.m	Isopar. quadratic triangular element
fem_isopaq.m	Isopar. quadratic quadrilateral element, Serendipity class
fem_isoraq.m	Isopar. quadratic boundary element
fem_pakell.m	Cubic parallogram element, Serendipity class
fem_rakell.m	Cubic hermitean Boundary element
fem_raqell.m	Quadratic boundary element
fem_ralell.m	Linear boundary element
fem_ffqdre.m	Shape functions for FEM_ISODRQ.M
fem_ffqbil.m	Shape functions for FEM_ISOBIL.M
fem_ffqpas.m	Shape functions for FEM_ISOPAQ.M

```
fem_ffquad.m  Shape functions for FEM_ISORAQ.M
myadapt.m     Simple adaptive mesh refinement
KAPITEL09/FEM_2, Discs and plates
bsp021g.m     Spanner, geometry data
bsp021h.m     Spanner, boundary data, loads
bsp022.m      Nine examples for plates
fem_batoz.m   Non-conforming quadratic triangular element
fem_batoz1.m  Auxiliary file for FEM_BATOZ.M
fem_drkpla.m  Non-conforming cubic triangular element
              after ZIENKIEWICZ
fem_drksch.m  Cubic disc element with condensation
              in triangle
fem_elstif.m  Non-conforming quadratic triangular element,
              other version
fem_pakpla.m  Non-conforming quadratic parallelogram element
              Serendipity class
fem_ripla.m   Conforming bicubic rectangular element
demo1.m       Masterfile for disc problems
demo2.m       Masterfile for plate problems after H.R.SCHWARZ
demo3.m       Masterfile for plate problems after BATOZ
scheibe3.m    Disc problem, cubic triangular element
spaqua1.m     Stress computation for cubic triangular element
KAPITEL03/FEM_3, Navier-Stokes Equations
Stream-function vorticity form
Time-dependent form after H.Ninomiya/K.Onishi; artificial
boundary conditions for vorticity automatically generated .
Time-independent form as elliptic system after Barragy-Carey.
demo1.m      lid driven cavity, time-dependent
demo2.m      flow past half cylinder, time-dependent,
demo3.m      flow past cylinder, time-dependent,
demo4.m      backfacing step, time-dependent
demo5.m      NS-part for transport problem, time-dependent
demo6.m      Example with exact solution, time-dependent
demo7.m      lid driven cavity, time-independent,
             Simple iteration
demo8.m      Example with exact solution, time-independent,
             Simple iteration
demo9.m      Example with exact solution, time-independent,
             Newton's method
demo10.m     Example with exact solution, time-dependent,
             with ode23.m
demo11.m     Coupled system, linear triangular elements
             simple Newton method, Example: lid driven cavity
demo12.m     Coupled system, linear parallelogram elements
             simple Newton method, Example: lid driven cavity
```

```
ellipt1.m:    Computes stream function by Poisson's equation
prepar.m      Mesh generation (with PDE TOOLBOX)
rside10.m     Right side for differential equation in ode23.m
velocity.m    Computes flow by stream function
vorticity.m   Computes vorticity
wbound.m      Computation of artificial boundary conditions
              for vorticity
KAPITEL09/FEM_4, Convection, Stream-function vorticity form
Time-dependent form after H.Ninomiya/K.Onishi; artificial
boundary conditions for vorticity automatically generated.
Time-independent form as elliptic system after W.N. Stevens
demo1.m         Thermal flow in a cup, time-dependent
demo2.m         Convection in a closed compartment, time-dependent
demo3.m         Convection in a square box, time-dependent
demo4.m         Thermal flow in a cup, time-independent
demo5.m         Convection in a unit square, time-independent
demo6.m         Example with exact solution, time-independent
convection.m    Computes temperature
vorticity_k.m   Computes vorticity for convection
lanscape.m      Neumann's boundary condition
matrizen.m      Matrices for coupled system
rightsides.m    Right sides for coupled system
KAPITEL09/STOKES, Navier-Stokes Problems in (u_1,u_2,p)-form
Fix one value of pressure p!
demo1.m: lid driven cavity with Taylor-Hood elements
         linear: without convection term
demo2.m: lid driven cavity with Mini elements
         linear: without convection term
demo3.m: lid driven cavity with Taylor-Hood elements
         nonlinear: with convection term, simple iteration
demo4.m: unit square with Taylor-Hood elements, example with
         exact solution, linear: without convection term
demo5.m: lid driven cavity with Taylor-Hood elements
         nonlinear: with convection term, NEWTON iteration
         simple continuation possible until NU = 0.002106
         Sequel for NU: [0.1,0.05,...,0.01,0.009,...0.003,
                        0.0029,..0.0022,0.00219,...0.002106]
demo6.m.M: Letters F E M with Taylor-Hood elements
         linear: without convection term
KAPITEL09/TIDAL, Shallow Water Equations
This directory contains MATLAB versions of BASIC programs
of H.Ninomiya/K.Onishi and further applications
demo1a.m    Island in a bay
demo1b.m    Island in a bay, different boundary computation
demo2.m     Finite channel with ode23.m
```

```
demo3.m      Long channel
demo4.m      Long wave on beach
flow_1.m     Velocity and water depth with lumped mass matrix
flow_2.m     As flow_1.m but with selective lumping
flow_3.m     As flow_1.m but with full mass matrix
lanscape.m   Island in a bay (geometry data, coast)
rside1.m     Right side of differential system
vnomal.m     Velocity at boundary (coast)
vnomal_n.m   Velocity at boundary (coast) (different way)
```

The representation of data in finite element methods follow to a large part as in MATLAB PDE TOOLBOX. For instance:

`[X,Y,P] = bspxxxf(segnr)`	boundary segments (often subfunction)
`[x,y] = bspxxxg(bs,s)`	geometry data
`[RD,RC,LOADS] = bspxxxh(p,e,t)`	boundary data

The external boundary has to be ordered counterclockwise, a possible internal boundary (cavity) clockwise.

Nodes: (Succession arbitrary fixed, never change during computation)
$p(1,:)$ x-components, $p(2,:)$ y-components.

Boundary: (ordered counterclockwise in simply connected domain)
$e(1,:)$ nr. of initial points of edges,
$e(2,:)$ nr. of end points of edges,
$e(3,:)$ initial values of line parameters,
$e(4,:)$ terminal values of line parameters,
$e(5,:)$ segment nrs.;
$e(6,:)$ etc., possibly additional characteristics

Linear triangular elements: (Succession arbitrary, may be changed during computation)
$t(1:3,:)$ Nrs. of vertices, possibly additional characteristics.

Quadratic triangular elements: (Succession arbitrary)
$t(1:3,:)$ Nrs. of vertices, $t(4:6,:)$ Nrs. of mid-points of edges, possibly additional characteristics in subsequent rows.

Parallelogram elements: (Succession arbitrary)
$t(1:4,:)$ Nrs. of vertices.

ATTENTION: Make MATLAB path to AAMESH permanent!

Frequently also the basic mesh of a problem is not built by using means of the MATLAB Toolbox but given directly and refined by programs of the folder AAMESH.

Example 9.28. Somewhat simplified mesh of Example 9.2:
Hint: the third and fourth row of the edge file e can be set equal to zero if not used in very simple examples.

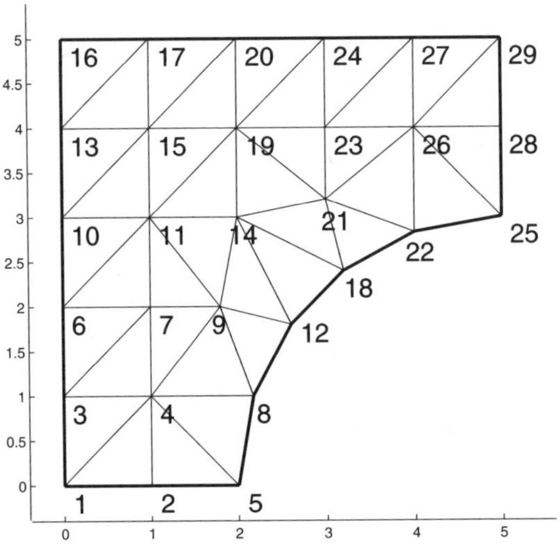

Figure 9.48. Ex. 9.28, mesh geometry

Geometry file for mesh of Figure 9.48:

```
function [p,e,t] = bsp01
% Nodes
p1 = [0 1 0 1 2 0   1 2.17157 1.8 0 1 2.6 0 2 1 0;
      0 0 1 1 0 2   2 1       2   3 3 1.8 4 3 4 5];
p2 = [1 3.2 2 2 3   4         3   3 5 4   4 5 5;
      5 2.4 4 5 3.2 2.82843   4   5 3 4   5 4 5];
p = [p1,p2];
% Triangles
t1 = [2  4  4  4  8  7   9   9 12  14  14  18 19 21 21 22 25 1;
      5  5  8  9 12  9  14  12 18  18  21  22 21 22 26 25 28 2;
      4  8  9  7  9 11  11  14 14  21  19  21 23 26 23 26 26 4];
t2 = [4  3  7  6  11 10 15 11 19 13 17 15 20 19 24 23 27 26 29;
      3  4  6  7  10 11 13 14 15 15 16 19 17 23 20 26 24 28 27;
      1  7  3 11  6  15 10 19 11 17 13 20 15 24 19 27 23 29 26];
t = [t1,t2];
% boundary
e = [1 2 5  8 12 18 22 25 28 29 27 24 20 17 16 13 10 6 3;
     2 5 8 12 18 22 25 28 29 27 24 20 17 16 13 10  6 3 1;
     0 0 0  0  0  0  0  0  0  0  0  0  0  0  0  0  0 0 0;
     0 0 0  0  0  0  0  0  0  0  0  0  0  0  0  0  0 0 0;
     1 1 2  2  2  2  2  3  3  4  4  4  4  4  5  5  5 5 5];
```

10

A Survey on Tensor Calculus

10.1 Tensor Algebra

As long as we work with a fixed coordinate system and with data fields of dimension not higher than two, the previously introduced formulas of matrix calculation and vector analysis are completely sufficient and result in a clear, concise picture. However, the TAYLOR expansion of a function $\underline{f} : \mathbb{R}^m \to \mathbb{R}^n$ shows already the limits of this representation. Let $\underline{x}_0 = \underline{0}$ be the expansion point then the TAYLOR polynomial of order two reads:

$$\begin{aligned} \underline{f}(\underline{x}) &\approx \underline{f}(\underline{0}) + \nabla \underline{f}(\underline{0})\underline{x} + \frac{1}{2}\nabla\nabla \underline{f}(\underline{0})(\underline{x}, \underline{x}) \\ f_i(\underline{x}) &\approx f_i(\underline{0}) + \sum_{j=1}^{n} \frac{\partial f_i}{\partial x_j}(\underline{0})x_j + \frac{1}{2}\sum_{j=1}^{n}\sum_{k=1}^{n} \frac{\partial^2 f_i}{\partial x_j\, \partial x_k}(\underline{0})x_j x_k\,, \; i = 1 : n\,. \end{aligned}$$

Already the next term of the expansion does no longer consist of a matrix in classical meaning but rather a tensor of third order. In *differential geometry* the defects become even more evident if we try to formulate, e.g., GAUSS' divergence theorem or STOKES' rotation theorem on a spherical surface or, to be more general, on a differentiable manifold. *Tensor Algebra* and its infinitisimal sister, *Tensor Analysis*, provide the calculus to overcome these deficiencies; however, the higher information content is at the cost of considerably more writing.

For a full development of this calculus, we have to work with *pairs* of vector spaces, namely with a basic (real) vector space $\mathcal{V}$ and its *dual space* $\mathcal{V}_d$ whose elements are called *covectors*. A covector $v_d \in \mathcal{V}_d$ is understood first of all as a linear bounded mapping $v_d : \mathcal{V} \ni v \mapsto v_d(v) \in \mathbb{R}$ according to Sect. 1.7**(e)**. However, since the sample space $\mathcal{V}$ is always *finite-dimensional* in tensor calculus, all these linear *functionals* are automatically bounded, and the second requirement is omitted, along with it the question of the underlying topology.

The *dual pairing* of V and $\mathcal{V}_d$ is emphasized in a special way by means of *angle brackets* $v_d(v) = \langle v_d, v\rangle$. On the other side, $v \in \mathcal{V}$ can also be interpreted as a mapping $v : \mathcal{V}_d \ni v_d \mapsto v(v_d) = \langle v, v_d\rangle := \langle v_d, v\rangle \in \mathbb{R}$. This *identification* of $\mathcal{V}$ with its bidual space $[\mathcal{V}_d]_d$ is called *canonical isomorphism*, and the fundamental relationship $\langle v, v_d\rangle = \langle v_d, v\rangle$ will be permanently used in this chapter; see also **(c)**.

Generically, a tensor is a multilinear mapping with arguments in $\mathcal{V}$ or $\mathcal{V}_d$ and with *scalar values* in $\mathbb{R}$ hence a *multilinear functional* (exact definition later). Tensor algrebra investigates the behavior of this *mapping* under transformations of the coordinate system. After having developed the transformation rules, the basis elements of $\mathcal{V}$ and $\mathcal{V}_d$ are frequently *cancelled* and one considers only the associated data fields of real numbers.

Data of an n-dimensional data field are fixed by a set of n indices. In an arbitrary field without symmetry, these indices cannot to be permutated or, in other words, their *horizontal* position is fixed. However, their *vertical* position, below or above, is free and remains at our disposal for further information.

As previously emphasized several times, it makes sense to interpret the elements of the *primal space* $\mathcal{V}$ as formal *column vectors* and the elements of the dual space $\mathcal{V}_d$ as formal *row vectors.* Then the *coordinate space* $\mathbb{R}^n$ of *column vectors* and the coordinate space $\mathbb{R}_n$ of *row vectors* are always illustrative examples of the vector spaces $\mathcal{V}$ and $\mathcal{V}_d$.

Rule for real and formal vectors:	
Index below (columns):	$\underline{v}_i \in \mathbb{R}^n\,,\quad \underline{v}_i \in \mathcal{V}$
Index above (rows):	$\underline{v}^i \in \mathbb{R}_n\,,\quad \underline{v}^i \in \mathcal{V}_d$

.

The elements of a column vector $\underline{x} \in {\mathbb{R}^n}_1 =: \mathbb{R}^n$ are *rows* with a single element hence their index is written above etc.; cf. Sect. 1.1**(a1)**.

Then, for instance, "lowering of indices" means transformation of rows into columns and "raising" has the converse meaning (two fundamental operations in MATLAB). Also, the correctness of large formulas can be checked easily by applying them to the model spaces $\mathbb{R}^n$ and $\mathbb{R}_n$. As a consequence, it makes sense to start by using the rules of matrix computation and to introduce the scalar product later.

Remember also EINSTEIN's convention which is used througout this chapter:

- If an index appears once, then the equation applies to all the values of this index.
- If an index appears twice, then it is to be sumed up over that index.
  In this case one index should stand *left below* and the other *right above if possible* to indicate row-column multiplication in application.

Exceptions to the summation rule cannot be avoided and must be marked in a special way.

Example 10.1. (1°) (Tensors in Component Form) (i) Let $\underline{a} \in \mathbb{R}^m$ be a column vector then a 1-tensor with data field $\underline{a}$ is $\mathsf{T} : \mathbb{R}_m \ni \underline{x} \mapsto \underline{x}\,\underline{a} \in \mathbb{R}$.

(ii) Let $A \in \mathbb{R}^m{}_n$ be a matrix then a 2-tensor with data field A is $\mathsf{T} : \mathbb{R}_m \times \mathbb{R}^n \ni (\underline{x}, \underline{y}) \mapsto \underline{x}A\underline{y} \in \mathbb{R}$; see **(e)**.

(2°) For a matrix $A = [a^i{}_k] \in \mathbb{R}^m{}_n$ the upper index i is the row index and the lower index k is the column index. The transposed matrix $A^T = [a_i{}^k]$ originates from A by *lowering* of the first index and *raising* of the second index (permutation of the two index letters i and k does not make any sense).

(3°) Multiplication of matrices is non-commutative. Consider two compatible matrices $A = [a^i{}_k]$ and $B = [b^i{}_k]$ then $A\,B \neq B\,A$ but according to EINSTEIN we can also write

$$\begin{aligned} C = [c^i{}_k] &:= A\,B &= a^i{}_j b^j{}_k = b^j{}_k a^i{}_j \\ C^T = [c_i{}^k] &:= B^T A^T &= b_j{}^k a_i{}^j = a_i{}^j b_j{}^k \end{aligned} \tag{10.1}$$

since *scalar* multiplication is commutative. Therefore the *position* of the components in the product does not play any role rather the position of the indices.

(4°) For $\underline{x} \in \mathbb{R}^n$ and $\underline{y} \in \mathbb{R}_n$ of course $\underline{y}A^T = (A\underline{y}^T)^T$. Moreover, by EINSTEIN's rules,

$$\begin{aligned} A\underline{x} &= \underline{a}_1 x^1 + \ldots + \underline{a}_n x^n = a^i{}_k x^k \; (= x^k a^i{}_k) = [b^i]_{i=1}^m \in \mathbb{R}^m \\ \underline{y}A &= y_1 \underline{a}^1 + \ldots + y_m \underline{a}^m = y_i a^i{}_k \; (= a^i{}_k y_i) = [b_k]_{k=1}^n \in \mathbb{R}_n \\ A^T\underline{x} &\; (= a^i{}_k x^i = x^i a_i{}^k) = a_i{}^k x^i = [b^k]_{k=1}^n \in \mathbb{R}^n \\ \underline{y}A^T &\; (= y_k a^i{}_k = a_i{}^k y_k) = y_k a_i{}^k = [b_i]_{i=1}^m \in \mathbb{R}_m\,. \end{aligned}$$

(a) Transformation of Basis and Components Let $\boldsymbol{\delta}$ be the unit tensor with components $\delta^i{}_k$ (KRONECKER *symbol*), i.e., $\delta^i{}_i = 1$ (without summation) and $\delta^i{}_k = 0$ for $i \neq k$, and let

$$\begin{aligned} \mathcal{E} &= \{\underline{e}_1, \ldots, \underline{e}_n\},\; \mathcal{F} = \{\underline{f}_1, \ldots, \underline{f}_n\} \text{ two bases in } \mathcal{V}, \\ \mathcal{E}_d &= \{\underline{e}^1, \ldots, \underline{e}^n\},\; \mathcal{F}_d = \{\underline{f}^1, \ldots, \underline{f}^n\} \text{ the } \textit{dual bases} \text{ in } \mathcal{V}_d\,, \end{aligned}$$

such that

$$\langle \underline{e}^i, \underline{e}_k \rangle := \underline{e}^i(\underline{e}_k) = \delta^i{}_k\,,\quad \langle \underline{f}^i, \underline{f}_k \rangle := \underline{f}^i(\underline{f}_k) = \delta^i{}_k\,; \tag{10.2}$$

for instance, the functional $\underline{e}^i$ is evaluated here at the point $\underline{e}_k$. The unique existence of the dual basis is warranted by the requirement (10.2). The basis $\{\underline{e}_1, \ldots, \underline{e}_n\}$ shall be called *reference basis* or *canonical basis* henceforth (for instance the unit vectors in $\mathbb{R}^n$). The elements of the basis $\mathcal{F}$ can be written as linear combination of the elements of $\mathcal{E}$, and the elements of the dual basis $\mathcal{F}_d$ as linear combination of the elements of $\mathcal{E}_d$, $\underline{f}_k = \underline{e}_i\, a^i{}_k$, $\underline{f}^i = b^i{}_k \underline{e}^k$, hence in *formal* writing

$$\mathcal{F} = \mathcal{E}A\,,\quad \mathcal{F}_d = B\mathcal{E}_d\,,\quad \mathcal{E}_d\mathcal{E} = \boldsymbol{\delta}\,,\quad \mathcal{F}_d\mathcal{F} = \boldsymbol{\delta} \tag{10.3}$$

with regular matrices $A = [a^i{}_k]$ and $B = [b^i{}_k]$ where "formal" means that we may handle the elements of $\mathcal{E}$, $\mathcal{F}$ as column vectors and the elements of $\mathcal{E}_d$, $\mathcal{F}_d$ as row vectors. Note that

$$\delta = \mathcal{F}_d\mathcal{F} = B\mathcal{E}_d\mathcal{E}A = BA \implies B = A^{-1}$$

by (10.3). *Scalar products* in $\mathcal{V}$ and $\mathcal{V}_d$ may be different, by the way, but they are denoted here *both* with the same point "$\cdot$". If, after introduction of a scalar product in $\mathcal{V}$, the basis $\mathcal{E}$ is a *cartesian* or *orthogonal* basis (actually orthonormal basis) with the property $\underline{e}_i \cdot \underline{e}_k = \delta^i{}_k$, then $\mathcal{F}$ is also cartesian if and only if A orthogonal, and $B = A^T$.

For two vectors,

$$\underline{v} = \underline{e}_1 x^1 + \ldots \underline{e}_n x^n =: \underline{e}_i x^i \in V\,,\ \ \underline{w} = y_1\underline{e}^1 + \ldots y_n\underline{e}^n =: y_i\underline{e}^i \in V_d\,,$$

we obtain immediately by formal writing

$$\begin{aligned}\underline{v} &= \mathcal{F}\widetilde{\underline{x}} = \mathcal{E}\underline{x} = \mathcal{E}AA^{-1}\underline{x} = \mathcal{F}A^{-1}\underline{x} &&\implies \widetilde{\underline{x}} = A^{-1}\underline{x} \equiv B\underline{x} \in \mathbb{R}^n\\ \underline{w} &= \widetilde{\underline{y}}\mathcal{F}_d = \underline{y}\mathcal{E}_d = \underline{y}B^{-1}B\mathcal{E}_d = \underline{y}B^{-1}\mathcal{F}_d &&\implies \widetilde{\underline{y}} = \underline{y}A \in \mathbb{R}_n\,.\end{aligned} \tag{10.4}$$

(Note that we make a strong difference between a vector $\underline{v} \in \mathcal{V}$ and its component vector $\underline{x} \in \mathbb{R}^n$).

Infinite-dimensional vector spaces are generically defined by properties of their elements ("the continuous function on $[a\,,\,b]$ form the vector space $\mathcal{C}[a,b]$") whereas finite-dimensional vector spaces are defined commonly by a basis of linearly independent elements, and components cannot exist without having declared a basis before. By this reason the following fundamental notations are primarily oriented by the *basis* of a vector space:

- A transformation is called *covariant* (with the transformation of a basis of $\mathcal{V}$) if it has the same form as $\mathcal{F} = \mathcal{E}A$.
- A transformation is called *contravariant* (to the transformation of a basis of $\mathcal{V}$) if it has the same form $\widetilde{\underline{x}} = A^{-1}\underline{x}$ as the transformation of the *component vectors* $\underline{x} \in \mathbb{R}^n$ of $\mathcal{V}$.
- Consequently, by (10.3), the basis of $\mathcal{V}_d$ is transformed *contravariant*, $\mathcal{F}_d = A^{-1}\mathcal{E}_d$, and the component vector $\underline{y} \in \mathbb{R}_n$ of $\underline{w} \in \mathcal{V}_d$ is transformed *covariant*, $\widetilde{\underline{y}} = \underline{y}A$.
- Regarding the customs in engineering sciences, the elements $\underline{v}$ of $\mathcal{V}$ are called *contravariant* elements and the vector space $\mathcal{V}$ is called *contravariant vector space* (since the behavior of the components is more important than that of the basis in applications). Accordingly, the elements of $\mathcal{V}_d$ and the vector space $\mathcal{V}_d$ are called *covariant.*
- *Contravariant* components are written with *index above* (as the scalar elements of a column vector in $\mathbb{R}^n$). *Covariant* components are written with *index below* (as the scalar elements of a row vector in $\mathbb{R}_n$).
- Relative to the transformation of components in $\mathcal{V}$, the transformation matrix A^T plays the major role and not the matrix A; it is then frequently A replaced by A^T and A^T by A but not in this volume.

Let us return to the above introduced notations. The transformation matrix A is orthogonal in transforming *cartesian coordinate systems* and thus $A^{-1} =$

A^T; then $\underset{\sim}{\tilde{x}} = A^{-1}\underset{\sim}{x} = A^T\underset{\sim}{x}$ holds in the first case of (10.4), and in the second case we have $\underset{\sim}{\tilde{y}}^T = A^T\underset{\sim}{y}^T$. If now the component vectors are considered as column vectors in both cases (yielding an error message in MATLAB), then we find

| no different transformational behavior in cartesian coordinate systems |,

and the distinction between covariant and contravariant tensors can be neglected. It is however maintained by theoretical and computational reasons.

Example 10.2. (1°) The space Π_n of polynomials of degree $\leq n$ is a vector space of dimension $n+1$. Choose $n+1$ mutually distinct numbers x_j, $j = 1:n+1$, then the basis $q_i \in \Pi_n$ of LAGRANGE polynomials has the property $q_i(x_j) = \delta^i{}_j$; see Sect. 2.1**(c)**. The dual basis of this basis consists of the *mappings* $\pi_j : \Pi_n \ni p \mapsto p(x_j) \in \mathbb{R}$ because $\pi_j(q_i) = q_i(x_j) = \delta^i{}_j$.
(2°) The polynomials of degree ≤ 2 on an arbitrary, non-degenerated triangle $T(x,y)$ form a vector space $\mathcal{Q}$ of dimension 6. Let $\underset{\sim}{x}_j$, $j = 1:3$ be the vertices , $\underset{\sim}{x}_j$, $j = 4:6$ the mid-points of the edges and $\varphi_i(x,y)$, $i = 1:6$ the *shape functions* from Sect. 11.3**(c)** having the unit value in precisely one of these points and zero else. Then the dual basis of this basis consists also of the *mappings* $\pi_j : Q \ni p \mapsto p(x_j) \in \mathbb{R}$, $j = 1:6$.

(b) Scalar Product Spaces Let $\mathcal{V}$ be a vector space with basis $\{\underset{\sim}{e}_1, \dots, \underset{\sim}{e}_n\}$ and let $M_e = [e_{ik}]$ be an arbitrary (real)-symmetric, positive definite matrix then a *scalar product* p is defined on $\mathcal{V}$ by $p(\underset{\sim}{e}_i, \underset{\sim}{e}_k) := e_{ik}$ for all basis vectors $\underset{\sim}{e}_j$ of the basis and thus also for all elements of $\mathcal{V}$. If conversely a scalar product $p(\cdot, \circ)$ is given, then the matrix $M_e = [e_{ik}]$ with $e_{ik} := p(\underset{\sim}{e}_i, \underset{\sim}{e}_k)$ is positive definite. This matrix resp. the corresponding bilinear mapping is called *covariant metric tensor* of the basis $\{\underset{\sim}{e}_1, \dots, \underset{\sim}{e}_n\}$. In consequence there exists always an arbitrary number of scalar products on $\mathcal{V}$ and we say that $\mathcal{V}$ is a *scalar product space* or *inner product space* if one of these scalar products is declared for *canonical* (natural) in a particular way. As customary, we write for the canonical scalar product $\underset{\sim}{v} \cdot \underset{\sim}{w} := p(\underset{\sim}{v}, \underset{\sim}{w})$. Then a *linear measure* or *metric* is defined on $\mathcal{V}$ by $[\underset{\sim}{v} \cdot \underset{\sim}{v}]^{1/2}$ and the length of $\underset{\sim}{v} = \underset{\sim}{e}_i v^i \in \mathcal{V}$ can be *declared* by means of the metric tensor:

$$|\underset{\sim}{v}|^2 = \underset{\sim}{v} \cdot \underset{\sim}{v} = (\underset{\sim}{e}_i v^i) \cdot (\underset{\sim}{e}_i v^i) = \underset{\sim}{x}^T M_e \underset{\sim}{x}\,, \quad \underset{\sim}{x} = [v^1, \dots, v^n]^T \in \mathbb{R}^n\,.$$

| Henceforth let $\mathcal{V}$ always be ascalar product space |.

A basis $\{\underset{\sim}{e}_1, \dots, \underset{\sim}{e}_n\}$ of $\mathcal{V}$ is *orthonormal*, i.e., a *normed orthogonal system (in short NOS)* if

$$\underset{\sim}{e}_i \cdot \underset{\sim}{e}_k = \delta^i{}_k \text{ (KRONECKER symbol).}$$

A basis $\{\underset{\sim}{r}^1, \dots, \underset{\sim}{r}^n\}$ of V is called *reciprocal* relative to a basis $\{\underset{\sim}{e}_1, \dots, \underset{\sim}{e}_n\}$ of V if

$$\underset{\sim}{r}^i \cdot \underset{\sim}{e}_k = \delta^i{}_k\,.$$

The reciprocal basis relative to a given basis exists uniquely (and is actually a basis). A NOS has always the metric tensor $M_e = I$ (unit matrix) w.r.t. the canonical basis. If $\{\underline{r}^i\}$ is reciprocal relative to $\{\underline{e}_k\}$ then obviously $\{\underline{e}_k\}$ is reciprocal relative to $\{\underline{r}^i\}$. In a scalar product space, the dual basis may be *identified canonically* with the reciprocal basis therefore the reciprocal bases is written with *upper index* by exception although it is not contained in $\mathcal{V}_d$. Orthonormal bases are reciprocal relative to themselves. They may be derived from an arbitrary basis by the GRAM-SCHMIDT method and thus we may work in scalar product spaces always with orthonormal bases.

Example 10.3. Let $A \in \mathbb{R}^n{}_n$ be regular then the column of A are a basis of $\mathbb{R}^n$ and the columns of $[A^{-1}]^T$, i.e., the transposed rows of A^{-1}, are reciprocal to that basis because $A^{-1}A = I$.

For a basis $\{\underline{e}_i\}$ and the reciprocal basis $\{\underline{r}^i\}$, we write

$$\boxed{e_{ik} = \underline{e}_i \cdot \underline{e}_k\,, \quad e^{ik} = \underline{r}^i \cdot \underline{r}^k\,, \quad \underline{v} = \underline{e}_i v^i = \underline{r}^k v_k \in \mathcal{V}}\,. \tag{10.5}$$

The components v^i are called *contravariant* components by the above convention and the components v_k are called *covariant* components of $\underline{v}$. They are obtained by multiplying $\underline{v}$ with $\underline{r}^k$ resp. with $\underline{e}_k$,

$$\underline{v} \cdot \underline{r}^k = (\underline{e}_i v^i) \cdot \underline{r}^k = v^i \underline{e}_i \cdot \underline{r}^k = v^k\,, \quad \underline{v} \cdot \underline{e}_k = (\underline{r}^i v_i) \cdot \underline{e}_k = \underline{v}_k\,; \tag{10.6}$$

the matrix $M_e^d := [e^{ik}]$ is called *contravariant metric tensor.* Using M_e and M_e^d we have the relation

$$v^i = \underline{v} \cdot \underline{r}^i = (\underline{r}^j v_j) \cdot \underline{r}^i = e^{ij} v_j\,, \quad v_j = \underline{v} \cdot \underline{e}_i = (\underline{e}_j v^j) \cdot \underline{e}_i = e_{ij} v^j \tag{10.7}$$

between the components v^i and v_k.

(c) Identifying $\mathcal{V}$ and $\mathcal{V}_d$. Let $\mathcal{V}$ be a scalar product space then $\mathcal{V}$ and $\mathcal{V}_d$ may be identified with each other which is customary in engineering sciences. This procedure is performed by using the metric tensors therefore we consider them more exactly and introduce to this end the RIESZ *mapping* $\mathcal{R} : \mathcal{V} \to \mathcal{V}_d$ being uniquely declared by pointwise definition for all $\underline{v}, \underline{w} \in \mathcal{V}$:

$$\boxed{(\mathcal{R}\underline{v})(\underline{w}) \equiv \langle \mathcal{R}(\underline{v}), \underline{w} \rangle := \underline{v} \cdot \underline{w} = \underline{w} \cdot \underline{v} =: \langle \mathcal{R}(\underline{w}), \underline{v} \rangle \equiv (\mathcal{R}\underline{w})(\underline{v})}\,.$$

This mapping is linear and bijective hence an isomorphism, frequently called likewise *canonical isomorphism.* Obviously it is the finite-dimensional analogue to the RIESZ mapping introduced in Sect. 1.11**(a)**. Since reciprocal basis and dual basis $\{\underline{e}_k\}$ are uniquely determined relative to a given basis and $\underline{r}^i \cdot \underline{e}_k = \delta^i{}_k$, we obtain directly

$$\langle \mathcal{R}(\underline{r}^i), \underline{e}_k \rangle \equiv \underline{r}^i \cdot \underline{e}_k = \delta^i{}_k \Longrightarrow \underline{e}^i = \mathcal{R}(\underline{r}^i) \tag{10.8}$$

from the definition. In other words, reciprocal basis in $\mathcal{V}$ and dual basis in $\mathcal{V}_d$ can be identified canonically by the *isomorphism* $\mathcal{R}$. But this property of bases holds also for the appertaining vector spaces, hence also $\mathcal{V}$ and $\mathcal{V}^d$ can be identified by means of the mapping $\mathcal{R}$. To emphasize this close relation, we write likewise $\langle \underline{v}, \underline{w} \rangle = \underline{v} \cdot \underline{w} \ \forall \ \underline{v}, \ \underline{w} \in V$. An identification infers always that *one* symbol has *two* meanings, namely here $\underline{v} \in V$ and $\underline{v} \in V_d$.

Example 10.4. Let $\mathcal{V} = \mathbb{R}^n$ and let A be an arbitrary symmetric and positive definite matrix, then a scalar product is defined by $p(\underline{x}, \underline{y}) := \underline{x}^T A \underline{y}$, $\forall \ \underline{x}, \ \underline{y} \in \mathbb{R}^n$, and we have $\mathcal{R}(\underline{x}) : \underline{y} \mapsto \underline{x}^T A \underline{y} \in \mathbb{R}$ as *mapping* and $\mathcal{R}(\underline{x}) = \underline{x}^T$ after identifying this mapping with an element $\underline{x}^T \in \mathcal{V}_d = \mathbb{R}_n$ (the canonical scalar product being naturally defined by the identity $A = I$).

The elements $\mathcal{R}(\underline{e}_i)$ in $\mathcal{V}_d$ may be written as linear combination of the dual basis, $\mathcal{R}(\underline{e}_i) = c^i{}_j \underline{e}^j$, which implies that

$$e_{ik} = \underline{e}_i \cdot \underline{e}_k =: (\mathcal{R}(\underline{e}_i))(\underline{e}_k) = c^i{}_j \langle e^j, \underline{e}_k \rangle = c^i{}_j \delta^j{}_k = c^i{}_k \implies \mathcal{R}(\underline{e}_i) = e_{ij} \underline{e}^j \,.$$

In other words, the images of the basis elements $\underline{e}_i$ w.r.t. $\mathcal{R}$ are written by using the components of the metric tensor M_e!

If $\mathcal{V}$ is a scalar product space, there exists precisely one scalar product $[\,\cdot\,, \circ\,]$ in $\mathcal{V}_d$ such that the canonical isomorphism $\mathcal{R}$ becomes an *isometry*, i.e.,

$$\forall \ \underline{v} \in \mathcal{V} : [\mathcal{R}(\underline{v}), \mathcal{R}(\underline{v})] = \underline{v} \cdot \underline{v} \,.$$

This scalar product is defined also by the metric tensor as

$$[\underline{e}^i, \underline{e}^j] = e^{ij} \,, \quad M_e^{-1} = [e^{ij}] \,.$$

Then we have on the one side

$$\underline{v} \cdot \underline{v} = (\underline{e}_i v^i) \cdot (\underline{e}_j v^j) = v^i e_{ij} v^j = \underline{x}^T M_e \underline{x} \,, \quad \underline{x} = [v^1, \cdots, v^n]^T$$

and on the other side, using $\mathcal{R}(\underline{e}_i) = e_{ij} \underline{e}^j$,

$$\forall \ \underline{x} \in \mathbb{R}^n : [\mathcal{R}(\underline{v}), \mathcal{R}(\underline{v})] = [e_{ij} \underline{e}^j v^i, e_{kl} \underline{e}^l v^k] = \underline{x}^T M_e M_e^{-1} M_e \underline{x} = \underline{x}^T M_e \underline{x} \,,$$

hence an isometry.

Finally, to establish a relation between the covariant and the contravariant metric tensor, we apply the last equation to $\underline{v} = \underline{r}^i$ and regard (10.5), $\underline{r}^i \cdot \underline{r}^k = [\mathcal{R}(\underline{r}^i), \mathcal{R}(\underline{r}^k)] = [\underline{e}^i, \underline{e}^k]$. Then, by Definition (10.5), $M_e^d = M_e^{-1}$.

If $\mathcal{V}$ is a scalar product space with metric tensor M_e, then $\mathcal{V}_d$ shall be always a vector space with *canonical* scalar product defined by the metric tensor M_e^d and this product is *likewise* denoted by a point "$\cdot$".

Of course, there exists also a reciprocal basis $\{\underline{r}_1, \ldots, \underline{r}_n\}$ relative to the dual basis in $\mathcal{V}_d$ with the property $\underline{r}_i \cdot \underline{e}^k = \delta^i{}_k$, of which the indices are consistently written *below* although this basis is contained in dual space. Then $\mathcal{R}(\underline{e}_i) = \underline{r}_i$

because of the isomorphism property of $\mathcal{R}$ and since $\{\underline{e}_i\}$ is reciprocal to $\{\underline{r}^k\}$. By (10.7), we obtain

$$\boxed{\underline{r}_i = \mathcal{R}(\underline{e}_i) = e_{ik}\,\underline{e}^k\,,\;\; \underline{r}^i = \mathcal{R}^{-1}(\underline{e}^i) = \underline{e}_k e^{ik}}\,. \tag{10.9}$$

These two formulas play a role in the operations "raising" and "lowering" of indices dealed with in subsection **(l)**.

(d) General Tensors Let $\mathcal{V}^q$ be the *vector space* of q-tuples consisting of elements $\underline{v} \in \mathcal{V}$ and let $(\mathcal{V}_d)^p$ be the *vector space* of p-tuples consisting of elements $\underline{w} \in \mathcal{V}_d$. A $(p+q)$-linear form $\mathsf{T}^p{}_q$ is a mapping

$$\mathsf{T}^p{}_q \in \mathcal{T}^p{}_q(\mathcal{V}) := \mathcal{L}((\mathcal{V}_d)^p \times \mathcal{V}^q; \mathbb{R})$$

$$\mathsf{T}^p{}_q : (\mathcal{V}_d)^p \times \mathcal{V}^q \ni (\underline{w}^1, \dots, \underline{w}^p, \underline{v}_1, \dots, \underline{v}_q) \mapsto \mathsf{T}^p{}_q(\underline{w}^1, \dots, \underline{w}^p, \underline{v}_1, \dots, \underline{v}_q) \in \mathbb{R}$$

which is *linear in each argument.*

$\mathsf{T}^p{}_q$ is called a p-fold contravariant and q-fold covariant tensor or briefly (p,q)-tensor.

This notation reflects the behavior in transforming the *components* of the tensor which is the crucial operation in computations!

Remember that $\mathcal{V}$ is called *contravariant* vector space and $\mathcal{V}_d$ *covariant* vector space because of the corresponding transformational behavior of the respective *components.* By consequence, relative to the transformational behavior of the respective components somewhat uncomfortably but correct:

A p-fold *contravariant* tensor T^p has p *covariant* arguments in $\mathcal{V}_d$. A q-fold *covariant* tensor T_q has q *contravariant* arguments in $\mathcal{V}$.

In particular, a contravariant vector $\underline{v} \in \mathcal{V}$ defines a contravariant 1-tensor (mapping) with elements of the covariant vector space $\mathcal{V}_d$ for arguments and a covariant vector $\underline{v}_d \in \mathcal{V}_d$ defines a covariant 1-tensor (mapping) with elements of the contravariant vector space $\underline{v}$ for arguments,

$$\begin{aligned} \mathcal{T}^1(\mathcal{V}) &= \mathcal{L}(\mathcal{V}_d; \mathbb{R}) = (\mathcal{V}_d)_d \sim \mathcal{V} \\ \mathcal{T}_1(\mathcal{V}) &= \mathcal{L}(\mathcal{V}; \mathbb{R}) \;= \mathcal{V}_d\,. \end{aligned}$$

(The sign "$\sim$" shall point to canonical identification.) See also **(f)**.

The succession of the p covariant and q contravariant arguments is *arbitrary but fixed,* e.g., as used here for simplicity. A $(0,0)$-form is a scalar, a $(p,0)$-form is a contravariant tensor of order p, a $(0,q)$-form is a covariant tensor of order q. The number $p+q$ is the total order of the tensor $T^p{}_q$. After having specified a basis, also the n^{p+q} components in $\mathbb{R}$ of the tensor are called (p,q)-tensor.

Example 10.5. A simple matrix $T \in \mathbb{R}^n{}_n$ defines (the field of numbers of) a tensor $\mathsf{T} \in \mathcal{T}^1{}_1(\mathcal{V})$ where $\mathcal{V} = \mathbb{R}^n$, $\mathbb{R}^n \times \mathbb{R}_n \ni (u,v) \mapsto \mathsf{T}(u,v) = v^T T u \in \mathbb{R}$. The transformational device for the matrix T in changing the system of coordinates is well-known and reads $T \mapsto A^{-1}TA$.

Let now $\mathcal{M}$ is an arbitrary set, then the *mapping*

$$\mathcal{M} \ni x \mapsto \mathsf{T}^p{}_q(x) \in \mathcal{L}((\mathcal{V}_d)^p \times \mathcal{V}^q; \mathbb{R})$$

is called *tensor field*, the *components* depending on the variable x. For instance,

$\mathcal{M} \ni x \mapsto \mathsf{T}^0{}_0(x) \in \mathbb{R}$		is a scalar field
$\mathcal{M} \ni x \mapsto \mathsf{T}_1(x)$	$\in \mathcal{L}(\mathcal{V}; \mathbb{R})$	is a covariant vector field (linear mapping "waiting" for arguments $\underline{v} \in \mathcal{V}$)
$\mathcal{M} \ni x \mapsto \mathsf{T}^1(x)$	$\in \mathcal{L}(\mathcal{V}_d; \mathbb{R})$	is a contravariant vector field (linear mapping "waiting" for arguments $\underline{w} \in \mathcal{V}_d$)

Of course, $\mathcal{T}^p{}_q(\mathcal{V})$, being a space of multilinear mappings, is a vector space. Addition und scalar multiplication are inherited from the image space $\mathbb{R}$:

For $\mathsf{S}, \mathsf{T} \in \mathcal{T}^p{}_q(\mathcal{V})$, we have

$$\begin{aligned}
(\mathsf{S}+\mathsf{T})(\underline{v}^1, \dots, \underline{v}^p, \underline{v}_1, \dots, \underline{v}_q) &:= \mathsf{S}(\underline{v}^1, \dots, \underline{v}^p, \underline{v}_1, \dots, \underline{v}_q) \\
&\quad + \mathsf{T}(\underline{v}^1, \dots, \underline{v}^p, \underline{v}_1, \dots, \underline{v}_q) \\
(\alpha\mathsf{S})(\underline{v}^1, \dots, \underline{v}^p, \underline{v}_1, \dots, \underline{v}_q) &:= \alpha\mathsf{S}(\underline{v}^1, \dots, \underline{v}^p, \underline{v}_1, \dots, \underline{v}_q) \\
0(\underline{v}^1, \dots, \underline{v}^p, \underline{v}_1, \dots, \underline{v}_q) &:= 0 \in \mathbb{R}\,.
\end{aligned}$$

(e) Representation and Transformation of Tensors

(e1) Tensors of Order One $\mathsf{T} \in \mathcal{L}(\mathcal{U}; \mathbb{R})$.

Case 1: $\mathcal{U} = \mathcal{V}$, $p = 0$, $q = 1$, $\mathsf{T} \equiv \mathsf{T}_1 \in \mathcal{L}(\mathcal{V}; \mathbb{R})$; $\underline{v} = \underline{e}_i v^i = \underline{f}_j \widetilde{v}^j \in \mathcal{V}$.

Case 2: $\mathcal{U} = \mathcal{V}_d$, $p = 1$, $q = 0$, $\mathsf{T} \equiv \mathsf{T}^1 \in \mathcal{L}(\mathcal{V}_d; \mathbb{R})$; $\underline{v} = v_i \underline{e}^i = \widetilde{v}_j \underline{f}^j \in \mathcal{V}_d$.

$$\boxed{\begin{aligned}
\mathsf{T}_1(\underline{v}) &= \mathsf{T}_1(\underline{f}_i \widetilde{v}^i) = T_1(\underline{f}_i)\widetilde{v}^i =: t_{(f)i}\widetilde{v}^i \\
&= \mathsf{T}_1(\underline{e}_j a^j{}_i)\widetilde{v}^i = \mathsf{T}_1(\underline{e}_j)a^j{}_i\widetilde{v}^i =: t_{(e)j}a^j{}_i\widetilde{v}^i \\
\mathsf{T}^1(\underline{v}) &= \mathsf{T}^1(\widetilde{v}_i \underline{f}^i) = \widetilde{v}_i \mathsf{T}^1(\underline{f}^i) =: \widetilde{v}_i t_{(f)}{}^i \\
&= \widetilde{v}_i \mathsf{T}^1(b^i{}_j \underline{e}^j) = \widetilde{v}_i b^i{}_j \mathsf{T}^1(\underline{e}^j) =: \widetilde{v}_i b^i{}_j t_{(e)}{}^j
\end{aligned}}\,.$$

Result for the components with matrix T:

$$\boxed{\begin{aligned}
t_{(f)i} = t_{(e)j}a^j{}_i &\Longrightarrow T_{(f)1} = T_{(e)1}A\,, \\
t_{(f)}{}^i = b^i{}_j t_{(e)}{}^j &\Longrightarrow T_{(f)}{}^1 = A^{-1}T_{(e)}{}^1
\end{aligned}}\,.$$

The components t_i of the *covariant* tensor T_1 are transformed *covariant*, hence the index is written *below*. The components t^i of the *contravariant* tensor T^1 are transformed *contravariant* hence they are written with *upper* index.

(e2) Tensors of Order Two $\mathsf{T} \in \mathcal{L}(\mathcal{U} \times \mathcal{W}; \mathbb{R})$.

Case 1: Twofold covariant tensor, $\mathcal{U} = \mathcal{W} = \mathcal{V}$, $p = 0$, $q = 2$, $\mathsf{T} = \mathsf{T}_2 \in \mathcal{L}(\mathcal{V} \times \mathcal{V}; \mathbb{R})$; $\underline{v} = \underline{e}_i v^i = \underline{f}_j \widetilde{v}^j$, $\underline{w} = \underline{e}_i w^i = \underline{f}_j \widetilde{w}^j$.

$$\begin{array}{|llll|}
\mathsf{T}_2(\underline{v},\underline{w}) & = \mathsf{T}_2(\underline{f}_i\widetilde{v}^i,\underline{f}_j\widetilde{w}^j) & = \mathsf{T}_2(\underline{f}_i,\underline{f}_j)\widetilde{v}^i\widetilde{w}^j & =: t_{(f)ij}\widetilde{v}^i\widetilde{w}^j \\
t_{(f)ij}\widetilde{v}^i\widetilde{v}^j & = \mathsf{T}_2(\underline{e}_k a^k{}_i,\underline{e}_l a^l{}_j)\widetilde{v}^i\widetilde{w}^j & = \mathsf{T}_2(\underline{e}_k,\underline{e}_l)a^k{}_i a^l{}_j\widetilde{v}^i\widetilde{w}^j & \\
t_{(f)ij}\widetilde{v}^i\widetilde{w}^j & = t_{(e)kl}a^k{}_i a^l{}_j\widetilde{v}^i\widetilde{w}^j & = t_{(e)kl}a^k{}_i\widetilde{v}^i a^l{}_j\widetilde{w}^j & \\
t_{(f)ij} & = t_{(e)kl}a^k{}_i a^l{}_j\,. & &
\end{array}$$

or

$$\begin{aligned}
t_{(e)kl}v^k w^l &= t_{(e)kl}a^k{}_i a^l{}_j\widetilde{v}^i\widetilde{w}^j = t_{(f)ij}\widetilde{v}^i\widetilde{w}^j \\
t_{(f)kl}\widetilde{v}^k\widetilde{w}^l &= t_{(f)kl}b^k{}_i b^l{}_j v^i w^j = t_{(e)ij}v^i w^j\,.
\end{aligned}$$

The covariant tensor T_2 is transformed double *covariant*; $T_{2(f)}$ is created from $\mathsf{T}_{2(e)}$ by replacing the *component vectors of the arguments* $\underline{x}$ by $A\underline{\widetilde{x}}$ and $\underline{y}$ by $A\underline{\widetilde{y}}$ (double *contravariant* (!) transformation of the components of the *arguments*); $\underline{x} = [v^1,\dots,v^n]^T$ etc.

Case 2: $\mathcal{U} = \mathcal{V}\,,\ \mathcal{W} = \mathcal{V}_d\,,\ \underline{v}\in\mathcal{V}\,,\ \underline{w}\in\mathcal{V}_d\,,\ \mathsf{T}_1{}^1\in\mathcal{L}(\mathcal{V}\times\mathcal{V}_d;\mathbb{R})\,.$

$$\begin{array}{|lll|}
T_1{}^1(\underline{v},\underline{w}) = & T_1{}^1(\underline{f}_i\widetilde{v}^i,\widetilde{w}_j\underline{f}^j) & = T_1{}^1(\underline{f}_i,\underline{f}^j)\widetilde{v}^i\widetilde{w}_j \\
t_{(f)i}{}^j & := T_1{}^1(\underline{e}_k a^k{}_i, b^j{}_l\underline{e}^l) & = t_{(e)k}{}^l a^k{}_i b^j{}_l
\end{array}\,.$$

Case 3: $\mathcal{U} = \mathcal{V}_d\,,\ \mathcal{W} = \mathcal{V}\,,\ \mathsf{T}^1{}_1\in\mathcal{L}(\mathcal{V}_d\times\mathcal{V};\mathbb{R})\,.$

$$\boxed{t_{(f)}{}^i{}_j := \mathsf{T}^1{}_1(\underline{f}^i,\underline{f}_j) = \mathsf{T}^1{}_1(b^i{}_k\underline{e}^k,\underline{e}_l a^l{}_j) = t_{(e)}{}^k{}_l b^i{}_k a^l{}_j}\,,$$

Case 4: $\mathcal{U} = \mathcal{W} = \mathcal{V}_d\,,\ \mathsf{T}^2\in\mathcal{L}(\mathcal{V}_d\times\mathcal{V}_d;\mathbb{R})\,.$

$$\boxed{t_{(f)}{}^{ij} := \mathsf{T}^2(\underline{f}^i,\underline{f}^j) = \mathsf{T}^2(b^i{}_k\underline{e}^k,b^j{}_l\underline{e}^l) = t_{(e)}{}^{kl}b^i{}_k b^j{}_l}\,.$$

(Compare with (10.1)). The transformation of tensors of order two can be written in matrix form when both component vectors are strictly written in column form, $\underline{x} = [v^1,\dots,v^n]^T\in\mathbb{R}^n$ and $\underline{y} = [w^1,\dots,w^n]^T\in\mathbb{R}^n$:

$$\begin{aligned}
T_{(f)2} &= A^T T_{(e)2}A\,, & T_{(f)1}{}^1 &= A^T T_{(e)1}{}^1 A^{-T}\,, \\
T_{(f)}{}^1{}_1 &= A^{-1}T^1_{(e)1}A\,, & T_{(f)}{}^2 &= A^{-1}T_{(e)}{}^2 e A^{-T}\,,
\end{aligned} \tag{10.10}$$

Then, e.g., $T_{(f)2}(\underline{\widetilde{x}},\underline{\widetilde{y}}) = \underline{\widetilde{x}}^T A^T T_{(e)2}A\underline{\widetilde{y}} = \underline{x}^T T_{(e)2}\underline{y}$ where $\underline{x} = A\underline{\widetilde{x}}$ and $\underline{y} = A\underline{\widetilde{y}}$.

Rule for tensors of order two with component vectors in column form:

If a contravariant vector $\underline{v}\in\mathcal{V}$ for argument is replaced by a covariant vector $\underline{w}\in\mathcal{V}_d$, then A is to be replaced by the *"contragredient"* A^{-T} in the corresponding place. Accordingly, if A orthogonal, i.e. $A = A^{-T}$, then both transformations are *identical.*

(f) Tensor Product As well known, there is a symmetric scalar product and a skew-symmetric vector product in coordinate space $\mathbb{R}^3$. The same holds in tensor spaces but the alternating product is postponed to the next section.

Cum grano salis we may state the following:
If an element $\underline{v}_d \in \mathcal{V}_d$ applies to an element $v \in \mathcal{V}$, then the result is a number $\langle v_d, v\rangle \in \mathbb{R}$, and the entire operation is called contraction. If conversely an element $v \in \mathcal{V}$ applies to an element $v_d \in \mathcal{V}_d$, then one obtains again a number $\langle v, v_d\rangle$ as contraction, also called rejunevation sometimes. By the canonical isomorphism, we have $\langle v_d, v\rangle = \langle v, v_d\rangle$, and it doesn't matter which applies to which.

In other words, a vector $v \in \mathcal{V}$ may be considered as contravariant tensor of order one (mapping) because its argument is an element from $\mathcal{V}_d$, and a vector $v_d \in \mathcal{V}_d$ as covariant tensor of order one because its argument is an element $v \in \mathcal{V}$; cf. **(e1)**. How does this matching now work in *two* and more vectors?

(f1) The **Tensor Product of Two Vectors** is defined by one of the following mappings

$$\left.\begin{array}{ll} \underline{u} \otimes \underline{v}_d : \mathcal{V}_d \times \mathcal{V} \ni (\underline{u}_d, \underline{v}) & \mapsto \underline{u}_d(\underline{u}) \cdot \underline{v}_d(\underline{v}) \\ \underline{u} \otimes \underline{v} : \mathcal{V}_d \times \mathcal{V}_d \ni (\underline{u}_d, \underline{v}_d) & \mapsto \underline{u}_d(\underline{u}) \cdot \underline{v}_d(\underline{v}) \\ \underline{u}_d \otimes \underline{v}_d : \mathcal{V} \times \mathcal{V} \ni (\underline{u}, \underline{v}) & \mapsto \underline{u}_d(\underline{u}) \cdot \underline{v}_d(\underline{v}) \end{array}\right\} = \langle \underline{u}_d, \underline{u}\rangle\langle \underline{v}_d, \underline{v}\rangle \in \mathbb{R}\,.$$

On the right side, we have always the product of two real numbers. The meaning is explained most suitable in coordinate space writing by exception $\underline{v}^T \in \mathbb{R}_n$ for row vectors to make the notation more apparent. Then, using temporarily a point for multiplication in $\mathbb{R}$,

$$< \underline{u}_d, \underline{u} >< \underline{v}_d, \underline{v} >= \underline{u}_d^T\underline{u} \cdot \underline{v}_d^T\underline{v} = \underline{v}_d^T\underline{v} \cdot \underline{u}_d^T\underline{u} = \underline{v}_d^T[\underline{v} \otimes \underline{u}_d^T]\underline{u}$$

with *dyadic product* $\underline{v} \otimes \underline{u}_d^T \in \mathbb{R}^n{}_n$ and, e.g., $\underline{u}$ and $\underline{v}$ are the independent variables in the third definition.

The general tensor product of vectors is a mapping

$$\bigotimes : \mathcal{V} \times \underbrace{\cdots}_{p-times} \times \mathcal{V} \times \mathcal{V}_d \times \underbrace{\cdots}_{q-times} \times \mathcal{V}_d \to \mathcal{T}^p{}_q(\mathcal{V})\,,$$

being pointwise defined by

$$\bigotimes : (\underline{v}_1, \ldots, \underline{v}_p, \underline{v}^1, \ldots, \underline{v}^q) \mapsto \underline{v}_1 \otimes \cdots \otimes \underline{v}_p \otimes \underline{v}^1 \otimes \cdots \otimes \underline{v}^q\,.$$

Of course, the spaces $\mathcal{V}$ and $\mathcal{V}_d$ may appear in every other succession. This mapping is likewise a multilinear mapping but the representation is not unique as in all multiplications; e.g., $\underline{v} \otimes \underline{v}_d = \left(2^{-1}\,\underline{v}\right) \otimes (2\,\underline{v}_d)$. A tensor is called *simple* if it can be written as tensor product of vectors. Let $\underline{u}_i, \underline{v}_i \in \mathcal{V}$, $\underline{u}^k, \underline{v}^k \in \mathcal{V}_d$ be arbitrary then

$$\mathsf{T}^p{}_q = \underline{v}_1 \otimes \cdots \otimes \underline{v}_p \otimes \underline{v}^1 \otimes \cdots \otimes \underline{v}^q$$

is defined pointwise by

$$\begin{aligned}&(\underline{v}_1 \otimes \cdots \otimes \underline{v}_p \otimes \underline{v}^1 \otimes \cdots \otimes \underline{v}^q)\,(\underline{u}^1, \ldots, \underline{u}^p, \underline{u}_1, \ldots, \underline{u}_q)\\ &:= \langle \underline{u}^1, \underline{v}_1\rangle \cdots \langle \underline{u}^p, \underline{v}_p\rangle\langle \underline{v}^1, \underline{u}_1\rangle \cdots \langle \underline{v}^q, \underline{u}_q\rangle \in \mathbb{R}\,.\end{aligned}$$

(f2) Tensor Product of Tensors For $\mathsf{S} \in \mathcal{T}^p{}_q(\mathcal{V})$, $\mathsf{T} \in \mathcal{T}^r{}_s(\mathcal{V})$, we have

$$\boxed{\begin{aligned}&(\mathsf{S} \otimes \mathsf{T})(\underline{v}^1, \ldots, \underline{v}^{p+r}, \underline{v}_1, \ldots, \underline{v}_{q+s})\\ &:= \mathsf{S}(\underline{v}^1, \ldots, \underline{v}^p, \underline{v}_1, \ldots, \underline{v}_q) \cdot \mathsf{T}(\underline{v}^{p+1}, \ldots, \underline{v}^{p+r}, \underline{v}_{q+1}, \ldots, \underline{v}_{q+s})\end{aligned}}$$

(again the point shall denote temporarily the multiplication of two real *numbers*). Accordingly, the tensor product of a (p, q)-tensor and a (r, s)-tensor is a $(p + r, q + s)$-tensor. The tensor product is not commutative, e.g., in $\mathbb{R}^n$,

$$\underline{x}^T A\underline{y} \cdot \underline{u}^T B\underline{v} = \underline{u}^T B\underline{v} \cdot \underline{x}^T A\underline{y} \neq \underline{x}^T B\underline{y} \cdot \underline{u}^T A\underline{v}\,.$$

(By the way, we have also $(g\otimes f)(x, y) := g(x)f(y) \neq f(x)g(y) = (f \otimes g)(x, y)$ for $f, g : \mathbb{R} \mapsto \mathbb{R}$ and $x \neq y$.)

(g) Vector Space of Tensors Let $\mathcal{V}$ be a vector space of dimension n then $\mathcal{T}^p{}_q(\mathcal{V})$ is a vector space of dimension n^{p+q}. If $\{\underline{e}_1, \ldots, \underline{e}_n\}$ denotes a basis of $\mathcal{V}$ and $\{\underline{e}^1, \ldots, \underline{e}^n\}$ the dual basis of $\mathcal{V}_d$ then the tensors

$$\mathsf{T}^p{}_q := \underline{e}_{i_1} \otimes \cdots \otimes \underline{e}_{i_p} \otimes \underline{e}^{j_1} \otimes \cdots \otimes \underline{e}^{j_q},$$

constitute for $i_1, \ldots, i_p, j_1, \ldots, j_q = 1 : n$ a basis of $\mathcal{T}^p{}_q(\mathcal{V})$ by *simple* tensors.

(h) Representation of General Tensors By using a basis, we can now write a tensor as a *mapping* in formal correct way. Let $\underline{v}_j = \underline{e}_i v^i{}_j \in \mathcal{V}$, $\underline{w}^i = w^i{}_j\underline{e}^j \in \mathcal{V}_d$ and let $\mathsf{T} \in \mathcal{T}^p_q(\mathcal{V})$ then

$$\boxed{\begin{aligned}&\mathsf{T}(\underline{w}^1, \ldots, \underline{w}^p, \underline{v}_1, \ldots, \underline{v}_q)\\ &:= t^{i_1\ldots i_p}{}_{j_1\ldots j_q}\underline{e}_{i_1} \otimes \cdots \otimes \underline{e}_{i_p} \otimes \underline{e}^{j_1} \otimes \cdots \otimes \underline{e}^{j_q}(\underline{w}^1, \ldots, \underline{w}^p, \underline{v}_1, \ldots, \underline{v}_q)\\ &= t^{i_1\ldots i_p}{}_{j_1\ldots j_q} w^1{}_{i_1} \cdots w^p{}_{i_p}\, v^{j_1}{}_1 \cdots v^{j_q}{}_q\end{aligned}}\,.$$

The components of a tensor are obtained by applying the tensor to the basis,

$$\begin{aligned}&\mathsf{T}(\underline{e}^{k_1}, \ldots, \underline{e}^{k_p}, \underline{e}_{l_1}, \ldots, \underline{e}_{l_q})\\ &:= t^{i_1\ldots i_p}{}_{j_1\ldots j_q}\underline{e}_{i_1} \otimes \cdots \otimes \underline{e}_{i_p} \otimes \underline{e}^{j_1} \otimes \cdots \otimes \underline{e}^{j_q}(\underline{e}^{k_1}, \ldots, \underline{e}^{k_p}, \underline{e}_{l_1}, \ldots, \underline{e}_{l_q})\\ &= t^{i_1\ldots i_p}{}_{j_1\ldots j_q}\langle \underline{e}^{k_1}, \underline{e}_{i_1}\rangle \cdots \langle \underline{e}^{k_p}, \underline{e}_{i_p}\rangle\langle \underline{e}^{j_1}, \underline{e}_{l_1}\rangle \cdots \langle \underline{e}^{j_q}, \underline{e}_{l_q}\rangle\\ &= t^{k_1\ldots k_p}{}_{l_1\ldots l_q}\,.\end{aligned}$$

Likewise, the components of a tensor product are computed by using a product basis,

$$(\mathsf{S}\otimes\mathsf{T})^{i_1\dots i_{p+r}}{}_{j_1\dots j_{q+s}} = S^{i_1\dots i_p}{}_{j_1\dots j_q}\, T^{i_{p+1}\dots i_{p+r}}{}_{j_{q+1}\dots j_{q+s}}\,.$$

(i) Transformation of General Tensors If the transformation of a basis is given by $\underline{f}_j = \underline{e}_i a^i{}_j$, $\underline{f}^j = b^j{}_k \underline{e}^k$, or *formal* $\mathcal{F} = \mathcal{E}A$, $\mathcal{F}_d = B\mathcal{E}_d$, then we obtain for the elements of a basis of $\mathcal{T}^p{}_q(\mathcal{V})$

$$\begin{aligned}&\underline{f}_{i_1}\otimes\cdots\otimes\underline{f}_{i_p}\otimes\underline{f}^{j_1}\otimes\cdots\otimes\underline{f}^{j_q}\\ &= a^{k_1}{}_{i_1}\cdots a^{k_p}{}_{i_p}\, b^{j_1}{}_{l_1}\cdots b^{j_q}{}_{l_q}\underline{e}_{k_1}\otimes\cdots\otimes\underline{e}_{k_p}\otimes\underline{e}^{l_1}\otimes\cdots\otimes\underline{e}^{l_q}\,.\end{aligned}$$

For the components of a tensor $\mathsf{T}\in\mathcal{T}^p{}_q(\mathcal{V})$ we obtain by direct computation using the linearity w.r.t. each argument

$$(t_f)^{i_1\dots i_p}{}_{j_1\dots j_q} = b^{i_1}{}_{k_1}\dots b^{i_p}{}_{k_p}\, a^{l_1}{}_{j_1}\cdots a^{l_q}{}_{j_q}\,(t_e)^{k_1\cdots k_p}{}_{l_1\dots l_q}\,. \tag{10.11}$$

The upper index indicates contravariant transformation and the lower index covariant transformation.

The converse is also correct: If two sets of respective n^{p+q} real numbers are given,

$$\left\{(t_f)^{i_1\dots i_p}{}_{j_1\dots j_q}\right\},\ \left\{(t_e)^{k_1\dots k_p}{}_{l_1\dots l_q}\right\}, \tag{10.12}$$

satisfying (10.11), then there exists a tensor $\mathsf{T}\in\mathcal{T}^p{}_q(\mathcal{V})$ having the components (10.12) w.r.t. the bases $\{\underline{e}_j\}$, $\{\underline{f}_j\}$ and the corresponding dual bases. For instance, let $\underline{u} = \underline{e}_i u^i \in \mathcal{V}$ etc. then a threefold covariant tensor is specified by

$$\begin{aligned}\mathsf{T} &:= t_{ijk}\underline{e}^i\otimes\underline{e}^j\otimes\underline{e}^k\ :\ \mathcal{V}\times\mathcal{V}\times\mathcal{V}\to\mathbb{R}\\ \mathsf{T}(\underline{u},\underline{v},\underline{w}) &= t_{ijk}\langle\underline{e}^i,\underline{u}\rangle\langle\underline{e}^j,\underline{v}\rangle\langle\underline{e}^k,\underline{w}\rangle = t_{ijk}u^iv^jw^k\,,\end{aligned}$$

and a twofold covariant and onefold contravariant tensor reads, e.g.:

$$\begin{aligned}\mathsf{T} &:= t_i{}^j{}_k\underline{e}^i\otimes\underline{e}_j\otimes\underline{e}^k\ :\ \mathcal{V}\times\mathcal{V}_d\times\mathcal{V}\to\mathbb{R}\\ \mathsf{T}(u,v^d,w) &= t_i{}^j{}_k\langle\underline{e}^i,\underline{u}\rangle\langle\underline{v}^d,\underline{e}_j\rangle\langle\underline{e}^k,\underline{w}\rangle = t_i{}^j{}_k u^iv_jw^k\,,\ \ \underline{v}^d = v_l\underline{e}^l\,.\end{aligned}$$

(j) Contraction (or Rejuvenation) If a contravariant index (above) and a covariant index (below) are denoted by the same letter and it is summed over this index (following EINSTEIN's convention), then we speak of a *contraction* or (translated from german) *rejuvenation* of a (p,q)-tensor. Of course, this operation may be applied repeatedly. If $p=q$ then the result after p contractions is a scalar. If $\mathsf{S}\in\mathcal{T}^p{}_q(\mathcal{V})$ and and $\mathsf{T}\in\mathcal{T}^q{}_p(\mathcal{V})$ then $\mathsf{S}\otimes\mathsf{T}$ is a $(p+q,p+q)$-tensor and a $p+q$-fold contraction yields a number called *scalar product of the tensors* S *and* T. This product is commutative after identification of $\mathcal{V}_d$ with $\mathcal{V}$ (where of course $\mathcal{V}$ must be a scalar product space). For instance, consider the simple tensor

$$\mathsf{T} := \underline{v}_1 \otimes \cdots \otimes \underline{v}_p \otimes \underline{v}^1 \otimes \cdots \otimes \underline{v}^q \in \mathcal{T}^p{}_q(\mathcal{V})\,,$$

then a single contraction is a linear mapping $C^i{}_j : \mathcal{T}^p{}_q(\mathcal{V}) \to \mathcal{T}^{p-1}{}_{q-1}(\mathcal{V})$ with the properties

$$C^i{}_j\mathsf{T} = \langle \underline{v}^j, \underline{v}_i\rangle \underline{v}_1 \otimes \cdots \otimes \widehat{\underline{v}}_i \otimes \underline{v}_{i+1} \otimes \cdots \otimes \underline{v}_p \otimes \\ \underline{v}^1 \otimes \cdots \otimes \underline{v}^{j-1} \otimes \widehat{\underline{v}}^j \otimes \underline{v}^{j+1} \otimes \cdots \otimes \underline{v}^q\,.$$

The symbol $\widehat{\ }$ means here "dropping" (the both vectors must take their hat). Especially we have $< \underline{w}, \underline{u} > = w_i u^i$ for $\underline{u} = \underline{e}_i u^i \in \mathcal{V}$ and $\underline{w} = w_i \underline{e}^i \in \mathcal{V}_d$. For the components of an arbitrary tensor $\mathsf{T} \in \mathcal{T}^p{}_q(\mathcal{V})$ we obtain likewise

$$(C^i{}_j\mathsf{T})^{k_1 \ldots k_p}{}_{l_1 \ldots l_q} = T^{k_1 \ldots k_{i-1}\sigma k_{i+1} \ldots k_p}{}_{l_1 \ldots l_{j-1}\sigma l_{j+1} \ldots l_q}$$

with summation over the index σ.

Examples (1°) If $\mathsf{T} \in \mathcal{T}^1{}_1(\mathcal{V}) = \mathcal{L}(\mathcal{V}_d, \mathcal{V}; \mathbb{R})$, then $C^1{}_1\mathsf{T} = T^i{}_i$ is the *trace operator.*
(2°) If $\mathcal{V} = \mathbb{R}^n$ then the divergence of $\underline{v}$ is the contraction of grad $\underline{v}$.
(3°) The ordinary matrix-vector multiplication may be understood as contraction of a tensor of order two with a tensor of order one to a tensor of order one.

(k) Scalar Product of Tensors Let $\mathsf{S} \in \mathcal{T}^p{}_q(\mathcal{V})$ be a (p, q)-tensor and $\mathsf{T} \in \mathcal{T}^q{}_p(\mathcal{V})$ a (q, p)-tensor then $\mathsf{S} \otimes \mathsf{T} \in \mathcal{T}^{p+q}{}_{p+q}(\mathcal{V})$ is a $(p + q, p + q)$-tensor of which the $(p + q)$-fold contraction yields a *number* as already mentioned. Defining the operator $\mathbf{C}$ by

$$\mathbf{C} := C^1{}_1 \circ \underbrace{\cdots}_{q-times} \circ C^1{}_1 \circ C^1{}_{q+1} \circ \underbrace{\cdots}_{p-times} \circ C^1{}_{q+1},$$

$$< \mathsf{S}, \mathsf{T} > := \mathbf{C}(\mathsf{S} \otimes \mathsf{T}) \in \mathbb{R}$$

is called *scalar product* of the tensors S and T. In the case where the bases are fixed and omitted, we obtain for the scalar product

$$< \mathsf{S}, \mathsf{T} > = s^{i_1 \ldots i_p}{}_{j_1 \ldots j_q}\, t^{j_1 \ldots j_q}{}_{i_1 \ldots i_p}\,.$$

(l) Raising and Lowering of Indices Let first $\mathsf{T} \in \mathcal{T}^p(\mathcal{V})$ be a simple tensor, i.e.,

$$\mathsf{T}(\underline{u}^1, \ldots, \underline{u}^p) = (\underline{v}_1 \otimes \cdots \otimes \underline{v}_p)(\underline{u}^1, \ldots, \underline{u}^p) = \langle \underline{u}^1, \underline{v}_1\rangle \cdots \langle \underline{u}^p, \underline{v}_p\rangle$$

with arguments $\underline{v}_i \in \mathcal{V}$. Then $\mathcal{R}(\underline{v}_i)$ is an element of the dual space $\mathcal{V}_d$ hence an application of the RIESZ mapping $\mathcal{R}$ to all arguments leads to an operation $\mathbf{R}^p$ of the form

$$\mathbf{R}^p : \mathsf{T} \mapsto \mathbf{R}^p\mathsf{T} := \underbrace{\mathcal{R}\underline{v}_1}_{\in \mathcal{V}_d} \otimes \cdots \otimes \underbrace{\mathcal{R}\underline{v}_p}_{\in \mathcal{V}_d} \in \mathcal{T}_p(\mathcal{V})\,,$$

and the application of the tensor $\mathbf{R}^p\mathsf{T}$ to the arguments $\underline{u}_i \in \mathcal{V}$ yields

$$\mathbf{R}^p T(\underline{u}_1, \ldots, \underline{u}_p) = \langle \mathcal{R}\underline{v}_1, \underline{u}_1\rangle \cdots \langle \mathcal{R}\underline{v}_p, \underline{u}_p\rangle = (\underline{v}_1 \cdot \underline{u}_1) \cdots (\underline{v}_p \cdot \underline{u}_p)\,.$$

By a suitable generalization to arbitrary $\mathsf{T} \in \mathcal{T}^p(\mathcal{V})$, the operation

$$\mathbf{R}^p : \mathcal{T}^p(\mathcal{V}) \to \mathcal{T}_p(\mathcal{V})$$

is defined uniquely, linear and bijective hence an isomorphism. But, as the tensor product is not an injective mapping, it has to be shown for the proof of the mentioned properties that the mapping $\mathbf{R}^p$ does not depend on the individual representation of T . In other words, it has to be shown that $\mathbf{R}^p$ does not depend of the individual basis of $\mathcal{V}$; cf. (Bowen), vol. I. Now the operator $\mathbf{R}^p$ "draws down" all indices and thus has the desired property (but $\mathbf{R}^p$ is not the same as $\mathcal{R} \circ \cdots \circ \mathcal{R}$ (p-times).) By this way, $\mathcal{T}^p(\mathcal{V})$ and $\mathcal{T}_p(\mathcal{V})$ are isomorph, but also the converse mapping exists,

$$(\mathbf{R}^p)^{-1} : \mathcal{T}_p(\mathcal{V}) \ni \underline{v}^1 \otimes \cdots \otimes \underline{v}^p \mapsto \mathcal{R}^{-1}\underline{v}^1 \otimes \cdots \otimes \mathcal{R}^{-1}\underline{v}^p \in \mathcal{T}^p(\mathcal{V})$$

as well as their generalization to arbitrary tensors $\mathsf{T} \in \mathcal{T}_p(\mathcal{V})$. The generalization of the "lowering" operation to mixed tensors is now obvious. Let us consider first a *simple* mixed tensor

$$\mathsf{T} = \underline{v}_1 \otimes \cdots \otimes \underline{v}_p \otimes \underline{u}^1 \otimes \cdots \otimes \underline{u}^q \in \mathcal{T}^p{}_q(\mathcal{V})\,,$$

then the operator $\mathbf{R}^p{}_q$ is defined by

$$\mathbf{R}^p{}_q T := \mathcal{R}\underline{v}_1 \otimes \cdots \otimes \mathcal{R}\underline{v}_p \otimes \underline{u}^1 \otimes \cdots \otimes \underline{u}^q \in \mathcal{T}_{p+q}(\mathcal{V})\,.$$

The generalization to arbitrary tensors $\mathsf{T} \in \mathcal{T}^p{}_q(\mathcal{V})$, $\mathbf{R}^p{}_q : \mathcal{T}^p{}_q(\mathcal{V}) \to \mathcal{T}_{p+q}(\mathcal{V})$, is then likewise an isomorphism with the inversion

$$(\mathbf{R}^p{}_q)^{-1}(\underline{v}^1 \otimes \cdots \otimes \underline{v}^p \otimes \underline{u}^1 \otimes \cdots \otimes \underline{u}^q) = \mathcal{R}^{-1}\underline{v}^1 \otimes \cdots \otimes \mathcal{R}^{-1}\underline{v}^p \otimes \underline{u}^1 \otimes \cdots \otimes \underline{u}^q\,.$$

The composition of two *compatible* isomorphisms is again an isomorphism hence also the mapping $(\mathbf{R}^{p_1}{}_{q_1})^{-1} \circ \mathbf{R}^p{}_q$ is an isomorphism in the case where $p_1 + q_1 = p + q$,

$$\mathcal{T}^p{}_q(\mathcal{V}) \overset{\mathbf{R}^p{}_q}{\longrightarrow} \mathcal{T}_{p+q}(\mathcal{V}) = \mathcal{T}_{p_1+q_1}(\mathcal{V}) \overset{(\mathbf{R}^{p_1}{}_{q_1})^{-1}}{\longrightarrow} \mathcal{T}^{p_1}{}_{q_1}(\mathcal{V})\,.$$

Fazit:

All tensor space of the same total order are isomorph in the case where the basis vector space $\mathcal{V}$ is a scalar product space.

Of course, all indices can also be drawn up instead down. This operation reads for simple tensors as follows:

$$(\mathbf{R}^p{}_q)^{-1}(\underline{v}^1 \otimes \cdots \otimes \underline{v}^p \otimes \underline{u}^1 \otimes \cdots \otimes \underline{u}^q) = \underbrace{\mathcal{R}^{-1}\underline{v}^1}_{\in\mathcal{V}} \otimes \cdots \otimes \underbrace{\mathcal{R}^{-1}\underline{v}^p}_{\in\mathcal{V}} \otimes \underbrace{\underline{u}^1}_{\in\mathcal{V}} \otimes \cdots \otimes \underbrace{\underline{u}^q}_{\in\mathcal{V}}\,.$$

Then, by using the inverse mapping $\mathcal{R}^{-1}$, a general tensor can be written in purely contravariant form,

$$\mathcal{T}^p{}_q(V) \overset{\mathbf{R}^p{}_q}{\longrightarrow} \mathcal{T}_{p+q}(\mathcal{V}) \overset{(\mathbf{R}^{p+q})^{-1}}{\longrightarrow} \mathcal{T}^{p+q}(\mathcal{V})\,.$$

The *representation* of the tensor $\mathbf{R}^p{}_q\,\mathsf{T}$ w.r.t. the reference basis $\{\underline{e}_1, \ldots, \underline{e}_n\}$ and the dual basis $\{\underline{e}^1, \ldots, \underline{e}^n\}$ may be verified easily by using (10.14). Recall that the components of the metric tensors satisfy

$$\begin{aligned}\underline{e}_i \cdot \underline{e}_k &=: (\mathcal{R}\underline{e}_i) \cdot \underline{e}_k &&= \langle e_{ij}\underline{e}^j, \underline{e}_k\rangle = e_{ij}\langle \underline{e}^j, \underline{e}_k\rangle = \underline{e}_{ik}\\ \underline{e}^i \cdot \underline{e}^k &=: \underline{e}^i \cdot (\mathcal{R}^{-1}\underline{e}^k) &&= \langle \underline{e}^i, \underline{e}_l e^{lk}\rangle = e^{lk}\langle \underline{e}^i, \underline{e}_l\rangle = e^{ik}\,.\end{aligned}$$

Let $\mathsf{T} \in \mathcal{T}^p{}_q(\mathcal{V})$ have the representation

$$\mathsf{T}_e = T^{i_1\ldots i_p}{}_{j_1\ldots j_q}\, \underline{e}_{i_1} \otimes \cdots \otimes \underline{e}_{i_p} \otimes \underline{e}^{j_1} \otimes \cdots \otimes \underline{e}^{j_q},$$

then lowering of the first p indices yields

$$\begin{aligned}\mathbf{R}^p{}_q\mathsf{T}_e &= T^{i_1\ldots i_p}{}_{j_1\ldots j_q}\, \mathcal{R}\underline{e}_{i_1} \otimes \cdots \otimes \mathcal{R}\underline{e}_{i_p} \otimes \underline{e}^{j_1} \otimes \cdots \otimes \underline{e}^{j_q}\\ &= T^{i_1\ldots i_p}{}_{j_1\ldots j_q}\, \underline{e}_{i_1k_1} \otimes \cdots \otimes \underline{e}_{i_pk_p} \otimes \underline{e}^{k_1} \otimes \cdots \otimes \underline{e}^{k_p} \otimes \underline{e}^{j_1} \otimes \cdots \otimes \underline{e}^{j_q}\\ &= T_{k_1,\ldots,k_p j_1,\ldots,j_q}\, e_{i_1k_1} \otimes \cdots \otimes e_{i_pk_p}\underline{e}^{k_1} \otimes \cdots \otimes \underline{e}^{k_p} \otimes \underline{e}^{j_1} \otimes \cdots \otimes \underline{e}^{j_q}\,.\end{aligned} \tag{10.13}$$

The raising of the last q indices by using $(\mathbf{R}^p{}_q)^{-1}$ is carried out in the same way applying $\mathcal{R}^{-1}$ instead of $\mathcal{R}$.

Let now finally $\{\underline{e}_1, \ldots, \underline{e}_n\}$ be an *normed orthogonal system*, then the metric tensors are the *unit tensors* $\boldsymbol{\delta}$,

$$e_{ij} = e^{ij} = \delta^i{}_j \quad \text{(Kronecker symbol)},$$

hence also

$$\mathcal{R}\underline{e}_j \cdot \underline{e}^k = e_{ij}\underline{e}^j \cdot \underline{e}^k = \delta^i{}_k\,,\quad \mathcal{R}^{-1}\underline{e}^j \cdot \underline{e}_k = e^{ij}\underline{e}^j \cdot \underline{e}_g k = \delta^i{}_k\,,$$

i.e., $\mathcal{R}\underline{e}_j := \underline{r}^j \in \mathcal{V}^d$ is reciprocal basis to $\{\underline{e}^j\}$ in $\mathcal{V}_d$, and $\mathcal{R}^{-1}\underline{e}^k := \underline{r}_k \in \mathcal{V}$ is reciprocal basis to $\{\underline{e}_k\}$ in $\mathcal{V}$. Accordingly, we *choose*

$$\begin{aligned}&\text{in } \mathcal{T}^p{}_q(\mathcal{V}) \text{ the reference basis } e^{i_1} \otimes \cdots \otimes \underline{e}^{i_p} \otimes \underline{e}_{j_1} \otimes \cdots \otimes \underline{e}_{j_q}\\ &\text{in } \mathcal{T}_{p+q}(\mathcal{V}) \text{ the reference basis } \underline{r}_{i_1} \otimes \cdots \otimes \underline{r}_{i_p} \otimes \underline{e}_{j_1} \otimes \cdots \otimes \underline{e}_{j_q}\,.\end{aligned}$$

Then the tensors $\mathsf{T} \in \mathcal{T}^p{}_q(\mathcal{V})$ and $\mathbf{R}^p{}_q\mathsf{T} \in \mathcal{T}_{p+q}(\mathcal{V})$ have the *same components* by the transformation rule (10.11) for tensors and the representation (10.13)! In the other cases, the behavior is readily described by (10.11).

(m) Examples

Example 10.6. in $\mathbb{R}^2$. Let $\boldsymbol{\varepsilon} = [\varepsilon^i{}_j]$ be the strain tensor and $\boldsymbol{\sigma} = [\sigma^i{}_j]$ the stress tensor of Sect. 8.5**(d)** (both symmetric) then the componentwise representation instead of (8.38)

$$\sigma^i{}_j = u^{im}{}_{jl}\varepsilon^l{}_m \text{ double contraction, } u^{im}{}_{jl} = \lambda\delta^i{}_j\delta^l{}_m + \mu(\delta^i{}_l\delta^m{}_j + \delta^m{}_i\delta^j{}_l)\,.$$

On the other side,

$$\delta^l{}_m\varepsilon^l{}_m = \mathrm{trace}(\boldsymbol{\varepsilon})\,,\quad \delta^i{}_l\delta^m{}_j\varepsilon^l{}_m = \delta^i{}_l\varepsilon^l{}_j = \varepsilon^i{}_j\,,\quad \delta^m{}_i\delta^j{}_l\varepsilon^l{}_m\delta^m{}_i\varepsilon^j{}_m = \varepsilon^j{}_i,$$

hence, using the unit tensor $\boldsymbol{\delta}$, we obtain (8.38) again

$$\boxed{\boldsymbol{\sigma} = \mu(\boldsymbol{\varepsilon} + \boldsymbol{\varepsilon}^T) + \lambda\,\mathrm{trace}(\boldsymbol{\varepsilon})\boldsymbol{\delta}}\,.$$

Examples with cartesian coordinates in $\mathbb{R}^n$. Let $\mathcal{V} = \mathbb{R}^n$ be again the coordinate space of column vectors equipped with canonical scalar product, and let $\mathcal{E} = \{\underline{e}_1, \dots, \underline{e}_n\}$ be a cartesian reference basis. Then $\mathcal{E}$ constitutes a real and orthogonal (n, n)-matrix with columns $\underline{e}_i$. The dual basis $\mathcal{E}_d$ consists of the rows $\underline{e}^k \in \mathbb{R}_n$ of $\mathcal{E}$, and the reciprocal basis is produced by transposing these rows into columns. All these basis vectors have length one. Hence the RIESZ mapping $\mathcal{R}$ transforms in this case each row vector into a column vector and thus identifies the row space with the column space. As a consequence, we may restrict ourselves to the vector space $\mathcal{V} = \mathbb{R}^n$ with its reference basis and the distinction between contravariant and covariant tensors may be dropped.

Example 10.7.

$$\begin{aligned}
&\text{0-tensor:} \quad \text{scalar,} \\
&\text{1-tensor: } \underline{v} = v_i \underline{e}_i\,, && v_i = \langle \underline{v}, \underline{e}_i \rangle\,, \\
&\text{2-tensor: } \mathsf{T} = t_{ij} \underline{e}_i \otimes \underline{e}_j\,, && t_{ij} = \mathsf{T}(\underline{e}_i, \underline{e}_j)\,, \\
&\text{3-tensor: } \mathsf{T} = t_{ijk} \underline{e}_i \otimes \underline{e}_j \otimes \underline{e}_k\,, && t_{ijk} = \mathsf{T}(\underline{e}_i, \underline{e}_j, \underline{e}_k)\,.
\end{aligned}$$

Evaluation in the arguments $\underline{u} = x_i \underline{e}_i$, $\underline{v} = y_i \underline{e}_i$, $\underline{w} = z_i \underline{e}_i$ yields

$$\begin{aligned}
(t_{ij} \underline{e}_i \otimes \underline{e}_j)(x_l \underline{e}_l) &= t_{ij} x_l \underline{e}_i \langle \underline{e}_j, \underline{e}_l \rangle = t_{ij} x_j \underline{e}_i \\
(t_{ij} \underline{e}_i \otimes \underline{e}_j)(x_l \underline{e}_l, y_m \underline{e}_m) &= t_{ij} x_l y_m \langle \underline{e}_i, \underline{e}_l \rangle \langle \underline{e}_j, \underline{e}_m \rangle = t_{ij} x_i y_j \\
(t_{ijk} \underline{e}_i \otimes \underline{e}_j \otimes \underline{e}_k)(x_l \underline{e}_l, y_m \underline{e}_m, z_n \underline{e}_n) &= t_{ijk} x_l y_m z_n \langle \underline{e}_i, \underline{e}_l \rangle \langle \underline{e}_j, \underline{e}_m \rangle \langle \underline{e}_k, \underline{e}_n \rangle \\
&= t_{ijk} x_i y_j z_k
\end{aligned} \tag{10.14}$$

since $\langle \underline{e}_i, \underline{e}_j \rangle = \underline{e}_i \cdot \underline{e}_j = \delta_{ij}$. In somewhat modified form, e.g., for a 3-tensor

$$\begin{aligned}
\mathsf{T}\underline{u}\,\underline{v}\,\underline{w} := \mathsf{T}(\underline{u}, \underline{v}, \underline{w}) &= T_{ijk} \underline{e}_i \otimes \underline{e}_j \otimes \underline{e}_k (\underline{u}, \underline{v}, \underline{w}) \\
&= T_{ijk} \langle \underline{e}_i, \underline{u} \rangle \langle \underline{e}_j, \underline{v} \rangle \langle \underline{e}_k, w \rangle = t_{ijk} x_i y_j z_k\,.
\end{aligned}$$

Example 10.8. Of course, the *scalar product* of, e.g., two 3-tensors S and T yields as above $\mathsf{S} : \mathsf{T} = \mathsf{T} : \mathsf{S} = t_{ijk} s_{ijk} \in \mathbb{R}$, and for the *tensor product* we obtain

$$(t_{ijk} \underline{e}_i \otimes \underline{e}_j \otimes \underline{e}_k) \otimes (s_{ijk} \underline{e}_i \otimes \underline{e}_j \otimes \underline{e}_k) = t_{ijk} s_{lmn} \underline{e}_i \otimes \underline{e}_j \otimes \underline{e}_k \otimes \underline{e}_l \otimes \underline{e}_m \otimes \underline{e}_n\,.$$

It is customary in technical applications to drop the basis entirely as in (10.14) quite right. One considers only the component vectors $\underline{x}$, $\underline{y}$, $\underline{z}$ and writes $T(\underline{x}_1, \dots, \underline{x}_n)$ instead $\mathsf{T}(\underline{v}_1, \dots, \underline{v}_n)$. However, this form is somewhat misleading. For instance, we obtain for the tensor product

$$\begin{aligned}
S = [s_{i_1} \cdots s_{i_p}]\,, \quad T &= [t_{i_1} \cdots t_{i_p}] \\
S \otimes T = [w_{i_1 \cdots i_p\, j_1 \cdots j_q}] &= [t_{i_1 \cdots i_p}\, s_{j_1 \cdots j_q}]
\end{aligned}$$

(especially, e.g., $[t_{ik}][s_{pqr}] = [t_{ik}\, s_{pqr}]$), which does weakly display that the tensor product is not commutative; cf. also the introduction to this chapter.

Finally, *arbitrary* vectors $\underline{e}, \underline{f}, \underline{g}, \underline{u}, \underline{v}, \underline{w}$ generate, e.g., *simple* tensors

$$(\underline{e} \otimes \underline{f})(\underline{v}) := \underline{e}\langle \underline{f}, \underline{v}\rangle\,, \qquad (\underline{e} \otimes \underline{f})(\underline{u}, \underline{v}) = \langle \underline{e}, \underline{u}\rangle\langle \underline{f}, \underline{v}\rangle$$
$$(\underline{e} \otimes \underline{f} \otimes \underline{g})(\underline{w}) = \underline{e} \otimes \underline{f}\langle \underline{g}, \underline{w}\rangle\,, \qquad (\underline{e} \otimes \underline{f} \otimes \underline{g})(\underline{v}, \underline{w}) = \underline{e}\langle \underline{f}, \underline{v}\rangle\langle \underline{g}, \underline{w}\rangle$$
$$(\underline{e} \otimes \underline{f} \otimes \underline{g})(\underline{u}, \underline{v}, \underline{w}) = \langle \underline{e}, \underline{u}\rangle\langle \underline{f}, \underline{v}\rangle\langle \underline{g}, \underline{w}\rangle$$

where again $\langle \underline{e}, \underline{u}\rangle = \underline{e} \cdot \underline{u}$ etc..

References: (Bowen).

10.2 Algebra of Alternating Tensors

An *algebra* is a vector space $\mathcal{V}$ equipped with a bilinear mapping $\mathcal{V} \times \mathcal{V} \to \mathcal{V}$ as multiplication; the mapping must have an unit element in V but is not necessarily commutative. Note that the result of this multiplication is again an element of the vector space. For instance the vector space of (n, n)-matrices is an algebra with the usual matrix multiplication, but also the vector space of mappings $f, g : [0\,,\,1] \mapsto [0\,,\,1]$ with the composition $f \circ g$ for multiplication constitutes an algebra.

We consider purely covariant tensors $T \in \mathcal{T}_p(\mathcal{V})$ over the n-dimensional *scalar product space* V since this type plays the major role in theory of *differential forms.*

(a) Alternating Tensors A tensor $\mathsf{T} \in \mathcal{T}_p(\mathcal{V})$ is called *skew-symmetric, antisymmetric* or *alternating, exterior p-form* if, for all $\underline{v}_i \in \mathcal{V}$,

$$\mathsf{T}(\underline{v}_1, \ldots, \underline{v}_p) = \varepsilon_{i_1 \ldots i_p}\, \mathsf{T}(\underline{v}_{i_1}, \ldots, \underline{v}_{i_p}) \quad \text{where}$$

$$\begin{aligned} \varepsilon_{i_1 \ldots i_p} &= 1 \text{ if } (i_1, \ldots, i_p) \text{ is an even permutation of } (1, \ldots, p),\\ \varepsilon_{i_1 \ldots i_p} &= -1 \text{ if } (i_1, \ldots, i_p) \text{ is an odd permutation of } (1, \ldots, p),\\ \varepsilon &= 0 \text{ else, i.e. if two indices are equal;} \end{aligned}$$

the tensor with components $\varepsilon_{i_1, \ldots, i_p}$ is called *ε-tensor.*

Remember: A permutation is *even* if it consists of an even number of pairwise inversions, and it is *odd* if that number is odd. (The *property* "even" or "odd" is unique but there may be different ways to attain the resulting permutation.) The permutation of *two* elements always needs an odd number of inversions:

An alternating tensor changes its sign if two arguments are permutated.

Example 10.9. $\{1\ 2\ 3\ 4\ 5\}$ has no inversion, $\{5\ 1\ 3\ 4\ 2\}$ has six inversions.

$\{1\ 2\ 3\}$ 0 inversions, $\{3\ 2\ 1\}$ 3 inversions
$\{2\ 3\ 1\}$ 2 inversions, $\{2\ 1\ 3\}$ 1 inversions
$\{3\ 1\ 2\}$ 2 inversions, $\{1\ 3\ 2\}$ 1 inversions.

The set of alternating tensors $\mathcal{A}_p(\mathcal{V}) := \{\mathsf{T} \in \mathcal{T}_p(\mathcal{V}),\ \mathsf{T}$ alternating$\}$ is a subspace of $\mathcal{T}_p(\mathcal{V})$ (of course with the scalar multiplication and the addition defined in $\mathcal{T}_p(\mathcal{V})$) with dimension $\binom{n}{p}$. Let $P_p : \mathcal{T}_p(\mathcal{V}) \to \mathcal{A}_p(V)$ be the projector of $\mathcal{T}_p(\mathcal{V})$ onto the linear subspace $\mathcal{A}_p(\mathcal{V})$ then $\mathrm{Ker}(P_p) = \{\mathsf{T} \in \mathcal{T}_p(\mathcal{V}),\ P_p\mathsf{T} = 0\}$ is the kernel (null space) of P_p, and
(1°) $\mathrm{Ker}(P_p) = \mathrm{span}\{\underline{v}^1 \otimes \cdots \otimes \underline{v}^p,\ \underline{v}^i \in \mathcal{V}^d$, at least two elements equal$\}$,
(2°) $\mathcal{T}_p(\mathcal{V}) = \mathcal{A}_p(\mathcal{V}) \oplus \mathrm{Ker}(P_p)$ (direct sum).

Accordingly, a tensor $\mathsf{T} \in \mathcal{A}_p(\mathcal{V})$ assumes the value zero, $\mathsf{T}(\underline{v}_1, \ldots, \underline{v}_p) = 0$, if two arguments $\underline{v}_i$, $\underline{v}_k$ coincide, or if the arguments $\underline{v}_1, \ldots, \underline{v}_p$ are linearly dependent, or if $p > \dim(\mathcal{V}) = n$ holds for the order of the tensor.

(b) Alternating Part of Tensors The *alternating* part

$$\mathrm{Alt}(\mathsf{T}) := P_p\,\mathsf{T} \in \mathcal{A}_p(\mathcal{V})$$

of a tensor $\mathsf{T} \in \mathcal{T}_p(\mathcal{V})$ is determined uniquely; for $p = 0$ we write $\mathrm{Alt}(\mathsf{T}) = \mathsf{T}$ and for $p = 1$ we have $\mathrm{Alt}(\mathsf{T}) = \mathsf{T}$ because the definition is empty.

Definition 10.1. *For* $p \geq 2$, *the* generalized KRONECKER symbols *are defined by* (1°)

$$\delta(i,j) = \begin{cases} 1 & i < j \\ 0 & i = j \\ -1 & i > j \end{cases},\quad \delta(i_1, \ldots, i_p) = \prod_{\mu,\nu=1,\,\mu<\nu}^{p} \delta(i_\mu, j_\nu)\,.$$

(2°)

$$\delta^{i_1\ldots i_p}{}_{j_1\ldots j_p} = \det\begin{bmatrix} \delta^{i_1}{}_{j_1} & \ldots & \delta^{i_1}{}_{j_p} \\ & \ldots & \\ \delta^{i_p}{}_{j_p} & \ldots & \delta^{i_p}{}_{j_p} \end{bmatrix},\quad (\delta^i{}_j \textit{ usual Kronecker symbol}).$$

Rules:

$$\begin{aligned} \delta(i_1, \ldots, i_\kappa, i_{\kappa+1}, \ldots, i_p) &= -\delta(i_1, \ldots, i_{\kappa+1}, i_\kappa, \ldots, i_p) \\ \delta(i_1, \ldots, i_p) &= \varepsilon_{i_1\ldots i_p} = \varepsilon^{i_1\ldots i_p} \\ \varepsilon_{i_1\ldots i_p} = \delta^{1\ldots p}{}_{j_1\ldots j_p} &= \varepsilon^{i_1\ldots i_p} = \delta^{i_1\ldots i_p}{}_{1\ldots p} \\ \varepsilon^{i_1\ldots i_p}{}_{j_1\ldots j_p} &= \delta^{i_1\ldots i_p}{}_{j_1\ldots j_p}\,; \end{aligned}$$

For instance, $\delta^{12}{}_{12} = 1$, $\delta^{12}{}_{21} = -1$, $\delta^{13}{}_{12} = 0$, $\delta^{13}{}_{21} = 0$, $\delta^{11}{}_{12} = 0$, and

$$\varepsilon^{123} = \varepsilon^{312} = \varepsilon^{231} = 1,\quad \varepsilon^{132} = \varepsilon^{321} = \varepsilon^{213} = -1,\quad \varepsilon^{112} = \varepsilon^{222} = \varepsilon^{233} = 0\,.$$

By using this notation, the alternating part $\mathrm{Alt}(\mathsf{T})$ of $\mathsf{T} \in \mathcal{T}_p(\mathcal{V})$ can be written explicitely (insignicant in later computation)

$$Alt(\mathsf{T}) = \frac{1}{p!}\,\delta(i_1, \ldots, i_p)\,\mathsf{T}(v_{i_1}, \ldots, v_{i_p})$$

(summation over all double occuring indices). If for instance $\mathsf{T} \in \mathcal{T}_2(\mathcal{V})$ then $\mathrm{Alt}(\mathsf{T})(\underline{v}, \underline{u}) = [\mathsf{T}(\underline{u}, \underline{v}) - \mathsf{T}(\underline{v}, \underline{u})]/2$.

(c) Exterior Product of Tensors

(c1) Example (Determinants). For $\mathcal{V} = \mathbb{R}^n$, the mapping "det" (determinant), det : $(\underline{v}_1, \dots, \underline{v}_n) \mapsto \det(\underline{v}_1, \dots, \underline{v}_n) =: \underline{v}_1 \wedge \dots \wedge \underline{v}_n \in \mathbb{R}$ is uniquely defined by

(1°) $\det \in \mathcal{A}_p(\mathcal{V})$, i.e., "det" is multilinear and alternating,

(2°) $\det(\underline{e}_1, \dots, \underline{e}_n) = \underline{e}_1 \wedge \dots \wedge \underline{e}_n = 1$ holds for the canonical basis $\underline{e}_i = [\delta^i{}_k]$, $i = 1 : n$.

For a matrix $A = [\underline{a}_1, \dots, \underline{a}_n]$ with columns $\underline{a}_i$ we have as well-known

$$\det(A) = \det(\underline{a}_1, \dots, \underline{a}_n) =: \underline{a}_1 \wedge \dots \wedge \underline{a}_n$$

and, for the "volume" of the parallelepiped with edges $\underline{a}_1, \dots, \underline{a}_n$,

$$\mathrm{Vol}(\underline{a}_1, \dots, \underline{a}_n) = |\det(\underline{a}_1, \dots, \underline{a}_n)| . \tag{10.15}$$

The computational rule $\det(A \cdot B) = \det(A) \cdot \det(B)$ for (n, n)-matrices A and $B = [\underline{b}_1, \dots, \underline{b}_n]$ is equivalent to $A\underline{b}_1 \wedge \dots \wedge A\underline{b}_n = \det(A)(\underline{b}_1 \wedge \dots \wedge \underline{b}_n)$.

(c2) Definition By and large, the exterior product is defined by the determinant. The following definition generalizes this product to arbitrary alternating tensors; cf. Sect. 10.1**(f)**.

Definition 10.2. *(Exterior product of tensors, wedge product) For* $\mathsf{S} \in \mathcal{T}_p(\mathcal{V})$ *and* $\mathsf{T} \in \mathcal{T}_q(\mathcal{V})$,

$$\mathsf{S} \wedge \mathsf{T} := \frac{(p+q)!}{p!q!} \mathrm{Alt}(\mathsf{S} \otimes \mathsf{T}) \in \mathcal{A}_{p+q}(\mathcal{V})$$

is the exterior product *of the tensors* S *and* T.

Of course, this product is compatible with the product "determinant" (nevertheless to be shown).

Some **Computational Rules** follow directly from the definition. Let $\mathsf{S} \in \mathcal{T}_p(\mathcal{V})$, $\mathsf{T} \in \mathcal{T}_q(\mathcal{V})$, $\mathsf{R} \in \mathcal{T}_r(\mathcal{V})$ then

$$\boxed{\begin{aligned}(\mathsf{S}+\mathsf{T}) \wedge \mathsf{R} &= \mathsf{S} \wedge \mathsf{R} + \mathsf{T} \wedge \mathsf{R}\\ \mathsf{S} \wedge (\mathsf{T}+\mathsf{R}) &= \mathsf{S} \wedge \mathsf{T} + \mathsf{S} \wedge \mathsf{R}\\ \alpha\mathsf{S} \wedge \mathsf{T} &= \mathsf{S} \wedge \alpha\mathsf{T} = \alpha(\mathsf{S} \wedge \mathsf{T})\\ \mathsf{S} \wedge \mathsf{T} &= (-1)^{p \cdot q}\mathsf{T} \wedge \mathsf{S}\end{aligned}},$$

$$(\mathsf{S} \wedge \mathsf{T}) \wedge \mathsf{R} = \mathsf{S} \wedge (\mathsf{T} \wedge \mathsf{R}) = \mathsf{S} \wedge \mathsf{T} \wedge \mathsf{R} = \frac{(p+q+r)!}{p!q!r!} \mathrm{Alt}(\mathsf{S} \otimes \mathsf{T} \otimes \mathsf{R}) .$$

In particular, for $\mathsf{T} \in \mathcal{T}_p(\mathcal{V})$ and odd p, $\mathsf{T} \wedge \mathsf{T} = (-1)^{p \cdot p}\mathsf{T} \wedge \mathsf{T} = -\mathsf{T} \wedge \mathsf{T}$, hence $\mathsf{T} \wedge \mathsf{T} = 0$, and, for all p,

$$(\underline{v}^1 \wedge \dots \wedge \underline{v}^p) \wedge (\underline{v}^{p+1} \wedge \dots \wedge \underline{v}^{p+r}) = \underline{v}^1 \wedge \dots \wedge \underline{v}^{p+r} .$$

(c3) Componentwise Representation

$$\mathsf{S} = S_{i_1 \dots i_p}\, \underline{e}^{i_1} \otimes \cdots \otimes \underline{e}^{i_p}\,, \quad \mathsf{T} = T_{j_1 \dots j_q}\, \underline{e}^{j_1} \otimes \cdots \otimes \underline{e}^{j_q}$$

$$\mathsf{S} \wedge \mathsf{T} = \frac{1}{p!q!} \delta^{i_1 \dots i_{p+q}}{}_{j_1 \dots j_{p+q}}\, S_{i_1 \dots i_p}\, T_{i_{p+1} \dots i_{p+q}}\, \underline{e}^{j_1} \otimes \cdots \otimes \underline{e}^{j_{p+q}}\,.$$

(d) Basis For *simple* alternating tensors $\mathsf{T} = \underline{v}^1 \wedge \cdots \wedge \underline{v}^p \in \mathcal{A}_p(\mathcal{V})$, $\underline{v}^i \in \mathcal{V}_d$, we have $\mathsf{T}(\underline{v}_1, \dots, \underline{v}_p) = \det(C)$, with (p,p)-matrix $C = [\langle \underline{v}^i, \underline{v}_k \rangle]$. If $\{\underline{e}_1, \dots, \underline{e}_n\}$ is a basis of $\mathcal{V}$, then $\{\underline{e}^{i_1} \wedge \cdots \wedge \underline{e}^{i_p}\,,\ 1 \le i_1 < \ldots < i_p \le n\}$ is a basis of $\mathcal{A}_p(\mathcal{V})$.

(e) Representation of Alternating Tensors Let $\mathsf{T} = T_{i_1 \dots i_p}\, \underline{e}^{i_1} \otimes \cdots \otimes \underline{e}^{i_p} \in \mathcal{A}_p(\mathcal{V})$ and let P_p be again the projection onto the alternating part then obviously $P_p \mathsf{T} = \mathsf{T}$ therefore

$$\begin{aligned} \mathsf{T} = P_p \mathsf{T} &= T_{i_1 \dots i_p}\, P_p(\underline{e}^{i_1} \otimes \cdots \otimes \underline{e}^{i_p}) \\ &= \frac{1}{p!} T_{i_1 \dots i_p}\, \underline{e}^{i_1} \wedge \cdots \wedge \underline{e}^{i_p} = \sum\nolimits_{i_1 < \ldots < i_p} T_{i_1 \dots i_p}\, \underline{e}^{i_1} \wedge \cdots \wedge \underline{e}^{i_p}\,. \end{aligned}$$

The last form is called tensor representation in *strict* components, the others being zero. Let $\mathsf{S} \in \mathcal{A}_p(\mathcal{V})$ and $\mathsf{T} \in \mathcal{A}_q(\mathcal{V})$ such that

$$\mathsf{S} = \sum_{i_1 < \ldots < i_p} S_{i_1 \dots i_p}\, \underline{e}^{i_1} \wedge \cdots \wedge \underline{e}^{i_p}\,, \quad \mathsf{T} = \sum_{i_1 < \ldots < i_q} T_{i_1 \dots i_q}\, \underline{e}^{i_1} \wedge \cdots \wedge \underline{e}^{i_q} \implies$$

$$\mathsf{S} \wedge \mathsf{T} = \sum_{\substack{1 \le i_1 < \ldots < i_p \le n, \\ 1 \le j_1 < \ldots < j_p \le n}} S_{i_1 \dots i_p}\, T_{j_1 \dots j_q}\, \underline{e}^{i_1} \wedge \cdots \wedge \underline{e}^{i_p} \wedge \underline{e}^{j_1} \wedge \cdots \wedge \underline{e}^{j_q}\,.$$

Evaluating T with arguments $\underline{v}_i = \underline{e}_j v^j{}_i$ yields

$$\mathsf{T}(\underline{v}_1, \dots, \underline{v}_q) = \sum_{1 \le i_1 < \ldots < i_q \le n} T_{i_1 \dots i_q} \det \begin{bmatrix} \langle \underline{e}^{i_1}, \underline{v}_1 \rangle & \dots & \langle \underline{e}^{i_1}, \underline{v}_q \rangle \\ & \dots & \\ \langle \underline{e}^{i_q}, \underline{v}_1 \rangle & \dots & \langle \underline{e}^{i_q}, \underline{v}_q \rangle \end{bmatrix}.$$

This form leads eventually to

$$\mathsf{T}(\underline{v}_1, \dots, \underline{v}_p) = \sum_{1 \le i_1 < \ldots < i_p \le n} T_{i_1 \dots i_p} \det \begin{bmatrix} v^{i_1}{}_1 & \dots & v^{i_1}{}_p \\ & \dots & \\ v^{i_p}{}_1 & \dots & v^{i_p}{}_p \end{bmatrix}$$

where $\langle \underline{e}^k, \underline{v}_i \rangle = \langle \underline{e}^k, \underline{e}_j v^j{}_i \rangle = v^k{}_i$ hence $\langle \underline{e}^{i_k}, \underline{v}_l \rangle = v^{i_k}{}_l$.

(f) Basis Transformation Let $\underline{f}^i = b^i{}_j \underline{e}^j$ be the rule for the dual basis then we obtain for simple tensors apparently

$$\underline{f}^1 \wedge \cdots \wedge \underline{f}^p = b^1 j_1 \dots b^p{}_{j_p}\, \underline{e}^{j_1} \wedge \cdots \wedge \underline{e}^{j_p}\,.$$

Here, some components are zero again hence the representation by *strict components* can be chosen,

$$\underline{f}^1 \wedge \cdots \wedge \underline{f}^p = \sum_{1 \le i_1 < \ldots < i_p \le n} b^1{}_{j_1} \cdots b^p{}_{j_p}\, \delta^{j_1 \dots j_p}{}_{k_1 \dots k_p}\, \underline{e}^{k_1} \wedge \cdots \wedge \underline{e}^{k_p}\,.$$

In general, only strict components are nonzero hence we have for an arbitrary tensor $\mathsf{T} \in \mathcal{A}_p(V)$

$$\mathsf{T} = T_{(f)\,i_1 \ldots i_p}\, \underline{f}^{i_1} \wedge \cdots \wedge \underline{f}^{i_p}, \quad T_{(f)\,i_1 \ldots i_p} = \mathsf{T}_{(f)}(\underline{f}_{i_1}, \ldots, \underline{f}_{i_p}),$$

the *strict* representation by **(e)**,

$$\begin{aligned}\mathsf{T}_{(f)}(\underline{f}_{i_1}, \ldots, \underline{f}_{i_p}) &= \sum_{j_1 < \ldots < j_p} T_{(e)\,j_1 \ldots j_p} \underline{e}^{j_1} \wedge \cdots \wedge \underline{e}^{j_p}(\underline{f}_{i_1}, \ldots, \underline{f}_{i_p}) \\ &= \sum_{i_1 < \ldots < i_p} T_{(e)\,j_1 \ldots i_p} \det \begin{bmatrix} \langle \underline{e}^{j_1}, \underline{f}_{i_1} \rangle & \ldots & \langle \underline{e}^{j_1}, \underline{f}_{i_p} \rangle \\ & \ldots & \\ \langle \underline{e}^{j_p}, \underline{f}_{i_1} \rangle & \ldots & \langle \underline{e}^{j_p}, \underline{f}_{i_p} \rangle \end{bmatrix}.\end{aligned}$$

Because $\langle \underline{e}^i, \underline{f}_j \rangle = \langle \underline{e}^i\,,\, \underline{e}_l a^l{}_j \rangle \;= a^i{}_j$, then also

$$\mathsf{T}_{(f)}(\underline{f}_{i_1}, \ldots, \underline{f}_{i_p}) = \sum_{j_1 < \ldots < j_p} T_{(e)\,j_1 \ldots j_p} \det \begin{bmatrix} a^{j_1}{}_{i_1} & \ldots & a^{j_1}{}_{i_p} \\ & \ldots & \\ a^{j_p}{}_{i_1} & \ldots & a^{j_p}{}_{i_p} \end{bmatrix}.$$

Suming up, the transformation of the *components* of an alternating tensor reads:

$$\begin{aligned} T_{(f)\,i_1 \ldots i_p} &= \sum_{j_1 < \ldots < j_p} T_{(e)\,j_1 \ldots j_p} \det \begin{bmatrix} a^{j_1}{}_{i_1} \ldots a^{j_1}{}_{i_p} \\ a^{j_p}{}_{i_1} \ldots a^{j_p}{}_{i_p} \end{bmatrix} && \text{(covariant tensor)} \\ T_{(f)}^{i_1 \ldots i_p} &= \sum_{j_1 < \ldots < j_p} T_{(e)}^{j_1 \ldots j_p} \det \begin{bmatrix} b^{j_1}{}_{i_1} \ldots b^{j_1}{}_{i_p} \\ b^{j_p}{}_{i_1} \ldots b^{j_p}{}_{i_p} \end{bmatrix} && \text{(contravariant tensor)} \end{aligned}$$

where $B = [b^i{}_k] = A^{-1}$.

(g) Scalar Product of Alternating Tensors Let $\mathcal{V}$ and thus also $\mathcal{V}_d$ be a scalar product space then a scalar product may be introduced for alternating tensors $\mathsf{S}\,,\, \mathsf{T} \in \mathcal{A}_p(\mathcal{V})$ by

$$\langle \mathsf{S}, \mathsf{T} \rangle = \frac{1}{p!} \; < \mathsf{S}, \mathsf{T} >,$$

where the product on the right side is the usual scalar product of tensors; cf. Sect. 10.1**(k)**.

Computational Rules:

$$\langle \underline{v}^1 \wedge \cdots \wedge \underline{v}^p, \underline{u}^1 \wedge \cdots \wedge \underline{u}^p \rangle = \det([\underline{v}^i \cdot \underline{u}^j]_{i,j=1}^p)$$

where the right side is to be evaluated using the canonical product in dual space.

Letting

$$\begin{aligned} \mathsf{S} &= \textstyle\sum_{i_1 < \ldots < i_p} S_{i_1 \ldots i_p}\, \underline{e}^{i_1} \wedge \cdots \wedge \underline{e}^{i_p} \;\in \mathcal{A}_p(\mathcal{V}), \\ \mathsf{T} &= \textstyle\sum_{i_1 < \ldots < i_p} T_{i_1 \ldots i_p}\, \underline{e}^{i_1} \wedge \cdots \wedge \underline{e}^{i_p} \;\in \mathcal{A}_p(\mathcal{V}) \end{aligned}$$

we have further

$$\boxed{\langle \mathsf{S},\mathsf{T}\rangle = \sum_{i_1<\ldots<i_p, j_1<\ldots<j_p} \det\begin{bmatrix} e^{i_1j_1} & \ldots & e^{i_1j_p} \\ & \ldots & \\ e^{i_pj_1} & \ldots & e^{i_pj_p}\end{bmatrix} S_{i_1\ldots i_p}\,T_{j_1\ldots j_p}}$$

where $e^{ij} = \underline{e}^i \cdot \underline{e}^j$ are the components of the contravariant metric M_e^d. In particular, if the basis $\{\underline{e}_1, \ldots, \underline{e}_n\}$ is an orthonormal system,

$$\boxed{\langle \mathsf{S},\mathsf{T}\rangle = \sum_{i_1<\ldots<i_p} S_{i_1\ldots i_p}\,T_{i_1\ldots i_p}}\,. \tag{10.16}$$

(h) Hodge-Star-Operator Because $\begin{pmatrix} n \\ p\end{pmatrix} = \begin{pmatrix} n \\ n-p\end{pmatrix}$ we have $\dim \mathcal{A}_p(\mathcal{V}) = \dim \mathcal{A}_{n-p}(\mathcal{V})$, hence the spaces $\mathcal{A}_p(\mathcal{V})$ and $\mathcal{A}_{n-p}(\mathcal{V})$ are isomorph. Accordingly, there exists an isomorphism $H_p : \mathcal{A}_p(\mathcal{V}) \to \mathcal{A}_{n-p}(\mathcal{V})$. This isomorphism may be chosen such that it becomes an isometry, i.e.,

$$\langle \mathsf{S},\mathsf{T}\rangle = \langle H_p\mathsf{S}, H_p\mathsf{T}\rangle\,,\quad \mathsf{S},\mathsf{T} \in \mathcal{A}_p(\mathcal{V})\,.$$

Then H_p is uniquely defined and is called HODGE-*star-operator* frequently denoted by "$*$" only. It will be described more thorougly in connection differential forms. For simple tensors and hence also for the basis and general tensors T, we obtain

$$\mathsf{T} = \underline{v}^{i_1} \wedge \cdots \wedge \underline{v}^{i_p}\,, \quad {}^*\mathsf{T} = \varepsilon_{i_1\ldots i_p\, j_1\ldots j_{n-p}}\, \underline{v}^{j_1} \wedge \cdots \wedge \underline{v}^{j_{n-p}}$$

with the components of the ε-tensor. If we suppose here that $i_1 < \ldots < i_p$ and $j_1 < \ldots < j_{n-p}$ then zero components are dropped again.

References: (Bowen), (Grauert).

10.3 Differential Forms in $\mathbb{R}^n$

(a) Abstract Tangential Space and Pfaffian Forms In this section, we reconsider the well-known directional derivative $\partial f(x; h)$ of Sect. 1.7**(a)** from an abstract point of view, namely as *operator* acting on a function space with elements f. More exactly, let $\emptyset \neq \mathcal{M} \subset \mathbb{R}^n$ be an open set and let f, g etc. be sufficiently smooth scalar-valued functions on $\mathcal{M}$, writing briefly $f, g \in F(\mathcal{M})$. The *abstract tangential space* at the point $\underline{x} \in \mathcal{M}$ is defined straightforward by means of the directional derivative. Consider the family of mappings

$$\varPhi_{\underline{a}} : F(\mathcal{M}) \ni f \mapsto \varPhi_{\underline{a}}(f) := \frac{d}{d\tau} f(\underline{x} + \tau\underline{a})|_{\tau=0} = (\operatorname{grad} f(\underline{x}))\underline{a} \in \mathbb{R}$$

for a *fixed* $\underline{x} \in \mathcal{M}$ and arbitrary $\underline{a} \in \mathbb{R}^n$ recalling that grad $f(\underline{x})$ is a row vector. For each $\underline{a}$, these mappings are linear and LEIBNIZ' product rule $\Phi_{\underline{a}}(f \cdot g) = \Phi_{\underline{a}}(f)g(\underline{x}) + \Phi_{\underline{a}}(g)f(\underline{x})$ does hold for $f \cdot g : \underline{x} \mapsto f(\underline{x}) \cdot g(\underline{x})$. Writing briefly

$$D_i|_{\underline{x}} : F(\mathcal{M}) \ni f \mapsto \frac{\partial f}{\partial x^i}(\underline{x}) \in \mathbb{R}\,, \quad D|_{\underline{x}} = [D_1|_{\underline{x}}, \ldots D_n|_{\underline{x}}] \qquad (10.17)$$

for the operators of partial derivation at the point $\underline{x}$, we obtain $\Phi_{\underline{a}} = a^i\, D_i|_{\underline{x}} = \underline{a} \cdot D|_{\underline{x}}$ (sometimes ∂_x is written for D and/or the evaluation point $\underline{x}$ is omitted). The operators $D_i|_{\underline{x}}$ are linearly independent, if namely $\pi_k : \underline{x} \mapsto x^k$ is the projection onto the k-th component of $\underline{x} \in \mathbb{R}^n$ then

$$\forall\, k : \;\; a^i D_i|_{\underline{x}} = 0 \implies 0 = (a^i D_i|_{\underline{x}})(\pi_k) = a^i D_i \pi_k(\underline{x}) = a^k\,.$$

Therefore the operators $D_i|_{\underline{x}}$ of partial differentiation define a vector space of *operators* $\mathcal{T}_{\underline{x}} := \{\Phi_{\underline{a}},\, \underline{a} \in \mathbb{R}^n\}$ called *(abstract) contravariant tangential space* at the point $\underline{x} \in \mathcal{M}$ and the elements of this space are called *abstract tangential vectors* or also *(abstract) vector fields* in case $\underline{a}$ is a classical vector field. This vector space plays the role of the former sample space $\mathcal{V}$ in this section. The dual space $\mathcal{V}_d = [\mathcal{T}_{\underline{x}}]_d = \mathcal{L}(\mathcal{V}; \mathbb{R})$ is called *covariant tangential space* at the point $\underline{x}$.

Remark 1. More genuinely, we should write $\Phi_{\underline{a}} = D|_{\underline{x}}\, \underline{a}$, $\underline{a} \in \mathbb{R}^n$, since the elements of the basic vector space $\mathcal{V}$ have been considered as formal column vectors up to now. This notation leads however to misinterpretations later on, therefore we prefer the above introduced form as scalar product.

Remark 2. Let a composition "$\circ$" be defined for some compatible functions or operators F and G then either $F \circ G = G \circ F$, i.e., both commute with each other, or a new function/operator F^* is created by $F^* \circ G := G \circ F$.

Accordingly, there exists for all scalar functions $f \in F(\mathcal{M})$ an element $df(\underline{x}) \in [\mathcal{T}_{\underline{x}}]_d$ defined by

$$\boxed{\forall\, \Phi_{\underline{a}} \in \mathcal{T}_{\underline{x}} : \;\; df(\underline{x}) \circ \Phi_{\underline{a}} = \Phi_{\underline{a}} \circ f \;(= \Phi_{\underline{a}}(f) = \operatorname{grad} f(\underline{x}) \cdot \underline{a})}\,. \qquad (10.18)$$

In particular, we obtain for the above projection $f = \pi_i$

$$d\pi_i(\underline{x}) \circ D_j|_{\underline{x}} := D_j(\pi_i)(\underline{x}) = \delta^i{}_j\,, \qquad (10.19)$$

thus the linear functionals $dx_i := d\pi_i(\underline{x})$, $i = 1 : n$, constitute the *dual basis* in $[\mathcal{T}_{\underline{x}}]_d$ relative to the basis $\{D_i|_x\} \in \mathcal{T}_{\underline{x}}$. Especially we have for arbitrary $a^j D_j|_{\underline{x}} \in \mathcal{T}_{\underline{x}}$ that $dx_i \circ (a^j D_j|_{\underline{x}}) = a^j D_j(\pi_i)(\underline{x}) = a^j \delta^i{}_j = a^i$. Therefore, the functional dx_i can be identified with the projection $\pi_i : \underline{a} \mapsto a^i$ for the vector of *components* $\underline{a} = [a^j]$ in $\mathcal{T}_{\underline{x}}$ by $dx_i \circ a^j D_j|_{\underline{x}} \sim \pi_i(\underline{a}) = a^i$.

The elements of $\mathcal{T}_{\underline{x}}$ are now briefly denoted by ξ , $\xi \in \mathcal{T}_{\underline{x}}$.

Theorem 10.1. *(*PFAFF*ian Forms) The element $df(\underline{x}) \in [\mathcal{T}_{\underline{x}}]_d$ defined by (10.18) satisfies*

$$df(\underline{x}) = D_i|_{\underline{x}}(f)dx_i \equiv \sum_{i=1}^{n} \frac{\partial f}{\partial x^i}(\underline{x})dx_i\,.$$

Proof. We have $df(\underline{x}) = a^i dx_i \in [\mathcal{T}_{\underline{x}}]_d$ because the functionals dx_i form the dual basis, hence $\forall\ \xi \in \mathcal{T}_{\underline{x}} : df(\underline{x}) \circ \xi = a^i(\xi \circ \pi_i)$. Choosing in particular $\xi = D_j|_{\underline{x}}$ yields $df(\underline{x}) \circ D_j|_{\underline{x}} = a^i(D_j|_{\underline{x}} \circ \pi_i) = a^j \cdot 1$. But, on the other side, $df(\underline{x}) \circ D_j|_{\underline{x}} = D_f|_{\underline{x}}(f) = (\partial f/\partial x^j)(\underline{x})$, which proves the assertion. □

(b) Differential Forms Alternating tensor fields on the tangential space $\mathcal{V} = \mathcal{T}_{\underline{x}}$ are called *p-differential forms* (here over the euclidean $\mathbb{R}^n$); they are customarily denoted by $\Omega^p(\mathcal{M})$ today. Frequently one speaks also simply of *p*-forms in a somewhat confusing way. An element $\omega \in \Omega^p(\mathcal{M})$ is thus a *mapping*

$$\boxed{\omega : \mathcal{M} \ni \underline{x} \mapsto \omega(\underline{x}) = \sum_{i_1<\cdots<i_p} a_{i_1\cdots i_p}(\underline{x})\, dx_{i_1} \wedge \cdots \wedge dx_{i_p} \in \mathcal{A}_p(\mathcal{T}_{\underline{x}})}\ . \tag{10.20}$$

The computational rules are carried over directly from Sect. 10.2(**c**).

Example 10.10. (1°), $n = 3$.
0-Form: $\underline{x} \mapsto a(\underline{x})$
1-Form: $\underline{x} \mapsto a_1(\underline{x})\, dx_1 + a_2(\underline{x})\, dx_2 + a_3(\underline{x})\, dx_3$
2-Form: $\underline{x} \mapsto a_{12}(\underline{x})\, dx_1 \wedge dx_2 + a_{23}(\underline{x})\, dx_2 \wedge dx_3 + a_{13}(\underline{x})\, dx_1 \wedge dx_3$
3-Form: $\underline{x} \mapsto a_{123}(\underline{x})\, dx_1 \wedge dx_2 \wedge dx_3$.
(2°) Vector product, $n = 3$:

$$\begin{aligned}&(a\, dx + b\, dy + c\, dz) \wedge (e\, dx + f\, dy + g\, dz)\\ &= (bg - cf)\, dy \wedge dz + (ce - ag)\, dz \wedge dx + (af - be)\, dx \wedge dy\,.\end{aligned}$$

(3°) Inner product, $n = 3$, e.g.:

$$\begin{aligned}&(a\, dx + b\, dy + c\, dz) \wedge (p\, dy \wedge dz + q\, dz \wedge dx + r dx \wedge dy)\\ &= (ap + bq + cr)\, dx \wedge dy \wedge dz\,.\end{aligned}$$

(4°) Let more generally $\xi_j = a^i_j D_i|_{\underline{x}} \in \mathcal{T}_{\underline{x}}$ then

$$dx_1 \wedge \cdots \wedge dx_p(\xi_1, \cdots, \xi_p) = \det\,[dx_i(\xi_j)] = \det\,[a^i{}_j]\,.$$

For $\omega_1,\ \omega_2 \in \Omega^p(\mathcal{M})$, i.e., more exactly for

$$\begin{aligned}\omega_1(\underline{x}) &= \sum_{i_1<\cdots<i_p} a_{i_1\cdots i_p}(\underline{x}) dx_{i_1} \wedge \cdots \wedge dx_{i_p}\\ \omega_2(\underline{x}) &= \sum_{j_1<\cdots<j_p} b_{j_1\cdots j_p}(\underline{x}) dx_{j_1} \wedge \cdots \wedge dx_{j_p}\,,\end{aligned}$$

the scalar product of tensors has the form

$$\omega_1(\underline{x}) \cdot \omega_2(\underline{x}) = \sum_{i_1<\ldots<i_p} a_{i_1\ldots i_p}(\underline{x})\, b_{i_1\ldots i_p}(\underline{x})$$

and $|\omega(\underline{x})| := (\omega(\underline{x}) \cdot \omega(\underline{x}))^{1/2}$ is the *norm* of $\omega(\underline{x})$.

(c) Exterior Derivatives If ω is a 0-form, then the PFAFFian form $d\omega\,(\underline{x}) := D_i|_{\underline{x}}\,\omega(\underline{x})\,dx_i$ is called *exterior derivative* of ω at the point $\underline{x} \in \mathcal{M}$. If ω is a p-form (10.20) and $p \geq 1$, then

$$d\omega\,(\underline{x}) := \sum_{i_1<\cdots<i_p} da_{i_1\ldots i_p}(\underline{x}) \wedge dx_{i_1} \wedge \cdots \wedge dx_{i_p} \tag{10.21}$$

is the *exterior derivative* of ω at the point $\underline{x} \in \mathcal{M}$ where $da_{i_1\cdots i_p}$ is the PFAFFian form of the scalar function $a_{i_1\ldots i_p}(\underline{x})$. Accordingly, if ω is a p-form, then $d\omega$ is a $(p+1)$-form.

Example 10.11. For $n = 3$.

$$\begin{aligned}
p = 0 \;&: da = D_i a\, dx_i\,, \text{ component vector: grad}\,\underline{a}.\\
p = 1 \;&: d(a_1\,dx_1 + a_2\,dx_2 + a_3\,dx_3)\\
&= D_i a_1\,dx_i \wedge dx_1 + D_i a_2\,dx_i \wedge dx_2 + D_i a_3\,dx_i \wedge dx_3\\
&= -D_2 a_1\,dx_1 \wedge dx_2 - D_3 a_1 dx_1 \wedge dx_3 + D_1 a_2\,dx_1 \wedge dx_3\\
&\quad -D_3 a_2 dx_2 \wedge dx_3 + D_1 a_3\,dx_1 \wedge dx_3 + D_2 a_3 dx_2 \wedge dx_3\\
&= (D_1 a_2 - D_2 a_1)\,dx_1 \wedge dx_2 + (D_1 a_3 - D_3 a_1)\,dx_1 \wedge dx_3\\
&\quad +(D_2 a_3 - D_3 a_2)\,dx_2 \wedge dx_3\\
&\quad \text{component vector: rot}\,\underline{a}\,;
\end{aligned}$$

$$\begin{aligned}
p = 2 \;&: d(a_{12}\,dx_1 \wedge dx_2 + a_{23}\,dx_2 \wedge dx_3 + a_{13}\,dx_1 \wedge dx_3)\\
&= \cdots = (D_1 a_{23} - D_2 a_{13} + D_3 a_{12})\,dx_1 \wedge dx_2 \wedge dx_3\,.
\end{aligned}$$

different notation for a 2-form:
$\omega = b_1\,dx_2 \wedge dx_3 + b_2\,dx_3 \wedge dx_1 + b_3\,dx_1 \wedge dx_2$, then
$d\omega = (D_1 b_1 + D_2 b_2 + D_3 b_3)\,dx_1 \wedge dx_2 \wedge dx_3$,
component vector is a scalar div $\underline{b}$.

Theorem 10.2. *Let ω_1, $\widetilde{\omega}_1$ be p-forms, ω_2 a q-form and f a 0-form (all sufficiently smooth) then*
(1°) $d(\omega_1 + \widetilde{\omega}_1) = d\omega_1 + d\widetilde{\omega}_1$.
(2°) $d(f\,\omega_1) = df \wedge \omega_1 + f\,d\omega_1$.
(3°) $d(\omega_1 \wedge \omega_2) = d\omega_1 \wedge \omega_2 + (-1)^p \omega_1 \wedge d\omega_2$.
(4°) $d(d\omega) = 0$.

The last rule constitutes the most important difference to the classical differential calculus.

Proof. (1°) Clear! Because (1°), the remaining part is proved only for simple tensors, $\omega_1 = a(\underline{x})\,dx_1 \wedge \cdots \wedge dx_p$.
(2°) Because $d(fa) = a\,df + f\,da$ and by the definition of the wedge product,

$$\begin{aligned}
d(f\omega_1) &= d(fa) \wedge dx_1 \wedge \cdots \wedge dx_p\\
&= (a\,df + f\,da) \wedge dx_1 \wedge \cdots \wedge dx_p\\
&= df \wedge (a\,dx_1 \wedge \cdots \wedge dx_p) + f(da \wedge dx_1 \wedge \cdots \wedge dx_p)\\
&= df \wedge \omega_1 + f\,d\omega_1\,.
\end{aligned}$$

(3°) Let $\omega_2 = b\, dx_{p+1} \wedge \cdots \wedge dx_{p+q}$ then

$$\begin{aligned}
\omega_1 \wedge \omega_2 &= (ab)\, dx_1 \wedge \cdots \wedge dx_{p+q} \\
d(\omega_1 \wedge \omega_2) &= (b\, da + a\, db) \wedge dx_1 \wedge \cdots \wedge dx_{p+q} \quad \text{(shifting of } db\text{)} \\
&= da \wedge dx_1 \wedge \cdots \wedge dx_p \wedge b\, dx_{p+1} \wedge \cdots \wedge dx_{p+q} \\
&+ a\, dx_1 \wedge \cdots \wedge dx_p \wedge db \wedge dx_{p+1} \wedge \cdots \wedge dx_{p+q} \cdot (-1)^p \\
&= d\omega_1 \wedge \omega_2 + (-1)^p \omega_1 \wedge d\omega_2 \,.
\end{aligned}$$

(4°) *Case 1:* If a is a 0-form, then, because of the symmetry of the JACOBI matrix,

$$d(da) = d(D_i a\, dx_i) + d(D_i a) \wedge dx_i = D_j D_i a\, dx_j \wedge dx_i = 0\,.$$

Case 2: If $\omega = a\, dx_1 \wedge \cdots \wedge dx_p$ is a monom, then, because (3°),

$$\begin{aligned}
d(d\omega) &= d(da \wedge dx_1 \wedge \cdots \wedge dx_p) \\
&= d(da) \wedge (dx_1 \wedge \cdots \wedge dx_p) - da \wedge d(dx_1 \wedge \cdots \wedge dx_p) = 0 + 0\,,
\end{aligned}$$

since $dx_1 \wedge \cdots \wedge dx_p = 1\, dx_1 \wedge \cdots \wedge dx_p$ hence $d(dx_1 \wedge \cdots \wedge dx_p) = 0$ by the definition of the exterior derivative (10.21). □

(The proof of (4°) follows also in direct way by using the definition (10.21).

Example 10.12. Let f be a 0-form and $\omega = a^i\, dx_i$ be a 1-form then

$$\begin{aligned}
d(df) = 0 &\Longrightarrow \operatorname{rot}(\operatorname{grad} f) = 0 \\
d(d\omega) = 0 &\Longrightarrow \operatorname{div}(\operatorname{rot} \underline{a}) = 0\,.
\end{aligned}$$

In functions of a single variable the interdependence of derivation and integration is managed via the primitive function as well-known. This interdependence can be carried over to functions of several variables by means of the rule $d(d\omega) = 0$. However, what are the conditions for the existence of a $(p-1)$-form π relative to a p-form ω with the property $d\pi = \omega$? By Theorem 10.2(4°) we obtain immediately that $d(d\pi) = d\omega = 0$ hence $d\omega = 0$ is a *necessary condition.* But, as well-known, the domain of definition must still suffice some requirements for sufficiency.

(d) Closed and Exact Forms If ω is a continuously differentiable p-form on $\mathcal{M} \subset \mathbb{R}^n$ and $d\omega = 0$ then ω is said to be *closed.* On the other side, if ω is a continuous p-form on $\mathcal{M} \subset \mathbb{R}^n$ and there exists a $(p-1)$-form π on $\mathcal{M}$ such that $d\pi = \omega$ then ω is called *exact.* If $\omega \in \Omega^p(\mathcal{M})$ is sufficiently smooth then

$$\boxed{\omega \text{ exact} \Longrightarrow \omega \text{ closed}}\,.$$

by *Example* 10.12. But the conversion is *not* true in general:

Theorem 10.3. (Lemma of POINCARÉ) *Let $\mathcal{M} \subset \mathbb{R}^n$ be star-shaped and let $\omega \in \Omega^p(\mathcal{M})$ closed then ω is exact.*

Recall that a set $\mathcal{M} \subset \mathbb{R}^n$ is *star-shaped* if $\exists\, p \in \mathcal{M}\, \forall\, x \in \mathcal{M} \Longrightarrow [p,x] \subset \mathcal{M}$, ($[p,x]$ straight line connecting p and x); cf. Sect. 1.2(**e**).

The computation of a *primitive function* π with $d\pi = \omega$ is equivalent to the solution of a system of partial differential equations of first order:

Example 10.13. For $n = 3$, $p = 2$. Let

$$\begin{aligned} \omega &= b_1\, dx_2 \wedge dx_3 + b_2\, dx_3 \wedge dx_1 + b_3 dx_1 \wedge dx_2\,, \\ \pi &= v_1\, dx_1 + v_2\, dx_2 + v_3 dx_3\,, \end{aligned}$$

then the relation $d\pi = \omega$ leads to the equations $D_2 v_3 - D_3 v_2 = 0$, etc. and the condition for integrability $d\omega = 0$ yields $D_1 b_1 + D_2 b_2 + D_3 b_3 = 0$, hence together

$$\boxed{d\pi = \omega \iff \operatorname{rot} v = b\,, \quad d\omega = 0 \iff \operatorname{div} b = 0}\,.$$

Example 10.14.

$$\omega(x) = \frac{x^1\, dx_2 - x^2\, dx_1}{(x^1)^2 + (x^2)^2}\,, \quad \mathcal{M} = \mathbb{R}^2 \setminus \{0\}\,.$$

$\mathcal{M}$ is not star-shaped in a neighborhood of zero!

$$\begin{aligned} d\omega &= d\Big(\frac{1}{(x^1)^2 + (x^2)^2}\Big) \wedge (x^1\, dx_2 - x^2\, dx_1) \\ &+ \frac{1}{(x^1)^2 + (x^2)^2} d(x^1\, dx_2 - x^2\, dx_1) \\ &= \frac{-2x^1\, dx_1 - 2x^2\, dx_2}{((x^1)^2 + (x^2)^2)^2} \wedge (x^1\, dx_2 - x^2\, dx_1) \\ &+ \frac{1}{(x^1)^2 + (x^2)^2} (dx_1 \wedge dx_2 - dx_2 \wedge dx_1) \\ &= \frac{-2((x^1)^2 + (x^2)^2)\, dx_1 \wedge dx_2}{((x^1)^2 + (x^2)^2)^2} + \frac{2}{(x^1)^2 + (x^2)^2} dx_1 \wedge dx_2 = 0\,. \end{aligned}$$

Accordingly, ω is closed. Let now $f : \mathbb{R} \longrightarrow \mathbb{R}^2 \setminus \{0\}$ be a function such that $df = \omega$ then ω would be exact and

$$g := \frac{\partial f}{\partial x^1} = -\frac{x^2}{(x^1)^2 + (x^2)^2}\,, \quad h := \frac{\partial f}{\partial x^2} = \frac{x^1}{(x^1)^2 + (x^2)^2}\,,$$

and also $g_{x^2} = h_{x^1}$. But, consider the function $\underline{x} : \mathbb{R} \ni t \mapsto (x^1, x^2) = (\cos t, \sin t) \in \mathbb{R}^2 \setminus \{0\}$, then the composed mapping $g := f \circ \underline{x} : \mathbb{R} \longrightarrow \mathbb{R}$ is continuous, periodic and smooth hence it has a maximum in some t_0 with $g_t(t_0) = 0$. On the other side, we have but

$$g'(t) = dg(t) = \frac{\partial f}{\partial x^1}(\underline{x}(t))\, \frac{dx^1}{dt}(t) + \frac{\partial f}{\partial x^2}(\underline{x}(t))\, \frac{dx^2}{dt}(t) = \frac{\sin^2 t + \cos^2 t}{\sin^2 t + \cos^2 t} = 1\,.$$

This is a contradiction hence ω cannot be exact. The assumption of star shape of the domain of definition $\mathcal{M}$ in POINCARÉ's Lemma cannot be dropped.

The constant of integration is now replaced by a more general differential form:

Theorem 10.4. *Let the assumption of Theorem 10.3 hold and let* $d\pi_1 = d\pi_2 = \omega$ *then* $\pi_1 - \pi_2$ *is a* $(p-1)$*-form.*

Definition 10.3. *(Operator of Integration)* (1°) *Let* $\mathcal{M} \subset \mathbb{R}^n$ *be a star-shaped set relative to the origin and let* $\omega(\underline{x}) = a(\underline{x})dx_{i_1} \wedge \cdots \wedge dx_{i_p}$ *,* $p \geq 1$ *be a simple p-form on* $\mathcal{M}$*.*

(2°) *Let* $\sigma_{i_1 \ldots i_p} = \sum_{k=1}^{p} (-1)^{k-1} x_{i_k} dx_{i_1} \wedge \cdots \wedge \widehat{dx_{i_k}} \wedge \cdots \wedge d_{i_p}$.

(The symbol $\widehat{\ }$ *means again "dropping".)*

(3°) *Let the operator* $\mathcal{J}$ *be defined by* $\mathcal{J}(\omega) = \left(\int_0^1 t^{p-1} a(t\,\underline{x})\, dt \right) \sigma_{i_1 \ldots i_p}$.

This integration operator $\mathcal{J}$ is linear hence it can be generalized immediately to arbitrary p-forms, moreover, it is a uniquely defined $p-1$-form. By and large, it describes the converse of the exterior derivation as the following theorem shows.

Theorem 10.5. *Adopting the assumption of Definition 10.3, let* $\omega \in \Omega^p(\mathcal{M})$ *be a continuously differentiable p-form,* $p \geq 1$ *, then*
(1°) $\mathcal{J}(d\omega) + d(\mathcal{J}\omega) = \omega$ *;*
(2°) *if* $p \geq 2$ *and* π *is a primitive form of* ω*, i.e.,* ω *is closed, then* $\pi = \mathcal{J}\omega + d\eta$ *where* $\eta \in \Omega^{p-2}(\mathcal{M})$.

Example 10.15. In $\mathbb{R}^3$, the 0-form $\mathcal{J}\omega$ originates from the 1-form ω, and the 1-form $\mathcal{J}\widetilde{\omega}$ from the 2-form $\widetilde{\omega}$:

$$\begin{aligned}
\omega &= a_1 dx + a_2 dy + a_3 dz \\
\mathcal{J}\omega &= x_1 \int_0^1 a_1(t\underline{x})\, dt + x_2 \int_0^1 a_2(t\underline{x})\, dt + x_3 \int_0^1 a_3(t\underline{x})\, dt\,, \\
\widetilde{\omega} &= a_1 dx_2 \wedge dx_3 + a_2 dx_3 \wedge dx_1 \wedge a_3 dx_1 \wedge dx_2 \\
\mathcal{J}\widetilde{\omega} &= b_1 dx_1 + b_2 dx_2 + b_3 dx_3 \\
b_i(\underline{x}) &= x_{2+i} \int_0^1 t a_{1+i}(t\underline{x})\, dt - x_{1+i} \int_0^1 t a_{2+i}(t\underline{x})\, dt \mod i\,.
\end{aligned}$$

(e) Hodge-Star-Operator and Integral Theorems

(e1) The following **Computational Rules** hold for the (linear) HODGE-operator "$*$" defined in Sect. 10.2**(h)**:

(1°) $^*(f\omega) = f(^*\omega) \quad (f \text{ 0-form})$,

(2°) $^*(\omega_1 + \omega_2) =^* \omega_1 +^* \omega_2$,

(3°) $(\omega_1 \cdot \omega_2)\, dx_1 \wedge \cdots \wedge dx_n = \omega_1 \wedge^* \omega_2$,

(4°) $^{**}\omega = (-1)^{p(n-p)}\omega$,

(5°) $\omega_1 \wedge^* \omega_2 = \omega_2 \wedge^* \omega_1$,

(6°) $|\omega| = |^*\omega|$.

For instance, let $\underline{x} = [x, y, z]^T \in \mathbb{R}^3$ then

$$\begin{aligned} {}^*1 &= dx \wedge dy \wedge dz \quad \text{(volume element)}, \\ {}^*dx &= dy \wedge dz\,, \quad {}^*dy = -(dx \wedge dz)\,, \quad {}^*dz = dx \wedge dy\,, \\ {}^*(dx \wedge dy) &= dz\,, \quad {}^*(dx \wedge dz) = -dy\,, \quad {}^*(dy \wedge dz) = dx\,, \\ {}^*(dx \wedge dy \wedge dz) &= 1\,. \end{aligned}$$

(e2) For the 1-form $\omega(\underline{x}) = v^i(\underline{x})\, dx_i$ we obtain

$${}^*\omega(\underline{x}) = \sum_{i=1}^{n} (-1)^{i-1}\, v^i(\underline{x})\, dx_1 \wedge \cdots \wedge dx_{i-1} \wedge dx_{i+1} \wedge \cdots \wedge dx_n\,.$$

In particular, for $\underline{x} = [x, y, z]^T \in \mathbb{R}^3$,

$${}^*\omega(\underline{x}) = v^1(\underline{x})\, dy \wedge dz + v^2(\underline{x})\, dz \wedge dx + v^3(\underline{x})\, dx \wedge dy\,.$$

Thus, with the notation introduced in Sect. 1.2, ${}^*\omega(\underline{x}) = \underline{v}(\underline{x}) \cdot \underline{n}\, dF$ (flux of $\underline{v}$ through the surface ∂F with normal vector $\underline{n}$).

(e3) The Theorem of GAUSS, $\displaystyle\int_{\partial G} \underline{v} \cdot \underline{n}\, dF = \int_G \operatorname{div} \underline{v}\, dV$, can now be written in $\mathbb{R}^3$ as

$$\boxed{\int_{\partial G} {}^*\omega(\underline{x}) = \int_G \operatorname{div} \underline{v}(\underline{x})\, dx \wedge dy \wedge dz\,, \quad \omega(\underline{x}) = v^i(\underline{x})\, dx_i}\,.$$

For the Theorems of STOKES and GREEN's formulas, let $D \subset \mathbb{R}^n$ be a domain (bounded open set) and $u, v \in C^\infty(D; \mathbb{R})$ scalar-valued functions. Then du is a 1-form and one verifies that

$$d({}^*du) = \left(\sum_{i=1}^{n} \frac{\partial^2 u}{(\partial x^i)^2} \right) dx_1 \wedge \cdots \wedge dx_n = \Delta u\, dx_1 \wedge \cdots \wedge dx_n\,.$$

Let now $G \subset \mathbb{R}^n$ be compact and let $D[u, v]$ be the DIRICHLET integral,

$$D[u, v] := \int_G du \wedge^* dv = \int_G dv \wedge^* du = \int_G \sum_i \left(\frac{\partial u}{\partial x^i} \frac{\partial v}{\partial x^i} \right) dx_1 \wedge \cdots \wedge dx_n\,.$$

The Theorem of STOKES, $\displaystyle\int_{\partial G} \omega = \int_G d\omega$ leads to $\displaystyle\int_{\partial G} u^* dv = \int_G d(u^* dv)$. Furthermore, we have

$$d({}^*dv) = du \wedge^* dv + u\, d({}^*dv) = du \wedge^* dv + u \Delta v\, dx_1 \wedge \cdots \wedge dx_n\,,$$

therefore GREEN's formula in $\mathbb{R}^n$ reads in the present context:

$$\int_{\partial G} u({}^*dv) = D[u, v] + \int_G u \Delta v\, dx_1 \wedge \cdots \wedge dx_n\,.$$

Permutation of u and v and subtraction yield GREEN's formula in symmetric form:

$$\boxed{\int_{\partial G} (u^* dv - v^* du) = \int_G (u\Delta v - v\Delta u)\, dx_1 \wedge \cdots \wedge dx_n} \,.$$

Here one writes customarily $dx_1 \cdots dx_n := dx_1 \wedge \cdots \wedge dx_n$ for the volume element.

(f) Transformations We now consider abstract tangential vectors under "transformations", i.e., the images under a mapping F which is denoted by a capital letter by exception to distinguish it from the arguments f, g etc. of the abstract tangential vectors $\xi \in \mathcal{T}_{\underline{x}}$. More exactly, let again $\mathcal{M} \subset \mathbb{R}^n$ be open and let $F : \mathbb{R}^n(\underline{x}) \supset \mathcal{M} \ni \underline{x} \mapsto F(\underline{x}) = \underline{y} \in \mathcal{N} \subset \mathbb{R}^m(\underline{y})$ be a continuously differentiable. Furthermore, let

$$\begin{aligned}
\mathcal{T}_{\underline{x}} &= \mathrm{span}\{D_1|_{\underline{x}}, \cdots, D_n|_{\underline{x}}\} && \text{the contravariant tangential space in } \underline{x} \in \mathcal{M}\\
\mathcal{T}_{\underline{x}}^* &= \mathrm{span}\{dx_1, \cdots, dx_n\} && \text{the covariant tangential space in } \underline{x} \in \mathcal{M}\\
\mathcal{T}_{\underline{y}} &= \mathrm{span}\{D_1|_{\underline{y}}, \cdots, D_m|_{\underline{y}}\} && \text{the contravariant tangential space in } \underline{y} \in \mathcal{N}\\
\mathcal{T}_{\underline{y}}^* &= \mathrm{span}\{dy_1, \cdots, dy_m\} && \text{the covariant tangential space in } \underline{y} \in \mathcal{N}\,.
\end{aligned}$$

(One writes $\mathcal{T}_x^*$ instead $[\mathcal{T}_x]_d$ etc. for the dual space in this context.) Then $f \circ F \in F(\mathcal{M})$ holds for a scalar function $f \in F(\mathcal{N}) = C^\infty(\mathcal{N};\mathbb{R})$ and a mapping $F_* : \mathcal{T}_x \to \mathcal{T}_y$ is defined between the both tangential spaces by

$$\boxed{\forall\, f \in F(\mathcal{N})\ \forall\, \xi \in \mathcal{T}_x :\ F_* \circ \xi \circ f := \xi \circ f \circ F} \,,$$

$$\begin{array}{ccccc}
\mathbb{R}^n(\underline{x}) \supset \mathcal{M} & \longrightarrow & F & \longrightarrow & \mathcal{N} \subset \mathbb{R}^m(\underline{y})\\
\uparrow & & & & \uparrow\\
\mathcal{T}_{\underline{x}} & \longrightarrow & F_* & \longrightarrow & \mathcal{T}_{\underline{y}}
\end{array} \,. \tag{10.22}$$

Theorem 10.6. (1°) $F_* \in \mathcal{L}(\mathcal{T}_{\underline{x}}, \mathcal{T}_{\underline{y}})$, *i.e.,* F_* *is linear and* $\forall\, \xi \in \mathcal{T}_{\underline{x}} : F_*\,\xi := F_*(\xi) = \zeta \in \mathcal{T}_{\underline{y}}$.
(2°) *If* $F : \mathcal{M} \to \mathcal{N}$ *and* $G : \mathcal{N} \to \mathcal{Q}$ *are continuously differentiable then* $(G \circ F)_* = G_* \circ F_*$.
(3°) $\mathcal{T}_{\underline{y}} \ni F_* D_k|_x = {a^i}_k(x) D_i|_{\underline{y}}$, $\underline{y} = F(\underline{x})$, *where*

$${a^i}_k(\underline{x}) = \frac{\partial F^i}{\partial x^k}(\underline{x}) \ \ \textit{(}[{a^i}_k(\underline{x})] = \mathrm{grad}\, F(\underline{x}) \ \text{JACOBI } \textit{matrix).}$$

Observe that it is sumed up over the *row index* i of the JACOBI matrix in (3°) on the right side.

For the *proof* of (3°) note that the functionals $D_i|_{\underline{y}}$ constitute a basis of $\mathcal{T}_{\underline{y}}$ whence at first $F_* D_k|_{\underline{x}} = {a^i}_k(\underline{x}) D_i|_{\underline{y}}$ for some ${a^i}_k$ which depend obviously on the point $\underline{x}$. Applying the projection $\pi_j : \underline{y} \mapsto y^j$ on both sides, we obtain on the one side for $F = [F^1, \ldots, F^m]^T$

$$F_* D_k|_{\underline{x}}(\pi_j) = D_k|_{\underline{x}}(\pi_j \circ F) = D_k(F^j)(\underline{x}) = \frac{\partial F^j}{\partial x^k}(\underline{x})$$

and on the other side $a^i{}_k(\underline{x}) D_i|_{\underline{y}}(\pi_j) = a^i{}_k(\underline{x})\delta^i j = a^j{}_k(\underline{x})$. □

In the case where $m = n$ and F is a C^1 diffeomorphism we can also insert $\underline{x} = F^{-1}(\underline{y})$ to obtain an image of F_* entirely in $\underline{y}$-coordinates.

Remember: For $A \in \mathcal{L}(\mathcal{X}, \mathcal{Y})$, the adjoint operator $A_d \in \mathcal{L}(\mathcal{Y}_d, \mathcal{X}_d)$ is defined by $A_d \circ \underline{y}_d := \underline{y}_d \circ A$, $\underline{y}_d \in \mathcal{Y}_d$ (dual space), i.e., by pointwise definition,

$$\forall\, \underline{x} \in \mathcal{X}\ \forall\, \underline{y}_d \in \mathcal{Y}_d : (A_d \circ \underline{y}_d)(\underline{x}) := (\underline{y}_d \circ A)(\underline{x}) \in \mathbb{R}\,.$$

The dual operator F^* relative to the operator F_* is also *pointwise* defined by this general device (again one writes F^* instead $[F_*]_d$ in this context):

$$\boxed{\forall\, \xi \in \mathcal{T}_{\underline{x}}\ \forall\, dy \in \mathcal{T}^*_{\underline{y}} :\ \ (F^* \circ dy)(\xi) = (dy \circ F_*)(\xi)}\,.$$

Theorem 10.7. (1°) $F^* \in \mathcal{L}(\mathcal{T}^*_{\underline{y}}, \mathcal{T}^*_{\underline{x}})$, *i.e.,* F^* *is linear and* $\forall\, dy \in \mathcal{T}^*_{\underline{y}} : F^*\, dy := F^* \circ dy \in \mathcal{T}^*_{\underline{x}}$.
(2°) *If* $F : \mathcal{M} \to \mathcal{N}$ *und* $G : \mathcal{N} \to \mathcal{Q}$ *are continuously differentiable then* $(G \circ F)^* = F^* \circ G^*$.
(3°) $F^*\, dy_i = a^i{}_k(\underline{x})\, dx_k$ *where* $a^i{}_k(\underline{x}) = \dfrac{\partial F^i}{\partial x^k}(\underline{x})$ *such that* $[a^i{}_k(\underline{x})] = \operatorname{grad} F(\underline{x})$ *is again the* JACOBI *matrix.*

Note that it is sumed up over the *column index* k of the JACOBI matrix in (3°).

For the *proof* of (3°) observe first that the operators dx_k constitute a basis of $\mathcal{T}^*_{\underline{x}}$ hence $F^*\, dy_i = b^i{}_k(\underline{x})\, dx_k$ with some $b^i{}_k$. Applying both sides to the basis element $D_j|_{\underline{x}} \in \mathcal{T}_{\underline{x}}$ yields, by definition of F^* and Theorem 10.6,

$$F^*\, dy_i(D_j|_{\underline{x}}) = dy_i(F_* D_j|_{\underline{x}}) = dy_i(a^l{}_j(\underline{x}) D_l|_{\underline{y}}) = a^l{}_j(\underline{x})\, dy_i\, D_l|_{\underline{y}} = a^i{}_j(\underline{x})\,,$$

because $\{dy_i\}$ is the dual basis relative to $\{D_k|\underline{y}\}$. On the other hand,

$$b^i{}_k(\underline{x})\, dx_k\, D_j|_{\underline{x}} = b^i{}_j(\underline{x}) \Longrightarrow b^i{}_j(\underline{x}) = a^i{}_j(\underline{x}) = \frac{\partial F^i}{\partial x^j}(\underline{x})\,. \quad \square$$

Example 10.16. Let $F : \mathbb{R}^n(\underline{x}) \supset \mathcal{M} \to \mathcal{N} \subset \mathbb{R}^m(\underline{y})$ be smooth and $\underline{x} : \mathbb{R} \supset \mathcal{I} \to \mathcal{M} \subset \mathbb{R}^n(\underline{x})$ be a smooth curve with $\underline{x}(t_0) = \underline{x}_0$. Then

$$\underline{x}_*(t_0) D_t|_{t_0} = \sum_{i=1}^n (x^i)'(t_0) D_i|_{\underline{x}_0}$$

does hold for the tangential space $\mathcal{T}_{t_0} = \{a D_t|_{t_0},\ a \in \mathbb{R}\}$. This is the (unnormed) tangential vector at the point t_0 and $(F \circ \underline{x})_*\, D_t|_{t_0} = F_*(\underline{x}_*(t_0) D_t|_{t_0})$ is its image in $\underline{y}$-space.

Example 10.17. (1.) Let $F : \mathbb{R} \ni t \mapsto (x^1, x^2) = (t^2, t^3) \in \mathbb{R}^2$ and $\omega(x) = x^1\, dx^2$, then $(F^*\omega)(t) = t^2 \dfrac{\partial x^2}{\partial t}(t)\, dt = 3t^4\, dt$.
(2.) Let $F : \mathbb{R}^2 \ni (x^1, x^2) \mapsto t = x^1 - x^2 \in \mathbb{R}$ and $\omega(t) = dt$, then $F^*\, dt = dx^1 - dx^2$.

Example 10.18. Let $\omega \in \mathcal{T}^*_{\underline{y}}$ be a 1-form, i.e., $\omega(\underline{y}) = a^i(\underline{y})\, d_i(\underline{y})$, then

$$(F^*\omega)(\underline{x}) = (\omega \circ F_*)(\underline{x}) = a^i(F(\underline{x}))\, \frac{\partial F^i}{\partial x^k}(\underline{x})\, d_k(\underline{x}) =: b^k(\underline{x}) d_k(\underline{x})\,.$$

Accordingly, the component vector $[a^i(\underline{y})]$ satisfies the transformation rules

$$b^k(\underline{x}) = a^i(\underline{y})\, \frac{\partial F^i}{\partial x^k}(\underline{x}) \implies \underline{b}(\underline{x}) = \underline{a}(\underline{y})\, \mathrm{grad}\, F(\underline{x})\,, \quad \underline{y} = F(\underline{x})\,.$$

Summary. Recall that $D|_{\underline{x}} = [D_1|_{\underline{x}}, \ldots D_n|_{\underline{x}}]$ by (10.17) where the operators $D_i|_{\underline{x}}$ of partial differentation are formal column vectors (as elements of the sample space $\mathcal{V}$) and suppose consistently that the elementary differentials dx_i are formal row vectors (as elements of the dual space $\mathcal{V}_d$). Then the assertions (3°) of Theorems 10.6 and 10.7 say that the homomorphisms F_* and F^* can be defined by

$$F_* D|_{\underline{x}} = D|_{\underline{y}}\, \mathrm{grad}\, F(\underline{x})\,, \quad F^* \begin{bmatrix} dy_1 \\ \vdots \\ dy_m \end{bmatrix} = \mathrm{grad}\, F(\underline{x}) \begin{bmatrix} dx_1 \\ \vdots \\ dx_n \end{bmatrix}.$$

In case F is a diffeomorphism, we can write $\underline{x} = F^{-1}(\underline{y})$ in the first equation. Also, verify, e.g., Theorem 10.6(2°) by direct computation,

$$(G \circ F)_* D|_{\underline{x}} = D|_{\underline{z}}\, \mathrm{grad}(G \circ F) = D|_{\underline{z}}\, \mathrm{grad}\, G(F(\underline{x}))\, \mathrm{grad}\, F(\underline{x}) \overset{(!)}{=} G_* \circ F_* D|_{\underline{x}}\,.$$

(g) "Push Forward" of (abstract) vector fields is readily explained by Theorem 10.6 and proof; cf. also (10.22). To deal with **"Pull Back"** of differential forms, Theorem 10.7 and the last example suggest to write $\omega \circ F = F^*\omega$; see also (Abraham), p. 265. In other words, by the linear mapping F^*, a differential form

$$\boxed{\omega \circ F := F^*\omega = \omega \circ F_*}$$

on the set $\mathcal{M}$ ($\underline{x}$-coordinates) is associated to the differential form ω on the set $\mathcal{N}$ ($\underline{y}$-coordinates) ("pull back"),

$$\begin{array}{ccccc} \mathcal{M} & \longrightarrow & F & \longrightarrow & \mathcal{N} \\ \uparrow & & & & \uparrow \\ \Omega^p(\mathcal{M}) & \longleftarrow & F^* & \longleftarrow & \Omega^p(\mathcal{N}) \end{array}.$$

In general, let $\omega(\underline{y}) = \sum_{i_1<\cdots<i_p} a_{i_1\ldots i_p}(\underline{y})\, dy_{i_1} \wedge \cdots \wedge dy_{i_p}$ then

$$(\omega \circ F)(\underline{x}) := [F^*(\underline{x})\, \omega(\underline{y})](\underline{x}) = \sum_{1\le i_1<\cdots<i_p\le n} a_{i_1\ldots i_p}(F(\underline{x}))\, dF_{i_1} \wedge \cdots \wedge dF_{i_p}\,,$$

where $dF_j = dF_j(\underline{x}) = D_k F^j(\underline{x})\, dx_k$. Even more general,

$$(\mathsf{T} \circ F)(\xi_1, \ldots, \xi_p) = \mathsf{T}(F_*\xi_1, \ldots, F_*\xi_p)\,, \quad \xi_i \in \mathcal{T}_x\,.$$

does hold for an arbitrary tensor field $\mathsf{T} \in \mathcal{T}_p(\mathcal{M})$.

Computational Rules for $\omega \circ F := F^*\omega \in \Omega^p(\mathcal{M})$:
(1°) $(\alpha\omega_1 + \beta\omega_2) \circ F = \alpha(\omega_1 \circ F) + \beta(\omega_2 \circ F)$.
(2°) $(\omega_1 \wedge \omega_2) \circ F = (\omega_1 \circ F) \wedge (\omega_2 \circ F)$.
(3°) If $p > n$ then $\omega \circ F \equiv 0$, if $p > m$ then $\omega \equiv 0$.

Example 10.19.

$$F^*\,(dy_1 \wedge dy_2) = F^*\, dy_1 \wedge F^*\, dy_2 = \left(\frac{\partial y^1}{\partial x^i}\, dx_i\right) \wedge \left(\frac{\partial y^2}{\partial x^j}\, dx_j\right)$$

$$\begin{aligned}
&= \sum_{i=1}^{n}\sum_{j=1}^{n} \frac{\partial y^1}{\partial x^i}\frac{\partial y^2}{\partial x^j}\, dx_i \wedge dx_j && \text{(all components)}\\
&= \sum_{1\le i<j\le n} \left[\frac{\partial y^1}{\partial x^i}\frac{\partial y^2}{\partial x^j} - \frac{\partial y^1}{\partial x^j}\frac{\partial y^2}{\partial x^i}\right] dx_i \wedge dx_j && \text{(strict components)}\\
&= \sum_{1\le i<j\le n} \frac{\partial(y^1, y^2)}{\partial(x^i, x^j)}\, dx_i \wedge dx_j && \text{(determinants).}
\end{aligned}$$

Theorem 10.8. *With the above notations,*

$$\boxed{d(\omega \circ F) \equiv d(F^*\omega) = F^* d\omega \equiv d\omega \circ F}\,.$$

Proof. (1°) Let φ be a 0-form on $\mathcal{N}$ then $d\varphi = \dfrac{\partial \varphi}{\partial y^i} dy_i$ hence

$$F^*\, d\varphi \overset{\text{“pull back”}}{=} \frac{\partial \varphi}{\partial y^i}\frac{\partial F^i}{\partial x^k}\, dx_k = \frac{(\partial\varphi \circ F)}{\partial x^k}\, dx_k = d(\varphi \circ F)\,.$$

(2°) Let $p > 0$ and $\omega(\underline{y}) = \sum_{[i]} a_{i_1\ldots i_p}(\underline{y})\, dy_{i_1} \wedge \cdots \wedge dy_{i_p}$ (index $[i]$ denoting summation over all strict components) then

$$\begin{aligned}
d(\omega \circ F)(\underline{x}) &= d\left(\sum_{[i]} (a_{i_1\ldots i_p} \circ F)(\underline{x})\, dF_{i_1} \wedge \cdots \wedge dF_{i_p}\right)\\
&= \sum_{[i]} d(a_{i_1\ldots i_p} \circ F)(\underline{x}) \wedge dF_{i_1} \wedge \cdots \wedge dF_{i_p}\\
&\quad + \sum_{[i]} (a_{i_1\ldots i_p} \circ F)(\underline{x})\, d(dF_{i_1} \wedge \cdots \wedge dF_{i_p})\,.
\end{aligned}$$

Because $dd\omega = 0$, we obtain by induction that the second sum disappears, moreover, $d(a_{\dots} \circ F) = da \circ F$, by the first part of the proof, therefore

$$d(\omega \circ F)(\underline{x}) = \sum_{[i]} da_{i_1 \dots i_p}(F(\underline{x}))\, dF_{i_1} \wedge \cdots \wedge dF_{i_p} = d\omega(F(\underline{x})) \equiv (d\omega \circ F)(\underline{x})\,. \quad \square$$

References: (Bowen), (Flanders), (Grauert).

10.4 Tensor Analysis

(a) Euklidian Manifolds The EUCLID*ian space* (also EUCLIDian point space) consists of points and vectors; more exactly, it is composed of three components $(\mathcal{M}, \mathcal{V}, \varphi)$ with the following properties:

(1°) $\emptyset \neq \mathcal{M}$ is a set consisting of "points" $x \in \mathcal{M}$.
(2°) $\mathcal{V}$ is a n-dimensional vector space (of translations) with canonical scalar product "$\cdot$" and dual vector space $\mathcal{V}_d$ of which the scalar product is defined in canonical way by the scalar product in $\mathcal{V}$; cf. § 10.1**(c)**.
(3°) There exists a mapping $\varphi : \mathcal{M} \times \mathcal{M} \ni (x, y) \to \varphi(x, y) =: x - y \in \mathcal{V}$ called *difference* such that

$$\forall\, x, y, z \in \mathcal{M} : \varphi(x, y) = \varphi(x, z) + \varphi(z, y)$$
$$\forall\, x \in \mathcal{M} \ \ \forall\, \underline{v} \in \mathcal{V} \ \ \exists!\, y \in \mathcal{M} : \varphi(x, y) = \underline{v}\,.$$

The rules $x - x = 0$, $y - x = -(x - y)$, $x - y = x' - y' \implies x - x' = y - y'$ follow directly from the definition, and the set $\mathcal{M}$ is a metric space (not vector space) with the (EUCLIDian) metric $d(x, y) = [(x - y) \cdot (x - y)]^{1/2} =: |x - y|$. The set $\mathcal{M}$ is also called EUKLIDian space of dimension n for brevity. But there is no addition defined on $\mathcal{M}$ but only the difference of two elements of $\mathcal{M}$ which is contained in the vector space $\mathcal{V}$. Also, the elements $x \in \mathcal{M}$ are denoted as *point vectors* and the elements $\underline{v} \in \mathcal{V}$ as *free vectors*. However, the different notation, namely $x \in \mathcal{M}$ and $\underline{v} \in \mathcal{V}$ cannot be keeped up throughout; besides, every vector space $\mathcal{V}$ with scalar product is made an EUCLIDian space by defining $\varphi(\underline{u}, \underline{v}) = \underline{u} - \underline{v}$.

Definition 10.4. *Let $(\mathcal{M}, \mathcal{V}, \varphi)$ (in short $\mathcal{M}$) be an* EUKLID*ian space.*

(1°) *A pair $(\mathcal{U}, \Psi)$ is a* chart *in $x \in \mathcal{M}$ if $\mathcal{U}$ is* open *in $\mathcal{M}$ with $x \in \mathcal{U}$ and $\Psi : \mathcal{U} \to \Psi(\mathcal{U}) \subset \mathbb{R}^n$ is a* diffeomorphism. *The* mapping *(!) Ψ is a* coordinate system *in $\mathcal{U}$.*
(2°) *If two charts $(\mathcal{U}_1, \Psi_1)$, $(\mathcal{U}_2, \Psi_2)$ are given and $\mathcal{W} := \mathcal{U}_1 \cap \mathcal{U}_2 \neq \emptyset$ then*

$$\Psi_2 \circ \Psi_1^{-1} : \mathbb{R}^n \supset \Psi_1(\mathcal{W}) \to \Psi_2(\mathcal{W}) \subset \mathbb{R}^n$$

is a coordinate transformation.

(3°) *Let* $\mathcal{I}$ *be an index set (not necessarily countable) and let* $\mathcal{A} = \{(\mathcal{U}_i, \Psi_i)\,,\ i \in \mathcal{I}\}$ *be a family of charts such that* $\mathcal{M} = \bigcup_{i\in\mathcal{I}} \mathcal{U}_i$ *, then* $\mathcal{A}$ *is an* atlas *on* $\mathcal{M}$ *. In this atlas, for every chart there must exist at least one other, overlapping chart such that the intersection, say* $\mathcal{W}$ *, has an open, non-empty interior,* $\operatorname{int}\mathcal{W} \neq \emptyset$ *.*

(4°) *If* $(\mathcal{M}, \mathcal{V}, \varphi)$ *is an* EUCLID*ian space and* $\mathcal{A}$ *an atlas on* $\mathcal{M}$ *, then the quintuple* $(\mathcal{M}, \mathcal{V}, \varphi, \mathcal{A})$ *is a* differentiable EUCLIDian manifold *(MF).*

Henceforth the set $\mathcal{M}$ of a MF $(\mathcal{M}, \mathcal{V}, \varphi, \mathcal{A})$ is also called MF for brevity.

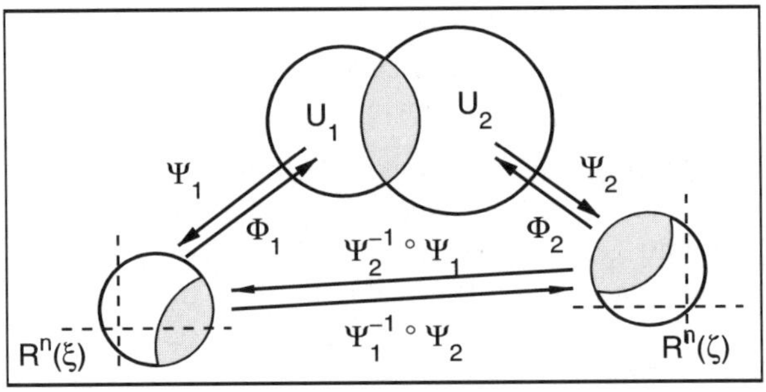

Figure 10.1. Two charts of a manifold

(b) Natural Coordinate Systems Let $\mathcal{M}$ be a MF and let $(\mathcal{U}, \Psi)$ be a chart on $\mathcal{M}$ with inverse mapping $\Phi = \Psi^{-1}$, i.e.,

$$\begin{aligned} \Psi = [\psi^1, \ldots, \psi^n]^T : \quad & \mathcal{M} \supset \mathcal{U} \ni x \mapsto \Psi(x) = \xi \in \mathbb{R}^n \\ & \Phi : \mathbb{R}^n \ni \Psi(\mathcal{U}) \ni \xi \mapsto \Phi(\xi) = x \in \mathcal{U} \subset \mathcal{M} \end{aligned}$$

where $\psi^i : \mathcal{U} \to \mathbb{R}$ are scalar fields. As the *parameter representation* Φ of a chart is commonly used for computation, all relations are to be "pulled back" into parameter space. Recall that

$$\operatorname{grad}\Psi(x) \circ \operatorname{grad}\Phi(\xi) = \operatorname{grad}\Phi(\xi) \circ \operatorname{grad}\Psi(x) = I\,,$$

because the inverse is determined uniquely.

We introduce two *natural coordinate systems* at the point $x = \Phi(\xi) \in \mathcal{M}$:

(1°) Let

$$\boxed{\underline{\widetilde{g}}^{i}(x) := \operatorname{grad}\psi^i(x) \in \mathcal{L}(\mathcal{V};\mathbb{R}) = \mathcal{V}_d\,,\ \ \underline{g}^{i}(\xi) := \underline{\widetilde{g}}^{i}(\Phi(\xi))\,,\quad i = 1:n}\,,$$

then $\underline{\widetilde{g}}^{i}(x_0) = \underline{g}^{i}(\xi_0) \in \mathcal{V}_d$ is the *normal vector* at the point x_0 on the i-th *coordinate surface* $\psi^i(x) = \psi^i(x_0)$.

(2°) Let

$$\boxed{\underline{g}_i(\xi) = D_i\Phi(\xi) := \frac{d}{dt}\Phi(\xi + t\underline{e}_i)|_{t=0}\,,\ \ \widetilde{\underline{g}}_i(x) = \underline{g}_i(\Psi(x)) \in \mathcal{V}\,,\ \ i = 1:n} \tag{10.23}$$

be the directional derivative w.r.t. to the i-th unit vector $\underline{e}_i = [\delta^k{}_i]_{k=1}^n \in \mathbb{R}^n$. The vector $\underline{g}_i(\xi) = \operatorname{grad}\Phi(\xi)\,\underline{e}_i$ is an element of the vector space $\mathcal{V}$ because $\operatorname{grad}\Phi(\xi) \in \mathcal{L}(\mathbb{R}^n;\mathcal{V})$. Moreover, $\underline{g}_i(\xi_0)$ is *tangential vector* at point $x_0 = \Phi(\xi_0)$ on the i-th *coordinate curve* in $\mathcal{U}$, i.e., of the curve

$$\mathbb{R} \ni \zeta \mapsto \Phi(\xi_0^1, \ldots, \xi_0^{i-1}, \zeta, \xi_0^{i+1}, \ldots, \xi_0^n)\,,\ \ x = \Phi(\xi)\,.$$

Because $\xi = \Psi \circ \Phi(\xi)$, $\xi^i = \psi^i \circ \Phi(\xi)$, we have in general

$$\frac{\partial}{\partial \xi^j}\xi^i = \delta^i{}_j = \operatorname{grad}\psi^i(x)\frac{\partial \Phi}{\partial \xi^j}(\xi) = \langle \widetilde{\underline{g}}^i(x), \underline{g}_j(\xi)\rangle\,,\ \ x = \Phi(\xi)\,,$$

hence $\langle \underline{g}^i(\xi), \underline{g}_j(\xi)\rangle = \delta^i{}_j$. (In numerical computations, we have to differ strongly between $\underline{g}_i$ and $\widetilde{\underline{g}}_i$ but in theoretical considerations both vectors are mostly identified with each other.) The vectors $\underline{g}_j(\xi) \in \mathcal{V}$ and the vectors $\underline{g}^i(\xi) \in \mathcal{V}_d$ are linearly independent, respectively, since Ψ shall be a diffeomorphism. Because (10.24), $\{\underline{g}^i(\xi)\}$ constitutes the dual basis in $\mathcal{V}_d$ relative to the basis $\{\underline{g}_i(\xi)\}$ in $\mathcal{V}$. But, as $\mathcal{V}$ is a scalar product space, the dual basis can be *identified canonically* with the *reciprocal* basis $\{\underline{r}^i(\xi)\} \subset \mathcal{V}$ relative to $\{\underline{g}_i(\xi)\} \subset \mathcal{V}$, i.e., $\underline{g}^i(\xi) \simeq \underline{r}^i(\xi)$ by writing $\langle \underline{v}, \underline{w}\rangle := \underline{v} \cdot \underline{w}$ again.

Then we obtain

$$\boxed{\langle \underline{g}^i(\xi), \underline{g}_j(\xi)\rangle = \underline{g}^i(\xi) \cdot \underline{g}_j(\xi) = \delta^i{}_j} \tag{10.24}$$

on the (fixed) chart $(\mathcal{U}, \Psi)$ and the vectors $\underline{g}^i(\xi)$ are uniquely defined by this relation, being important for applications. These vectors $\underline{g}^i(\xi)$ are the rows of the (temporarily) formal matrix $[\nabla\Phi(\xi)]^{-1}$ written as columns. The upper index and the *position* in the scalar product are however keeped up to remember their origin which is also of advantage in MATLAB implementations. Mostly, the computation of the inverse mapping Ψ by Φ is not necessary in applications and $[\nabla\Phi(\xi)]^{-1}$ is computed directly by using CRAMER's rule yielding acceptable results in $\mathbb{R}^3$.

Notations relative to an individual chart:

(1°) The basis $\{\underline{g}_1(\xi), \ldots, \underline{g}_n(\xi)\}$ is called *covariant basis* of $\mathcal{V}$ in $\mathcal{U} \subset \mathcal{M}$.
(2°) The basis $\{\underline{g}^1(\xi), \ldots, \underline{g}^n(\xi)\}$ is called *contravariant basis* of $\mathcal{V}$ in $\mathcal{U} \subset \mathcal{M}$.

(3°) The real matrix $M_g(\xi) := [g_{ij}(\xi)] := [\underline{g}_i(\xi) \cdot \underline{g}_j(\xi)]$ with scalar product in $\mathcal{V}$ resp. the tensor (mapping)

$$\mathsf{M}_g(\xi) = g_{ij}(\xi)\, \underline{g}^i(\xi) \otimes \underline{g}^j(\xi)$$

are called *covariant metric tensor* in $x \in \mathcal{U} \subset \mathcal{M}$; and $\mathsf{M}_g(\underline{u}, \underline{v}) = g_{ij} u^i v^j$ holds for $\underline{u} = \underline{g}_i u^i$, $\underline{v} = \underline{g}_i v^i$.

(4°) The real matrix $M_g{}^d(\xi) = [g^{ij}(\xi)] = [\underline{g}^i(\xi) \cdot \underline{g}^j(\xi)]$ with *canonical* scalar product in $\mathcal{V}_d$ resp. the tensor (mapping)

$$\mathsf{M}_g{}^d(\xi) = g^{ij}(\xi)\, \underline{g}_i(\xi) \otimes \underline{g}_j(\xi)$$

are called *contravariant metric tensor* in $x \in \mathcal{U} \subset \mathcal{M}$. Of course $M_g(\xi)\, M_g^d(\xi) = I \in \mathbb{R}^n{}_n$ for all $\xi \in \Psi(\mathcal{U})$ since the scalar product is just chosen in that way.

(5°) *Contrary* to the usual notation $g(\xi) = \det(M_g(\xi))$ let

$$g(\xi) := \det(M_g(\xi))^{1/2} = \det([g_{ij}(\xi)])^{1/2}\,.$$

(6°) For brevity, let

$$\boxed{\partial_i = \frac{\partial}{\partial \xi^i}\,,\quad \widehat{\partial}_i = \frac{\partial}{\partial x^i}\,,\ \text{if } \mathcal{M} = \mathcal{V} \text{ and } x = \underline{e}_i x^i}\,.$$

(7°) Let $\widehat{\underline{v}} : \mathcal{M} \ni x \mapsto \widehat{\underline{v}}(x) \in \mathcal{V}$ be a vector field living on $\mathcal{M}$. In the sequel, always the global-local transformation $\widehat{\underline{v}} \to \underline{v}$ resp. the local-global transformation $\underline{v} \to \widehat{\underline{v}}$ shall apply as follows

$$\boxed{\widehat{\underline{v}}(x) = \widehat{\underline{v}}(\Phi(\xi)) = \underline{v}(\xi) = \underline{v}(\Psi(x))}\,; \qquad (10.25)$$

$\underline{v}$ is also said to be the *coordinate representation* of $\widehat{\underline{v}}$.

Often x is written instead of ξ in literature if the notation is not related specifically to an individual chart.

The vectors

$$\widehat{\underline{g}}_i(\xi) = \underline{g}_i(\xi) g_{ii}(\xi)^{-1/2}\,,\quad \widehat{\underline{g}}^i(\xi) = \underline{g}^i(\xi) g_{ii}(\xi)^{1/2}\,,\quad i = 1:n\,,$$

(no summation) have unit length in *orthogonal* natural coordinate systems. The components v^i of the representation $\underline{v} = \widehat{\underline{g}}_i(\xi) v^i \in \mathcal{V}$ are then called *anholonomic* or *physical* (contravariant) components of $\underline{v}$ relative to the covariant basis $\{\underline{g}_i(\xi)\}$. The physical components relate to a *normed orthogonal* basis which however is no longer a genuine "natural" basis.

Simplification. For *model* of the physical situation in technical applications, $\mathcal{M} = \mathcal{V}$ is chosen frequently, i.e., the distinction between points and vectors is dropped. Then a cartesian coordinate system (NOS) $\{\mathcal{O}; \underline{e}_1, \ldots, \underline{e}_n\}$ with origin $\mathcal{O}$ is chosen for canonical basis such that $\underline{e}_i \cdot \underline{e}_j = \delta^i{}_j$. Mostly also $\underline{e}^i$ and $\underline{e}_i$ are identified, $\underline{e}^i = \underline{e}_i$, but the different position of the indices is keeped up. Let $\mathcal{E} = \{\underline{e}_1, \ldots, \underline{e}_n\}$ be the matrix of formal column vectors again then $\Phi : \mathbb{R}^n \ni \xi \mapsto \underline{e}_i \widetilde{\varphi}^i(\xi) = x \in \mathcal{V}$ on a chart $(\mathcal{U}, \Phi^{-1})$ or, formally written with components $\widetilde{\Phi}(\xi) \in \mathbb{R}^n$ and $\widetilde{\Psi}(x) \in \mathbb{R}_n$,

$$\Phi(\xi) = \mathcal{E}\widetilde{\Phi}(\xi), \quad \nabla\Phi(\xi) = \mathcal{E}\nabla\widetilde{\Phi}(\xi), \quad \nabla\widetilde{\Phi}(\xi) = \left[\frac{\partial\widetilde{\varphi}^i}{\partial\xi^j}(\xi)\right]$$

$$\Psi(x) = \widetilde{\Psi}(x)\mathcal{E}^T, \ \widehat{\nabla}\Psi(x) = \widehat{\nabla}\widetilde{\Psi}(x)\mathcal{E}^T, \ \widehat{\nabla}\widetilde{\Psi}(x) = \left[\frac{\partial\widetilde{\psi}^i}{\partial x^j}(x)\right]$$

(i row index). Then

$\underline{g}_i(\xi) = \underline{e}_k\, \partial_i \widetilde{\varphi}^k(\xi)$	covariant basis with contravariant components components are columns of $\nabla\widetilde{\Phi}(\xi)$
$\underline{g}^i(\xi) = \widehat{\partial}_k \widetilde{\psi}^i(x))\underline{e}^k$	contravariant basis with covariant components components are rows of $\widehat{\nabla}\widetilde{\Psi}(x)$

and, according to the above definition,

$$\boxed{M_g(\xi) = \nabla\Phi(\xi)^T \nabla\Phi(\xi)\,,\ M_g{}^d(\xi) = \widehat{\nabla}\Psi(x)\widehat{\nabla}\Psi(x)^T\,,\ g(\xi) = |\det \operatorname{grad} \Phi(\xi)|}$$

The remaining notations do not change. The *reciprocal* basis relative to $\{\underline{g}_i(\xi)\}$ consists here of the *column vectors* $\underline{g}^i(\xi)^T$.

Example 10.20. Shells consist frequently of a single chart of a three-dimensional MF in coordinate space $\mathbb{R}^3$. If first $\omega \subset \mathbb{R}^2(\xi_1, \xi_2)$ is a domain and $\underline{\varphi} : \omega \to \mathbb{R}^3$ is the parameter representation of a *surface*, then the mapping Φ is declared by

$$\Omega = \omega \times (-h, h) \ni \xi \mapsto \Phi(\xi) = \underline{\varphi}(\xi_1, \xi_2) + \xi_3\, \underline{n}(\xi_1, \xi_2) \in \mathbb{R}^3\,,\ h > 0\,,$$

where $\underline{n} = \underline{g}_1 \times \underline{g}_2$ is the *normal vector* of the surface. One chooses $\{\underline{g}_1(\xi_1, \xi_2), \underline{g}_2(\xi_1, \xi_2), \underline{g}_3(\xi_1, \xi_2) = \underline{n}(\xi_1, \xi_2)\}$ for covariant basis, and the contravariant basis follows from the relation $\underline{g}^i(\xi) \cdot \underline{g}_j(\xi) = \delta^i{}_j$.

Example 10.21. In a vector space $\mathcal{V}$ with cartesian coordinate system $\{\mathcal{O}; \underline{e}_1, \underline{e}_2, \underline{e}_3\}$, the components of the (linear) elasticity tensor and the (linear) strain tensor have the form

$$\begin{aligned} A^{ijkl}(x) &= \lambda\, \delta^{ij}\delta^{kl} + \mu\, [\delta^{ik}\delta^{jl} + \delta^{il}\delta^{jk}] \\ e_{ij}(\underline{u}(x)) &= \frac{1}{2}\left(\widehat{\partial}_j u^i + \widehat{\partial}_i u^j\right)(x)\,, \end{aligned}$$

$\lambda > 0$ and $\mu > 0$ being the LAMÉ constants. On a chart $(\mathcal{U}, \Psi)$ of a three-dimensional MF, the same components have the form, pulled back into parameter space,

$$\begin{aligned} A^{ijkl}(\xi) &= \lambda g^{ij}(\xi)g^{kl}(\xi) + \mu\left[g^{ik}(\xi)g^{jl}(\xi) + g^{il}(\xi)g^{jk}(\xi)\right] \\ e_{i\|j}(\underline{u}(\Phi(\xi))) &= \left[\frac{1}{2}\left(\partial_j(u^i\circ\Phi) + \partial_i(u^j\circ\Phi)\right) - \Gamma_{ij}^k(u_k\circ\Phi)\right](\xi) \end{aligned}$$

where Γ_{ij}^k are the CHRISTOFFEL symbols dealed with below.

(c) Representation and Transformation

(c1) Basis Transformation Let $(\mathcal{U}_1, \Psi_1)$ and $(\mathcal{U}_2, \Psi_2)$ be two (overlapping) charts with non-empty intersection $\mathcal{W} := \mathcal{U}_1 \cap \mathcal{U}_2 \neq \emptyset$, i.e.,

$$\begin{aligned} &\Psi_1 : \mathcal{U}_1 \ni x \mapsto \Psi_1(x) = \xi \in \mathbb{R}^n,\ \Phi_1 = \Psi_1^{-1} \\ &\Psi_2 : \mathcal{U}_2 \ni x \mapsto \Psi_2(x) = \zeta \in \mathbb{R}^n,\ \Phi_2 = \Psi_2^{-1}, \end{aligned}$$

and thus $\xi = (\Psi_1 \circ \Phi_2)(\zeta)$, $\zeta = (\Psi_2 \circ \Phi_1)(\xi)$. Let $\{\underline{g}_i(\xi)\}$, $\{\underline{g}^j(\xi)\}$ be the natural bases of the chart $(\mathcal{U}_1, \Psi_1)$ and $\{\underline{h}_i(\zeta)\}$, $\{\underline{h}^j(\zeta)\}$ the natural bases of the chart $(\mathcal{U}_2, \Psi_2)$, as well as

$$A(\zeta) = [a^i{}_k(\zeta)] := \nabla\Psi_1(x)\,\nabla\Phi_2(\zeta),\quad B(\xi) = [b^i{}_k(\xi)] := \nabla\Psi_2(x)\nabla\Phi_1(\xi),$$

then $B(\xi) = A(\zeta)^{-1} \in \mathbb{R}^n{}_n$. On $\mathcal{W}$ we have $\Phi_2(\zeta) = (\Phi_1 \circ \Psi_1 \circ \Phi_2)(\zeta)$, $\Psi_2(x) = (\Psi_2 \circ \Phi_1 \circ \Psi_1)(x)$, hence we obtain

$$\underline{h}_i(\zeta) = \underline{g}_j(\xi)a^j{}_i(\zeta),\quad \underline{h}^i(\zeta) = b^i{}_j(\xi)\underline{g}^j(\xi),\quad i = 1:n, \tag{10.26}$$

or formally by § 10.1**(b)**

$$\underline{\mathcal{H}}(\zeta) = \underline{\mathcal{G}}(\xi)A(\zeta),\quad \overline{\mathcal{H}}(\zeta) = B(\xi)\overline{\mathcal{G}}(\xi) = A(\zeta)^{-1}\overline{\mathcal{G}}(\xi)$$

where $\underline{\mathcal{G}}(\zeta) = [\underline{g}_1(\zeta), \ldots, \underline{g}_n(\zeta)]$ etc.. Accordingly, $A(\zeta)$ is the transformation matrix in transforming the covariant basis $\{\underline{g}_i\}$ of $\mathcal{V}$, and $A(\xi)^{-1}$ is the transformation matrix in transforming the contravariant basis $\{\underline{g}^i\}$ of $\mathcal{V}$. In orthogonal, natural coordinate systems, we have $\overline{\mathcal{H}}(\xi) = \underline{\mathcal{H}}(\zeta(\xi))^T$ as well as $A(\xi)^{-1} = A(\xi)^T$, and both transformations are equal.

(c2) Component Transformation Every vector field $\underline{v} : \mathcal{M} \ni x \mapsto \underline{v}(x) \in \mathcal{V}$ has two representations on a chart $(\mathcal{U}, \Psi)$, $\underline{v}(x) = \underline{g}_i(\xi)v^i(x) = v_j(x)\underline{g}^j(\xi)$, with *covariant* basis $\{\underline{g}_i\}$ and *contravariant* components v^i, or with *contravariant basis* $\{\underline{g}^i\}$ and *covariant* components v_j. Following (10.9) and (10.13), the indices may be lowered or raised using the components of the metric tensors,

$$\boxed{v^k(x) = v_i(x)g^{ik}(\xi),\quad v_k(x) = v^i(x)g_{ik}(\xi)}\,.$$

Frequently, the vector field $\underline{v}$ is given or sought in a global coordinate system $\mathcal{E} = \{\underline{e}_1, \ldots, \underline{e}_n\}$, i.e., $\underline{v}(x) = \underline{e}_i\widehat{v}^i(x) = \widehat{v}_j(x)\underline{e}^j$ because $\underline{e}^i = \underline{e}_i$. Then

$$\underline{v}(x) = \underline{e}_i\widehat{v}^i(x) = \underline{g}_i(\xi)v^i(\xi),\quad x = \Phi(\xi), \tag{10.27}$$

on a fixed chart $(\mathcal{U},\Psi)$ and the components $v^i(\xi)$ are to be found:

$$v^i(\xi) = \underline{g}^i(\xi)\cdot\underline{g}_j(\xi)v^j(\xi) = \underline{g}^i(\xi)\cdot\underline{e}_j\widehat{v}^j(x)$$
$$\widehat{v}^i(x) = \underline{e}^i\cdot\underline{e}_j\widehat{v}^j(x) = \underline{e}^i\cdot\underline{g}_j(\xi)v^j(\xi)\,.$$

On the intersection $\mathcal{W}\ni x = \Phi_1(\xi) = \Phi_2(\zeta)$ of two charts, the vector field has *four* representations

$$\underline{v}:\mathcal{U}_1\ni x\mapsto \underline{v}(x) = \underline{g}_i(\xi)v^i(x) = v_i(x)\underline{g}^i(\xi)\in\mathcal{V}\,,$$
$$\underline{v}:\mathcal{U}_2\ni x\mapsto \underline{v}(x) = \underline{h}_i(\zeta)w^i(x) = w_i(x)\underline{h}^i(\zeta)\in\mathcal{V}\,.$$

Changing the chart yields, by **(c1)** and § 10.1**(a)**,

$$w^i(x) = b^i{}_j(\xi)v^j(x)\,,\ \ w_i(x) = v_j(x)a^j{}_j(\zeta)\ \text{ or}$$

$$\boxed{\overline{[\underline{w}]}(x) = B(\xi)\,\overline{[\underline{v}]}(x)\,,\ \ \underline{[\underline{w}]}(x) = \underline{[\underline{v}]}(x)\,A(\zeta)}$$

for the

contravariant components $\overline{[\underline{v}]} = [v^1,\dots,v^n]^T\,,\ \overline{[\underline{w}]} = [w^1,\dots,w^n]^T$

covariant components $\underline{[\underline{v}]} = [v_1,\dots,\underline{v}_n]\,,\ \ \underline{[\underline{w}]} = [w_1,\dots,w_n]\,.$

Often one writes e.g. $\underline{v}(\xi) := \underline{v}(\Phi(\xi))$ instead e.g. $\widetilde{\underline{v}}(\xi) := \underline{v}(\Phi(\xi))$ for the coordinate representation which has to be regarded in *computation.*

(c3) Transformation of Tensor Fields A *covariant* tensor field $\widehat{\mathsf{T}}$ of order p on a MF $\mathcal{M}$ is a mapping $\widehat{\mathsf{T}}:\mathcal{M}\ni x\mapsto\widehat{\mathsf{T}}(x)\in\mathcal{T}_p(\mathcal{V})$ with *local* representation (on a fixed chart)

$$\begin{aligned}
\widehat{\mathsf{T}}(x) &= T_{i_1\dots i_p}(\xi)\,\underline{g}^{i_1}(\xi)\otimes\dots\otimes\underline{g}^{i_p}(\xi)\\
\widehat{\mathsf{T}}(x)(\underline{v}_1,\dots,\underline{v}_p) &= T_{i_1\dots i_p}(\xi)\,\langle\underline{g}^{i_1}(\xi),\underline{v}_1\rangle\cdots\langle\underline{g}^{i_p}(\xi),\underline{v}_p\rangle\,,\ \ \underline{v}_i\in\mathcal{V}\\
T_{i_1\dots i_p}(\xi) &= \widehat{T}(x)\,(\underline{g}_{i_1}(\xi),\dots,\underline{g}_{i_p}(\xi))\,,\ \ x=\Phi(\xi)
\end{aligned}$$

and *covariant* components $T_{i_1\dots i_p}(\xi)$; mostly both attributes are dropped. On the intersection of two charts, we have with the above notations

$$\widehat{\mathsf{T}}(x) = \big(T_{i_1\dots i_p}\,\underline{g}^{i_1}\otimes\dots\otimes\underline{g}^{i_p}\big)(\xi) = \big(\widetilde{T}_{i_1\dots i_p}\,\underline{h}^{i_1}\otimes\dots\otimes\underline{h}^{i_p}\big)(\zeta)\,.$$

In coordinate transformation (change of chart), we obtain the transformation rule

$$\widetilde{T}^{i_1\dots i_p}(\xi) = b^{i_1}{}_{k_1}(\xi)\dots b^{i_p}{}_{k_p}(\xi)\,T^{k_1\dots k_p}(\xi)$$

for *contravariant* components, and

$$\widetilde{T}_{i_1\dots i_p}(\zeta) = a^{i_1}{}_{k_1}(\zeta)\dots a^{i_p}{}_{k_p}(\zeta)\,T_{k_1\dots k_p}(\zeta)\,,\ \ x=\Phi_1(\xi)=\Phi_2(\zeta) \qquad (10.28)$$

for *covariant* components.

Example 10.22. On a chart $(\mathcal{U},\Psi)$, we have for the metric tensor

$$\begin{aligned}M_g(\xi) &= g_{ij}(\xi)\,\underline{g}^i(\xi)\otimes\underline{g}^j(\xi) = \underline{g}_j(\xi)\otimes\underline{g}^j(\xi) = \underline{g}^j(\xi)\otimes\underline{g}_j(\xi)\\ &= g^{ij}(\xi)\,\underline{g}_i(\xi)\otimes\underline{g}_j(\xi) = \delta^i{}_j\underline{g}_i(\xi)\otimes\underline{g}^j(\xi) = \boldsymbol{\delta}\end{aligned}$$

(unit tensor on the chart $(\mathcal{U},\Psi)$), although the components are not constants in normal case! For instance, for $\underline{v}\in\mathcal{V}$,

$$\mathsf{M}_g\underline{v} = \underline{g}_j\langle\underline{g}^j,\underline{v}\rangle = \underline{g}_j(\underline{g}^j\cdot\underline{v}) = \underline{g}_j v^j = \underline{v}\,,\quad \mathsf{M}_g(\underline{u},\underline{v}) = \underline{u}\cdot\underline{v}\,.$$

Example 10.23. If $\underline{e}_1,\ldots,\underline{e}_n$ is a NOS in $\mathcal{V}$, then the following representation holds for the *tensors* $\widehat{\mathsf{E}}$ *of volume unit* of § 10.2**(a)**

$$\widehat{\mathsf{E}}(x) = \underline{e}_1\wedge\cdots\wedge\underline{e}_n = \varepsilon_{i_1\ldots i_n}\,\underline{e}_{i_1}\otimes\cdots\otimes\underline{e}_{i_n}\,.$$

The appertaining tensor field $\mathcal{M}\ni x\mapsto\widehat{\mathsf{E}}(x) = \widehat{\mathsf{E}}$ is obviously constant. It follows by the theory of determinants that this tensor has, on a chart $(\mathcal{U},\Psi)$, the form

$$\widehat{\mathsf{E}} = E_{i_1\ldots i_n}(\xi)\,\underline{g}^{i_1}(\xi)\otimes\cdots\otimes\underline{g}^{i_n}(\xi) = E^{i_1\ldots i_n}(\xi)\,\underline{g}_{i_1}(\xi)\otimes\cdots\otimes\underline{g}_{i_n}(\xi)$$

$(\det(A\cdot B) = \det(A)\cdot\det(B))$ where

$$E_{i_1\ldots i_n}(\xi) = e\cdot g(\xi)^{1/2}\varepsilon_{i_1\ldots i_n}\,,\quad E^{i_1\ldots i_n}(\xi) = e\cdot g(\xi)^{-1/2}\varepsilon^{i_1\ldots i_n}\,.$$

Here, $e=1$ in positive orientation of $\underline{g}_i$ relative to $\{\underline{e}_i\}$, and $e=-1$ else, and $g(\xi) = \det([g_{ij}(\xi)])$. For conversion of the components into contravariant components, as always, $E^{i_1\cdots i_n}(\xi) = g^{i_1j_1}(\xi)\cdots g^{i_nj_n}(\xi)E_{j_1\cdots j_n}(\xi)$.

(c4) Tangents of a Curve on the MF $\mathcal{M}$ Let $\mathcal{I}=[a,b]$ be an interval and Ξ a curve in $\mathcal{M}$, $\Xi:\mathcal{I}\ni t\mapsto z(t)\in\mathcal{M}$. Let $t_0\in(a,b)$ and $(\mathcal{U},\Psi)$ be a chart on $\mathcal{M}$ with $\Phi=\Psi^{-1}$ and $z_0=z(t_0)\in\mathcal{U}$. Then

$$\frac{d}{dt}z(t_0) = \lim_{\tau\to 0}\frac{z(t_0+\tau)-z(t_0)}{\tau}\in\mathcal{V} \tag{10.29}$$

is the tangential vector in z_0 (if $x,\ y\in\mathcal{M}$ then $y-x\in\mathcal{V}$). We have $\Psi(z(t)) =: \zeta(t)\in\mathbb{R}^n$, $z(t)=\Phi(\zeta(t))\in\mathcal{U}\subset\mathcal{M}$, and $\mathcal{I}\ni t\mapsto\zeta(t)$ is the curve in $\mathbb{R}^n$ being "pulled back into parameter space",

$$\frac{d}{dt}z(t) = \partial_i\Phi(\zeta(t))\,\frac{d\zeta^i}{dt}(t) = \underline{g}_i(\zeta(t))v^i(t)\,,\quad v^i(t) = \frac{d}{dt}\zeta^i(t)\in\mathbb{R}\,,\quad i=1:n\,. \tag{10.30}$$

The components of the tangential vector relative to the covariant basis are the components of the tangential vector of the curve pulled back into parameter space.

As a direct conclusion from (10.29) we obtain

$$\Psi(z(t+\Delta t)) = (\zeta^1(t) + v^1\Delta t + o(\Delta t), \ldots, \zeta^n(t) + v^n\Delta t + o(\Delta t))$$

where $v^i = d\zeta^i(t)/dt$.

(c5) Gradient of a Scalar Function The gradient of a scalar function $\widehat{f}$ is defined by

$$\langle \widehat{\mathrm{grad}} \widehat{f}(x), \underline{v}\rangle = \widehat{\mathrm{grad}} \widehat{f}(x) \cdot \underline{v} = \frac{d}{d\tau} f(x + \tau\underline{v})\big|_{\tau=0}.$$

On a chart $(\mathcal{U}, \Psi)$, it follows by the Chain Rule, using $\underline{v} = \underline{g}_i(\xi)v^i$,

$$\begin{aligned}\frac{d}{d\tau}\widehat{f}(x+\tau\underline{v})|_{\tau=0} &= \frac{d}{d\tau}(\widehat{f}\circ\Phi\circ\Psi)(x+\tau\underline{v})\big|_{\tau=0} \\ &= \partial_i(\widehat{f}\circ\Phi)(\Psi(x))\, \widehat{\mathrm{grad}}\psi^i(x)\cdot\underline{v} \\ &= \partial_i(\widehat{f}\circ\Phi)(\Psi(x))\, \underline{g}^i(\xi)\cdot\underline{g}_j(\xi)v^j = \partial_i f(\xi)v^i,\end{aligned}$$

or

$$\boxed{\widehat{\mathrm{grad}}\widehat{f}(x) = \partial_i f(\xi)\, \underline{g}^i(\xi) = \partial_i f(\xi)\, g^{ik}(\xi)\, \underline{g}_k(\xi)}\,. \tag{10.31}$$

(d) Christoffel Symbols Let $(\mathcal{U}, \Psi)$ be a chart on the MF $\mathcal{M}$ and $\Phi = \Psi^{-1}$ again. One writes for brevity

$$\Gamma_{ij}^k(\xi) := \underline{g}^k(\xi)\partial_j\underline{g}_i(\xi) = \underline{g}^k(\xi)\frac{\partial^2\Phi}{\partial\xi^i\partial\xi^j}(\xi) \in \mathbb{R}. \tag{10.32}$$

The *mappings* $\Gamma_{ij}^k : \xi \mapsto \Gamma_{ij}^k(\xi) \in \mathbb{R}$ are called CHRISTOFFEL *symbols* (of second order). One finds by partial derivation of $\underline{g}^k\cdot\underline{g}_i = 0$ that

$$\partial_j\underline{g}^k\cdot\underline{g}_i + \underline{g}^k\cdot\partial_j\underline{g}_i = \partial_j\underline{g}^k\cdot\underline{g}_i + \Gamma_{ij}^k = 0\,.$$

By multiplication from left by $\underline{g}^i(\xi)$, resp. in (10.31) from right by $\underline{g}_k(\xi)$, we obtain

$$\boxed{\partial_j\underline{g}_i(\xi) = \underline{g}_k(\xi)\Gamma_{ij}^k(\xi)\,,\quad \partial_j\underline{g}^k(\xi) = -\Gamma_{ij}^k(\xi)\underline{g}^i(\xi)} \tag{10.33}$$

where permutation of the components is allowed as already mentioned. The CHRISTOFFEL symbols relate always to an individual chart. They are symmetric in indices i and j but they are not tensors but scalar functions. We obtain by partial derivation of $g_{ij} = \underline{g}_i\cdot\underline{g}_j$ w.r.t. ξ_k that

$$\partial_k g_{ij} = \partial_k\underline{g}_i\cdot\underline{g}_j + \underline{g}_i\cdot\partial_k\underline{g}_j = \Gamma_{ik}^l\underline{g}_l\cdot\underline{g}_j + \underline{g}_i\cdot\Gamma_{jk}^l\underline{g}_l = \Gamma_{ik}^l g_{lj} + \Gamma_{jk}^l g_{il}\,.$$

By utilizing the symmetry properties of the CHRISTOFFEL symbols, we derive from this relations the formula

$$\boxed{\Gamma_{ij}^k = \frac{1}{2} g^{kl}(\partial_j g_{il} + \partial_i g_{jl} + \partial_l g_{ij})}\,,$$

which allows to compute the CHRISTOFFEL symbols by means of the metric tensor.

The metric tensor is a diagonal matrix in *orthogonal* coordinate systems by (10.9), and $M_g M_g{}^d = I$ implies $g^{ii} = (g_{ii})^{-1}$ (no summation). In this situation we thus obtain by (10.31) (without summation)

$$\begin{aligned} \Gamma_{ij}^k &= 0\,, \qquad i \neq j\,,\, i \neq k\,,\, j \neq k\,, & \Gamma_{ii}^j &= -\frac{\partial_j g_{ii}}{2g_{jj}}\,,\, i \neq j\,,\\ \Gamma_{ij}^i = \Gamma_{ji}^i &= -\frac{\partial_j g_{ii}}{2g_{ii}}\,,\, i \neq j\,, & \Gamma_{ii}^i &= \frac{\partial_i g_{ii}}{2g_{ii}}\,. \end{aligned} \tag{10.34}$$

In a orthogonal and normed i.e. cartesian coordinate system, the metric tensor is the unit matrix and all CHRISTOFFEL symbols are zero by consequence. A MATLAB program for the computation of CHRISTOFFEL symbols in spherical coordinates is offered in KAPITEL10.

(e) Divergence of Gradient of a Scalar Field By **(c5)**, we have $\widehat{\text{grad}}\,\widehat{f}(x) = \partial_i f(\xi)\underline{g}^i(\xi)$ and hence

$$\begin{aligned} \widehat{\text{grad}}\,\widehat{\text{grad}}\,\widehat{f}(x) &= \partial_j(\partial_i f(\xi)\underline{g}^i(\xi)) \otimes \underline{g}^j = [\partial_i\partial_j f\underline{g}^i + \partial_i f \partial_j \underline{g}^i] \otimes \underline{g}^j\\ &= [\partial_i\partial_j f\underline{g}^i - \partial_i f\Gamma_{kj}^i \underline{g}^k] \otimes \underline{g}^j = [\partial_i\partial_j f - \partial_k f\Gamma_{ij}^k]\underline{g}^i \otimes \underline{g}^j\\ &= [\partial_i\partial_j f - \partial_k f\Gamma_{ij}^k] g^{ij} \underline{g}_i \otimes \underline{g}_j\,. \end{aligned}$$

Consequently,

$$\widehat{\text{div}}\,\widehat{\text{grad}}\,\widehat{f}(x) = C_{1,2}\widehat{\text{grad}}\,\widehat{\text{grad}}\,\widehat{f}(x) = [\partial_i\partial_j f - \partial_k f\Gamma_{ij}^k](\xi) g^{ij}(\xi)$$

where $C_{1,2}$ is the contraction operator. On the other side, partial derivation of $g^{ij} = \underline{g}^i \cdot \underline{g}^j$ w.r.t. ξ^k yields

$$\partial_k g^{ij} = \partial_k \underline{g}^i \cdot \underline{g}^j + \partial_k \underline{g}^j \cdot \underline{g}^i = -\Gamma_{lk}^i \underline{g}^l \cdot \underline{g}^j - \Gamma_{lk}^j \underline{g}^l \cdot \underline{g}^i = -\Gamma_{lk}^i g^{lj} - \Gamma_{lk}^j g^{li}\,.$$

Using this result and $g = \det[g_{ij}]^{1/2}$, one computes

$$\begin{aligned} g^{-1}\partial_i(g\, g^{ij}\partial_j f) &= g^{-1} g^{ij}\partial_j f \partial_i g + \partial_j f \partial_i g^{ij} + g^{ij}\partial_i\partial_j f\\ &= g^{ij}\partial_i\partial_j f + g^{ij}\partial_j f\Gamma_{ik}^k - \partial_j f[\Gamma_{li}^i g^{lj} + \Gamma_{li}^j g^{li}]\\ &= g^{ij}\partial_i\partial_j f + g^{ij}\partial_j f\Gamma_{ik}^k - \partial_j f[\Gamma_{ik}^k g^{ij} + \Gamma_{li}^j g^{li}]\\ &= g^{ij}\partial_i\partial_j f - g^{li}\Gamma_{li}^j \partial_j f\,. \end{aligned}$$

Therefore we obtain the relation

$$\boxed{\widehat{\Delta}\,\widehat{f}(x) = \left(\frac{1}{g}\partial_i\left(g\, g^{ij}\partial_j f\right)\right)(\xi)}\,.$$

In the same way, one obtains for vector fields, using (10.25),

$$\widehat{\Delta}\widehat{\underline{v}}(x) = \left(\frac{1}{g}\partial_i\left(g\, g^{ij}\partial_j \underline{v}\right)\right)(\xi)\,.$$

(f) The **Gradient of a Tensor** is computed in the same way as in **(c5)**. For example, let $\mathcal{M} \ni x \mapsto \widehat{\mathsf{T}}(x) \in \mathcal{T}_p(\mathcal{V})$ be a *covariant* tensor field (with *covariant* components). The gradient of $\widehat{\mathsf{T}}$ at the point $x \in \mathcal{M}$ is the derivation of the tensor $\widehat{\mathsf{T}}(x)$ as multilinear mapping,

$$\widehat{\mathsf{T}}(x+\underline{v}) = \widehat{\mathsf{T}}(x) + (\widehat{\operatorname{grad}}\widehat{\mathsf{T}}(x))\underline{v} + r(x, |\underline{v}|)\,,\ \ \underline{v} \in \mathcal{V}\,,\ \ \lim_{\underline{v}\to 0}\frac{r(x,|\underline{v}|)}{|\underline{v}|} = 0\,.$$

Because $\widehat{\mathsf{T}}(x) \in \mathcal{L}(\mathcal{V}^p;\mathbb{R})$ also $(\widehat{\operatorname{grad}}\widehat{\mathsf{T}}(x))\underline{v} \in \mathcal{L}(\mathcal{V}^p;\mathbb{R})$ hence

$$\widehat{\operatorname{grad}}\widehat{\mathsf{T}}(x) \in \mathcal{L}(\mathcal{V};\mathcal{L}(\mathcal{V}^p;\mathbb{R})) = \mathcal{L}(\mathcal{V}^{p+1};\mathbb{R})\,. \tag{10.35}$$

(More genuinely one should write $(\widehat{\operatorname{grad}}\widehat{\mathsf{T}})(x)$ instead $\widehat{\operatorname{grad}}\widehat{\mathsf{T}}(x)$.) By definition

$$(\widehat{\operatorname{grad}}\widehat{\mathsf{T}}(x))\underline{v} = \frac{d}{d\tau}\widehat{\mathsf{T}}(x+\tau\underline{v})\big|_{\tau=0}\,.$$

On an individual chart $(\mathcal{U},\Psi)$ there follows by (10.25)

$$\frac{d}{d\tau}\widehat{\mathsf{T}}(x+\tau\underline{v})\big|_{\tau=0} = \frac{d}{d\tau}(\widehat{\mathsf{T}}\circ\Phi\circ\Psi)(x+\tau\underline{v})|_{\tau=0} = \partial_j\mathsf{T}(\xi)\,\widehat{\operatorname{grad}}\psi^j(x)\cdot\underline{v}\,,$$

and

$$\boxed{\widehat{\operatorname{grad}}\widehat{\mathsf{T}}(x) = \partial_j\mathsf{T}(\xi)\otimes\underline{g}^j(\xi)\,,\ \ x = \Phi(\xi)} \tag{10.36}$$

must hold because (10.35). Customarily, one chooses here the contravariant representation of $\widehat{T}(x)$. Then we have on a chart $(\mathcal{U},\Psi)$

$$\mathsf{T}(\xi) = T^{i_1\ldots i_p}(\xi)\,\underline{g}_{i_1}(\xi)\otimes\cdots\otimes\underline{g}_{i_p}(\xi)\,. \tag{10.37}$$

Note that $\xi \mapsto \mathsf{T}(\xi)$ is a mapping $\mathbb{R}^n \supset \Psi(\mathcal{U}) \to \mathbb{R}$ for which the partial derivatives are defined in classical sense. Using (10.36) and (10.37) we then obtain the following representation for the tensor $\widehat{\operatorname{grad}}\,\widehat{\mathsf{T}}(x) \in \mathcal{T}_{p+1}(\mathcal{V})$,

$$\widehat{\operatorname{grad}}\widehat{\mathsf{T}}(x) = A(\xi)^{i_1\ldots i_p}{}_{|j}\,\underline{g}_{i_1}(\xi)\otimes\cdots\otimes\underline{g}_{i_p}(\xi)\otimes\underline{g}^j(\xi)\,,\ \ x = \Phi(\xi)\,, \tag{10.38}$$

where $A(\xi)^{i_1\ldots i_p}{}_{|j}$ shall be at first simply the components of $\widehat{\operatorname{grad}}\,\widehat{T}(x)$ in sense of (10.25), i.e. $\partial_j\mathsf{T}(\xi) = A(\xi)^{i_1\ldots i_p}{}_{|j}\,\underline{g}_{i_1}(\xi)\otimes\cdots\otimes\underline{g}_{i_p}(\xi)$. Let now $\widehat{\mathsf{T}}(x)$ be a tensor in contravariant representation, $\widehat{\mathsf{T}}(x) = T^{i_1\ldots i_p}(\xi)\,\underline{g}_{i_1}(\xi)\otimes\cdots\otimes\underline{g}_{i_p}(\xi)$ then, by the product rule,

$$\begin{aligned}\partial_j \mathsf{T}(\xi) &= \partial_j T^{i_1\dots i_p}(\xi)\, \underline{g}_{i_1}(\xi) \otimes \dots \otimes \underline{g}_{i_p}(\xi) \\ &\quad + T^{i_1\dots i_p}(\xi) \left[\partial_j \underline{g}_{i_1}(\xi) \otimes \dots \otimes \underline{g}_{i_p}(\xi) + \ldots + \underline{g}_{i_1}(\xi) \otimes \dots \otimes \partial_j \underline{g}_{i_p}(\xi)\right]\end{aligned} \tag{10.39}$$

and, because $\partial_j \underline{g}_i(\xi) = \Gamma^k_{ij}(\xi)\, \underline{g}_k(\xi)$, we obtain, e.g., with argument ξ,

$$\begin{aligned}T^{i_1\dots i_p} \partial_j \underline{g}_{i_1} \otimes \underline{g}_{i_2} \otimes \dots \otimes \underline{g}_{i_p} &= T^{i_1\dots i_p} \Gamma^k_{i_1 j}\, \underline{g}_k \otimes \underline{g}_{i_2} \otimes \dots \otimes \underline{g}_{i_p} \\ &= T^{k\, i_2\dots i_p} \Gamma^{i_1}_{kj}\, \underline{g}_{i_1} \otimes \underline{g}_{i_2} \otimes \dots \otimes \underline{g}_{i_p}\,.\end{aligned}$$

Altogether, it follows by (10.39)

$$\boxed{A(\xi)^{i_1\dots i_p}{}_{|j} = \partial_j T^{i_1\dots i_p} + T^{k\, i_2\dots i_p} \Gamma^{i_1}_{kj} + \ldots + T^{i_1\dots i_{p-1} k} \Gamma^{i_p}_{kj}} \tag{10.40}$$

for the components of $\widehat{\mathrm{grad}}\, \widehat{T}(x)$ where the argument is dropped.

Example 10.24. A vector field $\widehat{\underline{v}} : \mathcal{M} \ni x \mapsto \widehat{\underline{v}}(x) \in \mathcal{V}$ has the contravariant representation $\widehat{\underline{v}}(x) = \underline{g}_i(\xi) v^i(\xi)$ on a chart $(\mathcal{U}, \Psi)$ by (10.25). One obtains by (10.39) for the gradient

$$\begin{aligned}\widehat{\mathrm{grad}}\, \widehat{\underline{v}}(x) &= \left[\partial_j v^i(\xi) \underline{g}_i(\xi) + v^i(\xi) \partial_j \underline{g}_i(\xi)\right] \otimes \underline{g}^j(\xi) \\ &= \left[\partial_j v^i(\xi) \underline{g}_i(\xi) + v^i(\xi) \Gamma^k_{ij}(\xi) \underline{g}_k(\xi)\right] \otimes \underline{g}^j(\xi) \\ &= v^i{}_{\|j}(\xi)\, \underline{g}_i(\xi) \otimes \underline{g}^j(\xi) \\ v^i{}_{\|j}(\xi) &:= \partial_j v^i(\xi) + v^k(\xi) \Gamma^i_{kj}(\xi)\,,\end{aligned}$$

in the same way, for $\widehat{\underline{v}}(x) = v_i(\xi) \underline{g}^i(\xi)$ in covariant representation,

$$\begin{aligned}\widehat{\mathrm{grad}}\, \widehat{\underline{v}}(x) &= v_{i\|j}(\xi)\, \underline{g}^i(\xi) \otimes \underline{g}_j(\xi) \\ v_{i\|j}(\xi) &:= \partial_j v_i(\xi) - v_k(\xi) \Gamma^k_{ij}(\xi)\,.\end{aligned}$$

Example 10.25. The gradient of a 2-tensor $\widehat{\mathsf{T}}(x) = T^{rs}(\xi) \underline{g}_r(\xi) \otimes \underline{g}_s(\xi)$, is the 3-tensor

$$\begin{aligned}\widehat{\mathrm{grad}}\, \widehat{\mathsf{T}}(x) &= \left(\partial_j A^{rs} \underline{g}_r \otimes \underline{g}_s \otimes g^j\right)(\xi) \\ &= \left([\partial_j T^{rs} + T^{ks} \Gamma^r_{kj} + T^{rk} \Gamma^s_{kj}] \underline{g}_r \otimes \underline{g}_s \otimes \underline{g}^j\right)(\xi)\,.\end{aligned}$$

Example 10.26. The metric tensor ${\mathsf{M}_g}^d = g^{ij} \underline{g}_i \otimes \underline{g}_j$ is as well as M_g in *Example* 10.22 a constant tensor, ${\mathsf{M}_g}^d\, \underline{v} = \underline{v}$ because, e.g. for $\underline{v} = \underline{g}^k v_k$,

$${\mathsf{M}_g}^d \underline{v} = g^{ij} \underline{g}_i \langle \underline{g}^k v_k, \underline{g}_j \rangle = g^{ij} \underline{g}_i v_j = v^i \underline{g}_i = \underline{v}\,.$$

Thus $\mathrm{grad}\, {\mathsf{M}_g}^d = O$-tensor must hold by definition of the gradient, i.e., $g^{ij}{}_{|k} = 0$. On the other side, one computes by (10.39) for the gradient of the metric tensor

$$\operatorname{grad}\mathsf{M}_g{}^d = (\operatorname{grad}\mathsf{M}_g{}^d)^{i_1 i_2}{}_j \underline{g}_{i_1} \otimes \underline{g}_{i_2} \otimes \underline{g}^j$$

the components

$$\begin{aligned}(\operatorname{grad}\mathsf{M}_g{}^d)^{i_1 i_2}{}_j &= \partial_j g^{i_1 i_2} + g^{k i_2}\Gamma^{i_1}_{kj} + g^{i_1 k}\Gamma^{i_2}_{kj}\\ &= \partial_j g^{i_1}\underline{g}^{i_2} + \partial_j g^{i_2}\underline{g}^{i_1} + g^{k i_2}\Gamma^{i_1}_{kj} + g^{i_1 k}\Gamma^{i_2}_{kj} = 0\,.\end{aligned}$$

Of course, we have likewise $(\operatorname{grad}\mathsf{M}_g)_{ij,k} = 0$ for the components of the tensor $\operatorname{grad}\mathsf{M}_g(x)$.

Example 10.27. The tensor $\widehat{\mathsf{E}}$ of volume unit, cf. *Example* 10.23, $g = \det[g_{ij}]^{1/2}$,

$$\widehat{\mathsf{E}}(x) = E_{i_1\ldots i_n}(\xi)\,\underline{g}^{i_1}(\xi)\otimes\cdots\otimes\underline{g}^{i_n}(\xi) = E^{i_1\ldots i_n}(\xi)\,\underline{g}_{i_1}(\xi)\otimes\cdots\otimes\underline{g}_{i_n}(\xi)$$

$$\begin{aligned}&E_{i_1\ldots i_n}(\xi) = e\cdot g(\xi)\varepsilon_{i_1\ldots i_n}\,,\quad E^{i_1\ldots i_n}(\xi) = e\cdot g(\xi)^{-1}\varepsilon^{i_1\ldots i_n}\\ &E^{i_1\ldots i_n}(\xi) = g^{i_1 j_1}(\xi)\ldots g^{i_n j_n}(\xi)E_{j_1\ldots j_n}(\xi)\end{aligned}$$

is likewise a constant tensor. We have

$$O = e\,E^{i_1\ldots i_n}{}_j = \varepsilon^{i_1\ldots i_n}\partial_j g - g\,\varepsilon^{k i_2\ldots i_n}\Gamma^{i_1}_{kj} - g\,\varepsilon^{i_1\ldots i_{n-1}k}\Gamma^{i_n}_{kj}\,.$$

If for instance $\varepsilon^{i_1\ldots i_n} = 1$ then $\varepsilon^{k i_2\ldots i_n} = 0$ for $k \neq i_1$ by definition because two indices appear double. Thus $\partial_j g = g\,\Gamma^{i_1}_{i_1 j} + \ldots + g\,\Gamma^{i_n}_{i_n j}$ without summation over the indices i_k or

$$\boxed{\frac{\partial_j g}{g} = \Gamma^k_{jk}\,,\quad g = \det[g_{ij}]^{1/2}} \tag{10.41}$$

with summation over k.

(g) Divergence of a Tensor Field The divergence of a tensor field $\widehat{\mathsf{T}} : \mathcal{M} \ni x \mapsto \widehat{\mathsf{T}}(x) \in \mathcal{T}_p(\mathcal{V})$ is the contraction of the gradient w.r.t. the last component, cf. § 10.1**(j)**, $\widehat{\operatorname{div}}\,\widehat{\mathsf{T}}(x) = C_{p,p+1}(\widehat{\operatorname{grad}}\,\widehat{\mathsf{T}})$, hence

$$\boxed{\begin{aligned}\widehat{\operatorname{div}}\,\widehat{\mathsf{T}}(x) &=: A^{i_1\ldots i_{p-1}l}{}_{|l}(\xi)\underline{g}_{i_1}(\xi)\otimes\cdots\otimes\underline{g}_{i_{p-1}}(\xi)\\ A^{i_1\ldots i_{p-1}l}{}_{|l} &= \partial_l T^{i_1\ldots i_{p-1}l}\\ &\quad + T^{k\,i_2\ldots i_{p-1}l}\Gamma^{i_1}_{kl} + \ldots + T^{i_1\ldots i_{p-2}kl}\Gamma^{i_{p-1}}_{kl}\\ &\quad + T^{i_1\ldots i_{p-1}k}\Gamma^l_{kl}\end{aligned}}\,. \tag{10.42}$$

Example 10.28. We have $< g^j(\xi), v^i(\xi)\Gamma^k_{ij}(\xi)g_k(\xi) > = v^i(\xi)\Gamma^j_{ij}(\xi)$, hence it follows in particular for the divergence of a vector field $\widehat{\underline{v}}$ that

$$\widehat{\operatorname{div}}\,\widehat{\underline{v}}(x) = \partial_i v^i(\xi) + v^i(\xi)\Gamma^j_{ij}(\xi)\,,\quad x = \Phi(\xi)\,.$$

Using (10.41), we obtain from this relation that

$$\boxed{\widehat{\operatorname{div}}\, \widehat{\underline{v}}(x) = \left(\frac{1}{g}\partial_i(g\, v^i)\right)(\xi)}\,.$$

Example 10.29. The LAPLACE operator of a tensor field is the divergence of the gradient,

$$\boxed{\widehat{\Delta}\widehat{\mathsf{T}}(x) := \widehat{\operatorname{div}}\, \widehat{\operatorname{grad}}\, \widehat{\mathsf{T}}(x) = C_{p+1,p+2}(\widehat{\operatorname{grad}}\, \widehat{\operatorname{grad}}\, \widehat{\mathsf{T}})(x)}$$

where $C_{p+1,p+2}$ is the contraction operator of a tensor $\widehat{\mathsf{T}}(x) = T^{i_1 \dots i_p}(\xi)\, g_{i_1}(\xi) \otimes \cdots \otimes g_{i_p}(\xi)$ of order p. We obtain as above

$$\widehat{\Delta}\widehat{\mathsf{T}}(x) = \left(g^{kl}\, T^{i_1 \dots i_p}{}_{|kl}\, g_{i_1} \otimes \cdots \otimes g_{i_p}\right)(\xi) \text{ contravariant tensor}$$
$$\widehat{\Delta}\widehat{\mathsf{T}}(x) = \left(g_{kl}\, T_{i_1 \dots i_p|kl}\, g^{i_1} \otimes \cdots \otimes g^{i_p}\right)(\xi) \text{ covariant tensor}\,.$$

Observe however that the formula for the LAPLACE operator is frequently displayed for physical components and not as here for natural components.

(h) Rotation of a Vector Field In accordance with the formula in $\mathbb{R}^3$,

$$\operatorname{rot} \underline{w}(x) \times (y - x) = [\nabla \underline{w}(x) - \nabla \underline{w}(x)^T](y - x)\,, \tag{10.43}$$

the rotation of $\widehat{\underline{v}}$ may be defined as skew-symmetric part of the gradient hence, with the projection operator $P_p : \mathcal{T}_p(\mathcal{V}) \to \mathcal{A}_p(\mathcal{V})$ onto the alterating part, $\widehat{\operatorname{rot}}\, \widehat{\underline{v}}(x) = P_2(\widehat{\operatorname{grad}}\, \widehat{\underline{v}})(x)$ or

$$\widehat{\operatorname{rot}}\, \widehat{\underline{v}}(x) = \left(\frac{1}{2}(v_{i\|j} - v_{j\|i})\underline{g}_i \otimes \underline{g}^j\right)(\xi)\,. \tag{10.44}$$

On an oriented three-dimensional MF we then obtain

$$\widehat{\operatorname{rot}}\, \widehat{\underline{v}}(x) = -E^{ijk}(\xi) v_{j\|k}(\xi) \underline{g}_i(\xi)$$

with identification by (10.43); cf. *Example* 10.27. This relation (10.44) can be generalized to arbitrary *alternating tensors* by means of differential forms; cf. (Bowen).

References: (Barner), (Bowen), (Ciarlet00), (Flanders), (Grauert).

10.5 Examples

Without demand for completeness, this section contains a collection of curvilinear orthogonal coordinate systems in a EUCLIDian vector space $\mathbf{E}^3$ with fixed cartesian coordinate system $\{\mathcal{O}; \underline{e}_1, \underline{e}_2, \underline{e}_3\}$.

(a) Brief Recapitulation We identify the elements of this space with their coordinate vectors $\mathbf{E}^3 \ni x = \underline{e}_i x^i \simeq [x^1, x^2, x^3]^T = x \in \mathbb{R}^3$. If again $x = \Phi(\xi)$, $\xi = [\xi^1, \xi^2, \xi^3]^T$, then the columns of $\operatorname{grad}\Phi(\xi) = [\underline{g}_1(\xi), \underline{g}_2(\xi), \underline{g}_3(\xi)]$ form the natural covariant basis at point $x = \Phi(\xi) \in \mathbb{R}^3$ resp. $\mathbf{E}^3$. We consider manifolds of which the metric tensor $M(\xi) = [g_{ij}(\xi)] = [\operatorname{grad}\Phi(\xi)]^T \operatorname{grad}\Phi(\xi)$ is a **diagonal matrix**. If now $\Psi = \Phi^{-1}$ is the inversion then, by the chain rule, $\operatorname{grad}\Psi(x)\operatorname{grad}\Phi(\xi) = I$, $x = \Phi(\xi)$. In case we make *no* difference between row and column vectors, the columns of $\operatorname{grad}\Phi(\xi)$ form a basis and the rows of $\operatorname{grad}\Psi(x)$, $[\underline{g}^1(\xi), \underline{g}^2(\xi), \underline{g}^3(\xi)] := [\operatorname{grad}\Psi(x)]^T$, $\xi = \Psi(x)$, form a basis of $\mathbb{R}^3$ at the point x as well and

$$\underline{g}^i(x) \cdot \underline{g}_j(\xi) = \delta^i{}_j \text{ (KRONECKER symbol)}.$$

Therefore $\{\underline{g}^i\}$ is called dual basis relative to $\{\underline{g}_j\}$; reciprocal and dual basis coincide by canonical identification of rows and columns in $\mathbb{R}^n$.

Let again

$$\begin{aligned} g_i(\xi) &:= m_{ii}(\xi)^{1/2} = [\underline{g}_i(\xi) \cdot \underline{g}_i(\xi)]^{1/2}, \quad i = 1:3, \\ g(\xi) &= \det(M(\xi))^{1/2} = g_1(\xi) \cdot g_2(\xi) \cdot g_3(\xi). \end{aligned}$$

The columns

$$\begin{aligned} [\widetilde{\underline{g}}_1(\xi), \widetilde{\underline{g}}_2(\xi), \widetilde{\underline{g}}_3(\xi)] &:= [\operatorname{grad}\Phi(\xi)] M(\xi)^{-1/2} \\ [\widetilde{\underline{g}}^1(\xi), \widetilde{\underline{g}}^2(\xi), \widetilde{\underline{g}}^3(\xi)] &:= [\operatorname{grad}\Psi(x)]^T M(\xi)^{1/2} \end{aligned} \tag{10.45}$$

form respectively a *normed* orthogonal basis in $\mathbb{R}^3$ resp. $\mathbf{E}^3$.

The representation $\underline{v} = \underline{g}_i(\xi) v^i = v_j g^j(\xi)$, holds for a vector $\underline{v} \in \mathbb{R}^3$ where it is to be summed up over double appearing indices. If the basis vectors are normed to EUCLIDian length one as in (10.45) then the components of $\underline{v}$ are called *physical* components. Any of these bases defines a particular metric (length) in point x, e.g.,

$$|\underline{v}|^2 = (\underline{g}_i(\xi) v^i) \cdot (\underline{g}_j(\xi) v^j) = [v^1, v^2, v^3] \nabla\Phi(\xi)^T \nabla\Phi(\xi) \begin{bmatrix} v^1 \\ v^2 \\ v^3 \end{bmatrix},$$

and the matrix (resp. tensor field) $M(\xi) = [g_{ij}(\xi)] = [\operatorname{grad}\Phi(\xi)]^T \operatorname{grad}\Phi(\xi)$ is called metric tensor in $x = \Phi(\xi)$ rightly.

(b) Orthogonal Natural Coordinate Systems appear if $\operatorname{grad}\Phi(\xi)$ is an orthogonal but not necessarily orthonormal matrix (there is mostly no difference made between these two notations). Then primal and dual basis coincide up to normalization. If *in addition* the columns of $\operatorname{grad}\Phi(\xi)$ are normed to EUCLIDian ubit length then the metric tensor becomes the unit matrix obviously and the components are again the *physical* components.

In the subsequent examples only the diagonal of the metric tensor M is specified because the other elements are zero. It shows how the natural basis has to be normed if necessary.

Example 10.30. Cylinder coordinates. $\xi = (r, \varphi, \zeta)$.
Transformation: $x = \left[r\cos\varphi,\ r\sin\varphi,\ \zeta\right]^T$, $0 < r$, $0 \le \varphi < 2\pi$, $\zeta \in \mathbb{R}$.

$$\operatorname{grad}\Phi(\xi) = \begin{bmatrix} \cos\varphi & -r\sin\varphi & 0 \\ \sin\varphi & r\cos\varphi & 0 \\ 0 & 0 & 1 \end{bmatrix}, \quad \operatorname{diag}(M(\xi)) = [1,\ r^2,\ 1]\,.$$

Example 10.31. Spherical Coordinates; cf. also the geographic version in *Example* 6.17. $\xi = (r, \vartheta, \varphi)$ (!).
Transformation : $x = [r\,\sin\vartheta\,\cos\varphi,\ r\,\sin\vartheta\,\sin\varphi,\ r\cos\vartheta]^T$, $0 < r$, $0 \le \varphi < 2\pi$, $0 < \vartheta < \pi$.

$$\operatorname{grad}\Phi(\xi) = \begin{bmatrix} \sin\vartheta\,\cos\varphi & r\cos\vartheta\,\cos\varphi & -r\sin\vartheta\,\sin\varphi \\ \sin\vartheta\,\sin\varphi & r\cos\vartheta\,\sin\varphi & r\sin\vartheta\,\cos\varphi \\ \cos\vartheta & -r\sin\vartheta & 0 \end{bmatrix},$$

$$\operatorname{diag}(M(\xi)) = [1,\ r^2,\ (r\sin\vartheta)^2]\,.$$

Origin as well as north- and south pole ($\vartheta = 0$, $\vartheta = \pi$) are to be excluded.

Example 10.32. Paraboloid coordinates. $\xi = (u, v, w)$, $(u, v) \neq (0, 0)$, $\le u$, $0 \le v$, $0 \le w < 2\pi$.
Transformation: $x = \left[u\,v\,\cos w,\ u\,v\,\sin w,\ (u^2 - v^2)/2\right]^T$,

$$\operatorname{grad}\Phi(\xi) = \begin{bmatrix} v\cos w & u\cos w & -u\,v\,\sin w \\ v\sin w & u\sin w & u\,v\,\cos w \\ u & -v & 0 \end{bmatrix},$$

$$\operatorname{diag}(M(\xi)) = [u^2 + v^2,\ u^2 + v^2,\ u^2 v^2]\,.$$

Example 10.33. Elliptic cylinder coordinates, $a > 0$, $\xi = (u, v, w)$.
Transformation: $x = \left[a\cosh u\cos v\,,\ a\sinh u\sin v\,,\ w\right]^T$, $0 < u$, $0 \le v < 2\pi$.

$$\operatorname{grad}\Phi(\xi) = \begin{bmatrix} a\sinh u\cos v & -a\cosh u\sin v & 0 \\ a\cosh u\sin v & a\sinh u\cos v & 0 \\ 0 & 0 & 1 \end{bmatrix},$$

$$\operatorname{diag}(M(\xi)) = [a^2(\sinh^2 u + \sin^2 v),\ a^2(\sinh^2 u + \sin^2 v),\ 1]\,.$$

Example 10.34. Ellipsoid coordinates. $a > 0$, $\xi = (u, v, w)$, $0 \le u$, $0 < v \le \pi$, $0 \le w < 2\pi$.
Transformation: $x = a\left[\cosh u\sin v\cos w,\ \cosh u\sin v\sin w,\ \sinh u\cos v\right]^T$,

$$\operatorname{grad}\Phi(\xi) = a\begin{bmatrix} \sinh u\sin v\cos w & \cosh u\cos v\cos w & -\cosh u\sin v\sin w \\ \sinh u\sin v\sin w & \cosh u\cos v\sin w & \cosh u\sin v\cos w \\ \cosh u\cos v & -\sinh u\sin v & 0 \end{bmatrix},$$

$$\begin{aligned}\operatorname{diag}(M(\xi)) = a^2[&\cosh^2 u\cos^2 v + \sinh^2 u\sin^2 v,\\ &\cosh^2 u\cos^2 v + \sinh^2 u\sin^2 v,\ \cosh^2 u\sin^2 v]\,.\end{aligned}$$

Example 10.35. Torus coordinates, $a > 0$, $\xi = (u, v, w)$.

$$x = a\left[\frac{\sinh u \cos w}{\cosh u - \cos v}, \frac{\sinh u \sin w}{\cosh u - \cos v}, \frac{\sin v}{\cosh u - \cos v}\right]^T,$$

$$G(\xi) :=$$
$$\begin{bmatrix} (1-\cosh u \cos v)\cos w & -\sinh u \sin v \cos w & -\sinh u(\cosh u - \cos v)\sin w \\ (1-\cosh u \cos v)\sin w & -\sinh u \sin v \sin w & \sinh u(\cosh u - \cos v)\cos w \\ -\sinh u \sin v & \cosh u \cos v - 1 & 0 \end{bmatrix},$$

$$\operatorname{grad}\Phi(\xi) = \frac{a}{(\cosh u - \cos v)^2}\, G(\xi),$$

$$\operatorname{diag}(M(\xi)) = \frac{a^2}{(\cosh u - \cos v)^2}\,[1,\ 1,\ \sinh^2 u].$$

Example 10.36. Bispherical coordinates, $a > 0$, $\xi = (u, v, w)$.

$$x = a\left[\frac{\sin v \cos w}{\cosh u - \cos v}, \frac{\sin v \sin w}{\cosh u - \cos v}, \frac{\sinh u}{\cosh u - \cos v}\right]^T,$$

$$G(\xi) :=$$
$$\begin{bmatrix} -\sinh u \sin v \cos w & (\cosh u \cos v - 1)\cos w & -(\cosh u - \cos v) * \sin v \sin w \\ -\sinh u \sin v \sin w & (\cosh u \cos v - 1)\sin w & \cosh u - \cos v)\sin v \cos w \\ 1 - \cosh u \cos v & -\sinh u \sin v & 0 \end{bmatrix}$$

$$\operatorname{grad}\Phi(\xi) = \frac{a}{(\cosh u - \cos v)^2}\, G(\xi),$$

$$\operatorname{diag}(M(\xi)) = \frac{a^2}{(\cosh u - \cos v)^2}[1,\ 1,\ \sin^2 v].$$

(c) In the subsequent relations it is always summed up *once* over the index i following EINSTEIN's convention, and the indices are to be computed *modulo 3*, e.g., $i+3$ is to be replaced by i. In this context, remember once more the notation $\widehat{f}(x) = f(\xi)$, $\partial_i = \partial\xi^i$, $\widehat{\partial}_i = \partial x^i$ etc..

(c1) Using (10.45), we have

$$\widehat{\operatorname{grad}}\,\widehat{f}(x) = \widehat{\partial}_i \widehat{f}(x)\underline{e}_i = \partial_i f(\xi)\underline{g}^i(\xi) = \frac{1}{g_i(\xi)}\partial_i f(\xi)\underline{\widetilde{g}}^i(\xi)$$

for a scalar function $\widehat{f} : \mathbf{E}^3 \ni x \mapsto \widehat{f}(x) \in \mathbb{R}$.

In the remaining examples, let $\underline{v} : \mathbf{E}^3 \ni x \mapsto \underline{v}(x) \in \mathbf{E}^3$ be a vector field with representation

$$\widehat{\underline{v}}(x) = \underline{g}_i(\xi)v^i(\xi) = \underline{\widetilde{g}}_i(\xi)g_i(\xi)v^i(\xi) \equiv \underline{\widetilde{g}}_i(\xi)\widetilde{v}^i(\xi),\quad x = \Phi(\xi)$$

in contravariant components $v^i(\xi)$ resp. in physical components $\widetilde{v}^i(\xi)$.

(c2)

$$\widehat{\operatorname{div}}\,\widehat{\underline{v}}(x) = \frac{1}{g(\xi)}\frac{\partial}{\partial \xi^i}[g(\xi)v^i(\xi)] := \frac{1}{g(\xi)}\frac{\partial}{\partial \xi^i}\left[\frac{g(\xi)}{g_i(\xi)}\,\widetilde{v}(\xi)\right].$$

(c3) By formal expansion w.r.t. the first row, we have

$$\widehat{\operatorname{rot}}\,\widehat{\underline{v}}(x) = \frac{1}{g(\xi)}\begin{vmatrix} g_1(\xi)\widetilde{\underline{g}}_1(\xi) & g_2(\xi)\widetilde{\underline{g}}_2(\xi) & g_3(\xi)\widetilde{\underline{g}}_3(\xi) \\ \partial/\partial\xi^1 & \partial/\partial\xi^2 & \partial/\partial\xi^3 \\ g_1(\xi)\widetilde{v}^1(\xi) & g_2(\xi)\widetilde{v}^2(\xi) & g_3(\xi)\widetilde{v}^3(\xi) \end{vmatrix}$$

$$\widehat{\operatorname{rot}}\,\widehat{\underline{v}}(x) = \frac{g_i(\xi)}{g(\xi)}\left[\frac{\partial(g_{i+2}(\xi)\widetilde{v}^{i+2}(\xi))}{\partial\xi_{i+1}} - \frac{\partial(g_{i+1}(\xi)\widetilde{v}^{i+1}(\xi))}{\partial\xi_{i+2}}\right]\widetilde{\underline{g}}_i(\xi)$$

with simple summation over $i \bmod 3$.

(c4) $\widehat{\Delta}\,\widehat{f}(x) = \widehat{\operatorname{div}}\,\widehat{\operatorname{grad}}\,\widehat{f}(x) = \dfrac{1}{g(\xi)}\left[\dfrac{\partial}{\partial\xi^i}\left(\dfrac{g(\xi)}{g_i(\xi)^2}\dfrac{\partial f(\xi)}{\partial\xi^i}\right)\right].$

(c5) $\widehat{\Delta}\,\widehat{\underline{v}}(x) = \widehat{\operatorname{grad}}\,\widehat{\operatorname{div}}\,\widehat{\underline{v}}(x) - \widehat{\operatorname{rot}}\,\widehat{\operatorname{rot}}\,\widehat{\underline{v}}(x)$

$$= \left(\frac{1}{g_i(\xi)}\frac{\partial\Gamma(\xi)}{\partial\xi^i} + \frac{g_i(\xi)}{g(\xi)}\left[\frac{\partial\Gamma_{i+1}(\xi)}{\partial\xi^{i+2}} - \frac{\partial\Gamma_{i+2}(\xi)}{\partial\xi^{i+1}}\right]\right)\widetilde{\underline{g}}_i(\xi);$$

where

$$\Gamma_i = \frac{g_i(\xi)}{g(\xi)}\begin{vmatrix} \partial/\partial\xi^{i+1} & \partial/\partial\xi^{i+2} \\ g_{i+1}(\xi)v_{i+1}(\xi) & g_{i+2}(\xi)v_{i+2}(\xi) \end{vmatrix}$$

$$\Gamma = \frac{1}{g(\xi)}\sum_{i=1}^{3}\frac{\partial}{\partial\xi^i}\left[\frac{g(\xi)v_i(\xi)}{g_i(\xi)}\right]$$

with simple summation and expansion w.r.t. the first row.

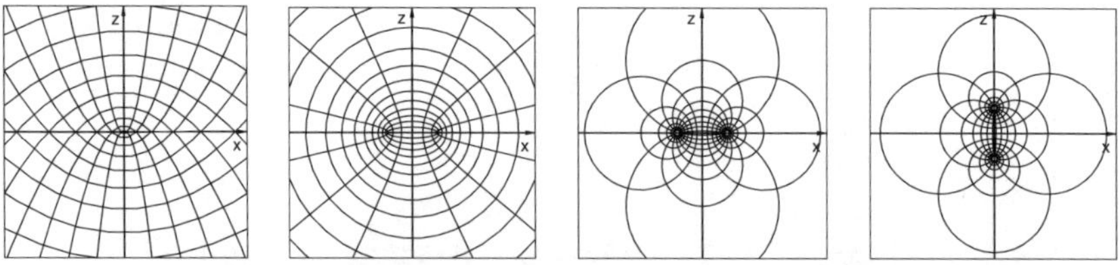

(a) (b) (c) (d)

Figure 10.2. Examples of natural coordinates

for $a = 1$, $w = 0$,
(a) paraboloid coordinates, (b) ellipsoid coordinates, (c) torus coordinates, (d) bispherical coordinates.

10.6 Transformation Groups

(a) Notations and Definitions Let $\mathcal{X}$ and $\mathcal{Y}$ be BANACH spaces. A set $\mathcal{G}$ with composition $\mathcal{G} \times \mathcal{G} \ni (a,b) \mapsto ab \in \mathcal{G}$ is called *group* if $\forall\ a,b,c \in \mathcal{G}$: $(ab)c = a(bc)$; $\exists\ e \in \mathcal{G}\ \ \forall\ a \in \mathcal{G} : ea = a$; $\forall\ a \in \mathcal{G}\ \ \exists\ a^{-1} \in \mathcal{G} : aa^{-1} = e$.

Notations:

(1°) Let $\mathcal{M}$ be an arbitrary set and $\mathcal{G}$ an arbitrary group then

$$\tau(\mathcal{M}) := \{f : \mathcal{M} \to \mathcal{M}\,,\ f \text{ bijective},\ f \cdot g := f \circ g\,,\ \mathbf{1} = \text{ identity}\}$$

is called *transformation group of* $\mathcal{M}$ and $f \in \tau(\mathcal{M})$ *transformation.*

(2°) Let $\pi : \mathcal{G} \to \tau(\mathcal{M})$ be an arbitrary group homomorphism then π is called *representation of* G *on* $\tau(\mathcal{M})$. The image (subgroup) $\pi(\mathcal{G}) \subset \tau(\mathcal{M})$ is called likewise representation of $\mathcal{G}$ as transformation group on $\tau(\mathcal{M})$; the representation is *exact* if π is an isomorphism.

(3°) The mapping $\varPhi : \mathcal{G} \times \mathcal{M} \ni (g,x) \mapsto \pi(g)(x) \in \mathcal{M}$ defined by π is called *action of the group* $\mathcal{G}$ *on* $\mathcal{M}$; it is not determined uniquely like π .

(4°) If $\mathcal{G} = \{\mathbb{R}\,,\ +\,\}$ and $\mathcal{M} = \mathcal{X}$ is a BANACH space then the representation $\pi : \mathbb{R} \to \tau(\mathcal{X})$ is called *locally exact* if $\exists\ \varepsilon > 0\ \forall\ t \in (-\varepsilon,\,\varepsilon) : \pi(t) = \mathrm{id}_X \iff t = 0$.

(5°) The mapping $\pi : \mathbb{R} \to \tau(x)$ (or $\pi(\mathbb{R})$) is called *one-parametric transformation group of* $\mathcal{X}$ if π is a locally exact representation of $\mathbb{R}$.

(6°) In analogeous way, $\pi : \mathbb{R}^r \to \tau(\mathcal{X})$ (or $\pi(\mathbb{R}^r)$) is called r-parametric transformation group of $\mathcal{X}$.

The *representation theory* has for subject the construction of representations $\pi(\mathcal{G}) \subset \mathcal{L}(\mathbb{R}^n, \mathbb{R}^n)$ for a group $\mathcal{G}$.

Addendum:

(1°) Note that $\forall\ g_1, g_2 \in G : \pi(g_1) \circ \pi(g_2) = \pi(g_1 \cdot g_2)\ \Big(= \pi(g_1 \circ g_2)\Big)$ because π is an homomorphism.

(2°) π exact $\iff (\pi(g) = \mathrm{id}_{\mathcal{M}} \iff g = \mathbf{1})$.

(3°) If $\pi : \mathbb{R} \to \mathcal{M}$ is a local exact representation of $\mathbb{R}$ then the actions

$$\varPhi(t,x) = \pi(t)(x) \tag{10.46}$$

satisfy

(1°) $\forall\ x \in \mathcal{X} : \varPhi(0,x) = x$

(2°) $\forall\ s,t \in \mathbb{R}\quad \forall\ x \in \mathcal{X} : \varPhi(t, \varPhi(s,x)) = \varPhi(s+t, x)$

(3°) $\exists\ \varepsilon > 0\quad \Big(\forall\ x \in \mathcal{X} : t \in (-\varepsilon,\,\varepsilon) \wedge \varPhi(t,x) = x\Big) \implies t = 0$.

(4°) Conversely, every mapping (action) $\varPhi$ with the properties (1°) – (3°) supplies by (10.46) a one-parametric transformation group $\mathcal{G}(\varPhi) := \varPhi(\mathbb{R}, \circ) \subset \tau(X)$; one writes $\mathcal{G}(\varPhi) \in C^k$, $0 \le k \le \infty$ if $\varPhi : \mathcal{X} \times \mathbb{R} \to \mathcal{X}$ is k-times continuously diffentiable; cf. Sect. 1.7.

(b) Examples Many groups $\mathcal{G}(\Phi)$ carry an own name:

Example 10.37. Translation group, notation $\mathcal{G}(\Phi) = T_t$.

$$\widetilde{x} = \Phi(t,x) := x + t\,x_0\,, \quad x \in \mathcal{X}\,, \quad t \in \mathbb{R} \quad 0 \neq x_0 \in X \text{ fixed.}$$

Example 10.38. Let $\mathcal{GL}(m) = \{A \in \mathbb{R}^m{}_m\,, \ \det(A) \neq 0\}$ be the group of invertible (m,m)-matrices (general linear) and $\mathcal{G}(\Phi) = \mathcal{GL}(m)$.

$$\widetilde{x} = \Phi(A,x) := Ax\,, \quad x \in \mathbb{R}^m\,, \quad A \in \mathcal{GL}(m)\,.$$

This m^2-parametric group is not connected because it contains reflections.

Example 10.39. Let $\mathcal{O}(m) = \{A \in \mathcal{GL}(m)\,, \ A^TA = I\}$ be the group of orthogonal (m,m)-matrices (rotations and reflections; dimension $m(m-1)/2$) and $\mathcal{G}(\Phi) = \mathcal{O}(m)$.

$$\widetilde{x} = \Phi(A,x) := Ax\,, \quad x \in \mathbb{R}^m\,, \quad A \in \mathcal{O}(m)\,.$$

Example 10.40. Let $\mathcal{SO}(m) = \{A \in \mathcal{O}(m)\,, \ \det(A) = +1\}$ (special orthogonal) and $G(\Phi) = \mathcal{SO}(m)$, $\widetilde{x} = Ax$. This $m(m-1)/2$-parametric group is connecting. For instance,

$$\mathcal{SO}(2) = \left\{\begin{bmatrix} \cos\varphi & -\sin\varphi \\ \sin\varphi & \cos\varphi \end{bmatrix}, \quad 0 \leq \varphi < 2\pi\right\}.$$

The action is a linear mapping in the groups of Examples 10.37 – 10.40 hence these groups are called also *linear groups.*

Example 10.41. The EUCLIDian group in $\mathbb{R}^3$ of dimension 3 + 3 consists of arbitrary translations and arbitrary rotations without reflections:

$$\widetilde{x} = \Phi(P,a,x) := Px + a\,, \quad x \in \mathbb{R}^3\,, \quad P \in \mathcal{SO}(3)\,, \quad a \in \mathbb{R}^3\,.$$

Example 10.42. The GALILEIan group of dimension $3 + 4 + 3 = 10$ in space $\mathbb{R}^4 = \{(t,\underline{x})\,, \ \underline{x} \in \mathbb{R}^3\}$ of "world points" consist of the following individual actions:

(1°) Uniform motion with velocity $\underline{v}$:

$$(\widetilde{t},\widetilde{\underline{x}}) = \Phi_1(t,\underline{x}) := (t, \underline{x} + \underline{v}\,t) \quad \text{(3 free parameter)}$$

(2°) Separated translation in time and space:

$$(\widetilde{t},\widetilde{\underline{x}}) = \Phi_2(t,\underline{x}) := (t + s, \underline{x} + \underline{s}) \quad \text{(4 free parameter)}$$

(3°) Rotation in $\mathbb{R}^3$:

$$(\widetilde{t},\widetilde{\underline{x}}) = \Phi_3(t,x) := Ax\,, \ A \in \mathcal{SO}(3) \quad \text{(3 free parameter)}\,.$$

Example 10.43. Scaling in $\mathbb{R}^2$. There are several possibilites here, e.g.,

$$(\widetilde{x}_1, \widetilde{x}_2) = \Phi(x_1, x_1, a) = (\varrho^a x_1, \sigma^a x_2)\,, \quad \varrho,\, \sigma \text{ fixed.}$$

Example 10.44. $\mathcal{G}(\Phi) := \{A : \mathbb{R} \ni a \mapsto A(a) \in SO(m)\}\,,\ \underline{x} \in \mathbb{R}^m\,.$

$$\widetilde{\underline{x}} = \Phi(\underline{x}, A(a)) := A(a)\underline{x}\,, \quad x \in X\,, \quad a \in \mathbb{R}\,.$$

We have $A(0) = \mathrm{id}_X$ and $A(a) \circ A(b) = A(a+b)$ by the group properties. If now A is differentiable and $B = A'(0)$ then

$$\begin{aligned}&\frac{\partial}{\partial b}\Big|_{b=0} A(a+b) = \frac{\partial}{\partial b} A(b) \circ A(a)\Big|_{b=0}\\ &\Longrightarrow\ \partial A(a) = BA(a) \ \Longrightarrow\ A(a) = \exp(aB)\,.\end{aligned}$$

Therefore B is called *generator* of $A : a \mapsto A(a)$. If $B = -B^T$ is skew-symmetric then

$$\begin{aligned}&A(a)^T = [\exp(aB)]^T = \exp(-aB)\\ &\Longrightarrow\ A(a)A(a)^T = I \ \Longrightarrow\ (A(a) \text{ orthogonal})\,.\end{aligned}$$

If in particular $A(\varphi) \in SO(2)$ then

$$\frac{d}{d\varphi}\Big|_{\varphi=0} \begin{bmatrix} \cos\varphi & -\sin\varphi \\ \sin\varphi & \cos\varphi \end{bmatrix} = \begin{bmatrix} 0 & -1 \\ 1 & 0 \end{bmatrix} = B\,,$$

hence $A(\varphi) = \exp(B\varphi)$.

(c) In the remaining part we confine ourselves to **One-Parametric Transformation Groups** and remember

Definition 10.5. *Let $\emptyset \neq \mathcal{U} \subset \mathcal{X}$ open, $\varepsilon > 0$, $\Delta := (-\varepsilon, \varepsilon)$, $\Phi : \Delta \times \mathcal{U} \to \mathcal{X}$ and $\Phi(t, \circ) : \mathcal{U} \to \Phi(t, \mathcal{U})$ bijective for all $t \in \Delta$. Then $\mathcal{G}(\Phi)$ is called* local one-parametric transformation group *(of local transformations of the space $\mathcal{X}$) (*Lie *group, briefly LTG) if* (1°) $\forall\, x \in \mathcal{X} : \Phi(0, x) = x$,
(2°) $\forall\, s, t \in \Delta\ \ \forall\, x \in \mathcal{X} : \Phi(t, \Phi(s, x)) = \Phi(s+t, x)$,
(3°) $\forall\, x \in \mathcal{X}\ \ \forall\, t \in \Delta : \Phi(t, x) = x \ \Longrightarrow\ t = 0$,
(4°) $\Phi \in C^\infty(\Delta \times \mathcal{U}; \mathcal{X})$.

Example 10.45. Projective transformation group: $\Delta = (-1,\, 1)\,,\ \mathcal{U} = \{(x,y) \in \mathbb{R}^2\,,\ |x| < 1\,,\ y \in \mathbb{R}\}$, $\widetilde{x} = x/(1-ax)\,,\ \widetilde{y} = y/(1-ay)\,,\ a \in \Delta$; the properties of Definition 10.5 are violated for larger Δ, $\mathcal{U}$.

Definition 10.6. *Let $\Phi_1 : \Delta_1 \times \mathcal{U}_1 \to \mathcal{X}$ action of a LTG $\mathcal{G}(\Phi_1)$,
$\Phi_2 : \Delta_2 \times \mathcal{U}_2 \to \mathcal{X}$ action of a LTG $\mathcal{G}(\Phi_2)$ and
$\mathcal{P} = \{p : \mathcal{U}_1 \to \mathcal{U}_2\,,\ p \text{ diffeomorphism}\}$. Then Φ_1 and Φ_2 are* similar, $\Phi_1 \sim \Phi_2$, *if*

$$\exists\, p \in \mathcal{P}\ \ \exists\, \Delta \subset \Delta_1 \cap \Delta_2\ \ \forall\, t \in \Delta : \Phi_2(t, \circ) = p \circ \Phi_1(t, \circ) \circ p^{-1}\,.$$

The groups $\mathcal{G}(\Phi_1)$ and $\mathcal{G}(\Phi_2)$ are similar, *if Φ_1 and Φ_2 are similar.*

Example 10.46. Dilatation in $\mathbb{R}^n$, $\mathcal{U} = \{x \in \mathbb{R}^n, x^i > 0\}$, $\widetilde{x}^i = \Phi_1^i(t, \underline{x}) = x^i \exp(t\lambda_i)$, $i = 1 : m$, $\underline{\lambda} \in \mathbb{R}^n$ fixed. For the diffeomorphism $p : \mathcal{U} \to \mathcal{X}$: $p_i(\underline{x}) = \ln x^i$, ${p^{-1}}_i(\underline{x}) = \exp(x^i)$ one computes

$$\begin{aligned}\widetilde{x} &= \Phi_2(t,\underline{x}) = p\circ\Phi_1(t,\cdot)\circ p^{-1}(\underline{x}) = p(\Phi_1(t,p^{-1}(\underline{x})))\\ \widetilde{x}^i &= \ln(\Phi_1(t,{p^{-1}}_i(\underline{x}))) = \ln(\Phi_1^i(p^{-1}(\underline{x})))\\ &= \ln({p^{-1}}_i(\underline{x})\exp(t\lambda_i)) = \ln(\exp(x^i)\exp(t\lambda_i)) = x^i + t\lambda_i\,.\end{aligned} \tag{10.47}$$

Therefore the dilatation group (in above notation) and the translation group in $\mathcal{X}$ are *similar*!

(d) Generator of a Group

Definition 10.7. *Let $\Phi : \Delta \times \mathcal{U} \to \mathcal{X}$ be the action of a LTG $\mathcal{G}$.*
(1°) *For fixed $x \in \mathcal{U}$, $\Phi(\Delta, x) = \{\Phi(t,x)\,,\ t \in \Delta\}$ is the* orbit *of $x \in \mathcal{X}$.*
(2°) *The* mapping

$$v : \mathcal{U} \ni x \mapsto \frac{d}{dt}\Phi(0,x) =: v(x) \in \mathcal{X}$$

is called tangential vector *of Φ in $x \in \mathcal{X}$; then $v(x)$ is tangent of the orbit of x in $t = 0$.*

If we differentiate $\Phi(t, \Phi(s,x)) = \Phi(s+t, x)$ w.r.t. t then, with $\Phi(0,x) = x$, we obtain directly the initial value problem for Φ

$$\frac{d}{dt}\Phi(t,x) = v(\Phi(t,x))\,,\quad \Phi(0,x) = x\,, \tag{10.48}$$

which is called LIE equation in this context.

Theorem 10.9. (LIE) *Let $v : \mathcal{U} \to \mathcal{X}$ be a vector field with $v \in C^\infty(\mathcal{U};\mathcal{X})$, $\mathcal{U}$ open and $v(x) \neq 0$ for $x \in \mathcal{U}$. Then the solution Φ of (10.47) is* action *of a LTG $\mathcal{G}(\Phi)$ with tangential vector v ; notation $\mathcal{G}(\Phi) = \mathcal{G}(v)$.*

Of course, the relation between $\mathcal{G}(\Phi)$ and v is unique up to a multiplicative factor only, the groups $\mathcal{G}(v)$ and $\mathcal{G}(\lambda v)$, $0 \neq \lambda \in \mathbb{R}$ are not discernible.

One computes $v(x) = x_0$ in Ex. 10.37, $v(\underline{x}) = B\underline{x}$ in Example 10.44, $v(\underline{x}) = (x^2, xy)$ in Example 10.45, $v(\underline{x}) = (\lambda_1 x^1, \ldots, \lambda_m x^m)$ in Example 10.46.

If Φ_1 and Φ_2 are similar — cf. (10.47) — and $\underline{y} = p(\underline{x})$ then

$$\begin{aligned}&v_2(\underline{y})\frac{\partial}{\partial t}\Big|_{t=0}\Phi_2(t,\underline{y})\\ &= \nabla p(\Phi_1(t,p^{-1}(\underline{y})D_t\Phi_1(t,p^{-1}(\underline{y}))\Big|_{t=0} = \nabla p(p^{-1}(\underline{y}))v_1(p^{-1}(\underline{y}))\end{aligned}$$

hence, using $\underline{x} = p^{-1}(\underline{y})$, we obtain

$$v_2(\underline{y}) = \nabla p(\underline{x})v_1(\underline{x})\,,\quad \underline{y} = p(\underline{x})\,. \tag{10.49}$$

Example 10.47. In Example 10.46, it follows by using $p^i(\underline{x}) = \ln x^i$ that $\nabla p(\underline{x}) = \operatorname{diag}\left(\dfrac{1}{x^1}, \dots, \dfrac{1}{x^m}\right)$ and (10.49) yields

$$v_2(\underline{x}) = \nabla p(\underline{x}) v_1(\underline{x}) = \nabla p(x)(\lambda_1 x^1, \dots, \lambda_m x^m)^T = (\lambda_1, \dots, \lambda_m)^T = \underline{\lambda}\,.$$

Let now $\Phi : \Delta \times \mathcal{U} \to \mathcal{X}$ be action of a LTG, $\mathcal{Y}$ a BANACH space, $F \in C^1(\mathcal{X};\mathcal{Y})$ and $x \in \mathcal{X}$ fixed, then $v(x)$ is tangent of the orbit of Φ in x and $t = 0$. How does the image of this tangent look like in $\mathcal{Y}$ under the mapping F? One computes directly with $\Phi(0, x) = x$

$$D_t(F \circ \Phi(t,x))\Big|_{t=0} = \nabla F(\Phi(t,x)) D_t \Phi(t,x)\Big|_{t=0} = \nabla F(x) v(x) =: (v \cdot \partial F)(x)\,; \tag{10.50}$$

the last notation has historical reasons.

Example 10.48. If $\mathcal{X} = \mathbb{R}^m$ and $\mathcal{Y} = \mathbb{R}$ then

$$(v \cdot \partial)F = \nabla F v = \sum_{i=1}^m v^i(\circ) \frac{\partial F}{\partial x^i}(\circ)$$

therefore $\partial = (\partial_1, \dots, \partial_m) = \left(\dfrac{\partial}{\partial x^1}, \dots, \dfrac{\partial}{\partial x^m}\right)$.

Definition 10.8. *The operator* $v \cdot \partial : F \mapsto \nabla F v$ *is the* (infinitisimal) generator *of the group* $\mathcal{G}(v)$.

We obtain $v \cdot \partial = \underline{x}_0 \cdot \partial$ for the translation $\widetilde{\underline{x}} = \underline{x} + a\underline{x}_0 \in \mathbb{R}^n$; further $(v \cdot \partial)F(\underline{x}) = \nabla F(\underline{x}) B \underline{x}$ for Example 10.44, and for Example 10.46

$$(v \cdot \partial)F(\underline{x}) = \nabla F(\underline{x}) \operatorname{diag}(\lambda)\underline{x} = \sum_{i=1}^m \lambda^i x^i \frac{\partial F}{\partial x^i}(\underline{x})\,.$$

For pure rotations in $\mathbb{R}^2$ — cf. Ex. 10.40 — we have

$$(v \cdot \partial)F(x) = -y \frac{\partial F}{\partial x}(\underline{x}) + x \frac{\partial F}{\partial y}(\underline{x}) \quad \Longrightarrow \quad v \cdot \partial = -y\partial_x + x\partial_y\,.$$

In the GALILEIan group we obtain three generators $t\partial_x$, $t\partial_y$, $t\partial_z$ for Φ_1, four generators ∂_t, ∂_x, ∂_y, ∂_z for Φ_2, and three generators $z\partial_y - y\partial_z$, $-z\partial_x + x\partial_z$, $y\partial_x - x\partial_y$ for Φ_3. In general, a r-parametric group has r generators.

If $\mathcal{G}(v)$ and $\mathcal{G}(w)$ are similar, i.e., $w(y) = \nabla F(x) v(x)$, $y = F(x)$, then the generators remain invariant: $v \cdot \partial_x = w \cdot \partial_y$. Of course, $v \cdot \partial$ is nothing else than the abstract tangential vector introduced in Sect. 10.4 with the properties

$$\begin{aligned} &(1^\circ)\ v \cdot \partial(\alpha f + \beta g) = \alpha v \cdot \partial f + \beta v \cdot \partial g && \text{linearity} \\ &(2^\circ)\ v \cdot \partial(f \cdot g) \quad\ = g \cdot (v \cdot \partial f) + f \cdot (v \cdot \partial g) && \text{LEIBNIZ rule}\,. \end{aligned} \tag{10.51}$$

Here, α, $\beta \in \mathbb{R}$ and f, g are scalar functions with the composition $(f \cdot g)(x) = f(x)g(x)$.

Further references to chapter 10: (Abraham), (Berger), (Bishop), (Lippmann), (Ovsiannikov) and many others.

11

Case Studies

11.1 An Example of Gas Dynamics

Consider an infinitely long tube with cross-section 1 $[L^2]$ being filled with an ideal gas (where $pV = \text{const}$ following BOYLE-MARIOTTE). The change of temperature is neglected. Let ϱ be the density, T the temperature, c the specific heat capacity, $\varepsilon = cT$ the specific interior energy, and v the scalar velocity. Moreover, let

$$p(t,x) \text{ pressure}, \quad m(t,x) = \varrho(t,x)v(t,x) \text{ momentum}$$
$$e(t,x) = \varepsilon(t,x) + \frac{v(t,x)^2}{2} \quad \text{specific interior plus kinetic energy}\,.$$

Then, writing $a = \Phi(A,t)$ and $b = \Phi(B,t)$,

$$M(t) = \int_a^b \varrho(t,x)dx \text{ total mass}, \quad I(t) = \int_a^b m(t,x)dx \text{ total momentum}$$
$$E(t) = \int_a^b \varrho(t,x)e(t,x)dx \quad \text{total energy}$$

in the section $(\Phi(A,t), \Phi(B,t))$ of the tube at time t. At time $t = 0$ let $\Omega = (A,B)$ be the reference section. The laws of physics yield directly the local balance theorems:

$$\begin{aligned}
\dot{I}(t) &= \frac{D}{Dt}I(t) = -p(\Phi(B,t),t) + p(\Phi(A,t),t)\\
\dot{E}(t) &= \frac{D}{Dt}E(t) = -p(\Phi(B,t),t)v(\Phi(B,t),t) + p(\Phi(A,t),t)v(\Phi(A,t),t)\,.
\end{aligned}$$

For instance, in the notation of Axiom 8.2 in Chap. 8, $\mathbf{t} = -p\boldsymbol{\delta} = -p$, $U = [A,B]$ and

$$\int_{\partial\Phi(t,U)} \mathbf{t}\underline{n}\,do = -p\Big|_a^b\,.$$

The integral form of the balance theorems has here the general form

$$\frac{D}{Dt}\int_a^b f(x,t)dx + g(x,t)\Big|_a^b = 0\,.$$

Applying REYNOLDS' transport theorem and $\mathrm{div}_x\, f = f_x$, we obtain the integral form

$$\int_a^b \left[\frac{\partial}{\partial t} f(x,t) + (vf+g)_x(x,t)\right] dx = 0$$

and the local form

$$\boxed{f_t + (vf+g)_x = 0}\,. \tag{11.1}$$

The integral form follows here also directly from LEIBNIZ' rule for parameter-dependent integrals. The balance theorem of mass supplies

$$\frac{\partial \varrho}{\partial t} + \mathrm{div}(\varrho\, v) = \varrho_t + (\varrho\, v)_x = 0\,, \tag{11.2}$$

which corresponds to the choice of $f = \varrho$ and $g = 0$ in (11.1). Because $m = \varrho\, v$, (11.2) is equivalent to

$$\varrho_t + m_x = 0\,. \tag{11.3}$$

The choice $f = m$ and $g = p$ leads to $vf + g = v\, m + p$. Because $\varrho\, v = m$ hence $v = m/\varrho$, the local balance theorem of moments follows by (11.1),

$$m_t + (v\, m + p)_x = m_t + \left[\frac{m^2}{\varrho} + p\right]_x = 0\,. \tag{11.4}$$

The choice $f = \varrho\, e$, $g = p\, v$ and $v = m/\varrho$ yields

$$vf + g = v\varrho\, e + p\, v = \frac{m}{\varrho}(\varrho\, e + p)$$

and thus, by (11.1),

$$(\varrho\, e)_t + \left[(\varrho\, e + p)\frac{m}{\varrho}\right]_x = 0\,. \tag{11.5}$$

Altogether, (11.3), (11.4) and (11.5) yields the system

$$\boxed{U_t + F(U)_x = 0} \tag{11.6}$$

where

$$U = \begin{bmatrix} \varrho \\ m \\ \varrho\, e \end{bmatrix}\,,\quad F(U) = \frac{1}{\varrho}\begin{bmatrix} \varrho\, m \\ m^2 + \varrho\, p \\ (\varrho\, e + p)m \end{bmatrix}\,.$$

The JACOBI matrix of F has real eigenvalues hence (11.6) is a *nonlinear hyperbolic system* of first order. It consists of three equations for four unknowns

ϱ, m, e, and p. The laws of thermodynamics supply the state equation $p = \psi(\varepsilon, \varrho)$. If, e.g., $p = (\gamma - 1)\,\varrho\,\varepsilon$ then, writing $v = m/\varrho$,

$$\varrho\, e = \varrho\,\varepsilon + m^2/2\varrho\,, \quad \varepsilon = (\varrho\, e - m^2/2\varrho)/\varrho\,, \quad p = (\gamma - 1)(\varrho\, e - m^2/2\varrho)\,. \tag{11.7}$$

Accordingly, (11.6) reads now

$$\begin{aligned} \varrho_t + m_x &= 0 \\ m_t + [m^2/\varrho + (\gamma-1)(\varrho\, e - m^2/2\varrho)]_x &= 0 \\ (\varrho\, e)_t + [m\{\varrho\, e + (\gamma-1)(\varrho\, e - m^2/2\varrho)\}/\varrho]_x &= 0\,. \end{aligned}$$

If ε is constant then the last equation may be replaced directly by $\varrho\, e = \varrho\,\varepsilon + m^2/2\varrho$ following (11.7). Cf. (Richtmyer).

In problems of this form, frequently *compression shocks* appear, then the system is now longer *simple* (i.e. smooth) and the balance theorems must be modified.

11.2 The Reissner-Mindlin Plate

The conditions (8.55) and (8.56) imposed on the *shear-rigid* KIRCHHOFF plate are dropped in the REISSNER-MINDLIN plate or *shear-soft* plate. But $\sigma_{33} = 0$ is still maintained because a plate of moderate thickness is supposed; moreover, some components of the stress tensor are equipped with a *shear-correction factor k*. We begin again with (8.54) and the first assumption of (8.55), i.e., we suppose that

$$\boxed{\begin{gathered} u_1 = -z\,\varphi_1(x,y)\,, \quad u_2 = -z\,\varphi_2(x,y)\,, \quad u_3 = w(x,y) \\ \sigma_z := \sigma_{33} = 0 \end{gathered}} \tag{11.8}$$

and obtain then with angles φ_1 and φ_2 as independent variables

$$\begin{aligned} \varepsilon_x &= u_{1,x} = -z\,\varphi_{1,x}(x,y)\,, & \varepsilon_y &= u_{2,y} = -z\,\varphi_{2,y}(x,y)\,, \\ \gamma_{xy} &= u_{1,y} + u_{2,x} = -z\varphi_{1,y} - z\varphi_{2,x}\,, & \gamma_{yz} &= u_{2,z} + u_{3,y} = -\varphi_2 + w_y\,, \\ \gamma_{zx} &= u_{3,x} + u_{1,z} = w_x - \varphi_1\,. \end{aligned}$$

For simplicity, we define in formal the same way as in plane stress of discs

$$\begin{aligned} \boldsymbol{\varepsilon}(\underline{\varphi}) &= \begin{bmatrix} \varphi_{1,x} & (\varphi_{1,y} + \varphi_{2,x})/2 \\ (\varphi_{1,y} + \varphi_{2,x})/2 & \varphi_{2,y} \end{bmatrix}, \\ \boldsymbol{\sigma}(\underline{\varphi}) &= 2\mu\boldsymbol{\varepsilon}(\underline{\varphi}) + 2\widetilde{\lambda}\,\mathrm{trace}(\boldsymbol{\varepsilon}(\underline{\varphi}))\boldsymbol{\delta} \end{aligned}$$

where $\underline{\varphi} = (\varphi_1\,,\ \varphi_2)$.

After having integrated over the z-variable, the extremal problem of the REISSNER-MINDLIN plate reads:

$$\begin{aligned}&\frac{h^3}{12}\int_{\Omega}\frac{1}{2}\varepsilon(\underline{\varphi}):\boldsymbol{\sigma}(\underline{\varphi})\,dF + h\mu k\int_{\Omega}(\operatorname{grad} w-\underline{\varphi})\cdot(\operatorname{grad} w-\underline{\varphi})\,dF\\&-h\int_{\Omega}\underline{\varphi}\cdot\underline{k}\,dF - h\oint_{\Gamma_C}\underline{\varphi}\cdot\underline{g}\,ds = \min!\,.\end{aligned}\tag{11.9}$$

The corresponding EULER equation, being derived in the usual way, reads:

$$\begin{aligned}&\frac{h^3}{12}\int_{\Omega}\varepsilon(\underline{\psi}):\boldsymbol{\sigma}(\underline{\varphi})\,dF + 2h\mu k\int_{\Omega}(\operatorname{grad} w-\underline{\psi})\cdot(\operatorname{grad} v-\underline{\varphi})\,dF\\&= h\int_{\Omega}\underline{\psi}\cdot\underline{k}\,dF - h\oint_{\Gamma_C}\underline{g}\cdot\underline{\psi}\,ds\end{aligned}\tag{11.10}$$

and has to be transformed again by means of GREEN's formula for a formulation of the problem as boundary value problem. We obtain

$$\begin{aligned}\int_{\Omega}(\operatorname{grad} w-\underline{\psi})(\operatorname{grad} v-\underline{\varphi})\,dF &= -\int_{\Omega}\underline{\psi}\cdot(\operatorname{grad} w-\underline{\varphi})\,dF\\&-\int_{\Omega}v\,(\Delta w-\operatorname{div}\underline{\varphi})\,dF + \oint_{\Gamma}v\,(\operatorname{grad} w-\underline{\varphi})\cdot\underline{n}\,ds\,.\end{aligned}$$

Applying (11.9) it then follows that

$$\begin{aligned}&-\frac{h^3}{12}\int_{\Omega}\underline{\psi}\cdot\left(\mu\Delta\underline{\varphi}+(2\widetilde{\lambda}+\mu)\operatorname{grad}\operatorname{div}\underline{\varphi}\right)\,dF + \frac{h^3}{12}\oint_{\Gamma}\underline{\psi}\cdot\boldsymbol{\sigma}(\underline{\varphi})\underline{n}\,ds\\&\qquad -2h\mu k\int_{\Omega}v\,\left[(\operatorname{grad} w-\underline{\varphi})+(\Delta w-\operatorname{div}\underline{\varphi})\right]\,dF\\&+2h\mu k\oint_{\Gamma}v\,\left(\operatorname{grad} w-\underline{\varphi}\right)\cdot\underline{n}\,ds = h\int_{\Omega}\underline{\psi}\cdot\underline{k}\,dF - h\oint_{\Gamma_C}\underline{\psi}\cdot\underline{g}\,ds\end{aligned}\tag{11.11}$$

where v is a test function relative to u and $\underline{\psi}$ is a test function relative to the vector field $\underline{\varphi}$. Now, an application of the fundamentallemma of variational calculus yields the desired system of elliptic equations for the REISSNER-MINDLIN plate

$$\boxed{\begin{aligned}-\frac{h^3}{12}\left(\mu\Delta\underline{\varphi}+(2\widetilde{\lambda}+\mu)\operatorname{grad}\operatorname{div}\underline{\varphi}\right) - 2h\mu k\operatorname{grad} w + 2h\mu k\underline{\varphi} &= h\underline{k}\ \text{in}\ \Omega\\ \operatorname{div}\underline{\varphi} - \qquad \Delta w \qquad &= 0\quad \text{in}\ \Omega\end{aligned}}\tag{11.12}$$

with (weak) boundary conditions

$$\underline{\psi}\cdot\left[\frac{h^3}{12}\boldsymbol{\sigma}(\underline{\varphi})\underline{n}-h\underline{g}\right]=0\ \text{ and }\ v(\operatorname{grad} w-\underline{\varphi})\cdot\underline{n}=0\ \text{ on }\Gamma\,.\tag{11.13}$$

These boundary conditions allow again four different combinations in the same way as in KIRCHHHOFF's plate; cf. § 8.7 **(c)**.

11.3 Examples of Multibody Problems

Every moving vehicle and, *cum grano salis* also human beings with its extremities, constitute a multibody problem. The overwhelming abundance of possibilities hardly allows a classification, cf., e.g., (Schiehlen90), and would be far beyound the scope of this volume. In this section, we consider three classical examples in the plane where each body K_i is a *rigid* disc $\Omega_i \subset \mathbb{R}^2$. Every disc is equipped with a local coordinate system (COS) (ξ_i, η_i) with gravity center S_i in origin and rotational angle $\varphi_i(t)$ relative to the global coordinate system. Moreover, the i-th disc has mass M_i and polar moment of inertia T_i relative to the origin of the local COS. In spatial systems the tensor of inertia takes the place of the scalar moment of inertia and three angles describe the position. The gravity center (origin) is given by the point vector $(x_i(t), y_i(t)) \in \mathbb{R}^2$ in global coordinates. Thus, altogether, the i-th disc is described by

$$K_i \simeq \{x_i(t), y_i(t), \varphi_i(t); \Omega_i,\ M_i, T_i\}\,, \quad i = 1 : I\,,$$

and the generalized state vector $\underline{q}(t)$ is a $3I$-dimensional vector

$$\underline{q}(t) = [x_1(t), \ldots, x_I(t), y_1(t), \ldots, y_I(t), \varphi_1(t), \ldots, \varphi_I(t)]^T$$

im *"configuration space"*. At least the kinetic energy of a multibody system in plane may now be displayed more precisely as

$$E_{\text{kin}}(\dot{\underline{q}}(t)) = \frac{1}{2}\left[\sum_{i=1}^{I} M_i\big(\dot{x}_i(t)^2 + \dot{y}_i(t)^2\big) + \sum_{i=1}^{I} T_i\,\dot{\varphi}_i(t)^2\right] \tag{11.14}$$

whereas potential energy and possible friction or other damping depend strongly on the individual problem.

Example 11.1. Double pendulum without moments of inertia as mechanical system; cf. (Hairer), II, p. 484. Let $\mathcal{O}$ be the origin of the COS, let $P_1(x_1, x_2)$ und $P_2(x_3, x_4)$ be points with masses m_1, m_2, and let $l_1 = |\overrightarrow{\mathcal{O}P_1}|$ and $l_2 = |\overrightarrow{P_1P_2}|$. In this example, the LAGRANGE multipliers z_i constitute stresses in rod i of length l_i which keep the mass points on their path.
Energy, constraints, and LAGRANGE function:

$$\begin{aligned}
E &= \frac{m}{2}\left(\dot{x}_1^2 + \dot{x}_2^2\right) + \frac{m}{2}\left(\dot{x}_3^2 + \dot{x}_4^2\right) + mg\,x_2 + mg\,x_4\,,\\
0 &= x_1^2 + x_2^2 - l_1^2\,,\ \ 0 = (x_3 - x_1)^2 + (x_4 - x_2)^2 - l_2^2\,,\\
L &= \frac{m}{2}\left(\dot{x}_1^2 + \dot{x}_2^2\right) + \frac{m}{2}\left(\dot{x}_3^2 + \dot{x}_4^2\right) - mg\,x_2 - mg\,x_4\\
&\quad -z_1\left(x_1^2 + x_2^2 - l_1^2\right) - z_1\left((x_3 - x_1)^2 + (x_4 - x_2)^2 - l_2^2\right)\,.
\end{aligned}$$

EULER equations:

$$\begin{aligned} m_1\ddot{x}_1 &= -2x_1z_1 & m_2\ddot{x}_3 &= -2x_3z_2 \\ m_1\ddot{x}_2 &= -2x_2z_1 - m_1\,g & m_2\ddot{x}_4 &= -2x_4z_2 - m_2\,g \\ 0 &= x_1^2 + x_2^2 - l_1^2\,, & 0 &= (x_3 - x_1)^2 + (x_4 - x_2)^2 - l_2^2\,. \end{aligned}$$

Example 11.2. Andrew's Squeezer (seven-body problem) is a well-known test problem. The complete numerical data of (Schiehlen90) are found also in (Hairer) II somewhat more easily attainable.
Data, computation and image sequence in `KAPITEL11\SECTION_3`.

Example 11.3. Roboter following (Schiehlen89); for the notations see Figure 11.4. Geometrical parameter:

$$C = 0.05\ [m]\,,\quad L = 0.50\ [m]\,,\quad T = 2\ [s]\,\text{runtime}\,.$$

The gravity center S_i of each body K_i, $i = 1 : 3$, lies in the origin of the body-fixed coordinate system. Force and turning moment have to be defined for every connection between two bodies. Generalized coordinates:

$$\underline{q} = [Z1\,,\ GA1\,,\ Y2\,,\ BE2\,,\ AL3]^T\,.$$

Equations of motion:

$$M(\underline{q})\underline{\ddot{q}} + K(t, \underline{q}, \underline{\dot{q}}) = Q(t, \underline{q}, \underline{\dot{q}})\,.$$

Masses and moments of inertia:

	Body		
	1	2	3
Mass $[kg]$	250	150	100
Moment of inertia $[kg \cdot m^2]$			
T_x	(90)	13	4
T_y	(10)	0.75	1
T_z	90	13	4.3

Initial conditions:

$$Z1 = 2.25\,[m]\,,\ GA1 = -0.5236\,[rad]\,,\ Y2 = 0.75\,[m]\,,\ BE2 = 0\,,\ AL3 = 0\,.$$

F1Z Force in direction Z1	F2Y Force in direction Y2
L1Z Torque in direction Z1	L3X Torque in direction X3

The turning moment in direction Y2 disappears.

Time t $[s]$	Data $[N]$, $[Nm]$
0 to 0.5	$\mathrm{F1Z} = 6348\,,\ F2Y = 36 \cdot t + 986$
	$\mathrm{L1Z} = 673 \cdot t - 508\,,\ \mathrm{L3X} = 64.5$
0.5 to 1.5	$\mathrm{F1Z} = 4905\,,\ \mathrm{F2Y} = \text{- } 2$
	$\mathrm{L1Z}\ 148\ \ exp(-5.5 \cdot (t - 0.5)) + 8\,,\ \mathrm{L3X} = 49{,}05$
1.5 to 2	$\mathrm{F1Z} = 3462\,,\ \mathrm{F2Y} = \text{- } 1019$
	$\mathrm{L1T} = 240\,,\ \mathrm{L3X} = 34.6$

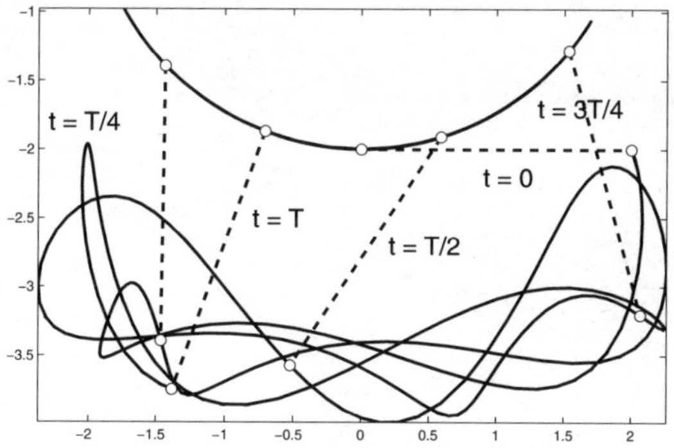

Figure 11.1. Double pendulum

```
KAPITEL11/SECTION_3, Multibody Problems
demo1.m        Masterfile for the multibody problems
demo2.m        Movie for Andrew's Squeezer
demo3.m        Andrew's Squeezer, fixed angle beta
demo4.m        Roboter following Schiehlen, with ode23t.m
hem4.m         Masterfile for semi-implicit Runge-Kutta methods
hem3_kern.m    Semi-implicit RKV of order 3 by Brasey
hem4_kern.m    Semi-implicit RKV of order 4 by Brasey
```

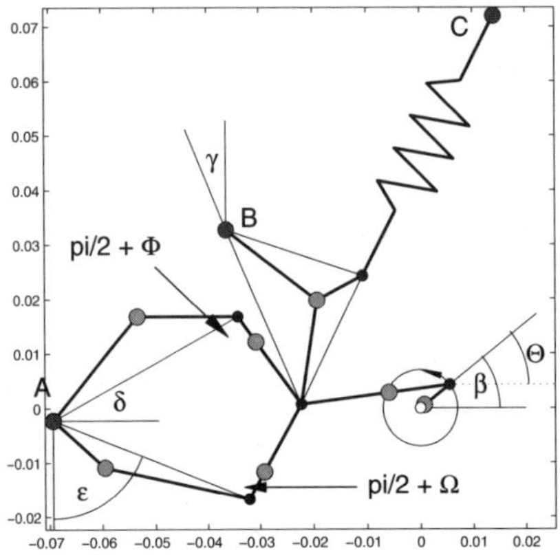

Figure 11.2. Seven-body problem

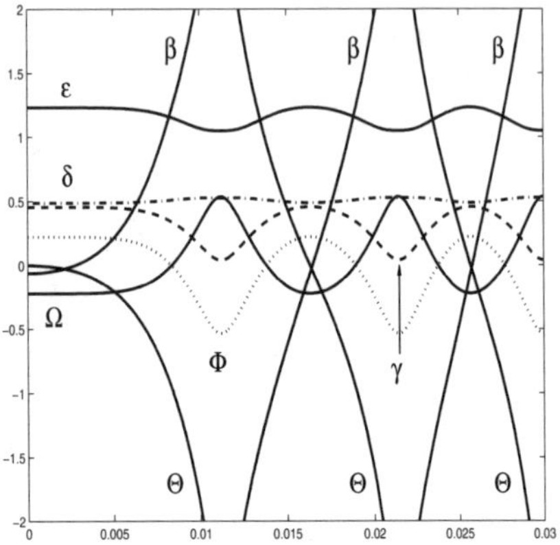

Figure 11.3. Angles mod 2π

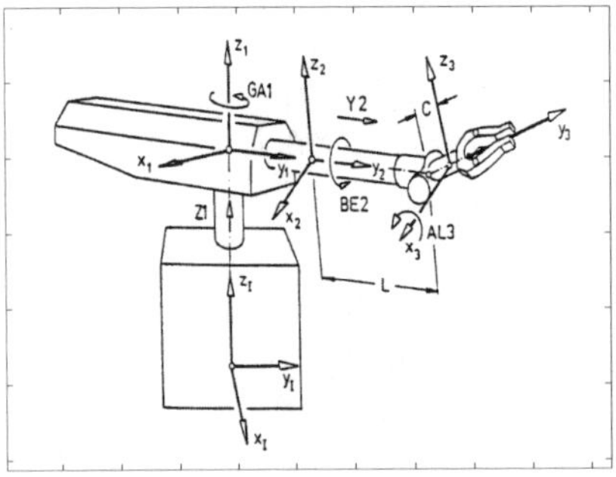

Figure 11.4. Roboter (Schiehlen89)

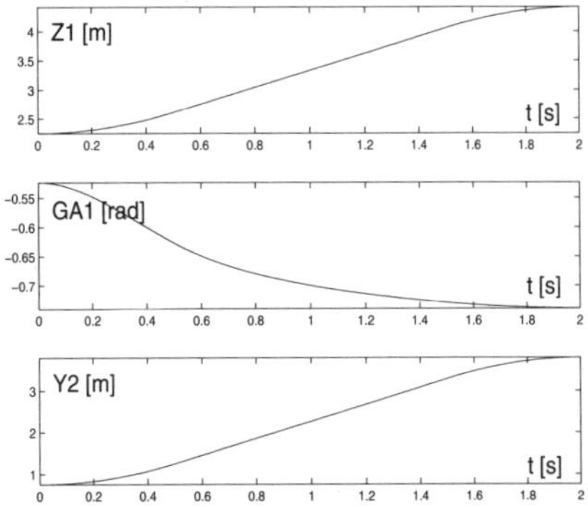

Figure 11.5. Curves of motion

11.4 Dancing Discs

Besides top and multibody systems, rotating cogwheels are a rather simple but not less striking example for working with different coordinate systems.

(a) General Discs as, e.g., *cogwheels* are designed as closed polygons for illustration. Four different cases are implemented in directory `KAPITEL11\SECTION_4`:

(1°) Disc $\mathcal{A}$ rolls on (or in) a fixed disc $\mathcal{B}$; cf. Figure 11.10, 1–4.

(2°) Disc $\mathcal{B}$ rolls backward with properly chosen velocity such that disc $\mathcal{A}$ moves only back and forth (optional).

(3°) The centers of both discs are fixed; cf. Figure 11.10, 5–7.

(4°) Centers and rotational velocity of both discs are fixed; cf. Figure 11.10, 8.

In the first two cases, the *instantaneous* center of rotation *on* the boundary of the fixed disc and the corresponding local rotation angle have to be calculated in each step of rotation. Both discs are *not necessarily convex.* Therefore "the range of sight" of the moving disc has to be fixed in advance. In other words, the number of polygonal points involved in the computation of a single step of rotation has to be adapted to the individual problem and must be given by the user perhaps after some preliminary trials. In the third case, disc $\mathcal{A}$ drives disc $\mathcal{B}$ whose unknown rotation angle is calculated in each step of motion by the method of *bisection* and applying the MATLAB command `inpolygon.m`. In every case the instantaneous point of rotation P must lie on *both discs* because they shall touch each other at that point. Thus P is either a polygonal point on both discs or it is a polygonal point of one disc (which?) and lies between two polygonal points of the other disc. Besides, the local process of rolling, e.g., on a single tooth flank may be visualized easily by the ZOOM function offered in the MATLAB suite.

(b) Cogwheels Cf., e.g., (Decker), (Wentzell). Let a large gear $\mathcal{A}$ and a smaller gear $\mathcal{B}$ be given in a global COS with centers $\underline{M}$ and $\underline{m}$ and the *pitch circles* W and w. Both pitch circles shall touch each other permanently at the *pitch point* $\underline{C}$.

All data of the large gear are written as capitals and all data of the smaller gear (or pinion) as small letters.

Notations:

R_W, r_w	radii of pitch circles
$\underline{M} = (-R_W, 0)$	center of gear $\mathcal{A}$ in global COS
$\underline{m} = (r_w, 0)$	center of gear $\mathcal{B}$ in global COS
N, n	number of revolutions
Z, z	number of thooths
Ω, ω	angular velocities
$\underline{B}(t)$, $\underline{b}(t)$	points of contact of tooth flanks in global COS
$\underline{T}(t)$, $\underline{t}(t)$	tangents of tooth flanks in $\underline{B}$, $\underline{b}$
$\underline{N}(t)$, $\underline{n}(t)$	normals of tooth flanks in $\underline{B}$, $\underline{b}$
$R_B(t)$, $r_b(t)$	radii of points of contact $\underline{B}$, $\underline{b}$
$\underline{V}(t)$, $\underline{v}(t)$	absolute velocity of $\underline{B}$, $\underline{b}$, $\dot{\underline{B}}(t) = \underline{V}(t)$ (perpendicular to radius vectors $\underline{R}(t)$, $\underline{r}(t)$)
$\underline{V}_T(t)$, $\underline{v}_t(t)$	projection of $\underline{V}$, $\underline{v}$ onto $\underline{T}$, $\underline{t}$
$\underline{V}_N(t)$, $\underline{v}_n(t)$	projection of $\underline{V}$, $\underline{v}$ onto $\underline{N}$, $\underline{n}$
$\lvert V_W\rvert = \lvert v_w\rvert$	$= 2\pi\, R_W\, N = 2\pi\, r_w\, n$ circumferential velocity of pitch circles

Translation:

$$1 \le u = \frac{n}{N} = \frac{R_W}{r_w} = \frac{\omega}{\Omega} = \frac{Z}{z}\,. \tag{11.15}$$

Remark on Geometry: Without loss of generality, let the pitch point $\underline{C}$ be the origin of the global COS and let both centers of the gears lie on the x-axis, $\underline{M} = (-M, 0)$, $M > 0$, $\underline{m} = (m, 0)$, $m > 0$. The larger gear $\mathcal{A}$ rotates counterclockwise.

Requirement 1: Both angular velocities Ω and ω are constant.

Then, for boundary curves of both gears,

$$\underline{X}(\varphi) = [X(\varphi), Y(\varphi)]^T, \quad \underline{x}(\varphi) = [x(\varphi), y(\varphi)]^T, \quad 0 \le \varphi \le 2\pi,$$

it follows that

$$\begin{aligned}
&\text{absolute motion of point } \underline{X}(\varphi) : \underline{Y}(t, \varphi) = \underline{M} + D(\Omega t)\underline{X}(\varphi) \\
&\text{absolute motion of point } \underline{x}(\varphi) : \ \underline{y}(t, \varphi) \ = \underline{m} + D(\pi - \omega t)\underline{x}(\varphi)
\end{aligned}$$

with rotation matrix

$$D(\alpha) = \begin{bmatrix} \cos(\alpha) & -\sin(\alpha) \\ \sin(\alpha) & \cos(\alpha) \end{bmatrix}, \quad C = \begin{bmatrix} 0 & -1 \\ 1 & 0 \end{bmatrix}.$$

Requirement 2: Both gears touch each other permanently:

$$\begin{aligned}
\underline{B}(t) &= \underline{Y}(t, \Omega t) && = \underline{M} + D(\Omega t)\underline{X}(\Omega t) \\
= \underline{b}(t) &= \underline{y}(t, \pi - \omega t) && = \underline{m} + D(\pi - \omega t)\underline{x}(\pi - \omega t), \\
\dot{\underline{B}}(t) &= \Omega\Big[CD(\Omega t)\underline{X}(\Omega t) && + D(\Omega t)\underline{X}'(\Omega t)\Big] \\
= \dot{\underline{b}}(t) &= -\omega\Big[CD(\pi - \omega t)\underline{x}(\pi - \omega t) + D(\pi - \omega t)\underline{x}'(\pi - \omega t)\Big].
\end{aligned}$$

Of course, the requirement $\underline{B}(t) = \underline{b}(t)$ implies that $\underline{N}(t) = \alpha\,\underline{n}(t)$ und $\underline{T}(t) = \beta\,\underline{t}(t)$ and the *law of touching*

$$\underline{V}_N(t) = \underline{v}_n(t), \tag{11.16}$$

for both velocity vectors, but not $\underline{V}_T(t) = \underline{v}_t(t)$.

Both velocity vectors in (11.16) may be considered as tangential vectors of some circles, namely in point P_A of the *base circle* A_G about $\underline{M}$ with radius R_G and in point P_B of the base circle B_G about $\underline{m}$ with radius r_G. The straight line connecting P_a and P_b intersects the straight line connecting $\underline{M}$ and $\underline{m}$ at a point $\underline{C}'$ with distance R' of $\underline{M}$ and r' of $\underline{m}$. Because

$$|\underline{V}_N(t)| = R_G(t)\,\Omega, \quad |\underline{v}_n(t)| = r_G(t)\,\omega,$$

(11.15) and $\omega/\Omega = u = constant$, also $R_G/r_G = u$ must be constant. It then follows that $R'/r' = u$ hence $\underline{C}' = \underline{C}$ and thus the *law of gearing*, namely that the individual normals in a touching point pass through the (constant) pitch point $\underline{C}$:

$$\boxed{[\underline{B}(t), \underline{C}] \equiv [\underline{b}(t), \underline{C}] = \gamma\,\underline{N}(t) = \delta\,\underline{n}(t)}$$

($[\underline{B}, \underline{C}]$ straight line for $\underline{B} \neq \underline{C}$).

Example 11.4. If both "gears" are pure circle discs then
$\underline{B}(t) = \underline{C}$ and in present COS $\underline{T}(t) = [0,\ \Omega]^T$, $\underline{N}(t) = [1,\ 0]^T$,
$\underline{t}(t) = [0,\ -\omega]^T$, $\underline{n}(t) = [-1,\ 0]^T$.

In general, the touching point $\underline{B}(t) = \underline{b}(t)$ (*action point*) describes locally

a curve in global COS (*action line*)
a relative curve on moving gear $\mathcal{A}$ (tooth flank A)
a relative curve on gear $\mathcal{B}$ (tooth flank B).

All three curves have to pass through the pitch point, and the different types of gears are classified by their action line in global COS.

Requirement 3: Let the action line be (locally) a straight line. Then this line has an angle α to the tangent of the pitch circles in pitch point which is prescribed by DIN to $\alpha = \pi/9$.

This condition leads to an *involute gear.* In an alternative case, the action line consists of two circle segments with equal radii but opposite curvature on both sides of the pitch point. This case, being not further prosecuted here, leads to a *cycloid gear*.

The angle α of the action line fixes base circle and outside circle on both gears.

Gear $\mathcal{A}$ base circle A_B with radius R_B,
outside circle A_O with radius $R_0 > R_B$
Gear $\mathcal{B}$ base circle B_b with radius r_b,
outside circle B_O with radius $r_O > r_b$.

External circle A_O intersects base circle B_b in initial point P_B of the action line,
External circle B_O intersects base circle A_B in terminal point P_A of the action line.

Then, by *Requirement 3*, α is the angle between $\overrightarrow{\underline{M}\,C}$ and $\overrightarrow{\underline{M}\,P_A}$ and likewise between $\overrightarrow{\underline{m}\,C}$ and $\overrightarrow{\underline{m}\,P_B}$ and moreover

$$
\begin{aligned}
&\text{radius of } A_G : R_G = R_W \cos\alpha\,, \quad \text{radius of } B_G : r_g = r_w \cos\alpha \\
&\text{radius of } A_O : R_O = \Big((R_W + r_w)^2 - R_W(R_W + 2r_W)\cos^2\alpha\Big)^{1/2} \\
&\text{radius of } B_O : r_O = \Big((R_W + r_w)^2 - r_w(r_w + 2R_W)\cos^2\alpha\Big)^{1/2}.
\end{aligned}
$$

The outside circles do not play a role in the sequel, and the above defined action lines contains segments being not involved. The *modul* m is a reference measure to which the other data refer. Modul values in mm by DIN 780:

0.05	0.80	0.10	0.12	0.16	0.20	0.25	0.3	0.4	0.6	0.8	1	1.25
1.5	2	2.5	3	4	5	6	8	10	12	16	20	60

(c) In a gear with **Zero-Gearing** the pitch circle is *circle of partition* to which the uniform circle partition refers. Aside from the *clearance* c, the pitch circle intersects the tooth flank in the mid-point.

Further data for gear $\mathcal{A}$ (likewise for gear $\mathcal{B}$ with same modul):

reference circle diameter radius (pitch circle radius)	$R_W = Z \cdot m/2$
Head circle radius $\leq$ (outside circle radius)	$R_K = R_W + h_K\,, \quad (\text{z.B. } h_K = m)$
root circle radius	$R_F = R_W - h_F \quad (h_F = h_K + c > 0)$
partition	$p = m \cdot \pi$
Circular pitch	$p_e = p \cdot \cos\alpha = m \cdot \pi \cdot \cos\alpha$
Zero-axes distance	$a_d = R_W + r_w = \dfrac{m}{2}(Z + z)\,.$

The modul m resp. the radius of the pitch circle is specified by the relation $R_W = Z \cdot m/2$; the data of wheel $\mathcal{B}$ then follow by the gear transmission ratio. The radius of the base circle R_G is larger than the radius of the foot circle R_F in normal case. Theoretically, the tooth flanks may be shaped arbitrarily out of base circle and head circle as far as they keep out of their way.

Example 11.5. (Decker) $Z = 81\,,\ z = 17\,,\ m = 4\,,\ R_W = 162\,[mm]\,,$ $r_w = 34\,[mm]\,.$

By the law of gearing, the normals of the curves in instantaneous touching point as well as the action line pass through the pitch point. Thus, for straight action line, the current normals are parallel and the tangents perpendicular to the action line. Then we obtain for both gear boundaries (discs) in touching point

$$
\begin{aligned}
&0 = (\underline{B}(t) - \underline{C})^T\dot{\underline{B}}(t) = \underline{B}(t)^T\dot{\underline{B}}(t) = \\
&\Big[\underline{M} + D(\Omega\, t)\underline{X}(\pi + \Omega t)\Big]^T \Big[\Omega C D(\Omega t)\underline{X}(\pi - \Omega t) + D(\Omega t)\underline{X}'(\pi - \Omega t)(-\Omega)\Big] \\
&0 = (\underline{b}(t) - \underline{C})^T\dot{\underline{b}}(t) = \underline{b}(t)^T\dot{\underline{b}}(t) \\
&= \Big[\underline{m} + D(\omega\, t)^T\underline{x}(\omega t)\Big]^T \Big[\omega C^T D(\omega t)^T\underline{x}(\omega t) + D(\omega t)^T\underline{x}'(\omega t)\omega\Big]\,.
\end{aligned}
$$

Evolvent for gear $\mathcal{A}$, $\text{inv}\varphi := \tan(\varphi) - \varphi$:

$$
\begin{aligned}
\underline{Z}(\xi) &= \underline{M} + R_G\begin{bmatrix}\cos\xi\\ \sin\xi\end{bmatrix} + R_G\,\xi\begin{bmatrix}\sin\xi\\ -\cos\xi\end{bmatrix}, \quad 0 \leq \xi \leq \xi_1\,,\\
\underline{Z}'(\xi) &= R_G\begin{bmatrix}-\sin\xi\\ \cos\xi\end{bmatrix} + R_G\begin{bmatrix}\sin\xi\\ -\cos\xi\end{bmatrix} + R_G\,\xi\begin{bmatrix}\cos\xi\\ \sin\xi\end{bmatrix} = R_G\,\xi\begin{bmatrix}\cos\xi\\ \sin\xi\end{bmatrix}.
\end{aligned}
$$

By this result, $\underline{n} = [\sin\xi\,,\ -\cos\xi]^T$ is the normal vector at curve point $\underline{Z}(\xi)$.

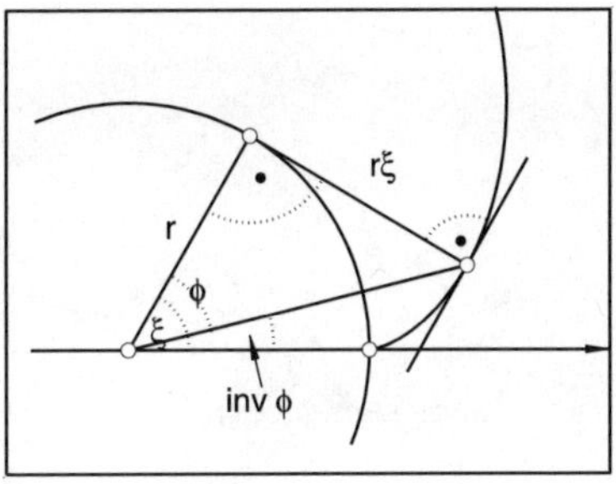

Figure 11.6. Evolvent

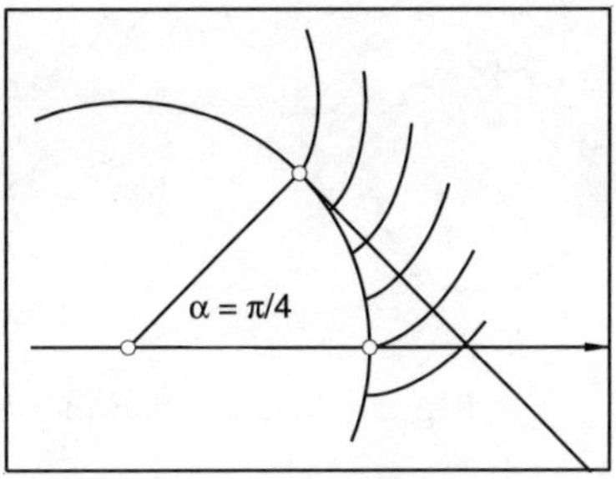

Figure 11.7. Rotated evolvents

In the following figure, $\alpha = \pi/4$ is chosen for more suitable visualizing.

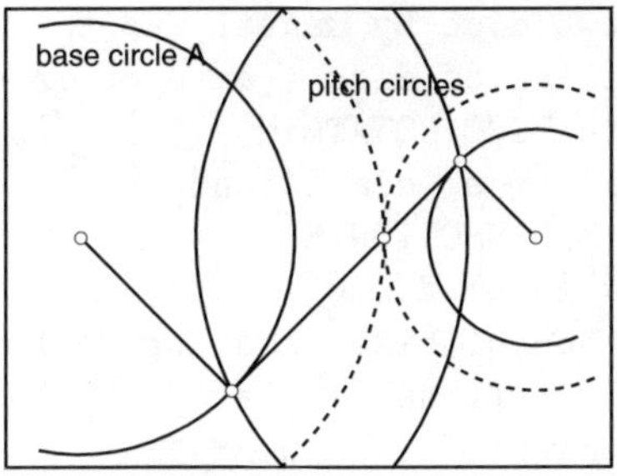

Figure 11.8. Pitch circles

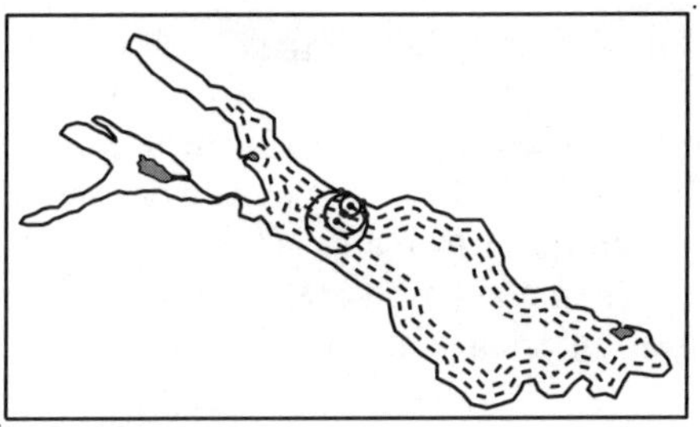

Figure 11.9. Lake Constance

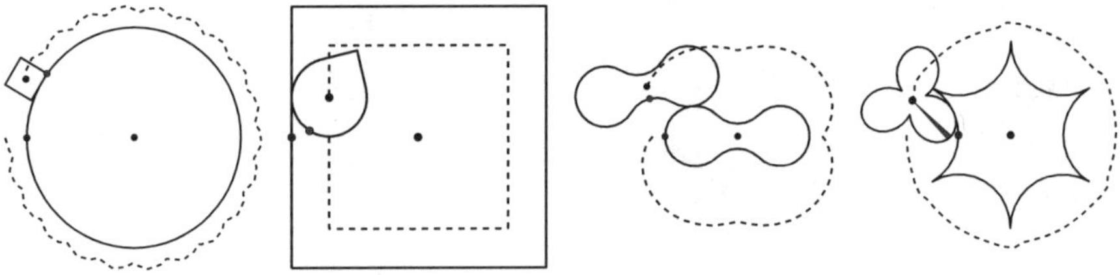

Figure 11.10. Examples 1–4 of discs (image sequences)

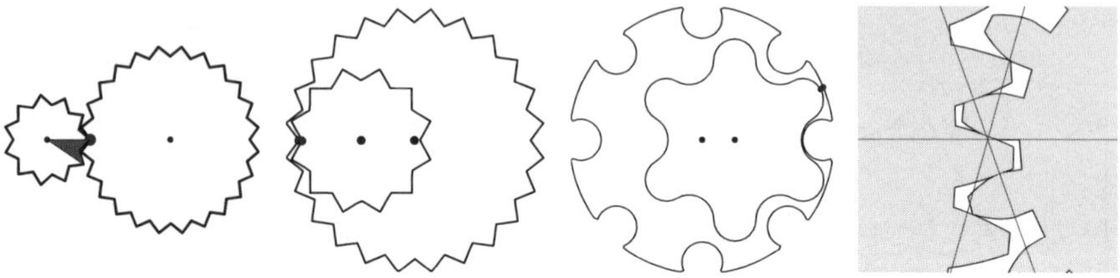

Figure 11.11. Examples 5–8 of discs (image sequences)

```
KAPITEL11/SECTION_4, Dancing Discs
At first both discs are to be constructed by
SCHEIBE01.M -- SCHEIBE24.M.
Both discs must touch each other at beginning.
demo1.m          Draws disc by manual input
demo2.m          Rolling of disc A onto or in disc B
                 with DISC_ROTATE.M
demo3.m          Rolling of disc A onto or in disc B
                 with BISECTION.M
demo4.m          Movie for discs
bisection.m      Method of bisection for computation
                 of rotational angle
disc_aendern.m   Geometry for DEMO1.M
disc-rotate.m    Geometry for DEMO2.M
```

11.5 Buckling of a Circular Plate

Consider a thin, circular, elastic plate of thickness h and radius R, subjected to a uniform lateral pressure p. Let $w(r)$ be the deflection of the plate normal to the unconstrained middle surface and $q(r)$ be the AIRY stress function with dimension of a force. Then VON KARMAN's equations (8.67) for rotationally symmetric deformation in polar coordinates are

$$\begin{aligned}\kappa\Delta^2 w(r) &= p + \frac{h}{r}\frac{d}{dr}\left(\frac{dq(r)}{dr}\frac{dw(r)}{dr}\right)\\ \Delta^2 q(r) &= -\frac{E}{2r}\frac{d}{dr}\left(\frac{dw(r)}{dr}\right)^2 .\end{aligned} \tag{11.17}$$

Here E is YOUNG's modulus, ν is POISSON's ratio and

$$\kappa = \frac{Eh^3}{12(1-\nu^2)}\,, \quad \Delta w = \frac{1}{r}\frac{d}{dr}[rw_r]_r\,.$$

The problem (11.17) is a nonlinear bending problem but spontane plate buckling is modelled by an eigenvalue problem as buckling of a beam in Sect. 7.2, where deflection jumps from zero in the new state and zero equilibrium exists further on but becomes unstable. Approximatively, we use the membrane equation (8.69), $-\mu\Delta w = p$, to replace the pressure p, and the bifurcation parameter μ becomes a boundary-force density $[N/L]$; see, e.g., (Machinek).

Then each of the equations (11.17) can be integrated once and the constants of integration can be evaluated by the condition of symmetry at the center, $r = 0$:

$$\kappa r\left[\frac{1}{r}[rw_r]_r\right]_r = -\mu r w_r + w_r q_r\,,\quad r\left[\frac{1}{r}[rq_r]_r\right]_r = -\frac{E}{2}(w_r)^2\,.$$

These equations can be transformed into dimensionless form

$$Lu(x) = -u(x)v(x) - \lambda x u(x)\,,\quad Lv(x) = \frac{1}{2}u^2(x) \tag{11.18}$$

where we have introduced the dimensionless quantities

$$\begin{aligned} x &= \frac{r}{R}\,,\quad v(x) = \frac{-(12(1-\nu^2))}{E\,R}\left(\frac{R}{h}\right)^2\frac{dq(r)}{dr} \\ u(x) &= -(12(1-\nu^2))^{1/2}\frac{R}{h}\frac{dw(r)}{dr}\,,\quad \lambda = \frac{(12(1-\nu^2))}{ER^2}\left(\frac{R}{h}\right)^2\mu \end{aligned} \tag{11.19}$$

and the linear differential operator

$$L(x): \varphi \mapsto L(x)\varphi = x\big[x^{-1}(x\varphi)_x\big]_x = x\varphi_{xx} + \varphi_x - x^{-1}\varphi$$

with fundamental system $\varphi_1(x) = x$ and $\varphi_2(x) = x^{-1}$. Boundary conditions at the center, $x = 0$, and edge of the plate, $x = 1$, must be specified to complete the formulation. From the assumed symmetry and regularity at the center we have $u(0) = v(0) = 0$. At the edge a variety of conditions may be imposed. Mainly, there are three conditions:
Case 1: Clamped with zero radial displacement, $u(1) = 0$, $v_x(1) = \nu v(1)$.
Case 2: Simply supported with zero radial displacement, $u_x(1) = -\nu u(1)$, $v_x(1) = \nu v(1)$.
Case 3: Simply supported with zero radial membrane stress, $u_x(1) = -\nu u(1)$, $v(1) = 0$.

A direct computation shows that

$$(L(x)\varphi,\psi) - (\varphi, L(x)\psi) = \varphi'(1)\psi(1) - \varphi(1)\psi'(1)\,,\quad \langle\varphi,\psi\rangle = \int_0^1 \varphi(x)\psi(x)\,dx\,;$$

cf. (5.60). Therefore L is selfadjoint with the same adjoint boundary conditions for u and likewise for v as above. Now

$$\begin{aligned} F(u,v,\lambda) &= \begin{bmatrix} L(x)u + \lambda x u + uv \\ L(x)v - u^2 \end{bmatrix} = \underline{0} \\ F_{(u,v)}(0,0,\lambda)\begin{bmatrix}\varphi\\ \psi\end{bmatrix} &= \begin{bmatrix} L(x) + \lambda x & 0 \\ 0 & L(x)\end{bmatrix}\begin{bmatrix}\varphi\\ \psi\end{bmatrix} = \underline{0}\,. \end{aligned} \tag{11.20}$$

Let $u_1(x) := J_1(x)$ be the BESSEL function of order 1 and degree 1, i.e., a solution of $x^2y'' + xy' + (x^2 - m)y = 0$ for $m = 1$ then $J_1(\sqrt{\lambda}x)$ is a solution of $x^2y'' + xy' + (\lambda x^2 - 1)y = 0$. Therefore $(\lambda_0, \varphi_0, \psi_0) = (\lambda_0, J_1(\sqrt{\lambda_0}x), 0)$ is a solution of the eigenvalue problem (11.20) where λ_0 follows from the boundary condition for $u(1)$, e.g.,

$$\begin{aligned} u(1) &= 0 & &\Longrightarrow J_1(\sqrt{\lambda_0}\cdot 1) = 0\,, & \lambda_0 &= 14.682\ldots \\ u'(1) &= -\nu u(1) & &\Longrightarrow J_1'(\sqrt{\lambda_0}x)\big|_{x=1} = -\nu J_1(\sqrt{\lambda_0}\cdot 1)\,, & \lambda_0 &= 4.198\ldots\,. \end{aligned}$$

To find the bifurcating solution let

$$u(\varepsilon, x) = \varepsilon u_1(x) + \varepsilon^2 U(\varepsilon, x)\,,\ v(\varepsilon, x) = \varepsilon V(\varepsilon, x)\,,\ \lambda(\varepsilon) = \lambda_0 + \varepsilon\xi(\varepsilon)\,.$$

Both differential equations in (11.20) are divided by x for adaption to the box scheme. Let $\widetilde{L}(x) = x^{-1}L(x)$ then solve iteratively the following system for fixed $\varepsilon > 0$ as large as possible that the iteration still converges (BC = boundary conditions):

$$\begin{aligned} [\widetilde{L}(x) + \lambda_0]U &= -\xi(u_1 + \varepsilon U) - x^{-1}(u_1 + \varepsilon U)V\,,\ \ \langle u_1, U\rangle = 0 & &\text{with BC for } u \\ \widetilde{L}(x)V &= 0.5\varepsilon\, x^{-1}(u_1 + \varepsilon U)^2 & &\text{with BC for } v \\ 0 &= \langle u_1, \xi x(u_1 + \varepsilon U) + (u_1 + \varepsilon U)V\rangle & &\langle u_1, xu_1\rangle \neq 0\,. \end{aligned}$$

The last equation is the branching equation. The optimal ε depends on the mesh width because the condition of the system varies with this parameter. Eventually, w and q must be recovered from u and v by means of (11.19). See also (Keller58), (Keener72). Figure 11.12 illustrates the non-dimensional results for $\nu = 0.3$ after some continuation w.r.t. λ. $w(x) = \int_0^x u(s)ds - w(1)$ is the non-dimensional deflection and $z(x) = -q(x)/x$ the non-dimensional radial membrane stress.

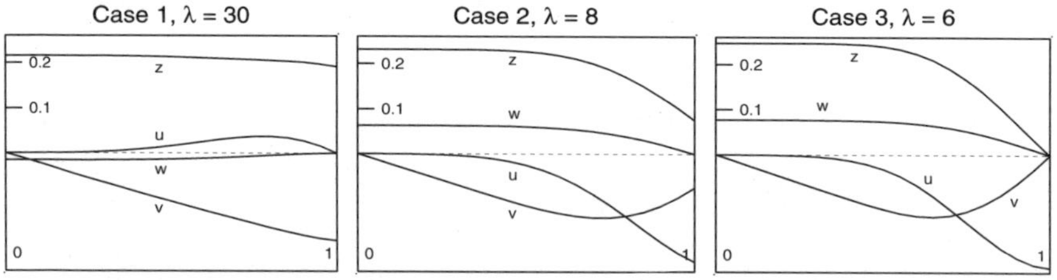

Figure 11.12. Buckling of circular plate

12

Appendix

12.1 Notations and Tables

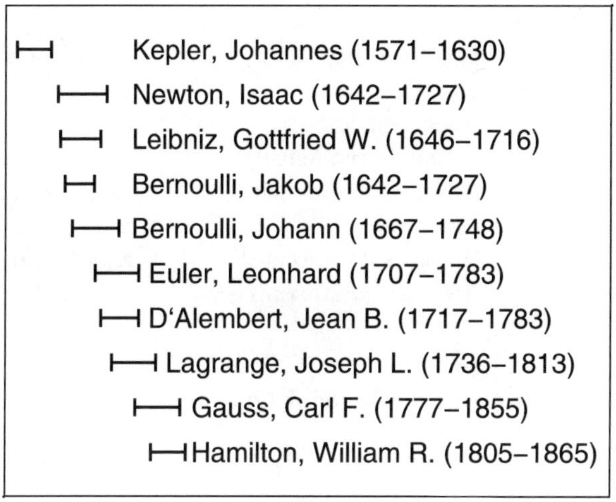
Kepler, Johannes (1571–1630)
Newton, Isaac (1642–1727)
Leibniz, Gottfried W. (1646–1716)
Bernoulli, Jakob (1642–1727)
Bernoulli, Johann (1667–1748)
Euler, Leonhard (1707–1783)
D'Alembert, Jean B. (1717–1783)
Lagrange, Joseph L. (1736–1813)
Gauss, Carl F. (1777–1855)
Hamilton, William R. (1805–1865)

Figure 12.1. Timetable

Remembering: (1°) Let A and B two sets. A "device" (mathematically *relation*) $f : A \to B$ is a *mapping* if $\forall : x, y \in A$: $f(x) \neq f(y) \Longrightarrow x \neq y$. A is the *domain* and $f(A) \subset B$ the *range* of f. The mapping f is *injective* if $x \neq y \Longrightarrow f(x) \neq f(y)$, *surjective* if $f(A) = B$, and *bijective* if it is both injective and surjective (hence invertible).

(2°) Let $\mathcal{U}$ and $\mathcal{V}$ be two vector spaces over $\mathbb{R}$ or $\mathbb{C}$ (where addition and scalar multiplication are denoted $\mathcal{U}$ and $\mathcal{V}$ wizh the same signs). Let $f : \mathcal{U} \to \mathcal{V}$ be a *linear* mapping, i.e.,

$$\forall\, u, v \in \mathcal{U} \;\; \forall\, \alpha, \beta \in \mathbb{R} : f(\alpha u + \beta v) = \alpha f(u) + \beta f(v)\,.$$

Then f is a *homomorphism* (*endomorphism* if $\mathcal{V} = \mathcal{U}$), f is a *isomorphism* (*automorphism* if $\mathcal{V} = \mathcal{U}$) if f bijective in addition.

(a) Notations All variable quantities have to be sufficiently smooth!
Elementary Notations:

$X \in \Omega$	material point, $x = \Phi(t, X)$ space point
$\underline{v}(t, x)$	velocity of material point X at time t
$\underline{b}(t, x)$	acceleration
$\underline{n}(t, x)$	normed normal vector at point x of the curve
$\underline{t}(t, x)$	tangential vector at point x
$\vartheta(t, x)$	temperature (> 0)

Spezific quantities (relative to mass unit):

$\varepsilon(t, x) \in \mathbb{R}$	spezific energy density	[energy/mass]
$c(t, x) \in \mathbb{R}$	spezific heat capacity	[heat/(mass · temperature)]
$\underline{f}(t, x) \in \mathbb{R}^3$	spezific volume force density	[force/mass]
$r(t, x) \in \mathbb{R}$	spezific thermal source density	[heat/(mass · time)]
$s(t, x) \in \mathbb{R}$	spezific entropy	[heat/(temperature · mass)]
$\psi = \varepsilon - \vartheta s$	free energy	[energy/mass]

Further quantities:

$\varrho(t, x) \in \mathbb{R}$	mass density	[mass/volume]
$\boldsymbol{\varepsilon}(t, x) \in {\mathbb{R}^3}_3$	strain tensor	[ ]
$\boldsymbol{\sigma}(t, x) \in {\mathbb{R}^3}_3$	stress tensor	[force/area]
$\underline{g}(t, x; \underline{n}(t, x)) \in \mathbb{R}^3$	surface force density	[force/area]
$\underline{k}(t, x) = \varrho(t, x)\underline{f}(t, x)$	volume force density	[force/volume]
$p(t, x) \in \mathbb{R}$	pressure	[force/area]
$\underline{q}(t, x) \in \mathbb{R}^3$	energy source vector	[energy/(area · time)]
	resp. thermal source vector	[heat/(area · time)]
$e = \varepsilon + \lvert\underline{v}\rvert^2/2$		

(b) Measure Units and Physical Quantities

Table 12.1. Simple SI-Units (System International):

Quantity	SI-Unit	Definition	Quantity	SI-Unit	Definition
length	meter	m	mass	kilogram	kg
time	second	s	temperature	Kelvin	K
angle	arc length	rad			

Table 12.2. Further SI-Units:

Physical Quantity	SI-Unit	Definition
force	Newton (N)	$N = kg \cdot m/s^2$
pressure	Pascal (Pa)	$N/m^2 = kg/(m \cdot s^2)$
energy	Joule (J)	$N \cdot m = kg \cdot m^2/s^2$
power	Watt (W)	$J/s = kg \cdot m^2/s^3$
pressure	stand. atmospere (atm)	$1\, atm = 101325\, Pa$
temperature	degree Celsius	$^oC = K - 273.15$
energy	calorie (cal)	$1\, cal\, (15^oC) = 4.1855\, J$

Table 12.3. Physical Constants:

Notation	Physical Quantity	Definition
$g = 9.81$	gravity acceleration (earth)	m/s^2
μ	viscosity	$Pa \cdot s = kg/(m \cdot s)$
$\nu = \mu/\varrho$	spezific or kinematic viscosity	m^2/s
η	mech. diffusion coefficient	m^2/s
κ	thermal conductivity	$J/(m \cdot s \cdot K)$
$\lambda = \kappa/(\varrho \cdot c)$	heat conduction coefficient	m^2/s
$\varrho \cdot c$	heat capacity	$J/(m^3 \cdot K)$
h	thermal transmission coefficent	$J/(m^2 \cdot s \cdot K)$
β	thermal expansion coefficient	$1/K$

Table 12.4. Non-Dimensional Quantities:

Quantity	Definition	Quantity	Definition
Froude number	$F_r = (R_e)^2/G_r$	Grashoff number	$G_r = g\,\beta \cdot \Delta\vartheta \cdot L^3/\nu^2$
Peclet number	$P_e = U \cdot L/\lambda$	Prandl number	$P_r = \nu/\lambda$
Rayleigh number	$R_a = G_r \cdot P_r$	Reynolds number	$R_e = U \cdot L/\nu$
Schmidt number	$S_c = \nu/\eta$		

(L, U characteristic length and velocity, $\Delta\vartheta$ characteristic temperature difference)

Table 12.5. Material Constants of Elastic Bodies and Fluids:

$E = \dfrac{\mu(3\lambda + 2\mu)}{\lambda + \mu} > 0$	Young's modulus (modulus of elasticity)
$G = \dfrac{E}{2(1+\mu)}$	shear modulus (modulus of rigidity, torsion modulus)
$K = \dfrac{E}{3(1-2\mu)}$	bulk modulus of elasticity (modulus of compression), bulk viscosity in fluids
$\nu = \dfrac{\lambda}{2(\lambda+\mu)}$	Poisson number
$\lambda = \dfrac{E\nu}{(1+\nu)(1-2\nu)} > 0$	Lamé constant, first coeff. of viscosity in fluids
$\mu = \dfrac{E}{2(1+\nu)} > 0$	Lamé constant (also shear modulus) second coefficient of viscosity in fluids
$K = \dfrac{3\lambda + 2\nu}{3}$	volumetric viscosity, bulk modulus
$\kappa = 1/K$	compressibility

$$\boxed{\lambda > 0 \text{ and } \mu > 0 \Longleftrightarrow 0 < \nu < \frac{1}{2} \text{ and } E > 0}$$

(c) Shape Functions for Complete Cubic Triangular Elements

$$\begin{aligned}
\psi_1 &= (1-\xi-\eta)[(1-\xi+2\eta)(1+2\xi-\eta)-16\xi\eta] &&= \zeta_1^2(3-2\zeta_1) - 7\zeta_1\zeta_2\zeta_3\\
\psi_2 &= \xi(1-\xi-2\eta)(1-\xi-\eta) &&= \zeta_1\zeta_2(\zeta_1-\zeta_3)\\
\psi_3 &= \eta(1-2\xi-\eta)(1-\xi-\eta) &&= \zeta_1\zeta_3(\zeta_1-\zeta_2)\\
\psi_4 &= \xi^2(3-2\xi) - 7\xi\eta(1-\xi-\eta) &&= \zeta_2^2(3-2\zeta_2) - 7\zeta_1\zeta_2\zeta_3\\
\psi_5 &= \xi^2(\xi-1) + 2\xi\eta(1-\xi-\eta) &&= \zeta_2^2(\zeta_2-1) + 2\zeta_1\zeta_2\zeta_3\\
\psi_6 &= -\xi\eta(1-2x-\eta) &&= -\zeta_2\zeta_3(\zeta_1-\zeta_2)\\
\psi_7 &= \eta^2(3-2\eta) - 7\xi\eta(1-\xi-\eta) &&= \zeta_3^2(3-2\zeta_3) - 7\zeta_1\zeta_2\zeta_3\\
\psi_8 &= -\xi\eta(1-\xi-2\eta) &&= -\zeta_2\zeta_3(\zeta_1-\zeta_3)\\
\psi_9 &= \eta^2(\eta-1) + 2\xi\eta(1-\xi-\eta) &&= \zeta_3^2(\zeta_3-1) + 2\zeta_1\zeta_2\zeta_3\\
\psi_{10} &= 27\xi\eta(1-\xi-\eta) &&= 27\zeta_1\zeta_2\zeta_3\,.
\end{aligned}$$

Succession of points in unit triangle and gravity center:

$$\begin{aligned}
&Q_1(0,0) : \psi_1\,,\ \psi_2\,,\ \psi_3\,,\quad Q_2(1,0) : \psi_4\,,\ \psi_5\,,\ \psi_6\,,\\
&Q_3(0,1) : \psi_7\,,\ \psi_8\,,\ \psi_9\,,\quad Q_4(1/3,1/3) : \psi_{10}.
\end{aligned}$$

Note that the representation of the complete cubic polynomial in Example 2.6 uses partly some *combinations* of shape functions; see also `SUPPLEMENT\chap09c`.

(d) Argyris' Triangle Element See Sect. 9.4(c). Let $\underline{x}_i$, $i = 1:3$, be the vertices of the triangle T and $\underline{x}_{12}$, $\underline{x}_{23}$, $\underline{x}_{31}$ the midpoints of the edges of which the lengths are denoted by ℓ_i. Let $C_i = (x_i, y_i, \mu_i, \zeta_i)$, $i = 1:3$, where

$$\mu_1 = (\ell_3^2 - \ell_2^2)/\ell_1^2\,,\ \mu_2 = (\ell_1^2 - \ell_3^2)/\ell_2^2\,,\ \mu_3 = (\ell_2^2 - \ell_1^2)/\ell_3^2\,,$$

are coefficients relating to the normal derivatives. Let also in barycentric coordiates ζ_i

$$\begin{aligned}
&\Psi(C_1, C_2, C_3; p(\underline{x}_1), \nabla p(\underline{x}_1), \nabla^2 p(\underline{x}_1), p_n(\underline{x}_{12}))\\
&= p(\underline{x}_1)\Big\{\zeta_1\big[1 + \zeta_2(1 + 3\zeta_1\zeta_2)(\zeta_1 - \zeta_2) + \zeta_3(1 + 3\zeta_1\zeta_3)(\zeta_1 - \zeta_3)\\
&\quad +2\zeta_2\zeta_3(3\zeta_1 - 1)\big] - 15\mu_1\zeta_1^2\zeta_2^2\zeta_3 + 15\mu_3\zeta_1^2\zeta_2\zeta_3^2\Big\}\\
&+ 2^{-1}\nabla p(\underline{x}_1)(\underline{x}_3 - \underline{x}_1)\cdot\\
&\quad\Big[\zeta_1\zeta_3\big[1 + \zeta_1 - \zeta_3 + \zeta_1(3\zeta_2 + \zeta_3) - \zeta_2(\zeta_2 + \zeta_3) + 3\zeta_1\zeta_3(\zeta_1 - \zeta_3)\big]\\
&\quad +(8 + 7\mu_3)\zeta_1^2\zeta_2\zeta_3^2 - 16\zeta_1^2\zeta_2^2\zeta_3\Big]\\
&+ 2^{-1}\nabla p(\underline{x}_1)(\underline{x}_2 - \underline{x}_1)\cdot\\
&\quad\Big[\zeta_1\zeta_2\big[1 + \zeta_1 - \zeta_2 + \zeta_1(3\zeta_3 + \zeta_2) - \zeta_3(\zeta_2 + \zeta_3) + 3\zeta_1\zeta_2(\zeta_1 - \zeta_2)\big]\\
&\quad +(8 - 7\mu_1)\zeta_1^2\zeta_2^2\zeta_3 - 16\zeta_1^2\zeta_2\zeta_3^2\Big]\\
&+ 4^{-1}(\underline{x}_3 - \underline{x}_1)^T\nabla^2 p(\underline{x}_1)(\underline{x}_3 - \underline{x}_1)\left[\zeta_3^2\zeta_1^2(1 + \zeta_1 - \zeta_3) + (4 + \mu_3)\zeta_1^2\zeta_2\zeta_3^2\right]\\
&+ 4^{-1}(\underline{x}_3 - \underline{x}_1)^T\nabla^2 p(\underline{x}_1)(\underline{x}_2 - \underline{x}_1)\left[4\zeta_1^2\zeta_2\zeta_3 - 8\zeta_1^2\zeta_2^2\zeta_3 - 8\zeta_1^2\zeta_2\zeta_3^2\right]\\
&+ 4^{-1}(\underline{x}_2 - \underline{x}_1)^T\nabla^2 p(\underline{x}_1)(\underline{x}_2 - \underline{x}_1)\left[\zeta_1^2\zeta_2^2(1 + \zeta_1 - \zeta_2) + (4 - \mu_1)\zeta_1^2\zeta_2^2\zeta_3\right]\\
&+ 2|T|16p_n((\underline{x}_1 + \underline{x}_2)/2)\zeta_1^2\zeta_2^2\zeta_3/\ell_1\,.
\end{aligned}$$

Then ARGYRIS' interpolating polynomial can be written in cyclic form and reads:

$$\begin{aligned}
\Phi_T(\zeta_1, \zeta_2, \zeta_3; p) &= \Psi(C_1, C_2, C_3; p(\underline{x}_1), \nabla p(\underline{x}_1), \nabla^2 p(\underline{x}_1), p_n(\underline{x}_{12}))\\
&+ \Psi(C_2, C_3, C_1; p(\underline{x}_2), \nabla p(\underline{x}_2), \nabla^2 p(\underline{x}_2), p_n(\underline{x}_{23}))\\
&+ \Psi(C_3, C_1, C_2; p(\underline{x}_3), \nabla p(\underline{x}_3), \nabla^2 p(\underline{x}_3), p_n(\underline{x}_{31}))\,.
\end{aligned}$$

The shape functions of the element can be found immediately by this representation in cyclic form; see also `SUPPLEMENT\chap09f`. MATLAB implementation in `KAPITEL02\TRIANGLES`.

12.2 Matrix Zoo

Let $A \in \mathbb{K}^n{}_n$ be a real or complex matrix, $\mathbb{K} \in \{\mathbb{R}, \mathbb{C}\}$. A^T denotes the transposed matrix and $A^H = \overline{A}^T$ the transposed conjugate matrix. Every vector norm $\|x\|$ defines a matrix norm (*operator norm*) by

$$\|A\| = \max\nolimits_{\|x\|=1}\|Ax\| \left(= \sup_{\|x\|<1}\|Ax\| = \sup_{x\neq 0}\frac{\|Ax\|}{\|x\|}\right)\,. \tag{12.1}$$

Three frequently used operator norms:

$\|A\|_\infty = \max_{1\le i\le n} \sum_{k=1}^{n} |a^i{}_k|$ row sum norm,

$\|A\|_1 = \max_{1\le k\le n} \sum_{i=1}^{n} |a^i{}_k|$ column sum norm,

$\|A\|_2 = (\varrho(A^H A))^{1/2}$ spectral norm or EUKLID norm.

Properties of general matrix norms:
(1°) $\|A\| \ge 0\,;\ \|A\| = 0 \Longleftrightarrow A = 0$ (positivity, else "semi-norm"),
(2°) $\forall\, \alpha \in \mathbb{R} : \|\alpha A\| = |\alpha|\|A\|$ (homogenity),
(3°) $A, B \in \mathbb{R}^m{}_n : \|A + B\| \le \|A\| + \|B\|$ (triangle inequality).
Moreover, *operator norms* have the crucial property
(4°) $\forall\, A\,,\, B \in \mathbb{C}^n{}_n : \|A \cdot B\| \le \|A\|\|B\|$ (submultiplicativity).

Operator norms are called also lub-norms ("least upper bound norm") because of the following property where $|||A|||$ denotes an arbitrary matrix norm:

$$\forall\, x : \|Ax\| \le |||A|||\, \|x\| \Longrightarrow \|A\| \le |||A|||\,, \quad \|A\| = \max_{\|x\|=1} \|Ax\|\,.$$

A similar to B $\Longleftrightarrow \exists\, X \in \mathbb{K}^n{}_n : B = XAX^{-1}$
A diagonally dominant $\Longleftrightarrow \forall\, k : |a^k{}_k| > \sum_{i\ne k} |a^i{}_k|$
or $\forall\, i : |a^i{}_i| > \sum_{k\ne i} |a^i{}_k|$
A diagonalizable $\Longleftrightarrow$ A similar to diagonal matrix
A Hermitian $\Longleftrightarrow A = A^H = \overline{A}^T$
A idempotent $\Longleftrightarrow A^2 = A$
A involutoric $\Longleftrightarrow A^2 = I$ (identity)
A non-negative $\Longleftrightarrow \forall\, i, k : a^i{}_k \ge 0$
A nilpotent $\Longleftrightarrow \exists\, k \in \mathbb{N} : A^k = 0$

A normal $\Longleftrightarrow A^H A = AA^H$
A normalizable $\Longleftrightarrow$ A similar to normal matrix
A orthogonal $\Longleftrightarrow$ A real and $A^T = A^{-1}$
A positive definite $\Longleftrightarrow$ A Hermitian and
$\forall\, 0 \ne x \in \mathbb{C}^n : x^H Ax > 0$
A permutation matrix $\Longleftrightarrow$ in every row and column precisely one element is one and all other zero
A reducibel $\Longleftrightarrow$ there exists a permutation matrix P and matrices B , C such that
$$PAP^T = \begin{bmatrix} B & D \\ 0 & C \end{bmatrix}$$
A real symmetric $\Longleftrightarrow$ A real and $A = A^T$
A stochastic $\Longleftrightarrow$ A non-negative and $\forall\, k : \sum_{i=1}^{n} a^i{}_k = 1$

A unitary $\Longleftrightarrow A^{-1} = A^H$
A upper triangular matrix $\Longleftrightarrow \forall\, i > k : a^i{}_k = 0$
A lower triangular matrix $\Longleftrightarrow \forall\, i < k : a^i{}_k = 0$
A projector $\Longleftrightarrow$ A Hermitian and idempotent
A reflector $\Longleftrightarrow$ A Hermitian and involutoric.

Some of these notations remain valid also for non-quadratic matrices.

$\sigma(A) := \{\lambda \in \mathbb{C}\,,\ \lambda \text{ eigenvalue of } A\}$	spectrum of A (quadratic),
$\varrho(A) = \max_{\lambda \in \mathrm{Sp}(A)} \vert\lambda\vert$	spectral radius of A (quadratic).

12.3 Translation and Rotation

It is easily shown by passing to polar or sperical coordinates that rigid motion consists locally of a translation and a rotation. We consider the rotation of a point of mass m about the *fixed* axis $\underline{a}$ of *length one* with radius vector $\underline{r}(t)$ of constant length $r := |\underline{r}(t)|$.

Linear Motion	Rotation
path $\underline{x}(t)$	angle of rotation $\underline{\varphi}(t) = \varphi(t)\underline{a}$ (vector)
velocity $\underline{v}(t) := \dot{\underline{x}}(t)$	angular velocity $\underline{\omega}(t) = \dot{\varphi}(t)\underline{a}$
	$\dot{\underline{r}}(t) = \underline{\omega}(t) \times \underline{r}(t)$
(linear) momentum	angular momentum $[ML^2T^{-1}]$
$\underline{\ell}(t) = m\underline{v}(t)$	$\underline{\ell}(t) = m\,\underline{r}(t) \times \dot{\underline{r}}(t) =: T\underline{\omega}(t)$
	$T = m\,r^2$ moment of inertia
force $\underline{k}(t) = m\,\ddot{\underline{x}}(t)$	$\underline{k}(t) = m\,\ddot{\underline{r}}(t)$ $[N]$
moment (of force)	torque $[NL]$
$\underline{p}(t) = \underline{x}(t) \times \underline{k}(t)$	$\underline{p}(t) = m\,\underline{r}(t) \times \ddot{r}(t)$

kinetic energy $[ML^2T^{-2}]$	rotational energy $[ML^2T^{-2}]$
$\dfrac{m\vert\underline{v}\vert^2}{2}$	$\dfrac{T\vert\underline{\omega}\vert^2}{2}$
work $[ML^2T^{-2}]$	
$\underline{k} \cdot \underline{x}$	$\underline{p} \cdot \underline{\varphi}$
power $[ML^2T^{-3}]$	
$\underline{k} \cdot \underline{v}$	$\underline{p} \cdot \underline{\omega}$

Rule:

angular velocity	replaces velocity
moment of inertia	replaces mass
torque	replaces force

.

For instance, NEWTON's axiom reads:

Force		Momentum	Torque		Angular Momentum
$\underline{k}$	$=$	$\dfrac{d}{dt}m\underline{v}(t)$	$\underline{p}(t)$	$=$	$\dfrac{d}{dt}T\underline{\omega}(t)$

Moment of Inertia: We have $\dot{\underline{r}} = \underline{\omega} \times \underline{r}$ hence, by the *representation formula* with constant $r = |\underline{r}|$

$$\underline{\ell} = m[\underline{r} \times \dot{\underline{r}}] = m[\underline{r} \times (\underline{\omega} \times \underline{r})] = m[r^2\underline{\omega} - (\underline{r} \cdot \underline{\omega})\underline{r}] = mr^2\underline{\omega} = T\underline{\omega}$$

because $\underline{r} \cdot \underline{\omega} = 0$.

Rotational Energy: We have $\dot{\underline{r}} = \underline{\omega} \times \underline{r}$ hence $\dot{r} = \omega r$. NEWTON's axiom $\underline{k} = m\ddot{\underline{r}}$ yields for the work (A)

$$A = \int_0^t \underline{k} \cdot \dot{\underline{r}}\, dt = m \int_0^t \ddot{\underline{r}}(t) \cdot \dot{\underline{r}}(t)\, dt = \frac{1}{2}m|\dot{\underline{r}}(t)|^2 - \frac{1}{2}m|\dot{\underline{r}}(0)|^2$$

hence, by $\dot{\underline{r}}(0) = 0$,

$$A = \frac{1}{2}mr^2|\underline{\omega}|^2 = \frac{1}{2}T|\underline{\omega}|^2 .$$

NEWTON*'s Equation:* We have

$$\frac{d}{dt}m(\underline{r} \times \dot{\underline{r}}) = m(\dot{\underline{r}} \times \dot{\underline{r}}) + m(\underline{r} \times \ddot{\underline{r}}) = \underline{r} \times (m \cdot \ddot{\underline{r}}) = \underline{r} \times \underline{k} = \underline{p} .$$

The *radius vector* $\underline{r}$ of the mass point may be replaced by its point vector, but then $\underline{x} \cdot \underline{\omega} \neq 0$ in the representation formula. Namely, by projection $\underline{x}_a$ of $\underline{x}$ onto $\underline{a}$,

$$\begin{aligned} \underline{r} &= \underline{x} - \underline{x}_a = \underline{x} - \frac{\underline{x} \cdot \underline{\omega}}{\underline{\omega} \cdot \underline{\omega}}\underline{\omega} \\ \dot{\underline{r}} &= \underline{\omega} \times \underline{r} = \underline{\omega} \times (\underline{x} - \frac{\underline{x} \cdot \underline{\omega}}{\underline{\omega} \cdot \underline{\omega}}\underline{\omega}) = \underline{\omega} \times \underline{x} . \end{aligned}$$

The projection $\underline{x}_a$ of $\underline{x}$ onto the axis of rotation is a constant vector. By $\underline{x} = \underline{x}_a + \underline{r}$ we then obtain $\dot{\underline{x}} = \dot{\underline{r}} = \underline{\omega} \times \underline{x}$.

Tensor of Inertia: By the representation formula we obtain for the angular momentum

$$\begin{aligned} \underline{\ell} &= m[\underline{r} \times \dot{\underline{r}}] = m[(\underline{x} - \kappa\underline{\omega}) \times (\underline{\omega} \times \underline{x})] = m[\underline{x} \times (\underline{\omega} \times \underline{x})] \\ &= m[(\underline{x}^T \underline{x})\underline{\omega} - (\underline{x}^T \underline{\omega})\underline{x}] = m[(\underline{x}^T \underline{x})\underline{\omega} - (\underline{x}\, \underline{x}^T)\underline{\omega}] \\ &= m[\underline{x}^T \underline{x}\, \boldsymbol{\delta} - \underline{x}\, \underline{x}^T]\underline{\omega} =: \underline{T}(\underline{x})\, \underline{\omega} . \end{aligned}$$

Accordingly, the moment of inertia T has now become a matrix $\underline{T}(\underline{x})$.

Let there n mass points $P(\underline{x}_i)$ be given with mass m_i then the *tensor of inertia* of the system of mass points is obtained by suming up all matrices $\underline{T}(\underline{x}_i)$ to

$$\mathbf{T} = \sum_{i=1}^{n} m_i[\underline{x}_i^T \underline{x}_i \boldsymbol{\delta} - \underline{x}_i\, \underline{x}_i^T] \in \mathbb{R}^3{}_3 .$$

ATTENTION: The point vectors $\underline{x}_i$ and the axis of rotation $\underline{a}$ on which the tensor applies relate to a *body fixed* (cartesian) coordinate system. In particular, the axis of rotation must pass through the origin of this coordinate system.

If the system is a rigid body with the geometric shape $\Omega \subset \mathbb{R}^3$ and mass desity ϱ, the sum is replaced by the integral and we obtain

$$\boxed{\mathbf{T} = \int_\Omega \varrho(\underline{x})[\underline{x}^T\underline{x}\,\boldsymbol{\delta} - \underline{x}\,\underline{x}^T]\,dV \in \mathbb{R}^3{}_3} \,.$$

12.4 Trigonometric Interpolation

In this section the letter i denotes always the *imaginary unit.*

(a) Fourier Series To begin with, we recall the main properties of FOURIER series.

Definition 12.1. *A function* $f : [a,b] \to \mathbb{C}$ *is* piecewise continuous *resp.* piecewise continuously differentible *if it is continuous resp. continuously differentiable in* $[a,b]$ *up to a finite number of points, and all possible one-sided limit values of* f *resp. of* f *and* df/dx *exist overall in* $[a,b]$ *.*

Let $f : [0,2\pi] \to \mathbb{C}$ be piecewise continuous.

(1°) *Complex* FOURIER *coefficients* of f:

$$c_j := \frac{1}{2\pi}\int_0^{2\pi} f(x)e^{-ijx}dx\,, \quad j \in \mathbb{Z}\,.$$

(2°) FOURIER *coefficients* of f in sinus-cosinus form:

$$a_n := \frac{1}{\pi}\int_0^{2\pi} f(x)\cos(nx)dx\,,\ n \in \mathbb{N}_0\,;\ b_n := \frac{1}{\pi}\int_0^{2\pi} f(x)\sin(nx)dx\,,\ n \in \mathbb{N}\,.$$

(3°) (Formal) FOURIER *series* of f:

$$S(x;f) := \sum_{j=-\infty}^{\infty} c_j e^{ijx} = \frac{a_0}{2} + \sum_{n=1}^{\infty}\big(a_n\cos(nx) + b_n\sin(nx)\big)\,.$$

(4°) *Transformation rules*:

$$c_0 = \frac{1}{2}a_0,\ c_n = \frac{1}{2}(a_n - ib_n)\,,\ c_{-n} = \frac{1}{2}(a_n + ib_n)\,,$$
$$a_0 = 2c_0,\ a_n = c_n + c_{-n}, \quad b_n = i(c_n - c_{-n})\,, \quad n \in \mathbb{N}\,.$$

If f is real-valued, $c_{-n} = \overline{c}_n$ and the coefficients a_n and b_n of the FOURIER series are likewise real. Using the notation $f(x\pm) = lim_{h\to 0, h>0} f(x \pm h)$ the following *Representation Theorem* collects the essential properties of FOURIER series $S(\,\cdot\,;f)$ being necessary for technical applications.

Theorem 12.1. *Let the 2π-periodic function $f : \mathbb{R} \to \mathbb{C}$ be piecewise continuously differentiable in $[0, 2\pi]$. Then*

(1°) *$S(\cdot\,; f)$ is uniformly convergent to f on $[a,b]$ if f is continuous on the finite intervall $[a,b] \subset \mathbb{R}$.*
(2°) *$S(x; f) = [f(x+) + f(x-)]/2$ for all $x \in \mathbb{R}$.*
(3°) GIBBS' *phenomenon occurs in every jumping point of f (overshooting of $\sim 18\ \%$ in the limit).*

Proof, e.g., (Meyberg). Obviously we may choose also the interval $[-\pi, \pi]$ for integration then

$$a_n = 0 \text{ if } f \text{ real and odd}, \quad b_n = 0 \text{ if } f \text{ real and even.}$$

Let now S_N be the N-th partial sum of the FOURIER series and T_N an arbitrary trigonometric polynomial of degree N,

$$S_N(x; f) = \sum_{j=-N}^{N} c_j e^{ijx}, \quad c_j = \frac{1}{2\pi} \int_0^{2\pi} f(x) e^{-ijx} dx, \quad T_N(x) = \sum_{j=-N}^{N} d_j e^{ijx}.$$

With the *mean energy of the signal* f for measure of distance,

$$||f||_2^2 = \frac{1}{2\pi} \int_0^{2\pi} f(x)^2 dx,$$

the *extremal property* of FOURIER series may be stated:

Theorem 12.2. *Let f be piecewise continuous in $[0, 2\pi]$ then*
(1°) $\lim_{N\to\infty} ||f - S_N||_2 = 0$, (2°) $||f - S_N||_2 \le ||f - T_N||_2$.

The next theorem provides details on the growing up of FOURIER coefficients; c.f., e.g., (Stoer):

Theorem 12.3. *Let the derivatives $f^{(i)}$, $i = 1, \ldots, m-1$, be continuous in $\mathbb{R}$ and let $f^{(m)}$ in $[0, 2\pi]$ be continuously differentiable. Then there exists a number $M > 0$ such that $|c_n| \le M/|n|^{m+1}$, $n \in \mathbb{Z}\backslash\{0\}$.*

Hence there exists a pointwise convergent majorant to the FOURIER series in case $m \ge 1$.

Example 12.1. (1°) (GIBBS' phenomenon.) Consider the *square-wave oscillation* with 2π-periodic function $f(x) = 1$ for $0 \le x < \pi$ and $f(x) = -1$ for $\pi \le x < 2\pi$. Direct computation yields the FOURIER coefficients

$$a_n = 0, \quad b_n = \begin{cases} 4/n\pi, & \text{for } n \text{ odd} \\ 0 & \text{for } n \text{ even}. \end{cases}$$

Accordingly, the FOURIER series reads:

$$S(x; f) = \frac{4}{\pi}\left(\frac{\sin x}{1} + \frac{\sin 3x}{3} + \frac{\sin 5x}{5} + \ldots\right) \sim \begin{cases} 1 & 0 < x < \pi \\ -1 & \pi < x < 2\pi \\ 0 & \text{else}. \end{cases}$$

In Figure 12.2 the FOURIER partial sums $S_N(x; f)$ are shown for $N = 1, 5, 11$.
(2°) (Fundamental example of FOURIER analysis.) The *trigonometric series* $\sum_{n=1}^{\infty}(\sin nx)/n$ represents a 2π-periodic *sawtooth oscillation*,

$$S(x; f) = \frac{\sin x}{1} + \frac{\sin 3x}{3} + \frac{\sin 5x}{5} + \ldots = \begin{cases} 0 & x = 0 \\ (\pi - x)/2 & 0 < x < 2 * \pi \,. \end{cases}$$

In Figure 12.3 the FOURIER partial sum $S_N(x; f)$ of the sawthooth oscillation is shown for $N = 7$.

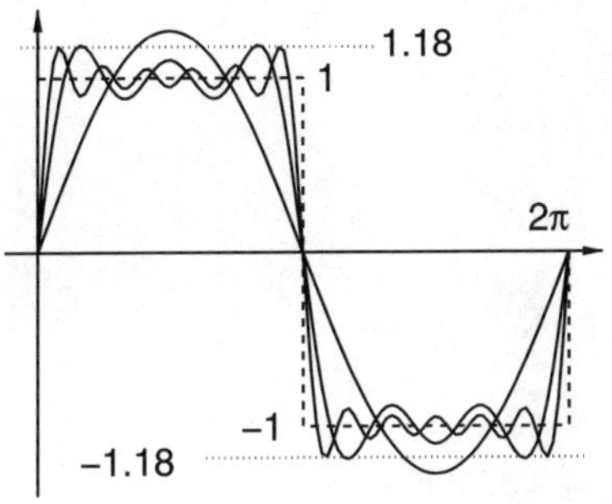

Figure 12.2. Example 12.1 (1°)

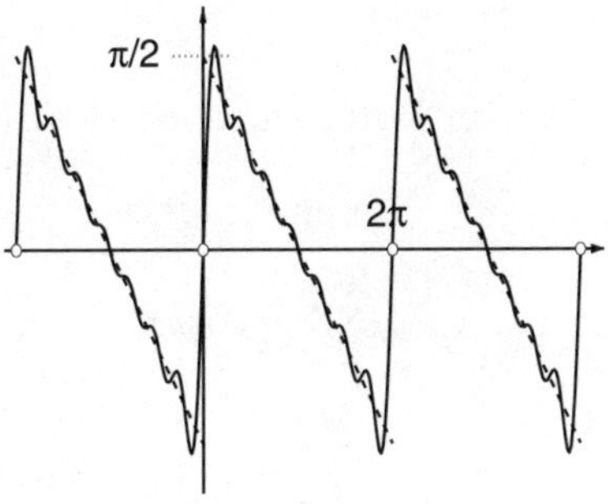

Figure 12.3. Example 12.1 (2°)

(b) Discrete Fourier Transformation The composite trapezoidal rule has a kind of *super-convergence* property in integration of smooth periodic functions by Lemma 2.3 therefore it is commonly used in integration of trigonometric series:

$$c_j = \frac{1}{2\pi}\int_0^{2\pi} f(x)e^{ijx}dx \simeq \frac{1}{2\pi}\frac{2\pi}{n}\sum_{k=0}^{n-1} f(x_k)e^{-ijx_k}\,, \quad x_k = 2\pi k/n\,. \tag{12.2}$$

In general,

$$y_j^* = \frac{1}{n}\sum_{k=0}^{n-1} y_k e^{-ijk2\pi/n}\,,\ j = 0 : n-1\,, \tag{12.3}$$

is called *discrete* FOURIER *transformation* (DFT) of the sequence $\{y_j\}_{j=0}^{\infty}$. ((IDFT) *inverse* discrete FOURIER transformation).

Let $\omega = e^{2\pi i/n}$ be the n-th complex root of unity then $\overline{\omega^k} = \overline{\omega}^k = \omega^{-k} = e^{-2k\pi i/n}$ and

$$y_j^* = \frac{1}{n}\sum_{k=0}^{n-1} y_k\omega^{-jk}\,,\ j = 0 : n-1\,,\quad F_n := [\omega^{jk}]_{j,k=0}^{n-1} \ \text{FOURIER matrix.}$$

The FOURIER matrix is complex but symmetric, $F_n = F_n^T$ (not Hermitian).

Lemma 12.1. $F_n\overline{F}_n = nI_n$.

By this result F_n is always invertible: $F_n^{-1} = \dfrac{1}{n}\overline{F}_n$. For $y = [y_0, \dots, y_{n-1}]^T$, $y^* = [y_0^*, \dots, y_{n-1}^*]^T$, we obtain

$$
\begin{aligned}
&\text{DFT} \qquad y^* = \frac{1}{n}\overline{F}_n y \Longleftrightarrow y_j^* = \frac{1}{n}\sum_{k=0}^{n-1} y_k\omega^{-jk} , \\
&\text{IDFT} \qquad y = F_n y^* \Longleftrightarrow y_k = \sum_{j=0}^{n-1} y_j^*\omega^{jk} .
\end{aligned}
\tag{12.4}
$$

Comparison with FOURIER series where $\varrho = 2\pi/n$:

finite	infinite
$y_j = \sum_{k=0}^{n-1} y_k^* e^{ijk\varrho}$	$f(t) = \sum_{k=-\infty}^{\infty} c_k e^{ikt},$
$y_k^* = \frac{1}{n}\sum_{j=0}^{n-1} y_j e^{-ijk\varrho}$	$c_k = \frac{1}{2\pi}\int_0^{2\pi} f(t)e^{-ikt}dt .$

Computational rules by n-periodic continuation $(y_{j+n} = y_j)$:

(1°) linearity: $\alpha y + \beta z \overset{DFT}{\longrightarrow} \alpha y^* + \beta z^*$,

(2°) translation: $r \in \mathbb{N}_0\ [y_{k+r}]_k \overset{DFT}{\longrightarrow} [\omega^{kr} y_k^*]_k$

(3°) periodic convolution: $y * z := \dfrac{1}{n}\left[\sum_{j=0}^{n-1} y_j z_{k-j}\right]_k \overset{DFT}{\longrightarrow} [y_k^* \cdot z_k^*]_k$,

(4°) PARSEVAL's equation: $\sum_{j=0}^{n-1} |y_j|^2 = \dfrac{1}{n}\sum_{k=0}^{n-1} |y_k^*|^2$.

Interpretation of (3°): n scalar products in time domain correspond to n multiplications in frequency domain.

(c) Trigonometric Interpolation (c1) Let

$$n = 2m \in \mathbb{N} ,\ \varrho = 2\pi/n ,\ \omega_n = e^{i\varrho} ,\ x_j = j\varrho ,\ j = 0 : n-1 .$$

Lemma 12.2. *There exists a unique complex trigonometric polynomial*

$$p(x; y) = \sum_{j=-m}^{m-1} y_j^* e^{ijx}$$

with the interpolation property

$$y_k = p(x_k; y) = \sum_{j=-m}^{m-1} y_j^* e^{ijx_k} = \sum_{j=-m}^{m-1} y_j^* e^{ijk\varrho}\,, \quad k = 0 : 2m-1\,. \tag{12.5}$$

The coefficients of this polynomial are

$$y_j^* = \frac{1}{2m} \sum_{k=0}^{2m-1} y_k e^{-ikx_j} = \frac{1}{2m} \sum_{k=0}^{2m-1} y_k e^{-ijk\varrho}\,, \quad j = -m : m-1\,. \tag{12.6}$$

Proof. Existence and Uniqueness follow from the corresponding result on complex interpolation polynomials. A simple computation shows that

$$\sum_{k=0}^{2m-1} e^{i(l-j)k\varrho} = y_j^* = \begin{cases} 1 \text{ for } l = j \\ 0 \text{ for } l \neq j \end{cases},$$

hence there follows for the representation of y_j^*

$$\begin{aligned} \frac{1}{2m} \sum_{k=0}^{2m-1} y_k e^{-ijk\varrho} &= \frac{1}{2m} \sum_{k=0}^{2m-1} \sum_{l=-m}^{m-1} y_l^* e^{ilx_k} e^{-ikx_j} \\ &= \sum_{l=-m}^{m-1} y_l^* \frac{1}{2m} \sum_{k=0}^{2m-1} e^{i(l-j)k\varrho} = y_j^*\,. \end{aligned}$$

(c2) Let $\varrho = 2\pi/n$ then

$$e^{-i(j-n)k\varrho} = e^{-ijk\varrho + ik2\pi} = e^{-i(-jk\varrho)}\,,$$

therefore we have $y_{j-n}^* = y_j^*\,,\ j = m : n-1$ hence

$$y_{-(m-k)}^* = y_{k-m}^* = y_{k+m-2m}^* = y_{k+m-n}^* = y_{m+k}^*\,, \quad k = 0 : m-1\,.$$

Instead of (12.5) and (12.6) we may thus use the pair

$$\begin{aligned} y_j^* &= \frac{1}{2m} \sum_{k=0}^{2m-1} y_k \omega_n^{-jk}\,, \; j = 0 : 2m-1\,, \\ y_k &= \sum_{j=0}^{2m-1} y_j^* \omega_n^{jk}\,, \qquad k = 0 : 2m-1\,, \end{aligned} \tag{12.7}$$

where $\underline{y}^* = [y_0^*, y_1^*, \ldots, y_{m-1}^*, y_{-m}^*, \ldots, y_{-1}^*]^T$, $\underline{y} = [y_0, y_1, \ldots, y_{2m-1}]^T$. By this equivalence, the *fast* FOURIER *transformation*, being usually implemented for the system (12.7), may be applied directly to the interpolation polynomial of Lemma 12.2.

(c3) Let now all ordinate values y_k be **real** then we obtain by (12.6) directly that $y_j^* = \overline{y_{-j}^*}$, $j = -m : m$; and y_m^* is real because

$$y_{-m}^* = \frac{1}{2m}\sum_{k=0}^{2m-1} y_k e^{-ik(-m)2\pi/2m} = \frac{1}{2m}\sum_{k=0}^{2m-1} y_k e^{ik\pi}$$
$$= \frac{1}{2m}\sum_{k=0}^{2m-1} y_k(-1)^k = \frac{1}{n}\sum_{k=0}^{2m-1} y_k e^{-ik\pi} = y_m^* .$$

Therefore, and by reasons of symmetry it is of advantage to apply the trigonometric interpolation polynomial in the following, slightly modified form,

$$p(x;\underline{y}) = \sum_{j=-m+1}^{m-1} y_j^* e^{ijx} + \frac{1}{2} y_m^* (e^{imx} + e^{-imx}) \qquad (12.8)$$

where the interpolation property of Lemma 12.2 is still fulfilled because

$$e^{imx_k} = e^{imk2\pi/2m} = e^{-imk2\pi/2m} = e^{-imx_k} , \quad k = 0 : 2m-1 .$$

For real ordinate values y_k, the interpolation polynomial (12.8) is real with complex-valued coefficients but it can be written also in entire real form for real y_k:

$$\begin{aligned} p(x;\underline{y}) &= y_0^* + \sum_{j=1}^{m-1}(y_j^* e^{ijx} + y_{-j}^* e^{-ijx}) + \frac{1}{2} y_m^*(e^{imx} + e^{-imx}) \\ &= y_0^* + 2\sum_{j=1}^{m-1}\left[(\mathrm{Re}\; y_j^*)\cos(jx) - (\mathrm{Im}\; y_j^*)\sin(jx)\right] + y_m^*\cos(mx) \\ y_0^* &= (2m)^{-1}\sum_{k=0}^{2m-1} y_k , \quad \text{and for } |j| \le m : \\ \mathrm{Re}\; y_j^* &= (2m)^{-1}\sum_{k=0}^{2m-1} y_k \cos(jk\varrho) , \quad \mathrm{Im}\; y_j^* = -(2m)^{-1}\sum_{k=0}^{2m-1} y_k \sin(jk\varrho) . \end{aligned}$$

Note that the trigonometric interpolation polynomial

$$\widetilde{p}(x;\underline{y}) = \sum_{j=0}^{2m-1} y_j^* e^{ijx} , \quad y_j^* = \frac{1}{2m}\sum_{k=0}^{2m-1} y_k e^{-ijk\varrho} , \quad j = 0 : 2m-1 , \qquad (12.9)$$

has the interpolation property, too, but it is not real for real values y_k in general. Instead, the real part has to be chosen for real interpolation in this case. Besides, this polynomial has worse stability properties because the sum is twice as large.

Example 12.2. $m = 1$:

$$m = 1 : y_0^* = \frac{1}{2}(y_0 + y_1)\,,\quad y_1^* = \frac{1}{2}(y_0 + y_1 e^{-i\pi}) = \frac{1}{2}(y_0 - y_1)$$

$$p(x; \underline{y}) = \frac{1}{2}(y_0 + y_1) + \frac{1}{2}(y_0 - y_1)\cos x$$

$$\widetilde{p}(x; \underline{y}) = \frac{1}{2}(y_0 + y_1) + \frac{1}{2}(y_0 - y_1)(\cos x + i \sin x)$$

$$m = 2 : p(x; \underline{y}) = \frac{1}{4}(y_0 + y_1 + y_2 + y_3) + \frac{1}{2}(y_0 - y_2)\cos x + \frac{1}{2}(y_1 - y_3)\sin x + \frac{1}{4}(y_0 - y_1 + y_2 - y_3)\cos 2x\,.$$

12.5 Further Properties of Vector Spaces

(a) Let $\mathcal{I}_k \subset \mathbb{R}$, $k \in \mathbb{N}$, be a *countable* set of open (or closed) intervals. A set $\mathcal{S} \subset \mathbb{R}$ is called *set of (*LEBESGUE*) measure zero* or briefly *null set* if

$$\forall\, \varepsilon > 0\ \exists\, \mathcal{I}_k\,,\ k \in \mathbb{N} : S = \bigcup_{k\in\mathbb{N}} I_k \ \text{ and } \ \sum_{k=1}^{\infty} |\mathcal{I}_k| \le \varepsilon\,.$$

A function $f : \mathbb{R} \supset \mathcal{I} \to \mathbb{R}$ has a property *almost everywhere* (a.e.) in $\mathcal{I}$ if it has that property in $\mathcal{I}$ up to a set of measure zero.

Let $\Delta_m = \{(x_0, x_1, \dots, x_m)\,,\ a = x_0 < x_1 < \dots < x_m = b\}$ be a partition of the interval $[a, b]$ then a function $f : [a, b] \to \mathbb{R}$ is called *absolutely continuous* if $\forall\, \varepsilon > 0\ \exists\, \delta > 0\ \forall\, m \in \mathbb{N}\ \forall\, \Delta_m$:

$$\sum_{j=1}^{m} |x_j - x_{j-1}| < \delta \implies \sum_{j=1}^{m} |\underline{f}(x_j) - \underline{f}(x_{j-1})| < \varepsilon\,.$$

In particular, absolutely continuous functions are continuous. They allow a generalization of the main theorem of differential and integral calculus to LEBESGUE-integrable functions:

Theorem 12.4. *Let $f : [a, b] \to \mathbb{R}$ be absolutely continuous then f is a.e. differentiable, the derivative is L-integrable and, by application of the* LEBESGUE-*Integral,*

$$f(x) = f(a) + \int_a^x f'(t)\,dt\,.$$

Proof see e.g. (Heuser80), Theorem 131.3.

(b) The number

$$\kappa_f := \int_a^b |df(x)| := \sup_{m\in\mathbb{N}} \sup_{\Delta_m} \sum_{i=1}^{m} |f(x_i) - f(x_{i-1})|$$

is called *variation* of f. If e.g. $f(x) = c$ is constant then $\kappa_f = c$, and if a function increases monotonically then $\kappa_f = f(b) - f(a)$. We now introduce two normed vector spaces namely the space of functions of **b**ounded **v**ariation and the subspace of **n**ormed functions of **b**ounded **v**ariation below:

$$\boxed{\begin{array}{l} \mathrm{BV}[a,b] = \{f : [a,b] \to \mathbb{R}\,,\ \kappa_f < \infty\}\,,\ \ \|f\| = \|f(a)\| + \kappa_f \\ \mathrm{NBV}[a,b] = \{f \in \mathrm{BV}[a,b]\,,\ f(b) = 0\,,\ \ f \text{ right-continuous in } [a,b]\} \end{array}}\ .$$

Some further Properties of functions $f : [a,b] \to \mathbb{R}$:

(1°) $f \in \mathrm{BV}[a,b]$ if and only if it is the difference $f = g - h$ of two monotonically increasing functions g and h; (Heuser80), I, Sect. 91.
(2°) A *continuous* function f is of bounded variation if and only if it is the difference $f = g - h$ of two continuous and monotonically increasing functions g and h; (Heuser80), I, Sect. 91.
(3°) f absolutely continuous then $f \in \mathrm{BV}[a,b]$, (Heuser80), II, Sect. 131, but the converse is not true.
(4°) The functions of $\mathrm{BV}[a,b]$ are continuous up to a countable set of points in which the one-sided limit values do exist, they are a.e. differentiable and R-integrable on $[a,b]$; (Heuser80), I, Sect. 91, II, SEct. 131.

Now we arrive at the main subject of this section and define the RIEMANN-STIELTJES integral as limiting value of RIEMANN sums for $f \in C[a,b]$ and $g \in \mathrm{BV}[a,b]$:

$$\int_a^b f(x)dg(x) = \lim_{m\to\infty} \sum_{j=1}^m f(\widetilde{x}_j)\big[g(x_j) - g(x_{j-1})\big]\,,\ \ \widetilde{x}_j \in [x_{j-1}\,,\,x_j]$$

where f is the *integrand* and g the *integrator*. This integral counts the jumpings of the *derivatives* of g multiplying them by the value of f at that points.

Example 12.3.(1°) If f scalar and $g(x) = |x|$ then

$$\int_{-1}^1 f(x)dg(x) = -\int_{-1}^0 f(x)\,dx + \int_0^1 f(x)\,dx + 2\,f(0)\,.$$

(2°) (Heuser80) I, Sect. 92. Let f and the derivative g' of g be R-integrable on $[a,b]$ then $\displaystyle\int_a^b f(x)dg(x)$ exists and and is equal to $\displaystyle\int_a^b f(x)g'(x)\,dx$.
(3°) (Heuser80) I, Sect. 92. Let $f \in C[a,b]$ and let g be a step function having exactly jumps of height $g_1, \ldots, g_m$ at the points $x_1, \ldots, x_m$, then

$$\int_a^b f(x)dg(x) = \sum_{i=1}^m f(x_i)g_i\,.$$

Conversely, if $f \in \mathrm{BV}[a,b]$ and g is continuous then the RS-integral $\int_a^b f(x)dg(x)$ exists always; (Heuser80). The following result shows that the vector space $\mathrm{BV}[a,b]$ is the dual space $(C[a,b])_d$ relative to the vector space $C[a,b]$ after canonical identification.

Lemma 12.3. *For $y \in (C[a,b])_d$ there exists an integrator $g \in BV[a,b]$ such that* $y(f) = \displaystyle\int_a^b f(x)dg(x)\,,\ \ \|y\| = \kappa_g\,.$

If we now require in addition that $g \in \mathrm{NBV}[a,b]$ then g is uniquely determined for every $y \in (C[a,b])_d$ and vice versa, and $\mathrm{BV}[a,b]$ can be replaced by $\mathrm{NBV}[a,b]$ in Lemma 12.3; cf. (Taylor), pp. 198–200.

(c) A normed vector space $\mathcal{X}$ is *separable* if there exists a finite or countable set $\mathcal{S}$ such that $\overline{\mathcal{S}} = \mathcal{X}$ holds for the closure of $\overline{\mathcal{S}}$ of $\mathcal{S}$, i.e. $\mathcal{S}$ is *dense* in $\mathcal{X}$.

Example 12.4. Let $\mathcal{I} = [a,b]$ be a comnpact interval with non-empty interior.

(1°) The vector space $\mathcal{B}(\mathcal{I})$ of bounded functions with maximum norm $\|f\|_\infty = \max_{x\in\mathcal{I}}|f(x)|$ is *not separable* but closed hence a BANACH space; (Taylor), pp. 89, 102.

(2°) The vector space $\mathcal{C}(\mathcal{I})$ of continuous functions with maximum norm $\|f\|_\infty = \max_{x\in\mathcal{I}}|f(t)|$ is separable since WEIERSTRASS' theorem 2.3 holds also for polynomials with rational numbers for coefficients, and it is closed hence a BANACH space because $\mathcal{C}(\mathcal{I})$ is closed in $\mathcal{B}(\mathcal{I})$; (Taylor), p. 103.

(3°) The space $\mathcal{C}^1(\mathcal{I})$ of continuously differentiable functions with the norm $\|f\| = \|f\|_\infty + \|f'\|_\infty$ is separable and closed as well (Amann).

(4°) Ignoring the distinction between $\widetilde{\mathcal{L}}$ and $\mathcal{L}$, cf. Sect. 1.7**(f)**, the vector space $\mathcal{L}_p(\mathcal{I})\,,\ 1 \le p < \infty$ consists of all *measurable* functions defined a.e. in $\mathcal{I}$ where $|f(x)|^p$ is L-integrable; (Taylor), pp. 16, 90, 372. Equipped with the norm $[\int_{\mathcal{I}}|f(x)|^p\,dx]^{1/p}$ (L-integral), these spaces are separable and closed hence BANACH spaces; (Taylor), p. 90.

(5°) The vector space $\mathcal{L}_\infty(\mathcal{I})$ consists of all measurable functions defined a.e. on $\mathcal{I}$ with the supremum norm $\|f\|_\infty = \inf\{\gamma\,,\ |f(t)| \le \gamma\,,\ \text{a.e. in } \mathcal{I}\}$. This space is not separable but closed; (Taylor), pp. 91, 104. A function $f \in \mathcal{L}_\infty(\mathcal{I})$ is L-integrable by LEBESGUE's theorem on dominated convergence.

(6°) The vector space $\mathcal{W}^1_\infty[a,b]$ consists of all continuous functions x where $x(t) = x(a) + \int_a^t y(\tau)\,d\tau$ in L-sense with $y \in \mathcal{L}_\infty[a,b]$.

12.6 Cycloids

(a) Orthocycloids Suppose that a disc with center M and radius r rolls on a straight line without friction in *positive* direction, then any fixed point P on the rolling disc describes an *orthocycloid* (Fig. 12.4). Choosing the angle φ between the straight line MP and the *negative* y-axis for independent variable such that $\varphi = 0$ at beginning, one obtains the representation

$$x = r\varphi - c\sin(\varphi + \alpha), \quad y = r - c\cos\varphi$$

where c denotes the distance between M and P, and α is the initial angle between the line MP and the negative y-axis.

(b) Epicycloids Suppose that a disc with center M and radius r rolls without friction *on* a fixed disc with center O and radius R then a fixed point P on the rolling disc describes an *epicycloid* (Figs. 12.5 and 12.6). Choosing the angle φ between the line OM and the x-axis for independent variable one obtains the representation

$$x = (R+r)\cos\varphi + c\cos\left(\frac{R+r}{r}\varphi + \alpha\right),$$
$$y = (R+r)\sin\varphi + c\sin\left(\frac{R+r}{r}\varphi + \alpha\right), \quad \varphi \geq 0,$$

where c is the distance between M and P again, and α is the initial angle between MP and the x-axis. The epizycloid is a closed curve for $R/r \in \mathbb{N}$.

(c) Hypocycloids Suppose that a disc with center M and radius r rolls without friction *in* a fixed disc with center O and radius R then a fixed point P on the rolling disc describes a *hypocycloid* (Fig. 12.7). Choosing the angle φ between OM and the x-axis for independent variable, one obtains the representation

$$x = (R-r)\cos\varphi + c\cos\left(\frac{R-r}{r}\varphi + \alpha\right),$$
$$y = (R-r)\sin\varphi - c\sin\left(\frac{R-r}{r}\varphi + \alpha\right),$$

where c is the distance between M and P, and α is the initial angle between MP and the x-axis. The hypocycloid is a closed curve for $R/r \in \mathbb{N}$ again.

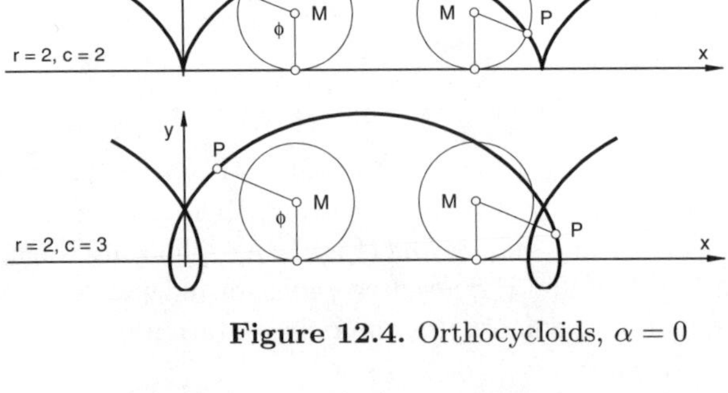

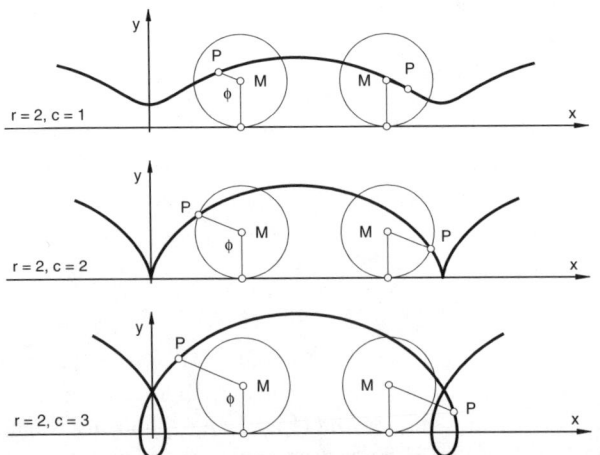

Figure 12.4. Orthocycloids, $\alpha = 0$

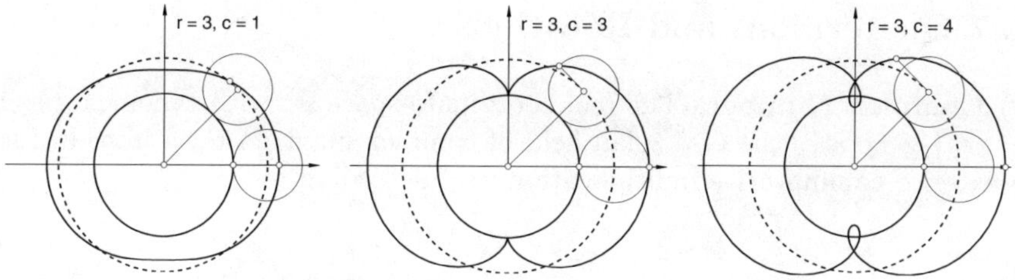

Figure 12.5. Epicycloids, $\alpha = 0\,,\ R = 6$

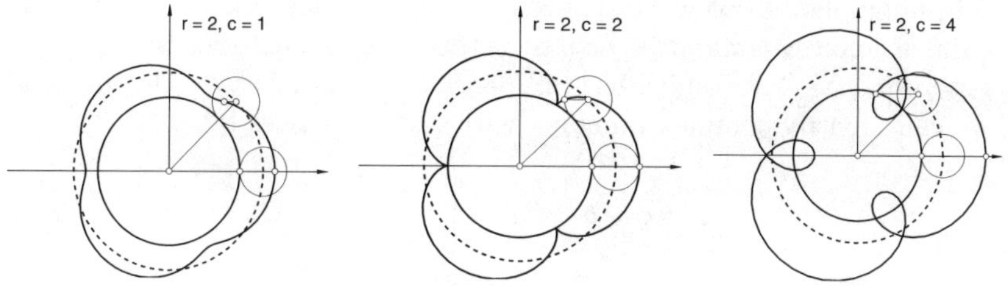

Figure 12.6. Epicycloids, $\alpha = 0\,,\ R = 6$

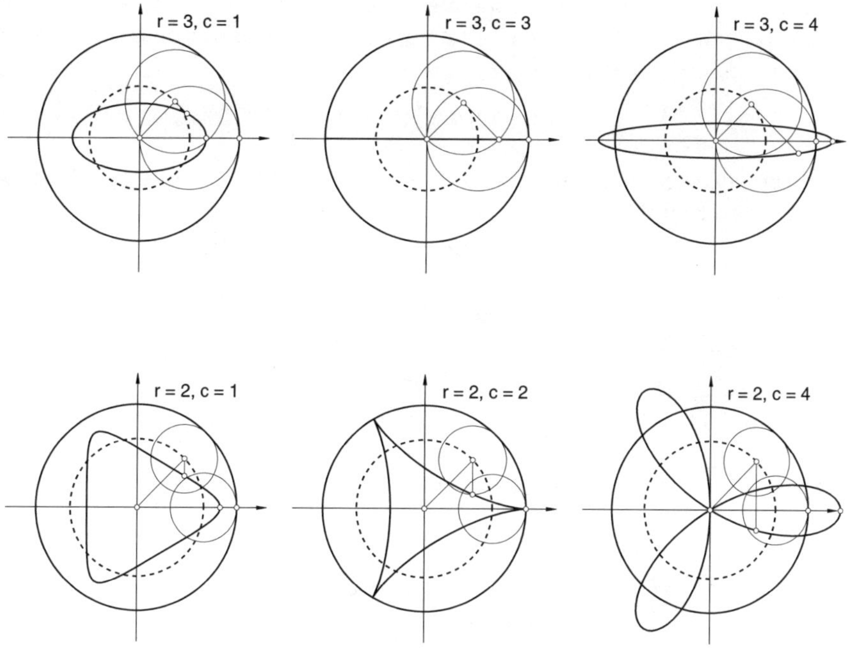

Figure 12.7. Hypozycloids, $\alpha = 0\,,\ R = 6$

12.7 Quaternions and Rotations

(a) Complex Numbers The real coordinate space $\mathbb{R}^2$ with canonical basis $\{\underline{e}_1, \underline{e}_2\}$ generates the GAUSSian field of complex numbers $\mathbb{C}$ by introducing an exterior commutative multiplication "$*$" according to

$$\underline{e}_1 * \underline{e}_1 = \underline{e}_1 \,,\; \underline{e}_1 * \underline{e}_2 = \underline{e}_2 * \underline{e}_1 = \underline{e}_2 \,,\; \underline{e}_2 * \underline{e}_2 = -\underline{e}_1 \,. \tag{12.10}$$

Then

$$(a\underline{e}_1 + b\underline{e}_2) * (c\underline{e}_1 + d\underline{e}_2) = (ac - bd)\underline{e}_1 + (ad + bc)\underline{e}_2$$

and we write simply $\underline{e}_1 = (1\,,\,0)\,,\, \underline{e}_2 = (0\,,\,i)$ or also $a + b\,i$ instead $a(1\,,\,0) + b(0\,,\,i)$ in unambiguous way. Mathematically spoken, the real $\mathbb{R}^2$ has become a two-dimensional *commutative field* by establishing an exterior product $\underline{w} = \underline{u} * \underline{v} \in \mathbb{R}^2$ for $\underline{u}\,,\,\underline{v} \in \mathbb{R}^2$ which has an identity and an inverse. On the other side, we can identify complex numbers with (2, 2)-matrices by

$$z = a + b\,i \sim Z = \begin{bmatrix} a & b \\ -b & a \end{bmatrix} \tag{12.11}$$

and then use the (perhaps more) familiar matrix multiplication instead of (12.10) which is also *commutative* for matrices of type (12.11).

(b) Quaternions Also the coordinate space $\mathbb{R}^4$ can be equipped with a suitable exterior product by an ingenious idea of HAMILTON, yielding however a *non-commutative* field of numbers (skew field). Consider two complex numbers $u = a + i\,b\,,\; v = c + i\,d$ and associate to that pair a matrix similarly built as in (12.11),

$$(u\,,\,v) \sim Q(u,v) = \begin{bmatrix} u & v \\ -\overline{v} & \overline{u} \end{bmatrix} ,$$

where the corresponding matrices (12.11) are inserted and overbar denotes the conjugate complex number. The matrix Q then reads explicitely

$$Q(a,b,c,d) = \left[\begin{array}{cc|cc} a & b & c & d \\ -b & a & -d & c \\ \hline -c & d & a & -b \\ -d & -c & b & a \end{array}\right] \tag{12.12}$$

and the product of two such matrices yields again a matrix of the same form. For the definition of quaternions relative to multiplication of basis elements we write briefly as in complex numbers

$$q = q_0 + q_1\,i + q_2\,j + q_3\,k \;\text{ then}$$
$$i^2 = j^2 = k^2 = -1\,,\; ij = -ji = k\,,\; ki = -ik = j\,.$$

The elements q of the skew field $\mathcal{Q}$ of quaternions are usually *not* considered as vectors but as "numbers" like the elements of complex field $\mathbb{C}$; therefore they are written as ordinary small letters not being underlined but with *non-commutative multiplication.* If we consider the quaternion as *vector* in $\mathbb{R}^4$, we write $\underline{q}$ instead q.

Note that the data of the quaternion are contained in the first row of the matrix Q in (12.12) thus the first row of the *matrix product* $Q(a,b,c,d)Q(e,f,g,h)$ contains the components of the vector $\underline{r}$ of the quaternion product $r = pq$. Using `KAPITEL12\quaternion.m` we obtain explicitely

$$\underline{p} = \begin{bmatrix} a \\ b \\ c \\ d \end{bmatrix}, \quad \underline{q} = \begin{bmatrix} e \\ f \\ g \\ h \end{bmatrix}, \quad \underline{r} = \begin{bmatrix} a\,e - b\,f - c\,g - d\,h \\ a\,f + b\,e + c\,h - d\,g \\ a\,g - b\,h + c\,e + d\,f \\ a\,h + b\,g - c\,f + d\,e \end{bmatrix}. \tag{12.13}$$

Further notations and properties:

$$\begin{array}{lll} \text{Conjugate } \overline{q} & & = q_0 - q_1 i - q_2 j - q_3 k \\ \text{Norm} & |q| & = \left(q_0^2 + q_1^2 + q_2^2 + q_3^2\right)^{1/2} \\ \text{Inverse} & q^{-1} & = \overline{q}/|q|^2 . \end{array}$$

$s(q) := q_0$ is the *scalar* of the quaternion q and $\mathrm{vec}(q) := q_1\underline{e}_1 + q_2\underline{e}_2 + q_3\underline{e}_3$ is the *vector* of q therefore $q = q_0 + \mathrm{vec}(q)$.

Let $\mathrm{vec}(p)$ and $\mathrm{vec}(q)$ be two *vector quaternions* then $s(\mathrm{vec}(p)\,\mathrm{vec}(q)) = -\,\mathrm{vec}(p)\cdot\mathrm{vec}(q)$ and $\mathrm{vec}(p\,q) = \mathrm{vec}(p)\times\mathrm{vec}(q)$ hence

$$\boxed{\mathrm{vec}(p)\,\mathrm{vec}(q) = -\,\mathrm{vec}(p)\cdot\mathrm{vec}(q) + \mathrm{vec}(p)\times\mathrm{vec}(q)}\,. \tag{12.14}$$

Of course a quaternion q can also be written with normed vector component $\widehat{\mathrm{vec}}(q)$

$$q = |q|\left(\frac{q_0}{|q|} + \frac{|\,\mathrm{vec}(q)|}{|q|}\frac{\mathrm{vec}(q)}{|\,\mathrm{vec}(q)|}\right) =: |q|\big(\cos\varphi + \sin\varphi\,\widehat{\mathrm{vec}}(q)\big)$$

where the *angle* φ is defined by

$$\cos\frac{\varphi}{2} := \frac{q_0}{|q|}, \quad \sin\frac{\varphi}{2} := \frac{|\mathrm{vec}(q)|}{|q|} \implies \cos^2\frac{\varphi}{2} + \sin^2\frac{\varphi}{2} = 1 \tag{12.15}$$

($\varphi/2$ is here chosen instead of φ because φ represents the rotation angle below).

(c) Composed Rotations Recall formula (1.11) for rotation about the axis $\underline{a} \in \mathbb{R}^3$, $|\underline{a}| = 1$, with angle φ,

$$\underline{y} = \underline{x}_a + \cos(\varphi)(\underline{x} - \underline{x}_a) + \sin(\varphi)(\underline{a}\times\underline{x})$$

where $\underline{x}_a = (\underline{x} \cdot \underline{a})\underline{a}$ is the projection of the position vector $\underline{x}$ onto the rotation axis $\underline{a}$. Recall also the representation (1.3),

$$\underline{a} \times (\underline{b} \times \underline{x}) = (\underline{a} \cdot \underline{x})\underline{b} - (\underline{a} \cdot \underline{b})\underline{x} = \left(\underline{b}\,\underline{a}^T - \underline{a}^T \underline{b}\, I\right) \underline{x}\,,$$

(I identity) therefore

$$\underline{y} = \underline{x} + (1 - \cos(\varphi))\underline{a} \times (\underline{a} \times \underline{x}) + \sin(\varphi)(\underline{a} \times \underline{x})\,.$$

Remember $1 - \cos\varphi = 2\sin^2(\varphi/2)$ and $\sin\varphi = 2\sin(\varphi/2)\cos(\varphi/2)$, assign a quaternion q of unit length to this rotation by

$$\boxed{q = q_0 + \mathrm{vec}(q) = \cos\left(\frac{\varphi}{2}\right) + \sin\left(\frac{\varphi}{2}\right) \underline{a}}\,, \tag{12.16}$$

and let C_q be the skew-symmetric matrix defined by $C_q\underline{x} = \mathrm{vec}(q) \times \underline{x}$ as in Sect. 1.1(**i**). Then

$$\begin{aligned}\underline{y} &= \underline{x} + 2\,\mathrm{vec}(q) \times (\mathrm{vec}(q) \times \underline{x}) + 2q_0\ \mathrm{vec}(q) \times \underline{x} \\ &= \left[I + 2\big(\mathrm{vec}(q)\,\mathrm{vec}(q)^T - \mathrm{vec}(q)^T\,\mathrm{vec}(q) I\big) + 2q_0 C_q\right]\underline{x}\,.\end{aligned}$$

Suppose now that we have two rotations with rotation matrices $D(\underline{a}, \varphi)$, $D(\underline{b}, \psi)$, assign the corresponding quaternions p and q by (12.16) and form the product $r = p\,q$ then the rotation matrix is

$$D(\underline{c}, \vartheta) := D(\underline{b}, \psi)D(\underline{a}, \varphi) = I + 2\big(\mathrm{vec}(r)\,\mathrm{vec}(r)^T - \mathrm{vec}(r)^T\,\mathrm{vec}(r) I\big) + 2r_0 C_r\,.$$

The scalar r_0 and vector $\mathrm{vec}(r)$ of r are found by (12.14), and the associated rotation angle ϑ and the new axis by normalization and (12.15).

Example 12.5. First rotation with axis $\underline{a} = \underline{e}_2$ and angle $\varphi = \pi/2$, second rotation with axis $\underline{b} = \underline{e}_1$ and angle $\psi = \pi/2$. Rotation matrix $D(\underline{c}, \vartheta) = D(\underline{e}_1, \psi)D(\underline{e}_2, \varphi)$. Assigned quaternions are p for the first rotation and q for the second rotation,

$$\begin{aligned}p &= \cos(\varphi/2) + \sin(\varphi/2)\,\underline{a} = \cos(\pi/4) + \sin(\pi/4)\,\underline{e}_2 = \frac{\sqrt{2}}{2}(1 + j) \\ q &= \cos(\psi/2) + \sin(\psi/2)\,\underline{b} = \cos(\pi/4) + \sin(\pi/4)\,\underline{e}_1 = \frac{\sqrt{2}}{2}(1 + i)\,.\end{aligned}$$

Let $r = q\,p$ be the quaternion of the composed rotation then we obtain $\underline{r} = 0.5\,[1\,,\,1\,,\,1\,,\,1]$ by applying (12.13). Therefore the new normed rotation axis is $\underline{c} = [1\,,\,1\,,\,1]/\sqrt{3}$ and by $r_0 = \cos(\vartheta/2) = 1/2$ we find also the rotation angle $\vartheta = 2\pi/3$.

References

Abraham. Abraham, R., et al.: Manifolds, Tensor Analysis, and Applications. Springer, Berlin Heidelberg New York (1983)

Acheson. Acheson, D.: Vom Calculus zum Chaos, An Introduction to Dynamics. University Press, Oxford (1997)

Agmon. Agmon, Sh.: Elliptic Boundary Value Problems. Van Nostrand, New York (1965)

Akhiezer. Akhiezer, N.I.: The Calculus of Variations. Blaisdell, New York (1962)

Allgower89. Allgower, E.L., et al.: Large sparse continuation problems. J. Comp. Appl. Math. **26**, 3–21 (1989)

Allgower90. Allgower, E.L., Georg, K.: Numerical Continuation Methods. Springer, Berlin Heidelberg New York (1990)

Altenbach. Altenbach, J., Altenbach, H.: Einführung in die Kontinuumsmechanik. Teubner, Stuttgart (1994)

Amann. Amann, H.: Gewöhnliche Differentialgleichungen. De Gruyter, Berlin (1983)

Arenstorf. Arenstorf, R.F.: Periodic solutions of the restricted three body problem representing analytic continuations of Keplerian elliptic motions. Amer. J. Math., **85**, 27–35 (1963)

Argyris. Argyris, J.H, et al.: The TUBA family of plate elements for the matrix displacement method. The Aeronautical J. of the Royal Aeronautical Society, **72**, 701–709 (1968).

Arnold78. Arnold, V.I.: Mathematical Methods of Classical Mechanics. Springer, Berlin Heidelberg New York (1978)

Arnold80. Arnold, V.I.: Gewöhnliche Differentialgleichungen. Springer, Berlin Heidelberg New York (1980)

Arnold90. Arnold, V.I.: Huygens and Barrow, Newton and Hooke. Birkhäuser, Basel (1990)

Babuska63. Babuska, I.: The theory of small changes in the domain of existence of partial differential equations and its applications. In: Differential Equations and their Applications. Academic Press, New York (1963)

Babuska90. Babuska, I., Pitkaeranta, J.: The plate paradox for hard and simple soft support. SIAM J. Math. Anal. **21**, 551–576 (1990)

Bank. Bank, R.E., Mittelmann, H.D.: Stepsize selection in continuation problems. J. Comp. Appl. Math. **26**, 67–77 (1989)

Barner. Barner, M., Flohr, F.: Analysis II. De Gruyter, Berlin (1983)

Barragy. Barragy, E., Carey, G.F.: Stream function vorticity solutions using high-p element-by-element techniques. Comm. Numer. Meth. Eng. **9**, 387–395 (1993)

Barton. Barton, I.E.: The entrance effect of laminar flow over a backward-facing step geometry. Int. J. Numer. Meth. Fluids **25**, 633–644 (1997).

Batoz. Batoz, J-L., et al.: A study of three-node triangular plate bending elementes. Int. J. Num. Meth. Eng. **15**, 1771–1812 (1980)

Becker75. Becker, E., Bürger, W.: Kontinuumsmechanik. Teubner, Stuttgart (1975)

Becker80. Becker, K.H., Seydel, R.: A Duffing equation with more than 20 branching points. In: E.L. Allgower et al. (eds.) Lecture Notes Bd. 878. Springer, Berlin Heidelberg New York (1980)

Bell. Bell, K.: A refined triangular plate bending finite element. Internat. J. Numer. Meth. Eng. **1**, 101–122 (1969)

Ben-Israel. Ben-Israel, A.: Linear equations and inequalities in finite dimensional, real or complex, vector spaces: a unified theory. J. Math. Anal. Appl. **27**, 367–389 (1969)

Berger. Berger, M., Gostiaux, B.: Differential Geometry: Manifolds, Curves, and Surfaces. Springer, Berlin Heidelberg New York (1988)

Berkowitz. Berkowitz, L.D.: Optimal Control Theory. Springer, Berlin Heidelberg New York (1974)

Bertram. Bertram, A.: Axiomatische Einführung in die Kontinuumsmechanik. Wissenschaftsverlag, Mannheim (1989)

Bertrand. Bertrand, J.: Théorème relatif au movement d'un point attiré vers un centre fixe. Compt. Rend. **77**, 849–853 (1873)

Best. Best, M.J., Ritter, K.: Linear Programming. Prentice Hall, Englewood Cliffs, N.Y. (1985)

Betten. Betten, J.: Tensorrechnung für Ingenieure. Teubner, Stuttgart (1987)

Bhatia. Bhatia, A.B., Singh, R.N.: Mechanics of Deformable Media. Adam Hilger, Bristol Boston (1986)

Birkhoff. Birkhoff, G.D., Kellogg, O.D.: Invariant points in function space. Trans. Amer. Math. Soc. **23**, 96–115 (1922).

Bishop. Bishop, R.L., Goldberg, S.: Tensor Analysis on Manifolds. Macmillan, New York (1968)

Blanchard. Blanchard, Ph., Brüning, E.: Direkte Methoden der Variationsrechnung. Springer, Berlin Heidelberg New York (1982)

Bogacki. Bogacki, P., Shampine, L.F.: A 3(2) pair of Runge-Kutta formulas. Appl. Math. Lett. **2**, 1–9 (1989)

Bourne. Bourne, D.E., Kendall, P.C.: Vektoranalysis. Teubner, Stuttgart (1973)

Bowen. Bowen, R.M., Wang, C.C.: Introduction to Vectors and Tensors. Plenum Press, New York London (1976)

Braess. Braess, D.: Finite Elemente. Springer, Berlin Heidelberg New York (1997)

Brasey92. Brasey, V.: A half-explicit Runge-Kutta method of order 5 for solving constrained mechanical systems. Computing **48**, 191–201 (1992)

Brasey93. Brasey, V., Hairer, E.: Half-explicit Runge-Kutta methods for differential-algebraic systems of index 2. SIAM J. Numer. Anal. **30**, 538–552 (1993)

Brenner. Brenner, S.C., Scott, L.R.: The Mathematical Theory of Finite Element Methods. Springer, Berlin Heidelberg New York (1994)

Brezzi. Brezzi, F., Fortin, M.: Mixed and Hybrid Finite Element Methods. Springer, Berlin Heidelberg New York (1991)

Bryson. Bryson, A.E., Ho, J.C.: Applied Optimal Control. Wiley, New York (1975)

Budden. Budden, P.J., Norbury, J.: A non-linear elliptic eigenvalue problem. J. Inst. Math. Appl. **24**, 9–33 (1979)

Burg. Burg, Kl., Haf, H, Wille, F.: Höhere Mathematik für Ingenieure, I–V. Teubner, Stuttgart (1991)

Burges. Burges, D., Graham, A.: Introduction to Control Including Optimal Control. Horwood, Chichester (1980)

Butcher. Butcher, J.C.: On the attainable Order of Runge-Kutta methods. Math. Computation **19**, 408–417 (1965)

Byrd. Byrd, R.H., et al.: An interior point algorithm for large-scale nonlinear programming. SIAM J. Optim. **9**, 877–900 (1999)

Calahan. Calahan, D.A.: A stable, accurate method of numerical integration for nonlinear systems. Proc. IEEE **56**, 744 (1968)

M.Chen. Chen, M.: On the solution of circulant linear systems. SIAM J. NumerȦnal. **24**, 668–683 (1987)

X.Chen. Chen, X.-N.: Hydrodynamics of Wave-Making in Shallow Water. Thesis, Universität Stuttgart, Stuttgart (1999)

Cheng. Cheng, M., et al.: A hybrid vortex method for flows over a bluff body. Int. J. Num. Meth. Fluids **24**, 253–274 (1997)

Chorin. Chorin, A.J., Marsden, J.E.: A Mathematical Introduction to Fluid Mechanics. Springer, Berlin Heidelberg New York (1979)

Chow. Chow, Sh.-N., Hale, J.H.: Methods of Bifurcation Theory. Springer, Berlin Heidelberg New York (1982)

Ciarlet79. Ciarlet, Ph. G.: The Finite Element Method for Elliptic Problems. North–Holland, Amsterdam (1979)

Ciarlet93. Ciarlet, Ph. G.: Mathematical Elasticity I, Three-Dimensional Elasticity. North–Holland, Amsterdam (1993)

Ciarlet96. Ciarlet, Ph.G., Lions, J.L.: Handbook of Numerical Analysis. North–Holland, Amsterdam (1996)

Ciarlet97. Ciarlet, Ph. G.: Mathematical Elasticity II, Theory of Plates. North Holland, Amsterdam (1997)

Ciarlet00. Ciarlet, Ph. G.: Mathematical Elasticity III, Theory of Shells. Elsevier, Amsterdam (2000)

Clarke. Clarke, F.H.: Optimization and Nonsmooth Analysis. Wiley, New York (1983)

Clegg. Clegg, J.C.: Calculus of Variations. Oliver & Boyd, Edinburgh (1968)

Cole. Cole, R.H.: Theory of Ordinary Differential Equations. McGraw-Hill, New York (1955)

Collatz60. Collatz, L.: The Numerical Treatment of differential Equations. Springer, Berlin Heidelberg New York (1960)

Collatz63. Collatz, L.: Eigenwertaufgaben mit technischen Anwendungen. Teubner, Leipzig (1963)

Courant. Courant, R., Hilbert, D.: Methoden der Mathematischen Physik, I. Springer, Berlin Heidelberg New York (1968)

Crandall. Crandall, M.G., Rabinowitz, P.H.: Bifurcation from simple eigenvalues. J. Functional Anal. **8**, 321–340 (1971)

Craven78. Craven, B.D.: Mathematical Programming and Control Theory. Chapman and Hall, London (1978)

Craven95. Craven, B.D.: Control and Optimization. Chapman and Hall, London (1995)

Crouzeix73. Crouzeix, M., Raviart, P.A.: Conforming and non-conforming finite element methods for solving stationary Stokes equations. R.A.I.R.O. **R-3**, 77–104 (1973)

Crouzeix75. Crouzeix, M.: Sur l'approximation des equations differentielles operationelles lineaires par des methods de Runge-Kutta. Thesis. Universite de Paris (1975)

Crouzeix80. Crouzeix, M., Raviart, P.-A.: Approximation des Problemes d'Evolution. Unpublished manuskript. Universite de Paris (1980).

Cuvelier. Cuvelier, C., et al.: Finite Element Methods and Navier Stokes Equations. Reidel, Boston (1986)

DeBoor. DeBoor, C.: A Practical Guide to Splines. Springer, Berlin Heidelberg New York (1978)

Decker. Decker, K.H.: Maschinenelemente. Hanser, München (1997)

Decker/Keller. Decker, D.W., Keller, H.B.: Multiple limit point bifurcation. J. Math. Anal. Appl. **75**, 417–430 (1980)

Dekker. Dekker, K., Verwer, J.G.: Stability of Runge-Kutta Methods for stiff nonlinear differential equations. North Holland, Amsterdam (1984)

Dellnitz. Dellnitz, M.: Computational bifurcation of periodic solutions in systems with symmetry. IMA J. Numer. Anal. **12**, 429–455 (1992)

Demoulin. Demoulin, Y.-M. J., Chen, Y.M.: An iteration method for solving nonlinear eigenvalue problems. SIAM J. Appl. Math. **28**, 588–595 (1975)

Denninger. Denninger, G.: Univ. Stuttgart, Private Mitteilung. (2005)

Dennis. Dennis, J.E., Schnabel, R.B.: Numerical Methods for Unconstrained Optimization and Nonlinear Equations. Prentice Hall, Englewood Cliffs, N.J. (1983)

Deuflhard84. Deuflhard, P.: Computation of periodic solutions of nonlinear ODEs. BIT **24**, 456–466 (1984)

Deuflhard87. Deuflhard, P.: Efficient numerical pathfollowing beyound critical points. SIAM J. Numer. Anal. **24**, 912–927 (1987)

Dieudonne. Dieudonné, J.: Foundations of Modern Analysis. Academic Press, New York (1960)

DIN84. DIN–Taschenbuch 202. Beuth, Berlin (1984)

Dormand. Dormand, J.R., Prince, P.J.: A family of imbedded Runge-Kutta formulae. J. Comp. Appl. Math. **6**, 223–232 (1980)

Dyer. Dyer, P., McReynolds, S.R.: The Computation and Theory of Optimal Control. Academic Press, New York (1970)

Ehle. Ehle, B.L.: High order A-stable methods for the numerical soluiton of systems of. D.E.'s. BIT **8**, 276-278 (1968)

Ekeland. Ekeland, I., Temam, R.: Convex Analysis and Variational Problems. North–Holland, Amsterdam (1976)

Ergatoudis. Ergatoudis, J.G. et al.: Curved isoparametric quadrilateral elements for finite element analysis. Int. J. Solids Struct. **4**, 31–42 (1968)

Evans. Evans, L.C.: Partial Differential Equations. AMS, Rhode Island (1998)

Fiedler. Fiedler, B.: Global Hopf bifurcation of two-parametric flows. Arch. Rat. Mech. Anal. **94**, 59–81 (1986)

Flanders. Flanders, H.: Differential Forms. Academic Press, New York (1963)

French. French, A.P.: Newtonsche Mechanik. De Gruyter, Berlin (1996)

Gekeler84. Gekeler, E.W.: Discretization Methods for Stable Initial Value Problems. Lecture Notes Math. Bd. 1044, Springer, Berlin Heidelberg New York (1984)

Gekeler86. Gekeler, E.W., Widmann, R.: On the order conditions of Runge-Kutta methods with higher derivatives. Numer. Math. **50**, 183–203 (1986)

Gekeler89. Gekeler, E.W.: On the numerical solution of double-periodic elliptic eigenvalue problems. Computing **43**, 97–114 (1989)

Gekeler92. Gekeler, E.W.: On trigonometric collocation and Hopf bifurcation. In: E.Allgower et al. (ed) Bifurcation and Symmetry, ISNM vol.104. Birkhäuser, Basel (1992)

Gekeler95. Gekeler, E.W.: On the perturbed eigenvalue problem. J. Math. Anal. Appl. **191**, 540–546 (1995)

Gelfand. Gelfand, I.M., Fomin, S.V.: Calculus of Variations. Prentice–Hall, Englewood Cliffs, N.J. (1963)

Gervais. Gervais, J.J, et al.: Some experiments with stability analysis of discrete incompressible flows in the lid-driven cavity. Int. J. Num. Meth. Fluids **24**, 477–492 (1997)

Girault. Girault, V., Raviart, P.-A.: Finite Element Methods for Navier-Stokes Equations. Springer, Berlin Heidelberg New York (1986)

Glass. Glass, L.: Combinatorial and topological methods in nonlinear chemical kinetics. J. Chem. Phys. **63**, 1325–1335 (1975)

Goldfarb. Goldfarb, D., Idnani, A.: A numerically stable dual method for solving strictly convex quadratic programs. Math. Programming **27**, 1–33 (1983)

Golub. Golub, G.H., VanLoan, Ch. F.: Matrix Computations. John Hopkins University Press, Baltimore London (1989)

Golubitsky. Golubitsky, M., Schaeffer, D.G.: Singularities and Groups in Bifurcation Theory, I, II. Springer, Berlin Heidelberg New York (1985) (1987)

Grauert. Grauert, H., Lieb, I.: Differential- und Integralrechnung III. Springer, Berlin Heidelberg New York (1968)

Gregory. Gregory, J., Lin, C.: Constrained Optimization in the Calculus of Variations and Optimal Control Theory. Van Nostrand, New York (1992)

Gresho. Gresho, P.M., Sani, R.L.: Incompressible Flow and the Finite Element Method, I,II. Wiley, New York (2000)

Grimm. Grimm, W.: Private Mitteilung. Univ. Stuttgart, Inst. of Flight Mechanics and Control (2004)

Gross A. Gross, D., et al.: Technische Mechanik I-III. Springer, Berlin Heidelberg New York (1998) (1999)

Gross B. Gross, D., et al.: Formeln und Aufgaben zur Technischen Mechanik I-III. Springer, Berlin Heidelberg New York (1998) (1999)

Gummert. Gummert, P., Reckling, K.-A.: Mechanik. Vieweg, Braunschweig (1987)

Haemmerlin. Hämmerlin, G., Hoffmann, K.H.: Numerische Mathematik. Springer, Berlin Heidelberg New York (1994).

Hairer. Hairer, E., et al.: Solving Ordinary Differential Equations I,II. Springer, Berlin Heidelberg New York (1987)

Hajek. Hajek, B.: Cooling schedules for optimal annealing. Mathematics of Operation Research, **13**, 311–329 (1988)

Hale. ale J.K.: Ordinary differential Equations. Wiley, New York (1969)

Halmos. Halmos, P.R.: Measure Theory. Van Nostrand, New York (1950)

Hamel. Hamel,G.: Theoretische Mechanik. Springer, Berlin Heidelberg New York (1978)

Hartl. Hartl,R.F., et al.: A survey of the maximum principle for optimal control problems with state constraints. SIAM Review **37**, 181–218 (1995)

Hartmann. Hartmann, F., Katz, C.: Structural Analysis with Finite Elements. Springer, Berlin London (2004)

Hassard. Hassard, B., et al.: Theory and Applications of Hopf Bifurcation. Cambridge University Press, Cambridge 1981.

Heil. Heil, M., Kitzka, F.: Grundkurs Theoretische Mechanik. Teubner, Stuttgart (1984)

Heuser80. Heuser, H.: Lehrbuch der Analysis I,II. Teubner, Stuttgart (1980) (1981)

Heuser86. Heuser, H.: Funktionalanalysis. Teubner, Stuttgart (1986)

Himmelblau. Himmelblau, D.M.: Applied Nonlinear Programming. McGraw–Hill, New York (1972)

C.Hirsch. Hirsch, C.: Numerical Computation of Internal and External Flows, I,II. Wiley, New York (1988)

M.Hirsch. Hirsch, M.W.: Differential Topology. Springer, Berlin Heidelberg New York (1976)

Hirzebruch. Hirzebruch, F., Scharlau, W.: Einführung in die Funktionalanalysis. Bibliograph. Institut, Mannheim (1971)

Hoellig. Höllig, K.: Grundlagen der Numerik. Klaus Höllig, Zavelstein (1998)

Holzmann. Holzmann, G., et al.: Technische Mechanik. Teubner, Stuttgart (1990)

Hopf. Hopf, E.: Abzweigung einer periodischen Lösung von einer stationären Lösung eines Differentialsystems. Ber. math.-phys. Klasse Sächs. Akad. Wiss. **94**, 1–22 (1942)

Householder. Householder, A.S.: The Theory of Matrices in Numerical Analysis. Dover Publications, New York (1964).

Irons. Irons, B., Ahmad, S.: Techniques of Finite Elements. Wiley, New York (1980)

Jeyachandrabose. Jeyachandrabose, C., et al.: An alternative explicit formulation for the DKT plate bending element. Int. J. Num. Meth. Eng. **21**, 1289-1293 (1985)

F.John. John, F.: Extremum problems with inequalities as subsidiary conditions. In: Studies and Essays, Courant Anniversary Volume. New York, 187–204 (1948)

Kardestuncer. Kardestuncer, H. (Ed.): FEM Handbook. McGraw-Hill, New York (1987)

Kawahara. Kawahara, M. et al.: Selective lumping finite element method for shallow water flow. Int. J. Num. Methods in Fluids **2**, 89–112 (1982)

Keener72. Keener, J.B., Keller, H.B.: Perturbed bifurcation and buckling of circular plates. Conf. on the Theory of Ordinary and Partial Differential Equations. Springer Lecture Notes, vol. 208, Dundee/Scotland (1972).

Keener73. Keener, J.P., Keller, H.B.: Perturbed bifurcation theory. Arch. Rat. Mech. Anal. **(50)**, 159–175 (1973)

Keener74. Keener, J.P.: Perturbed bifurcation theory at multiple eigenvalues. Arch. Rat. Mech. Anal. **(56)**, 348–366 (1974)

Keller58. Keller, H.B., Reiss, E.L.: Iterative solutions for the non-linear bending of circular plates. Comm. Pure Appl. Math. **11**, 273-292 (1958).

Keller72. Keller, H.B., Langford, W.F.: Iterations, perturbations and multiplicities for nonlinear bifurcation problems. Arch. Rational Mech. Anal. **48**, 83–108 (1972)

Keller87. Keller, H.B.: Numerical Methods in Bifurcation Problems. Springer, Berlin Heidelberg New York (1987)

Kielhoefer. Kielhöfer, H.: Hopf bifurcation at multiple eigenvalues. Arch. Rat. Mech. Anal. **69**, 53–83 (1979)

Kirchgaessner. Kirchgässner, K.: Exotische Lösungen des Bénardschen Problems. Math. Mech. in Appl. Sci. **(1)**, 453–467 (1997)

Kirsch. Kirsch, A., et al.: Notwendige Optimierungsbedingungen und ihre Anwendung. Springer, Berlin Heidelberg New York (1978)

Kosmol. Kosmol, P.: Optimierung und Approximation. De Gruyter, Berlin (1991)

Krasnoselki. Krasnoselki, M.A. Topological Methods in the Theory of Nonlinear Integral Equations. Macmillan, New York (1964)

Kroener. Kroener, D.: Numerical Schemes for Conservation Laws. Wiley und Teubner, Stutgart (1997)

Krabs. Krabs, W.: Optimierung und Approximation. Teubner, Stuttgart (1975)

Kubicek. Kubicek, M.: Algorithm 502. Dependence od solution of nonlinear systems on a parameter. JACM Transactions Math. Software **2**, 98–107 (1976)

Kuznetsov. Kutnetsov, Y.A.: Elements of Applied Bifurcation Theory. Springer, Berlin Heidelberg New York (1995)

Lanczos. Lanczos, C.: The Variational Principles of Mechanics. Toronto (1970)

Landau. Landau, L.D., Lifschitz, E.M.: Lehrbuch der theoretischen Physik. Akademie–Verlag, Berlin (1989)

Landman. Landman, K.A, Rosenblat, S.: Bifurcation from a multiple eigenvalue and stability of solutions. SIAM J. Appl. Math. **34**, 743–759 (1978)

Langford77a. Langford, W.F.: Numerical solution of bifurcation problems for ordinary differential equations. Numer. Math. bf 28, 171–190 (1977)

Langford77b. Langford, W.F.: A shooting algorithm for the best least squares solution of two-point boundary value problems. SIAM J. Numer. Anal. bf 14, 527–542 (1977)

Langford78. Langford, W.F.: The generalized inverse of an unbounded linear operator with unbounded constraints. SIAM J. Math. Anal. bf 9, 1038–1095 (1978)

Liao. Liao, S.-J., Zhu, J.-M.: A short note on high-order streamfunction-vorticity formulations of 2D steady state Navier-Stokes equations. Int. J. Num. Meth. Eng. **22**, 1–9 (1996)

Lippmann. Lippmann, H.: Angewandte Tensorrechnung. Springer, Berlin Heidelberg New York (1993)

Ljusternik. Ljusternik, L.A.: Conditional extrema of functionals. Mat. Sbornik **41**, 390–401 (1934)

Luenberger. Luenberger, D.G.: Optimization by Vector Space Methods. Wiley, New York (1969)

Machinek. Machinek, A.K., Troger, H.: Post-buckling of elastic annular plates at multiple eigenvalues. Dynamics and Stability of Systems, **3**, 79–98 (1988)

Magnus. Magnus, K., Müller, H.H.: Grundlagen der Technischen Mechanik. Teubner, Stuttgart (1982)

Marsden76. Marsden, J.E., McCracken, M.: The Hopf Bifurcation and Its Applications. Springer Appl. Math. Sci. Lecture Notes Series, Vol. 19, Springer, Berlin (1976)

Marsden. Marsden, J.E., Hughes, Th.J.: Mathematical Foundations of Elasticity. Prentice–Hall, Englewood Cliffs, N.J. (1983)

Marti. Marti, J.T.: Konvexe Analysis. Birkhäuser, Stuttgart (1977)

McLeod. McLeod, J.B. and Sattinger, D.H.: Loss of Stability and bifurcation at a double eigenvalue. J. Functional Analysis **14**, 62–84 (1973)

Meschkowski. Meschkowski, H.: Mathematiker-Lexikon. Bibliographisches Institut, Mannheim (1964)

Meyberg. Meyberg, K., Vachenauer, P.: Höhere Mathematik I,II. Springer, Berlin Heidelberg New York (1991)

Meyers. Meyers, N.G, Serrin, J.: H = W. Proc. Nat. Acad. Sci. USA **51**, 1055–1056 (1964)

Michlin. Michlin, S.G, Smolitzky, Ch.L.: Näherungsmethoden zur Lösung von Differential- und Integralgleichungen. Teubner, Leipzig (1969)

Mindlin. Mindlin, R.D.: Influence of rotary inertia and shear on flexural motion of elastic plates. Trans. ASME Ser. E, J. Appl. Mech. **18**, 31–38 (1951)

Mittelmann. Mittelmann, H.D., Weber, H.: Multigrid Solution of Bifurcation Problems. SIAM J. Sci. Stat. Comput. **6**, 49–60 (1985)

MooreA. Moore, G.: The numerical treatment of non-trivial bifurcation points. Numer. Funct. Anal. Optim. **2**, 441–472 (1980)

MooreB. Moore, G., Spence, A.: The calculation of turning points of nonlinear equations. SIAM J. Numer. Anal. **17**, 567–576 (1980)

Morley. Morley, L.S.: On the constant moment plate bending element. J. Strain Anal. **6**, 20–24 (1971).

Ninomiya. Ninomiya, H., Onishi, K.: Flow Analysis Using a PC. Southampton-Boston: Computational Mechanics Publications, Southampton Boston (1991)

Orlt. Orlt, M.: Regularitätsuntersuchungen und FEM-Fehlerabschätzungen ... . Dissertation Univ. Rostock, Rostock (1998)

Ortega. Ortega, J.M., Rheinboldt, W.C.: Iterative Solution of Nonlinear Equations in Several Variables. Academic Press, New York (1970)

Ovsiannikov. Ovsiannikov, L.V.: Group Analysis of Differential Equations. Academic Press, New York (1982)

Peraire. Peraire, J., et al.: Shallow water equations: A general explizit formulation. Int. J. Numer. Meth. Eng. **22**, 547–574 (1986)

Peregrine. Peregrine, D.H.: Long waves on a beach. J. Fluid Mech. **27**, 815–827 (1967)

Perelomov. Perlomov, A.M.: Integrable Systems of Classical Mechanics and Lie Algebras. Birkhäuser, Basel (1990)

Petera. Petera, J., Nassehi, V.: A new two-dimensional finite element model for the shallow water equations using a Lagrangian framework constructed along fluid particle trajectories. Int. J. Num. Mech. Eng. **39**, 4159–4182 (1996)

Peterson. Peterson, D.W.: A review of constraint qualifications in finite-dimensional spaces. SIAM Review **15**, 639–654 (1973)

Petrov. Petrov, I.P.: Variational Methods in Optimum Control Theory. Academic Press, New York (1968)

Polak. Polak, E.: Optimization. Springer, Berlin Heidelberg New York (1997)

Quartapelle. Quartapelle, L.: Numerical Solution of the Incompressible Navier-Stokes Equations. Birkhäuser, Basel (1993)

Quarteroni. Quarteroni, A., Valli, A.: Numerical Approximation of Partial Differential Equations. Springer, Berlin Heidelberg New York (1997)

Rabinowitz71. Rabinowitz, P.H.: Bifuraction from simple eigenvalues. J. Fuctional Analysis (**8**), 321–340 (1971)

Rabinowitz73. Rabinowitz, P.H.: Some aspects of nonlinear eigenvalue problems. Rocky Mountain J. Math. **(3)**, 161–202 (1973)

Rade. Rade, L., Westergren, B.: Mathematische Formeln. Springer, Berlin Heidelberg New York (1997)

Rannacher. Rannacher, R., Turek, S.: Simple nonconforming quadrilateral Stokes elements. Numer. Meth. Part. Diff. Equations **8**, 97–111 (1992).

Reid. Reid, W.T.: Generalized Greens' matrices for two-point boundary value problems. SIAM J. Appl. Math. **15**, 856–870 (1967)

Reissner. Reissner, E.: On the theory of bending of elastic plates. J. Math. Phys. **23**, 184–191 (1944)

Rheinboldt70. Rheinboldt, W.C.: Iterative Solution of Nonlinear Equations in Several Variables. Academic Press, New York (1970)

Rheinboldt86. Rheinboldt, W.C.: Numerical Analysis of Parametrized Nonlinear Equations. Wiley, New York (1986)

Richtmyer. Richtmyer, R.D.: Principles of Advanced Mathematical Physics. Springer, Berlin Heidelberg New York (1981)

Robinson. Robinson, S.M.: Stability theory for systems of inequalities, Part II. SIAM J. Numer. Analysis **13**, 497–513 (1976)

Roose. Roose, D., Hlavacek, V.: A direct method for the computation of Hopf bifurcation points. SIAM J. Appl. Math. **45**, 879–894 (1985)

Rouche. Rouche, N., Mauwhin, J.:, Ordinary Differential Equations. Pitman, Boston (1973)

Rutishauser. Rutishauser H.: Ausdehnung des Rombergschen Prinzips. Numer. Math. **5**, 48–54 (1963)

Sather. Sather, D.: Branching solutions of nonlinear equations. Rocky Mountain J. Math. **(3)**, 203–250 (1973)

Schaeffer. Schaeffer, H.H.: Topological Vector Spaces. Macmillan, New York (1966)

Schiehlen86. Schiehlen, W.: Technische Dynamik. Teubner, Stuttgart (1986)

Schiehlen89. Daberkov, A., Eismann, W., Schiehlen, W.: Test Examples for Multibody Systems. Universität Stuttgart, Inst. B für Mechanik, Inst.-bericht IB-13. Stuttgart (1989)

Schiehlen90. Schiehlen, W.: Multibody systems handbook. Springer, Berlin Heidelberg New York (1990)

Schneider. Schneider, M.: Himmelsmechanik. Bibliographisches Institut, Mannheim (1981)

Schneiders. Schneiders, R.: Refinig quadrilateral and hexahedral element meshes. Proc. NUMIGRID 96, Eds. Soni, B.K. et al., 679-689 (1996)

Schwarz80. Schwarz, H.R.: Methode der Finiten Elemente. Teubner, Stuttgart (1980)

Schwarz91. Schwarz, H.R.: FORTRAN-Programme zur Methode der Finiten Elemente. Teubner, Stuttgart (1991)

Seydel79. Seydel, R.: Numerical computation of branch points in nonlinear equations. Numer. Math. **33**, 339–352 (1979)

Seydel83. Seydel, R.: Branch switching in bifurcation problems for ordinary differential equations. Numer. Math. **41**, 93–116 (1983)

Seydel94. Seydel, R.: Practical Bifurcation and Stability Analysis. Springer, Berlin Heidelberg New York (1994)

Shampine82. Shampine, L.F.: Implementation of Rosenbrock methods. ACM Trans. on Math. Soft. **8**, 93–113 (1982)

Shampine97. Shampine, L.F, Reichelt, M.W.: The Matlab ODE suite. SIAM J. Sci. Computing **18**, 1–22 (1997)

Sohn. Sohn, J.L., Heinrich, J.C.: A Poisson equation formulation for pressure calculations in penalty finite element models for viscous incompressible flows. Int. J. Numer. Meth. Eng. **30**, 349–361 (1990)

Specht. Specht, W.: Modified shape functions for the three-node plate bending element passing the patch test. Int. J. Num. Meth. Eng. **26**, 705–715 (1968)

Spellucci. Spellucci, P.: Numerische Verfahren der Nichtlinearen Optimierung. Birkhäuser, Basel (1992)

Spotz. Spotz, W.F., Carey, G.F.: High-order compact scheme for the steady streamfunction vorticity equations. Int. J. Num. Meth. Eng. **38**, 3497–3512 (1995)

Stakgold. Stakgold, I.: Branching of solutions of nonlinear equations. SIAM Review **13**, 289–329 (1971)

Stevens. Stevens, W.N.: Finite Element, stream function-vorticity solution of steady laminar natural convection. Int. J. Numer. Methods in Fluids, **2**, 349–366 (1982)

Stoer. Stoer,J., Bulirsch, R.: Introduction to Numerical Analysis. Springer, Berlin Heidelberg New York (1980)

Strang. Strang, G., Fix, G.J.: An Analysis of the Finite Element Method. Prentice–Hall, Englewood Cliffs, N.J. (1973)

Stroud. Stroud, A.H.: Approximate Calculation of Multiple Integrals. Prentice–Hall, Englewood Cliffs, N.J. (1971)

Szabo76. Szabo, I.: Geschichte der Mechanischen Prinzipien. Birkhäuser, Basel 1976

Szabo77. Szabo, I.: Höhere Technische Mechanik. Springer, Berlin Heidelberg New York (1977)

Szegoe. Szegö, G.: Orthogonal Polynomials. AMS Soc. Coll. Publ., New York **23** (1939).

Tanahashi. Tanahashi, T., et al.: GSMAC finite element method for unsteady incompressible Navier-Stokes equations at high Reynolds numbers. Int. J. Num. Methods in Fluids **11**, 479–499 (1990)

Taylor. Taylor, A.E.: Introduction to Functional Analysis. John Wiley, New York (1958)

Teo89. Teo, K.L., Goh, C.J.: A computational method for combined optimal parameter selection and optimal control problems with general constraints. J. Austral. Math. Soc., Ser. B, **30**, 350–364 (1989)

Teo91. Teo, K.L., et.al.: A Unified Computational Approach to Optimal Control Problems. Longman, London (1991)

Timoshenko. Timoshenko, S., et al.: Vibration Problems in Engineering. Wiley, New York (1974)

Turek. Turek, St.: Efficient Solvers for Incompressible Flow Problems. Springer, Berlin Heidelberg New York (1999)

VanderBauwhede. VanderBauwhede, A.: Local Bifurcation and Symmetry. Pitman, Boston (1982)

VanLaarhoven. VanLaarhoven, P.J., Aarts, E.H.: Simulated Annealing: Theory and Applications. Reidel Publishing Company, Dordrecht Boston (1987)

Velte. Velte, W.: Direkte Methoden der Variationsrechnung. Teubner, Stuttgart (1976)

Voigt. Voigt, W.: Lehrbuch der Kristallphysik. Teubner, Leipzig (1910).

Weber79. Weber, H.: Numerische Behandlung von Verzweigungsproblemen bei gewöhnlichen Differentialgleichungen. Numer. Math. **32**, 17–29 (1979)

Weber85. Weber, H.: Multigrid bifurcation iteration. SIAM J. Numer. Anal. **22**, 262–279 (1985)

Wentzell. Wentzell, T.H.: Machine Design. Delmar Cengage Learning (2003)

Werner. Werner, J.: Optimization Theory and Applications. Vieweg, Braunschweig (1984)

Wilkinson. Wilkinson, J.H.: The Algebraic Eigenvalue Problem. Wiley, New York (1965)

Wloka. Wloka, J.: Funktionalanalysis und Anwendungen. De Gruyter, Berlin (1971)

Yosida. Yosida, K.: Functional Analysis. Springer, Berlin Heidelberg New York (1968)

Zangwill. Zangwill, B.I.: Nonlinear Programming. Prentice–Hall, Englewood Cliffs, N.J. (1969)

Zienkiewicz. Zienkiewicz, O.C., Taylor, R.L.: The Finite Element Method I,II. McGraw–Hill, London (1991)

Index